T0181541

Handbook of
Behavioral Neurobiology

Volume 10
Neurobiology of
Food and
Fluid Intake

HANDBOOK OF BEHAVIORAL NEUROBIOLOGY

General Editor:
Norman T. Adler
University of Pennsylvania, Philadelphia, Pennsylvania

Volume 1   Sensory Integration
Edited by R. Bruce Masterton

Volume 2   Neuropsychology
Edited by Michael S. Gazzaniga

Volume 3   Social Behavior and Communication
Edited by Peter Marler and J. G. Vandenbergh

Volume 4   Biological Rhythms
Edited by Jürgen Aschoff

Volume 5   Motor Coordination
Edited by Arnold L. Towe and Erich S. Luschei

Volume 6   Motivation
Edited by Evelyn Satinoff and Philip Teitelbaum

Volume 7   Reproduction
Edited by Norman T. Adler, Donald Pfaff, and Robert W. Goy

Volume 8   Developmental Psychobiology and Developmental Neurobiology
Edited by Elliott M. Blass

Volume 9   Developmental Psychobiology and Behavioral Ecology
Edited by Elliott M. Blass

Volume 10   Neurobiology of Food and Fluid Intake
Edited by Edward M. Stricker

Handbook of
Behavioral Neurobiology

Volume 10
Neurobiology of Food and Fluid Intake

Edited by

Edward M. Stricker

University of Pittsburgh
Pittsburgh, Pennsylvania

Springer Science+Business Media, LLC

Library of Congress Cataloging-in-Publication Data

Neurobiology of food and fluid intake / edited by Edward M. Stricker.
       p.   cm. -- (Handbook of behavioral neurobiology ; v. 10)
    Includes bibliographical references.
    Includes index.
    ISBN 978-1-4612-7874-0   ISBN 978-1-4613-0577-4 (eBook)
    DOI 10.1007/978-1-4613-0577-4
    1. Hunger--Physiological aspects. 2. Thirst--Physiological
 aspects. 3. Neurobiology.   I. Stricker, E. II. Series.
    [DNLM: 1. Drinking Behavior--physiology. 2. Feeding Behavior-
 -physiology. 3. Neurobiology.   W1 HA51I v. 10 / WL 102 N4945208]
 QP147.N48 1990
 612.3'91--dc20
 DNLM/DLC
 for Library of Congress   90-7651
                                                               CIP

This Handbook is dedicated to the memory of

Claude Bernard, Walter Cannon, and Curt Richter,

whose pioneering work provided the foundation for the
modern behavioral neurobiology of food and fluid ingestion.

Contributors

MATTHEW BECK, *Department of Psychology, McMaster University, Hamilton, Ontario L8S 4K1, Canada*

DAVID A. BOOTH, *School of Psychology, University of Birmingham, Birmingham B15 2TT, England*

L. ARTHUR CAMPFIELD, *Neurobiology and Obesity Research, Hoffmann-La Roche, Inc., Nutley, New Jersey 07110*

J. A. DEUTSCH, *Department of Psychology, University of California San Diego, LaJolla, California 92093*

ALAN N. EPSTEIN, *Department of Biology, University of Pennsylvania, Philadelphia, Pennsylvania 19104*

LEONARD H. EPSTEIN, *Western Psychiatric Institute and Clinic, University of Pittsburgh School of Medicine, Pittsburgh, Pennsylvania 15213*

J. T. FITZSIMONS, *The Physiological Laboratory, University of Cambridge, Cambridge CB2 3EG, England*

MARK I. FRIEDMAN, *Monell Chemical Senses Center, Philadelphia, Pennsylvania 19104*

BENNETT G. GALEF, JR., *Department of Psychology, McMaster University, Hamilton, Ontario L8S 4K1, Canada*

HARVEY J. GRILL, *Department of Psychology and Institute of Neurological Sciences, University of Pennsylvania, Philadelphia, Pennsylvania 19104*

F. REED HAINSWORTH, *Department of Biology, Syracuse University, Syracuse, New York 13244*

W. G. HALL, *Department of Psychology, Duke University, Durham, North Carolina 27706*

JOEL M. KAPLAN, *Department of Psychology and Institute of Neurological Sciences, University of Pennsylvania, Philadelphia, Pennsylvania 19104*

HARRY R. KISSILEFF, *Departments of Medicine and Psychiatry, Columbia University College of Physicians and Surgeons at St. Luke's-Roosevelt Hospital Center, New York, New York 10025*

PAUL R. MCHUGH, *Department of Psychiatry and Behavioral Sciences, The Johns Hopkins University School of Medicine, Baltimore, Maryland 21205*

DAVID J. RAMSAY, *Department of Physiology, University of California, San Francisco, California 94143*

PAUL N. ROZIN, *Department of Psychology, University of Pennsylvania, Philadelphia, Pennsylvania 19104*

JAY SCHULKIN, *Department of Anatomy, University of Pennsylvania, Philadelphia, Pennsylvania 19104*

THOMAS R. SCOTT, *Department of Psychology, University of Delaware, Newark, Delaware 19716*

FRANCOISE J. SMITH, *Neurobiology and Obesity Research, Hoffmann-La Roche, Inc., Nutley, New Jersey 07110*

ELIOT STELLAR, *Department of Anatomy and Institute of Neurological Sciences, University of Pennsylvania, Philadelphia, Pennsylvania 19104*

EDWARD M. STRICKER, *Department of Behavioral Neuroscience, University of Pittsburgh, Pittsburgh, Pennsylvania 15260*

TERRY N. THRASHER, *Department of Physiology, University of California, San Francisco, California 94143*

THEODORE B. VANITALLIE, *Department of Medicine, Columbia University College of Physicians and Surgeons at St. Luke's-Roosevelt Hospital Center, New York, New York 10025*

JOSEPH G. VERBALIS, *Departments of Medicine and Behavioral Neuroscience, University of Pittsburgh, Pittsburgh, Pennsylvania 15261*

LARRY L. WOLF, *Department of Biology, Syracuse University, Syracuse, New York 13244*

Preface

When I began graduate school in 1961, Physiological Psychology was alive with adventure and opportunity. It seemed possible, indeed easy, to determine which part of the brain influenced which aspect of behavior, and the relative absence of technical hurdles encouraged neophytes into the laboratory. New theories of brain function based on a wealth of reliable and provocative findings also stimulated further laboratory investigation. And the results obtained in studies of food and fluid ingestion certainly were exciting, albeit perplexing. For example, eating could be stimulated by injecting one chemical agent into the rat brain, whereas drinking was stimulated by injecting a different chemical through the same hypothalamic cannula. After focal brain lesions rats would overeat but not work harder to obtain food. After other brain lesions in adjacent sites, rats would stop eating and drinking altogether, but ingestive behaviors would return gradually over a period of weeks or months despite permanent brain injury. Although some of these observations and related findings may provide less insight into the central control of ingestive behavior than had been believed initially, there was a strong impression then that much more was known about eating and drinking than other behaviors, and they became models of motivated activities in addition to being of interest in their own right.

Twenty-two years ago, the American Physiological Society published the first handbook devoted exclusively to the subject of alimentary behavior. The volume contained 32 chapters that admirably summarized the findings to date over a broad range of topics. Each chapter represented an especially important contribution to the literature because in the 1960s few comprehensive reviews were available. Thus, the *Handbook of Physiology*, with its diverse coverage and large-page format, was able to provide in a single volume a unique and substantial source of information and perspective. Perhaps of equal significance was the usefulness of the volume itself in defining as a field of inquiry the physiology of alimentary behavior, which was not (and still is not) in the mainstream of either Physiology or Psychology. Although there had been relatively little interest in food and fluid ingestion then, the impressive collection of material contained in that *Handbook*, and its publication by the prestigious American Physiological Society, brought attention and legitimacy to the subject matter and thereby helped to create the much larger audience and group of participants that exists today.

The milieu for this second volume on food and fluid ingestion is different from that which existed 20 years ago: relevant review articles now are abundant, and the edited proceedings of numerous workshops and symposia on the topic also are readily available to concerned investigators and their students. Because today's readers, therefore, are much more likely to be familiar with the facts before coming to this volume than were early readers of the first *Handbook*, I saw no need for the present *Handbook* to provide summaries of subject matter that in many cases had been adequately summarized recently. Furthermore, I had to restrict the number of chapters that might be included in the present volume so that the length of each could be reasonable and the total size could fall within practical limits. Consequently, I did not seek thorough coverage of the field in planning this *Handbook*. Instead, I identified major focal issues in contemporary research and theory concerning the behavioral neurobiology of food and fluid consumption and then asked selected authors to write either a comprehensive chapter (if none was available elsewhere), a chapter focusing on designated issues of particular significance, or an essay that retrospectively evaluates how the field has moved forward during the past two decades or prospectively considers where it should be heading.

There were two general points that I wanted to have emphasized throughout the volume, and I am pleased to believe that both points have been made. First, I wanted it to be clear that aspects of ingestive behavior can be best understood in a biological context, and that physiological responses complementing ingestive behavior are not totally independent of behavior. This is the contemporary view of behavioral neurobiology: according to it, discussions of thirst, for example, are incomplete without considering renal contributions to body fluid homeostasis, while aspects of kidney function might be appreciated only when fluid ingestion also is considered. Second, I wanted it to be clear that these multidisciplinary considerations of eating and drinking permit insights into various dysfunctions in human ingestive behavior, and vice versa. This is the traditional view of experimental medicine: according to it, an appropriate dialogue between laboratory investigators and their colleagues in the clinics enables each group to accomplish together what they could not do separately.

To make these points, I selected as contributors to the *Handbook* investigators who were distinguished by their unique research accomplishments and biological perspective on behavior, and I asked some of them to give their presentations a definite clinical orientation. I chose not to include several prominent investigators with much to say who recently had written lengthy reviews or entire volumes on their subjects of expertise because I did not believe that a condensed version of those writings would be as useful to the field as certain other points of view, that had not yet received the attention that inclusion in the *Handbook* might provide.

It will not take long for the reader to note that some views stated in one chapter or essay seem to contradict those proposed elsewhere in the *Handbook*. Such contradictions are real. Indeed, I confess that I do not agree with all that each author has written. However, I opted to edit this book rather than prepare a book of my own largely because I thought this unsettled field would be much better served by the presentation of multiple viewpoints, articulated clearly and forcefully by some of their best advocates, than by a single perspective. Put another way, I believe that such diversity of views energizes the field and generates light as well as heat. Thus, in my role as editor, I did not wish to blunt, much less conceal, such differences of opinion, and I hope their presence in this volume will stimulate thought and provoke the research that ultimately clarifies existing controversies.

Concerns about homeostasis understandably pervade most of this volume. Such concerns had their principal origins in the work of Claude Bernard and then Walter Cannon, who defined the phenomenon and began to describe its physiological bases. Equally important to the modern behavioral neurobiology of food and fluid ingestion, however, is the subsequent work of Curt Richter. His broad vision and seminal research established as significant the behavioral contributions to organismal homeostasis, and by innumerable observations of lasting significance he enriched the Biological Sciences and Psychology and prepared those disciplines for a later alliance. Bengt Andersson, Derek Denton, and Vincent Dethier were singular leaders in the next generation who consistently saw individual issues as examples of larger themes, who were able to communicate those themes in elegant prose, and who developed research programs 30 years ago that provided exemplary models of an integrated psychobiological approach to homeostasis, which then was novel but now prevails. This *Handbook* is dedicated to the memory of Professors Bernard, Cannon, and Richter and also acknowledges, with respect and gratitude, the later contributions of Professors Andersson, Denton, and Dethier. I hope that the research described in these pages will stimulate a new generation of investigators as their foundational work has inspired ours, so that future scientists will continue to find, as we do, that studies on the neurobiology of ingestive behavior increasingly provide excitement, surprise, and satisfaction in ample measure.

I enjoyed editing this volume, and I thank Norman Adler, the general editor of this series of handbooks, for the invitation to do so. I also thank Mary Born, Senior Editor at Plenum Press, and Shuli Traub, Production Editor, for their able assistance in transforming a collection of manuscripts into a handsome book. Finally, I thank my wife, Marcy, and children, Judd and Emily, for their patience and support during the months when it must have seemed that I was giving this project the highest priority. It only seemed that way.

EDWARD M. STRICKER

Contents

CHAPTER 6

Caudal Brainstem Participates in the Distributed Neural Control
    of Feeding

Harvey J. Grill and Joel M. Kaplan

PART III FOOD SELECTION

CHAPTER 10

Gustatory Control of Food Selection

Thomas R. Scott

CHAPTER 11

Comparative Studies of Feeding

F. Reed Hainsworth and Larry L. Wolf

CHAPTER 12

Food Selection

Paul N. Rozin and Jay Schulkin

CHAPTER 13

Diet Selection and Poison Avoidance by Mammals Individually
    and in Social Groups

Bennett G. Galef, Jr. and Matthew Beck

PART IV THIRST, SODIUM APPETITE, AND FLUID HOMEOSTASIS

CHAPTER 14

Thirst and Water Balance

David J. Ramsay and Terry N. Thrasher

CHAPTER 15

Sodium Appetite

Edward M. Stricker and Joseph G. Verbalis

CHAPTER 16

Clinical Aspects of Body Fluid Homeostasis in Humans

Joseph G. Verbalis

PART V PROSPECTIVE ESSAYS

CHAPTER 17

The Behavioral and Neural Sciences of Ingestion

David A. Booth

CHAPTER 18

Prospectus: Thirst and Salt Appetite

Alan N. Epstein

CHAPTER 19

Making Sense Out of Calories

Mark I. Friedman

CHAPTER 20

Clinical Issues in Food Ingestion and Body Weight Maintenance

Paul R. McHugh

PART I
Retrospective Essays

Brain and Behavior

ELIOT STELLAR

RETROSPECTIVE VIEW OF BRAIN AND BEHAVIOR

The question of how the brain generates behavioral control of food and water intake, motivation for food and water reward, and the experience of hunger and thirst was not framed until Lashley (1938) published his classical paper. Until that time, the concern was primarily with the peripheral physiology underlying hunger and thirst sensation in humans, particularly the role of gastric contractions and dry mouth (Cannon, 1932). The behavioristic framework was just being assembled by Skinner (1938) and Hull (1943), who used hunger and thirst motivation and food and water reward as model systems of operant or instrumental conditioning. Hull believed he was dealing with motivation and reward as important intervening variables. Skinner was just as certain that they did not exist and that specifying changes in the stimulus–response relationship was all that was needed. Neither Skinner nor Hull, however, was concerned with the role of the brain in behavior. Both thought that behavioral analysis alone could yield a complete explanation and that when the physiology and the neurology were eventually worked out, it might be interesting to make "coordinating definitions" between the behavior and the brain. As a consequence, neither man anticipated the explosive rise of physiological psychology or the later development of behavioral neuroscience, which Lashley's work had foretold. Their contribution was to frame the question in behavioristic terms, which allowed the use of animals to answer research questions and gave a methodology that would permit fine-grained analysis of the behavior that, of course, reflected the functioning of the brain.

Lashley's specific contribution was to ask how the state of the internal environment, when combined with internal and external sensory stimulation, could yield motivated behavior (he called it "instinctive behavior") ranging from

ELIOT STELLAR   Department of Anatomy and Institute of Neurological Sciences, University of Pennsylvania, Philadelphia, Pennsylvania 19104.

a spider spinning its web to sexual behavior in the female rat. His model physiological system was the invertebrate microstoma's search for the nettles or stinging cells of hydra: microstoma does not grow its own stinging cells but acquires them by ingesting hydra; when it has shot all its nettles, it is in a depleted state and becomes ravenous in its search and ingestion of hydra, whereas once it has a full complement of nettles, it is in a satiated state and selectively avoids eating hydra even when it has no other source of food. His neural model, as he transferred his thinking to mammals, was Sherrington's (1906) central excitatory state, but he had no idea of the specific structures in the brain that might be involved, because he wrote just before Ranson's group reported their explorations of the hypothalamus with the newly rediscovered stereotaxic instrument (Hetherington & Ranson, 1942).

It is the purpose of this retrospective view of research on the role of the brain in the control of food intake to trace the several lines of historical development that led to our present state of knowledge and conceptualization. Of special concern is the development of our ideas about how the brain is involved in physiological regulation and how it yields the behavior that contributes to physiological regulation, that is, the role the brain plays in the physiology of motivated behavior. Because the taste and ingestion of food and water are rewarding, we also are interested in the neural basis of the hedonic processes of reward and reinforcement and in hedonic experience in humans. All of these behavioral and psychological processes are involved in hunger and the control of food intake, and all involve parts of the same widespread systems of the brain, particularly the limbic structures. On the behavioral side, these processes are of increasing complexity. On the neural side, they are organized in hierarchical fashion along the neuraxis. The scientific problem of behavioral and systems neuroscience is to relate the two. In addition, there is the exciting, new possibility of relating the behavioral and systems neuroscience to the cellular and molecular level of description, to the special properties of neuronal cells, their neurotransmitter and neurosecretory properties, and their receptor and metabolic properties.

Before coming to grips with the issues, it is important to note that scientific progress depends on repeated cycles of progress in techniques and concepts. This certainly is true in the study of hunger and the control of food intake. The earliest of the modern concepts was based on hunger as a sensation, and this led to the use of available peripheral techniques (e.g., recording gastric contractions) to study the phenomenon in humans. As behavioristic techniques for studying animals developed, and more particularly as stereotaxic surgery was reintroduced into animal research, the depths of the brain could be explored, and emphasis shifted from peripheral concepts to central neural concepts. Clearly, it was recognized that both were involved, but it was not until modern methods for anatomic tracing and physiological recording were used that some of the specific peripheral–central–peripheral systems were identified, involving the limbic system from the brainstem to the hypothalamus and amygdala. Now we have the challenge to apply the new techniques of cell and molecular biology to reveal the great specificity these systems must have in yielding not only the control of food intake and the sensation of hunger and satiation but also specific hungers, food rewards, and hedonic experience in humans. Through the use of these new techniques, new concepts will emerge, and old concepts will be refined.

The role of behavior in physiological regulation was recognized by Darwin (1873), who envisioned a role of adaptive behavior in evolution. It gained direct

physiological significance with Claude Bernard's (1861/1957) conception of the
constancy of the "milieu interieur," in which behavior participated. It was not
until the work of Andre Mayer (1900), who recognized that thirst and water
consumption were essential components of body fluid homeostasis, that the first
explicit statement was made of the role of behavior in regulating the internal
environment. It was Richter (1942–1943), however, who developed the more
general concept of self-regulatory behavior and applied it not only to thirst but
also to the hunger for calories and to specific hungers, including most notably
the appetite for sodium.

The modern history of the study of salt appetite is instructive. Richter
(1936) first identified the role of the adrenal gland in behavior, which was known
to play a role in salt conservation at the kidney, in the generation of salt appetite.
He showed that bilateral extirpation of the adrenal glands of rats led to increased intake of 3% NaCl solution, enough to keep the animals alive despite
uncontrolled sodium loss in the urine. Desoxycorticosterone replacement therapy, in small doses, restored the control of sodium excretion and eliminated the
excess appetite. To explain these effects, Richter (1939) invoked a peripheral
mechanism, and his idea was that adrenalectomy lowered the taste threshold for
sodium on the tongue and that this accounted for greater ingestion. Pfaffmann
and Bare (1950), however, could find no difference between adrenalectomized
and normal rats when they investigated electrophysiological taste thresholds,
recording from the chorda tympani nerve. Shortly thereafter, Carr (1952) confirmed these findings and reported no difference in the rat's threshold for salt in
a behavioral discrimination test. More recently, however, Contreras (1977) has
shown that following adrenalectomy, the discharge rate of the chorda tympani to
stimulation of the tongue with sodium chloride is reduced, and he has suggested
that this may be a peripheral mechanism involved in the increased salt appetite.

Nevertheless, attention turned centrally when it was discovered that the
peptide angiotensin could elicit salt appetite when applied directly to the brain
and that it exerted some of its effects through synergistic action with aldosterone
(Epstein, 1982a). As the work reported in Chapter 18 makes clear, the search is
now on for aldosterone receptors in the brain and for the sites of action of brain
angiotensin in the roles that the two hormones play as synergistic generators of
salt appetite. So it appears that the sodium levels of the internal environment are
regulated by a mechanism that involves, in part, complex hormonal actions in
the brain to raise the salt intake that are coordinated with the actions of aldosterone at the kidney to conserve salt and angiotensin at peripheral resistance
vessels to support blood pressure.

Theoretically, similar arguments can be made for thirst and the behavioral
regulation of body water. The original emphasis was on dry mouth and the thirst
sensation in humans (Cannon, 1932). As attention turned to central neural
mechanisms controlling water intake, we learned that there is a dual regulation
(Epstein, Kissileff, & Stellar, 1973)—an osmotic regulation and a volumetric or
volemic regulation—and that these involve somewhat different mechanisms in
the brain. Osmotic regulation depends on osmoreceptors in the preoptic area of
the brain and basal forebrain structures (Andersson, 1953; Blass & Epstein,
1971; Johnson & Buggy, 1978). Volemic regulation, it has been proposed, depends in part on the renin–angiotensin system in both the periphery and the
brain, with its focus on the anterior wall of the third ventricle, particularly the
angiotensin-sensitive subfornical organ, the nucleus medianus, and the organum

vasculosum of the lamina terminalis (Epstein, 1982b). Anatomic studies by Miselis (1981) show that these regions of the brain are richly interconnected with each other and with the paraventricular and supraoptic nuclei (the sources of vasopressin, the antidiuretic hormone), the posterior pituitary, and the lateral hypothalamus. So here again is a case of integration of physiology and behavior to promote homeostasis where water deficiency simultaneously stimulates a conserving hormone and elicits intake.

Other specific hungers are more complicated. Whereas the appetites for salt and water, and also sugars, appear to be innate, a great deal of learning appears to be involved in the specific appetites for vitamins and other minerals (Rozin, 1982). Although the early cafeteria-feeding studies of Richter (Richter, Holt, & Barelare, 1938) showed that rats could self-select a balanced diet, later studies by Pilgrim and Patton (1947) showed considerable individual variation in success, and P. T. Young's (1948) work clearly demonstrated the importance of individual learning and experience (see Chapter 12).

The problem in understanding the hunger for calories is that we still do not know what is being regulated (Brobeck, 1965), as we do in our studies of salt appetite, thirst, and, to pick a distant example, thermoregulatory behavior. It sometimes has been suggested that this hunger is the sum total of all specific hungers, and there have been theories based on the separate regulations of glucose utilization, protein, and fat as well as more indirect theories of body weight regulation and set point and energy balance as determinants of food intake. Be this as it may, the search for neural and humoral signals controlling food intake is an active and productive one, and neurotransmitters (norepinephrine and serotonin), insulin, glucagon, cholecystokinin, opiates, and other peptides all have been suggested to be significant in this regard.

From the foregoing, we can conclude that whereas motivated behavior serves in the regulation of the internal environment, its contribution is not always a simple or direct one in terms of what is being regulated. In addition, we know that some motivated ingestive behaviors are not homeostatic, such as the adjunctive behavior seen in the excessive drinking of schedule-induced polydipsia (Falk, 1964) and the excessive ingestion of tasty solutions in the absence of hunger or thirst (Ernits & Corbit, 1973). Despite these exceptions, the main line of evidence is that ingestive behavior often contributes to the regulation of the internal environment, that both peripheral physiological and central neural mechanisms are involved, and that the same neuroendocrine mechanisms that control physiological regulation also may control behavior, including the attributes of motivation, reward, and hedonic process. In our present state of knowledge, we have much better understanding of the mechanisms involved in salt and water ingestion and the regulation of fluid balance than we do of food intake and food selection and the physiological and metabolic control of nutrition and energy balance. To repeat what we have said earlier, this is because we know what is regulated in salt and water ingestion, and the variables are much simpler than in the case of food ingestion.

THE PHYSIOLOGY OF MOTIVATION

Motivation has not always been a recognized area of research in psychology. Boring never mentions the term in his *History of Experimental Psychology* (1929), although it clearly was a concern of Freud and McDougall (who used the term instinct) and was brought into experimental psychology by Woodworth (1918) with his concept of drive. An early ethologist, Craig (1918), had already made the sophisticated distinction between appetitive and consummatory aspects of instincts, and today these are useful concepts in the study of motivation. With regard to ingestive behavior, the appetitive motivation is the approach to the food or fluid, whether it is natural foraging behavior or a learned operant act such as pressing a lever for food, whereas consummatory behavior is, of course, the act of ingestion. Other European ethologists, including von Frisch, Lorenz, and Tinbergen (1951), studied adaptive behavior that they called instinctive behavior and put emphasis on the unlearned aspects of the behavior. This stance served to isolate them from American behaviorists, who had banished the term instinct because in America it was typically used only as an explanatory concept and was not the subject of investigation. Furthermore, the behaviorists took the view that most behavior involved a great deal of learning, and the idea of innate behavior was anathema. Even Richter did not often use the term "motivated behavior." He preferred the term "self-regulatory behavior" for food and water motivation and confined himself mainly to the consummatory acts.

It was Lashley who accepted the concept of goal-directed behavior. He was a student of J. B. Watson's and a behaviorist who became more and more influenced by gestalt psychology with its emphasis on innate patterns of behavior dependent on a hard-wired nervous system. A broad scholar, he was familiar with the work of the European ethologists and drew many of his models from their studies of invertebrates, fish, and birds. At the same time, he was a neurologist, greatly influenced by his mentor, Shepherd Ivory Franz, and by Pavlov and Sherrington. He was a centralist who believed in the integrative action of the brain, bringing together peripheral influences of the humoral environment and sensory input. Finally, his interest was less in localizing function in the brain than in investigating patterns of input to brain and how they led to patterns of output or behavior.

C. T. Morgan, in the first edition of *Physiological Psychology* (1943), formalized Lashley's views of the central neural mechanisms involved in adaptive behavior, and he spoke directly of motivation. His earlier work on how insulin stimulated eating in the rat (Morgan & Morgan, 1940) led to his concept of a humoral motive factor whereby changes in the internal environment caused arousal of the brain mechanisms for motivated behavior, contributing along with sensory stimulation to a central motive state, a priming of the brain to specific action such as eating or drinking. Astutely, he recognized that the satiation as well as the arousal of behavior had to be considered. He conceived of satiation as a reduction of the central motive state caused by eliminating the humoral motive factor through ingestion of food or water, by providing a different humoral signal from the gut to the brain, or by afferent neural input produced by the acts of ingestion.

It was not, however, until 1950 that the full impact of the work of the Ranson group at Northwestern was felt (Morgan & Stellar, 1950). By this time, it became evident that the hypothalamus played an important role in the control of food and water intake. Stereotaxic surgery, reintroduced by Ranson, became generally available, and the new technique allowed investigators to place discrete lesions within the depths of the brain. Taken together with the development in the 1930s by W. R. Hess (1954) of implanting electrodes in the diencephalon of the cat, this methodology also provided the means to stimulate deep structures electrically or chemically through implanted cannulae as well as to ablate them.

With this background, it was possible to construct a general scheme of the physiological control of motivated behavior and of food intake in particular (Stellar, 1954). There were several components to the conception (Figure 1). The central theme was that hunger and food intake were under multifactorial control, involving external sensory input (taste and smell), internal sensory stimulation (gastric contractions and gastric distension), internal environment factors (insulin, nutrients), and modification by learning and experience.

The question of how the brain integrated all this peripheral information into a single act of food intake seemed straightforward enough in light of the fact that lesions of the ventromedial nucleus of the hypothalamus led to overeating (Hetherington & Ranson, 1942; Teitelbaum, 1955; Reeves & Plum, 1969) and lesions of the lateral hypothalamus led to starvation (Anand & Brobeck, 1951; Teitelbaum & Stellar, 1954). The fact that these lesions produced other changes in behavior and in physiological functions was no surprise, for electrolytic lesions were crude interventions in a closely packed hypothalamus where many structures and functions are woven together. The general idea was that, under natural conditions, the afferent neural input and the specific changes in the internal environment would find their ways selectively to the hypothalamic structures involved in food intake.

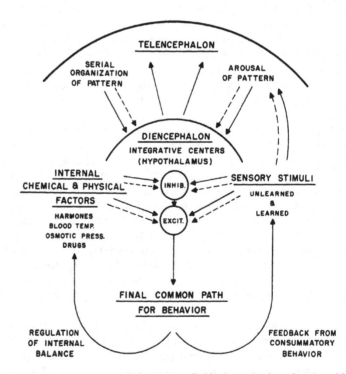

Figure 1. The 1954 conception of a multifactor-controlled brain mechanism of motivated behavior with input from peripheral sensory stimuli, events in the internal environment, and forebrain influences; excitatory and inhibitory systems in the hypothalamus were seen as the major integrators, controlling a response mechanism in the lower neuraxis (Stellar, 1954).

Given these observations on brain-damaged animals and the fact that hunger and satiation are two of the prominent features of the motivation for food, a dual control mechanism was envisioned: a satiation mechanism involving the ventromedial hypothalamus and a hunger mechanism involving the lateral hypothalamus. Both were thought to be under the combined control of sensory and humoral peripheral input and would function together like Sherringtonian excitatory and inhibitory mechanisms. But it was not clear whether their reciprocal influences were on each other or on a final common path to the behavior in response mechanisms lower in the neuraxis.

In addition to the recognition that the hypothalamus exerted its effects through somatic and visceral influences on brainstem and spinal cord response mechanisms, it also was recognized that forebrain structures (e.g., infraorbital frontal cortex, amygdala) had controlling input to the hypothalamus. Thus, the central neural mechanism envisioned placed the hypothalamus at the center of an extensive system involving the whole neuraxis. It was seen as a mechanism responsive to and controlling both the autonomic and somatic aspects of the regulation of food intake, digestion, and the metabolism of nutrients.

In the years that followed soon thereafter, the excitement in research focused on the exploration of the properties of the central neural mechanism in the hypothalamus with the newfound stereotaxic techniques. With a few exceptions, investigations of the periphery were neglected, and the study of possible humoral factors proved quite refractory. In contrast, major advances were made in the study of the hypothalamus. Among other things, the experiments of W. R. Hess were rediscovered by the American workers with the publication of his work in English in 1954, showing that stimulation of the hypothalamus led to a variety of autonomic and somatic responses and could evoke eating behavior in the satiated cat. Smith (1961) showed the same thing in the chronic waking rat, with lateral hypothalamic stimulation evoking eating and ventromedial hypothalamic stimulation inhibiting it. In a similar fashion, Epstein (1960) injected the local anesthetic procaine and a stimulating solution of hypertonic saline into the lateral and ventromedial hypothalamus and found the predicted increases and decreases in food intake in all four possible combinations. These were impressive findings, complementing the effects of electrolytic lesions of the same areas.

Efforts to be more specific about the naturally occurring chemicals that might influence these brain structures (e.g., insulin, glucose) were not successful in Epstein's experiment. However, in 1960, Grossman reported highly specific effects of norepinephrine and the cholinergic agonist carbachol, which, when applied through the same cannula in the lateral hypothalamus, selectively aroused eating and drinking, respectively. Since that time much elegant work has been done by Leibowitz (1985) and Hoebel (1984), showing that norepinephrine, epinephrine, and serotonin play significant roles in the medial and lateral hypothalamus when injected into the chronic, waking rat. Depending on where these neurotransmitters are injected, they increase or decrease eating behavior (see Figure 6; Hoebel, 1984) and alter food selection and meal patterns (Leibowitz, 1984), providing data for some very interesting speculations as to how they may operate in the control of food intake.

In 1954, Olds and Milner published their classical paper on electrical self-stimulation of the brain, and the reward mechanism eventually revealed fit in with the concept of a hypothalamic mechanism for the motivation to ingest food.

Lateral hypothalamic stimulation turned out to be excitatory and positively rewarding, whereas medial hypothalamic stimulation was negative. Somewhat later, Hoebel and Teitelbaum (1962) and Margules and Olds (1962) showed that stimulation of lateral hypothalamic electrodes elicited both self-stimulation and feeding behavior and that variables that influenced feeding such as gastric loading and ventromedial hypothalamic lesions also influenced self-stimulation in the same way.

Further research led to a clarification of the ventromedial and lateral hypothalamic syndromes and made it clear that the conceptualization of 1954 was an oversimplification. In regard to the ventromedial mechanism, it was shown that a lesion of the ventromedial nucleus itself was not critical for hyperphagia; rather, destruction of the fiber paths leading through it, from the brainstem to the paraventricular nucleus lying anterior to it, produced the hyperphagia (Gold, 1973; Kapatos & Gold, 1973). Secondly, despite earlier evidence suggesting that ventromedial hypothalamic lesions produced a direct behavioral change and not a metabolic change, work by Powley and Opsahl (1974) and others showed that the lesions produced a marked hyperinsulinemia, which could be abolished by vagotomy along with the hyperphagia and obesity. Clearly, ventromedial hypothalamic lesions produced neurohumoral and autonomic effects as well as behavioral changes, and the suggestion was made that the hyperphagia and obesity might be secondary to the endocrine and autonomic changes. However, King, Carpenter, Stamoutsos, Frohman, & Grossman (1978) later showed that when the vagotomy was done before the ventromedial hypothalamic lesions and the animal was allowed to recover from the acute adverse effects of the vagotomy, both hyperphagia and obesity were produced, even though insulin levels did not increase. Furthermore, it is interesting that rats with ventromedial hypothalamic lesions show hyperphagia after vagotomy when they are fed high-fat diets (Powley & Opsahl, 1974) or highly palatable supermarket diets (Sclafani, Aravich, & Landman, 1981). So electrolytic lesions of the medial hypothalamus are, indeed, crude interventions that produce multiple somatic, autonomic, endocrine, and metabolic effects. More refined techniques should be able to separate these different physiological and behavioral functions.

In regard to the lateral hypothalamic mechanism, it became clear that lesions here produced a wide range of symptoms, including adipsia (Anand & Brobeck, 1951), deficits in behavioral thermoregulation (Satinoff & Shan, 1971), sleep (Nauta, 1946), self-stimulation (Janas & Stellar, 1987), and sensory neglect, akinesia, and catalepsy (Marshall, Turner, & Teitelbaum, 1971). As Stricker (1983) has pointed out, rats with lateral hypothalamic lesions also have a defect in arousal and are slow to respond, but when stimulated sufficiently or given enough time, they can make many of the adaptive responses thought to be lost to them. Furthermore, it was shown by Ungerstedt (1971) that injection of the neurotoxin 6-hydroxydopamine (6-OHDA) directly into the medial forebrain bundle of the lateral hypothalamus destroys the catecholamine fibers of the nigrostriatal pathway and produces much of the lateral hypothalamic syndrome. The picture is complicated, however, for to be effective, 90–95% of the striatal dopamine must be depleted (Stricker & Zigmond, 1976), an interesting finding in light of the fact that electrolytic lesions in the lateral hypothalamus cause only a 40–60% depletion of striatal dopamine (Oltmans & Harvey, 1972). Furthermore, using the new technique of transplanting fetal substantia nigral tissue into the striatum reduces the deficits of sensory neglect and akinesia and restores self-stimulation, depending on the locus in the striatum, but so far the deficits in eating and drinking have not been corrected by such transplants (Dunnett, Bjorklund, & Stenevi, 1983; Fray, Dunnett, Iverson, Bjorklund, & Stenevi, 1983). So the multiple deficits produced by lateral hypothalamic lesions can be partially untangled with more sophisticated methods.

In addition to the changes in motivated behavior induced, lateral hypothalamic lesions and stimulation clearly have widespread visceral and metabolic effects, leading to changes in gastric motility and secretion, oxygen consumption, body temperature, and glucagon release and blood sugar levels, to name a few (Nicolaidis, 1981; Stricker, 1983). These changes, of course, become part of the sensory and humoral input to the brain, especially the hypothalamus. On the other hand, recording from the lateral hypothalamus of waking monkeys shows a great deal of specificity, for individual neurons there are responsive to food stimuli only when the animal is hungry, and they also respond to neutral stimuli

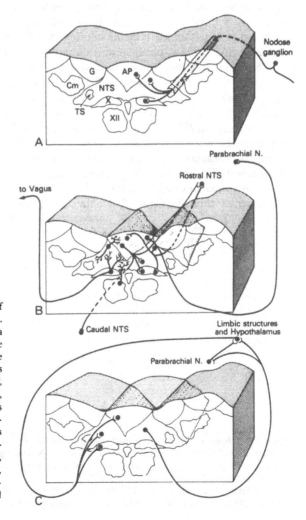

Figure 2. Central connections of the vagus in the caudal medulla. (A) Synapses in the area postrema (AP), caudal medial nucleus of the solitary tract (cmNTS), and the dorsal motor nucleus of the vagus (X). (B) Connectivity between AP, cmNTS, rostral and caudal NTS, X, and parabrachial nucleus (PBN). (C) Ascending connections to PBN and hypothalamus and other limbic structures, descending connection to AP, NTS, and X. G, gracile nucleus; Cm, medial cuneate nucleus; TS, tractus solitarius; XII, hypoglossal nucleus (Hyde & Miselis, 1983).

that have been associated with food (Rolls, 1982). Interestingly, stimulating these cells evokes both eating behavior and self-stimulation. Again, more specific techniques yield more specific results.

That more than the hypothalamus is involved in food motivation has been clear from the beginning. Lesions of the medial and lateral amygdala, for example, produce decreases and increases, respectively, in the eating behavior of dogs and in their responses to social stimulation and petting (Fonberg, 1974). More recently the striking experiments of Norgren and Grill (1982) with the decerebrate rat identified a caudal brainstem mechanism at work. They succeeded in studying eating in the decerebrate, where others before them failed (Bard & Macht, 1958), by virtue of the intraoral cannulae they used to present liquid food stimuli to the mouth. The decerebrate rat will not eat voluntarily, but it will ingest or reject food put into the mouth, depending on the taste of the food and on the state of the animal. The hungry decerebrate, for example, accepts sucrose solutions introduced into its mouth but rejects quinine solutions. When the animal had been fed recently, however, it also will reject sucrose solutions. At the physiological level, there is a cephalic insulin response to the taste of sweet solutions. Yet, there is no comparable acceptance of water by the thirsty decerebrate rat or of salt solutions by the salt-depleted rat. So the brainstem mechanism revealed thus far is a specific brainstem integrator for feeding behavior.

Some ideas of the circuits involved, connecting brainstem and hypothalamic integrators, have been worked out by Hyde and Miselis (1983), who used modern anatomic tracing methods to follow the vagus into the caudal brainstem, particularly to the area postrema (AP) and the caudal medial nucleus of the solitary tract (cmNTS) (Figure 2). From here, the fibers project forward to the parabrachial nucleus (PB) of the pons and then branch to the classical dorsal thalamocortical taste pathway and to the ventral limbic taste system, described by Norgren (1977), involving the lateral hypothalamus, the paraventricular nucleus of the hypothalamus, the central nucleus of the amygdala, and the bed nucleus of the stria terminalis (Figure 3). Furthermore, there are descending projections, as Figure 2 indicates, from the paraventricular nucleus, the ventromedial nu-

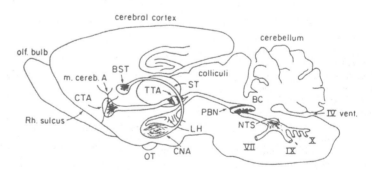

Figure 3. Scheme of the two taste pathways in a lateral view of the rat's brain. (1) The dorsal thalamocortical path from cranial nerves VII, IX, and X, making synapses in the rostral nucleus tractus solitarius (NTS) and the parabrachial nucleus (PBN) of the pons, the thalamic taste area (TTA), and the cortical taste area (CTA). (2) The ventral limbic taste pathway from PBN to the lateral hypothalamus (LH), central nucleus of the amygdala (CNA), and the stria terminals (ST) and its bed nucleus (BST). BC, brachium conjunctivum; OT, optic tract; Rh, rhinal sulcus (Norgren, 1977).

cleus, and the lateral hypothalamus to the caudal medial NTS, the area post-rema, and the dorsal motor nucleus of the vagus.

Quite clearly, the brainstem and hypothalamic integrators involve the autonomic as well as the somatic nervous systems. W. R. Hess recognized this in his early experiments stimulating the hypothalamus of the waking cat, and it also was evident in the early lesion experiments of Ranson's group at Northwestern. Only now, however, are we beginning to learn of the specific pathways involved, both afferent and efferent. In addition, both the hypothalamus and the brain-stem are exposed to the internal environment via the circumventricular organs (Figure 4), which have reduced blood–brain barriers (Phillips, 1978), so humoral control of feeding behavior is possible by direct influences on the brain. The hypothalamus, moreover, has neurosecretory functions that allow it to control the pituitary gland and, thus, the other endocrine glands and the internal environment. The elegant studies of Nicolaidis (1981), in fact, show that there are complex peripheral–central–peripheral neural and humoral loops that could be involved in both the arousal and satiation of feeding behavior.

One of the earliest proponents of peripheral, autonomic control mechanisms in feeding behavior was Russek (1971). He recognized the significance of postabsorptive metabolic events in the liver, activated by glucose entering the hepatic portal vein and conveyed to the brain mainly over the vagus nerve. Support for this view and extension of it comes from the reviews of Friedman and Stricker (1976) and of Novin and Vander Weele (1977), who emphasize the importance of a hepatic neural signal to the brain in the short-term control of food intake.

Another example of peripheral input comes from the work on the role of the peptide cholecystokinin (CCK) in the satiation of feeding behavior, started by Gibbs, Young, and Smith (1973). This hormone, secreted by intestinal cells, is considered by some to be a satiety factor because intraperitoneal injection of CCK not only brings a meal to an early end but also induces the satiation syndrome in which the rat grooms, explores, and sleeps postprandially. Cutting the vagus nerve eliminates this peripheral CCK effect, so its afferent neural

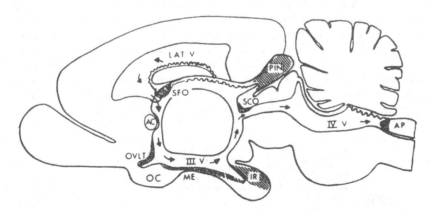

Figure 4. The seven circumventricular organs, seen in a lateral view of the rat's brain: SFO, subfor-nical organ; OVLT, organum vasculosum of the lamina terminalis; ME, median eminence; IR, infundibular recess and neurophyphophysis; SCO, subcommissural organ; PIN, pineal gland; AP, area postrema (Phillips, 1978).

pathway to the brain has been identified. Because CCK also occurs in the brain, the question arises as to whether CCK of central origin may play a role in satiation. To test this idea, CCK has been injected into brain parenchyma or into the cerebral ventricles with variable results. Some investigators have reported reductions in food-motivated behavior (Della-Fera & Baile, 1979; Maddison, 1977; Zhang, Bula, & Stellar, 1986), whereas others have not been able to confirm these findings or obtain consistent effects. To complicate matters, the question arises as to whether CCK produces the reduction of food intake by inducing nausea the way lithium chloride does (Deutsch & Hardy, 1977). This is not a logically easy question to control for or to answer. However, it is important to note that it has been shown that intravenous administration of CCK to humans reduces food intake without producing reports of nausea (Kissileff, Pi-Sunyer, Thornton, & Smith, 1981). It is still possible, however, that there may be other negative effects of CCK as yet unreported.

Thus, the effects of CCK on food intake are not as clear as the effects of angiotensin on thirst, where both peripheral and central administration of the peptide produce robust dipsogenic effects (Epstein, 1983). Nor is it like the more complicated case of the role of insulin, which appears to arouse eating behavior when injected peripherally (Morgan & Morgan, 1940) and to act as a satiety peptide when injected into the ventricles (Nicolaidis, 1981; Woods, Lotter, McKay, & Porte, 1979), although some of this difference may be the distinction between pharmacological and physiological effects of insulin.

A quite different analysis of the physiological mechanisms controlling food intake comes from Stricker (1983). His view recalls the general arousal theory of motivation (Hebb, 1955; Lindsley, 1960), and he sees the motivation to eat as part of a general arousal mechanism, involving the contribution of eating to regulation of the activity of the reticular formation or "reticular homeostasis" in addition to subserving caloric homeostasis. His scheme (Figure 5) takes into account many of the facts we have been trying to deal with. Like the dual-center concept, it proposes central excitatory and central inhibitory systems in the brain, although their anatomic locations are not specified. Hunger develops when the central excitatory system is aroused by metabolic signals caused by the decrease in calories coming from the intestine and the liver (1) and from afferent input over taste and smell pathways (2) as the organism begins to eat. Activity of the catecholamines in the reticular formation arouses the organism generally and allows the feeding responses (3). The resulting calories from the intestine and liver along with insulin release contribute to satiation, inhibiting the metabolic arousal signals (4). This may occur, in part, because insulin favors the uptake into brain of tryptophan, the precursor of serotonin, which may play an important role in the central inhibitory system (6). The central inhibitory (sero-

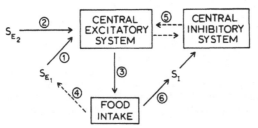

Figure 5. Scheme of central excitatory and central inhibitory mechanisms controlling food intake. Numbers explained in text (Stricker, 1983).

tonergic) and central excitatory (catecholaminergic) systems act reciprocally on each other (5). The conception has the strength that it directly addresses the starting and the stopping of eating and adds a rich account of the peripheral metabolic, endocrine, and sensory input to the brain. But there is nothing to say that this system does not function to maintain caloric homeostasis as well as reticular homeostasis. The concepts are not mutually exclusive.

In a subsequent essay (Stricker, 1984), Stricker goes even further in his provocative thinking and makes three key points: (1) "hunger does not reflect a biological need for food as thirst does for water" (metabolic fuels are continuously available to the organism, whether from intestinal absorption of nutrients or from the mobilization of liver and adipose tissue storage), (2) "there may be no stimulus for hunger" (rather, hunger and the sensations associated with it may be traced to a reduction of the inhibitory stimuli of satiety), and (3) "stimuli for satiety originate in the periphery, not in the brain" (in the liver and the gut but not in the brain, which is so well protected from fluctuations in energy sources). These are challenging ideas and help to redress the balance between peripheral and central neural factors controlling food intake. But we must not let the pendulum swing too far and recognize that in the control of food intake the brain is responsive to changes in energy metabolism (Nicolaidis, 1980), to humoral influences such as insulin (Woods *et al.*, 1979), and to afferent neural input, both learned and unlearned, depending on the state of the animal (Rolls, 1982).

We have come a long way in our understanding of the physiology of motivation, particularly as it applies to hunger and the control of food intake. At the same time, we have a long way to go, for there are many unresolved issues. On the positive side, we have learned that there are central neural mechanisms that operate in the arousal and satiation of feeding behavior and that these same brain mechanisms are involved in the autonomic and endocrine control of the viscera and the gastrointestinal system in particular. On the behavioral side, the motivation for food is controlled by oropharyngeal, gastric, and hepatic signals, both neural and humoral. On the physiological side, food intake has been most often related to energy balance and caloric needs, but it may also be driven by a more general arousal mechanism. The two kinds of mechanisms are not mutually exclusive, for both do what all physiological theories are trying to do, and that is to account for the arousal and satiation of eating behavior.

HEDONIC PROCESSES

Hull and Skinner took a chapter from Thorndike (1898) and focused attention on the importance of reward and reinforcement in instrumental and operant behavior, but we had no clue to the brain mechanisms that might be involved until the classical paper of Olds and Milner (1954). The implication of their work and subsequent studies (Gallistel, Shizgal, & Yeomans, 1981; Commons, Church, Stellar, & Wagner, 1988) was that rewarding electrical self-stimulation of the brain was a way to tap into the brain circuits involved in natural reward such as that provided by food for the hungry animal. As already mentioned, Hoebel and Teitelbaum (1962) directly demonstrated this likelihood by showing that physiological interventions that increased or decreased eating also increased and decreased electrical self-stimulation of the brain. Using the method of self-stimulation, numerous experiments have demonstrated that central dopaminergic neurons play an important role in reward, because drugs that block dopamine receptors also reduce the rewarding effects of self-stimulation (Stellar & Stellar, 1985). Careful behavioral analysis has shown that whereas a portion of this effect could be attributed to an impairment of motor function, much of it was a direct inhibitory effect on reward. Recently, Wise (1984) advanced the hypothesis that dopaminergic neurons also play a role in food reward, as judged by the effects of the dopamine receptor blocker pimozide in reducing food-rewarded operant behavior. He envisions the same circuit involved in self-stimulation: neurons of the lateral hypothalamus projecting to the ventral tegmental area and, from there, dopamine neurons projecting to the nucleus accumbens (Wise & Bozarth, 1984).

Perhaps the most comprehensive conception to date in the effort to bring together our understanding of the brain mechanisms involved in the control of food intake and the generation of reward has been presented by Hoebel (1984) as a speculative set of hypotheses. He adopts Wise's circuit for reward but points out that the lateral hypothalamus has both excitatory and inhibitory elements, the latter being mediated, in part, by epinephrine and norepinephrine, which reduce both reward and feeding when injected locally (Figure 6). In the same way, the medial hypothalamus has both inhibitory and excitatory elements; taste and visceral input via the vagus are inhibitory, and so is norepinephrine, leading to increased feeding and reward; serotonin is thought to be excitatory in the medial hypothalamus, inhibiting food intake. Although this scheme is still speculative, it summarizes a good deal of existing data and gives a good idea of the level of complexity we must learn to deal with. It also addresses directly the dual nature of hunger and the control of food intake by taking into account the fact

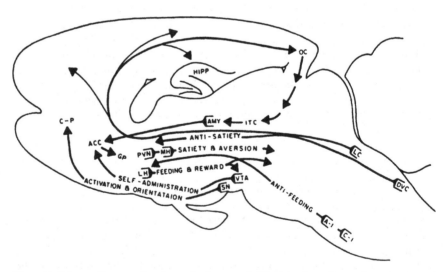

Figure 6. Scheme of areas and pathways in the control of food intake and reward seen in a lateral view of the rat's brain. ACC, nucleus accumbens; AMY, amygdala; A-1, norepinephrine cell group; C-1, epinephrine cell group; C–P, caudate–putamen; DVC, dorsal vagal complex; FC, frontal cortex; GP, globus pallidus; HIPP, hippocampus; ITC, infratemporal cortex; LC, locus coeruleus; LH, lateral hypothalamus; MH, medial hypothalamus; OC, optic cortex; PVN, paraventricular nucleus; SN, substantia nigra; VTA, ventral tegmental area (Hoebel, 1984).

that there have to be brain mechanisms for both starting and stopping eating in any system we envisage.

Only in humans can we directly address the question of hedonic experience as it relates to hunger, satiation, and the control of food intake. Behavioristic measures so successful in the investigation of animals now have been established in the study of humans. Consequently, it is possible to measure objectively the microstructure of single meals (Jordan, Wieland, Zebley, Stellar, & Stunkard, 1966; Kissileff, Klingsberg, & Van Itallie, 1980) and the daily pattern of meals (Hashim & Van Itallie, 1965; Bernstein, Zimmerman, Czeisler, & Weitzman, 1981). At the same time, by the use of analogue scales, it is possible to obtain ratings of hunger, palatability, and satiation as a function of deprivation, type of food, and the effects of eating. For example, as food intake declines during the course of a meal, not only do hunger decrease and satiation increase as expected but, in addition, the perceived palatability of a test food decreases (Spiegel *et al.*, 1987). Thus, hedonic rating is a function not only of taste but also of the internal state of the organism. Taste also is important in the reduction of hunger during a meal; when subjects were fed intragastrically by stomach tube, bypassing taste or oral factors completely, the hunger ratings did not decline over the course of the meal as they did with normal oral eating (Stellar & Jordan, 1970).

So we have come full circle in the study of hunger, once again looking at sensation in human subjects, albeit with different techniques that parallel those used in animal studies. What has not been possible with human subjects, however, is to ask what brain mechanisms are involved in these sensory experiences. However, with the development of higher-resolution noninvasive techniques for imaging the activity of the brain and for doing *in vivo* biochemistry, we soon should be able to assess human brain function as it is affected by hunger, satiation, and food palatability.

LOOKING BACK

We have come a long way in the study of the brain mechanisms involved in hunger and the control of food intake. From an initial emphasis on the role of peripheral factors involved in the sensation of hunger in humans, we have proceeded to the investigation of food motivation in animals as a self-regulatory behavior contributing to the control of the internal environment. This work, pioneered by Richter, applied the technique of endocrine extirpation and replacement therapy and led to the identification of a role of hormones in ingestive behavior. With the advent of stereotaxic technology, the depths of the brain involved in hunger were explored by lesion, electrical stimulation, chemical stimulation, and electrophysiological recording from individual neurons in the behaving animal.

Many suggestions have been made about what is regulated (if anything) in the control of food intake and what signals arouse and satiate food-motivated behavior. Carbohydrate, fat, and protein and their utilization have been implicated. Physiological signals from the periphery, both neural and hormonal, appear to be important in this regard. So are taste and smell signals. What role do catecholaminergic and serotonergic signals from the brainstem play? Is energy metabolism sensed in the hypothalamus? Is body weight regulated? Or is it a brain state that is regulated? These and many other issues appear to be involved in the multifactor control of the central brain system involved in hunger and satiation.

An initial conception of a dual central neural control mechanism, with the ventromedial hypothalamus involved in satiation and the lateral hypothalamus involved in hunger, gave way to the recognition of a more complicated system in which medial and lateral hypothalamic structures were shown to be influenced by neurotransmitters and neuromodulators, suggesting that both regions may have inhibitory and excitatory functions. Furthermore, not only was it shown that forebrain structures were involved in hunger and satiation but so also were brainstem structures. In fact, modern anatomic tracing studies revealed a complex system of ascending and descending pathways linking hypothalamic and brainstem structures and, very importantly, providing a basis for many peripheral autonomic and metabolic inputs to the brain and for peripheral somatic and autonomic outputs to the viscera. Clearly, the multifactor control system involved in hunger and the control of food intake involves both the periphery and the central nervous system, and a great deal of information evidently is processed in the periphery and the brainstem as well as the hypothalamus and forebrain structures.

Such a complex physiological system yields not only control of food intake but also the appetitive search for food and the rewarding and reinforcing effects of food stimuli. In addition, it provides a neural basis for the hedonic experiences of hunger and satiation that can be reported by human beings.

Looking Ahead

We have a long way to go, but the value of reviewing our past history is that it gives us a trajectory into the future. Friedman (Chapter 19) provides a detailed prospective view. But suffice it to say here that our past tells us that we need another leap forward in the technique–concept cycle. We need more refined anatomic methods for tracing the relevant circuits in the brain. We need more refined neurochemical and neuropharmacological methods for identifying the neurotransmitters and neuromodulators involved in these behavioral and motivational systems and for using them to separate discrete functions in the brain. We need to identify both the peripheral and central signals of hunger and satiation generated by energy metabolism, the flux of calories, and hormonal influences, and we must uncover their cellular mechanism of action. We need to continue our efforts to record in the behaving animal from neuronal units along the pathways that are identified, for we can learn many of the physiological and behavioral properties of the system from that kind of mapping.

To accomplish these ends and go beyond them, we need to understand the cellular mechanism at work, the receptors, the ion channels, the second messengers, and so on. We also need to avail ourselves of molecular probes to manipulate the cellular processes identified. Finally, we need to improve our behavioral techniques so that we can examine the appetitive as well as the consummatory aspects of the motivation for food, so we can analyze the microstructure of a single meal or the dynamics of daily food intake patterns in both animals and humans, so that we can look at taste reactivity and the reward properties of food and the brain mechanisms involved, and so that we can deal with the human subjective experiences of hunger, satiation, and food palatability reliably and quantitatively. In short, the task ahead is to deal with the problem of hunger and the control of food intake at all levels of analysis: psychological, behavioral, at the level of the neural systems involved, and at the cellular and molecular levels as well. We also need to have prepared minds for the unexpected finding, for the new windows that new techniques will open, and not least of all, for challenging and provocative new conceptualizations. Then we should be able to reach a better understanding of how the brain works to handle the multiple peripheral inputs that control food intake and produce the somatic, autonomic, endocrine, and metabolic responses involved in eating and the experiences of hunger, satiation, and food reward.

References

Anand, B. K., & Brobeck, J. R. (1951). Hypothalamic control of food intake in rats and cats. *Yale Journal of Biology and Medicine, 24,* 123–140.

Andersson, B. (1953). The effect of injections of hypertonic NaCl-solutions into different parts of the hypothalamus of goats. *Acta Physiologiae Scandinavia, 28,* 188–201.

Bernard, C. (1957). *An introduction to the study of experimental medicine.* New York: Dover Publications. (Original work published 1865).

Bard, P., & Macht, M. B. (1958). The behavior of chronically decerebrate cats. In G. E. W. Wolstenholm & C. M. O'Connor (Eds.), *Ciba Foundation symposium on the neurological bases of behavior* (pp. 55–75). Boston: Little, Brown.

Bernstein, I. L., Zimmerman, J. C., Czeisler, C. A., & Weitzman, E. D. (1981). Meal patterns in "free-running" humans. *Physiology and Behavior, 27,* 621–623.

Blass, E. M., & Epstein, A. N. (1971). A lateral preoptic osmosensitive zone for thirst in the rat. *Journal of Comparative and Physiological Psychology, 76,* 378–394.

Boring, E. G. (1929). *A history of experimental psychology.* New York: Appleton Century.

Brobeck, J. R. (1965). Exchange, control and regulation. In W. S. Yamamoto & J. R. Brobeck (Eds.), *Physiological controls and regulations* (pp. 1–13). Philadelphia: W. B. Saunders.

Cannon, W. B. (1932). *The wisdom of the body.* New York: Norton.

Carr, W. J. (1952). The effect of adrenalectomy upon the NaCl taste threshold in the rat. *Journal of Comparative and Physiological Psychology, 45,* 377–380.

Commons, M. L., Church, R. M., Stellar, J. R., & Wagner, A. R. (1988). *Quantitative analyses of behavior: Biological determinants of reinforcement* (Vol. VII). Hillsdale, NJ: Lawrence Erlbaum Associates.

Contreras, P. J. (1977). Changes in gustatory nerve discharges with sodium deficiency: A single unit analysis. *Brain Research, 121,* 373–378.

Craig, W. (1918). Appetites and aversions as constituents of instincts. *Biological Bulletin, 34,* 91–107.

Darwin, C. (1873). *The expression of emotions in man and animals.* New York: D. Appleton and Co.

Della-Fera, M. A., & Baile, C. A. (1979). Cholecystokinin octapeptide: Continuous picomole injections into the cerebral ventricles of sheep suppress feeding. *Science, 206,* 471–473.

Deutsch, J. A., & Hardy, W. T. (1977). Cholecystokinin produces bait shyness in rats. *Nature, 266,* 196.

Dunnett, S. B., Bjorklund, A., & Stenevi, U. (1983). Transplant-induced recovery from brain lesions: A review of the nigrostriatal model. In R. B. Wallace & G. P. Das (Eds.), *Neural transplantation research* (pp. 191–216). New York: Springer-Verlag.

Epstein, A. N. (1960). Reciprocal changes in feeding behavior produced by intrahypothalamic chemical injection. *American Journal of Physiology, 199,* 969–974.

Epstein, A. N. (1982a). Mineralocorticoids and cerebral angiotensin may act together to produce sodium appetite. *Peptides, 3,* 493–494.

Epstein, A. N. (1982b). The physiology of thirst. In D. W. Pfaff (Ed.), *The physiological mechanisms of motivation* (pp. 165–214). New York: Springer-Verlag.

Epstein, A. N. (1983). The neuropsychology of drinking behavior. In E. Satinoff & P. Teitelbaum (Eds.), *Handbook of behavioral neurobiology. Vol. 6, Motivation* (pp. 367–423) New York: Plenum Press.

Epstein, A. N., Kissileff, H. R., & Stellar, E. (Eds.). (1973). *The neuropsychology of thirst.* Washington, DC: V. H. Winston.

Ernits, T., & Corbit, J. D. (1973). Taste as a dipogenic stimulus. *Journal of Comparative and Physiological Psychology, 83*, 27–31.

Falk, J. L. (1964). Studies on schedule-induced polydipsia. In M. J. Wayer (Ed.), *Thirst* (pp. 95–116).

Fonberg, E. (1974). Amygdala functions within the alimentary system. *Acts Neurobiologiae Experimentalis, 34*, 435–466.

Fray, P., Dunnett, S., Iverson, S., Bjorklund, A., & Stenevi, U. (1983). Nigral transplants reinervating the dopamine-depleted neostriatum can sustain intracranial self-stimulation. *Science, 219*, 416–419.

Friedman, M. I., & Stricker, E. M. (1976). The physiological psychology of hunger: A physiological perspective. *Physiological Review, 83*, 409–431.

Gallistel, C. R., Shizgal, P., & Yeomans, J. S. (1981). A portrait of the substrate for self-stimulation. *Psychological Review, 88*, 228–273.

Gibbs, J., Young, R. C., & Smith, G. P. (1973). Cholecystokinin decreases food intake in rats. *Journal of Comparative and Physiological Psychology, 84*, 488–495.

Gold, R. M. (1973). Hypothalamic obesity: The myth of the ventromedial nucleus. *Science, 182*, 488–490.

Grossman, S. P. (1960). Eating or drinking by direct adrenergic or cholinergic stimulation of the hypothalamus. *Science, 132*, 301–302.

Hashim, S. A., & Van Itallie, T. B. (1965). Studies in normal and obese subjects with a monitored food dispensing device. *Annals New York Academy of Sciences, 131*, 654–661.

Hebb, D. O. (1955). Drives and the CNS (conceptual nervous system). *Psychological Review, 62*, 243–254.

Hess, W. R. (1954). *Diencephalon: Autonomic and extrapyramidal functions.* New York: Grune & Stratton.

Hetherington, A. W., Ranson, S. W. (1942). The spontaneous activity and food intake of rats with hypothalamic lesions. *American Journal of Physiology, 136*, 609–617.

Hoebel, B. G. (1984). Neurotransmitters and the control of feeding and its rewards: Monoamines, opiates, and brain–gut peptides. In A. J. Stunkard & E. Stellar (Eds.), *Eating and its disorders* (pp. 15–38). New York: Raven Press.

Hoebel, B. G., & Teitelbaum, P. (1962). Hypothalamic control of feeding and self-stimulation. *Science, 135*, 375–377.

Hull, C. L. (1943). *Principles of behavior.* New York: Appleton-Century-Crofts.

Hyde, T. M., & Miselis, R. R. (1983). Effects of area postrema/caudal medial nucleus of solitary tract lesions on food intake and body weight. *American Journal of Physiology, 244*, R577–R587.

Janas, J. D., & Stellar, J. R. (1987). Effects of knife-cut lesions of the medial forebrain bundle in self-stimulating rats. *Behavioral Neuroscience, 101*, 832–845.

Johnson, A. K., & Buggy, J. (1978). Periventricular preoptic–hypothalamus is vital for thirst and normal water economy. *American Journal of Physiology, 234*, R122–R129.

Jordan, H. A., Wieland, W. F., Zebley, S. P., Stellar, E., & Stunkard, A. J. (1966). Direct measurement of food intake in man: A method for the objective study of eating behavior. *Psychosomatic Medicine, 28*, 836–842.

Kapatos, G., & Gold, R. M. (1973). Evidence for ascending noradrenergic mediation of hypothalamic hyperphagia. *Pharmacology, Biochemistry, and Behavior, 1*, 81–87.

King, B. M., Carpenter, R. G., Stamoutsos, B. A., Frohman, L. A., & Grossman, S. P. (1978). Hyperphagia and obesity following ventromedial hypothalamic lesions in rats with subdiaphragmatic vagotomy. *Physiology and Behavior, 20*, 643–651.

Kissileff, H. R., Klingsberg, G., & Van Itallie, T. B. (1980). Universal eating monitor for continuous recording of solid or liquid consumption in man. *American Journal of Physiology, 238*, 14–22.

Kissileff, H. R., Pi-Sunyer, F. X., Thornton, J., & Smith, G. P. (1981). Cholecystokinin-octapeptide (CCK-8) decreases food intake in man. *American Journal of Clinical Nutrition, 34*, 154–160.

Lashley, K. S. (1938). An experimental analysis of instinctive behavior. *Psychological Review, 45*, 445–471.

Leibowitz, S. F. (1984). Brain monoamine projections and receptor systems in relation to food intake, diet preference, meal patterns, and body weight. In G. M. Brown, S. H. Koslow, & S. Reichlin (Eds.), *Neuroendocrinology of psychiatric disorders* (pp. 383–399). New York: Raven Press.

Leibowitz, S. F. (1985). Brain neurotransmitters and appetite regulation. *Psychopharmacological Bulletin, 21*, 412–418.

Lindsley, D. B. (1960). Attention, consciousness, sleep, and wakefulness. In J. Field, H. W. Magoun, & V. E. Hall (Eds.), *Handbook of physiology. Section 1: Neurophysiology. Vol III* (pp. 1553–1593). Washington, DC: American Physiological Society.

Maddison, S. (1977). Intraperitoneal and intracranial cholecystokinin depress operant responding for food. *Physiology and Behavior, 19*, 819–824.

Margules, D. L., & Olds, J. (1962). Identical "feeding" and "rewarding" systems in the lateral hypothalamus of rats. *Science, 135*, 374–375.

Marshall, J. F., Turner, B. H., & Teitelbaum, P. (1971). Sensory neglect produced by lateral hypothalamic damage. *Science, 174*, 523–525.

Mayer, A. (1900). Variations de la tension osmotique due sang chez les animoux prives de liquides. *Compte Rendu des Seances de la Societe le Biologie, 52*, 153–155.

Miselis, R. R. (1981). The efferent projections of the subfornical organ of the rat: A circumventricular organ with a neural network subserving water balance. *Brain Research, 230*, 1–23.

Morgan, C. T. (1943). *Physiological psychology.* New York: McGraw-Hill.

Morgan, C. T., & Morgan, J. D. (1940). Studies in hunger: I. The effects of insulin upon the rat's rate of eating. *Journal of Genetic Psychology, 56*, 137–147.

Morgan, C. T., & Stellar, E. (1950). *Physiological psychology.* New York: McGraw-Hill.

Nauta, W. J. H. (1946). Hypothalamic regulation of sleep in rats: An experimental study. *Journal of Neurophysiology, 9*, 285–316.

Nicolaidis, S. (1980). Hypothalamic convergence of external and internal stimulation leading to early ingestive and metabolic responses. *Brain Research Bulletin, 5*(Suppl. 4), 97–101.

Nicolaidis, S. (1981). Lateral hypothalamic control of metabolic factors related to feeding. *Diabetologia, 20*, 426–434.

Norgren, R. (1977). Synopsis of gustatory neuroanatomy. In J. Le Magnen & P. McLeod (Eds.), *Olfaction and taste* (Vol. VI, pp. 225–232). Washington, DC: IRL Publishers.

Norgren, R., & Grill, H. (1982). Brainstem control of ingestive behavior. In D. W. Pfaff (Ed.), *The physiological mechanisms of motivation* (pp. 99–131). New York: Springer-Verlag.

Novin, D., & Vander Weele, D. A. (1977). Visceral involvement in feeding: There is more to regulation than the hypothalamus. In J. M. Sprague & A. N. Epstein (Eds.), *Progress in psychobiology and physiological psychology* (Vol. 7, pp. 193–241). New York: Academic Press.

Olds, J., & Milner, P. (1954). Positive reinforcement produced by electrical stimulation of septal area and other regions of the rat brain. *Journal of Comparative and Physiological Psychology, 47*, 419–427.

Oltmans, G. A., & Harvey, J. A. (1972). LH syndrome and brain catecholamine levels after lesions of the nigrostriatal bundle. *Physiology and Behavior, 8*, 69–78.

Pfaffmann, C., & Bare, J. K. (1950). Gustatory nerve discharges in normal and adrenalectomized rat. *Journal of Comparative and Physiological Psychology, 40*, 320–324.

Phillips, M. I. (1978). Angiotensin in the brain. *Neuroendocrinology, 25*, 354–377.

Pilgrim, F. J., & Patton, R. A. (1947). Patterns of self-selection of purified dietary components by the rat. *Journal of Comparative and Physiological Psychology, 40*, 343–348.

Powley, T. L., & Opsahl, C. A. (1974). Ventromedial hypothalamic obesity abolished by subdiaphragmatic vagotomy. *American Journal of Physiology, 226*, 25–33.

Reeves, A. G., & Plum, F. (1969). Hyperphagia, rage, and dementia accompanying a ventromedial hypothalamic neoplasm area. *Archives of Neurology, 20*, 616–624.

Richter, C. P. (1936). Increased salt appetite in adrenalectomized rats. *American Journal of Physiology, 115*, 155–161.

Richter, C. P. (1939). Salt taste thresholds of normal and adrenalectomized rats. *Endocrinology, 24*, 367–371.

Richter, C. P. (1942–1943). Total self regulatory functions in animals and human beings. *Harvey Lectures Series, 38*, 63–103.

Richter, C. P. Holt, L. E., Jr., & Barelare, B., Jr. (1938). Nutritional requirements for normal growth and reproduction in rats studied by the self-selection method. *American Journal of Physiology, 122*, 734–744.

Rolls, E. T. (1982). Feeding and reward. In B. G. Hoebel & D. Novin (Eds.), *The neural basis of feeding and reward* (pp. 323–327). Brunswick, ME: Haer Institute.

Rozin, P. (1982). Human food selection: The interaction of biology, culture, and individual experience. In L. M. Barker (Ed.), *The psychobiology of human food selection* (pp. 225–254). Westport, CT: AVI Publishing.

Russek, M. (1971). Hepatic receptors and the neurophysiological mechanisms controlling feeding behavior. In S. Ehrenpreis (Ed.), *Neuroscience research* (pp. 213–282). New York: Academic Press.

Satinoff, E., & Shan, S. (1971). Loss of behavioral thermoregulation after lateral hypothalamic lesions in rats. *Journal of Comparative and Physiological Psychology, 77*, 302–312.

Sclafani, A., Aravich, P. F., & Landman, M. (1981). Vagotomy blocks hypothalamic hyperphagia in rats on a chow diet and sucrose solution, but not on a palatable mixed diet. *Journal of Comparative and Physiological Psychology, 95,* 720–734.

Sherrington, D. (1906). *The integrative action of the nervous system.* New York: Scribners.

Skinner, B. F. (1938). *The behavior of organisms.* New York: Appleton-Century.

Smith, O. A. (1961). Food intake and hypothalamic stimulation. In D. E. Sheer (Ed.), *Electrical stimulation of the brain* (pp. 367–370). Austin, Texas: University of Texas Press.

Spiegel, T. A., Stunkard, A. J., Shrager, E. E., O'Brien, C. P., Morrison, M. F., & Stellar, E. (1987). Effect of naltrexone on food intake, hunger, and satiety in obese men. *Physiology and Behavior, 40,* 135–141.

Stellar, E. (1954). The physiology of motivation. *Psychological Reviews, 61,* 5–22.

Stellar, E., & Jordan, H. A. (1970). Perception of satiety. In D. A. Hamburg, K. H. Pribram, & A. J. Stunkard (Eds.), *Perception and its disorders* (pp. 298–317). Baltimore: William & Wilkins.

Stellar, J. R., & Stellar, E. (1985). *The neurobiology of motivation and reward.* New York: Springer-Verlag.

Stricker, E. M. (1983). Brain neurochemistry and the control of food intake. In E. Satinoff & P. Teitelbaum (Eds.), *Handbook of Behavioral Neurobiology, Vol. 6. Motivation* (pp. 329–366). New York: Plenum Press.

Stricker, E. M. (1984). Biological bases of hunger and satiety: Therapeutic implications. *Nutrition Reviews, 42,* 333–340.

Stricker, E. M., & Zigmond, J. J. (1976). Recovery of function after damage to central catecholamine-containing neurons: A neurochemical model for the lateral hypothalamic syndrome. In J. M. Sprague & A. N. Epstein (Eds.), *Progress in psychobiology and physiological psychology* (Vol. 6, pp. 121–188). New York: Academic Press.

Teitelbaum, P. (1955). Sensory control of hypothalamic hyperphagia. *Journal of Comparative and Physiological Psychology, 48,* 156–163.

Teitelbaum, P., & Stellar, E. (1954). Recovery from failure to eat produced by hypothalamic lesions. *Science, 120,* 894–895.

Thorndike, E. L. (1898). Animal intelligence: An experimental study of associative processes in animals. *Psychological Review Monograph, Suppl. 2,* 1–109.

Tinbergen, N. (1951). *The study of instinct.* New York: Oxford University Press.

Ungerstedt, U. (1971). Stereotaxic mapping of the monoamine pathways in the rat brain. *Acta Physiologiae Scandinavia Supplementum, 367,* 1–48.

Wise, R. A. (1984). Neuroleptics and operant behavior: The anhedonia hypothesis. *Behavior and Brain Sciences, 5,* 39–87.

Wise, R. A., & Bozarth, M. (1984). Brain reward circuitry: Four circuit elements "wired" in opponent series. *Brain Research Bulletin, 12,* 203–208.

Woods, S. C., Lotter, E. C., McKay, L. D., & Porte, D. (1979). Chronic intracerebroventricular infusion of insulin reduces food intake and body weight of baboons. *Nature, 282,* 503–505.

Woodworth, R. S. (1918). *Dynamic psychology.* New York: Columbia University Press.

Young, P. T. (1948). Appetite, palatability, and feeding habit: A critical review. *Psychological Bulletin, 45,* 289–320.

Zhang, D.-M., Bula, W., & Stellar, E. (1986). Brain cholecystokinin as a satiety peptide. *Physiology and Behavior, 36,* 1183–1186.

2

Thirst and Sodium Appetite

J. T. FITZSIMONS

INTRODUCTION

Retrospection allows the identification of certain themes in the literature that can help us in our thinking about current research problems. In this chapter I do not attempt a comprehensive description of past work on thirst and sodium appetite such as I have undertaken before (Fitzsimons, 1979). Instead I present a pastiche of a few well-known historical events and pioneering experiments in the literature of thirst and sodium appetite, chosen because they interest me and because they highlight some present-day experimental endeavors. It alludes to only a few of the issues raised by Epstein in his far-ranging chapter (Chapter 18), but it may give a foretaste of some of the intellectual pleasures to be found in the scientific study of thirst and sodium appetite.

1758: THE BLACK HOLE OF CALCUTTA

George Henry Lewes (1817–1878) gives an account of the suffering caused by severe thirst in Chapter 1 of his *Physiology of Common Life*, which was first serialized in the *Cornhill Magazine* and therefore aimed at a lay readership, and then published in two volumes in 1859 and 1860. Lewes was the friend of George Eliot (Marian Evans), with whom he lived for many years after his first wife had deserted him. *Adam Bede*, George Eliot's first full-length novel published in 1858, was dedicated, "To my dear husband, George Henry Lewes." She was devoted to him and established the G. H. Lewes studentship in his memory after his death. The studentship has been held by many famous physiologists, including A. V. Hill, Henry Dale, and Charles Sherrington.

J. T. FITZSIMONS   The Physiological Laboratory, University of Cambridge, Cambridge CB2 3EG, England.

The *Physiology of Common Life* contains descriptions of historical events that Lewes used to illustrate various physiological principles. The torment of thirst is vividly demonstrated by a famous incident known as the *Black Hole of Calcutta*, which had occurred in 1758 and had made a considerable impression on the public throughout the late 18th and 19th centuries. The event as described by Lewes was as follows. The Nawab of Murishidabad, Suraj-ud-Dowlah, had marched against old Fort-William with a considerable force of armed men because the Governor of Fort-William had imprisoned a merchant called Omychund. The Nawab had besieged and taken the fort and had imprisoned the 146 survivors of the garrison in the barrack room named the Black Hole under such appalling conditions that only 23 men emerged alive next morning; the rest were dead.

Mr. Holwell, the officer in command, described the horrors of this imprisonment in the *Annual Register* for 1758, from which Lewes drew his account.

> Figure to yourself the situation of a hundred and forty-six wretches, exhausted by continual fatigue and action, crammed together in a cube of eighteen feet, in a close sultry night in Bengal, shut up to the eastward and southward (the only quarter whence air could reach us) by dead walls, and by a wall and door to the north, open only to the westward by two windows strongly barred by iron, from which we could receive scarce any the least circulation of fresh air. . . . . . We had been but a few minutes confined before every one fell into a perspiration so profuse, you can form no idea of it. This brought on a raging thirst, which increased in proportion as the body was drained of its moisture. . . . . . . Water! Water! became the general cry. An old Jemmantdaar, taking pity on us, ordered the people to bring us some skins of water. . . . . . Until the water came I had not myself suffered much from thirst, which instantly grew excessive. . . . . . But from the water I had no relief; my thirst was rather increased by it; so I determined to drink no more, but patiently wait the event. I kept my mouth moist from time to time by sucking the perspiration out of my shirt-sleeves, and catching the drops as they fell like heavy rain from my head and face; you can hardly imagine how unhappy I was if any of them escaped my mouth. . . . . . No Bristol water could be more soft or pleasant than what arose from perspiration. (Lewis, 1859, pp. 38–43)

The reinforcement of thirst as soon as water arrived is noteworthy; the prospect of water was enough to remind the victims of their need for it. Indeed, it is common experience that when relief is imminent the sensation of need becomes more imperious. Animals drink in order to obtain oropharyngeal (and other pregastric) sensations such as the taste of water, and even a small quantity of water in the mouth reinforces drinking (see Rolls & Rolls, 1981). Nothing surprising here. How different life would be if opportunity were not reinforcing!

The severe thirst experienced by the victims was certainly of complex causation with an element of hypovolemia as well as cellular dehydration. The rate of sweating would presumably have been at or near maximum. The fluid losses produced by high rates of sweating are enormous. Schmidt-Nielsen (1964) gives values of rates of sweating as high as 4 L per hour for 2 to 3 hours, a loss of fluid comparable to the plasma volume each hour, though a more usual value for heavy sweating is about a liter per hour, still a heavy drain on the body fluid reserves of about 42 L in the standard 70-kg man. Sweat is a hypotonic solution of sodium chloride with small amounts of potassium, urea, and lactic acid. At high rates of sweating the concentration of NaCl may reach 100 mmol/L, so that severe sodium deficiency quickly develops.

The dehydration produced was therefore a mixed one of sodium and water. Since the fluid loss in heavy sweating is hypotonic, cellular dehydration thirst would have occurred early on. The sodium loss would have caused hypovolemia with a rise in plasma protein concentration and hematocrit. This would have been reflected in inadequate venous filling, diminished cardiac output, tachycardia, selective vasoconstriction with pale cold extremities, oliguria, orthostatic hypotension, and eventual circulatory shock. It is likely that the victims would have been subjected to such rapid dehydration that the rate as well as the absolute amount of fluid loss would have been a significant factor in their early collapse. The cause of death would have been circulatory collapse brought on by hyperpyrexia and excessive loss of fluid and electrolytes.

The greater satisfaction afforded by drinking sweat rather than water is interesting. Perhaps the victims were already sodium deficient owing to the prevailing climatic conditions in Calcutta at that time of the year so that the salt content of the sweat would have satisfied a need. The awful conditions of their imprisonment would undoubtedly have heightened the need for sodium, and this would have been manifest as a craving for salt similar to that shown by miners and others engaged in heavy physical work who lose large amounts of sodium in their sweat. The development of a sodium appetite is an appropriate response to this type of dehydration because sodium is, of course, needed in addition to water to restore the extracellular fluid and plasma volumes to their predehydration states.

The expression, *the Black Hole of Calcutta*, has entered the language. Lewes concludes the chapter on a not unjustifiable hyperbole.

> Such are Hunger and Thirst, two mighty impulses, beneficent and terrible, monitors ever vigilant, warning us of the need there is for Food and Drink—sources of exquisite pleasure and of exquisite pains, motives to strenuous endeavour, and servants to our higher aims. We are all familiar with them in their gentler aspects; may we never know them in their dreadful importunities! (p. 50)

1816: THE SHIPWRECK OF LA MÉDUSE

In his classic book, *Thirst*, A. V. Wolf gives an account of this famous shipwreck. The French frigate, *La Méduse*, with three other vessels sailed from the Isle d'Aix on the west coast of France on the 17th of June 1816 for Senegal on the west coast of Africa. On the second of July, near its destination, the port of St. Louis, the ship ran aground on a dangerous reef. After efforts to free the ship had failed, the chief officers and some of the crew took to the six boats on the fifth of July, while the remaining 147–150 persons—officers, soldiers, and crew—were placed on a large raft provided with six casks of wine but only two of water, and 25 pounds of biscuits and flour. It was intended that the raft should be towed by the boats, but the raft was soon abandoned by its boats. The boats reached the shore safely, but the raft drifted helplessly for 13 days, during which time the soldiers became enraged with wine and dehydration and started fighting among themselves and throwing survivors overboard. On the morning of the 13th day the remaining survivors were saved by one of the ships that had been sent out for them from St. Louis. Of the 147–150 persons on the raft, 15 only were saved, and of these five died shortly afterwards.

In the account of the suffering endured, the dreadful thirst experienced by the victims and their attempts to assuage it by drinking urine and sea water were uppermost in their thoughts. The following passage is taken from Wolf's book.

"A raging thirst, which was redoubled in the daytime, by the beams of a burning sun, consumed us; it was such, that we eagerly moistened our parched lips with urine, which was cooled in little tin-cups—we put the cup in a place where there was a little water, that it might cool the sooner; it often happened that these cups were stolen from those who had prepared them—the cup was returned to him to whom it belonged, but not till the liquid it contained was drank." M. Savigny, a young Swiss surgeon, and one of the narrators, observed that the "urine of some of us was more agreeable than that of others. There was a passenger who could never prevail on himself to swallow it; in reality, it had not a disagreeable taste; but in some of us it became thick and extraordinarily acrid; it produced an effect truly worthy of remark: namely, that it was scarcely swallowed when it excited an inclination to urine anew. We also tried to quench our thirst by drinking sea-water—M. Griffin used it continually; he drank ten or twelve glasses in succession. But all these means failed, or diminished our thirst only to render it more severe a moment afterwards." . . . . . "Three days passed in inexpressible anguish; we despised life to such a degree, that many of us did not fear to bathe in sight of the sharks, many of which swam about the raft; others placed themselves naked in the front of the machine which was still submerged; these means diminished, a little, our burning thirst." (Wolf, 1958, pp. 247–248)

The increasing thirst experienced by the victims on the raft presumably arose mainly from cellular dehydration as a result of lack of water to drink and excessive sweating caused by the sun. Those who drank the wine would have suffered further water losses because of alcohol diuresis. What of those who drank their own urine or sea water? On the basis of experiments in which they gave various solutes to dehydrated men, McCance, Young, and Black (1944) concluded:

The results of the present experiments, and a general consideration of the known facts about the specific gravity and other characteristics of the urine passed during a period of water deprivation, suggest that a dehydrated man can do himself no good by drinking his own urine. By so doing he is merely asking the kidneys to repeat work which they have already done and cannot be expected to do better. (p. 427)

The truth of this passage is self-evident when it is a question of drinking the maximally concentrated urine of dehydration, but more dilute urine such as that produced in diabetes insipidus or renal failure presumably could be drunk with some benefit. The question of drinking sea water is more complicated and is one that has excited controversy, not to say passion. Whether or not osmotically free water can be gained by drinking sea water depends first on its salinity. This varies from about 28% in the Dead Sea to less than 1% in the Baltic and the Gulf of Finland; the Atlantic, Pacific, and Mediterranean are about 3.6%, i.e. about 1,200 mosmol/kg water. Secondly, it depends on how much is drunk and whether there is fresh water available to mix with it. Of course, it also depends on whether you are a man or a small desert rodent!

In dehydrated man, the urinary concentration may reach 1,200–1,500 mosmol/kg water, but only about two-thirds of this consists of electrolytes, mainly NaCl and KCl. Taking values of 1,200 mosmol/kg water for sea water (consisting entirely of electrolytes) and 800 mosmol/kg for maximal urinary electrolyte concentration, this would mean that for every 100 ml of sea water ingested, a minimum of 150 ml of urine would have to be excreted to rid the body of all the extra salt, a net loss of 50 ml of body water. But in this calculation it is assumed that the renal concentrating ability is unaffected by the extra salt ingested, in other words, that maximal vasopressin secretion always means that the urine is maximally concentrated. This assumption is incorrect. The urinary concentration also depends on the rate of solute excretion.

Ingestion of a hypertonic salt solution such as sea water causes an osmotic diuresis because the increased amounts of unreabsorbed solute passing along the renal tubule slow down the rate at which water is removed from the filtrate. In extreme cases, the increased flow of filtrate results in a urine osmolality approaching that of plasma. In *Reflections on Renal Function*, one of the best books on renal physiology ever written, Robinson (1954, p. 76) describes an experiment on a normal student who was producing 0.5 ml/min of concentrated urine after a period of water deprivation. This student responded to an intravenous infusion of 250 ml of 10% NaCl by increasing his urine flow to 24 ml/min, a greater diuresis than could be obtained by drinking water. At the same time the concentration of the urine dropped, and at the height of the diuresis it was little greater than that of plasma. As Robinson later pointed out (1988, p. 172), "vasopressin is strictly an anti-*water*-diuretic hormone." It cannot prevent an osmotic diuresis such as is caused by ingestion of sea water.

Whether or not ingestion of some sea water by itself might be temporarily helpful is disputed. It is obvious that where the salinity is low the water is potable. But unfortunately, in those circumstances where the sea water is dilute enough to drink, e.g., near the icecaps, death is likely to occur rapidly from exposure, whereas in warmer climates, e.g., the Pacific Ocean, the water is too concentrated to profit from drinking it. Ladell (see Wolf, 1958, pp. 266 and 276; Ladell, 1965) suggested that small amounts of even hypertonic sea water might be beneficial because the extracellular hypertonicity serves to maintain the extracellular and plasma volumes, though at the expense of the cellular water. However, the benefit is short-lasting, and it is paid for by earlier eventual collapse when nervous symptoms and respiratory paralysis occur as the cells reach their limiting osmotic pressure.

Hervey and McCance (1952) and many others have repeatedly pointed out the dangers of drinking sea water. Whatever the possible gain in the early stages, wartime records show that the mortality of castaways who drank sea water was 10 times that of those who kept strictly to their limited supplies of fresh water (see Ladell, 1965). Hervey and McCance (1952) argued against the drinking of sea water in the following terms:

> . . . . . ."addition of sea-water to fresh-water rations is actually deleterious and likely to shorten survival. From the first, it considerably accelerates the increase in the total osmotic pressure of the body; the improved water balance which it does initially produce adds fluid only to the extracellular space and is accompanied by the withdrawal of water from the cells; and, finally, over a longer period the gain in water balance is replaced by a loss which, once started, will continue at an increasing rate. (p. 544)

Drinking sea water exacerbates the thirst that it is meant to relieve, and the victim enters a vicious circle of increasing dehydration from osmotic diuresis and worsening thirst. The magnesium and sulfate ions in sea water contribute to the fluid and electrolyte disturbances by causing vomiting and diarrhea so that there may also be a degree of hypovolemia as well as severe cellular dehydration.

The renal response to hyperosmotic solutions is a factor that must be taken into account when looking at stimulus–drinking response relations in osmometric thirst. It may also explain why sodium-depleted animals tend to drink water before they take salt, as is discussed below.

1936: McCance's Experimental Sodium Chloride Deficiency in Man

The regulation of cell volume and blood volume are interdependent. Both depend on the control of fluid and electrolyte intake and excretion; inevitably the controlling mechanisms overlap in their functions. As far as the effect of cellular dehydration on thirst is concerned, many experiments since Gilman's (1937) demonstration that hypertonic NaCl causes more drinking than the same osmotic load of hyperosmotic urea have confirmed the importance of cellular dehydration as a stimulus to thirst. In normal hydration, a diminution in cellular water of 1–2% is enough to initiate drinking. Diencephalic osmoreceptors share in the general cellular dehydration, and they ensure that, in conjunction with renal osmoregulation, enough water is drunk to restore the cellular water to normal. Electrolyte depletion is not a feature of pure cellular dehydration, and any loss is generally replaced at leisure in the course of feeding.

Extracellular dehydration also arouses thirst, but the experimental validation of this has only come in the past 30 years, though the thirst of severe blood loss or cholera has been known for a very long time. Other instances where thirst of extracellular origin occurs are severe sodium deficiency, sequestration of extracellular fluid produced by intraperitoneal or subcutaneous injection of hyperoncotic colloid, interference with venous return to the heart by obstructing the inferior vena cava, and congestive heart failure. The common factor in all of these is hypovolemia, actual or simulated, signaled by vascular stretch receptors. Since the vascular and interstitial fluid compartments are directly coupled through the Starling capillary filtration–reabsorption system, control of the vascular compartment ensures control of the whole extracellular compartment. We can therefore describe this type of thirst as hypovolemic thirst. The renal renin–angiotensin system contributes to hypovolemic thirst, but its participation is not essential.

The fall in plasma volume required to arouse hypovolemic thirst approaches 10%, a considerably greater percentage change than for initiating cellular dehydration thirst. The threshold for getting rid of extracellular fluid is similarly about a 10% increase. There are good reasons for this. First, the circulation requires such varied rates and distribution of blood flow it would be an embarrassment if vascular stretch receptors did not have an ample margin within which to control the circulation before bringing into play intake and renal mechanisms. Second, there is an ample reserve of fluid in the interstitial fluid compartment, which can be quickly moved into the circulation when the need arises through adjustments in the Starling filtration–reabsorption mechanism. Conversely, in hypervolemia plasma is made to leave the circulation.

Invariably in extracellular dehydration, there is a delayed increase in sodium intake after the initial increase in water intake. This second behavioral response to hypovolemia is an appropriate response because sodium is needed as well as water to restore the extracellular fluid volume to normal. In animal experiments the sequence of early water intake/delayed sodium intake seems clear enough, but in man neither the behavior pattern nor the sensation responsible is straightforward. Unfortunately, only man can share his experience of the sensation with others.

The complexity of the sensation aroused in extracellular dehydration is well seen in a classical study on experimental sodium deficiency performed by R. A.

McCance on himself and three other subjects, described in *The Proceedings of the Royal Society* (McCance, 1936a) and in the Goulstonian lectures published in the *Lancet* (McCance, 1936b). Sodium deficiency was brought about by a combination of sweating induced by radiant heat baths and the taking of a salt-free diet. At least 25–30% of the body's extracellular NaCl was removed in this way, and the blood became dark and viscous, and the hematocrit and plasma proteins rose, indicating hypovolemia. McCance described the sensations as follows:

> As the deficiency developed all three subjects lost weight. . . . . . Their sense of flavour and taste was affected. E. interpreted this aberration or lack of sensation as thirst. She complained of it constantly and drank freely but without obtaining any relief. R.A.M. recognized the feeling as distinct from thirst. His mouth was not unduly dry but food was tasteless, even highly flavoured food, and this was the more noticeable because such foods were eagerly sought to make the meals more appetizing. . . . . . R.B.N. was not so much troubled by this symptom but felt it from time to time. He noted once that he was "Thirsty all morning—drank a lot but water seems to make little difference" and on another day reported that he had a "funny feeling in the mouth." (1936a, pp. 249–250)

There was no doubt that the "peculiar sensation in the mouth" was near enough to the sensation of thirst to cause some of the subjects to become polydipsic. Subsequent animal experimentation has fully confirmed that sodium-deficient animals become polydipsic (e.g., Cizek, Semple, Huang, & Gregersen, 1951), but the failure of water to relieve symptoms in McCance's experiment suggests that the sensations experienced during sodium deficiency were more complicated than thirst alone. The victims of the Black Hole of Calcutta also seemed to sense a need for something more than water, finding satisfaction in drinking the salty sweat that was being produced in abundance. However, a little surprisingly, only one of McCance's subjects complained of increased sodium appetite.

> R.B.N. longed for salt and often went to sleep thinking about it. R.A.M. felt no specific craving for salt and had difficulty in convincing himself that taking salt would at once make him feel all right again. (1936a, p. 251)

Although many clinicians may have personal knowledge of patients with adrenal insufficiency who show unusual craving for salt, Richter (1956) points out that it remains a problem why all patients with Addison's disease do not show such craving. Denton (1982) provides an explanation for the capriciousness of sodium appetite in man. If as is likely salt appetite were highly developed in early herbivorous man, it could follow from the advent of carnivorous behavior that the selection pressure favouring salt appetite may have dwindled to a low level because of adequate sodium intake in meat. Even so, in McCance's experiments NaCl was found to be refreshing in a way that water was not, as the following passages indicate.

> At the close of my second experiment I tried the effect of washing out my mouth with salt and water and found it very refreshing, and thought my sense of flavour was thereby restored. . . . . . I believe these observations are of clinical interest, for I think it quite possible that the "thirst" so often complained of by patients with intestinal obstruction may really be this curious loss of taste brought about by salt deficiency. At all events Mr. J. B. Hunter has informed me that a patient's "thirst" is often relieved by hypertonic saline in amounts too small to relieve the general dehydration. (1936b, p. 824)

> Recovery was quite dramatic. Half an hour after eating 15 gm of NaCl with bread, butter, and an egg E.'s sense of flavour and taste had returned, although no fluid had

been taken. This she spoke of as a quenching of her thirst. Genuine and almost unbearable thirst supervened later and was only satisfied by copious draughts of water. (1936a, pp. 251–252)

Thirst was apparently experienced while they were sodium-deficient by at least two of the subjects (E. and R.B.N.), so that rapid recovery after taking NaCl implies satisfaction of thirst as well as relief of other symptoms, including increased sodium appetite when this occurred. Relief of this thirst could be explained by the NaCl-induced shift of water from the cellular space to the extracellular space, leading to restoration of plasma volume and suppression of renin secretion. The "genuine" thirst—the qualification is significant—that supervened was, of course, the classical cellular dehydration thirst that would have been absent up to then because the cells were overhydrated. Since the subjects were doctors or medical students, it may have been that they were only too aware of current orthodox opinion that increased osmolality caused thirst whereas sodium deficiency did not.

The water drunk in response to the hypovolemia of severe sodium deficiency does not restore the blood volume to normal, though it will help toward restoration. As the water enters the body it is shared between the cellular and extracellular compartments according to their respective solute contents, and, because of the reduced extracellular sodium content, proportionately more water enters the cellular space than the extracellular space compared with what happens in the nondeficient state. The cellular space therefore becomes even more overhydrated. The overall concentration of the body fluids is already lower than normal because, as McCance describes, there has been relative preservation of volume at the expense of concentration.

It will be noted that at first a removal of sodium was followed *pari passu* by a loss of an equivalent amount of water. So long as this was so, red cell counts may have risen and other signs of anhydraemia appeared, but *no change in the osmotic pressure* of the body fluids had taken place. Further forced removal of sodium was *not* followed by an equivalent loss of water. The volume of the body fluids was relatively well maintained but *only at the expense of the osmotic pressure*, which fell. (1936a, pp. 262–263)

Decreased water excretion in sodium deficiency depends on vasopressin and vasopressin-independent mechanisms (see Shapiro & Anderson, 1987). Although the threshold for vasopressin release in response to hypovolemia is higher than it is for increases in osmolality, the hypovolemia—vasopressin relationship is steep, and the amounts of hormone released are much more than required for maximum antidiuresis. Reduced delivery of filtrate to the distal renal tubular segments as a result of avid proximal tubular sodium conservation also limits free water excretion independently of vasopressin. An inability to excrete a water load is therefore a feature of sodium deficiency and is the basis of a well-known test for adrenal insufficiency. The assumption has always been that limiting renal water losses in this way protects the extracellular volume, which is partly true, but it is also true that the hypovolemia has to be considerable before this happens and that some of the volume conserved is responsible for cellular overhydration.

Increased water intake is the only way in which volume can be restored in sodium deficiency; renal conservation of water without water intake only slows the rate of fluid loss. But of course, any ingested water is distributed throughout the body water, and only part of it is going to the extracellular compartment where it is required. Following absorption of water, the concentration of the body fluids falls still further, and drinking stops before the extracellular fluid volume has been brought back to its predeficiency state. It is now necessary to restore the extracellular sodium content, overall osmolality, and extracellular water to normal and to remove excess cellular water. Restoration of the extracellular sodium content goes a long way to achieving these objectives. The increase in extracellular sodium brings the osmolality up to normal and allows water to be finally shifted out of the overhydrated cells into the extracellular space where it is needed. It seems to make good physiological sense to phase in sodium appetite and increase the electrolyte intake once the required volume of water has been ingested. If sodium were taken prior to water, there is a risk of too much solute being introduced into the body relative to the reduced body water. With the shift of water out of the cells this might be enough to cause an osmotic diuresis and worsening of the dehydration.

1821–1832: Dupuytren, Magendie, O'Shaughnessy, and Latta and the Relief of Thirst by Intravenous Fluid

The relief of thirst by the use of intravenous fluids goes back to the famous experiment of Guillaume Dupuytren, described by Rullier in 1821 in the *Dictionaire des Sciences Médicales:*

> ...... M. le professeur Dupuytren a souvent apaisé la soif d'animaux soumis à ses expériences, et exposés plus ou moins longtemps à l'ardeur du soleil, en leur injectant de l'eau, du lait, du petit-lait et divers autres liquides dans les veines. (p. 469)

> [Professor Dupuytren has often assuaged the thirst of animals subjected to his experiments and exposed for shorter or longer times to the force of the sun, by injecting water, milk, whey, and various other liquids into the veins.]

François Magendie (1821), who was Dupuytren's contemporary in Paris, describes in the first number of the *Journal de Physiologie Expérimentale*, which he had founded, how he had collaborated with Dupuytren in the past in trying to relieve the symptoms of rabies in a young man by injecting a gummy extract of opium into the veins. The treatment failed, as it had in rabid dogs. But in the same article, Magendie reported some success in calming a rabid dog by injecting about 60 oz of water intravenously. Magendie pointed out that the blood of the enraged animal had become thick and lacking in serum because the animal had been unable to drink since the onset of rabies, and at the same time its pulmonary and cutaneous water losses had been more abundant than usual because of the agitation. Magendie had previously observed that bodily functions are depressed when there was "pléthore aqueuse artificielle." From this he argued that intravenous water would have a calming effect on the dog's symptoms. This in fact is what happened for a time. The dog remained calm, it did not bark, and it lay down apparently peacefully asleep until about 5 hours after the injection, when it died from some respiratory embarrassment.

Magendie's approach to the problem of treating a human patient with rabies was similar when 2 years later he was called to a consultation at the Hôtel-Dieu de Paris. In his description of the case (Magendie, 1823), he referred to his experience with the rabid dog, recollecting the beneficial effect of intravenous injections of warm water. He observed that the patient before him was unable to swallow because of violent convulsions—the mere sight of liquid was enough to excite the most violent agitation—and he decided to try the intravenous water treatment. To his gratification the patient was relieved of his symptoms by intravenous injection of 2 lb of water (about a liter) over a period of 25 minutes, although unfortunately the patient later died. Within less than an hour after the injection had been completed, the pulse rate had fallen from 150 to 80 beats per minute. Magendie (1823) described his patient in the following terms.

> Le malade reprit l'usage de ses sens et de sa raison, le calme de l'esprit remplaça la fureur, les yeux reprirent une expression naturelle, les mouvements convulsifs s'arrêtèrent, et, chose merveilleuse! il put boire sans aucune difficulté un verre de liquide qu'on lui présenta; enfin le changement d'état fut complet. (p. 386)

> [The patient regained the use of his senses and his reason, calmness of spirit replaced the fury, the eyes recovered a natural expression, the convulsive movements stopped, and, wonderfully, he was able to drink without any difficulty a glass of water that was presented to him; in short, the change in condition was complete.]

Magendie did not explicitly state that the victim's thirst was relieved by the intravenous water, although others assumed reasonably enough that Magendie's treatment did just that. For example, in his book, Wolf (1958, p. 41) stated that Magendie relieved thirst in hydrophobia by injecting water intravenously, and Schiff (1867) wrote:

> Les injections d'eau dans les veines ont été employées avec succès chez l'homme, dans un cas d'hydrophobie rapporté par Magendie. Le malade, ne pouvant satisfaire la soif violente qui le tourmentait, à cause du spasme des muscles de la déglutition que produisaient la vue et le contact de l'eau, fut très-promptement soulagé par l'injection dans les veines de ce liquide. (p. 41)

> [Injections of water into the veins have been successfully employed in man, in a case of rabies reported by Magendie. The patient, not being able to satisfy the severe thirst that was tormenting him because of spasms of the muscles of deglutition that the sight and contact of water produced, was very promptly relieved by injection of this liquid into the veins.]

Magendie's research interests and activity were devoted mainly to the central nervous system. It seems reasonably clear that his therapeutic goal was to give water not in order to restore the blood volume and maintain the circulation but in order to quiet down the overactive nervous system. The absence of any statement about the relief of thirst is therefore understandable.

A more striking example of relief of thirst by intravenous fluids was provided by a new treatment for cholera victims that found temporary favor shortly after this. During the course of the 19th century a number of cholera epidemics had spread across Asia and Europe and decimated the local populations. In the great epidemic of 1832, which broke out in Sunderland in England in October 1831, a bewildering variety of unsuitable treatments were in vogue, but particularly unsuitable were the practices of bleeding the victims and administering purges to them. The theory of bleeding was based on the great systems of humoral pathology that had been developed from the Middle Ages. However, the practitioners of bleeding came up against the irrefutable practical obstacle that the patients were so severely dehydrated that it was very difficult to bleed them.

More rational was the approach of the medical men of the Broussian or pathological school who traced symptoms to their source in the body and tried to combat them as they arose (Morris, 1976). This analytical approach led the nearest to modern treatment in the work of William O'Shaughnessy of Limerick and Thomas Latta of Leith, who introduced the practice of infusing saline into the veins in order to relieve symptoms in cholera. In December 1831, O'Shaughnessy published in the *Lancet* the results carried out in Newcastle of his analysis of the blood of cholera victims, which he found to be deficient in water and certain salts. He suggested that in desperate cases of cholera, tepid water holding a solution of the normal salts of the blood should be injected into the veins. His report, *The Chemical Pathology of Malignant Cholera,* was received by the Central Board of Health in December and under their authority given publicity in the medical journals.

The idea of giving medicinal remedies into the veins was recent at the time and very much in the air. A translation of Magendie's rabies report had appeared in the *Lancet* in 1824 and had received widespread publicity in the newspapers, and his animal work was well known and referred to in a leading article in the same issue of the *Lancet* in which Latta's intravenous treatment of cholera appeared. Magendie had also visited England in 1824 and had obtained first-hand impressions of the cholera epidemic when he visited Sunderland in December, 1831. When cholera broke out a little later in Paris, Magendie was constantly at the Hôtel-Dieu treating his patients. He was well aware of the depressed state of the circulation and of the difficulty in obtaining blood from the victims. But his treatment consisted of "punch" and skin frictions. Replacement of lost fluid by intravenous fluids did not find a place in his practice.

On the basis of O'Shaughnessy's work, Latta decided to try to restore the lost fluids and salts of cholera victims. Administration by mouth or *per anum* having failed, in one patient he injected about 3½ L of a rather hypotonic mixture of muriate of soda (0.22–0.34% sodium chloride) and subcarbonate of soda (0.08% sodium bicarbonate) into the veins with successful results. Latta's description of the effect of this treatment was published in the *Lancet* of 1832 under the heading, "Letter from Dr. Latta to the Secretary of the Central Board of Health, London, Affording a View of the Rationale and Results of his Practice in the Treatment of Cholera by Aqueous and Saline Injections," from which the following short excerpt is taken:

> At first there is but little felt by the patient, and symptoms continue unaltered, until the blood, mingled with the injected liquid, becomes warm and fluid; the improvement in the pulse and countenance is almost simultaneous, the cadaverous expression gradually gives place to appearances of returning animation, the horrid oppression at the praecordia goes off, the sunken turned-up eye, half covered by the palpebrae, becomes gradually fuller, till it sparkles with the brilliance of health, the livid hue disappears, the warmth of the body returns, and it regains its natural colour—words are no more uttered in whispers, the voice first acquires its true cholera tone, and ultimately its wonted energy, and the poor patient, who but a few minutes before was oppressed with sickness, vomiting, and burning thirst, is suddenly relieved from every distressing symptom; blood now drawn exhibits on exposure to air its natural florid hue. (Latta, 1832, p. 275)

The new treatment faced many difficulties. Critics and supporters could not have known of the risks of bacterial contamination, which led to septicemia and probably accounted for the varying success of the treatment. There was a willingness to try different treatments at the start of an epidemic, the results of which were difficult to assess because the disease always seemed to be less virulent at the end than at the beginning. Chemical pathology was a relatively new discipline, and other more firmly entrenched traditions dominated medical practice, particularly those of the medical establishment in London. Unfortunately, as the cholera epidemic subsided, the use of intravenous fluids fell out of use. As a treatment of severe dehydration it was ahead of its time, before the importance of the composition of the infused fluid and the necessity for sterility were understood.

The description by Latta and others of severe thirst associated with the impending circulatory collapse in cholera remained, until fairly recently, one of the few accounts of what we would now call hypovolemic thirst. Equally impressive was the relief of this "burning thirst" by intravenous fluid, but here recent work has shown that the position is more complicated than these early observations suggest, although rather few experiments have been performed, and few if any on the effect of intravenous fluid on hypovolemic thirst.

Normally the relief of thirst depends on the sequence of events that accompany drinking. Adolph, who did so much of the classical quantitative work on thirst and satiety, talked of "two satieties: (1) alimentary satiety, which leads to cessation of ingestion, and (2) tissue satiety, which comes later and erases the deficit" (Adolph, 1980, p. 339). In other words, the restoration of the body fluids to the predehydration state is the end result, but this is preceded by oropharyngeal sensations such as the taste of water and the muscular effort of swallowing, gastric distension, and gut and hepatic portal stimulation by water, all of which contribute to the termination of drinking (see Rolls & Rolls, 1981). Bypassing the preabsorptive signals by infusing water or other fluid directly into the circulation removes the precision of control. After the infusion, the animal takes some time to become aware that it is no longer thirsty. Holmes and Montgomery (1960) found that injection of fluids intravenously, starting 40 minutes before inducing thirst with intravenous hypertonic NaCl, did not inhibit drinking, whereas the same solutions put directly into the stomach through a fistula had a significant inhibitory effect on drinking. Intravenous infusion of water seems less effective at suppressing spontaneous drinking than water given at the same rate by intragastric infusion (Rowland & Nicolaïdis, 1976).

Preabsorptive sensations associated with taking water therefore play a privileged role in the relief of thirst caused by fluid deficit, and they prevent overdrinking. Equally important, their importuning provides the incentive to drink in the absence of any need for water, and they therefore ensure a fail-safe intake of water, i.e., an intake of water that is not dependent on an existing fluid deficit. This type of drinking is well described as *prophylactic* (Peters, 1980). Beyond the fact that prophylactic drinking must be of the utmost biological importance for a terrestrial organism, we know little of the underlying mechanisms. This is a field where fresh work and insights are required.

1954: THE KIDNEY AND THIRST REGULATION

The idea that the kidney may regulate thirst came from work by Linazasoro, Jiménez Díaz, and Castro Mendoza published in the little-known *Bulletin of the Institute of Medical Research (Madrid)* in 1954. Since this publication is not easily obtained, a summary of some of the results obtained by the Spanish workers is given in Tables 1 and 2. Linazasoro and his colleagues found that bilaterally

TABLE 1. THE EFFECTS OF NEPHRECTOMY AND RENAL EXTRACT (SAR) ON WATER INTAKE AND CHANGE IN BODY WEIGHT[a]

	n	Water drunk (mL per 100 g body wt.)	Change in body wt. (g per 100 g body wt.)
Normal rats	20	8.59 ± 0.28	−0.55 ± 0.10
Nephrectomized	24	5.13 ± 0.13	−3.00 ± 0.19
Nephrectomized + SAR	25	7.09 ± 0.20	2.73 ± 0.26

[a]Water intake was assessed at 24 hours and body weight at 48 hours in fasting rats with water available to drink. Results are mean values ± S.E.M. and are recalculated from the results of Linazasoro et al. (1954).

neprectomized rats kept without food for 48 hours but allowed free access to water lost more weight and drank less water than normal rats,

> . . . . . . in spite of the absence of salt output, which ought to result in an increase in extracellular osmotic tonicity and, therefore, in the water requirements, they drank almost half as much as normal animals. Such a lack of thirst appeared to be conditioned by the absence of the kidneys in such animals. (pp. 55–56)

These effects of nephrectomy could be prevented by injecting "active renal extract from pig (SAR)" (Table 1), for which, unfortunately, no details of the preparation are given.

> Instead of the weight loss and decreased water intake seen in merely nephrectomised animals, there was found an increase in thirst superior to extrarenal output which caused the animals to gain weight. (pp. 57–58)

On the basis of these findings, Linazasoro and his colleagues suggested that

> . . . . . . water balance and osmosis are regulated through the action of hormones but when such an action is not enough to prevent the concentration of intracellular fluids some substance is liberated from the kidney itself that conditions the sensation of thirst by acting on the centres. This would be a further example of the self-regulating capacity of the organs: the kidney, whose margin of safety is reduced to a minimum, contributes to osmotic regulation by liberating a thirst hormone. (p. 59)

In a further series of experiments, Jiménez Díaz, Linazasoro and Merchante (1959) confirmed the thirst-reducing effects of nephrectomy in rats and reported similar results in dogs. They also confirmed that the effects of nephrec-

TABLE 2. THE EFFECT OF RENAL EXTRACT (SAR) OR RENIN ON NaCl INTAKE AND CHANGE IN BODY WEIGHT[a]

	n	NaCl drunk (mL per 100 g body wt.)	Change in body wt. (g per 100 g body wt.)
Nephrectomized	12	7.74 ± 0.33	4.91 ± 0.70
Nephrectomized + SAR	12	13.38 ± 0.53	11.40 ± 0.87
Nephrectomized + renin	15	8.07 ± 0.22	4.44 ± 0.69

[a]Changes were assessed after 24 hours in male and female rats with physiological saline available to drink. Results are mean values ± S.E.M. and are recalculated from the results of Jiménez Díaz et al. (1959).

tomy could be reversed by injection of the renal extract. Surprisingly, however, they found that renin, prepared by the method of Braun Menéndez, was ineffective in nephrectomized rats (Table 2). They wrote:

> The thirst-stimulating action of renal extracts is not due to their renin content, for if the latter is injected instead of the whole extract, it has practically no effect on voluntary ingestion of liquid. . . . . . The intense thirst-reducing effect of nephrectomy in rats and dogs is confirmed by this new series of experiments. . . . . . The thirst-regulating effect and the effect on the permeability of cells, are regularly canceled out by injection of kidney extracts. On the other hand this canceling out effect is not achieved by injection of renin, which seems to show that the substance produced by the kidneys, whose function it is to do this, is not renin. (pp. 66–67)

In view of subsequent work that has established renin as an extremely potent dipsogenic substance, it is puzzling that Jiménez Díaz, Linazasoro, and Merchante were so dismissive of the possibility that their SAR contained renin. A possible reason for their failure to obtain any fluid intake after renin, as opposed to the large intakes after SAR, is that in this series of experiments the only solution offered to the animals to drink was "physiological saline serum"; water was apparently not offered. If SAR possessed much greater renin activity than the semipurified renin, and/or it possessed some additional components that affected NaCl intake, an explanation on the following lines might apply.

The animals of Jiménez Díaz, Linazasoro, and Merchante were made anuric by bilateral nephrectomy. It is known that eliminating renal function greatly reduces formalin-induced intake of 0.9% or 2.7% NaCl in the rat offered the choice of water and NaCl (Fitzsimons & Stricker, 1971) and also the NaCl intake that follows adrenalectomy (Fitzsimons & Wirth, 1978). On the face of it this might be taken to mean that renin secreted by the kidney is responsible for the appetite. Certainly, when components of the renin–angiotensin system are injected directly into the brain, NaCl intake is stimulated as well as water intake (Chiaraviglio, 1976; Avrith & Fitzsimons, 1980). However, eliminating renal excretory function by bilateral ureteric ligation, which results in increased activation of the renin–angiotensin system, also greatly reduces sodium appetite after formalin or adrenalectomy. Attempts to restore the sodium appetite of anuric animals with systemic renin have met with mixed success. Chiaraviglio (1976) succeeded in reinstating about one-third of the expected 1% NaCl intake in sodium-depleted nephrectomized rats by injecting renin intraperitoneally. But in the work referred to above, the effects of injecting renin or angiotensin systemically, or of stimulating renin release with isoproterenol in animals made anuric by ureteric ligation, were on the whole unsuccessful, though these procedures caused increased water intake. Even in normal animals, systemic renin or pharmacological activation of the renin–angiotensin system has rather little effect on NaCl intake, though both will stimulate water intake.

Therefore, in anuric animals, for reasons unknown, some consequence of uremia causes a sodium aversion that cannot easily be overcome with systemically administered renin or by activation of the animal's own renal renin–angiotensin system. The amount of semipurified renin administered to nephrectomized rats by Jiménez Díaz and his colleagues may have been insufficient to overcome this aversion, whereas nephrectomized SAR-treated animals drank the "physiological saline serum" because the crude SAR contained more renin activity. We also know that crude renal extracts or very large doses of renin cause increased vascular permeability, resulting in edema, ascites, and pleural effusions. The resulting hypovolemia, when added to the direct stimulating effect of renin on sodium appetite, may be sufficient to overcome the sodium aversion of uremia. Had Jiménez Díaz and his colleagues increased the dose of renin or offered their animals water to drink as well as NaCl, they might have found an increase in water intake after renin, perhaps less than after SAR, and they would presumably have concluded that at least some of the effects of SAR were caused by the renin it contained. There is obviously some unfinished work here. The kidney may yet yield surprises in its possible role in the regulation of sodium appetite.

In contrast to the negative findings of Jiménez Díaz, Linazasoro, and Merchante with renin, others who were working on renal extracts in relation to hypertensive vascular disease at this time were impressed by the effects of renal extracts containing renin on drinking behavior. Describing experiments on nephrectomized, renin-injected dogs allowed either water or 1% NaCl to drink, Masson, Plahl, Corcoran, and Page wrote in 1953:

> The syndrome elicited by bilateral nephrectomy in dogs is familiar; it was intensified by administration of renin. Especially noteworthy was the thirst manifested by renin-treated animals; most of these drank and vomited in alternation until they went into coma. (p. 87)

Masson and his colleagues gave no figures for fluid intake in this paper, and their interpretation of the mechanism of this effect was quite different from that of the Spanish workers. There was no suggestion of a specific humoral thirst factor; rather, the thirst was regarded as secondary to an effect of accelerated vascular damage and increased permeability on tissue electrolytes. They continued:

> However, we were impressed by the thirst elicited by renin in nephrectomized dogs; this thirst can best be attributed to an extrarenal action of renin on the distribution and concentration of electrolytes. (p. 96)

In further work, this time on nephrectomized rats, Masson and his colleagues (Masson, del Greco, Corcoran, & Page, 1956) reported that preparations of renin caused serous effusions and increased drinking of either water or NaCl; in this paper values for intakes of water or NaCl are given. They concluded that the formation of serous transudates was the consequence of a nephrectomy-potentiated effect of renin on vascular permeability and that

> . . . . . . the thirst induced by renin in nephrectomized rats is not the result of primary stimulation of thirst centers, nor of increased serum non-urea osmolarity. Rather it is due to the oligemia consequent on transudate formation. (p. 199)

Asscher and Anson (1963) obtained similar results with renal extracts on the water intake of nephrectomized rats. They described their extract as a "vascular permeability factor of renal origin," and they concluded that increased thirst was attributable to the fall in plasma volume caused by leakage of fluid into the tissues.

Except for the work of Linazasoro, Jiménez Díaz, and Castro Mendoza, none of the earlier work on renal extracts was regarded as indicating a role for the kidney in the control of drinking behavior. Renin (or some other permeability factor) was thought to cause thirst indirectly, most likely by inducing hypovolemia, and not by direct action of angiotensin on thirst centers. But most if not all of the early work on renal extracts was aimed at the elucidation of the vascular pathology of hypertensive disease and not to throw light on mechanisms of thirst. In contrast, Linazasoro, Jiménez Díaz, and Castro Mendoza were primarily interested in thirst and were certainly thinking in terms of a hormonal mechanism that would enable the kidney to safeguard the body against dehydration. They deserve the greatest credit for this important idea, even though the thirst hormone was regarded as contributing to the regulation of osmotic thirst rather than hypovolemic thirst. It is unfortunate that in the second paper, for reasons that I have tried to analyze, Jiménez Díaz, Linazasoro, and Merchante excluded renin as the renal thirst hormone.

It was another 10 years before the links missing in the 1954 idea of Linazasoro, Jiménez Díaz, and Castro Mendoza began to be sorted out. In experiments in which it was established that caval obstruction caused drinking in the rat (Fitzsimons, 1964), this stimulus to thirst was found to be relatively ineffective after nephrectomy, though still effective in the rat made anuric by ureteric ligation, leading to the suggestion that ". . . . . . the kidneys play a part in the response, possibly by release of renin or some other renal factor . . ." (p. 480). It was becoming evident at this time that caval obstruction, like a number of other procedures such as hemorrhage, sodium depletion, and sequestration of extracellular fluid by hyperoncotic dialysis, caused thirst, probably by altering the sensory discharge reaching the central nervous system from cardiovascular stretch receptors. Since actual or simulated hypovolemia was known to release renin, it was reasonable to explore the possibility that the renin–angiotensin system plays a role in hypovolemic thirst. Support for this view came with the finding that the renal dipsogen is identical with renin (Fitzsimons, 1969). Its effects are mediated by angiotensin II, which was shown to be dipsogenic when infused intravenously in water-replete rats (Fitzsimons & Simons, 1969). Angiotensin II was established as the most potent dipsogenic substance known (and has remained so to this day) when it was injected directly into the rat's diencephalon (Epstein, Fitzsimons, & Rolls, 1970).

1990: Renin–Angiotensin Systems, Thirst, and Sodium Appetite

The present view on how the renal renin–angiotensin system is involved in thirst is essentially the same as that originally put forward 20 years ago. Through a reduction in stretch receptor information from the underfilled vasculature, hypovolemia deinhibits the neural structures in the diencephalon responsible for drinking behavior. This arouses thirst and sodium appetite and results in increased intakes of water and sodium salts. It is possible that severe hypovolemia activates other types of receptors that stimulate drinking through an increase in nerve impulse traffic to thirst neurons. Hypovolemia also causes increased renal renin secretion and generation of angiotensin II, which acts directly on diencephalic structures, notably the subfornical organ and the vascular organ of the lamina terminalis, increasing the responsiveness of neural drinking systems to the altered receptor inputs from the vasculature. It was originally suggested that hypovolemic thirst was the outcome of synergistic action between nonendocrine and endocrine mechanisms. Later the idea of synergy between nonendocrine and endocrine mechanisms was extended to the sodium appetite of hypovolemia. This simple view of the genesis of hypovolemic thirst and sodium appetite lacks quantitative precision and has been rightly subjected to vigorous and skeptical scrutiny (e.g., Stricker, 1978; Denton, 1982).

Some of the questions that Stricker posed about angiotensin-induced drinking behavior have only been partly answered. We are uncertain of all the circumstances in which angiotensin-induced drinking is physiologically significant. We have very few clear ideas on the magnitude and time course of the contribution of angiotensin in those types of drinking behavior where it appears to be involved. There are other unresolved problems. Among the most pressing is the role, if any, of cerebral renin in thirst and sodium appetite and, if it is involved, its relation to renal renin in causing increased intakes of water and NaCl. Is one renin–angiotensin system more important than the other for a particular behavior or for responding to a particular pattern of fluid and electrolyte imbalance? What are the differences in mechanism between angiotensin-induced intake of water and intake of NaCl, and why is the increase in intake of NaCl slower in onset and longer lasting? Is the commonly used requirement that an animal drinks more of an aversive concentration of NaCl solution a satisfactory method of demonstrating an increase in sodium appetite?

Let us now summarize what is known about angiotensin-induced drinking and first water intake. Angiotensin-induced water intake is a short-latency response the characteristics of which suggest that it is caused by direct stimulation of neurons by the peptide. It is clear that in many circumstances where drinking behavior has been induced in the rat by various thirst stimuli, the plasma angiotensin II levels are well above the 200 fmol/mL necessary to initiate drinking when angiotensin II is infused intravenously (Mann, Johnson, Ganten, & Ritz, 1987). The response to circulating angiotensin could be a useful emergency backup to the basic nonrenin drinking response caused by acute hypovolemia and would result in extra water intake over and above that caused by the nonrenin component of the stimulus. It has been known for a long time that when angiotensin is infused intravenously into animals made thirsty by a variety of procedures, it simply adds its effect to that of the other thirst stimulus. Furthermore, angiotensin released in hypovolemia may be dipsogenically more effective than exogenously administered angiotensin. One reason for this is that any rise in blood pressure produced by endogenously generated hormone in response to hypovolemia is unlikely to reach the levels that have been shown to inhibit drinking when angiotensin is infused into normovolemic animals (Robinson & Evered, 1986).

The contribution of endogenously generated angiotensin to drinking in response to different thirst stimuli has been assessed by removing the source of renin by nephrectomy or by using antagonists of the renin–angiotensin system to prevent the formation or action of angiotensin II. Neither approach is entirely satisfactory, because eliminating one mechanism may cause the remaining mechanisms to become more strongly activated and therefore make greater contributions to the response than they normally do. Nevertheless, there is now good evidence that in some circumstances endogenous renin makes an important contribution to hypovolemic thirst. It seems that the more rapid the onset of circulatory changes, the greater is the contribution from blood-borne angiotensin. But this belief is based on rather limited data. For example, in caval obstruction, where the circulatory changes are maximal when the obstruction is first applied, angiotensin contributes in a major way to drinking. This may be contrasted with the apparently negligible contribution to drinking following hyperoncotic dialysis, where the circulatory changes develop more gradually.

Angiotensin-induced sodium appetite seems to be different from angiotensin-induced thirst; it starts more slowly, lasts longer, and the response is enhanced by previous experience. These characteristics, the multiplicity of hormones involved in the striking sodium appetite of pregnancy and lactation (Denton, 1982), and the synergy between angiotensin and mineralocorticoid action (Epstein, 1986) in arousing sodium appetite suggest that something more than straightforward depolarization of neurons is responsible. Perhaps synthesis of a neurotransmitter or neuromodulator, or a change in the morphology of the neural network, is a necessary prelude to the expression of sodium appetite. An attractive hypothesis needing a great deal more experimental work is that the cerebral renin–angiotensin system plays a more important role than the renal system in sodium appetite. For example, the exaggerated NaCl intake of spontaneously hypertensive rats (SHR) is attenuated by cerebroventricular infusion of captopril (Di Nicolantonio, Hutchinson, & Mendelsohn, 1982). The methods of molecular biology could help resolve this question, because it may be possible to use labeled complementary DNA probes to determine whether there are "switched on" renin genes in brain tissue when sodium appetite has been aroused.

Some of the differences between angiotensin-induced thirst and sodium appetite, as well as some of the problems requiring further investigation, are well illustrated by the results of experiments using the angiotensin-converting enzyme inhibitor captopril. By preventing conversion of angiotensin I to angiotensin II, captopril releases renal renin secretion from negative feedback inhibition by angiotensin II, with the result that renin secretion and angiotensin I formation increase. After moderate doses of captopril, peripheral block of converting enzyme is complete, but there is enough unblocked converting enzyme in the brain to convert the increased amounts of blood-borne angiotensin I reaching the brain to angiotensin II. The immediate effect of the angiotensin II formed in the brain is to stimulate water intake, but there is little immediate effect on NaCl intake (Elfont & Fitzsimons, 1985). With higher captopril dosage, the cerebral enzyme is also blocked, so that angiotensin I conversion ceases in the brain as well as in the periphery. Therefore, with no possibility of angiotensin I conversion anywhere in the body, there is no angiotensin II stimulation of brain structures. The absence of increases in intake of either water or NaCl is entirely to be expected.

However, when the higher captopril dosage is continued for some days, there is a delayed and developing increase in sodium appetite (Fregly, 1980; Elfont, Epstein, & Fitzsimons, 1984). This is thought to be because chronic captopril administration induces converting enzyme biosynthesis in the brain, overcoming the initial conversion block and allowing renewed formation of angiotensin II from the high levels of circulating angiotensin I. Continued captopril dosage now leads to the sustained exposure of structures in the brain to increases in angiotensin II, which seems necessary for the development of sodium appetite. This supports the view that some change in neuronal organization or synthesizing ability underlies sodium appetite. However, despite the increasing intake of NaCl, there is no long-term effect on water intake, a surprising finding for which there is no explanation.

In contrast to the absence of any continued increase in water intake with long-term administration of captopril is the striking polydipsia shown by two-kidney Goldblatt hypertensive rats during the first 2 weeks after operation, during which circulating renin is elevated (Costales, Fitzsimons, & Vijande, 1984). Except for the long-term polydipsia, the pattern of drinking shown by the two-kidney Goldblatt rat is similar to that of the rat subjected to chronic captopril treatment. The immediate postoperative response is increased water intake with no increase in NaCl intake. But this is followed by increasing intakes of both water and NaCl, which seem to depend partly on renal renin because both intakes are attenuated by removing the renin-producing kidney or preventing the formation of angiotensin II. The differences in drinking behavior between the two-kidney Goldblatt rat and the rat maintained on captopril are unexplained and demonstrate how complex angiotensin-induced thirst and sodium appetite are. Release of other renal factors and differences in hemodynamics and fluid and electrolyte balances are possibilities that require further experimental investigation.

CONCLUSION

One theme of this chapter is that accounts by victims of their personal experiences of severe thirst and sodium appetite often provide important insights into present experimental problems that are unobtainable in animal experimentation. A. V. Wolf's *Thirst* (1958) is particularly valuable in this respect because in addition to its more formal coverage of the scientific and clinical literature, it contains general accounts of incidents in which thirst and lack of water have come to dominate the minds of those affected by the events described. However, the choice available in the literature is enormous. The selection of events and experiments presented here overlaps Wolf's selection to a very limited extent, and the analysis is entirely personal. In these accounts, some of which go back a very long way, we can catch vivid glimpses of what is important about thirst and sodium appetite and unearth clues about unsuspected physiological processes that may have been overlooked in the laboratory.

As we saw, Mr. Holwell's "raging thirst" during his ordeal in the Black Hole of Calcutta, and his observations that his thirst was made worse by the prospect of water, whereas drinking the sweat that poured off him afforded some relief, tell us a good deal more than we might have gleaned from any animal experiment. How revealing also is the description of the dreadful thirst experienced by the victims of the shipwreck of *La Méduse!* After reading what the survivors recounted, there can be little doubt about the catastrophic effects of drinking urine or sea water, whereas perusal of some of the more recent scientific literature might leave us in a state of uncertainty about the benefits or otherwise of drinking such fluids. We are bound to conclude that it is scientific myopia to confine our observations to the laboratory and ignore what is happening in the world at large. Cannon in his Croonian lecture to the Royal Society in 1918, argued powerfully in favor of not ignoring what ordinary people have to say about thirst when he wrote

. . . . . . that the attitude of physiologist with reference to thirst has been much as it was with reference to hunger. In each condition a general bodily need has arisen from a lack of essential bodily material and is signalled by a well-defined sensation. In each the testimony of ingenuous persons regarding their feelings has been carefully set down, and then explained away. (Cannon, 1919, p. 292)

A clear first-hand account of events by the victim can be as important in understanding aspects of the physiology of thirst and sodium appetite as a good medical history taken from the patient is in making a correct diagnosis.

The second theme in this chapter is an extension of the first. In addition to what is found in the general literature, clinical observations may provide important information about drinking behavior, and, although experiments on thirst and sodium appetite are not often carried out on man, such experiments likewise provide first-hand personal accounts of the nature of the sensations. An instance of an early clinical observation on thirst is found in Latta's report on the use of intravenous fluid to treat cholera. The occurrence of "burning thirst," as he called it, and its relief by intravenous fluid highlight the urgency of hypovolemic thirst, even now often regarded as the poor relation of cellular thirst. Again, we saw how the sensations experienced by the sodium-deficient subjects of McCance's experiment indicate the complexity of the behavioral response to extracellular dehydration. Some subjects complained of thirst and became polydipsic, but at the same time hypertonic NaCl, which might have been expected to exacerbate the thirst, was found to be refreshing, suggesting that increased sodium appetite may have been an important component of the sensations at this stage in the deficiency.

Lest it be imagined otherwise from what has been written here, the essence of physiological investigation is animal experimentation, and there is a moral obligation to extract the maximum of information from such investigations. Study of papers on the kidney and thirst written before general awareness of the involvement of renal renin in drinking behavior remind us of some interesting questions as yet unanswered. Why did Linazasoro and his colleagues discount renin as the active principle of their dipsogenic renal extract? Was it tied in with the fact that in the second series of experiments the fluid offered to drink was "physiological saline serum" which for some reason the nephrectomized rats were less willing to drink than water? Was it simply that the activity of the renin sample was too low, or, a tantalising possibility, did the SAR of Linazasoro and his colleagues contain substances other than renin that affected drinking behavior?

Therefore, in some respects the most important theme of this chapter is that it is desirable and wholly worthwhile to scrutinize the raw data from past experiments and see what else can be learned from them. In current work, there is a natural tendency to select those past results that best accord with the theory that is being advanced. Results that are compatible with current thinking are therefore preserved, whereas other observations that do not fit in with current dogma may be forgotten. These forgotten results may be valuable indicators for future research.

REFERENCES

Adolph, E. F. (1980). Intakes are limited: Satieties. *Appetite, 1,* 337–342.

Asscher, A. W., & Anson, S. G. (1963). A vascular permeability factor of renal origin. *Nature, 198,* 1097–1099.

Avrith, D. B., & Fitzsimons, J. T. (1980). Increased sodium appetite in the rat induced by intracranial administration of components of the renin–angiotensin system. *Journal of Physiology, 301,* 349–364.

Cannon, W. B. (1919). The physiological basis of thirst. *Proceedings of the Royal Society, B90,* 283–301.

Chiaraviglio, E. (1976). Effect of renin–angiotensin system on sodium intake. *Journal of Physiology, 255,* 57–66.

Cizek, L. J., Semple, R. E., Huang, K. C., & Gregersen, M. I. (1951). Effects of extracellular electrolyte depletion on water intake in dogs. *American Journal of Physiology, 164,* 415–422.

Costales, M., Fitzsimons, J. T., & Vijande, M. (1984). Increased sodium appetite and polydipsia induced by partial aortic occlusion in the rat. *Journal of Physiology, 352,* 467–481.

Denton, D. A. (1982). *The hunger for salt.* Berlin, Heidelberg, New York: Springer Verlag.

Di Nicolantonio, R., Hutchinson, J. S., & Mendelsohn, F. A. O. (1982). Exaggerated salt appetite of spontaneously hypertensive rats is decreased by central angiotensin-converting enzyme blockade. *Nature, 298,* 846–848.

Elfont, R. M., Epstein, A. N., & Fitzsimons, J. T. (1984). Involvement of the renin–angiotensin system in captopril-induced sodium appetite in the rat. *Journal of Physiology, 354,* 11–27.

Elfont, R. M., & Fitzsimons, J. T. (1985). The effect of captopril on sodium appetite in adrenalectomized and deoxycorticosterone-treated rats. *Journal of Physiology, 365,* 1–12.

Epstein, A. N. (1986). Hormonal synergy as the cause of salt appetite. In G. de Caro, A. N. Epstein, & M. Massi (Eds.), *The physiology of thirst and sodium appetite* (pp. 395–404). New York: Plenum Press.

Epstein, A. N., Fitzsimons, J. T., & Rolls, B. J. (1970). Drinking induced by injection of angiotensin into the brain of the rat. *Journal of Physiology, 213,* 457–474.

Fitzsimons, J. T. (1964). Drinking caused by constriction of the inferior vena cava in the rat. *Nature, 204,* 479–480.

Fitzsimons, J. T. (1969). The role of a renal thirst factor in drinking induced by extracellular stimuli. *Journal of Physiology, 201,* 349–368.

Fitzsimons, J. T. (1979). *The physiology of thirst and sodium appetite,* Monographs of the Physiological Society, No. 35. Cambridge: Cambridge University Press.

Fitzsimons, J. T., & Simons, B. J. (1969). The effect on drinking in the rat of intravenous angiotensin, given alone or in combination with other stimuli of thirst. *Journal of Physiology, 203,* 45–57.

Fitzsimons, J. T., & Stricker, E. (1971). Sodium appetite and the renin–angiotensin system. *Nature, 231,* 58–60.

Fitzsimons, J. T., & Wirth, J. B. (1978). The renin–angiotensin system and sodium appetite. *Journal of Physiology, 274,* 63–80.

Fregly, M. J. (1980). Effect of the angiotensin converting enzyme inhibitor, captopril, on NaCl appetite of rats. *Journal of Pharmacology and Experimental Therapeutics, 215,* 407–412.

Gilman, A. (1937). The relation between blood osmotic pressure, fluid distribution and voluntary water intake. *American Journal of Physiology, 120,* 323–328.

Hervey, G. R., & McCance, R. A. (1952). The effects of carbohydrate and sea water on the metabolism of men without food or sufficient water. *Proceedings of the Royal Society, B139,* 527–545.

Holmes, J. H., & Montgomery, V. (1960). Relation of route of administration and types of fluid to satisfy thirst in the dog. *American Journal of Physiology, 199,* 907–911.

Jiménez Díaz, C., Linazasoro, J. M., & Merchante, A. (1959). Further study of the part played by the kidneys in the regulation of thirst. *Bulletin of the Institute of Medical Research (Madrid), 12,* 60–67.

Ladell, W. S. S. (1965). Water and salt (sodium chloride) intakes. In O. G. Edholm & A. L. Bacharach (Eds.), *The physiology of human survival* (pp. 235–299). London: Academic Press.

Latta, T. (1832). Letter. *The Lancet, 2,* 274–277.

Lewes, G. H. (1859). *The physiology of common life* (Vol. 1, pp. 1–50). Edinburgh: Blackwood.

Linazasoro, J. M., Jiménez Díaz, C., & Castro Mendoza, H. (1954). The kidney and thirst regulation. *Bulletin of the Institute of Medical Research (Madrid), 7,* 53–61.

Magendie, F. (1821). Expérience sur la rage, par M. Magendie. *Journal de Physiologie Expérimentale, 1,* 40–46.

Magendie, F. (1823). Histoire d'un hydrophobe, traité à l'Hôtel-Dieu de Paris, au moyen de l'injection de l'eau dans des veines. *Journal de Physiologie Expérimentale et Pathologique, 3,* 382–392.

Mann, J. F. E., Johnson, A. K., Ganten, D., & Ritz, E. (1987). Thirst and the renin–angiotensin system. *Kidney International, 32,* Suppl. 21, S27–S34.

Masson, G. M. C., del Greco, F., Corcoran, A. C., & Page, I. H. (1956). Metabolic and pathological effects of renin in nephrectomized rats. *American Journal of the Medical Sciences, 231,* 198–204.

Masson, G. M. C., Plahl, G., Corcoran, A. C., & Page, I. H. (1953). Accelerated hypertensive vascular disease from saline and renin in nephrectomized dogs. *American Medical Association Archives of Pathology, 55,* 85–97.

McCance, R. A. (1936a). Experimental sodium chloride deficiency in man. *Proceedings of the Royal Society, B119,* 245–268.

McCance, R. A. (1936b). Medical Problems in mineral metabolism. III. Experimental human salt deficiency. *Lancet, 1*, 823–830.

McCance, R. A., Young, W. F., & Black, D. A. K. (1944). The secretion of urine during dehydration and rehydration. *Journal of Physiology, 102*, 415–428.

Morris, R. J. (1976). *Cholera 1832. The social response to an epidemic.* London: Croom Helm.

O'Shaughnessy, W. B. (1831). *The Lancet, 1*, 490.

Peters, G. (1980). Mécanismes de réglage de l'ingestion d'eau. *Journale de Physiologie, 76*, 295–322.

Richter, C. P. (1956). Salt appetite of mammals: Its dependence on instinct and metabolism. In *L'Instinct dans le comportement des animaux et de l'homme* (pp. 577–629). Paris: Masson.

Robinson, J. R. (1954). *Reflections on renal function.* Oxford: Blackwell.

Robinson, J. R. (1988). *Reflections on renal function* (ed. 2). Oxford: Blackwell.

Robinson, M. M., & Evered, M. (1986). Angiotensin II and arterial pressure in the control of thirst. In G. de Caro, A. N. Epstein, & M. Massi (Eds.), *The physiology of thirst and sodium appetite* (pp. 193–198). New York: Plenum Press.

Rolls, B. J., & Rolls, E. T. (1981). The control of drinking. *British Medical Bulletin, 37*, 127–130.

Rowland, N., & Nicolaïdis, S. (1976). Metering of fluid intake and determinants of *ad libitum* drinking in rats. *American Journal of Physiology, 231*, 1–8.

Rullier. (1821). Soif. In *Dictionnaire des sciences médicales, par une société de médecins et de chirugiens* (Vol. 51, pp. 448–490). Paris: Panckoucke.

Schiff, M. (1867). *Leçons sur la physiologie de la digestion, faites au muséum d'histoire naturelle de florence* (Vol. 1, pp. 29–42). Florence: Loescher.

Schmidt-Nielsen, K. (1964). *Desert animals. Physiological problems of heat and water.* Oxford: Clarendon Press.

Shapiro, J. I., & Anderson, R. J. (1987). Sodium depletion states. In B. M. Brenner & J. H. Stein (Eds.), *Body Fluid Homeostasis* (pp. 245–276). New York: Churchill Livingstone.

Stricker, E. M. (1978). The renin–angiotensin system and thirst: Some unanswered questions. *Federation Proceedings, 37*, 2704–2710.

Wolf, A. V. (1958). *Thirst. Physiology of the urge to drink and problems of water lack.* Springfield, IL: Charles C. Thomas.

3

Homeostatic Origins of Ingestive Behavior

Edward M. Stricker

Introduction

Beginning in the mid-1960s, extraordinary developments in the biological sciences have permitted studies of the biological bases and consequences of behavior, once the province of physiological psychology, to evolve into the modern investigations of brain function that characterize the emerging discipline of behavioral neuroscience. To describe this change in more operational terms, there has been a notable transition in scientific approach from behavioral experiments that inferred putative biological principles to biological investigations that examined those principles directly. For example, we now can make chemically specific brain lesions in animals and, in minute pieces of brain tissue, extract and analyze the content of neurotransmitters and their metabolites, the activity of their biosynthetic enzymes, and the density and binding of their receptors. Moreover, we can deliver pharmacological agents into the brain or cerebrospinal fluid of animals by way of implanted miniature pumps, uninterruptedly and for weeks at a time, and we can extract small amounts of blood and determine the concentration of circulating hormones by sensitive radioimmunoassays. Such procedures as these have improved the conduct of our science in ways that still seem magical to those of us whose work has spanned the past three decades, and have liberated thought by allowing new proposals about brain biology and function to be formulated and then considered experimentally. In consequence, the present ideas about the brain's control of ingestive behavior are in many cases richer in detail and more tangible than the older ones, because the original data

Edward M. Stricker   Department of Behavioral Neuroscience, University of Pittsburgh, Pittsburgh, Pennsylvania 15260.

on which they were based have been reinterpreted in light of new findings and new perspectives.

This chapter considers some of the changes in perspective that have resulted from the remarkable progress in the laboratory. The points considered may seem obvious now, but it is instructive to note that they were much less clear 25 years ago. Then, physiological psychologists had segregated motivated behavior into two categories, those that were considered to be "homeostatic" and those that were not. The homeostatic behaviors included food intake motivated by hunger, water intake motivated by thirst, NaCl intake motivated by sodium appetite, and various behaviors subserving thermoregulation. They were perceived as adaptive behavioral responses to nutrient deficiencies, metabolic imbalances, and other threats to the internal milieu, and they were believed to contribute importantly to the maintenance of certain regulated variables that were critical to the organism's survival, such as body calories, water, sodium, and temperature. Other motivated activities that were not directed at stabilizing those variables and thereby promoting the development and health of the organism, such as reproductive and maternal behaviors, were considered to be "nonhomeostatic."

It is now plain that the distinction between these categories is blurred. In a broader conceptualization of homeostasis that includes stability of brain activational systems, discussed below, there can be recognized homeostatic bases of motivated behavior that do not reflect nutrient deficiencies or metabolic imbalances. It also is clear that ingestive behaviors do not invariably reflect a need state. Food or drink may be consumed in anticipation of hunger or thirst, or simply because they taste good, or for social reasons, or to obtain a desired chemical agent in the ingesta such as caffeine or alcohol despite an adequate caloric or hydrational state. Indeed, consumption of food or fluid might continue for such reasons even after satiety or overhydration has developed; witness the universal inclination to eat palatable deserts at the end of a filling meal, long after hunger has been relieved.

This chapter does not consider the bases of such intakes, not because they are unimportant—in fact, they probably account for much of what we consume—but because I wish to focus instead on three general issues concerning the homeostatic origins of ingestive behavior that have emerged during the past 25 years and have affected the design of our experiments and the interpretation of our results. The three issues involve the physiological and behavioral contributions to homeostasis, the multiple stimuli that control ingestive behavior, and the nonspecific activational component of motivated behavior. Each is considered in turn, after which some of the basic principles that underlie the control of food and fluid intake are discussed.

PHYSIOLOGICAL AND BEHAVIORAL CONTRIBUTIONS TO HOMEOSTASIS

Claude Bernard (1878/1974) and Walter Cannon (1932), while emphasizing reflex physiological processes, both recognized that appropriate behaviors also could contribute importantly to organismal homeostasis. However, it was largely the work of Curt Richter (1942) that established the significance of those behavioral contributions to homeostasis: the specific appetites for water and salt, for example, which appear to reflect bodily deficiencies and prompt appropriate intakes to correct those nutrient needs.

By now the joint roles of physiology and behavior in homeostasis are well accepted: renal water conservation accompanies thirst during dehydration, sodium conservation accompanies sodium appetite during sodium deficiency, shivering and peripheral vasoconstriction accompany nest building during cold stress, and so on. In addition, it is generally appreciated that the physiological and behavioral responses are plastic, and either can be increased when the other is constrained. For example, animals drink more water after receiving a salt load when their ability to excrete concentrated urine is eliminated; conversely, their urine is more concentrated when drinking water is withheld. Such observations make it evident that focused studies of either the physiological or the behavioral responses to homeostatic challenges are incomplete without full knowledge of the other adaptive ways in which the animal is or is not responding.

Whereas the presence of ingestive behavior might reflect some underlying stimulus, perhaps a nutrient deficiency or an alteration in the flux of metabolic fuels (see below), the absence of drinking or eating does not necessarily imply that there is no signal for thirst or hunger. Behavior may be signaled but precluded when certain vital physiological functions are compromised. For example, an organism that has become incapacitated by hemorrhagic hypotension does not drink despite the presence of more hypovolemia than is necessary to stimulate thirst in an alert and mobile animal. Thus, when subjects exposed to severe homeostatic imbalances do not eat or drink as expected, it has become routine for investigators to reject the trivial explanation of behavioral incompetence before proposing a more interesting biological interpretation of the results (e.g., Stricker, 1977).

It clearly is important to the animal that the complementary physiological and behavioral responses subserving the same general function occur simultaneously, but what are the biological mechanisms by which this integration is accomplished? One possibility is that the stimulus of ingestive behavior is derived from one of the physiological responses to the homeostatic imbalance, much as gonadal hormones are known to affect mating behavior in addition to influencing reproductive physiology. Even when these responses occur in the same time frame, however, the question remains as to whether they are causally related. For example, it is quite clear that vasopressin is not a stimulus of thirst despite its common appearance during dehydration and general association with thirst (Adolph, Barker, & Hoy, 1954). In this case, the physiological and behavioral responses are coincident because they are stimulated by the same sensory signal, and this organizational feature insures their integration.

Other examples involving the putative roles of angiotensin in thirst, cholecystokinin in satiety, and aldosterone in sodium appetite are less clear. In each case the administered hormone is known to have a marked effect on ingestive behavior. In each case the hormone also has a recognized biological function that complements the ingestive behavior: angiotensin is a potent pressor agent and stimulus of aldosterone secretion during hypovolemia, cholecystokinin inhibits gastric motility and emptying, and aldosterone promotes renal sodium conservation. These features are not disputed, nor is the reliability of the induced behavioral effects in doubt. Instead, the present controversies concern the physiological significance of those effects; the fact that exogenous hormone can affect behavior does not demand that it does so normally. For example, insulin given to rats in pharmacological doses elicits eating, but this finding has no apparent relevance to its actions as a hormone secreted in response to a meal. Whether angiotensin, cholecystokinin, and aldosterone similarly act as drugs when their administration alters food or fluid intake remains to be resolved. Consequently, as yet there is no definite example of a hormone whose physiological functions include a direct and substantial influence on ingestive behavior.

In addition to the prominent physiological and behavioral contributions to homeostasis, there are more subtle, intrinsic features of the mammalian organism that provide stability without requiring either regulatory hormones, neural reflexes, or ingestive behaviors. For example, the impact of water loss on plasma osmolality is blunted by the process of osmosis, which spreads the effects of the water lost from the extracellular fluid to the larger volumes of cellular water. Therefore, the fact that osmoregulatory thirst and vasopressin secretion result when increases in plasma osmolality exceed threshold values of 1–3% does not mean that they are easily provoked; it takes a relatively substantial water loss to affect plasma osmolality at all.

Similarly, the cardiovascular and renal systems contain features that resist disturbance of the circulatory flow. For example, the distensibility of veins permits excess blood to be retained in these capacitance vessels without influencing arterial blood pressure. Conversely, after hemorrhage the transient decrease in arterial pressure and unaltered plasma protein concentration permit a net movement of interstitial fluid into local capillaries and thereby help to restore lost circulatory volume rapidly. A parallel decrease in filtration from glomerular capillaries in the kidney reduces urine volume independently of vasopressin secretion. It takes a fairly sizable treatment to overcome these buffers and trigger the overt volume-regulatory responses that come under study.

There is intrinsic stability of body weight too, resulting from storage of excess calories as triglycerides. Triglycerides, when burned, yield 9 kcal per gram, as compared with only 4 kcal per gram when glycogen is oxidized. Triglycerides also require much less water for storage in white adipose tissue than glycogen requires for storage in liver. Thus, whereas approximately 80 kcal is stored as triglyceride in 10 g of adipose tissue in adult rats (the equivalent of their daily food intake), 80 g of liver would be needed to store that amount of energy as glycogen. Consequently, it takes relatively large amounts of extra calories to increase body fat appreciably, and it takes a larger negative calorie balance than results even from 24 hours of food deprivation to cause the loss of 10 g of body fat in rats. The efficient storage of calories, together with the apparent autoregulation of adipose tissue stores of triglyceride, can account for the general long-term maintenance of body fat without postulating a control of food intake according to some "set-point" arrangement (see Chapter 9).

Of course, it is easy for investigators to overwhelm the intrinsic stability of regulated variables by administering various chemical agents in massive doses. When they do so, the treatments used present substantial challenges to homeostasis. For example, systemic injection of 2 ml of 2 M NaCl solution is standard in studies of osmoregulatory thirst in adult rats, but the induced increase in plasma osmolality of 40 mOsm/kg is entirely unnatural and represents an extraordinary stimulus of thirst. It is impressive that the animals drink at all under these circumstances of acute and severe dehydration, and the fact that they drink promptly and in appropriate amounts is truly remarkable. A similar perspective can be applied to the drinking elicited by abrupt reductions in arterial blood pressure and to the eating induced by acute hypoglycemia. These common treatments stimulate uncommon responses. Under investigation then are the capabilities of behavioral systems at their outer limits, not their ordinary functions, and it should not be surprising or necessarily meaningful that animals do not behave so well in response to such severe challenges soon after undergoing brain damage or abdominal surgery (see below).

MULTIPLE STIMULI IN THE CONTROL OF INGESTIVE BEHAVIOR

A great many investigations during the past two decades have been concerned with the specific mechanisms by which ingestive behavior is controlled under various conditions. In considering why animals drink water after a period of forced water deprivation, for example, it is plain that they are not monitoring body water *per se* but become aware of dehydration because appropriate sensors detect specific consequences of the induced water loss. Thus, the search for the biological bases of thirst has been translated into a search for specialized sensors that detect the relevant manifestations of dehydration.

There are many ways of inducing thirst in animals, many more ways than there are stimuli for thirst. Thus, dissimilar treatments share a common stimulus for thirst. For example, the consumption of salted food elicits thirst without producing negative water balance, by raising plasma osmolality and thereby stimulating cerebral osmoreceptors (Gilman, 1937). The thirst that appears after water deprivation (Ramsay, Rolls, & Wood, 1977) or heat stress (Hainsworth, Stricker, & Epstein, 1968) also is predominantly osmoregulatory in nature, although both of these treatments are associated with actual water loss.

Although the concept of volume regulation still was controversial in the early 1960s, soon thereafter it became clear that thirst could be stimulated by an isosmotic loss of plasma volume. Thus, there are two dimensions in which dehydration can be detected, one related to the regulation of plasma osmolality and the other to the regulation of blood volume. The latter has been shown most directly in studies of rats given extravascular injections of colloidal solutions, which elicit thirst in proportion to the local sequestration of isosmotic extracellular fluid and the induced hypovolemia (Fitzsimons, 1961; Stricker, 1966, 1968). The decrease in plasma volume apparently is signaled by stretch receptors in the distensible walls of the great veins and right heart, which detect the contraction of volume (e.g., Kaufman, 1984). Thirst also can be elicited by the administration of β-adrenergic agonists, which reduce peripheral vascular resistance and thereby precipitate marked arterial hypotension without causing hypovolemia (Lehr, Mallow, & Krukowski, 1967). Thirst under these circumstances presumably is mediated by receptors known to be located on the high-pressure side of the circulation, in the carotid sinus and aortic arch (Hosutt, Roland, & Stricker, 1978), although there is little direct evidence in support of this conjecture. A third nonosmotic stimulus for thirst results from the hormone angiotensin (Fitzsimons, 1969), which normally is formed in blood from the proteolytic action of renin on a circulating plasma protein. Renin is an enzyme that is secreted from the kidneys in response to decreases in the renal blood supply and/or increases in sympathetic input to the kidneys, as occur during hypovolemia. Exogenous angiotensin is known to act in the brain to stimulate thirst, evidently at the subfornical organ (Simpson, Epstein, & Camardo,

1978), thus raising the possibility that endogenous angiotensin normally does so as well.

In short, there now appear to be four distinct stimuli of thirst, each of which results when an individual aspect of dehydration is detected by appropriate sensors. However, each is not equally likely to occur under normal circumstances. For example, osmotic dehydration seems to provide the stimulus of thirst that occurs commonly after solute gain or water loss. In contrast, sufficient volume depletion to cause arterial hypotension and raise blood angiotensin to dipsogenic levels probably is such an unusual event in the lifetime of an animal that those signals might best be considered as eliciting emergency responses.

Although food intake often is presented in apposition to water intake, the obvious similarities between eating and drinking behaviors distract attention from the fundamental differences between caloric homeostasis and body fluid homeostasis. Briefly, the cellular need for metabolic fuels is continuous, as it is for oxygen, and like oxygen the fuels are supplied to the cells automatically. The requisite delay for digestive processing of food makes it impossible for eating to provide these metabolic fuels in a timely basis, as breathing provides oxygen. Instead, the fuels are derived from the excess calories consumed in meals taken more remotely, and they are obtained either from the intestines or from endogenous depots. In contrast, water is not a nutrient that is continually utilized in cellular metabolism but a solvent that is slowly lost from the body by respiration, evaporation, and urinary excretion. The consequences of its loss are buffered by osmotic movement of water from large cellular stores, as mentioned, and thus the concentration of cellular solutes rises only gradually and is not immediately life-threatening. Hence, animals can afford to replace lost water by episodic drinking behavior.

These considerations do not deny the possibility that caloric deficiency and tissue metabolic deficits can be produced under certain experimental circumstances. Such a crisis can be created readily by administering insulin in pharmacological amounts and thereby promoting the uptake of circulating glucose into adipose and muscle cells. After blood glucose levels have been reduced sufficiently to compromise glucose utilization in the brain, cerebral chemoreceptors are activated, and the animals initiate both a sympathoadrenal response and eating (Steffens, 1969; Stricker, Rowland, Saller, & Friedman, 1977). Cerebral glucoprivation also can be produced by systemic or intracranial administration of 2-deoxyglucose, a glucose analogue that cannot be metabolized but instead blocks glycolysis in all body cells including brain (Miselis & Epstein, 1975; Stricker & Rowland, 1978). This metabolic emergency triggers appropriate counterregulatory responses, including ingestive behavior, but it does not in any obvious way reproduce the conditions that obtain normally at the onset of eating.

A more appropriate perspective of the principles by which food intake is controlled derives from the research of Jacques Le Magnen (1969), who showed that the intermeal intervals of free-feeding rats are proportional to the amounts consumed at the preceding meals, as though satiety signals generated by the meals had to be dissipated before the next meal could be initiated. In contrast, there was no evidence that food intake at each meal was related to the duration of the interval preceding the meal. Thus, the onset and termination of eating seem to be controlled by the disappearance and appearance of satiety signals caused by food consumption, respectively, rather than by inverse alterations in signals for hunger. My own views of these matters have been presented elsewhere (Stricker, 1984; Stricker & Verbalis, 1990) and are discussed briefly in a later section of this chapter.

During the past 20 years, multiple signals of satiety have been proposed on the basis of reports that food intake is inhibited by diverse treatments affecting the gastrointestinal tract, endocrine secretions, and/or circulating nutrient levels. However, because eating is a very labile behavior, observations that food intake can be reduced or abolished altogether are difficult to interpret, and the induced changes may not be associated with alterations in caloric homeostasis and the sensation of satiety. Indeed, dehydration and nausea each have much more potent inhibitory effects on feeding than satiety, and each inhibits gastric function as well (Stricker, McCann, Flanagan, & Verbalis, 1988). Thus, after observing that a given treatment decreases food intake, investigators have learned to examine the great variety of alternative possibilities before drawing inferences and to use human subjects to determine directly which sensation has been produced.

BEHAVIORAL AROUSAL AND STRESS

Among the physiological and behavioral contributions to homeostasis, some are specifically related to a particular regulatory function, and some are not. For example, aldosterone secretion from the adrenal cortex during hypovolemia permits sodium conservation in urine while promoting renal loss of potentially harmful potassium ions that accumulate during oliguria. This adaptive response to hypovolemia is not seen during homeostatic challenges that do not involve volume regulation. In contrast, secretion of glucocorticoids from the adrenal cortex and release of catecholamines from the adrenal medulla and sympathetic nerves occur not only during hypovolemia but also in response to a great variety of other stressors. Thus, animals are well known to become fragile after surgical disruption of either the pituitary–adrenal axis or the sympathoadrenal system, and their failure to deal adequately with stress never is interpreted as a specific vulnerability to any one treatment.

It is now clear that some of the same considerations apply to behavior. A nonspecific component of behavioral arousal has long been associated with motivated activities, and activity in the brainstem reticular formation is disturbed when stimulation is excessive or relatively low. In either case, specific behaviors are learned and performed in order to bring these central systems to some intermediate level of optimal arousal, and by doing so they may promote organismal homeostasis. For example, water-deprived animals drink water in order to eliminate the disquieting sensation of thirst, and usually they pursue that goal until plasma osmolality is restored. Similarly, eating may be used as a calming agent when the general level of activation is too high (because of hunger or any other source of excitement), although it also may be used as a source of stimulation when the general level of activation is too low. This concept of "reticular homeostasis" is consistent with the well-known formulations of Hebb (1955), Lindsley (1951), Dell (1958), and many others who have proposed a nonspecific component of excitation common to all motivated behaviors and who have suggested that much of behavior can be understood as an attempt to attain an optimal level of arousal.

As might be expected, surgical or pharmacological treatments that eliminate cerebral arousal abolish all voluntary behavior. With this in mind, it is now apparent that bilateral destruction of the lateral hypothalamus did not destroy feeding and drinking "centers," as first was proposed, but instead left animals akinetic and with profound impairments in sensorimotor integration that disrupted all motivated activities (Marshall, Turner, & Teitelbaum, 1971). Nor did the induced dysfunctions necessarily result from damage to the hypothalamus *per se*. Instead, the lesions seemed to be effective because they destroyed dopamine-containing fibers traversing the ventral diencephalon en route to the striatum from their origin in the midbrain, and damage anywhere along their length produced comparable behavioral impairments even when hypothalamic tissue was spared (Ungerstedt, 1971). The extensive destruction of central dopaminergic fibers in rats thus produced an animal model of human Parkinson's disease with characteristic dysfunctions in initiating movement (Zigmond & Stricker, 1989).

Other features of the lateral hypothalamic syndrome, such as the progressive recovery of function, similarly have been related to improved dopaminergic neurotransmission in the damaged system (Stricker & Zigmond, 1976, 1986). More specifically, recovery of function appears to result in part from an increase in dopamine release from residual undamaged neurons, a decrease in the inactivation of dopamine as a result of the loss of presynaptic uptake sites in association with the degeneration of axon terminals, and a further increase in the efficacy of released dopamine because of the proliferation of postsynaptic receptors. Collectively, these compensatory neurochemical responses to subtotal destruction of dopaminergic neurons may be viewed as homeostasis operating at the synaptic level. The adaptive changes are not unique to central dopaminergic neurons but also have been observed to occur after comparable damage in other monoaminergic neural systems including serotonin- and norepinephrine-containing neurons in the brain and the sympathetic noradrenergic nerves in the periphery (Zigmond & Stricker, 1985).

The loss of eating and drinking in response to acute homeostatic challenges after the apparent recovery of rats from the general impairments produced by large dopamine-depleting brain lesions also has been reevaluated in light of evidence that the animals then become akinetic and unresponsive again (Snyder, Stricker, & Zigmond, 1985; Stricker, Cooper, Marshall, & Zigmond, 1979). Thus, their failure to respond appropriately was not modality specific but instead reflected the general inability of brain-damaged rats to initiate voluntary movements during the intense physiological stimulation associated with those challenges (Stricker, 1976; Stricker, Friedman, & Zigmond, 1975). These results are reminiscent of Cannon's classic findings that extensive removal of the sympathetic ganglionic chain in cats produced little or no physiological change under basal laboratory conditions, whereas such lesions severely impaired the normal physiological responses to acute homeostatic challenges (Cannon, Newton, Bright, Menkin, & Moore, 1929). In retrospect, it seems probable that synaptic homeostasis permitted preservation of function under nonstressful conditions, but, with the residual neurons already having increased norepinephrine biosynthesis in response to partial denervation, a further increase in response to an environmental stimulus was no longer possible; in consequence, the animal was unable to emit an appropriate physiological response to stress (Fluharty, Rabow, Zigmond, & Stricker, 1985). Similarly, large dopamine-depleting brain lesions may be viewed as a "central sympathectomy," and the residual behavioral deficits observed after apparent recovery of function may reflect the limits of the compensatory process in the brain-damaged animals (Stachowiak, Keller, Stricker, & Zigmond, 1987).

In contrast to these investigations of the activational component of ingestive behavior, other more recent studies have found that focal brain lesions disrupt eating or drinking by destroying receptor systems or neural pathways that mediated specific regulatory responses. For example, damage to the basal forebrain eliminates the drinking response to osmotic dehydration but not to hypovolemia (Johnson & Buggy, 1978; Gardiner & Stricker, 1985), and damage to the subfornical organ eliminates the drinking response to angiotensin but not to osmotic dehydration or hypovolemia (Simpson *et al.*, 1978; Hosutt, Rowland, & Stricker, 1981). Consequently, by now investigators have learned to study brain-damaged animals in a wide variety of conditions when determining the specificity of induced behavioral impairments.

In addition to using multiple treatments to stimulate ingestive behavior, the provoked responses also have been subject to greater scrutiny. It has long been known that different responses are obtained when animals eat freely or work to obtain food (Miller, Bailey, & Stevenson, 1950) and when the food they eat is tasty or unpalatable (Teitelbaum & Stellar, 1954). It later became clear that different responses to acute homeostatic imbalance are observed when the animals are tested briefly or for prolonged periods (Stricker *et al.*, 1975) and when they are tested at night or in the day (Rowland, 1976). Each difference represents a significant variable that, once appreciated, makes the ingestive behavior richer in meaning and its control more complex.

MODELS OF INGESTIVE BEHAVIOR

The biological basis of alimentary behavior traditionally has been considered in terms of a depletion–repletion model. According to this perspective, shown in Figure 1A, a bodily deficiency of some nutrient, whether resulting from dietary deprivation or some other imbalance, somehow triggers a specific appetite for the needed nutrient, the ingestion of which leads to nutrient repletion and satiation. One of the most important developments in perspective during the past 25 years has been the replacement of this familiar model with

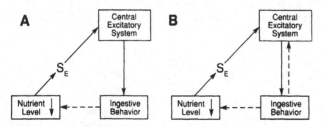

Figure 1. Schematic representation of possible central mechanisms by which ingestive behavior may be controlled. In this single-loop negative feedback system, ingestive behavior corrects the nutrient deficiency responsible for excitatory stimuli (S_E) that activate the behavior (A). This schema had been proposed for the control of osmoregulatory drinking and other behaviors. However, a more rapid and direct inhibitory effect of drinking on central excitatory systems appears to be involved as well (B). Unbroken arrows denote stimulatory influences, broken arrows inhibitory influences.

alternative views of the general mechanisms by which appetitive behaviors are stimulated and terminated.

Consider, for example, the depletion–repletion model as it may be applied to the control of water intake that results from water deprivation. With reference to Figure 1A, the "nutrient deficiency" is cellular water, the excitatory stimulus is a signal from cerebral osmoreceptors that activates some central system mediating thirst, and the consequence of water consumption is cellular rehydration, satiation, and the termination of drinking. The model actually seemed quite reasonable and was accepted widely until it was recognized that the termination of drinking normally precedes rehydration rather than results from it. Thus, there must be another, more rapid mechanism of negative feedback to the brain, indicated in Figure 1B, that accounts for the prompt inhibition of thirst before absorption of an ingested water load. Ramsay and his colleagues recently have associated this variable with the swallowing reflex, allowing an oropharyngeal metering of fluid consumption that provides inhibitory signals to the brain despite continued dehydration of bodily fluids. It is only later, after the ingested water is absorbed, that satiation accompanies rehydration (see Chapter 14).

The single-loop negative feedback model shown in Figure 1A is even less useful when applied to other alimentary behaviors than osmoregulatory drinking. During isosmotic hypovolemia, the neural signals generated by vascular baroreceptors stimulate thirst, but the water consumed does not remain intravascular and repair the "nutrient deficiency." Instead, ingested water is distributed by diffusion and osmosis throughout body water, with approximately two-thirds of it entering cells and swelling them purposelessly. At this rate, huge quantities of water would have to be consumed before sufficient amounts could repair the hypovolemia and eliminate its stimulus for thirst. But animals do not drink water until normal plasma volume is restored; instead, they drink only until body fluids are diluted by 3–5% (Stricker, 1969).

The termination of water intake then despite continued volume depletion reflects behavioral inhibition, not satiation. It has been associated with osmotic dilution rather than some other consequence of drinking because it also can be obtained by gastric water loads, which bypass the oropharynx. Thus, as modeled generally in Figure 2, a signal from vascular baroreceptors stimulates a central excitatory system that controls thirst, presumably the same system that is activated by osmoreceptors in association with cellular dehydration; however, the induced water intake does not correct the "nutrient deficiency," unlike drinking provoked by osmotic dehydration, but instead it generates a second stimulus associated with osmotic dilution, which inhibits the central excitatory system and

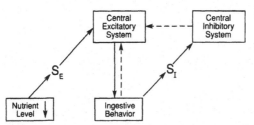

Figure 2. Schematic representation of possible central mechanisms by which ingestive behavior may be controlled. This schema, which appears to be most appropriate for volume-regulatory drinking, is more complex than that presented in Figure 1B because the ingestive behavior does not repair the nutrient deficiency that provoked this response. Instead, the intake generates inhibitory stimuli (S_I), which activate central inhibitory systems and thereby suppress further consumption despite the continued presence of excitatory stimuli (S_E). Arrows as in Figure 1.

thereby eliminates thirst and water consumption despite continued hypovolemia. Clearly, osmo- and volume-regulatory thirsts are associated with rather
different schematic models describing their control, as might be expected given
the different impact of water on plasma osmolality and volume.

It is of interest to note that the same factors involved in the control of thirst
also influence the control of vasopressin secretion from the posterior pituitary.
Specifically, water consumption by dehydrated animals rapidly inhibits vasopressin secretion too (Thrasher, Nistal-Herrera, Keil, & Ramsay, 1981), as does
osmotic dilution in hypovolemic rats (Stricker & Verbalis, 1986). Thus, the
prompt inhibitory feedback control of drinking is not restricted to behavior but
includes other responses to dehydration. As mentioned earlier, these physiological and behavioral responses occur in parallel because the same sensory signals
control each response.

The control of food intake appears to follow a third set of rules. The
concept of "nutrient deficiency" is not relevant in this case, as it is for water
intake, because normally there are no shortages of metabolic fuels. Instead, the
key variable in controlling food intake seems to be the alteration between the two
phases of caloric homeostasis, in which metabolic fuels are supplied either from
recently consumed food that still is in the alimentary canal and being assimilated
or from accumulated triglyceride, glycogen, and protein stores that resulted
from food consumed more remotely in time. Thus, according to this perspective,
the shift from satiety to hunger is not associated with some metabolic deficit but
instead occurs when animals enter the postabsorptive state and begin to mobilize
calories from stores (Friedman & Stricker, 1976). Nor does eating correct a
deficit or restore some aspect of tissue function; instead it simply reestablishes
the predominant supply of utilizable calories from the gastrointestinal tract
rather than from stores in adipose tissue, liver, or muscle.

A schematic representation of the factors that might control food intake
according to this perspective is displayed in Figure 3. In this arrangement,
inhibitory stimuli are generated by eating and result at first from gastric chem-
oreception of ingested calories and/or gastric distention (see Chapter 7). Hyperosmolality associated with the ingestion of dry or salted food may provide an
additional preabsorptive signal for satiety. Although the measured delivery of
calories from the stomach to the intestines gradually diminishes gastric contents,
the continued action of absorbed calories and insulin in the liver maintains this
phase of fuel storage and satiety until the intestines have emptied as well. By
then, all satiety signals have disappeared, and calories must begin to be mobilized
from liver cells and adipocytes. Hunger begins in consequence, although the
exact nature and timing of the factors signaling the end of satiety remain uncer-

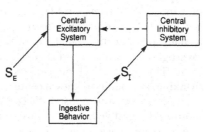

Figure 3. Schematic representation of possible
central mechanisms by which ingestive behavior
may be controlled. This schema, which appears to
be most appropriate for food intake, differs from
that presented in Figure 2 because there are no
nutrient deficiencies; thus, excitatory stimuli (S_E)
derive from exteroceptors detecting salient properties of the food, not from interoceptors (as in
Figures 1 and 2). However, ingestive behavior does not rely on such excitatory signals and may begin
when inhibitory stimuli (S_I) generated by a previous meal dissipate. Arrows as in Figure 1.

tain. An excitatory signal exists, but it derives from the stimulating taste of palatable food rather than from some bodily nutrient deficiency.

Under this schema, data relevant to the maintenance of glucose homeostasis and stable body fat stores can be reconsidered from the perspective of peripheral events. For example, evidence that gastric emptying is accelerated markedly in diabetic rats because of a tremendous increase in intestinal absorption of carbohydrate may explain why these animals are hyperphagic (Granneman & Stricker, 1984): the gastric phase of satiety terminates quickly and is not followed by a normal postabsorptive phase because the liver is unable to utilize calories efficiently in the absence of insulin (Friedman, 1978). The hyperphagia that follows ventromedial hypothalamic lesions similarly appears to result from an increase in gastric emptying (Duggan & Booth, 1986) plus a primary increase in lipogenesis in adipose tissue (Powley, 1977; Le Magnen, 1983), which reduce both the gastric and the postabsorptive satiety effects of a meal. In either case, a complex interaction among the stomach, liver, and adipose tissue could provide a mechanism by which food intake is controlled under diverse conditions, as described elsewhere (Stricker, 1984).

Note that an analogous schematic model can be used to describe the control of food intake by blowflies. As Dethier and his colleagues have shown (Dethier, 1976), these animals consume food in proportion to the gustatory stimulation associated with sweet taste until such excitation is counterbalanced by inhibition resulting from distention of the crop and foregut (i.e., "gastric" signals). Thus, the principal difference between the controls of eating in flies and rats is that there is no inhibitory signal associated with alterations in caloric flux in flies, either gastric or postgastric, because evidently they cannot detect calories directly (see Chapter 11).

Summary

Considerations of homeostasis in studies of brain and alimentary behavior have become much more sophisticated during the past 25 years, in parallel with, and no doubt in consequence of, comparable developments in laboratory techniques. Increasingly, ingestive behavior is described in biological terms and understood from a biological perspective. Thus, inevitably, studies of behavior require complementary studies of neurobiology to integrate biology and behavior in our theories as they are integrated in the animals we study.

In the 1960s, concerned investigators were preoccupied with the brain and its control of ingestive behavior. Le Magnen (1971) termed this a "bodiless psychology"; the body below the neck was considered of significance to behavior only insofar as it transported the brain from place to place. Nevertheless, we did not know then the central neural mechanisms by which the putative stimuli actually provoke motivated ingestive behaviors, despite the great number of investigations devoted to this goal, and today those mechanisms still are mysterious. The concept of brain centers that provide specific controls of eating and drinking, analogous to known centers that control autonomic functions, has long been discarded, and nonspecific components of arousal in the performance of motivated ingestive behaviors now are appreciated. However, new comprehensive concepts providing a description of cerebral events that account for feeding and drinking behaviors have not yet emerged, nor have specific pathways for these behaviors been identified. It seems clear that more incisive research is urgently needed to determine how the brain integrates specific input from relevant exteroceptors and interoceptors with nonspecific activational input and produces the appropriate motor activities involved in the acquisition and consumption of food and fluid.

One of the most significant changes in perspective during the past two decades has been the renewal of interest in physiological processes and peripheral signals. In this regard, studies of the hormones insulin, angiotensin, cholecystokinin, and aldosterone have been especially important in focusing attention to the periphery, and they already have served a useful purpose regardless of whether they ultimately are shown to have significant physiological roles in the control of food and fluid ingestion. When these are combined with the increasing number of studies focusing on the liver, stomach, adipose tissue, cardiovascular system, and kidneys in addition to the brain, there certainly is much more attention paid to basic physiological systems now than years ago.

The models of hunger and thirst that were described in this chapter and the comparable consideration of sodium appetite discussed elsewhere (see Chapter 15; Stricker & Verbalis, 1988) represent significant progress in my own understanding of these issues and how they might be approached. When I began work 25 years ago, drinking and eating were thought to be controlled by a simple depletion–repletion model, and in that fundamental respect they were viewed as generally similar to one another. Multiple stimuli in the control of thirst and satiety were not fully appreciated, and if the specific excitatory and inhibitory stimuli that seem to be involved in mediating hunger and thirst remain unsettled as yet, at least a great number of reasonable candidates now are available for consideration. In consequence of this progress, we can more adequately describe complex behaviors in terms that are consistent with the known biology of the organism and put what is known about the behavioral and biological contributions to homeostasis to much better use in illuminating clinical dysfunctions in human subjects. Indeed, the enormous breadth of contemporary investigations, ranging from molecules to behavior in both basic and clinical studies, represents the most significant development of all. It is a sound basis for deriving much satisfaction with the scientific research of the recent past and for optimism in considering our prospects for the future.

Acknowledgments

This chapter was based in part on a series of lectures given in May 1989 while I was a Visiting Professor at the University of Washington. I am grateful to Stephen C. Woods and his students and colleagues in the Department of Psychology at that institution for their stimulating discussion of these issues. I also thank Loretta M. Flanagan for her constructive comments on an early draft of this chapter.

REFERENCES

Adolph, E. F., Barker, J. P., & Hoy, P. A. (1954). Multiple factors in thirst. *American Journal of Physiology, 178*, 538–562.

Bernard, C. (1974). *Leçons sur les phenomenes de la vie, communs aux animaux et aux vegetaux*, translated to *Lectures on the phenomena of life common to animals and plants* (pp. 46–91). Springfield, IL: Charles C. Thomas. (Original work published 1878.)

Cannon, W. B. (1932). *The wisdom of the body.* New York: Norton.

Cannon, W. B., Newton, H. F., Bright, E. M., Menkin, V., & Moore, R. M. (1929). Some aspects of the physiology of animals surviving complete exclusion of sympathetic nerve impulses. *American Journal of Physiology, 89*, 84–107.

Dell, P. (1958). Some basic mechanisms of the translation of bodily needs into behavior. In G. E. W. Wolstenhome & C. M. O'Connor (Eds.), *Neurological Basis of Behaviour* (pp. 187–201). Boston: Little, Brown.

Dethier, V. G. (1976). *The hungry fly*. Cambridge, MA: Harvard University Press.

Duggan, J. P., & Booth, D. A. (1986). Obesity, overeating, and rapid gastric emptying in rats with ventromedial hypothalamic lesions. *Science, 231*, 609–611.

Fitzsimons, J. T. (1961). Drinking by rats depleted of body fluid without increase in osmotic pressure. *Journal of Physiology (London), 159*, 297–309.

Fitzsimons, J. T. (1969). The role of a renal thirst factor in drinking induced by extracellular stimuli. *Journal of Physiology* (London), 210, 349–368.

Fluharty, S. J., Rabow, L. E., Zigmond, M. J., & Stricker, E. M. (1985). Tyrosine hydroxylase activity in the sympathoadrenal system under basal and stressful conditions: Effect of 6-hydroxy-dopamine. *Journal of Pharmacology and Experimental Therapeutics, 235*, 354–360.

Friedman, M. I. (1978). Hyperphagia in rats with experimental diabetes mellitus: A response to a decreased supply of utilizable fuels. *Journal of Comparative and Physiological Psychology, 92*, 109–117.

Friedman, M. I., & Stricker, E. M. (1976). The physiological psychology of hunger: A physiological perspective. *Psychological Review, 83*, 409–431.

Gardiner, T. W., & Stricker, E. M. (1985). Impaired drinking responses of rats with lesions of nucleus medianus: Circadian dependence. *American Journal of Physiology, 248*, R224–R230.

Gilman, A. (1937). The relation between blood osmotic pressure, fluid distribution and voluntary water intake. *American Journal of Physiology, 120*, 323–328.

Granneman, J. G., & Stricker, E. M. (1984). Food intake and gastric emptying in rats with strep-tozotocin-induced diabetes. *American Journal of Physiology, 247*, R1054–R1061.

Hainsworth, F. R., Stricker, E. M., & Epstein, A. N. (1968). Water metabolism of rats in the heat: Dehydration and drinking. *American Journal of Physiology, 214*, 983–989.

Hebb, D. O. (1955). Drives and the C.N.S. (conceptual nervous system). *Psychological Review, 62*, 243–254.

Hosutt, J. A., Rowland, N., & Stricker, E. M. (1978). Hypotension and thirst in rats after iso-proterenol treatment. *Physiology and Behavior, 21*, 593–598.

Hosutt, J. A., Rowland, N., & Stricker, E. M. (1981). Impaired drinking responses of rats with lesions of the subfornical organ. *Journal of Comparative and Physiological Psychology, 95*, 104–113.

Johnson, A. K., & Buggy, J. (1978). Periventricular preoptic-hypothalamus is vital for thirst and normal water economy. *American Journal of Physiology, 234*, R122–R129.

Kaufman, S. (1984). Role of right atrial receptors in the control of drinking in the rat. *Journal of Physiology (London), 349*, 389–396.

Lehr, D., Mallow, J., & Krukowski, M. (1967). Copious drinking and simultaneous inhibition of urine flow elicited by beta-adrenergic stimulation and contrary effect of alpha-adrenergic stimulation. *Journal of Pharmacology and Experimental Therapeutics, 158*, 150–163.

Le Magnen, J. (1969). Peripheral and systemic actions of food in the caloric regulation of intake. *Annals of the New York Academy of Science, 157*, 1126–1157.

Le Magnen, J. (1971). Advances in studies on the physiological control and regulation of food intake. In E. Stellar & J. M. Sprague (Eds.), *Progress in Psychobiology and Physiological Psychology* (vol. 4, pp. 203–261). New York: Academic Press.

Le Magnen, J. (1983). Body energy balance and food intake: A neuroendocrine regulatory mechanism. *Physiological Reviews, 63*, 314–386.

Lindsley, D. B. Emotion. In S. S. Stevens (Ed.), *Handbook of Experimental Psychology* (pp. 473–516). New York: John Wiley & Sons.

Marshall, J. F., Turner, B. H., & Teitelbaum, P. (1971). Sensory neglect produced by lateral hypothalamic damage. *Science, 174*, 523–525.

Miller, N. E., Bailey, C. J., & Stevenson, J. A. F. (1950). Decreased "hunger" but increased food intake resulting from hypothalamic lesions. *Science, 112*, 256–259.

Miselis, R. R., & Epstein, A. N. (1975). Feeding induced by intracerebroventricular 2-deoxy-D-glucose in the rat. *American Journal of Physiology, 229*, 1438–1447.

Powley, T. L. (1977). The ventromedial hypothalamic syndrome, satiety, and a cephalic phase hypothesis. *Psychological Review, 84*, 89–126.

Ramsay, D. J., Rolls, B. J., & Wood, R. J. (1977). Thirst following water deprivation in dogs. *American Journal of Physiology, 232*, R93–R100.

Richter, C. P. (1942). Total self regulatory functions in animals and human beings. *Harvey Lectures, 38*, 63–103.

Rowland, N. (1976). Circadian rhythms and partial recovery of regulatory drinking in rats after lateral hypothalamic lesions. *Journal of Comparative and Physiological Psychology, 90*, 382–393.

Simpson, J. B., Epstein, A. N., & Camardo, J. S., Jr. (1978). Localization of receptors for the dipsogenic action of angiotensin II in the subfornical organ of rat. *Journal of Comparative and Physiological Psychology, 92*, 581–608.

Snyder, A. M., Stricker, E. M., & Zigmond, M. J. (1985). Stress-induced neurological impairments in an animal model of Parkinsonism. *Annals of Neurology, 18*, 544–551.

Stachowiak, M. K., Keller, R. W., Jr., Stricker, E. M., & Zigmond, M. J. (1987). Increased dopamine efflux from striatal slices during development and after nigrostriatal bundle damage. *Journal of Neuroscience, 7*, 1648–1654.

Steffens, A. B. (1969). The influence of insulin injections and infusions on eating and blood glucose level in the rat. *Physiology and Behavior, 4*, 823–828.

Stricker, E. M. (1966). Extracellular fluid volume and thirst. *American Journal of Physiology, 211*, 232–238.

Stricker, E. M. (1968). Some physiological and motivational properties of the hypovolemic stimulus for thirst. *Physiology and Behavior, 3*, 379–385.

Stricker, E. M. (1969). Osmoregulation and volume regulation in rats: Inhibition of hypovolemic thirst by water. *American Journal of Physiology, 217*, 98–105.

Stricker, E. M. (1976). Drinking by rats after lateral hypothalamic lesions: A new look at the lateral hypothalamic syndrome. *Journal of Comparative and Physiological Psychology, 90*, 127–143.

Stricker, E. M. (1977). The renin–angiotensin system and thirst: A reevaluation. II. Drinking elicited by caval ligation or isoproterenol. *Journal of Comparative and Physiological Psychology, 91*, 1220–1231.

Stricker, E. M. (1984). Biological bases of hunger and satiety: Therapeutic implications. *Nutrition Reviews, 42*, 333–340.

Stricker, E. M., Cooper, P. H., Marshall, J. F., & Zigmond, M. J. (1979). Acute homeostatic imbalances reinstate sensorimotor dysfunctions in rats with lateral hypothalamic lesions. *Journal of Comparative and Physiological Psychology, 93*, 512–521.

Stricker, E. M., Friedman, M. I., & Zigmond, M. J. Glucoregulatory feeding by rats after intraventricular 6-hydroxydopamine or lateral hypothalamic lesions. *Science, 189*, 895–897.

Stricker, E. M., McCann, M. J., Flanagan, L. M., & Verbalis, J. G. (1988). Neurohypophyseal secretion and gastric function: Biological correlates of nausea. In H. Takagi, Y. Oomura, M. Ito, & M. Otsuka (Eds.), *Biowarning System in the Brain* (pp. 295–307). Tokyo: University of Tokyo Press.

Stricker, E. M., and Rowland, N. (1978). Hepatic versus cerebral origin of stimulus of feeding induced by 2-deoxy-D-glucose in rats. *Journal of Comparative and Physiological Psychology, 92*, 126–132.

Stricker, E. M., Rowland, N., Saller, C. F., & Friedman, M. I. (1977). Homeostasis during hypoglycemia: Central control of adrenal secretion and peripheral control of feeding. *Science, 196*, 79–81.

Stricker, E. M., & Verbalis, J. G. (1986). Interaction of osmotic and volume stimuli in regulation of neurohypophyseal secretion in rats. *American Journal of Physiology, 250*, R267–R275.

Stricker, E. M., & Verbalis, J. G. (1988). Hormones and behavior: The biology of thirst and sodium appetite. *American Scientist, 76*, 261–267.

Stricker, E. M., & Verbalis, J. G. (1990). Control of appetite and satiety: Insights from biologic and behavioral studies. *Nutrition Reviews 48*, 49–56.

Stricker, E. M., & Zigmond, M. J. (1976). Recovery of function following damage to central catecholamine-containing neurons: A neurochemical model for the lateral hypothalamic syndrome. In J. M. Sprague & A. N. Epstein (Eds.), *Progress in Psychobiology and Physiological Psychology* (vol. 6, pp. 121–188). New York: Academic Press.

Stricker, E. M., & Zigmond, M. J. (1986). Brain monoamines, homeostasis, and adaptive behavior. In F. E. Bloom (Ed.), *Handbook of Physiology*, Section 1: *The Nervous System*, Volume IV, *Intrinsic Regulatory Systems of the Brain* (pp. 677–700). Bethesda: American Physiological Society.

Teitelbaum, P., & Stellar, E. (1954). Recovery from the failure to eat produced by hypothalamic lesions. *Science, 120*, 894–895.

Thrasher, T. N., Nistal-Herrera, J. F., Keil, L. C., & Ramsay, D. J. (1981). Satiety and inhibition of vasopressin secretion after drinking in dehydrated dogs. *American Journal of Physiology, 240*, E394–E401.

EDWARD M.
STRICKER

Ungerstedt, U. (1971). Adipsia and aphagia after 6-hydroxydopamine induced degeneration of the nigro-striatal dopamine system. *Acta Physiologica Scandinavica Supplementum, 367*, 95–122.

Zigmond, M. J., & Stricker, E. M. (1985). Adaptive properties of monoaminergic neurons. In A. Lajtha (Ed.), *Handbook of neurochemistry* (vol. 9, pp. 87–102). New York: Plenum Press.

Zigmond, M. J., & Stricker, E. M. (1989). Animal models of parkinsonism using selective neurotoxins: Clinical and basic implications. *International Review of Neurobiology, 31*, 1–79.

4

Behavioral Treatment of Obesity

Leonard H. Epstein

Introduction

This chapter provides an overview of the behavioral treatment of obesity. Obesity usually is defined as having a body weight greater than 20% over the ideal weight for a given height, with the ideal weights established by the life insurance norms. This excess body weight is associated with increased storage of fat relative to lean body mass. Over 25% of the adult population older than 30 years of age is obese (USDEW, 1979), and almost half of these obese adults are under a doctor's care to lose weight or are attempting to lose weight on their own (Stewart, Brook, & Kane, 1980).

Obese persons want to lose weight and to keep it off, not just for cosmetic reasons but because obesity is now recognized to be a major health risk (Simopoulos & Van Itallie, 1984). There are various causes of disability and death that are associated with obesity, including diabetes (Bonham & Brock, 1985), hypertension (Messerli, 1984), and stroke (Heyden *et al.*, 1971). The majority of type II diabetics are obese (National Diabetes Data Group, 1979) and experience rapid and significant improvements in metabolic control when they lose weight (Wing, Epstein, Nowalk, Koeske, & Hagg, 1985). Likewise, decreases in body weight reduce blood pressure in hypertensive obese persons (Reisen & Frohlich, 1978). It is more difficult to identify psychological problems that result from excess body weight (Wadden & Stunkard, 1987), although improvements in mood have been reported to occur with weight loss (Wing, Epstein, Marcus, & Kupfer, 1984).

Prior to 1960, obesity had been treated with various diets and exercise

Leonard H. Epstein　Western Psychiatric Institute and Clinic, University of Pittsburgh School of Medicine, Pittsburgh, Pennsylvania 15213.

programs with little systematic success (Stunkard, 1958). As obesity has become recognized as a major health problem, however, there has been an increase in the sophistication of therapists in developing methods for modifying the motivation of their clients, in targeting specific foods as problems, and in changing the client's environment and the attitudes and behaviors of significant people in his/her environment.

Such behavior modification programs grew from laboratory research to become the dominant approach to treating obesity. Behavior therapy now is included as the major treatment or as an adjunct in almost all current obesity programs (Brownell & Wadden, 1986). Behavioral approaches represent a very powerful and flexible technology for behavior change. They have been extensively researched, and they can incorporate methods and findings from other approaches. Research on the use of behavior modification for the treatment of obesity has been going on for over 25 years, which is long enough to characterize its success and to anticipate new developments. That is the general goal of this chapter.

This chapter is divided into four sections. The first section discusses briefly the rationale for including behavior therapy in the treatment of obesity. This is followed by an overview of clinic-based treatment procedures for adult obesity. As behavioral research has matured, treatment methods have improved and led to greater losses of weight than before. Next, the treatment of childhood obesity is reviewed. Conceptual differences in adult and child treatment are considered, as well as empirical differences in treatment outcome between adults and children. Finally, the potential for interfacing behavioral treatment approaches with research findings based on other methodological approaches is discussed. The incorporation of research in eating behavior, genetics, and the linking of specific treatments with specific etiologies may hold promise for further improvements in weight control and maintenance.

WHY BEHAVIOR THERAPY?

The treatment of obesity always involves changing the energy balance equation to one in which the person consumes fewer calories than he/she expends. One of the most direct ways to change energy balance is to alter behavior, such as by decreasing food intake and/or by increasing exercise. Behavioral treatments are well suited to this task because they focus on understanding the conditions under which behaviors develop and the methods for changing these behaviors. It is possible to target a general decrease in eating or to be more specific in altering the consumption of one particular type of food (Morganstern, 1974; Tondo, Lane, & Gill, 1975) or class of foods (Buxton, Williamson, Absher, Warner, & Moody, 1985; Epstein, Wing, & Valoski, 1985).

The goal of behavioral procedures is not only to shape and change eating and exercise, and thus produce weight change, but also to maintain the improved eating and exercise behaviors. Reinforcement contingencies can facilitate self-control (Cohen, Gelfand, Dodd, Jensen, & Turner, 1980; Epstein, Wing, Koeske, & Valoski, 1986) and incorporate social support to enhance motivation for behavior change (Brownell, Heckerman, Westlake, Hayes, & Monti, 1978; Rosenthal, Allen, & Winter, 1980). Reinforcement contingencies usually are used to change the behaviors of obese persons, but they also can be used to change the behaviors of family members (Epstein, Wing, Koeske, Andrasik, & Ossip, 1981) and co-workers (Brownell, Stunkard, & McKeon, 1985).

Perhaps most importantly, behavioral treatment procedures are based on a strong tradition of laboratory research focusing on the conditions in which behavior is learned and maintained (Leitenberg, 1976). This tradition is reflected in the empirical development and rigorous evaluation of treatment programs, which is quite unique in psychological therapies. For example, by 1981 there had been over 20 articles on behavioral approaches to weight loss (Foreyt, Goodrick, & Gotto, 1981). This empirical tradition is important in providing an accurate assessment of the strengths (Brownell & Wadden, 1986) and limitations (Foreyt *et al.*, 1981) of behavioral procedures.

This empirical tradition is compatible with the integration of other empirically based approaches to obesity, such as those based on pharmacology and genetics. As the procedures of behavior therapy have become more standard, and its strengths and limitations have been documented, specific behavioral techniques have been combined with specific biological regimens in order to increase the effectiveness of programs designed to change various behaviors associated with obesity. Behavior therapists also have included methodologies that may improve behavior change, which traditionally has been the focus of nonbehavioral approaches, as well as searching for new mechanisms to improve weight control.

TREATMENT OF ADULT OBESITY

The initial discussion of behavioral treatments of obesity usually is credited to Ferster and colleagues in 1962 (Ferster, Nurnberger, & Levitt, 1962). These investigators hypothesized that obese and nonobese persons differed in their eating behaviors and that behavioral procedures could be used to control body weight by making the eating behavior of the obese like that of nonobese persons. The first behavioral treatment of obesity reported by Stuart (1967) was a series of 10 cases treated over a 12-month period, with an average weight loss of 17.1 kg. He subsequently presented a comprehensive behavioral program that included nutrition, exercise, and behavioral management and published a popular treatment manual that provided further structure to behavioral treatment programs (Stuart & Davis, 1972).

Most behavioral programs still attend to the same three elements of treatment that Stuart used. Variations in treatments for nutrition, exercise, and behavioral management have emerged, but the similarities in effectiveness of the treatment packages across studies are in many ways more impressive than the differences across treatments (Wilson & Brownell, 1980; Wing & Jeffery, 1979). The average behavioral program reviewed in 1979 (Wing & Jeffery, 1979) obtained weight losses of about 5 kg across a wide variety of treatment approaches. These similarities suggested that elements common to each of the behavioral packages, such as self-monitoring and stimulus control, were among the most effective components of treatment (Brownell & Wadden, 1986).

Brownell and Wadden (1986) provide an interesting historical perspective on the development of behavioral treatments through 1984. The average weight loss for participants in behavioral programs in 1974 was 3.9 kg, whereas in 1984 the average loss was 7.0 kg. The factor that best characterized the improvement in weight loss was length of treatment. Although there has been no change over the years in the average weight loss of 0.5 kg per week, the programs have increased in length from 2 to 4 months, and thus weight loss has doubled during treatment.

The development of longer treatment protocols was in part a response to the recognition that obesity was a more complex and resistant problem than previously recognized by behavior therapists, and clients would not continue to lose weight after formal treatment sessions were completed. Longer programs could produce weight loss that was more likely to be clinically relevant.

In addition, longer programs could influence maintenance of new behavior change (Brownell & Wadden, 1986) because of improved learning of new habits. The maintenance of weight change is one of the major problems in treating obesity as well as many other habit change problems (Brownell, Marlatt, Lichenstein, & Wilson, 1986). Many obese persons lose large amounts of weight but over time return to old habits and regain weight, often returning to pretreatment levels of obesity. For this reason, it is important both to evaluate treatment effects at the end of intensive treatment and to assess the durability of treatment effects during follow-up.

Weight losses of 12 kg or more are common in these longer treatment programs (Jeffery, Gerber, Rosenthal, & Lindquist, 1983). The magnitude of these changes is sufficient to normalize the weight of many mildly overweight persons. For example, a woman who is 20% overweight (the minimal criterion for obesity) and is 60 in. in height and weighs 60 kg is 9.6 kg over her ideal weight, and thus she would be at her ideal weight if she had average success in a longer behavioral program.

In order for behavioral procedures to produce even larger weight losses that are clinically relevant for more obese persons, behavior therapists have begun to incorporate new and innovative procedures that increase weekly weight loss and thus the total amount of weight change. These newer approaches have included pharmacotherapy and very-low-calorie diets (VLCDs). Both pharmacotherapy and VLCD treatments were developed and used by nonbehavioral practitioners before they were incorporated into behavioral programs. Both of these procedures can produce weight losses that are superior to those produced by behavior therapy alone. However, the effectiveness of these nonbehavioral approaches has been limited because they typically are used only for short durations, they are not designed to change eating or exercise behaviors, and clients often return to their previous habits when treatment is discontinued. For this reason, it has been considered potentially advantageous to combine behavioral treatments and pharmacological and VLCD programs to shape new and more adaptive behaviors and thereby improve long-term weight control.

In one extensive test of this hypothesis, Craighead, Stunkard, and O'Brien (1981) compared the treatment effects of fenfluramine hydrochloride, behavior therapy, and the combination of pharmacotherapy and behavior therapy on separate groups of obese subjects. After 6 months, subjects in the groups given pharmacotherapy or pharmacotherapy plus behavior therapy had lost approximately 15 kg in body weight, significantly more than subjects receiving behavior therapy alone (10.9 kg). However, after a 12-month follow-up, the subjects given behavior therapy had maintained almost all the weight loss, whereas subjects in the other two groups regained most of their lost weight. The authors suggest that pharmacotherapy may have impaired the effects of behavior therapy rather than improving them, perhaps because subjects attribute the weight loss to the drug.

The combination of behavior therapy with VLCDs has been evaluated with a similar design in which the effects of behavior therapy alone were compared with the effects of VLCD alone and behavior therapy plus VLCD (Wadden & Stunkard, 1986). Results showed weight losses for the combined treatment (19.3 kg) to be superior to those for either the VLCD or behavior therapy alone (14.1, 14.3 kg). At 1-year follow-up, subjects given VLCD alone regained 9.4 kg and weighed more than subjects given behavior therapy plus VLCD, who regained only 6.4 kg. The smallest regain (4.8 kg) was in the group given behavior therapy alone. These data suggest that the combination of behavior therapy plus VLCD was better than either behavior therapy or VLCD alone in producing large weight losses that are maintained.

It is possible that other approaches, such as surgical procedures, may be combined with behavior therapy to improve weight control in the future. In addition, other pharmacological approaches may be better suited to combination with behavior therapy than fenfluramine. However, on the basis of the studies involving pharmacotherapy or VLCD and behavior therapy, it is clear that separate programs cannot simply be combined to produce additive effects, and these adjunctive treatments may limit rather than enhance the long-term effectiveness of behavior therapy.

TREATMENT OF CHILDHOOD OBESITY

The behavioral treatment of obese children has received much less research attention than the treatment of obese adults. However, there are several reasons to focus on the behavioral treatment of obese children. First, obese children are at a greater risk of becoming obese adults than their nonobese peers (Abraham, Collins, & Nordsieck, 1971; Abraham & Nordsieck, 1960), and thus treatment of obesity at an early age may prevent adult obesity. Second, children have a shorter history of eating and exercise behaviors that are associated with obesity, so it may be easier for them to change those behaviors and maintain the change than for adults. Third, it may be easier to mobilize social support for such changes in child behavior than in adult behavior, because parents usually are interested in and able to help their children change (Epstein, Koeske, & Wing, 1986). Fourth, there are important differences in the process by which adults and children become less overweight: adults do so only by losing weight, whereas children can become slimmer by maintaining their weight and growing taller. For this reason, the unit of change in children is the percentage that they are overweight, which takes into account changes in both height and weight rather than only the loss of weight in kilograms.

The major difference in the behavioral treatments of adults and children is that parent management is included with social support procedures for children, whereas many adult obese patients are treated independently of other family members. Because of the unusual degree of interaction and reciprocal determination of behavior between parents and children, numerous studies with children have focused on parenting strategies (Epstein & Wing, 1987) and found them to be among the most powerful determinants of outcome.

Treatment of children also may involve different diets from those used in treating adults. Children require a balanced diet for growth, and thus there have been no trials using VLCD with children. Likewise, there is much less use of pharmacotherapy with children and no controlled trials of pharmacotherapy in combination with behavior therapy. For these reasons, it might be expected that children would not do as well as adults in the typical behavior therapy program. However, the largest reported weight losses in children (−25.5% overweight; Epstein, Wing, Penner, & Kress, 1985) are within the range of percentage overweight changes produced in adults by longer behavior therapy programs (−24.1%; Wadden & Stunkard, 1986) and by more powerful programs such as combined behavior therapy plus VLCD (−33.6%; Wadden & Stunkard, 1986).

Perhaps the most important difference between the treatment of obese adults and children is in the maintenance of treatment effects. As mentioned, adults usually show poor maintenance of treatment effects across a variety of treatment programs (Brownell et al., 1986). Although there are examples of obesity treatment in children in which there similarly is a failure to maintain treatment effects over the long term (Epstein & Wing, 1987), we have shown in several studies that substantial changes in percentage overweight can be maintained over a 5-year period. In one study, we compared obese children who were targeted and reinforced for habit change and weight loss along with their parents to children in two control groups: obese children who were targeted and reinforced for habit change without their parents and obese children who were not reinforced for habit change or weight loss but rather for attendance at treatment meetings. As shown in Figure 1, children who worked on their weight problem along with their parents maintained their weight loss over 5 years, whereas children in the other two groups returned to base-line percentage overweight (Epstein, Wing, Koeske, & Valoski, 1987). These results suggest that there may be some reciprocal benefit to treating several obese persons within a family (Bandura, 1978).

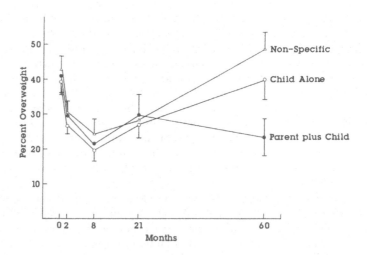

Figure 1. Average percentage overweight in obese children at 0, 2, 8, 21, and 60 months for children in groups in which parent and child were both targeted and reinforced for weight loss, the child alone was targeted for weight loss, or neither parent nor child was targeted or reinforced for weight loss. (From Epstein et al., 1987)

We have demonstrated long-term control of percentage overweight in obese children in a second study evaluating the effects of three types of exercise programs. In this study obese children with obese parents were randomized to three groups who were placed on the same diet but given different exercise programs. A lifestyle exercise program previously tested over a 17-month period (Epstein, Wing, Koeske, Ossip, & Beck, 1982) was compared to an isocaloric aerobic program and a calisthenics exercise group that was similar in amount of exercise time, goal setting, and feedback but that required considerably less caloric expenditure than the aerobic or lifestyle groups. As shown in Figure 2, beginning at 2 years (Epstein, Wing, Koeske, & Valoski, 1985) and extending to 5 years, children given lifestyle exercise had a significantly greater percentage overweight change than children given aerobic or calisthenics exercise programs. Parents assigned to the lifestyle group (−10.3%) also showed better long-term effects than parents assigned to the calisthenics group (−1.3%). This study is one of the few studies we are aware of showing differential maintenance of adult percentage overweight changes over time.

As yet, attempts have not been made to improve the effectiveness of behavioral treatments for children by increasing the length of treatment programs, as has become standard in treating adults. One reason for this difference may be that the treatment effects in children already are greater than those in adults. Another is the fact that many other variables that may contribute to child obesity remain unstudied. However, it seems likely that greater weight loss in children also will be produced by longer programs, and this modification should some day become common in child treatment programs.

One of the most interesting aspects of the treatment of childhood obesity is the possibility of preventing children who are at risk from becoming obese. Although behavioral programs to date always treated children after they became

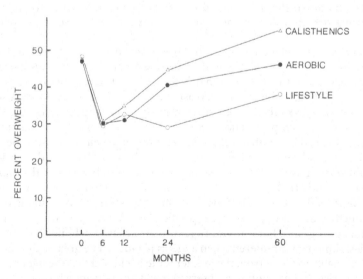

Figure 2. Average percentage overweight in obese children at 0, 6, 12, 24, and 60 months for children in groups in which they dieted plus engaged in lifestyle, aerobic, or calisthenic exercise. (Unpublished data, University of Pittsburgh, 1988)

obese, it may be possible to prevent obesity in children, particularly if those who are at risk can be identified and factors important in the etiology of obesity are known. The risk factor that has generated most interest has been parent weight and the familial aspects of obesity (Garn, Bailey, Solomon, & Hopkins, 1981; Garn & Clark, 1976; Garn, Cole, & Bailey, 1980); obese children are likely to be in families with obese parents and obese siblings. In those children caloric intakes and expenditures should be balanced so as to promote growth but not the accumulation of excess body fat. This could be done simply by monitoring body weight continuously and adjusting caloric intake and output in accordance with changes in height. The mere provision of information on appropriate nutrition to parents was sufficient to reduce the incidence of obesity in a series of young children followed for 3 years (Piscano, Lichter, Ritter, & Siegal, 1978).

INTERFACE OF BEHAVIORAL TREATMENTS WITH BASIC BEHAVIORAL, METABOLIC, AND GENETIC RESEARCH

BEHAVIORAL RESEARCH

Many findings of basic research on eating behaviors have not yet been incorporated into the behavioral treatment of obesity. Two potentially important examples are given, focusing on food preference and sensory-specific satiety.

Food preference is not typically considered in behavioral treatment programs. However, when obese people begin to diet, they often are asked to eat small amounts of low-calorie foods, which they do not prefer as much as the higher-calorie foods they had been eating. It is our experience that obese subjects do not develop preferences for the new foods they are given during a diet, even though these foods are consumed and lead to weight loss (Epstein, Wing, Valoski, & Penner, 1987). Because the subjects do not prefer these foods, it is not surprising that they do not continue to eat them once the treatment period is over.

The goal is to change food preference so that obese people will prefer low-calorie foods over high-calorie foods. One possible means to attain this goal is to use the new foods to reinforce other behavior, which Birch, Birch, Marlin, and Kramer (1982) have shown will increase the preference for that food. A second possibility is to make these foods very palatable. Behavioral investigators should attend more to food preference, because there is no question that maintenance of weight loss will be difficult if subjects do not enjoy eating low-calorie foods.

The second behavioral factor that could be considered is the variety of available food. Basic research has shown that satiety is a function of the availability of new foods (Rolls, 1985); thus, the amount of food consumed is greater when new foods are given to a person who has shown satiety to foods recently consumed. This concept of sensory-specific satiety has not yet influenced obesity treatment programs, but it would suggest that variety, in addition to number of calories, should be considered when a diet is planned. The greater the variety of foods available to the person, the greater is the likelihood that the person will eat more calories. In fact, it often is observed that many persons who go on fad diets that use very-high-calorie foods, such as the ice cream diet, lose weight as long as they eat only one food because they soon tire of it.

Genetic factors typically are not emphasized in the behavioral treatment of obesity. However, there is little doubt that genetics plays a role in the etiology of obesity (Foch & McLearn, 1980), and genetic factors may be important in its treatment.

Genetic influences may help us to understand who is at risk for obesity and how this risk influences the development of obesity. For example, parent obesity may not necessarily cause children to become obese, but rather it may increase the probability of obesity by exposing children to the high-calorie foods eaten by their parents or to family members not engaging in any exercise. If genetic factors predispose the child to obesity, then behavioral investigators may make major contributions to understanding the importance of how behavioral risk factors interact with these predispositions. Genetic factors that predispose a child toward obesity will operate by showing an increased energy intake over energy expenditure and positive energy balance, just as nongenetic factors would be responsible for obesity. In a previous review of this field, we found that there was slightly stronger evidence for a role of genetics in reducing energy expenditure than in increasing food intake but that both sides of the equation were likely to be affected (Epstein & Cluss, 1986).

Twin studies have not shown similarities in caloric intake or food preference beyond that attributable to a common shared environment (Fabsitz, Garrison, Feinleib, & Hjortland, 1978). On the other hand, they have provided evidence that activity level has a strong genetic component (Engstrom & Fischbein, 1977). Other components of expenditure, such as metabolic rate and diet-induced thermogenesis, also may be determined in part by genetic factors (Griffiths & Payne, 1976).

A second important goal for behavioral investigators interfacing with genetic research is to establish the extent to which genetic factors interact with appropriate treatment. It is common to think that when a disorder is identified as having a genetic component, it is impossible to treat because the genes cannot be modified. This shows a poor understanding of the role of genetics in outcome. For example, phenylketonuria (PKU) is a genetic disorder that is expressed when afflicted children eat diets that contain phenylalanine, an amino acid that they cannot metabolize and that leads to mental retardation when it accumulates. Simply maintaining the children in a diet low in phenylalanine during development will prevent retardation. A similar case may be made for obesity. Appropriate diets and exercise regimens may prevent obesity even in individuals with a genetic predisposition to become obese.

LINK SPECIFIC TREATMENTS WITH SPECIFIC ETIOLOGIES

Participants in obesity treatment programs usually are provided similar treatments, as if the problem of obesity were similar for all the participants. In the simplest sense this is true: all obese persons have become obese by consuming more calories then they expended. However, the causes for the positive energy balance are not likely to be the same across all participants. For example, some obese persons may eat to excess during stressful situations. Others may have long periods of inactivity when they cannot easily reduce caloric intake sufficiently to compensate for the lowered expenditure. It seems likely that the reasons for getting obese, or for regaining lost weight, differ across individuals. These specific differences should be linked with the development of individually tailored treatment programs.

This does not imply that the treatments will be totally different from one another. For example, it is likely that most programs for weight loss will emphasize dieting rather than exercising, because dieting produces significantly greater weight change than exercising (Wing & Jeffery, 1979). The need for individualized treatment protocols also does not imply that the relative degrees to which eating and exercise behavior are part of the protocol, once determined, are held constant. For example, a person may become obese because he has very low exercise levels, he may lose weight by dieting without exercising, he may regain weight by eating in stressful situations, and he may lose weight after a short relapse by beginning to exercise. It is difficult to imagine a program for successful long-term weight management that did not attend to these differences in circumstances. Finally, it seems likely that many of the treatment procedures that are currently being used are appropriate for incorporation in any protocol, such as self-monitoring, contracting, and goal setting. It is in the behavioral targets, the goals, and the methods for attaining the goals that specificity is introduced and individualized protocols are established.

CONCLUSIONS

The behavioral treatment of obesity has made substantial progress since its inception over 25 years ago in 1962. Soon thereafter, investigators found that the behavioral programs for obesity were reliable but often produced only small losses in body weight. The amount of loss was later shown to be related to the relatively short duration of the treatments. Now that the treatments have become longer, the average weight losses have doubled so that they now amount to 7 to 12 kg or so (Jeffery *et al.*, 1983).

Nevertheless, behavioral therapists working with obese adults have not been satisfied with the magnitude of these weight losses, and they have incorporated techniques other than typical dieting and exercising to change energy balance. These methods have included both pharmacological agents and VLCDs. However, research on pharmacological agents has not been promising, because the use of the agents either alone or together with behavior therapy has not been associated with better long-term follow-up than was obtained with behavior therapy alone. On the other hand, the use of VLCDs may be a better adjunctive treatment, because behavior therapy plus VLCDs seem to be superior to behavior therapy or to VLCDs alone in producing and maintaining body weight loss.

Research combining behavioral and nonbehavioral treatments is important to understanding more about the strengths of behavioral treatment. Simply adding two treatments together does not necessarily result in better treatment effects, and it certainly does not result in better maintenance effects. In fact, doing things in addition to behavior therapy may compromise the positive long-term effects that behavior therapy alone usually shows.

During the same 25-year period, the behavioral treatment of childhood obesity has made even greater strides than the treatment of obese adults, with research documenting and then replicating effective treatments for long-term control of body weight. Still greater increases in weight loss, perhaps by lengthening the duration of treatments and by using behavioral treatments to prevent obesity in high-risk children, seem to be reachable goals.

Much remains to be accomplished in the development of effective treatments for obese adults and children. Behavioral treatments have been derived in large part from learning theory, which provides an important empirical background for the critical evaluation of current behavioral programs. However, other areas in behavioral science also should influence the assessment and treatment of obesity, such as basic research in eating and genetics. Moreover, behavioral treatments of obesity should become more sensitive to individual differences that determine eating and activity for obese individuals. As discussed, there are a wide variety of possible reasons why people are obese and why they have trouble losing and maintaining their weight losses. It is likely that once we have obtained a better understanding of these mechanisms and developed appropriate methods to treat these problems, there will be much better weight losses and maintenance of weight loss than are commonplace now. Although it remains to be seen whether some day we will see a "cure" for obesity, it already is clear that behavioral treatments have taught us methods important in modifying eating and exercise and thereby affecting body weight in the obese.

Acknowledgments

Research reported here was supported in part by grants from National Institute of Child Health and Human Development.

REFERENCES

Abraham, S., Collins, G., & Nordsieck, M. (1971). Relationship of childhood weight status to morbidity in adults. *Public Health Reports, 85,* 273–284.

Abraham, S., & Nordsieck, M. (1960). Relationship of excess weight in children and adults. *Public Health Reports, 75,* 263–273.

Bandura, A. (1978). The self-system in reciprocal determinism. *American Psychologist, 33,* 344–358.

Birch, L. L., Birch, D., Marlin, D. W., & Kramer, L. (1982). Effects of instrumental consumption on children's food preference. *Appetite: Journal of Intake Research, 3,* 125–134.

Bonham, G. S., & Brock, D. B. (1985). The relationship of diabetes with race, sex and obesity. *American Journal of Clinical Nutrition, 41,* 776–783.

Brownell, K. D., Heckerman, C. L., Westlake, R. J., Hayes, S. C., & Monti, P. M. (1978). The effect of couples training and partner co-operativeness in the behavioral treatment of obesity. *Behaviour Research and Therapy, 16,* 323–334.

Brownell, K. D., Marlatt, G. A., Lichenstein, E., & Wilson, G. T. (1986). Understanding and preventing relapse. *American Psychologist, 41,* 756–782.

Brownell, K. D. Stunkard, A. J., & McKeon, P. E. (1985). Weight reduction at the work site: A promise partially fulfilled. *American Journal of Psychiatry, 141,* 47–51.

Brownell, K. B., & Wadden, T. A. (1986). Behavior therapy for obesity: Modern approaches and better results. In K. D. Brownell & J. P. Foreyt (Eds.), *Handbook of eating disorders* (pp. 180–197). New York: Basic Books.

Buxton, A., Williamson, D. A., Absher, N., Warner, M., & Moody, S. C. (1985). Self-management of nutrition. *Addictive Behaviors, 10,* 383–394.

Cohen, E. A., Gelfand, D. M., Dodd, D. K., Jensen, J., & Turner, C. (1980). Self-control practices associated with weight loss maintenance in children and adolescents. *Behavior Therapy, 11,* 26–37.

Craighead, L. W., Stunkard, A. J., & O'Brien, R. (1981). Behavior therapy and pharmacotherapy for obesity. Archives of General Psychiatry, 38, 763–768.

Engstrom, L. M., & Fischbein, S. (1977). Physical capacity in twins. *Acta Geneticae Medicae et Gemellologiae (Roma), 26,* 159–165.

Epstein, L. H., & Cluss, P. A. (1986). Behavioral genetics of childhood obesity. *Behavior Therapy, 17,* 324–334.

Epstein, L. H., Koeske, R., & Wing, R. R. (1986). The effect of family variables on child weight loss. *Health Psychology, 5,* 1–12.

Epstein, L. H., & Wing, R. R. (1987). Behavioral treatment of childhood obesity. *Psychological Bulletin, 101,* 331–342.

Epstein, L. H., Wing, R. R., Koeske, R., Andrasik, F., & Ossip, D. J. (1981). Child and parent weight loss in family-based behavioral modification programs. *Journal of Consulting and Clinical Psychology, 49,* 674–685.

Epstein, L. H., Wing, R. R., Koeske, R., Ossip, D. J., & Beck. S. (1982). A comparison of lifestyle change and programmed aerobic exercise on weight and fitness changes in obese children. *Behavior Therapy, 13,* 651–665.

Epstein, L. H., Wing, R. R., Koeske, R., & Valoski, A. (1985). A comparison of lifestyle exercise, aerobic exercise, and calisthenics on weight loss in obese children. *Behavior Therapy, 16,* 345–356.

Epstein, L. H., Wing, R. R., Koeske, R., & Valoski, A. (1986). Effects of parent weight on weight loss in obese children. *Journal of Consulting and Clinical Psychology, 54,* 400–401.

Epstein, L. H., Wing, R. R., Koeske, R., & Valoski, A. (1987). Long-term effects of family-based treatment of childhood obesity. *Journal of Consulting and Clinical Psychology, 55,* 91–95.

Epstein, L. H., Wing, R. R., Penner, B., & Kress, M. J. (1985). The effect of diet and controlled exercise on weight loss in obese children. *Journal of Pediatrics, 107,* 358–361.

Epstein, L. H., Wing, R. R., & Valoski, A. (1985). Childhood obesity. *Pediatric Clinics of North America, 32,* 363–379.

Epstein, L. H., Wing, R. R., Valoski, A., & Penner, B. C. (1987). Stability of food preferences during weight control: A study with 8- to 12-year-old children and their parents. *Behavior Modification, 11,* 87–101.

Fabsitz, R. R., Garrison, R. J., Feinleib, M., & Hjortland, M. (1978). A twin analysis of dietary intake: Evidence for a need to control for possible environmental differences in MZ and DZ twins. *Behavior Genetics, 8,* 15–25.

Ferster, C. B., Nurnberger, J. I., & Levitt, E. B. (1962). The control of eating. *Journal of Mathetics, 1,* 87–109.

Foch, T. T., & McLearn, G. E. (1980). Genetics, body weight, and obesity. In A. J. Stunkard (Ed.), *Obesity* (pp. 48–71). Philadelphia: W. B. Saunders.

Foreyt, J. P., Goodrick, K., & Gotto, A. M. (1981). Limitations of behavioral treatment of obesity: Review and analysis. *Journal of Behavioral Medicine, 4,* 159–174.

Garn, S. M., Bailey, S. M., Solomon, M. A., & Hopkins, P. J. (1981). Effect of remaining family members on fatness prediction. *American Journal of Clinical Nutrition, 34,* 148–153.

Garn, S. M., & Clark, D. C. (1976). Trends in fatness and the origins of obesity. *Pediatrics, 57,* 443–456.

Garn, S. M., Cole, P. E., & Bailey, S. M. (1980). Effect of parental fatness levels on the fatness of biological and adoptive children. *Biology of Food and Nutrition, 7,* 91–93.

Griffiths, M., & Payne, P. R. (1976). Energy expenditure in small children of obese and non-obese parents. *Nature, 260,* 698–700.

Heyden, S., Hames, C. G., Bartel, A., Cassel, J. C., Tyroler, H. A., & Cornoni, J. C. (1971). Weight and weight history in relation to cerebrovascular and ischemic heart disease. *Archives of Internal Medicine, 128,* 956–960.

Jeffery, R. W., Gerber, W. M., Rosenthal, B. S., & Lindquist, R. A. (1983). Monetary contracts in weight control: Effectiveness of group and individual contracts of varying size. *Journal of Consulting and Clinical Psychology, 51,* 242–248.

Leitenberg, H. (1976). *Handbook of behavior modification and behavior therapy.* Englewood Cliffs, NJ: Prentice Hall.

Messerli, F. H. (1984). Obesity in hypertension: How innocent a bystander? *American Journal of Medicine, 77,* 1077–1082.

Morganstern, K. P. (1974). Cigarette smoke as a noxious stimulus in self-managed aversion therapy for compulsive eating: Technique and case illustration. *Behavior Therapy, 5,* 255–260.

National Diabetes Data Group. (1979). Classification and diagnosis of diabetes mellitus and other categories of glucose intolerance. *Diabetes, 28,* 1039–1057.

Piscano, J. C., Lichter, H., Ritter, J., & Siegal, A. P. (1978). An attempt at prevention of obesity in infancy. *Pediatrics, 61,* 360–364.

Reisen, E., & Frohlich, E. D. (1978). Effects of weight without salt restriction on the reduction of blood pressure in overweight hypertensive patients. *New England Journal of Medicine, 298,* 1–5.

Rolls, B. J. (1985). Experimental analyses of the effects of variety in a meal on human feeding. *American Journal of Clinical Nutrition, 42*, 932–939.

Rosenthal, B., Allen, G. J., & Winter, C. (1980). Husband involvement in the behavioral treatment of overweight women: Initial effects and long-term follow-up. *International Journal of Obesity, 4*, 165–173.

Simopoulos, A. P., & Van Itallie, T. B. (1984). Body weight, health, and longevity. *Annals of Internal Medicine, 100*, 285–295.

Stewart, A. L., Brook, R. H., & Kane, R. L. (1980). *Conceptualization and measurement of health habits for adults in the health insurance study: Vol. II, Overweight.* Santa Monica, CA: Rand Corporation.

Stuart, R. B. (1967). Behavioral control of overeating. *Behaviour Research and Therapy, 5*, 357–365.

Stuart, R. B., & Davis, B. (1972). *Slim chance in a fat world: Behavioral control of overeating.* Champaign, IL: Research Press.

Stunkard, A. J. (1958). The management of obesity. *New York Journal of Medicine, 58*, 79–87.

Tondo, T. R., Lane, J. R., & Gill, K., Jr. (1975). Suppression of specific eating behaviors by covert response cost: An experimental analysis. *Psychological Record, 25*, 187–196.

U.S. Department of Health Education and Welfare. (1979). *Overweight adults in the United States. Advance data from Vital and Health Statistics of the National Center for Health Statistics, no. 51.* Rockville, MD: USDHEW.

Wadden, T. A., & Stunkard, A. J. (1986). Controlled trial of very low calorie diet, behavior therapy, and their combination in the treatment of obesity. *Journal of Consulting and Clinical Psychology, 54*, 482–488.

Wadden, T. A., & Stunkard, A. J. (1987). Psychopathology and obesity. In R. J. Wurtman & J. J. Wurtman (Eds.), *Human obesity* (pp. 55–65). New York: New York Academy of Sciences.

Wilson, G. T., & Brownell, K. B. (1980). Behavior therapy for obesity: An evaluation of treatment outcome. *Advances in Behavior Research and Therapy, 3*, 49–86.

Wing, R. R., Epstein, L. H., Marcus, M. D., & Kupfer, D. J. (1984). Mood changes in behavioral weight loss programs. *Journal of Psychosomatic Research, 28*, 189–196.

Wing, R. R., Epstein, L. H., Nowalk, M. P., Koeske, R., & Hagg, S. (1985). Behavior change, weight loss, and physiological improvements in type II diabetic patients. *Journal of Consulting and Clinical Psychology, 53*, 111–122.

Wing, R. R., & Jeffery, R. J. (1979). Outpatient treatments of obesity: A comparison of methodology and results. *International Journal of Obesity, 3*, 261–279.

Part II
Food Intake and
Caloric Homeostasis

5

The Ontogeny of Ingestive Behavior
Changing Control of Components in the Feeding Sequence

W. G. HALL

THE SEQUENTIAL AND COMPONENT NATURE OF EARLY APPETITIVE BEHAVIOR

At the beginning of postnatal life, in their first expression of appetitive behavior, young mammals suckle. A newborn foal, barely able to stand, will orient to the shape of the mare, approach, and make tentative probing movements below the udder and teat region. Then, guided by tactile cues to its muzzle and mouth and by the proximity of odor signals, it grasps and engulfs the teat with its mouth (Waring, 1983). Neonatal rabbits resting in their fur-lined nest begin excitedly searching for their mother when alerted by vestibular and tactile signals of her daily return (Distel & Hudson, 1985; Hudson & Distel, 1982). Then, as the dam crouches over them, the young contact her fur and, aroused by odor cues, are stimulated into a vigorous probing and searching of her ventrum. Eventual contact with the nipple elicits nipple-grasping responses followed by attachment, sucking, and milk withdrawal.

Mammalian appetitive behavior in general is epitomized by such suckling sequences and by the adaptive ordering of individual responses into an effective chain of component behavioral elements or modules (Hinde, 1970; Tinbergen, 1951; Figure 1A). In the suckling sequence there is an obvious progression of responses leading from one to the next, with each response positioning the infant to respond to cues for the next response (Figures 1B and 2A): presence of the mother elicits activity that leads to contact with the mother's fur or odor,

W. G. HALL   Department of Psychology, Duke University, Durham, North Carolina 27706.

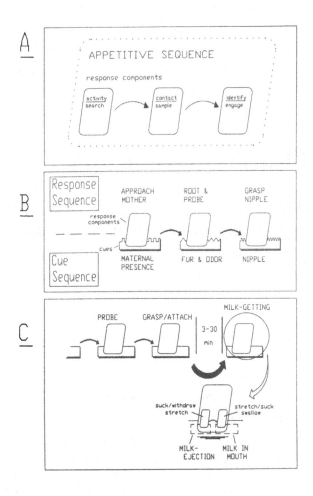

Figure 1. Diagram of the sequential nature of appetitive response chains. (A) A general framework representing the ordering of appetitive responses. This and similar schemes are elaborations of the ordering of behavioral elements or components as conceptualized by early physiologists and ethologists (e.g., Craig, 1918; Sherrington, 1906/1947). (B) The sequential organization of suckling. The topography and texture of the environmental cue sequence organize the response sequence—a cue elicits a response component, which leads the infant to encounter the next environmental cue, which organizes the next response component, and so on. Infants typically first respond to the mother's presence with approach, which leads to contact with her ventrum and elicits the rooting and probing of nipple search, which leads to contact with a nipple. (C) Several of the specific components of suckling in infant rodents are readily identified and can be isolated for study. These include probing into the fur, or nipple search, grasping the nipple on contact with it, and, after a period of time determined by maternal responsiveness, showing a milk-getting response in reaction to milk letdown. The milk-getting response can itself be divided into subcomponent responses, with each being subject to separate control.

which then elicits nipple search; eventual contact with a nipple elicits oromotor nipple-grasping responses; mechanical and suction attachment occur after the nipple has been grasped; and attachment quiets the neonate and terminates the appetitive sequence (Blass, Hall, & Teicher, 1979; Blass & Teicher, 1980; Hudson & Distel, 1982; for terminology see Hall, Hudson, & Brake, 1988). Even

A

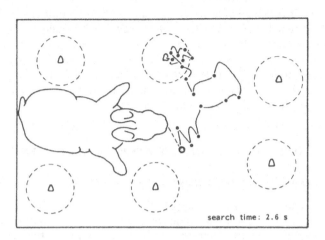

search time: 2.6 s

B

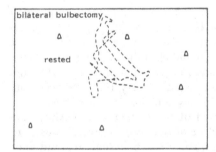

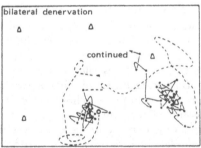

Figure 2. (A) Typical nipple-search sequence of a rabbit pup tested on the ventrum of a restrained doe. Broken line indicates locomotion without nipple search. Pup initiated search at the star. Dots indicate probes into fur. Note the shift in probing behavior as search brings pup within the (arbitrarily defined) nipple region, where it identifies the nipple and then attaches. (B) Search pattern of surgically manipulated rabbit pups. Left panel shows that probing and search are not initiated in pups lacking olfactory input after bilateral olfactory bulbectomy. Right panel shows the effect of snout denervation on pups and reveals the inability to shift from probing to nipple grasping. The pup identifies the nipple region but is unable to attach and moves on to another region. (Modified from Distel & Hudson, 1985, with permission.)

though the specific components of suckling and their eliciting cues may differ between species, this method of sequential environmental cuing provides a universal means of organizing a series of independent response components into an adaptive chain (e.g., Hinde, 1970). Behavioral patterning based on such direct responses to orderly features of environmental topography is perhaps less essential for more mature and flexible appetitive responses. Yet suckling instructively exaggerates the manner in which appetitive repertoires invariably are composed of sequences of distinct behavioral elements, partially patterned and guided by the structure of the environment and by the cascade of sensory–perceptual events in the normal chain of action (e.g., Fentress, 1976; Hinde, 1970; Norgren & Grill, 1982; Teitelbaum, 1977). Developmentally, such sequences may become increasingly hierarchically embedded, dynamically interdependent, and influenced by experience (Fentress, 1981, 1983; Gallistel, 1980; Tinbergen, 1951), but this maturation does not deny their essential multicomponent nature or reduce the need to understand the individual elements and their control as well as the processes that orchestrate the full sequence.

Separate Control of Appetitive Response Components

The Sequence of Suckling Components in Rodents

The consequence of viewing appetitive sequences as made up of distinct components is a recognition that these components may be modulated or controlled individually and by different mechanisms. I illustrate this principle of separate control with data from studies of suckling behavior in infant rodents. As with the horse and rabbit neonates described above, the suckling behavior of young rat pups exhibits an obvious progression of response components (Figure 1C), which can be elicited and studied individually.

Nipple Search. For rat pups, the nipple-search response, composed of head sweeping and probing, can be elicited by simply placing pups on fur, either in maternal pelage or on a synthetic fur surface. On synthetic fur, without nipples to terminate the search component and elicit the nipple-grasping component, young pups will probe at high rates for over an hour compared to their typical preattachment probing of less than a minute (Terry, 1988). The vigor and duration of pups' searching behavior in this situation is determined by the amount of time they have been deprived of the mother and suckling. This control by deprivation potentially could be attributable to several causes, for example, nutritive privation, dehydration, absence of maternal care, and so on. But one factor appears to be the primary determinant of the vigor of the search response, and that is the purely experiential loss of maternal interaction. When pups are left with nonlactating foster mothers and are thereby deprived of food and fluid but not maternal interaction, they subsequently show a low responsiveness when placed on synthetic fur, responsiveness similar to nondeprived pups (Figure 3). Experience with maternally behaving adults that lack nipples— therefore, the opportunity for nipple search without attachment—is equally effective in reducing subsequent probing. Exposure to synthetic fur itself also reduces the vigor of subsequent probing. These data suggest that the experience of, or the stimulation from, engaging in nipple-search behavior modulates the future elicitability of the behavior, a "use-dependent" form of intrinsic modulation of a response component, a subject to which I return.

Nipple Grasping. Control of the nipple-search response can be contrasted to nipple grasping, the succeeding component in the suckling sequence (Figure 1C). Nipple grasping can be separated for analysis from nipple search because it can be triggered by simply holding pups at the nipple (Drewett & Cordall, 1976;

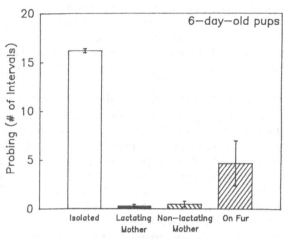

Figure 3. Amount of probing (nipple search) shown by 6-day-old rat pups in a 6-minute test on synthetic fur after various deprivation manipulations (mean ± SEM). Probing score reflects the number of 15-second intervals (of 24) in which pups were observed actively probing into fur. Note that the high levels of probing stimulated by deprivation are reduced not only by nutritive suckling but also by nonnutritive suckling and by mere experience on fur. (Modified from Terry & Hall, in press-a.)

Drewett, Statham, & Wakerley, 1974; Teicher & Blass, 1976). Elicitation of the grasping response requires appropriate tactile and olfactory stimuli from the nipple and its surround (Hofer, Fisher, & Shair, 1981; Hofer, Shair, & Singh, 1976; Teicher & Blass, 1976) and is influenced by time of day, though I do not emphasize this circadian effect here (Henning & Gisel, 1980; Holloway, Dollinger, & Denenberg, 1979). But, in contrast to the search component, in young pups the grasping response occurs independently of pups' experiential, nutritional, or hydrational state. Young pups, 10 days of age or less, typically grasp the nipple and become attached whether deprived or not. The lack of state dependence of this terminal nipple-grasping response accounts for a well-established finding that, early in the preweaning period, nipple attachment typically occurs regardless of pups' deprivational state, even when pups are simply put to suckle with the mother and must both search and attach (Hall, Cramer, & Blass, 1977; Henning, Chang, & Gisel, 1979; though see Cramer & Blass, 1982). This latter fact holds because, though modulated by experience, the preceding appetitive component of nipple search is expressed to some degree even in nondeprived pups (e.g., Hall *et al.*, 1977).

The absence of control of nipple attachment is made particularly obvious by the finding that gastric loads do not suppress attachment in young pups (Hall & Rosenblatt, 1978; Lorenz, 1983; Lorenz, Ellis, & Epstein, 1982). Moreover, even when pups receive large volumes of milk by oral infusion while attached to the nipple, they still persist in nipple attachment (Hall & Rosenblatt, 1977). Thus, pairing nipple attachment with the development of excessive gastric fill does not discourage pups from attaching or even reattaching to nipples. This persistence of attachment is consistent with an absence of deprivation-related controls of nipple grasping and further shows that young pups are not likely to use attaching to or detaching from nipples as a way to control intake. Later, in pups older than 10 days of age, nipple grasping comes to be controlled by pups' deprivation state (Drewett & Cordall, 1976), and so too does their overall attachment behavior when they are placed with a mother (Hall *et al.*, 1977; Henning *et al.*, 1979).

MILK CONSUMPTION. For those studying the development of rodent suckling behavior, it initially was puzzling that nipple-grasping and attachment responses in young pups seemed to be under little control other than that related to sensory factors. These findings were at apparent odds with the earlier demonstrations that the milk intake of young pups during normal suckling was modulated by deprivation and by gastric loads (Friedman, 1975; Houpt & Epstein, 1973; Houpt & Houpt, 1975; Lytle, Moorcroft, & Campbell, 1971). The expected results would have been findings of parallel modulation of all the appetitive components of suckling, effects that might have accounted for, or correlated with, the effects on milk consumption. In the present context of a component-by-component analysis of suckling, these findings are more readily interpreted. They emphasize the point that different response elements can be under very different control and lead to the specific understanding that the final component in the suckling sequence, milk consumption, is controlled independently of search and attachment.

To understand fully the control of milk consumption in rat pups, it is necessary to know something about how they acquire milk. Infant rodents actually are attached to the nipple for a considerable portion of the day, but milk consumption occurs only during brief episodes of maternal milk ejection lasting about 10 seconds each and happening many minutes after nipple attachment (Lincoln, Hill, & Wakerley, 1973). During the short episodes of milk ejection a pup rapidly whisks diet down its throat while engaging in a unique and stereotypic "stretch" response (Drewett *et al.*, 1974). These stretch responses and subsequent swallowing are triggered by fluid or tactile stimulation of the back of the mouth (Hall & Rosenblatt, 1978; Lau & Henning, 1985) at the occurrence of the mother's periodic milk ejections. Thus, milk intake is not closely linked in time to nipple-search, nipple-grasp, or attachment responses but occurs at irregular intervals. In addition, even though pups do make several types of sucking and mouthing movements during the extensive time they are attached (Brake, Wolfson, & Hofer, 1979), most of this sucking is nonnutritive in the sense that no milk is obtained.

For rodents, the act of milk getting during suckling can itself be viewed as a multicomponent response system made up of two elements (Brake, Tavana, & Myers, 1986): (1) a milk-detection and initial sucking and withdrawal phase, during which the "stretch" is initiated and milk is first drawn into the mouth, and (2) a swallowing component of the stretch response by which, while the "stretch" and sucking are continued, milk is passed from the back of the mouth into the esophagus and stomach (Figure 1C). The latter phase is the simplest and is perhaps helpfully conceptualized as a reflex. I consider it first.

Once milk has arrived at the back of the mouth and a stretch response has been triggered, swallowing invariably occurs, and the "stretch" is maintained. There appears to be little control or modulation of this final swallowing component of the ingestive sequence in pups less than 2 weeks of age. Rosenblatt and I (Hall & Rosenblatt, 1977) found that when fluids were provided by infusions through oral cannulae into the mouths of young pups attached to a nipple, the fluid stimulation invariably triggered a stretch response, and the solutions always were swallowed. Young pups that were receiving such infusions during suckling continued to show stretch responses and consume milk until large volumes were ingested and extreme gastric filling developed. These experiments indicated that, in young pups, control of intake did not occur once milk was in the mouth. The findings suggested to us that the initial milk detection/withdrawal component of milk getting was the only behavioral component at which the amount of milk ingested during normal suckling could be influenced. Our subsequent analyses confirmed that this initial portion of the milk-consumption response, along with the sucking associated with it, is modulated to provide a control over milk intake during suckling (described in Blass *et al.*, 1979). Specifically, the critical features of this control are the vigor in exertion of negative pressure at the nipple as well as in pups' vigilance in detecting milk availability. These response features are modulated (Brake & Hofer, 1980), and such modulation contributes to differential ingestion by pups in different test conditions, even though neither nipple grasping, attachment, nor swallowing of solutions is state dependent.

The details of the modulation of this first component of milk getting and the manner in which intake is affected are complex but instructive. We are largely indebted to Brake and his collaborators for our current, well-developed understanding of this process (recently reviewed by Brake, Shair, & Hofer, 1988). Brake's group first carefully analyzed the type of sucking behavior used by pups to generate negative pressure at the nipple. They identified two basic types: arrhythmic sucking and rhythmic sucking. Either could occur in the absence of milk. In fact, even sleeping pups frequently suck (Brake *et al.*, 1979). But only a rapid form of rhythmic sucking occurs when pups are ingesting milk during the stretch at a milk ejection (Brake *et al.*, 1986). It is this rhythmic sucking at the milk ejection that makes milk available for swallowing. Rhythmic sucking was shown to be influenced by a number of factors including sucking experience, stomach load, taste, and milk deliveries themselves (Brake, Sager, Sullivan, & Hofer, 1982; Pelchat & Brake, 1987). These findings are consistent with demonstrations that a number of factors, both nutritive and nonnutritive, can influence intake in young pups (e.g., deprivation and gastric fill, Houpt & Epstein, 1973; dehydration, Friedman, 1975; Friedman & Campbell, 1974; suckling experience with nonlactating mothers, Cramer & Blass, 1985).

A particularly helpful aspect of the work of Brake and his co-workers was resolving how the control over rhythmic sucking is effected. As noted above, pups were found to show sucking movements when awake and in both paradoxical sleep (PS) and slow-wave sleep (SWS). However, the distribution of sucking across these states was not equal. Pups did 53% of their sucking while awake, 33% in SWS, and 14% in PS. The important finding was that deprivation did not increase sucking by altering the rate of sucking in any state but simply increased the amount of time pups spent awake, and in this way increased the overall amount of sucking. Thus, according to Brake *et al.* (1988), effects on intake appear to be mediated through changes in sleep–wake behavior. They point out, however, that control is not simply a matter of a sleep/satiety occurring after milk ejections; pups actually are awakened after milk delivery. Rather, the effect of changes in sleeping and waking states is distributed across a longer period of time and over pups' behavior in general. In this way they have a cumulative effect on intake by influencing the reactions to episodic milk deliveries. This finding that intake is influenced by the amount of waking activity is consistent with other reports of nonnutritive and nutritive influences on sleep (e.g., Lorenz, 1986) and with earlier arguments that behavioral state has an important influence on suckling intake (e.g., Hall & Williams, 1983). Although it remains possible that the various effects on suckling intake may be mediated through different mechanisms or through different combinations of mechanisms (e.g., gastric fill might have a direct oromotor effect), mediation of the effects of all manipulations through a common effect on state seems at the moment to provide a consistent and parsimonious explanation for the modulation of suckling intake.

WHAT SUCKLING TELLS US ABOUT FEEDING SYSTEMS

This review of the three behavioral components in the suckling sequence has depicted their individual and differential control. More importantly, I have tried to indicate how understanding the expression of the overall sequence of an appetitive behavior requires an appreciation of the importance of individual controls in influencing the appetitive sequence and in determining its consummatory outcome. Thus, this analysis of a mammal's most primitive appetitive response system argues that appetitive behavior is best conceived in terms of multiple and sequential components, each of which can be subject to multiple and differential control. Because control or modulation at any component in the sequence has the potential to affect action, or inaction, at all subsequent components, any apparent modulation of an endpoint measure of appetitive behavior, such as intake volume, may be the result of a controlling or limiting influence elsewhere in the sequence (e.g., Hinde, 1970). Nipple grasping, for example, is not a behavioral site that influences intake in young pups. However, the modulation of grasping does influence intake in older pups. Further, even though two different manipulations may have similar effects on measures such as intake volume, these effects need not be based on the same modulatory mechanism or, more importantly, reflect controls exerted at the same neurobehavioral site. Elucidation of the control of appetitive behavior will require a concern for such sites of modulation in the response element sequence (Figure 4).

With respect to the nature of the controls of response elements, note that the analysis of appetitive aspects of suckling revealed a close relationship between experiential and physiological factors as modulators of response components. Psychobiologists' traditional emphasis on the homeostatic facet of feeding predisposes an appreciation of the physiological modulation of appetitive responses, control that can be conceived as an extrinsic input to a response component or system (Figure 4A). But analysis of nipple-search and milk-withdrawal behaviors called attention to the equally fundamental contribution of experience or "use" to the modulation of responding, control that can be conceived as an intrinsic property of the response element. I return to this aspect of appetitive control in a later section.

Finally, this analysis of suckling suggests that despite its overall parallel to adult feeding and drinking as an appetitive sequence and in having food ingestion as an endpoint, suckling is not homologous to feeding. There is little apparent similarity between the specific components of the suckling sequence and those of the adult feeding and drinking. Indeed, they have little resemblance

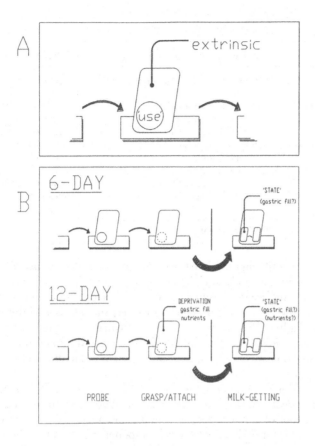

Figure 4. Sources of control for response components in the suckling sequence. (A) Any response component may be subject to control that is extrinsic to it (e.g., physiological modulation, modulation by activity or arousal states, learned influences), but it also is subject to a fundamental modulation by its own "use" or activation. (B) Some major sources of modulation of components in the suckling sequence of 6- and 12-day-old rat pups. In the suckling response sequence of 6-day-old rat pups, probing behavior is clearly modulated by "use" and seems to have little other control. Nipple grasping has little apparent control; "use" has not been carefully studied. For the milk-getting response component, sleep–wake state (and perhaps stomach fill directly) influences the first subcomponent, though there is little control of intake once diet is in the mouth and the stretch response has been triggered. By 12 days of age, a new extrinsic control has developed over nipple attachment, and this change reorganizes the expression and characteristics of suckling.

either in their sensory–motor features or in their control and modulation (Table 1; see review by Hall & Williams, 1983; Drewett, 1978; Epstein, 1984, 1986). Brake's finding that changes in sucking are a consequence of changes in sleep–wake state provides a good example and confirms previous suggestions that although intake was controlled during normal suckling, such control was indirect. Thus, suckling is not just an immature version of feeding. Indeed, the pattern of milk availability is so different from the typical form of food presentation to adult animals that it would seem unlikely that controls of intake could have much in common. Suckling was initially studied by psychobiologists as a

TABLE 1. SOME DIFFERENCES IN THE FEATURES OF COMPONENTS OF SUCKLING (FOR PUPS YOUNGER THAN 12 DAYS) AND ADULT FEEDING[a]

Feature	Suckling	Adult feeding
Ingestive response	'Stretch' response with guzzling of milk	Complex movements of mouth to lick, chew, and swallow
External control	Eliminated by anosmia	Limited effects of anosmia
Internal control	Attachment not affected by deprivation or dehydration	Deprivation-dependent; dehydration-inhibited
Neural substrate	Intake not affected by amphetamine or norepinephrine	Inhibited by amphetamine; stimulated by norepinephrine
Necessity for experience	Prevention of suckling disrupts later suckling	Prevention of suckling or early feeding does not affect adult feeding

[a]Summarized from Hall & Williams (1983).

precursor to adult ingestive systems, but the lack of evidence for continuity between this infant mode and the adult modes of ingestion is now recognized.

Suckling and weaning remain fascinating biological systems worthy of considerable study in their own right. Moreover, the analysis of the origins and ontogeny of adult ingestive systems can be instructively guided by our analysis of suckling's appetitive organization, and the two may overlap in some ways. There may even be some mechanisms, such as those involved in the modulation of sleep state (e.g., Lorenz, 1986), that are common between pups and adults. But study of the specific development of feeding response components can better proceed in a different fashion. In the following sections I describe studies of the ontogeny of components of adult ingestive systems and their control from this other perspective. I refer to these behavioral components, as they have been studied in developing rodents, as components of "independent ingestion," since they represent features of a response system that exist independent of the mother and suckling.

THE SIMPLE INDEPENDENT INGESTIVE RESPONSES OF YOUNG RATS

Infants of altricial rodent species like rats are born with immature sensory and motor capabilities. Much morphological, physiological, and neural maturation occurs postnatally (Figure 5). Even if, for argument's sake, all central sub-

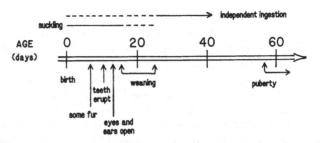

Figure 5. Timing of some of the major events in the development of rats. (From Hall, 1985.)

strates of an appetitive system for adultlike ingestion were present in these immature animals, such animals probably could not express ingestive behavior independent of the mother in any adultlike manner. Such apparent lack of feeding would result simply because pups could not respond to the flow of stimuli that lead from one component in the response chain to the next; they would be physically unable to get themselves from one trigger point to the next trigger point in the behavior sequence. Therefore, in order to identify and study those components of the ingestive sequence that might be present early in life, the components must be elicited in a fashion that obviates pups' sensorimotor limitations. Young pups must be tricked into showing the components that they possess. The demonstration of a response component then may be taken as an indication that one has presented to the neonate an age-appropriate stimulus for eliciting that component.

Although such methods for extracting early behavior might at first appear highly artificial, there is a good deal of sensitivity to the natural in the coaxing of behavior from immature animals. Such coaxing requires understanding the range in which signals are processed by a young animal and the degree to which other stimuli or conditions that may be irrelevant to older animals can interfere with the expression of behavior in the young. As experimenters have become increasingly sensitive to the world of infant animals, young animals have begun to exhibit capabilities not previously imagined. Illustrations include the demonstration of increasingly younger learning capabilities in infants of a number of species (e.g., Johanson & Hall, 1979; Hudson, 1985; Rudy & Cheatle, 1977; Smotherman & Robinson, 1985), including humans (e.g., DeCasper & Fifer, 1980; Lipsitt, 1977), and the discovery of the precocial presence of components of other behaviors, for example, in neonatal rodents precocial grooming behavior (Fentress, 1978) and precocial sexual responses (Williams & Lorang, 1985).

A particularly instructive example is Thelen's (1984; Thelen & Fisher, 1982) analysis of "stepping" behavior in human infants. For the first month or so after birth, human infants will make stepping movements when held erect, seeming to possess the neural organization for patterning this response and producing its alternating sequence of limb activation. Then this stepping behavior is lost, not to reappear until after 10 months. An earlier interpretation of these developmental changes was in terms of changing neural organization of limb control, with early stepping viewed as having only a fortuitous resemblance to later stepping, which required the development of new neural circuits. Thelen has shown, however, that infants continue to possess patterned stepping even during the period when it is not apparent. During this time the expression of stepping is normally overwhelmed by the peripheral biomechanical factors of added limb fat and weight. When Thelen and her co-workers held 7-month-old neonates with their legs suspended in water, reducing the effect of the added weight, stepping was shown. Thus, while these components of coordinated leg movement await muscle development for normal expression in stepping and walking, they are continuously present in infancy even though they are not normally apparent. When the situation is appropriately arranged, these components can be made apparent, and their characteristics and controls can be studied.

For developmental analysis of feeding and drinking, a basic behavioral component is the mouthing and swallowing of food or fluid once it is in the mouth. By identifying the appearance of this oromotor component in development and then determining how its controls mature, we can provide a foundation for an understanding of the origins and organization of adult ingestive systems. The analysis starts with the straightforward question: What does a newborn rat pup do with food or fluid placed in its mouth? Wirth and Epstein (1976) first addressed this question by holding rat neonates to a flowing water spout and monitoring their ingestion after different treatments. In such tests, pups were not required to produce all the preceding components of the appetitive response sequence (e.g., locating, approaching, and identifying food) in order to indicate whether they possessed the substrates of the response component for lapping and swallowing. This procedure for studying the final component of ingestion was later extended in my laboratory by making use of fine intraoral cannulas that permit controlled deliveries of test fluids to different positions in pups' mouths (Figure 6; Hall, 1979; see Phifer & Hall, 1987, for details of techniques).

THE FINAL INGESTIVE RESPONSE COMPONENT IS PRESENT AT BIRTH

Based on experiments studying the oromotor component of early ingestion using orally infused diets, we now know that, from birth, rats can actively lap, mouth, and swallow (e.g., Wirth & Epstein, 1976). Indeed, such responses have been observed in prematurely delivered pups (Terry & Hall, unpublished results) and in fetuses as young as embryonic day 18 (Smotherman & Robinson, 1985, 1987). Unlike the stretch response used by suckling pups to consume milk, these responses resemble the consummatory responses of adults in their motor topography (Hall, 1979). Thus, young pups possess neural substrates and effector systems needed to produce the final response component for adultlike ingestion, even though they do not normally feed until their third postnatal week.

As in the example from Thelen's work described above, a special context was required for eliciting this component of ingestion in young pups. Ingestive

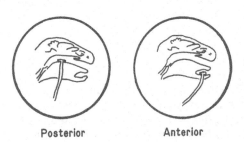

Posterior   Anterior

Figure 6. Diagram of position of oral cannulas in the mouth of a neonatal rat (top). Infusions at posterior locations stimulate swallowing that is reflexive in nature; intake of infusions at anterior locations is modulated by pups' deprivational condition. Other cannula locations also have been explored (e.g., Pelchat & Brake, 1987; Kehoe & Blass, 1985; Spear, Specht, Kirstein, & Kuhn, 1989). Using oral cannulas allows test infusions to be made remotely, without handling a pup (bottom). Pups typically behave as though they are ingesting a diet from the floor in front of them. (From Hall, 1979.)

responses did not occur unless testing was conducted in an incubator warmed to the temperature of the nest (also see Almli, 1973). When pups were tested at room temperature, even when their core temperature was still high, they showed little ingestive responding to oral infusions of milk (Johanson & Hall, 1980). This temperature constraint no doubt reflects the thermal vulnerability of pups, but it also reveals that even at primitive stages of organization, the expression of appetitive response components depends on external controlling factors in addition to the oral stimuli for ingestion. The number and variety of factors that constrain appetitive response expression increase as sensory systems mature and contextual influences on ingestion expand from thermal cues to include an influence of tactile, olfactory, and social cues (Johanson & Hall, 1981; Galef, 1982; Chapter 13).

Early Ingestive Responses Are Influenced by the Sensory Properties of Infused Diets

In addition to having a motor topography similar to ingestion in adult rats, ingestion in response to oral infusions also is influenced in an adultlike manner by sensory properties of food. Newborn pups are responsive to the odor characteristics of infused diets (Hall, 1979; Terry & Johanson, 1987) and show indications of taste responsivity in their oral behavior from the time of birth and before (Ganchrow, Steiner, & Canetto, 1986; Smotherman & Robinson, 1985). Yet, despite these general capabilities for discrimination, sensory systems subserving ingestion in neonates still are immature peripherally and centrally. Olfactory system maturation, for example, goes on for a considerable period in postnatal life (see review by Brunjes & Frazier, 1986). And for taste, electrophysiological studies in the chorda tympani nerve and in the brainstem of neonates indicate that gustatory responses are immature for several weeks (Ferrell, Mistretta, & Bradley, 1981; Hill & Almli, 1980; Hill, Bradley, & Mistretta, 1983; Hill, 1987). Thus, it is not surprising that in tests with oral infusions, behavioral responsivity to tastes and smells is only partially complete at birth and gradually develops postnatally. With respect to rats' taste reactions, the full expression of sucrose or polycose preference and quinine aversion is not apparent in rats until 15 days of age (Figure 7; Hall & Bryan, 1981; Johanson & Shapiro, 1986; Kehoe & Blass, 1985; Virgorito & Sclafani, 1988). Responsiveness to sodium salts also is poorly developed in the first week, not becoming completely adultlike until after weaning (Bernstein & Courtney, 1987; Midkiff & Bernstein, 1983; Moe, 1986). Finding developmental change in chemosensory systems means that although pups are capable of making ingestive discriminations, sensations underlying such discriminations are likely to be different quantitatively, if not qualitatively, from those of adult animals.

For rodents, tactile sensations from the mouth, face, and snout are also critical for the normal expression of appetitive behavior (Zeigler, 1983). After perioral tactile denervation, pups are not able to nipple attach, although they do show some nipple-search responses (Hofer et al., 1981; also Figure 2B). Perioral tactile information is carried in the trigeminal nerve, and there is evidence for marked change in the function of trigeminal systems during development (Verney & Axelrad, 1977). Although these changes have received little experimental attention with respect to feeding development, alterations in trigeminal function, and hence pups' orofacial perceptions, are likely to play a role in the changing nature of ingestive behavior and perhaps in the poorly understood

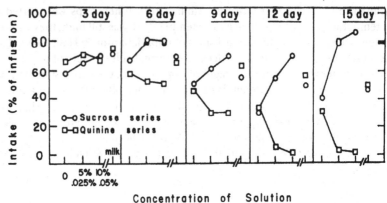

Figure 7. The developmental emergence of preference for sucrose and aversion for quinine as revealed by rat pups in short oral-infusion tests. For each age, mean percentage intake for an ascending series of sucrose (0, 5, and 10%) and quinine (0, 0.025, and 0.05%) concentrations is presented. Separate pups were tested at each data point. The unconnected circles and squares are intakes of milk at each age and are presented for comparison. (From Hall & Bryan, 1981.)

shift from suckling to feeding at normal weaning (Williams, Hall, & Rosenblatt, 1980).

Immaturity such as that seen in taste, smell, and orofacial tactile systems is even more obvious in other sensory systems in altricial infants (Gottlieb, 1971). Vision and audition do not become functional until about 2 weeks of age, and thus the young rat's ability to perceive its environment and to respond to food cues is limited. Maturation of motor capabilities also contributes to the way that developing animals interact with the array of stimuli in their environment and thus influences the organization of ingestive behavior at all ages. In short, despite the presence of a functioning final ingestive response component in young pups, limitations in pups' sensitivity, perception, and locomotor competence represent an important interpretive consideration for thinking about ingestive development.

Given a cautious recognition of these potential interpretive difficulties, developmental analysis still provides special opportunities for insight into the factors that modulate and control components of ingestive behavior. At early stages in development, because of physiological and neural immaturity, controls may be fewer or more simple and thus more easily identified and isolated for analysis. From such starting conditions, additional controls and complexity that appear during development then can be identified and studied sequentially. Indeed, application of this strategy has shown that there is a restricted set of physiological determinants of independent ingestion in rat neonates and an additional set of controls for ingestion that appears only in older pups.

EARLY INGESTIVE RESPONSES ARE MODULATED BY CHANGES IN PHYSIOLOGICAL STATE

Studies in several laboratories have revealed that changes in physiological state consistently modulate simple ingestive responses to oral infusions both in the youngest rat pups and in adult animals (Grill & Norgren, 1978). Effective manipulations in pups have included deprivation, dehydration, cholecystokinin (CCK) administration, and gastric loading with nutritive and nonnutritive solutions (e.g., Bruno, 1981; Hall, 1979; Hall & Bruno, 1984; Robinson, Moran, & McHugh, 1988; Wirth & Epstein, 1976). The question for developmental analysis has thus become whether this modulation of ingestion in young pups is produced by fewer, simpler, or more easily understood controls than in adult animals.

DEHYDRATION MAY BE THE ONLY PHYSIOLOGICAL STIMULUS FOR INGESTIVE RESPONDING IN RAT NEONATES. Newborn rat pups initially seemed a bit like blowflies (Dethier, 1976), always willing to ingest unless inhibited by a signal from stomach fill (e.g., Hall, 1985). However, recently collected data argue against the presence of an intrinsic feeding signal in young pups. Instead, it now appears that dehydration may be the primary stimulatory signal for all instances of experimentally enhanced ingestion in pups 6 days of age and younger, including ingestion elicited by deprivation of maternal milk.

It already had been well established that acute, experimentally induced dehydration is a potent stimulus for ingestion in neonates (Wirth & Epstein, 1976; Bruno, 1981; Bruno, Blass, & Amin, 1983). However, we had not viewed dehydration as an essential stimulus for early independent ingestion after deprivation of milk, because an empty stomach seemed at least as prominent a potential signal. But deprivation of milk results in both cellular and extracellular dehydration (Bruno, 1981; Friedman, 1979) as well as caloric privation. This means that in previous studies, dehydration was confounded with manipulations producing nutritional deficits or empty stomachs, and thus the stimulus for enhanced ingestion after deprivation actually had not been identified. In a recent study (Phifer, Ladd, & Hall, 1988), we removed the confound of dehydration in young food-deprived pups and found that when pups were not dehydrated, they showed no increase in ingestion after deprivation.

In this experiment, 6-day-old pups' hydration was maintained during a 22-hour "deprivation" period by a continuous infusion of isotonic saline through chronic gastric cannulas. Comparison groups received infusions of milk or no infusions. Two hours before testing, the infusions were stopped to allow time for pups' stomachs to empty. Then ingestive responses to oral infusions were tested. We found that, as expected, deprived pups receiving no infusions during the deprivation period ingested large volumes of the diet (Figure 8A). In contrast, pups that had been maintained in hydrational balance with isotonic saline infusions consumed little of the diet. Their intake was similar to that of milk-loaded control pups and equivalent to that of nondeprived pups in other experiments. In addition, the hydrated pups terminated their ingestion with very little stomach fill. This finding provided the important demonstration that, in the absence of dehydration, ingestion did not proceed to a high level of gastric fill. Thus, these hydrated pups showed little willingness to ingest despite being nutritively and maternally deprived and having empty stomachs. Like pups that received gastric infusions of milk, the hydrated pups simply appeared unresponsive to oral milk infusions.

To the degree that we distinguish hunger from thirst on the basis of the responsiveness of behavior to specific physiological stimuli, these findings suggest that rat pups are born without a hunger system. As is reviewed below, this

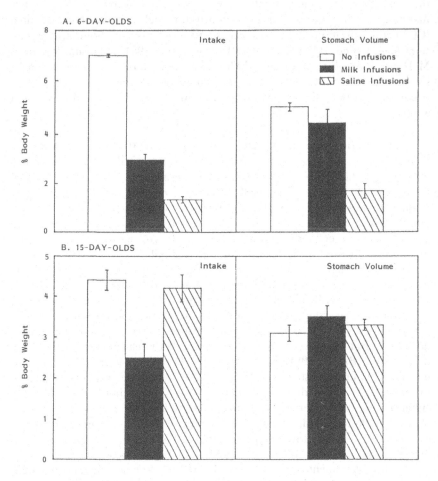

Figure 8. Analysis of pups' ingestion and terminal stomach volume for oral infusion tests after overnight deprivation or after a deprivation period during which they were continuously infused intragastrically with isotonic saline or milk. (A) Data from 6-day-old pups (mean ± SEM). Note that saline infusion reduces intake to the same or a greater degree than milk infusion (left panel) and that in this condition pups terminate their ingestion with low stomach volumes (right panel). (B) Data from 15-day-old pups. Note that saline infusions no longer reduce deprivation-induced intake. Thus, by this age there appears to be a stimulus for ingestion in addition to dehydration. (Modified from Phifer et al., in press.)

conclusion is strengthened by the fact that no manipulation other than dehydration has been found that potentiates ingestion of oral infusions in pups younger than 6 days of age. Nor has any manipulation been found that modulates ingestion at all except via an inhibitory signal related to gastric fill.

Confidence in this conclusion must be tempered by the possibility that the continuous intragastric delivery of isotonic saline to these young rats may have had some nonphysiological consequence, perhaps as a result of immature renal (Falk, 1955; Heller, 1949) or gastrointestinal function (Henning, 1987). Blood measures indicative of fluid balance indicate that while saline infusions maintained blood volume at near normal values, plasma osmolality was somewhat reduced (4–5%). Such osmotic dilution raises the possibility that saline-infused pups could have been debilitated by cellular overhydration. However, because these pups appeared behaviorally competent and did consume some diet, a general debilitation seems unlikely. A more specific inhibition of ingestion by overhydration also is a possible explanation for a failure of intake after overnight saline infusion (Stricker, 1969; Blass & Hall, 1976). However, inhibition by overhydration would only be expected for a hydrationally controlled thirst system, and thus inhibition by overhydration is consistent with the idea of an absent hunger system. In contrast to the case for these young pups, overhydration in adults releases or stimulates feeding (Gutman & Krausz, 1969; Hsiao & Trankina, 1969; Hsiao & Langenes, 1971; Schwartzbaum & Ward, 1958).

THE STOMACH MAKES AN INHIBITORY CONTRIBUTION TO EARLY INGESTION. Whereas the ingestive responses of young rats appear to be activated only by the stimulus of dehydration, a second factor, gastric fill, appears to be the sole inhibitory control of early ingestion. Young rats that are dehydrated and consuming orally infused solutions stop ingesting while dehydration probably still persists. Considerable evidence showing that this inhibition results from gastric fill has been collected by Phifer and collaborators. Much of this evidence relies on an adaptation of the technique of pyloric occlusion and on the development of gastric-fistula and sham-feeding techniques for pups (see Phifer & Hall, 1987, for methodological details). In particular, occluding the pylorus with a noose and preventing movement of ingested solutions from the stomach to the intestines (Deutsch & Wang, 1977; Kraly & Smith, 1978; Hall, 1973; Hall & Blass, 1977) has provided a means of evaluating the modulatory role of inhibitory signals from the stomach. Ingestion with and without pyloric occlusion can be compared to provide a contrast between intake with and without a postgastric signal.

Deprived 6-day-old rats with closed pyloric nooses ingested volumes comparable to those of pups ingesting normally; stomach fill alone was sufficient to stop ingestion (Figure 9A). Stomach volumes measured at the termination of ingestion revealed that pups' ingestion stopped at the same degree of gastric fill whether they were ingesting normally or with closed nooses; this is particularly apparent when viewed across load condition (Figure 9A, right panel). If postgastric nutritive signals had been required to inhibit ingestion, then pups should consume more with pyloric occlusion, and their terminal level of stomach fill should have been higher to create additional inhibition needed to replace the absent postgastric signal. In contrast to these findings from 6-day-olds, adult rats (Kraly & Smith, 1978) and older pups (see below) ingesting in the absence of postgastric nutritive signals terminate ingestion only after reaching greater than normal stomach volumes.

Inhibition resulting from gastric fill is a product of the amount rather than the chemical characteristics of the solution that fills the stomach. When pups' pyloruses are occluded, milk, isotonic saline, and glucose preloads all result in intake termination at an equivalent point of gastric fill. Thus, in contrast to feeding in adults (e.g., Deutsch, 1985), there seems to be no gastric chemoreceptor contribution to intake termination in young pups receiving oral infusions. The degree of gastric filling completely accounts for the inhibition of ingestive responses to orally infused solutions.

Robinson et al. (1988) have found that CCK supresses independent ingestion in pups from shortly after birth. But they also report that CCK inhibits

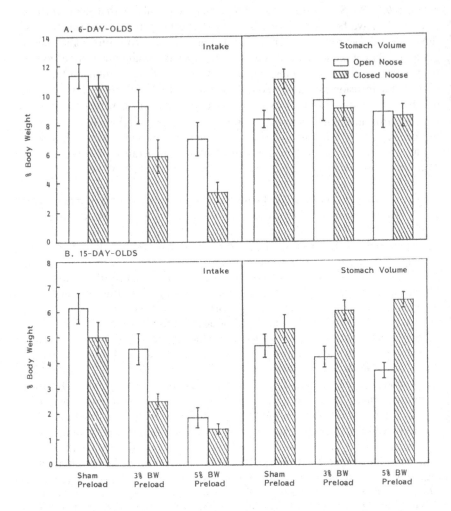

Figure 9. Intake and stomach volume (mean ± SEM) at the termination of ingestion for pups consuming oral infusions with either open or closed pyloric nooses and after having received sham, 3%, or 5% body weight gastric preloads of milk 2 hours earlier. (A) Data for 6-day-old pups. Note that stomach volumes are the same after various preloads, irrespective of whether noose is open or closed. Although stomach volumes in the closed-noose condition are slightly higher than those in the open-noose condition after the sham preload, in other studies we have found these two values to be equivalent (e.g., Phifer, Sikes, & Hall, 1986). (B) Data for 15-day-old pups. Note that pups receiving gastric preloads of milk terminate ingestion with smaller volumes of stomach fill in the open-noose condition, the condition in which the loads could have activated postgastric control mechanisms. (From Phifer & Hall, 1988.)

gastric emptying from the same age and thus suggest that the CCK effect on intake in pups is secondary to its influence on the gastric fill signal.

The gastric inhibitory signal appears to be carried by the vagus nerve. Lorenz *et al.* (1982) vagotomized pups at 7–9 days of age and found their intake to be markedly enhanced in oral infusion tests (Figure 10). Splanchnic denervation increased intake further when added to vagotomy but had little effect by itself. These findings are unusual in that they provide a demonstration of enhanced ingestion following vagotomy—an outcome not observed in studies with

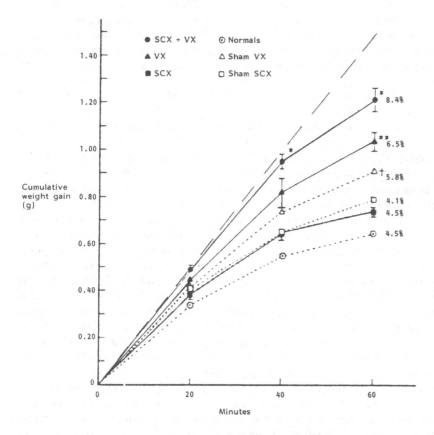

Figure 10. Intake in an oral infusion test for unoperated normal 7 to 9-day-old pups and pups that received vagotomy (VX), superior cervical cordotomy (SCX), cordotomy and vagotomy (SCX + VX), or sham surgery. Note that intake is increased after vagotomy and vagotomy plus cordotomy, though cordotomy alone has little effect. (From Lorenz *et al.*, 1982, with permission.)

adult animals, even though an inhibitory afferent contribution of the vagus is often presumed. In addition, the findings indicate that normal termination of intake in young rats with an intact vagus results from an inhibitory modulation of ingestion rather than reflecting a simple mechanical limit of abdominal capacity.

The modulatory effect of gastric fill is further demonstrated by findings that the final termination of intake varies according to pups' hydrational state. Recall that low levels of gastric fill terminate ingestion in hydrationally replete pups (e.g., Figure 8A). In contrast, termination of intake can be delayed until very high levels of fill are reached when pups are dehydrated. When non-deprived pups, which already had full stomachs (approximately 7.5% body weight), were acutely dehydrated by injection of 1 M NaCl (2% body weight, s.c.), they consumed as much as another 6% of their body weight. When these animals stopped ingesting they were found to have extremely large stomach volumes amounting to 11–12% body weight (Hall, Denzinger, & Phifer, 1988). Thus, the significance of the gastric fill signal is modulated by hydrational state and does not appear to be based on an absolute physical limit.

FURTHER EVIDENCE FOR AN ABSENCE OF POSTGASTRIC NUTRITIVE OR META-
BOLIC MODULATION OF EARLY INGESTIVE RESPONDING. Although gastric factors,
modulated by hydrational state, may completely account for intake termination
in the typical oral-infusion test of young pups, it still is conceivable that some
postgastric nutrient or metabolic sensor might exist that could contribute to the
modulation of ingestion. To enhance the likelihood of detecting any postgastric
nutritive or metabolic influence on ingestion, Phifer, Browde, and Hall (1986)
evaluated the effects of nutritive loads that were delivered 2 hours before testing, thereby allowing time for intestinal transport and absorption. In these tests,
we found no additional inhibition beyond the effects of gastric fill. Further, it is
unlikely that gastric signals were just overwhelming more subtle postgastric or
postabsorptive signals, because nutritive loads failed to affect sham feeding in
young pups, a situation in which there is no gastric fill signal because all ingested
diet spills from an open gastric fistula (Phifer, Sikes, & Hall, 1986).

These data suggest that for rats 6 days of age and younger, the ingestive
response component elicited by oral infusions is completely controlled by hydrational status and gastric fill. There has been no demonstration that an enhancement of independent ingestion can be produced by any factor other than dehydration. Even an empty stomach is not itself a stimulus for ingestion. No
postgastric feedback signal, other than rehydration, during or subsequent to
ingestion appears to influence behavior. This strong argument is one with major
implications for the analysis of the development of feeding in rats. It indicates
that in young rat pups a relatively simple ingestive control system is present, one
that should be quite amenable to the analysis of central and peripheral interactions in ingestive control. More importantly, it emphasizes that there is an entirely postnatal ontogeny for caloric control of feeding systems: nutritively elicited or modulated ingestion does not develop until some time after 6 days of age.
Although an ingestive response system is present in young rats, to the degree
that we consider "feeding" systems as responsive to some feature of caloric
depletion, a feeding system subserving independent ingestion is not present in
the first week of life.

THE ONSET OF FEEDING IN OLDER PUPS

The state of affairs just summarized for independent ingestion is one entirely to our experimental advantage because it places the ontogeny of feeding
controls, other than gastric-fill inhibition, in the postnatal period sometime after
6 days of age. It thus allows convenient study of their emerging physiological
mechanisms and neural substrates. In fact, after just 9 more days of development, by 15 days of age, ingestion in deprived pups has changed and depends on
more than dehydration. When, as described for 6-day-olds above, hydration was
maintained in 15-day-old pups during overnight "deprivation" by intragastric
infusions of isotonic saline, subsequent ingestion of oral infusions was not reduced (Phifer et al., 1988); saline-infused pups ingested just as much as completely deprived pups and stopped ingesting with large volumes in their stomachs (Figure 8B). Consistent with this finding, and in contrast to younger pups,
rehydrating deprived 15-day-old pups with gastric preloads of water had little
effect on ingestion of oral infusions (Phifer et al., 1986). Thus, by 15 days there
appears to be a cause other than dehydration for ingestion after deprivation.
The stimulus or stimuli for this intake remain unknown.

NEW NUTRITIVE CONTROLS OF INGESTION. Manipulating pups' nutritive state with caloric gastric loads also reveals new influences on the ingestive responses of 15-day-olds. Preloads of milk and glucose reduce intake relative to saline loads. More importantly, they result in a termination of ingestion at lower volumes of gastric fill (Figures 9B and 11; Phifer *et al.*, 1986; Phifer & Hall, 1988). This finding indicates that by 15 days of age there is a modulatory signal in addition to the one produced by gastric fill. This signal is postgastric, and, because nutritive loads also reduce sham feeding at 15 days of age, this signal has a direct effect on ingestion, as opposed to being simply mediated by an indirect

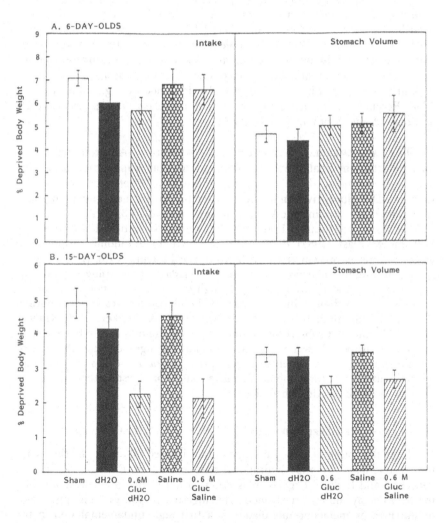

Figure 11. Intake and stomach volumes (mean ± SEM) at the termination of ingestion for pups that received vehicle or glucose gastric loads 2 hours before an oral-infusion test. (A) Data for 6-day-old pups. (B) Data for 15-day-old pups. Note that only in 15-day-olds does glucose have a specific inhibitory effect and that stomach volumes at the termination of ingestion are decreased after glucose loads then. (Modified from Phifer *et al.*, 1986.)

effect on gastric emptying or on the efficacy of the gastric signal. Nonetheless, this emergent postgastric signal probably does not make much of a contribution to the termination of ingestion in our standard oral-infusion tests.

In studies with pyloric nooses, terminal gastric volumes in the sham-load condition after milk intake were comparable regardless of whether or not the milk was allowed to enter the intestine (Figure 9B). It seems that only when nutrients have entered the intestine well before ingestion begins do these nutrients reduce the required terminal levels of gastric fill. This delayed inhibition is the case for milk diets; it remains possible that diets that more rapidly empty the stomach could have more immediate effects (Phifer & Hall, 1988). The postgastric contribution to milk intake thus is likely to influence the permanence of the inhibition of ingestion or subsequent ingestive episodes and thus fits with a number of demonstrations in adults that there is little effective postgastric contribution in the early phase of satiety (Davis & Campbell, 1973; Deutsch, 1983, 1985; Snowdon, 1970; though see Booth & Jarman, 1976). Note further that the gastric inhibition of ingestion at 15 days of age still is independent of gastric chemoreceptor stimulation, since isotonic saline loads are as effective as milk or glucose loads in pups with closed pyloric nooses (e.g., Phifer & Hall, 1988). Recent findings suggest that the shift in control becomes apparent as early as 9 days of age (Swithers & Hall, 1989).

THE LATE-EMERGING RESPONSE TO GLUCOPRIVATION. An additional modulatory effect on the final component of ingestion becomes apparent considerably later in development, the response to glucoprivation. When produced with insulin or 2-DG, glucoprivation has been shown to increase independent ingestive responding and suckling intake only in pups older than 3 weeks of age (e.g., Lytle et al., 1971; Houpt & Epstein, 1973; Drewett & Cordall, 1976; review by Williams & Blass, 1987) despite the fact that an autonomic response to glucoprivation is present from birth (Houpt & Epstein, 1973). A feeding response to insulin is marginally detectable at 21 days of age using measures of "nibbling" (Gisel & Henning, 1980), though one to 2-DG is not apparent for another week. Although the relevance of the glucoprivic response to normal feeding in adult animals is questionable (Friedman & Stricker, 1976; Kraly & Blass, 1974; Smith, Gibbs, & Stromayer, 1972), we also now can be fairly certain that glucoprivation does not become a stimulus for ingestion until the time of weaning or later in rats. Therefore, glucoprivic responsiveness can not be related to the initial and fundamental nutritive control of ingestion that emerges at 15 days or earlier, as described above.

SUMMARY OF THE DEVELOPMENT OF RATS' FINAL INGESTIVE COMPONENT

From the youngest ages, pups ingest orally infused solutions, and this ingestion is modulated by pups' hydrational state and inhibited by gastric fill. But between 6 and 15 days of age, there is a transition from hydrationally modulated to hydrationally and nutritively modulated ingestive responses. This appearance of nutritive or metabolic modulation is a first and fundamental step in the emergence of a "feeding" system.

We thus have a rough outline of significant transitional events in the appearance of controls of the final consummatory response (Table 2), but important questions about the controls remain unanswered. (1) When in this period do

TABLE 2. SUMMARY OF CURRENT UNDERSTANDING OF THE DEVELOPMENT OF
MODULATORY CONTROL OF INGESTIVE RESPONSES TO ORAL INFUSIONS

Stage	Age	Stimulates intake	Inhibits intake
Prefeeding, hydrational-only period	0–6 days	Dehydration (cellular and extracellular)	Gastric fill (modulated by hydration or rehydration)
Feeding and hydrational period	By 15 days[a]	Dehydration and some effect of nutrient deprivation	Gastric fill[b] and some postgastric nutrient effect[c]

[a]Transition probably occurs about 9 days.
[b]Gastric-fill inhibition may work by different mechanisms for dehydration- versus deprivation-induced ingestion.
[c]Nutrient inhibition may arise from a different mechanism than the nutrient-deprivation effect.

the changes actually occur? (2) What are the mechanisms, detector and effector, of the nutritive or metabolic modulatory system that becomes operative? (3) Are the modulatory effects of caloric gastric loads that emerge by 15 days achieved as a result of reduction of the new, deprivation-related nutritive stimulus for ingestion, or do they reflect an additional and separate short-term postabsorptive inhibitory control? Finally and most importantly, note that to this point we have considered only the final component of ingestion and its control. To appreciate the assembly of the whole appetitive sequence for independent ingestion, we need to consider other components in the sequence, their individual ontogeny and control, and the manner in which they are organized to achieve adaptive results.

BEYOND THE FINAL INGESTIVE RESPONSE COMPONENT: EXTENDING INGESTION INTO THE ENVIRONMENT

INGESTING FROM THE FLOOR

To this point I have focused on the ingestive responses pups make to orally infused solutions. We are fortunate to be able to utilize oral infusion as a way of studying this component in both pups and adult rats (e.g., Grill & Norgren, 1978). But it also is important to consider the preceding components in the ingestive sequence and their development and control, since in older animals it is ultimately the control and expression of the whole sequence that is feeding.

A component of ingestive behavior that precedes the swallowing of food is an animal's contact with and identification of ingesta in the environment and its subsequent movement of the ingesta into its mouth. We have been able to study this ingestive component in young rats by altering their environment to insure that pups contact and can respond to test solutions. We spread liquid diets on the floor of pups' test containers where the diets always were available at or near the pups' mouths. When deprived or dehydrated pups were tested in such situations, they were found to actively lap and swallow diets and to do so shortly after birth (Figure 12; Hall & Bryan, 1980). With this procedure, pups did not just passively accept or reject diets already in their mouths but contacted and actively

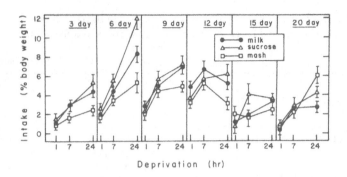

Figure 12. Ingestion of diet spread on the floor of pups' test container. Intakes (mean ± SEM) of various milk, sucrose solutions, or mash at different ages and after 1-, 7-, or 24-hour deprivation. (From Hall & Bryan, 1980.)

lapped solutions from the floor, setting their own pace for consumption and continued ingestion. Thus, when appropriately triggered, a preceding component of the ingestive sequence appeared to be present even in very young pups.

For pups 6 days of age and younger, adding this response component to tests of ingestion produced few differences from the results of oral infusion tests, either after manipulations of the sensory characteristics of ingesta or after manipulations of pups' physiological state. Sweetening the diet increased intake (Hall & Bryan, 1980), dehydration increased intake (Bruno, 1981), and gastric fill inhibited intake. Such findings indicate that this component either is subject to little control in young pups or is subject to the same control as the final oral component.

In older pups, additional controls of this component were found to develop, and these partially paralleled the emergent nutritive controls seen with oral infusions. For example, by 2 weeks of age, with tests of ingestion from the floor, nutritive preloads suppressed intake as they did in tests with oral infusions (Swithers & Hall, 1989). In short, the control of this earlier component in the sequence was generally similar to the control of the final oromotor component stimulated by oral infusions. However, we also found an illuminating difference in the control of this earlier component compared with pups' responses to oral infusions, and I return to this difference shortly.

Tests of the ontogeny of independent ingestive behavior in mice, using tests of lapping from the floor or oral infusions, provide some support for the idea that a calorically responsive feeding system may be late to develop in other altricial rodents (Hall & Browde, 1986). Mice do not feed in any type of independent ingestive test until after 10 days of age and show no responsiveness to deprivation. This finding of no ingestion in young mice initially was quite surprising. But mice, unlike rats, are not responsive to dehydration from birth. Thus, absence of early ingestion in mice may result from differences in thirst systems rather than feeding systems. Young mice may not ingest simply because they lack the dehydration-driven responsiveness that is present from birth in rats. Mice, in fact, begin to evidence nutritively related ingestion at the same time that it is observed in rats.

PATTERNS OF SPONTANEOUS INDEPENDENT INGESTION CHANGE DURING
DEVELOPMENT

101

ONTOGENY OF
INGESTIVE
BEHAVIOR

One feature of ingestion that cannot be studied with oral infusions is the spontaneous initiation of ingestion. Tests of feeding from the floor provide a means to assess the patterning of ingestion in a situation in which food is continuously available. An initial comparison between the patterning of ingestion in 6- and 15-day-old rat pups has revealed differences in behavior that may reflect factors involved in developmental transitions in ingestion.

When ingestion in 6-day-old pups was sampled on a minute-by-minute basis, deprived pups were found to ingest frequently and actively for several minutes (Figure 13A; Hall, Swithers, & McCall, 1988). They then essentially stopped ingesting. This termination of ingestion was relatively permanent. Six-day-olds showed little ingestion in the second half-hour of an hour-long test. In contrast, deprived and nondeprived 15-day-old pups were observed to feed periodically during an entire hour-long test, initiating numerous bouts of ingestion (Figure 13B). Measures of bout duration indicated that 6-day-olds ingested continuously for long durations, whereas the bouts of 15-day-olds were short, and their patterning was suggestive of a different reaction to the ingestive situation than for 6-day-olds.

A factor in this change in the spontaneous patterning of ingestion may be pups' increased responsiveness to the sensory qualities of diets. A more salient

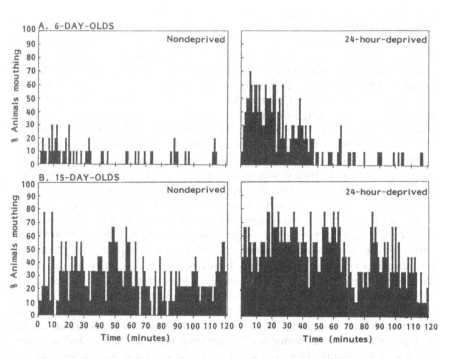

Figure 13. Patterns of ingestion for nondeprived and deprived pups feeding from the floor of their test containers. Bars represent percentage of pups mouthing at each 1-minute sampling point; (A) 6-day-old pups; (B) 15-day-old pups. (Modified from Hall, Swithers, & McCall, 1988.)

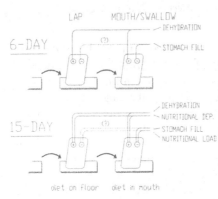

Figure 14. A summary of the suggested extrinsic and intrinsic controls for response components in the appetitive sequence for independent ingestion and their change from 6 to 15 days of age. In young pups, both lapping and mouthing/swallowing are enhanced by dehydration. Probably lapping is inhibited by gastric fill, as is certainly the case for the mouthing/swallowing component. There is little evidence for intrinsic control, though this has not been carefully studied. By 15 days of age, deprivation also enhances both components of ingestion, and an intrinsic use-dependent control now appears to modulate mouthing/swallowing.

taste stimulus, for example, could cause nondeprived pups to continue to sample diet throughout the test and, in addition, might enhance the oral contribution to satiety in deprived pups, producing short, small bouts of ingestion. As well, the altered nature of the emergent "feeding" signal between these two ages may contribute in some manner to the differential pattern of feeding. Determining the cause for the changing patterns of ingestion remains an important question for the developmental analysis of ingestion (Figure 14).

DIRECTING INGESTIVE BEHAVIOR

Our early experiments with pups suggested that whereas young pups would avidly contact and lap up liquid diets spread on the floor, they could not search out or maintain contact with a topographically restricted source of diet. For example, when 6-day-old pups were required to direct their ingestion to diet restricted to a limited area on the floor of the test container, they were very inefficient in maintaining contact with the solution and consumed relatively little (Hall & Bryan, 1980). Older pups (12 days) consumed similar amounts whether diet was available from the whole floor or from a restricted area. We thus initially believed that young pups either could not direct their behavior to, or maintain their ingestion at, restricted sources of ingesta in their environment. We have learned recently, however, that the apparent inability of young pups to maintain contact with a food source arises from the activation of competing appetitive responses that interfere with pups' guidance of their ingestion. When not distracted by these responses, even 3-day-old pups showed efficient ingestion from restricted diet sources and consumed amounts similar to pups ingesting from the entire surface (Hall *et al.*, 1988). Additional evidence that young pups can direct their appetitive responding comes from the demonstration that 1-day-old pups can be trained to press a paddle to receive oral infusions of milk and are capable of learning to discriminate a paddle that rewards them from one that does not (Johanson & Hall, 1979). Thus, although the ability to orient, direct, and guide appetitive behavior is often believed to be a complex and acquired feature of mammalian behavior, when young rat pups are appropriately tested such abilities can be readily found.

One of the first components in the ingestive sequence is food search or exploration, a behavioral component that is generally believed to reflect an increased locomotor activity as well as an increased responsiveness to potential food signals. This component of the appetitive sequence has not been studied for independent ingestion. We do not know when or how pups begin to seek out food in their environment or when physiological manipulations come to modulate such behavior. Although there is some information available on the effects of deprivation on locomotor activity in developing rats (Moorcroft, Lytle, & Campbell, 1973; Hofer, 1975, 1978), the similarity of the activity change at different ages is unclear, and the mechanisms or substrates of these effects are poorly understood (Campbell & Randall, 1977; Campbell & Raskin, 1978). A first step in the analysis of food-seeking behavior has been made by studying factors that contribute to sleep-state cycling in pups (e.g., Hofer & Shair, 1982; Lorenz, 1986; Shair, Brake, & Hofer, 1984), since an animal must be awake and active to explore for food. These studies indicate that time in various sleep states is modulated by a number of factors from early ages. At this early point in the analysis, however, such studies may be most important in emphasizing that neonatal rodents spend a large portion of their time asleep. Investigators of early behavior in human infants have paid considerable attention to interactions between testing conditions and state, but consideration of state has received limited and probably inadequate attention in studies of behavioral development in animals.

DRINKING VERSUS FEEDING: AN EXAMPLE OF CONTROL ONLY AT AN INITIAL COMPONENT IN THE APPETITIVE SEQUENCE

Despite many parallels between ingestive responses elicited by oral infusions and those found for ingestion from the floor, there was one telling difference that emerged between these two components. This difference illustrates particularly well the importance of determining on which components in an appetitive sequence the control or modulatory effects are exerted.

The difference was seen with respect to the differentiation of feeding and drinking. Differentiated responses are readily apparent in adult animals, for which the response to dehydration differs from that to nutritive deprivation, not just in the nature of the stimulating event but also in the animal's discrimination and choice of deficit-appropriate ingesta. For example, when they are dehydrated, adult animals that are otherwise hungry do not eat. Indeed, one of the most common features of the mammalian response to dehydration is an inhibition of food ingestion; this inhibition is known as "dehydration anorexia" (Lepkovsky, Lyman, Fleming, Nagumo, & Dimick, 1957). We analyzed how such dehydration inhibition of ingestive behavior emerges during development and where in the response sequence it is achieved.

In young pups, dehydration enhances the intake of water as well as food. For example, in oral infusion tests dehydration increased intake of milk and concentrated sucrose solutions that were themselves dehydrating; in fact, milk and water intakes were equivalently increased in young dehydrated pups (Bruno, 1981; or following angiotensin or renin administration, Leshem, Boggan, & Epstein, 1988; Misantone, Ellis, & Epstein, 1980). Thus, pups' discrimination of ingesta was not yet appropriate to their hydrational state. For young pups, this lack of discrimination might be attributable to an immature gustatory system. However, even later in development, as long as oral-infusion tests were used, milk intake was found to be stimulated rather than reduced by cellular dehydration. In 2-week-old animals with more mature taste responses, for example, although dehydration enhances water intake somewhat more than milk, intake of both solutions is increased (Bruno, 1981). Thus, if an assessment were based on tests of the final component in the ingestive sequence, it would appear that full differentiation of hunger and thirst has not yet occurred.

However, consider the effects of dehydration on preceding components in the ingestive sequence. The responses of pups to milk and water in tests of lapping from the floor and oral-infusion tests were similar before 15 days. In both types of test, intake of water and milk was increased by cellular dehydration. However, by 20 days of age, the tests gave different results (Bruno, 1981). Dehydration increased the ingestion of orally infused milk, but dehydration inhibited the intake of milk when animals were ingesting from the floor of their test containers. In other words, when food was placed in the mouth, ingestion was enhanced by dehydration; when milk had to be acquired from the environment, intake was blocked by dehydration. The final oral response component in the ingestive sequence thus appeared to be stimulated by dehydration irrespective of diet or age, although responses of the preceding approach and contact component in the sequence were inhibited.

This control exerted at the food-contact component of ingestion appears to be based on a hydrational modulation of a pup's response to the odor of food. The evidence is as follows: when 20-day-old pups were dehydrated and offered a diet with little or no odor, such as sucrose, their intake was enhanced as it was with water, even though the intake of such a solution can produce further dehydration (Bruno & Hall, 1982). However, when odor was added to the sucrose, ingestion was inhibited by dehydration. In fact, when odor was placed in or near water, dehydration reduced intake of the scented water (Figure 15). Thus, reaction to an olfactory signal appears to cause reduction of intake after dehydration, probably by limiting pups' approach to or contact with the solution. Orally infused solutions are ingested irrespective of whether they are food or water because the olfactory contribution to maintenance of contact is bypassed. Consistent with this hypothesis is the finding that dehydration anorexia is eliminated by rendering 20-day-old pups anosmic. The ontogeny of differential responses described here is for the thirst of cellular dehydration produced by hypertonic saline injections. Although shifts in taste preference for water occur at about the same age following extracellular dehydration and renin and angiotensin injection (Leshem et al., 1988), it is interesting that administration of renin or angiotensin in dipsogenic doses does not appear to produce anorexia, even in adults (Leshem & Epstein, 1988). This latter finding fits well with earlier demonstrations that the appetitive expression of the two thirsts is different (Burke, Mook, & Blass, 1972), and with the general idea expressed here that individual components of the response sequence can be differentially modulated by manipulations that share other effects in common.

The finding of the behavioral site of action of dehydration anorexia provides a clear demonstration of the manner in which controls of ingestion can operate on individual components in an appetitive sequence (Figure 16). Identifying where in the stream of behavior a control or modulatory influence is expressed contributes to fully understanding its actions and provides information on its neural mechanism. Thus, a differential response to dehydration in weanling rats occurs because of control exerted at an early component in the appetitive sequence and is based on an altered olfactory response during ap-

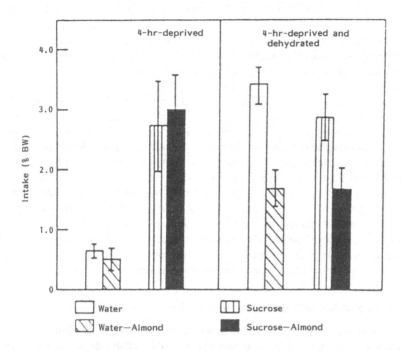

Figure 15. Intake (mean ± SEM) of 20-day-old pups ingesting from a cup in their test container. In dehydrated pups (right panel), adding the odor of almond to either water or sucrose depresses intake and thus results in the expression of dehydration anorexia. (From Bruno & Hall, 1982.)

roach to food. Although the dehydration anorexia of adult rats may be somewhat more complex, adult responses to oral infusions of milk are also relatively unaffected by dehydration (Bruno, Hall, & Grill, 1980). These findings exemplify how developmental analysis contributes to simplifying the identification of components and controls in the ingestive sequence.

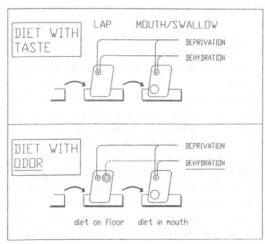

Figure 16. Hypothesized difference in the organization of response components for ingestion following dehydration when the characteristics of diet differ. For a liquid diet with taste but no odor, deprivation and dehydration both stimulate the lapping and mouthing components of ingestion. When diet has an odor, however, the lapping (or approach) response component is inhibited by dehydration, even though the subsequent mouthing component is not.

In addition to the importance of identifying developmental changes in the patterning and control of ingestive behavior, it is useful to consider their cause and source. How many developmental changes in behavior depend simply on animals' new physical features and motoric abilities, how many on new sensitivities, and how many on changes in more central processes? The breadth of potential candidates argues for an open mind in studying emerging sources of behavioral control and apetitive organization. I already have emphasized how changes in morphology and sensory–motor function during development can, in themselves, contribute to changes in the conditions in which behavior is expressed and to developmental reorganizations of the patterns of behavior. In fact, new patterns of overall responding may be established or unmasked by subtle and easily overlooked changes in response capabilities of animals. But it is likely that central changes also make contributions to the development of appetitive systems, particularly to the emerging adaptive modulation, inhibition, and potentiation of appetitive controls. I now consider specific internal and central sources of potential new controls for developing ingestive responses.

ALTERED SIGNALS OF PHYSIOLOGICAL STATE?

Maturation of gastrointestinal function and nutrient metabolism may be of relevance to change in ingestive controls. Marked changes in gastric secretions, in intestinal enzymes, and in intestinal absorption and breakdown of protein all occur during development. These maturational events have been reviewed recently by Henning (1981, 1987). In addition, and in contrast to adults, suckling animals utilize ketone bodies as a major metabolic substrate for brain metabolism rather than glucose (Hawkins, Williamson, & Krebs, 1971; Thurston & Hauhart, 1985), with ketone utilization decreasing at the time of weaning. The influence of this shift in metabolic substrate on feeding is unknown.

One feature of early gastrointestinal function is worth particular comment. An active transport mechanism for glucose absorption from the intestine is present from birth, though its capacity is somewhat less than in adults (e.g., Ghishan & Wilson, 1985). Thus, simple lack of absorption of glucose is not a likely explanation for insensitivity of young pups to glucose loads. Nonetheless, significant changes in the nature of absorption of other nutrients do occur. In young pups, protein and other large molecules, which arrive in the intestine largely intact because of low enzymatic activities in the mouth and stomach, are absorbed by active pinocytotic processes and either digested in the cells of the intestinal wall or passed as macromolecules to the blood. Only after 3 weeks of age do increases in gastric, pancreatic, and intestinal enzyme secretions result in efficient digestion in the intestinal lumen and in adultlike absorption (see Henning, 1987). Thus, physiological maturation alters nutrient uptake and permits the use of new sources of food and fluid. These changes in gastrointestinal function occur at the same time that we see new signals of nutritional or energetic status begin to influence ingestion. An additional development is an increased insulin release in response to food absorption as weaning time approaches (Blazquez, Montoya, & Quijada, 1970). Such release might figure indirectly in the emergent nutritive modulation of ingestive responses.

Gastrointestinal and metabolic maturational events could be secondary to the onset of ingestion of carbohydrate-rich food during weaning. As such they might be less likely to be indicative of fundamental change in ingestive control systems, since the onset of nutritive modulation of ingestion, for example, seems slightly to precede pups' dietary transitions. However, for cases that have been studied, an internal timing of the developmental changes in gastrointestinal physiology is indicated (e.g., Martin & Henning, 1984).

New Neural Controls of Ingestion

Considerable maturation of CNS function occurs during the postnatal period in altricial mammals. Although only a few areas of the mammalian brain actually undergo postnatal cell division, impressive development and differentiation occur throughout the brain (e.g., Lund, 1978; Purves & Lichtman, 1985). In particular, a number of maturational events have been described for integrative systems and structures believed to be related to ingestive behavior (e.g., monoaminergic system development, Breese & Traylor, 1972; Coyle & Axelrod, 1972; Horowitz, Heller, & Hoffman, 1982; Johnston, 1985; Lanier, Dunn, & Van Hartesveldt, 1976; Loizou, 1972; Mabry & Campbell, 1977; hypothalamic electrophysiological response maturation, Almli & McMullen, 1979). The CNS maturation, of course, offers one of the most intriguing forms of explanation for developmental changes in behavior and in particular for the emergence of new controls of behavior.

As yet there is little specific information regarding correlations of ingestive response development and neural maturation. Indeed, attempts at such developmental correlations have been limited and indirect and primarily have involved study of monoaminergic systems. From this work to date, what is most impressive is the extent to which massive damage to monoamine systems very early in development only minimally affects ingestive responses at the time of the damage or later in adulthood (Almli, 1984; Bruno, Snyder, & Stricker, 1984; Bruno, Zigmond, & Stricker, 1986).

Acute pharmacological manipulations of monoamine systems have provided stronger indications of relationships between neural maturation and appetitive behavior. Spear and co-workers have shown developmental changes in serotonergic modulation of mouthing in oral infusion tests (Caza & Spear, 1982; Ristine & Spear, 1985). Considerable differentiation of serotonergic neurons occurs in the preweaning period (Lidov & Molliver, 1982) and thus has a potential role in changes in ingestion.

Indeed, a role for maturation of serotonergic transmitter systems in changes in suckling behavior has been identified. Williams, Rosenblatt, and Hall (1979), Spear and Ristine (1982), and Stoloff and Supinski (1985) have reported pharmacological effects on behavior suggesting that maturation of serotonergic neurons is partially responsible for the normal loss of suckling at weaning. After 2 weeks of age, serotonin (5-HT) antagonists markedly enhance attachment in nondeprived pups whereas 5-HT agonists block attachment. A few days earlier these same manipulations have little (Williams et al., 1979) or opposite (Spear & Ristine, 1982) effects on suckling. Treatment with 5-HT blockers will reinstate suckling in pups up to 40 days of age, long after suckling is normally lost, although serotonergic effects on suckling are highly dependent on pups' suckling experience (Lichtman & Cramer, 1988a). Serotonergic manipulations at the time of weaning also modulate pups' interest in nonmaternal food sources and influence the nibbling of food (Williams *et al.*, 1979). Thus, the maturation of some aspects of serotonergic function may be contributing to the decline in suckling at weaning and to pups' shift to new food sources. However, we know little about whether serotonergic maturation may be contributing to the nutritive control of independent ingestion that is emerging in parallel to these changes in suckling.

To date, the only information regarding potential neural substrates of the emerging nutritive control of independent ingestion is suggestive but indirect. Ellis, Axt, and Epstein (1984) have shown that central injection of norepinephrine, which is a potent stimulant of food intake in adult rats when injected into the paraventricular nucleus (e.g., Leibowitz, 1976), does not enhance the intake of orally infused solutions until 9–10 days of age. At that time, it becomes markedly orexogenic, with pups showing enhanced intakes of 50–80%. This timing is interesting because it corresponds to the time at which ingestive responses become driven by factors other that dehydration. Thus, although we know little about the emergent nutritive or metabolic signal, a noradrenergic contribution to the stimulation or control of the oromotor response component is a possible substrate.

CHANGING CONTRIBUTIONS OF EXPERIENCE TO THE CONTROL OF INGESTION?

A primary feature of appetitive behavior is its capacity for adjustment as a result of experience (Teitelbaum, 1966). Since effects of experience may occur on any of the elements in a response sequence, experience effects can be multiple, independent, and complex. Thus, developmental changes in mechanisms for the experiential modulation of ingestion could figure prominently in the reorganization of ingestive systems during development. Analysis of these mechanisms during ontogeny permits an assessment of their potential roles and will contribute to the identification of their mechanisms and sites of action in appetitive behavior. I consider here the developmental aspects of two types of experiential contribution to ingestion: short-lasting effects related to ongoing ingestion and longer-lasting effects related to the consequences of ingestion and to early experiences in general.

USE-DEPENDENT MODULATION OF A RESPONSE COMPONENT. Consider again the suckling sequence I described in the first section of this chapter. The influence of preceding nipple-search behavior on the subsequent rooting and probing behavior illustrates how a basic appetitive response can be modulated by preceding experience (cf. Hinde, 1970). Nipple search reduces subsequent nipple search. Indeed, one of the most straightforward, though overlooked, contributions that experience can make to the control of ingestion is to modulate an ongoing response based on the previous stimulation or activation of that same response (Figure 4). Thus, many highly organized responses such as those that characterize suckling seem readily elicitable unless they have been activated recently. Stated in another manner, these responses habituate with use, and the degree of their habituation influences their subsequent expression. Thus habituation potentially becomes a fundamental characteristic of all component responses in appetitive systems.

Experience-based modulatory functions have been conceptually central to ethological theories of motivation (e.g., Lorenz, 1950; Tinbergen, 1951), yet they have not been directly incorporated into the psychobiological analysis of ingestive behavior. Nonetheless, although not usually viewed from a habituation perspective (though see Thorpe, 1966), considerable data from studies of feeding in adult animals support the importance of the previous activation of responses, or their "use," in modulating ingestive behavior. A classic example is the demonstration that "food by mouth is more effective than food by stomach" in reducing subsequent ingestion (e.g., Kohn, 1951; Miller & Kessen, 1952). That is, feeding must be experienced for satiety to occur. That the habituation process is specific and not secondary to a factor such as fatigue is supported by numerous additional demonstrations that satiety is specific to a diet that has been ingested. Ingestive responding typically is reinstated when a diet is replaced with a new one (i.e. "sensory-specific satieties"; Booth, 1985; Holland, 1988; Rolls, Rolls, Rowe, & Sweeney, 1981; Treit, Specht, & Deutsch, 1983). Thus, the use or activation of ingestive response components may be required for an animal to terminate ingestion simply because response components need an opportunity to habituate (see Gallistel, 1980, for an instructive incorporation of this principle into the organization of motivational control).

Of course, habituation cannot be all that is important in stopping or modulating ingestion, because sham feeding in deprived adult rats is relatively persistent (Davis & Campbell, 1973; Young, Gibbs, Antin, Holt, & Smith, 1974). However, sham feeding does not proceed uninterrupted. Habituationlike processes may contribute to these interruptions. A parsimonious conception of control of an ingestive response element is one that incorporates two general sources of control: (1) an intrinsic source arising from the consequences of ongoing ingestion that acts by a habituationlike process (2) that can itself be modulated in efficacy by an extrinsic signal (e.g. a gastric fill signal). A habituation-oriented notion of modulatory control provides a way of understanding what is meant by the "oral phase" in feeding control, a control frequently termed "oral metering." Given a habituating, use-dependent process intrinsic to the response system that also is modulated by systemic physiological signals, an animal is given a real-time interface with its physiology, an automatic means of integrating its state and recent history and expressing the result in ongoing behavior. From this perspective, the contribution of physiological state to the modulation of ingestion can be viewed much as a sensitizing or dishabituating input signal would be in a more typical habituation situation (e.g., Groves & Thompson, 1970). Only for ingestion, the situation is somewhat more complicated than a simple habituation experiment, because during consumption the accumulation of an ingested diet can desensitize or depotentiate (e.g., Gallistel, 1980) one or more components in the ingestive response sequence at the same time that the decremental process of response habituation is occurring.

A habituationlike process certainly characterizes the nipple-search component of suckling in young pups, but we have only suggestive information with respect to the presence or development of such processes for components of independent ingestion (Figure 14). In young pups, gavage loads are as effective or more effective at suppressing intake with closed pyloric nooses than the same volume of milk actively ingested (e.g., Figure 9A), suggesting an irrelevance of preceding ingestive experience. In older pups, a contribution of previous ingestion becomes apparent as the suppressive effects of gastric loads become less complete (Figure 9B; Phifer & Hall, 1988). An emerging importance of use dependency could also be involved in the shift in the patterning of ingestion between 6 and 15 days of age (e.g., Figure 13).

Further information on this experiential type of modulation of ingestive responses could come from comparisons of sham feeding in pups of different ages. We already know that rate of intake during sham feeding is reduced over time at most ages, but this reduction could be partly related either to fatigue or to more central characteristics of use dependency and habituation. The question of the specificity of inhibition of ingestion for a given diet, which directly tests fatigue versus habituation, has not been explored developmentally either for pups ingesting normally or during sham feeding.

I have emphasized the potential role of these short-term contributions of experience to ingestion because understanding how pups come to appreciate their own ongoing ingestive behavior may contribute considerably to understanding the control of feeding. Even though this type of learning about ongoing ingestion may be short-lived and usually is categorized as nonassociative, such nonassociative forms of learning are often overlooked in appetitive situations where they may, in fact, make a major contribution to the control of behavior.

LEARNING ABOUT INGESTIVE CONSEQUENCES. Potentially, any component of ingestion also is vulnerable to modulation by more specific forms of associative learning. Such associations may be related to short- or long-term consequences of ingestion. Further, they may occur with respect to any response component from the beginning to the end of ingestion. Such learning within the context of ingestive systems is usually referred to in terms of learned preference or aversion, learned appetite, or learned satiety (see Weingarten, 1985). Even though the richness of learned effects on feeding has been recognized, the site of action in the behavior sequence rarely is analyzed in either adult or developmental studies (although see Garcia, Lasiter, Bermudez-Rattoni, & Deems, 1985). Thus, we have little idea about where in the sequence of appetitive behavior the learning occurs and even less information about the ways in which ingestion-related associative capabilities of pups may change during development. However, there is considerable evidence that even newborns possess some of these capabilities.

From birth and before, rat pups are capable of considerable consequence-dependent appetitive learning. This learning has been demonstrated in the context of feeding and suckling and with odors, tastes, and tactile stimuli (see reviews by Hall & Oppenheim, 1987; Johanson & Terry, 1988). Oral milk infusions produce learned preferences for odors paired with them (Johanson & Hall, 1982) as well as conditioned mouthing responses to those odors (Figure 17; Johanson, Hall, & Polefrone, 1984). Thus, both approach and consummatory components of early ingestive sequences show plasticity, although the neural substrate for each learned component may be quite different (Kucharski, Johanson, & Hall, 1986). It has been further demonstrated that such learning can influence ongoing independent ingestion (Sullivan, Brake, Hofer, & Williams, 1986). Similar appetitive learning has been shown within the context of suckling (Brake, 1981; Terry, Craft, & Johanson, 1983). Moreover, pups can associate negative consequences with ingestion from early in development (Kucharski & Spear, 1984; Rudy & Cheatle, 1977), and these associations can influence appetitive behavior. Clearly, early appetitive response components, whether occur-

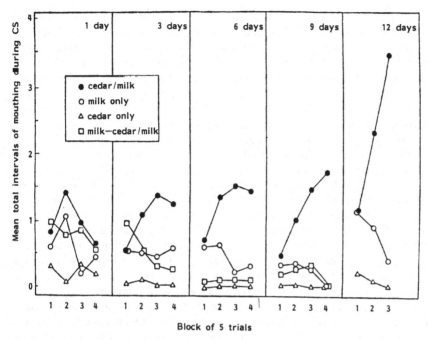

Figure 17. The potential for learned modulation of response elements is shown in conditioned mouthing responses of pups of different ages trained with pairings of cedar odor and oral milk infusions. Data are the mean number of intervals (of five per block) in which pups showed mouthing. The interval was the 5-second odor delivery period before milk infusion for the cedar/milk group or the appropriate control interval for other groups. (From Johanson *et al.*, 1984.)

ring normally or induced experimentally, have considerable vulnerability to the modulatory power of previous contingencies. Knowing that pups can learn about events related to ingestion is helpful, but it brings us back to the question of the site in the behavioral sequence at which learning occurs and raises two more specific questions with respect to feeding development: What is it about food, or events or consequences related to food, that is responsible for the learning? And does such ingestion-related learning contribute to the development of ingestive systems and influence later behavior?

Although it is clear that food- and suckling-related stimulation has rewarding effects in young pups, the cause of the reinforcement is not known and may be very different than it is for adults. For example, because young pups have no apparent sensitivity to the nutritive consequences of food with respect to ingestion, it seems unlikely that the postingestive consequences of food are initially important to them. For suckling, some studies indicate that although nipple attachment is rewarding (Amsel, Burdette, & Letz, 1976), getting milk does not add further reward value if pups are younger than 2 weeks of age (Blass *et al.*, 1979; Kenny & Blass, 1977; Letz, Burdette, Gregg, Kitrell, & Amsel, 1978). Similarly, for independent ingestion, some evidence suggests that even the immediate consequences of milk ingestion are important only insofar as they induce other reactions. For instance, in most paradigms involving food rewards, the reward itself creates a general locomotor activation. It is possible that it is this activation or excitement, rather than the direct action of the food taste or absorption, that reinforces learning. Camp and Rudy (1988) have argued that even foot shock leads to appetitive learning in young animals as long as it induces excitement. However, our understanding of these situations is still poor, and we cannot be certain whether the excitement has a direct effect or whether it is just an externalization of a more central process that supports learning. In this regard, Moran and co-workers (Moran, Lew, & Blass, 1981; Moran, Schwartz, & Blass, 1983) have shown that brain stimulation induces rewarding behavioral excitement similar to that produced by appetitive rewards.

We have found that although behavioral activation may be sufficient to produce learning in young animals, it is not a necessary process. In situations where there is little evidence of overt activation, both conditioned preference and mouthing still can be produced. Pups rewarded with oral infusions of milk after deprivation with foster mothers or after dehydration show little increase in activity yet show good evidence of learned preference and conditioned ingestive responding (Terry, 1988). Thus, it is not clear what feature or features of early reinforcers act to produce learning. The promise of developmental analysis is that comparison of the rewarding potencies of particular aspects of ingesta at different ages may allow us to dissect essential features of reinforcement and to determine when they become operational. Two important questions in this regard are how and when the postingestive consequences of feeding come to influence subsequent feeding. Recent emphasis on the importance of learned contributions to subsequent reactions to food has produced a new interest in this type of question (Booth, 1985, 1987; Weingarten, 1985).

It would seem reasonable that mechanisms for learning about the specific postingestive consequences of feeding are present as pups begin to sample food at the time of weaning and that these mechanisms may contribute to weaning (e.g., Rozin & Pelchat, 1988; see Blake & Henning, 1983; Blake, Okuhara, & Henning, 1984; Lichtman & Cramer, 1989b; Thiels, Cramer, & Alberts, 1988, for other influences). Learning about postingestive effects of foods may also influence pups' postweaning patterns of ingestion (Blake & Henning, 1985; Melcer & Alberts, 1989; Booth, Stoloff, & Nicholls, 1974). But in addition to these potential contributions of ingestion-related learning to weaning, normal suckling experiences also may be contributing to later feeding or to later behavior in general (see review by Johanson & Terry, 1988). This is a particularly relevant consideration given the fact that suckling appears to be becoming more like feeding in the week before weaning (Hall & Williams, 1983). Further, there is a clear indication that the experiences of suckling have a representation in later behavior. Taste and/or olfactory experiences during suckling can make at least transient contributions to later food preferences (Galef & Clark, 1972; Galef & Henderson, 1972; Galef & Sherry, 1973) as well as affect olfactory preferences during and beyond suckling (Coopersmith & Leon, 1986; Terry & Johanson, 1987). How these effects are achieved is not clear. They may reflect the pairing of relevant cues with the mother (Alberts, 1984) or motherlike stimulation (Sullivan & Hall, 1988), or they could be more specific to ingestive consequences, pre- or postabsorptive. They may or may not play a significant role in switching pups' ingestive preferences away from the mother and to solid food. But it does seem to be the case that such experiences can have long-lasting effects on many forms of appetitive behavior (Fillion & Blass, 1986; Leon, Galef, & Behse, 1977).

CHANGING ORGANIZATION, HIERARCHIES, AND THE DYNAMICS OF APPETITIVE
EXPRESSION AND CONTROL

113

ONTOGENY OF
INGESTIVE
BEHAVIOR

The eventual expression of adultlike ingestive response patterns and the emergence of the control processes for them must depend on the maturation of pups' morphological characteristics, response capabilities, and peripheral and central physiological and neural processes. These maturational events can alter production and modulation of all components of ingestive sequences. Thus, as a result of its continually changing reaction to the patterns and structures of its environment, a developing organism passes through many phases of appetitive organization. Indeed, there is evidence that such dynamic processes persist well past weaning (Wade, 1972) and past the time we typically deem feeding behavior to have reached a static form (Kolb & Nonneman, 1976), thus emphasizing the life-span nature of developmental processes.

I have called attention here to the manner in which the expression of appetitive behavior reflects the production of an organized sequence of response components. Although I have shown that many of these components can be studied individually during the course of ontogeny to reveal their control, these components are not normally expressed in organized behavior until late in the preweaning period. The adultlike pattern of feeding appears normally as the changing features of many components become integrated and coalesce into organized sequences. These dynamics of feeding ontogeny, as well as those of many other developmental processes, have recently been placed within the context of systems theory and synergistic analysis by Thelen (1988). Her review makes useful conceptual and empirical suggestions and emphasizes how the normal expression of the full sequence of ingestive behavior, or any behavior, is self-organized and depends on factors related to an animal's ontogenetic status, its immediate internal and external environment, and to the dynamic interaction between these factors.

These dynamics also may result in changing sites and shifting levels of control (Gallistel, 1980). Although individual and differential control applies to components in the suckling sequence in rat pups and has been increasingly recognized in the appetitive behavior of adult animals, be it for ingestion (Blundell, 1983; Bludell & Latham, 1980; Norgren & Grill, 1982; Zeigler, 1983), aggression (Bernston, Hughes, & Beattie, 1976; Flynn, 1972) or reproduction (Pfaff, 1980), appetitive sequences can become embedded in more global response hierarchies. Then, critical control of the sequence may or may not be lost to higher elements and may or may not involve integration of information from multiple levels (e.g., Gallistel's, 1980, "lattice" hierarchy). Such shifting control of organization of response sequences has been suggested as a central feature of ontogeny and experience (Fentress, 1983). Thus, adult appetitive behavior may be more complex because of the possibility of shifting sites of control. Nonetheless, we can begin its analysis from a perspective open to the identification of multiple behavioral components and of their potentially separable forms of modulation. In this regard, note that emphasis on behavioral components expands the current synthetic concept of "multiple controls of ingestion," emphasizing that such controls may operate at multiple behavioral and neural sites and that level of control may shift during ontogeny. Thus, understanding how ingestion is determined will require not simply the identification of what factors can affect volume of intake or body weight but an appreciation of where and when in the neurobehavioral stream of responding control is exerted.

SUMMARY AND CONCLUSIONS

Ingestive behavior, like other appetitive behaviors, is composed of a sequence of response elements or components. This sequential organization is particularly obvious during early postnatal development, a time when a number of these ingestive components can be conveniently isolated for study. Developmental analysis of such response components can reveal features of their patterning and control and suggest how these change with maturation.

In the normal ingestive sequence, the final component is oromotor, involving the lapping, or biting and chewing, then swallowing of diet. This final component can be studied in infant rats by making infusions of diet into pups' mouths. Young pups' responses to such oral infusions appear to be exclusively modulated by hydrational and mechanical signals and not at all by metabolic or nutritive signals. Nutritive or metabolic control appears only after the first week of life, in a context of morphological, physiological, and neural maturation. Thus, to the extent that "feeding" is defined in terms of a primary and direct nutrient-related control of intake, feeding does not develop until well after birth.

Controls of other ingestive components also emerge during the postnatal period and reveal that not all control of ingestion need be achieved by actions on a common component in the appetitive sequence. For example, as weaning approaches, the differentiation of feeding and drinking is not achieved by modulation of the final oral component but rather by the differential modulation of early components of the response sequence such as food approach and contact.

In this chapter I have reviewed a portion of the information available on the development of ingestion in rodents with the hope of indicating how such analysis can contribute to unraveling the structures and substrates of ingestive and appetitive behavior. A primary goal has been to illustrate the developmental opportunities for understanding behavior systems and to highlight some of the lessons learned about the organization of ingestive behaviors through developmental analysis. These lessons emphasize the multicomponent nature of appetitive response sequences and the necessity of considering where in this sequence modulatory controls exert their action. Ingestive controls have intrigued and frustrated psychobiologists for the last 50 years, and ontogenetic analysis remains a relatively unexplored means of determining the emergent control and organization of this fundamental response system.

I also wished to convey something of the current specific opportunities that the developmental approach has to offer in the analysis of feeding. These opportunities are of at least four types. First, at early stages of development, ingestive systems may be immature, simplified, and thus more easily understood. As such, they are available for study in isolation from factors that normally confound a dissection of their mechanism. Thus, 6-day-old rat pups provide an excellent preparation in which to study neural mechanisms for the production of dehydration-stimulated ingestion (Hall, 1989) or of the gastric inhibition of ingestive responses (Phifer & Hall, 1988). Second, complexities that are added during development can be appreciated one by one, and newly appearing substrates can be analyzed and contrasted to their absence at earlier stages. In this fashion the appearance of nutritive modulation of ingestion around 2 weeks of age provides an opportunity for identification of a primary determinant of feeding behavior. Third, developmental analysis permits control, manipulation, and evaluation of the significance of the history of an animal's ingestion-related experience. Only by such analysis can we hope to identify early events of major relevance in shaping later behavior. Finally, to the extent that aspects of the organization of ingestion and its control change independently of morphological and systemic physiological development, we may be able to associate change with events in neural maturation and thus establish brain– behavior relationships in ways not possible in the absence of the natural experiment of development. Although it still is early in the process of making use of these opportunities, the new perspectives on behavioral organization created by observing the dynamics of developmental change promise to make ontogenetic analysis one of our most powerful tools for understanding the biology of appetitive behavior.

Acknowledgments

Portions of the research described in this chapter were supported by NICHD grants RO1-HD17457 and RO1-HD17458 and by NIMH Research Scientist Development Award K01-MH00571. I thank C. M. Anderson, D. A. Booth, C. O. Eckerman, P. C. Holland, D. N. Lorenz, C. B. Phifer, S. M. Specht, S. E. Swithers, and L. M. Terry for comments on an earlier version of the manuscript, and E. M. Stricker for his thoughtful and considerable editorial contribution to this chapter.

REFERENCES

Alberts, J. R. (1984). Sensory–perceptual development in the Norway rat: a view toward comparative studies. In R. Kail & N. E. Spear (Eds.), *Comparative perspectives on memory development* (pp. 65–101). Hillsdale, NJ: Lawrence Erlbaum.

Almli, C. R. (1973). The ontogeny of the onset of drinking and plasma osmotic pressure regulation. *Developmental Psychobiology, 6,* 147–158.

Almli, C. R. (1984). Early brain damage and time course of behavioral dysfunction: Parallels with neural maturation. In S. Finger & C. R. Almli (Eds.), *Early brain damage, Vol. 2, Neurobiology and behavior* (pp. 99–116). Orlando, FL: Academic Press.

Almli, C. R., & McMullen, N. T. (1979). Ontogeny of lateral preoptic unit activity in rats. *Brain Research Bulletin, 4,* 773–781.

Amsel, A., Burdette, D. R., & Letz, R. (1976). Appetitive learning, patterned alteration, and extinction in 10-d-old rats with non-lactating suckling as reward. *Nature, 262,* 816–818.

Bernstein, I. L., & Courtney, L. (1987). Salt preference in the preweaning rat. *Developmental Psychobiology, 20,* 443–453.

Bernston, G. G., Hughes, H. C., & Beattie, M. S. (1976). A comparison of hypothalamically induced biting attack with natural predatory behavior in the cat. *Journal of Comparative and Physiological Psychology, 90,* 167–178.

Blake, H. H., & Henning, S. J. (1983). Weaning in the rat: A study of hormonal influences. *American Journal of Physiology, 244,* R537–R543.

Blake, H. H., & Henning, S. J. (1985). Basis for lactose aversion in the weanling rat *Physiology and Behavior, 35,* 313–316.

Blake, H. H., & Henning, S. J. (1986). Control of protein intake in the young rat. *Physiology and Behavior, 38,* 607–611.

Blake, H. H., Okuhara, C. T., & Henning, S. J. (1984). Role of dietary preferences in the weaning pattern of the rat. *American Journal of Physiology, 246,* R96–R101.

Blass, E. M., & Hall, W. G. (1976). Drinking termination: Interactions among hydrational, orogastric, and behavioral controls in rats. *Psychological Review, 83,* 356–374.

Blass, E. M., Hall, W. G., & Teicher, M. H. (1979). The ontogeny of suckling and ingestive behaviors. *Progress in Psychobiology and Physiological Psychology, 8,* 243–299.

Blass, E. M., & Teicher, M. H. (1980). Suckling. *Science, 210,* 15–22.

Blazquez, E., Montoya, E., & Quijada, C. L. (1970). Relationship between insulin concentrations in plasma and pancreas of foetal and weanling rats. *Journal of Endocrinology, 48,* 553–561.

Blundell, J. E. (1983). Problems and processes underlying the control of food selection and nutrient intake. In R. J. Wurtman & J. J. Wurtman (Eds.), *Nutrition and the brain* (vol. 6, pp. 163–221). New York: Raven Press.

Blundell, J. E., & Latham, C. J. (1980). Characterization of adjustments to the structure of feeding behavior following pharmacological treatment: Effects of amphetamine and fenfluramine and the antagonism produced by pimozide and methergoline. *Pharmacology and Biochemistry, 12,* 717–722.

Booth, D. A. (1985). Food-conditioned eating preferences and aversions with interoceptive elements: Learned appetites and satieties. *Annals of the New York Academy of Sciences, 443,* 22–37.

Booth, D. A. (1987). How to measure learned control of food or water intake. In F. M. Toates & N. E. Rowland (Eds.), *Techniques in the behavioral sciences (Vol. 1): Feeding and drinking* (pp. 111–149). Amsterdam: Elsevier.

Booth, D. A., & Jarman, S. P. (1976). Inhibition of food intake in the rat following complete absorption of glucose delivered into the stomach, intestine or liver. *Journal of Physiology (London), 259,* 501–522.

Booth, D. A., Stoloff, R., & Nicholls, J. (1974). Dietary flavor acceptance in infant rats established by association with effects of nutrient composition. *Physiological Psychology, 2,* 313–319.

Brake, S. C. (1981). Suckling infant rats learn a preference for a novel olfactory stimulus paired with milk delivery. *Science, 211,* 506–508.

Brake, S. C., & Hofer, M. A. (1980). Maternal deprivation and prolonged suckling in the absence of milk alter the frequency and intensity of suckling responses in neonatal rat pups. *Physiology and Behavior, 24,* 185–189.

Brake, S. C., Sager, D. J., Sullivan, R., & Hofer, M. A. (1982). The role of intraoral and gastrointestinal cues in the control of sucking and milk consumption in rat pups. *Developmental Psychobiology, 15,* 529–541.

Brake, S. C., Shair, H., & Hofer, M. A. (1988). Exploiting the nursing niche: The infant's sucking and feeding in the context of the mother–infant interaction. In E. M. Blass (Ed.), *Handbook of behavioral neurobiology* (pp. 347–388). New York: Plenum Press.

Brake, S. C., Tavana, S., & Myers, M. M. (1986). A method for recording and analyzing intra-oral negative pressure in suckling rat pups. *Physiology and Behavior, 36,* 575–578.

Brake, S. C., Wolfson, V., & Hofer, M. A. (1979). Electrophysiological patterns associated with nonnutritive sucking in 11–13-day old rat pups. *Journal of Comparative and Physiological Psychology, 93,* 760–770.

Breese, G. R., & Traylor, T. D. (1972). Developmental characteristics of main catecholamines and tyrosine hydroxylase in the rat: Effects of 6-hydroxydopamine. *British Journal of Pharmacology, 4P,* 210–222.

Brunjes, P. C., & Frazier, L. L. (1986). Maturation and plasticity in the olfactory system of vertebrates. *Brain Research Review, 11,* 1–45.

Bruno, J. P. (1981). Development of drinking behavior in preweanling rats. *Journal of Comparative and Physiological Psychology, 95,* 1016–1027.

Bruno, J. P., Blass, E. M., & Amin, F. (1983). Determinants of suckling versus drinking in weanling albino rats: Influence of hydrational state and maternal contact. *Developmental Psychobiology, 16,* 177–184.

Bruno, J. P., & Hall, W. G. (1982). Olfactory contributions to dehydration-induced anorexia in weanling rats. *Developmental Psychobiology, 15,* 493–505.

Bruno, J. P., Hall, W. G., & Grill, H. J. (1980). Dehydration-induced anorexia: Olfactory contributions. *Neuroscience Abstracts, 6,* 517.

Bruno, J. P., Snyder, A. M., & Stricker, E. M. (1984). Effect of dopamine-depleting brain lesions on suckling and weaning in rats. *Behavioral Neuroscience, 98,* 156–161.

Bruno, J. P., Zigmond, M. J., & Stricker, E. M. (1986). Rats given dopamine-depleting brain lesions as neonates do not respond to acute homeostatic imbalances as adults. *Behavioral Neuroscience, 100,* 125–128.

Burke, G. H., Mook, D. G., & Blass, E. M. (1972). Hyperactivity to quinine associated with osmotic thirst in the rat. *Journal of Comparative and Physiological Psychology, 78,* 32–39.

Camp, L. L., & Rudy, J. W. (1988). Changes in the categorization of appetitive and aversive events during postnatal development of the rat. *Developmental Psychobiology, 21,* 25–42.

Campbell, B. A., & Randall, P. J. (1977). Paradoxical effects of amphetamine on preweanling and postweanling rats. *Science, 195*, 888–890.

Campbell, B. A., & Raskin, L. A. (1978). Ontogeny of behavioral arousal: The role of environmental stimuli. *Journal of Comparative and Physiological Psychology, 92*, 176–184.

Caza, P., & Spear, L. P. (1982). Pharmacological manipulations of milk-induced behaviors in 3-day-old rat pups. *Pharmacology, Biochemistry and Behavior, 16*, 481–486.

Coopersmith, R., & Leon, M. (1986). Enhanced neural response by adult rats to odors experienced early in life. *Brain Research, 371*, 400–403.

Coyle, J. T., & Axelrod, J. (1972). Dopamine-β-hydroxylase in the rat brain: Developmental characteristics. *Journal of Neurochemistry, 19*, 449–459.

Craig, W. (1918). Appetites and aversions as constituents of instincts. *Biological Bulletin, 34*, 91–107.

Cramer, C. P., & Blass, E. M. (1982). The contribution of ambient temperature to suckling behavior in rats 3–20 days of age. *Developmental Psychobiology, 15*, 339–348.

Cramer, C. P., & Blass, E. M. (1983). Mechanisms of control of milk intake in suckling rats. *American Journal of Physiology, 245*, R154–R157.

Cramer, C. P., & Blass, E. M. (1985). Nutritive and nonnutritive determinants of milk intake of suckling rats. *Behavioral Neuroscience, 99*, 578–592.

Davis, J. D., & Campbell, C. S. (1973). Peripheral control of meal size in the rat: Effect of sham feeding on meal size and drinking rate. *Journal of Comparative and Physiological Psychology, 83*, 379–387.

DeCasper, A. H., & Fifer, W. P. (1980). Of human bonding: Newborns prefer their mothers' voices. *Science, 208*, 1174–1176.

Dethier, V. G. (1976). *The hungry fly: A physiological study of the behavior associated with feeding.* Cambridge: Harvard University Press.

Deutsch, J. A. (1983). Dietary control and the stomach. *Progress in Neurobiology, 20*, 313–332.

Deutsch, J. A. (1985). The role of the stomach in eating. *American Journal of Clinical Nutrition, 42*, 1040–1043.

Deutsch, J. A., & Wang, M.-L. (1977). The stomach as a site for rapid nutrient reinforcement sensors. *Science, 195*, 89–90.

Distel, H., & Hudson, R. (1985). The contribution of the olfactory and tactile modalities to the nipple-search behaviour of newborn rabbits. *Journal of Comparative Physiology A, 158*, 599–605.

Drewett, R. F. (1978). The development of motivational systems. *Progress in Brain Research, 48*, 407–417.

Drewett, R. F., & Cordall, K. M. (1976). Control of feeding in suckling rats: Effects of glucose and osmotic stimuli. *Physiology and Behavior, 16*, 711–718.

Drewett, R. F., Statham, C., & Wakerley, J. B. (1974). A quantitative analysis of the feeding behavior of suckling rats. *Animal Behavior, 22*, 907–913.

Ellis, S., Axt, K., & Epstein, A. N. (1984). The arousal of ingestive behaviors by the injection of chemical substances into the brain of the suckling rat. *Journal of Neuroscience, 4*, 945–955.

Epstein, A. N. (1984). The ontogeny of neurochemical systems for control of feeding and drinking. *Proceedings of the Society for Experimental Biology and Medicine, 175*, 127–134.

Epstein, A. N. (1986). The ontogeny of ingestive behaviors: Control of milk intake by suckling rats and the emergence of feeding and drinking at weaning. In R. Ritter, S. Ritter, & C. D. Barnes (Eds.), *Neural and humoral controls of food intake* (pp. 1–25). New York: Academic Press.

Falk, G. (1955). Maturation of renal function in infant rats. *American Journal of Physiology, 181*, 157–170.

Fentress, J. C. (1976). Dynamic boundaries of patterned behaviour: Interaction and self-organization. In P. P. G. Bateson & R. A. Hinde (Eds.), *Growing points in ethology* (pp. 135–169). London: Cambridge University Press.

Fentress, J. C. (1978). *Mus musicus.* The developmental orchestration of selected movement patterns in mice. In M. Bekoff & G. Burghardt (Eds.), *The development of behaviour: Comparative and evolutionary aspects* (pp. 321–342). New York: Garland Publishing.

Fentress, J. C. (1981). Order in ontogeny: Relational dynamics. In K. Immelmann, G. Barlow, M. Main, & L. Petrinovich (Eds.), *Behavioral development* (pp. 338–371). New York: Cambridge University Press.

Fentress, J. C. (1983). Ethological models of hierarchy and patterning of species-specific behavior. In E. Satinoff & P. Teitelbaum (Eds.), *Handbook of behavioral neurobiology, Vol.6, Motivation* (pp. 185–228). New York: Plenum Press.

Ferrell, M. F., Mistretta, C. M., & Bradley, R. M. (1981). Chorda tympani taste responses during development in the rat. *Journal of Comparative Neurology, 198*, 37–44.

Fillion, T. J., & Blass, E. M. (1986). Infantile experience with suckling odors determines adult sexual behavior in male rats. *Science, 231,* 729–731.

Flynn, J. P. (1972). Patterning mechanisms, patterned reflexes, and attach behavior in cats. *Nebraska Symposium on Motivation, 20,* 125–153.

Friedman, M. I. (1975). Some determinants of milk ingestion in suckling rats. *Journal of Comparative and Physiological Psychology, 89,* 636–647.

Friedman, M. I. (1979). Effects of milk consumption and deprivation on body fluids of suckling rats. *Physiology and Behavior, 23,* 1029–1034.

Friedman, M. I., & Campbell, B. A. (1974). Ontogeny of thirst in the rat: Effects of hypertonic saline, polyethylene glycol, and vena cava ligation. *Journal of Comparative and Physiological Psychology, 87* 37–46.

Friedman, M. I., & Stricker, E. M. (1976). The physiological psychology of hunger: A physiological perspective. *Psychological Review, 83,* 409–431.

Galef, B. G., Jr. (1982). The development of flavor preferences in man and animals: The role of social and nonsocial factors. In R. N. Aslin, J. R. Alberts, & M. R. Peterson (Eds.), *Development of perception: Psychobiological perspectives* (pp. 411–431). New York: Academic Press.

Galef, B. G., Jr., & Clark, M. M. (1972). Mother's milk and adult presence: Two factors determining initial dietary selection by weanling rats. *Journal of Comparative and Physiological Psychology, 78,* 220–225.

Galef, B. G., Jr., & Henderson, P. W. (1972). Mother's milk: A determinant of the feeding preferences of rat pups. *Journal of Comparative and Physiological Psychology, 78,* 213–219.

Galef, B. G., Jr., & Sherry, D. F. (1973). Mother's milk: A medium for the transmission of cues reflecting the flavor of mother's diet. *Journal of Comparative and Physiological Psychology, 83,* 374–378.

Gallistel, C. R. (1980). *The organization of action: A new synthesis.* New York: Lawrence Erlbaum.

Ganchrow, J. R., Steiner, J. E., & Canetto, S. (1986). Behavioral displays to gustatory stimuli in newborn rat pups. *Developmental Psychobiology, 19,* 163–174.

Garcia, J., Lasiter, P. S., Bermudez-Rattoni, F., & Deems, D. A. (1985). A general theory of aversion learning *Annals of the New York Academy of Sciences, 443,* 8–21.

Ghishan, F. K., & Wilson, F. A. (1985). Developmental maturation of D-glucose transport by rat jejunal brush-border membrane vesicles. *American Journal of Physiology, 248,* G87–G92.

Gisel, E. G., & Henning, S. J. (1980). Appearance of glucoprivic control of feeding behavior in the developing rat. *Physiology and Behavior, 24,* 313–318.

Gottlieb, G. (1971). Ontogenesis of sensory function in birds and mammals. In E. Toback, L. Avonson, & E. Shaw (Eds.), *The biopsychology of development* (pp. 67–128). New York: Academic Press.

Grill, H. J., & Norgren, R. (1978). The taste reactivity test. I. Mimetic responses to gustatory stimuli in neurologically normal rats. *Brain Research, 143,* 263–279.

Groves, P. M., & Thompson, R. F. (1970). Habituation: A dual-process theory. *Psychological Review, 77,* 419–450.

Gutman, Y., & Krausz, M. (1969). Regulation of food and water intake in rats as related to plasma osmolarity and volume. *Physiology and Behavior, 4,* 311–313.

Hall, W. G. (1973). A remote stomach clamp to evaluate oral and gastric controls of drinking in the rat. *Physiology and Behavior, 11,* 897–901.

Hall, W. G. (1979). Feeding and behavioral activation in infant rats. *Science, 190,* 1313–1315.

Hall, W. G. (1985). What we know and don't know about the development of independent ingestion in rats. *Appetite, 6,* 333–356.

Hall, W. G. (1989). Neural systems for early independent ingestion: Regional metabolic changes during ingestive responding. *Behavioral Neuroscience, 103* 386–411.

Hall, W. G., & Blass, E. M. (1977). Orogastric determinants of drinking in rats: Interaction between absorptive and peripheral controls. *Journal of Comparative and Physiological Psychology, 91,* 365–373.

Hall, W. G., & Browde, J. A., Jr. (1986). The ontogeny of independent ingestion in mice: Or, why won't infant mice feed? *Developmental Psychobiology, 18,* 211–222.

Hall, W. G., & Bruno, J. P. (1984). Inhibitory controls of ingestion in 6-day-old rat pups. *Physiology and Behavior, 32,* 831–841.

Hall, W. G., & Bryan, T. E. (1980). The ontogeny of feeding in rats. II. Independent ingestive behavior. *Journal of Comparative and Physiological Psychology, 94,* 746–756.

Hall, W. G., & Bryan, T. E. (1981). The ontogeny of feeding in rats. IV. Taste development as measured by intake and behavioral responses to oral infusions of sucrose and quinine. *Journal of Comparative and Physiological Psychology, 95,* 240–251.

Hall, W. G., Cramer, C. P., & Blass, E. M. (1977). The ontogeny of suckling in rats. *Journal of Comparative and Psychological Psychology, 91,* 1141–1155.

Hall, W. G., Denzinger, A., & Phifer, C. B. (1988). Behavioral activation and the guidance of early feeding. Unpublished manuscript.

Hall, W. G., Hudson, R., & Brake, S. C. (1988). Terminology for use in investigations of nursing and suckling. *Developmental Psychobiology, 21,* 89–91.

Hall, W. G., & Oppenheim, R. W. (1987). Developmental psychobiology: Prenatal, perinatal, and early postnatal aspects of behavioral development. *Annual Review of Psychology, 38,* 91–128.

Hall, W. G., & Rosenblatt, J. S. (1977). Suckling behavior and intake control in the developing rat pup. *Journal of Comparative and Physiological Psychology, 91,* 1232–1247.

Hall, W. G., & Rosenblatt, J. S. (1978). Development of nutritional controls of food intake in suckling rat pups. *Behavioral Biology, 24,* 412–427.

Hall, W. G., Swithers, S., & McCall, K. (1988). Developmental changes in feeding patterns for 6- and 15-day old rat pups. Unpublished manuscript.

Hall, W. G., & Williams, C. L. (1983). Suckling isn't feeding, or is it? A search for developmental continuities. In J. S. Rosenblatt, R. A. Hinde, R. A. Beer, & M. Busnell (Eds.), *Advances in the study of behavior* (pp. 219–254). New York: Academic Press.

Hawkins, R. A., Williamson, D. H., & Krebs, H. A. (1971). Ketone-body utilization by adult and suckling rat brain *in vivo. Biochemical Journal, 122,* 13–18.

Heller, H. (1949). Effect of dehydration on adult and newborn rats. *Journal of Physiology, 108,* 303–314.

Henning, S. J. (1981). Postnatal development: Coordination of feeding, digestion and metabolism. *American Journal of Physiology, 241,* G199–G214.

Henning, S. J. (1987). Functional development of the gastrointestinal tract. In L. R. Johnson (Ed.), *Physiology of the gastrointestinal tract* (pp. 285–300). New York: Raven Press.

Henning, S. J., Chang, S.-S. P., & Gisel, E. G. (1979). Ontogeny of feeding controls in suckling and weanling rats. *American Journal of Physiology, 237,* R187–R191.

Henning, S. J., & Gisel, E. G. (1980). Nocturnal feeding behavior in the neonatal rat. *Physiology and Behavior, 25,* 603–605.

Hill, D. L. (1987). Development of taste responses in the rat parabrachial nucleus. *Journal of Neurophysiology, 57,* 481–495.

Hill, D. L., & Almli, C. R. (1980). Ontogeny of chorda tympani nerve responses to gustatory stimuli in the rat. *Brain Research, 197,* 27–38.

Hill, D. L., Bradley, R. M., & Mistretta, C. M. (1983). Development of taste responses in rat nucleus of solitary tract. *Journal of Neurophysiology, 50,* 879–895.

Hinde, R. A. (1970). *Animal behavior: A synthesis of ethology and comparative psychology.* New York: McGraw-Hill.

Hofer, M. A. (1975). Studies on how early maternal separation produces behavioral change in young rats. *Psychosomatic Medicine, 37,* 245–264.

Hofer, M. A. (1978). Hidden regulatory processes in early social relationships. In P. P. G. Bateson & P. H. Klopfer (Eds.), *Perspectives in ethology* (pp. 135–163). New York: Plenum Press.

Hofer, M. A., Fisher, A., & Shair, H. (1981). Effects of infraorbital nerve section on survival, growth, and suckling behaviors of developing rats. *Journal of Comparative and Physiological Psychology, 95,* 123–133.

Hofer, M. A., & Shair, H. (1982). Control of sleep–wake states in the infant rat by features of the mother–infant relationship. *Developmental Psychobiology, 15,* 229–243.

Hofer, M. A., Shair, H., & Singh, P. (1976). Evidence that maternal ventral skin substances promote suckling in infant rats. *Physiology and Behavior, 17,* 131–136.

Holland, P. C. (1988). Excitation and inhibition in unblocking. *Journal of Experimental Psychology: Animal Behavior Processes, 14,* 261–279.

Holloway, W. R., Dollinger, M. J., & Denenberg, V. H. (1979). Nipple attachment in the neonatal rat: Evidence for a daily rhythm. *Biology and Behavior, 4,* 351–362.

Horowitz, J., Heller, A., & Hoffman, P. C. (1982). The effect of development of thermoregulatory function on the biochemical assessment of the ontogeny of neonatal dopaminergic neuronal activity. *Brain Research, 235,* 245–252.

Houpt, K. A., & Epstein, A. N. (1973). Ontogeny of controls of food intake in the rat: GI fill and glucoprivation. *American Journal of Physiology, 225,* 58–66.

Houpt, K. A., & Houpt, T. R. (1975). Effects of gastric loads and food deprivation on subsequent food intake in suckling rats. *Journal of Comparative and Physiological Psychology, 88,* 764–772.

Hsiao, S., & Langenes, D. J. (1971). Liquid intake, initiation and amount of eating as determined by osmolality of drinking liquids. *Physiology and Behavior, 7,* 233–237.

Hsiao, S., & Trankina, F. (1969). Thirst–hunger interactions. I. Effects of body fluid restoration on food and water intake in water deprived rats. *Journal of Comparative and Physiological Psychology, 69,* 448–453.

Hudson, R. (1985). Do newborn rabbits learn the odor stimuli releasing nipple-search behavior? *Developmental Psychobiology, 18,* 575–586.

Hudson, R., & Distal, H. (1982). The pattern of behavior of rabbit pups in the nest. *Behavior, 79,* 255–271.

Johanson, I. B., & Hall, W. G. (1979). Appetitive learning in 1-day-old rat pups. *Science, 205,* 419–421.

Johanson, I. B., & Hall, W. G. (1980). The ontogeny of feeding in rats. III. Thermal determinants of early ingestive responding. *Journal of Comparative and Physiological Psychology, 94,* 977–992.

Johanson, I. B., & Hall, W. G. (1981). The ontogeny of feeding in rats. V. Influence of texture, home odor, and sibling presence on ingestive behavior. *Journal of Comparative and Physiological Psychology, 95,* 837–847.

Johanson, I. B., & Hall, W. G. (1982). Appetitive conditioning in neonatal rats: Conditioned orientation to a novel odor. *Developmental Psychobiology, 15,* 379–397.

Johanson, I. B., Hall, W. G., & Polefrone, J. M. (1984). Appetitive conditioning in neonatal rats: Conditioned ingestive responding to stimuli paired with oral infusions of milk. *Developmental Psychobiology, 17,* 357–381.

Johanson, I. B., & Shapiro, E. C. (1986). Intake and behavioral responsiveness to taste stimuli in infant rats from 1 to 15 days of age. *Developmental Psychobiology, 19,* 593–606.

Johanson, I. B., & Terry, L. M. (1988). Learning in infancy: A mechanism for behavioral change during development. In E. M. Blass (Ed.), *Handbook of behavioral neurobiology* (pp. 245–281). New York: Plenum Press.

Johnston, M. V. (1985). Neurotransmitters. In R. C. Wiggins, D. W. McCandless, & S. J. Enna (Eds.), *Developmental neurochemistry* (pp. 193–224). Austin: University of Texas Press.

Kehoe, P., & Blass, E. M. (1985). Gustatory determinants of suckling in albino rats 5–20 days of age. *Developmental Psychobiology, 18,* 67–82.

Kenny, J. T., & Blass, E. M. (1977). Suckling as incentive to instrumental learning in preweanling rats. *Science, 196,* 898–899.

Kohn, M. (1951). Satiation of hunger from food injected directly into the stomach versus food ingested by mouth. *Journal of Comparative and Physiological Psychology, 44,* 412–422.

Kolb, B., & Nonneman, A. J. (1976). Functional development of prefrontal cortex in rats continues into adolescence. *Science, 193,* 335–336.

Kraly, F. S., & Blass, E. M. (1974). Motivated feeding in the absence of glucoprivic control of feeding in rats. *Journal of Comparative and Physiological Psychology, 87,* 801–807.

Kraly, F. S., & Smith, G. P. (1978). Combined pregastric and gastric stimulation by food is sufficient for normal meal size. *Physiology and Behavior, 21,* 405–409.

Kucharski, D., Johanson, I. B., & Hall, W. G. (1986). Unilateral olfactory conditioning in 6-day-old rat pups. *Behavioral Neural Biology, 46,* 472–490.

Kucharski, D., & Spear, N. E. (1984). Conditioning of aversion to an odor paired with peripheral shock in the developing rat. *Developmental Psychobiology, 17,* 465–479.

Lanier, L. P., Dunn, A. J., & Van Hartesveldt, C. (1976). Development of neurotransmitters and their function in brain. In S. Ehrenpreis & I. J. Kopin (Eds.), *Review of neurosciences* (vol. 2, pp. 195–256). New York: Raven Press.

Lau, C., & Henning, S. J. (1985). Investigation of the nature of the "stretch response" in suckling rats. *Physiology and Behavior, 34,* 649–651.

Leibowitz, S. F. (1976). Brain catecholaminergic mechanisms for control of hunger. In D. Novin, W. Wyricka, & G. Bray (Eds.), *Hunger: Basic mechanisms and clinical implications* (pp. 1–18). New York: Raven Press.

Leon, M., Galef, B. G., & Behse, J. H. (1977). Establishment of phermonal bonds and diet choice in young rats by odor pre-exposure. *Physiology and Behavior, 18,* 387–391.

Lepkovsky, S., Lyman, R., Fleming, D., Nagumo, M., & Dimick, M. M. (1957). Gastrointestinal regulation of water and its effect on food intake and rate of digestion. *American Journal of Physiology, 188,* 327–331.

Leshem, M., Boggan, B., & Epstein, A. N. (1988). The ontogeny of drinking evoked by activation of brain angiotensin in the rat pup. *Developmental Psychobiology, 21,* 73–76.

Leshem, M., & Epstein, A. N. (1988). Thirst induced anorexias and the ontogeny of thirst in the rat. *Developmental Psychobiology, 21*, 651–662.

Letz, R., Burdette, D. R., Gregg, B., Kittrell, M. E., & Amsel, A. (1978). Evidence for a transitional period for the development of persistence in infant rats. *Journal of Comparative and Physiological Psychology, 92*, 856–866.

Lichtman, A. H., & Cramer, C. P. (1989a). Methysergide-induced nipple attachment depends upon suckling experience in juvenile rats. *Behavioral Neuroscience, 103*, 254–261.

Lichtman, A. H., & Cramer, C. P. (1989a). Relative importance of experience, social facilitation, and availability of milk in weaning of rats. *Developmental Psychobiology, 22*, 347–356.

Lidov, H. G. W., & Molliver, M. E. (1982). An imunohistochemical study of serotonin neuron development in the rat: Ascending pathways and terminal fields. *Brain Research Bulletin, 8*, 389–430.

Lincoln, D. W., Hill, A., & Wakerley, J. B. (1973). The milk-ejection reflex of the rat: An intermittent function not abolished by surgical levels of anesthesia. *Journal of Endocrinology, 57*, 459–476.

Lipsitt, L. P. (1977). The study of sensory and learning processes of the newborn. *Clinics in Perinatology, 4*, 163–186.

Loizou, L. A. (1972). The postnatal ontogeny of monoamine-containing neurons in the central nervous system of the rat. *Brain Research, 40*, 300–311.

Lorenz, D. N. (1983). Effects of gastric filling and vagotomy on ingestion, nipple attachment and weight gain by suckling rats. *Developmental Psychobiology, 16*, 469–483.

Lorenz, D. N. (1986). Alimentary sleep satiety in suckling rats. *Physiology and Behavior, 38*, 557–562.

Lorenz, D. N., Ellis, S. B., & Epstein, A. N. (1982). Differential effects of upper gastrointestinal fill on milk ingestion and nipple attachment in the suckling rat. *Developmental Psychobiology, 15*, 309–330.

Lorenz, K. (1950). The comparative method for studying innate behavior patterns. *Symposium of the Society for Experimental Biology, 4*, 221–268.

Lund, R. D. (1978). *Development and plasticity of the brain.* New York: Oxford University Press.

Lytle, L. D., Moorcroft, W. H., & Campbell, B. A. (1971). Ontogeny of amphetamine anorexia and insulin hyperphagia in the rat. *Journal of Comparative and Physiological Psychology, 77*, 388–393.

Mabry, P. D., & Campbell, B. A. (1977). Developmental psychopharmacology. In L. L. Iversen, S. D. Iversen, & S. H. Iversen (Eds.), *Handbook of psychopharmacology* (vol. 7, pp. 393–444). New York: Plenum Press.

Martin, G. R., & Henning, S. J. (1984). Enzymic development of the small intestine: Are glucocorticoids necessary? *American Journal of Physiology, 246*, G695–G699.

Melcer, T., & Alberts, J. R. (1989). Recognition of food by individual, food-naive, weaning rats. *Journal of Comparative Psychology, 103*, 243–251.

Midkiff, E. E., & Bernstein, I. L. (1983). The influence of age and experience on salt preference of the rat. *Developmental Psychobiology, 16*, 385–394.

Miller, N. E., & Kessen, M. L. (1952). Reward effect of food via stomach compared with those of food via mouth. *Journal of Comparative and Physiological Psychology, 45*, 550–564.

Misantone, L. J., Ellis, S., & Epstein, A. N. (1980). Development of angiotensin-induced drinking in the rat. *Brain Research, 186*, 195–202.

Moe, K. E. (1986). The ontogeny of salt preference in rats. *Developmental Psychobiology, 19*, 185–196.

Moorcroft, W. H., Lytle, L. D. & Campbell, B. A. (1973). Ontogeny of starvation-induced behavioral arousal in the rat. *Journal of Comparative and Physiological Psychology, 75*, 50–67.

Moran, T. H., Lew, M. F., & Blass, E. M. (1981). Intracranial self-stimulation in 3-day-old rats. *Science, 214*, 1366–1368.

Moran, T. H., Schwartz, G. J., & Blass, E. M. (1983). Organized behavioral responses to lateral hypothalamic electrical stimulation in infant rats. *Journal of Neuroscience, 3*, 10–19.

Norgren, R., & Grill, H. (1982). Brain-stem control of ingestive behavior. In D. W. Pfaff (Eds.), *The physiological mechanisms of motivation* (pp. 99–131). New York: Springer-Verlag.

Pelchat, M. L., & Brake, S. C. (1987). Sapid savvy in sucklings: The effect of quinine hydrochloride on intra-oral negative pressure and intake by 11–13 day old rat pups. *Developmental Psychobiology, 20*, 261–276.

Pfaff, D. W. (1980). *Estrogens and brain function: Neural analysis of a hormone-controlled mammalian reproductive behavior.* New York: Springer-Verlag.

Phifer, C. B., Browde, J. A., Jr., & Hall, W. G. (1986). Ontogeny of glucose inhibition of independent ingestion in preweanling rats. *Brain Research Bulletin, 17*, 673–679.

Phifer, C. B., & Hall, W. G. (1987). Development of feeding behavior. In F. M. Toates & N. E.

Rowland (Eds.), *Techniques in the behavioral sciences (Vol. 1): Feeding and drinking* (pp. 189–230). Amsterdam: Elsevier.

Phifer, C. B., & Hall, W. G. (1988). Ingestive behavior in preweanling rats: Emergence of postgastric controls. *American Journal of Physiology, 255,* R191–R199.

Phifer, C. B., Ladd, M., & Hall, W. G. (1988). Feeding only emerges after 6 days of age. Unpublished manuscript.

Phifer, C. B., Sikes, C. R., & Hall, W. G. (1986). Ingestive controls in 6-day-old rat pups: Termination of intake by gastric fill alone. *American Journal of Physiology, 250,* R807–R814.

Purves, D., & Lichtman, J. W. (1985). *Principles of neural development.* Sunderland, MA: Sinauer Associates.

Raskin, L. A., & Campbell, B. A. (1981). The ontogeny of amphetamine anorexia: A behavioral analysis. *Journal of Comparative and Physiological Psychology, 95,* 425–435.

Ristine, L. A., & Spear, L. P. (1985). Is there a "serotonergic syndrome" in neonatal rat pups? *Pharmacology Biochemistry and Behavior, 22,* 265–269.

Robinson, P. H., Moran, T. H., & McHugh, P. R. (1988). Cholecystokinin inhibits independent ingestion in neonatal rats. *American Journal of Physiology, 255,* R14–R20.

Rolls, B. J., Rolls, E. T., Rowe, E. A., & Sweeney, K. (1981). Sensory specific satiety in man. *Physiology and Behavior, 27,* 137–142.

Rozin, P., & Pelchat, M. L. (1988). Memories of mammaries: Adaptations to weaning from milk. *Progress in Psychobiology and Physiological Psychology, 13,* 1–29.

Rudy, J. W., & Cheatle, M. D. (1977). Odor-aversion learning in neonatal rats. *Science, 198,* 845–846.

Schwartzbaum, J. S., & Ward, H. P. (1958). An osmotic factor in the regulation of food intake in the rat. *Journal of Comparative and Physiological Psychology, 51,* 555.

Shair, H., Brake, S., & Hofer, M. A. (1984). Suckling in the rat: Evidence for patterned behavior during sleep. *Behavioral Neuroscience, 98,* 366–370.

Sherrington, C. S. (1947). *The integrative action of the nervous system.* New Haven: Yale University Press. (Original work published 1906).

Smith, G. P., Gibbs, J., & Stromayer, A. J. (1972). Threshold doses of 2-deoxy-D-glucose for hyperglycemia and feeding in rats and monkeys. *American Journal of Physiology, 222,* 77–81.

Smotherman, W. P., & Robinson, S. R. (1985). The rat fetus in its environment: Behavioral adjustments to novel, familiar aversive and conditioned stimuli presented *in utero. Behavioral Neuroscience, 99,* 521–530.

Smotherman, W. P., & Robinson, S. R. (1987). Prenatal expression of species-typical action patterns in the rat fetus (*Rattus norvegicus*). *Journal of Comparative Psychology, 101,* 190–196.

Snowdon, C. T. (1970). Gastrointestinal sensory and motor control of food intake. *Journal of Comparative and Physiological Psychology, 71,* 68–76.

Spear, L. P., & Ristine, L. A. (1982). Suckling behavior in neonatal rats: Psychopharmacological investigations. *Journal of Comparative and Physiological Psychology, 96,* 244–255.

Spear, L. P., Specht, S. M., Kirstein, C. L., & Kuhn, C. M. (1989). Anterior and posterior, but not cheek, intraoral cannulation procedures elevate serum corticosterone levels in neonatal rat pups. *Developmental Psychobiology, 22,* 401–412.

Stoloff, M. L., & Supinski, D. M. (1985). Control of suckling and feeding by methysergide in weaning albino rats: A determination of Y-maze preferences. *Developmental Psychobiology, 18,* 273–285.

Stricker, E. M. (1969). Osmoregulation and volume regulation in rats: Inhibition of hypovolemic thirst by water. *American Journal of Physiology, 217,* 98–105.

Sullivan, R. M., Brake, S. C., Hofer, M. A., & Williams, C. L. (1986). Huddling and independent feeding of neonatal rats can be facilitated by a conditioned change in behavioral state. *Developmental Psychobiology, 19,* 625–635.

Sullivan, R. M., & Hall, W. G. (1988). Reinforcers in infancy: Classical conditioning using stroking or intra-oral infusions of milk as UCS. *Developmental Psychobiology, 21,* 215–224.

Swithers, S., & Hall, W. G. (1989). A nutritive control of independent ingestion in rat pups emerges by nine days of age. *Physiology and Behavior, 46,* 873–879.

Teicher, M. H., & Blass, E. M. (1976). Suckling in newborn rats: Eliminated by nipple lavage, reinstated by pup saliva. *Science, 193,* 422–425.

Teitelbaum, P. (1966). The use of operant methods in the assessment and control of motivational states. In W. K. Hoing, (Ed.), *Operant behavior: Areas of research and application* (pp. 565–608). New York: Appleton-Century-Crofts.

Teitelbaum, P. (1977). Levels of integration of the operant. In W. K. Honig & J. E. R. Staddon (Eds.), *Handbook of operant behavior* (pp. 7–27). New York: Prentice-Hall.

Terry, L. M. (1988). *Neurobehavioral organization of an early mammalian appetitive behavior.* Durham, NC: Duke University.

Terry, L. M., Craft, G. F., & Johanson, I. B. (1983). Early olfactory experiences modulate infant rats' ingestive behaviors. *Abstracts Annual Meeting International Society for Developmental Psychobiology, 1983,* 79.

Terry, L. M., & Johanson, I. B. (1987). Olfactory influences on the ingestive behavior of infant rats. *Developmental Psychobiology, 20,* 313–332.

Thelen, E. (1984). Learning to walk: Ecological demands and phylogenetic constraints. In L. P. Lipsitt & C. Rovee-Collier (Eds.), *Advances in infancy research* (vol. 3, pp. 213–250). Norwood, NJ: Ablex.

Thelen, E. (1988). Self-organization in developmental processes: Can systems approaches work? In M. Gunnar (Eds.), *Systems in development: The Minnesota Symposium in Child Psychology* (Vol. 22, pp. 77–117). Hillsdale, NJ: Lawrence Erlbaum.

Thelen, E., & Fisher, D. M. (1982). Newborn stepping: An explanation for a "disappearing reflex." *Developmental Psychology, 18,* 760–775.

Thiels, E., Cramer, C. P., & Alberts, J. R. (1988). Behavioral interactions rather than milk availability determine decline in milk intake of weaning rats. *Physiology and Behavior, 42,* 507–516.

Thorpe, W. H. (1966). *Learning and instinct in animals.* Cambridge, MA: Harvard University Press.

Thurston, J. H., & Hauhart, R. E. (1985). Ketone bodies, lactate, glucose: Metabolic fuels for developing brain. In R. C. Wiggins, D. W. McCandless, & S. J. Enna (Eds.), *Developmental neurochemistry* (pp. 100–126). Austin: University of Texas Press.

Tinbergen, N. (1951). *The study of instinct.* Oxford: Clarendon Press.

Treit, D., Specht, M. L., & Deutsch, J. A. (1983). Variety in the flavor of food enhances eating in the rat: A controlled demonstration. *Physiology and Behavior, 30,* 207–211.

Verney, R., & Axelrad, H. (1977). Functional maturation of the rat trigeminal nerve. *Neuroscience Letters, 5,* 133–139.

Vigorito, M., & Sclafani, A. (1988). Ontogeny of polycose and sucrose appetite in neonatal rats. *Developmental Psychobiology, 21,* 457–466.

Wade, G. N. (1972). Gonadal hormones and behavioral regulation of body weight. *Physiology and Behavior, 8,* 523–534.

Waring, G. H. (1983). *Horse behavior.* Park Ridge, NJ: Noyes Publications.

Weingarten, H. P. (1985). Stimulus control of eating: Implications for a two-factor theory of hunger. *Appetite, 6,* 387–401.

Williams, C. L., & Blass, E. M. (1987). Development of postglucoprivic insulin-induced suckling and feeding in rats. *American Journal of Physiology, 253,* R121–R127.

Williams, C. L., Hall, W. G., & Rosenblatt, J. S. (1980). Changing oral cues in suckling of weaning-age rats: Possible contributions to weaning. *Journal of Comparative and Physiological Psychology, 94,* 472–483.

Williams, C. L., & Lorang, D. (1985). Neural control of lordosis and ear wiggling in 6-day-old rats. *Society of Neuroscience Abstracts, 15,* 161.12.

Williams, C. L., Rosenblatt, J. S., & Hall, W. G. (1979). Inhibition of suckling in weaning-age rats: A possible serotonergic mechanism. *Journal of Comparative and Physiological Psychology, 93,* 414–429.

Wirth, J. B., & Epstein, A. N. (1976). The ontogeny of thirst in the infant rat. *American Journal of Physiology, 230,* 188–198.

Young, R. C., Gibbs, J., Antin, J., Holt, J., & Smith, G. P. (1974). Absence of satiety during sham feeding in the rat. *Journal of Comparative and Physiological Psychology, 87,* 795–800.

Zeigler, H. P. (1983). The trigeminal system and ingestive behavior. In E. Satinoff & P. Teitelbaum (Eds.), *Handbook of behavioral neurobiology. Vol. 6: Motivation* (pp. 265–327). New York: Plenum Press.

6

Caudal Brainstem Participates in the Distributed Neural Control of Feeding

HARVEY J. GRILL AND JOEL M. KAPLAN

INTRODUCTION

The notion that the neural control of feeding behavior is localized to a few brain sites has persisted for some time. During the past 15 years, we have begun to move away from this view toward a model encompassing more widespread control. This distributed control system involves critical afferent and efferent links with the digestive organs and glands via the autonomic nervous system as well as communication among a variety of sites within different levels of the central nervous system. Of the central locations involved in feeding control, the integrative substrates of the caudal brainstem, the principal focus of this chapter, are likely to receive increasing attention in the future.

Our discussion begins with a consideration of the neural basis of movement. We feel that lessons learned from the history of work in that area should be applied to the neural analysis of feeding. Figure 1 leads naturally to a discussion of *hierarchy* in the organization of the nervous system. In this case, stimulation of a site within the motor cortex gives rise to a very specific act, leg movement. Early studies using electrical stimulation of the neocortex repeatedly confirmed that discrete portions of the body musculature were represented at specific places within neocortex. Such results, however, did not rule out the possibility that stimulation of other sites within different levels of the nervous system can give rise to the same act, a fact abundantly demonstrated in subsequent work.

HARVEY J. GRILL AND JOEL M. KAPLAN   Department of Psychology and Institute of Neurological Sciences, University of Pennsylvania, Philadelphia, Pennsylvania 19104.

This apparent redundancy is known as *rerepresentation of function* and is a concept of the 19th century neurologist Hughlings Jackson.

On the basis of a series of studies in which the neuraxis was transected at different levels, Sherrington was led to the conclusion, foreshadowed by Jackson, that the movement system was built from the bottom up. For example, primary integrative processes that apply to all uses of a leg take place at the neuroanatomical level from which its motor and sensory innervation derives. Thus, the rhythmic bipedal stepping movements that survive midspinal transection and yet resemble the pattern of hindlimb movement during walking in the intact animal argue for a local spinal control of limb movement (Grillner, 1985). The similarity in movement topography of intact and transected animals led to inferences about (1) how higher anatomical levels of the control system in intact animals engage spinal mechanisms to initiate isolated movements of that musculature and (2) how the contribution of the spinal level is reflected in the control of complex movements.

Another strength of the Sherringtonian approach was in the identification of exteroceptive stimuli capable of eliciting movement. Although it is now generally agreed that central mechanisms alone can generate the basic sequence and timing of muscle activation for rhythmic movements such as locomotion and chewing, it also is clear that sensory input can initiate and sustain movement as well as provide critical feedback for its modification. In fact, attention to effective stimuli has led to a characterization of a number of fundamental reflexes whose neural substrates represent part of the movement control network. Accordingly, before it was possible to measure neural activity directly within the central ner-

THE FAR SIDE   By GARY LARSON

"Whoa! *That* was a good one! Try it, Hobbs —
just poke his brain right where my finger is."

Figure 1. *THE FAR SIDE*, copyright 1986 Universal Press Syndicate. Reprinted with permission. All rights reserved.

vous system, 19th and early 20th century neurologists were able to infer a great deal about the central integration underlying movement. These insights served as guides for later experimenters who were able to apply new methods to approach more directly the physiology of and interactions among the specific neurons involved in a particular response.

The neural control of feeding behavior, like that of walking, involves central patterning of rhythmic behavior that can be initiated and modulated by external stimuli. In addition to the general somatic afferents entering the CNS via the trigeminal nerve, feeding involves input from the chemical senses. Also, efferent output to the striated musculature is complemented by motor output to the viscera. Most importantly, feeding is a regulated behavior; that is, the ability of the exteroceptive stimulus to trigger a response varies as a function of the animal's internal state. This regulatory aspect brings into the analysis of feeding the *interoceptors* that monitor internal state as well as the integrative process by which this information alters the system's output. Despite the fact that feeding behavior involves these specialized control elements, we argue that its analysis will benefit from lessons taken from the classical approach to movement control.

Until very recently, research on the neural control of feeding has ignored these lessons. Instead, the notion of a *feeding center*, the unique localization of interoceptive regulatory and even somatic integrative function to a particular brain region, captured the imagination of researchers in the field and persists to this day. The view was supported initially by the correspondence of clinical and experimental observations associating distinct feeding "syndromes" with damage to or excitation of hypothalamic structures. The experimental results were often quite dramatic. For example, well-organized approach and consummatory responses to food could be elicited by a simple train of electrical pulses applied to the lateral hypothalamic region of nondeprived rats (Coons, Levak, & Miller, 1965); destruction of the same region produced a syndrome of aphagia and adipsia (Anand & Brobeck, 1951). The hypothalamic model also directed attention to the lateral and ventromedial hypothalamic regions as the *exclusive* sites of interoceptors relevant to behavioral as well as autonomic compensatory responses. The caudal brainstem, containing final common path neurons for ingestive consummatory responses and the peripheral autonomic system, was seen as part of the feeding control system, but its functional role was restricted to the production of somatic and autonomic responses.

The applicability of Jackson's concept of rerepresentation of function to feeding behavior can be surmised from the "hypothalamic" feeding syndromes that result from manipulation of CNS regions other than the hypothalamus. For example, lesions of amygdala, globus pallidus, or midbrain tegmentum produce an aphagia as intense and long lasting as that produced by lateral hypothalamic lesions (Morgane, 1961; Fonberg, 1974; Parker & Feldman, 1967; see also Gold, 1973; Sclafani & Kirchgessner, 1986). Similarly, electrically elicited feeding can be obtained from extrahypothalamic electrodes placed, for example, in the midbrain, cerebellum, and pons (Waldbillig, 1975). The fact that the lateral hypothalamus is a "hot spot" for electrical elicitation of feeding reveals no more about the neural organization underlying feeding behavior than leg movement evoked by cortical stimulation reveals about localization of function in locomotion (see Figure 1).

The location of interoceptors of relevance to feeding is one neural control issue to which Jackson's principle applies. The case for the importance of peripheral interoception in the initiation and satiation of feeding is now compelling (e.g., Deutsch, 1983; McHugh & Moran, 1985; Stricker, Rowland, Saller, & Friedman, 1977). Indeed, the feeding relevance of hypothalamic interoception was supported by data that now appear to have been overinterpreted (e.g., Oomura, Ono, Ooyama, & Wayner, 1969; Berthoud & Mogenson, 1977; Miselis & Epstein, 1975). Interoceptors located in the caudal brainstem may be the best candidates within the CNS for a critical role in metabolic homeostasis. Work on the caudal brainstem provides evidence against holding *any* key element of the feeding control system as the exclusive purview of the hypothalamus. On the output side, in addition to providing the motor neurons that activate feeding musculature, caudal brainstem substrates orchestrate the repeated activation of many muscles to yield efficient consummatory sequences. Finally, the caudal brainstem, isolated from hypothalamic and other forebrain influences by decerebration, modifies the consummatory pattern as a function of regulatory variables. Evidence supporting each of these points is discussed below.

To challenge the hypothalamus-as-feeding-center model is not to discount the relevance of hypothalamic function to feeding (see Chapter 1). Our argument is that a more balanced approach, entailing study of the forebrain, peripheral, and caudal brainstem mechanisms, will facilitate identification and analysis of relevant substrates at each level and will lead to effective experimentation focused on their interactions.

An Approach to the Neural Basis of Feeding Behavior

For an effective long-term approach to the distributed neural control of feeding behavior, it is important that a common testing and behavioral measurement framework link studies of different brain regions. It also is important to select for initial study regions that promise to yield the most useful information. Toward this end, we outline strategies, derived directly from the neurological approaches of Jackson and Sherrington, that direct attention to (1) response topography, (2) effective stimuli, and (3) independent processing capabilities of the lowest level of the system.

Response Topography

Goal-directed feeding behavior may be divided into two phases: appetitive and consummatory (Craig, 1918). *Appetitive* behaviors are those that direct the animal to food sources. The *consummatory* phase of feeding refers to the oral and pharyngeal actions that transport food from the environment to the esophagus. Although most manipulations that affect feeding influence both behavioral categories, we have chosen to focus almost exclusively on consummatory mechanisms for a number of practical reasons. It is difficult to identify and characterize neural elements mediating appetitive behaviors because the form of the movements can be quite variable. In addition, they are not constrained by sensory aspects of the food stimulus. They can involve complex perceptions and cognitive strategies such as those involved in foraging. By contrast, the task of identifying the neural hardware underlying consummatory action is more tractable. The stereotyped movement topography implicates particular motor neurons and premotor networks that pattern the behavior (Luschei & Goldberg, 1981; Chandler & Tal, 1986; Miller, 1982; Lowe, 1981, etc.).

Any regulatory influence that affects feeding must engage these movement-patterning mechanisms. Moreover, such influences evoke changes along relatively few dimensions of ongoing consummatory behavior. They may affect the frequency or pattern of stereotyped oromotor actions that result in swallowing or rejection of the food stimulus. Within meals, sensitive aspects of feeding "microstructure" include the ingestion rate while feeding, the duration of pauses in ingestive behavior, and the duration of "bursts" of ingestive behavior (e.g., McCleery, 1977). These consummatory response dimensions assay the balance of excitatory and inhibitory influences that affect such whole-meal descriptors as meal size and meal duration. Progress in characterizing the neural networks underlying the production of consummatory responses and their sensory–motor integration identifies both candidate sites for regulatory integration and sites of potential convergence of modulatory inputs from different anatomical levels of the feeding system.

A consummatory response analysis is particularly appropriate in studies that involve experimental destruction of CNS tissue. For a variety of neurological preparations, appetitive behavior is either eliminated or profoundly disrupted. However, the conclusion that critical regulatory substrates have been destroyed by such lesions is often erroneous. Under proper testing conditions, a physiologically regulated consummatory control system often can be demonstrated in otherwise aphagic preparations (Grill & Norgren, 1978b; Flynn & Grill, 1983; Fluharty & Grill, 1981). Analysis of consummatory behavior, therefore, is necessary to avoid overattribution of function to the substrates affected by the lesion.

Stimuli That Elicit, Sustain, and Terminate Consummatory Behavior

To account for the ability of exteroceptive and interoceptive stimuli to elicit or modify consummatory responses, the neural system controlling feeding must encompass sensory processes as well as the anatomical and physiological links between relevant inputs and mechanisms of consummatory response production.

Exteroceptive. Once food is at hand, taste receptors can trigger either consummatory behavior or rejection of the stimulus. These sensors assess the physical features of an available food to permit responding in relation to such biologically relevant variables as its sugar or cation content. If the decision is to ingest, then aspects of the gustatory stimulus, such as its chemistry and concentration, can affect each of the consummatory dimensions listed above (Davis & Levine, 1977; Grill, Spector, Schwartz, Kaplan, & Flynn, 1987).

Interoceptive. Input from interoceptors (e.g., gastric mechanoreceptors, intestinal chemoreceptors, metabolically sensitive cells near the fourth ventricle or in the liver) signals the relative availability or lack of necessary nutrients. Traditionally, input from these sensors was considered the principal determinant of the initiation and termination of feeding. However, "deficits" as such may not arise within normal intermeal intervals (e.g., LeMagnen & Tallon, 1966; Campfield & Smith, 1986; Stricker, 1984); meal initiation may "anticipate" deficits or be influenced by conditioned stimuli (e.g., Weingarten, 1984). Similarly, meal termination need not strictly correlate with the output of postingestive receptors that signal nutrient availability. For example, environmental variables such as spatial distribution of food or acquisition and consumption "costs" affect meal duration and size (Collier, 1985). These important insights, discussed in other chapters in this volume, do not contravert the many reliable effects of interoceptive stimulation on feeding. These effects include the stimulation of feeding by treatments such as insulin or glucose-antimetabolite injection and food deprivation. Manipulations that affect postingestive receptors, such as nutrient preloads and varying caloric density of foods, yield inhibitory effects on meal duration, meal size, and the microstructure of ongoing ingestions.

We do not wish to debate the relative importance of homeostatic versus nonhomeostatic mechanisms under various natural or experimental conditions. We embrace the homeostatic paradigm for its heuristic value to an explicitly neural analysis. A neural approach can be served by recording behavior that is "time-locked" to stimulus events. In addition, an appreciation of the manner by which consummatory behavior is influenced by graded, potentially measurable signals can lead to inferences about the neural processes underlying feeding regulation. To the extent that progress is made in the identification of interoceptive channels (see below) relevant to meal initiation and termination, inroads will be charted into the neural control system for feeding.

INTEGRATION AT LOWEST ANATOMICAL LEVEL IN THE NEURAL HIERARCHY

The feeding control system undoubtedly involves integrative neural processes seated at various levels of the neuraxis. We believe, as did Sherrington, that the most fruitful approach to the system as a whole begins at the anatomical level(s) of the relevant sensory inputs and motor outputs. The substrate mediating the most direct influence of sensory input (i.e., involving the fewest synapses) on the system's output defines the "lowest integrative level." Studying "primary" circuits both reduces the number of neurons to consider and lays the groundwork for an approach to the "higher" integrative substrates.

For the feeding system, many of the relevant inputs enter, and all of the consummatory motor outputs emerge, at the level of the caudal brainstem. The most direct neural pathway between sensory and motor structures involves one or more synaptic links within the brainstem. Anatomical and physiological studies have begun to highlight particular pathways and groups of neurons in the reticular formation that may represent part of the lowest integrative level for sensory and regulatory modulation of consummatory motor behavior (e.g., Travers & Norgren, 1983; Jean & Car, 1979; Kawamura & Yamamoto, 1978; Yamamoto, Fujiwara, Matsuo, & Kawamura, 1982; see Travers, Travers, & Norgren, 1987). However, it is also possible that these reticular connections play no role or a very minor one in the control of ongoing ingestion in the intact animal. From this perspective, the effect of taste on ingestion may depend more critically on primary integrative processes performed at another level of the neuraxis. Following battle lines drawn earlier, we might consider the hypothalamic substrates that receive sensory information as a leading alternative candidate (Norgren, 1976).

In this chapter, we argue that many of the lowest-level circuits relevant to feeding do indeed reside in the caudal brainstem. Specifically, we hypothesize that much of the neural circuitry that mediates the influence of exteroceptive and interoceptive stimulation on the consummatory pattern lies caudal to the diencephalon. Furthermore, this circuitry appears to function normally independent of the influence of anatomically higher levels. The most compelling evidence supporting this hypothesis comes from experiments in which the neuraxis is transected at the mesodiencephalic juncture, neurally isolating the brainstem from the hypothalamus and other forebrain circuits.

AN INVESTIGATION OF THE CONTRIBUTION OF THE CAUDAL BRAINSTEM TO FOOD INTAKE CONTROL

The caudal brainstem contains all the elements of a consummatory motor control system. It contains somatic motor neurons for tongue, jaw, and pharyngeal movements as well as autonomic outputs to salivary glands and other viscera. This level of the nervous system receives and processes cephalic sensory information from the mouth (taste and trigeminal). In addition, feeding-relevant interoceptors either reside in or report to the caudal brainstem. The issue at hand is the extent to which the caudal brainstem substrate that receives these critical inputs is actually capable of mediating their effects on the consummatory pattern. The decerebrate preparation provides evidence for the sufficiency of caudal brainstem integrators to mediate a variety of sensory effects on autonomic and somatic actions that serve feeding, including chewing movements (Lund & Dellow, 1971), salivation (Kawamura & Yamamoto, 1978), swallowing (Miller & Sherrington, 1915), preabsorptive insulin response (Grill, Berridge, & Ganster, 1984), and the sympathoadrenal response (DiRocco & Grill, 1979). The extensive integrative capability of brainstem circuits is uncovered by experiments outlined below, which show that the caudal brainstem, isolated from the forebrain, coordinates fractional ingestive components into a continuous, regulated consummatory pattern resembling that of the intact rat.

To examine the behavioral capacities of the isolated caudal brainstem, we used a classic but neglected neurological preparation, the chronic supracollicular decerebrate, developed by Bard and his colleagues (Bard & Macht, 1958; Macht, 1951; Woods, 1964). These investigators found that by allowing brainstem systems to recover, many complex, stimulus-guided behavioral sequences, including aspects of food and fluid ingestion, could be observed. The preparation fell into disuse shortly after these studies in part because of the difficulty of maintaining the chronic decerebrate and in part, we think, because of excitement attending the rediscovery of stereotaxic techniques that permitted focal lesion, stimulation, and recording studies. Grill and Norgren (1978b) revived the chronic decerebrate preparation, improved its longevity, and applied it to the behavioral analysis of ingestion. Appetitive ingestive behavior cannot be readily studied in this preparation. The chronic decerebrate is aphagic and adipsic and requires intragastric intubation of nutrients for survival. The decerebrate is, however, responsive to food stimuli provided that the stimulus is introduced directly into its mouth. Intraoral delivery was accomplished in the experiments described via indwelling oral cannulae. The remote delivery of fluid taste stimuli driven by an infusion pump allows a high degree of stimulus control without interfering with the preparation. Lack of appetitive behavior notwithstanding, the consummatory behavior of the chronic decerebrate receiving intraoral infusions provides an effective behavior assay for the integrative competence of the caudal brainstem. The richness of the consummatory response domain is studied by videotape and electromyographic recording of oropharyngeal actions during feeding and also by recording the duration of the ingestive bout and the amount consumed.

Once continuous bouts of efficient ingestive behavior are produced in response to a given nutritive stimulus, the research strategies outlined above can be applied. We pose a series of experimental questions: Does the chronic decerebrate demonstrate discriminative consummatory responses to sucrose, a prototypical taste stimulus whose digestion yields the metabolic fuels glucose and fructose? Do various internal state manipulations such as food deprivation, gastric load, or gastrointestinal hormone administration modulate the decerebrate's sucrose-elicited consummatory responses? If these questions are answered in the affirmative, as they are for the intact rat, then we can conclude that the caudal brainstem, in isolation from forebrain mechanisms, is sufficient for both production and regulation of sucrose ingestion. The same questions are also asked of other regulatory ingestive systems, such as those involved in sodium and water balance, where the relevant taste and internal state variables are different.

The Caudal Brainstem Produces Consummatory Behavior

A prolonged bout of ingestive consummatory behavior is observed when the *intact* rat receives a continuous intraoral infusion of a sucrose solution. The rat is not simply swallowing the fluid reflexively. The infusions are, in fact, delivered toward the front of the oral cavity and must be transported intraorally into position for swallowing. Intraoral transport is accomplished by rhythmic coupled movements of the jaw and tongue emitted at a frequency of approximately 6 Hz, which is indistinguishable from that seen during voluntary spout licking (Kaplan & Grill, 1989; Stellar & Hill, 1952). If the rhythmic oromotor behavior ceases before the infusion is discontinued by the experimenter, then the fluid drips passively from the mouth. Thus, infusion-accommodative behavior in the intact rat merits the designation "active ingestion."

From previous work on acutely decerebrated preparations, it was known that the caudal brainstem contains circuitry sufficient for the production of rhythmic oromotor and swallowing actions in response to various arbitrary forms of oral stimulation such as electrical stimulation of sensory nerves, tactile probes of the oral and pharyngeal mucosa, or squirts of fluid aimed at particular receptive surfaces (Miller & Sherrington, 1915; Lund & Dellow, 1971; Doty & Bosma, 1956; Thexton & Griffiths, 1979; reviews by Miller, 1982; Luschei & Goldberg, 1981). Our work on the chronically maintained decerebrate rat receiving intraoral infusions shows further that rhythmic oromotor behavior resembling that of the intact rat can be elicited *and sustained* by food stimuli. Active ingestion can continue for substantial durations in this preparation, resulting in intake of metabolically meaningful volumes of nutritive fluid (see below).

Ingestion might proceed in the chronic decerebrate, but with a consummatory topography that is anomalous or somehow disorganized when compared to that of the intact rat. To address this issue, two aspects of the consummatory pattern were examined: the pattern of swallowing when the rate of intraoral infusion was varied and the segmentation of the ingestive pattern into behavioral bursts separated by pauses in ingestive behavior. Having established the normative pattern for the intact rat (Kaplan & Grill, 1989), we describe results of our preliminary study on the chronic decerebrate rat.

Swallowing during intraoral infusions of sucrose solutions delivered across sessions at different rates was studied using electromyographic recording. To adjust its rate of ingestion to accommodate the experimental adjustment of infusion rate, the rat may vary either swallow frequency or swallow volume (infusion rate divided by the swallow frequency) while holding the other constant. Alternatively, adjustments in both swallow frequency and volume may share in mediating adjustments in the ingestion rate. We found that the latter mode of adjustment, as a rule, applies to the intact rat ingesting at different rates (Kaplan & Grill, 1989). The swallowing behavior of the chronic decerebrates studied thus far cannot be distinguished from that of intact rats studied under the same infusion rate conditions (0.5, 1.0, and 1.5 cc/min). No distinction between the intact and decerebrate rats could be drawn in terms of (1) the coadjustment of both swallow frequency and swallow volume across conditions, (2) absolute values for swallow frequency and volume at any ingestion rate, or (3) the temporal pattern of swallowing as reflected in the form of the interswallow interval histogram. The intact rat, during sustained intraoral infusions of sucrose solutions, displays pauses between bursts of rhythmic oromotor behavior. Swallows are emitted only within bursts of active ingestion (Kaplan & Grill, 1989). The initial segment of the intraoral intake interval is often largely uninterrupted. The pauses in behavior typically become pronounced as the intraoral intake test proceeds. Chronic decerebrate rats also show a burst–pause pattern with pauses increasing in duration over the course of the intraoral intake test.

The consummatory topography of chronic decerebrate rats receiving intraoral infusions, therefore, is remarkably similar to that of their intact controls. The significance of this result would be diminished if the similarity reflects the constraints imposed by the arbitrary intraoral infusions on the behavior of the intact rat more than it reflects a "normal" consummatory repertoire of the decerebrate. This is a weak objection, however, given the very similar oropharyngeal organization of voluntary and intraoral ingestion in the intact rat. First, the burst–pause patterning of voluntary meals of solid food or liquid test stimuli (e.g., Stellar & Hill, 1952; McCleery, 1977) is qualitatively analogous to that of intraoral intake. These pauses in voluntary consummatory behavior are quantitatively important in decreasing the average rate of ingestion over the course of the meal and become windows of initiative for switching into behavior modes other than feeding. In addition, a similar covariation of swallow frequency and swallow volume in the mediation of ingestion rate adjustments has been shown in intact rats drinking water delivered at different rates from a drinking spout (Weijnen, Wouters, & van Hest, 1984), in intact rats ingesting a meal of semisolid mash within which ingestion rate declines as the postingestive load accumulates (Kaplan & Grill, 1987), and intact and decerebrate rats ingesting intraoral infusions at different rates.

The qualitatively similar consummatory topographies of the intact and decerebrate rats suggest that the same brainstem integrative substrates that coordinate the consummatory pattern in the decerebrate may also account for the performance of the intact rat. Although not dismissed on logical grounds, the notion that the forebrain controls consumption on a movement-to-movement basis in the intact rat is unparsimonious. The burst–pause patterning would appear to represent a "higher" level of organization of consummatory behavior than the level of the individual movement cycle for swallowing or rhythmic oral behavior. Yet, data from the decerebrate suggest that the caudal brainstem may be at least partly responsible for this ingestion characteristic in the intact rat as well. Brainstem substrates, then, may pattern a discontinuous behavior stream in the intact animal as a function of regulatory influences or of descending inputs from the forebrain.

THE CAUDAL BRAINSTEM IS SUFFICIENT FOR THE PRODUCTION OF DISCRIMINATIVE RESPONSES TO TASTE

The intact rat's decision to ingest an intraorally infused fluid depends critically on its taste properties. A "bitter" quinine solution introduced into the mouth usually results in the expulsion of the stimulus. In intact rats, normally avoided tastes produce a stereotyped rejection response sequence that includes gapes, chin rubs, head shakes, face washes, forelimb shakes, and paw rubs as well as locomotion (Grill & Norgren, 1978a). The frequency of emitted rejection responses is concentration dependent (Schwartz & Grill, 1984). When tastes that are normally avoided by the intact rat are intraorally delivered to the chronic decerebrate, the same rejection response sequence is observed. Moreover, decerebrate and control rats show similar concentration–response relationships.

As described in the previous section, intact and decerebrate rats show prolonged bouts of rhythmic oromotor behavior in response to the intraoral infusion of sucrose and other normally preferred taste stimuli. The number of ingestion responses during a given interval of sucrose infusion increases with concentration for intact as well as decerebrate rats (Flynn & Grill, 1988). The similarities in threshold, dynamic range, and slope of concentration–response function between the taste reactivity of the decerebrate and the intact rat not only includes sucrose and quinine solutions but extends to every compound examined to date (e.g., NaCl, HCl, various sugars, and certain amino acids). Therefore, the caudal brainstem is sufficient for the production of discriminative consummatory responses to taste stimulation in the rat in isolation from other integrative substrates such as the hypothalamus and the amygdala that also receive taste information. Similarly the oral–facial responses of anencephalic and intact human newborns are virtually identical (Steiner, 1973). Taken together, these results suggest that the neural modulation of the consummatory response as a function of taste input in the intact animal reflects a primary contribution of caudal brainstem integration.

THE CAUDAL BRAINSTEM IS SUFFICIENT FOR INTEGRATION OF TASTE AFFERENT INPUT AND INTEROCEPTIVE SIGNALS TO REGULATE INGESTIVE CONSUMMATORY BEHAVIOR

Various physiological influences regulate the consumption of a given taste stimulus. These may be grouped into two categories: factors that promote intake and inhibitory postingestive factors that terminate the ingestive bout. Underlying these influences on intraoral intake are interoceptive processes whose identity and location are addressed with the decerebrate preparation. The following two sections deal, respectively, with the satiation and stimulation of intraoral intake in intact and decerebrate rats.

135

CAUDAL
BRAINSTEM ROLE

SATIATION OF INTRAORAL INTAKE. Rats do not continue to ingest even the most highly preferred substances indefinitely. Influences from the accumulating load act to terminate the ingestive bout. The influence of postingestive factors are greatly reduced in the "sham-feeding" preparation in which ingested fluids drain from the stomach through a surgically implanted fistula. For intact rats, the amount of sham-ingested sucrose solution consumed is a monotonically increasing (often linear) function of its intensity over a broad range of concentrations. This relationship is observed in the rat drinking the solution from a spout (Weingarten & Watson, 1982) and, as shown recently (Grill & Kaplan, in press), also holds true for intraoral intake tests.

In the presence of normal postingestive feedback, sucrose concentration/intraoral intake functions for chronic decerebrate and pair-fed intact rats are parallel (Flynn & Grill, 1988). This result has been extended to several other sugar solutions (S. Lee & H. J. Grill, unpublished results). Intake of a given sucrose solution in both control and chronic decerebrate rats is reduced in equal proportion by a pretest gastric load (Grill, 1980). In the absence of postingestive feedback resulting from an open gastric fistula, sucrose intake is a linear function of its concentration in both chronic decerebrate and intact control rats (Grill & Kaplan, in press). We conclude from these data that decerebrates and intact rats demonstrate a similar postingestive inhibition of taste-driven consummatory behavior. The argument that the postingestive regulation in intact rats reflects the action of caudal brainstem mechanisms is strengthened because the effects described for the decerebrate were in the same direction and of a similar magnitude as they were for the intact controls.

The catalogue of postingestive suppressive effects on intake has grown faster than an appreciation of the type and location of interoceptive elements that mediate the experimental effects. Candidate stimuli for satiety-relevant interoception include gastric distension, the action of nutrients on the stomach, the intestines, or the liver, and the release of gastrointestinal hormones (Deutsch, 1983; Houpt, 1982; Tordoff & Friedman, 1986; McHugh & Moran, 1985; Smith, Gibbs, & Kulkosky, 1982). Separate experiments are required to isolate the contributions of these and other putative satiety signals. Priorities for future research are to specify the afferent channels over which this information travels to the brain (e.g., Smith, Jerome, & Norgren, 1985) and to identify the integrative substrates that receive each type of input and mediate the behavioral effects originating from the successive digestive organs. The chronic decerebrate preparation may be useful in this work.

Gastrointestinal hormones have received a great deal of attention as putative satiety signals (Smith *et al.*, 1982). Of these hormones, cholecystokinin (CCK) has been most actively pursued. Cholecystokinin is released in the course of normal digestion. When exogenously administered to intact rats, CCK reduces intake under all meal-taking paradigms examined, including the intraoral intake test. When CCK-8 is administered to 24-hour food-deprived chronic decerebrate rats, they, like their pair-fed controls, reduced their intraoral intake and showed fewer ingestive oral motor responses to sucrose (Grill & Smith, 1988). Bombesin, another gastrointestinal hormone thought to play a role in postingestive satiety, also reduces intraoral intake of both intact and decerebrate rats (Flynn, 1987).

The intake-suppressive effects of these substances, therefore, do not require the forebrain.

INTEROCEPTIVE FACILITATION OF INTRAORAL INTAKE. Meal size is, in part, a function of the physiological state of the animal at the time the meal is initiated. Insulin administration is a prototypical state manipulation that enhances food intake at the first feeding opportunity. Flynn and Grill (1983) found that insulin injection increased to a comparable degree the sucrose intake of sated decerebrate and control rats. This treatment failed to increase the water intake in either preparation. Another manipulation broadly affecting physiological state in a manner that promotes food intake is simple food deprivation. Twenty-four-hour food-deprived chronic decerebrates, like control rats, ingested two to three times the volume of sucrose they had consumed in the sated condition after tube feeding (Grill & Norgren, 1978d; Grill, 1980). Water intake of both groups was unaffected by food deprivation. In this experiment, however, it is difficult to unambiguously attribute results to a facilitatory effect(s) of deprivation versus an inhibitory effect of the nutrient preload (the tube-fed meal). Nevertheless, the results of the experiment with decerebrate rats suggest that the metabolic consequences of food deprivation and repletion act on caudal brainstem mechanisms to affect sucrose intake. A goal for future research is to develop paradigms that more readily decipher the contribution of metabolic need from that of metabolic adequacy or surfeit.

The treatments discussed above that enhance or suppress consummatory responding are global physiological manipulations in that they affect a variety of theoretically relevant interoceptors. Nonetheless, the experiments described indicate that the critical receptors, whatever stimuli they respond to, report to caudal brainstem integrative mechanisms that mediate the regulatory effects on consummatory responses. The interoceptors themselves may reside in the periphery or in the parenchyma of the caudal brainstem. Because the adjustments of consummatory responding of the decerebrate under these disparate conditions strongly resemble those of their intact controls, the same caudal brainstem substrate(s) may perform primary regulatory integration for the intact rat. The forebrain substrates that may receive interoceptive information and perform integrative function of some relevance to normal feeding would, under this scenario, affect behavior via outputs to these caudal brainstem circuits.

CAUDAL BRAINSTEM ORCHESTRATES AUTONOMIC AS WELL AS BEHAVIORAL COMPENSATORY RESPONSES TO METABOLIC CHALLENGE

In the previous sections, we have considered only behavioral regulatory mechanisms. Modern homeostatic theory emphasizes both autonomic and behavioral compensatory responses to metabolic challenge. At some level, the neural control of the respective compensations is divergent; certainly the final common path to the organs of behavior and of autonomic action are different. However, it is entirely possible that the behavioral and autonomic responses are triggered by the same set of interoceptors that, moreover, engage the same or overlapping integrative substrates. Having considered caudal brainstem mechanisms for behavioral adjustments in the previous sections, we now address autonomic regulation at this neural level. Autonomic adjustments to depletion involve the sympathetic nervous innervation of various tissues (fat, liver, pancreas, etc.) as well as actions of blood-borne factors on the same tissues. In the following section, we focus particularly on the sympathoadrenal response (Cannon, McIver, & Bliss, 1924).

Compromises to glucose metabolism such as reductions of blood levels, rate of transport, or rate of glycolysis evoke a neurally mediated release of epinephrine from the adrenal medulla. This blood-borne epinephrine acts on the liver to promote glycogenolysis and on the pancreas to inhibit insulin secretion. The net effect of these actions is to enhance glucose availability to the CNS. DiRocco and Grill (1979) showed that the sympathoadrenal hyperglycemic responses induced by 2-deoxy-D-glucose (2-DG) in chronic decerebrate and pair-fed intact controls were similar.

The integrity of the sympathoadrenal response in chronic decerebrates implies that its afferent limb lies caudal to the forebrain—in the periphery and/or in the caudal brainstem itself. The experiment of Stricker et al. (1977) strongly supported the idea that the metabolic interoceptors for this response reside in the CNS and not in the periphery. The experiment assessed the ability of organ-specific fuels to counteract insulin-induced sympathoadrenal response. They found that systemic injection of fructose, a metabolic fuel not utilized by the brain but readily oxidized by the liver, failed to block the sympathoadrenal response. The converse experiment, using ketones, which are used by the brain but not by the liver, yielded the opposite result—inhibition of the sympathoadrenal response. The decerebrate data coupled with Stricker et al.'s (1977) finding (as well as other data discussed below) suggest a role for the brain caudal to the hypothalamus in the integration and interoception for the sympathoadrenal response. Identifying the interoceptors for the sympathoadrenal response and their adequate stimuli is a priority for future research. Insights gained in this endeavor could facilitate discoveries on the location of and adequate stimuli for interoceptors whose action evokes behavioral compensation.

CAUDAL BRAINSTEM LOCATION OF METABOLIC INTEROCEPTORS STIMULATING BOTH BEHAVIORAL AND AUTONOMIC COMPENSATION

A more direct demonstration of CNS metabolic interoceptors can be provided by biochemical manipulation of selected brain regions. On the assumption that what is being detected by these receptors is some correlate of energy availability, it should be possible to evoke compensatory response(s) as a function of direct application of agents that interfere with energy availability.

Glucose is the principal metabolic fuel of the central nervous system. When this substrate is insufficient for the metabolic needs of CNS neurons, compensatory behavioral, autonomic, and autonomic–endocrine responses are elicited to elevate circulating levels of glucose and other metabolic fuels (e.g., Smith & Epstein, 1969; Himsworth, 1970). Support for the existence of CNS metabolic interoceptors was provided by studies that introduced phlorizin, a glucose transport inhibitor, into the lateral ventricles. Intracerebroventricular (ICV) infusions of phlorizin in amounts too small to affect peripheral receptors elicited compensatory feeding behavior; sympathoadrenal responses were not measured in this study (Glick & Mayer, 1968). Subsequently, inhibition of brain glycolysis by lateral ICV injections of 2-DG, an inhibitor of intracellular glycolysis, was shown to evoke both behavioral and autonomic compensatory responses (Miselis & Epstein, 1975; Berthoud & Mogenson, 1977; Coimbra, Gross, & Migliorini, 1979).

HARVEY J. GRILL
AND JOEL M.
KAPLAN

Which regions of the CNS are exposed to a drug injected into the lateral ventricle? Although the ventricular application of drug doses used in these studies was unlikely to raise peripheral concentrations significantly, it is likely because of the normal caudad movement of cerebrospinal fluid, that receptors at many levels of the central nervous system are affected (Bradbury, 1979). The *Zeitgeist* of the 1960s and 1970s directed interpretation of the site of drug action to the hypothalamic nuclei proximal to the third ventricle. The presence of hypothalamic interoceptors had been inferred based on electrophysiological demonstrations of glucoresponsive neurons in this region (Chhina, Anand, & Rao, 1971; Desiraju, Benerjee, & Anand, 1968; Oomura *et al.*, 1969). The existence of hypothalamic neurons responsive to glucose, however, neither demonstrated that these cells were the exclusive site of interoception nor proved that they alone are responsible for eliciting compensatory behavioral and autonomic responses in the lateral ICV studies.

One of the regions affected by lateral ICV injections of glucodynamic drugs is the caudal brainstem surrounding the fourth ventricle. It has recently been shown that metabolically sensitive cells in the caudal brainstem do provide an afferent limb for compensatory behavioral and autonomic responses. R. Ritter and colleagues showed that injections of 5-thioglucose (5-TG), another inhibitor of intracellular glycolysis, restricted to the fourth ventricle by means of a cerebral aqueduct plug, produced both feeding and sympathoadrenal responses (Ritter, Slusser, & Stone, 1981). Lateral ICV injections of 5-TG restricted to the forebrain ventricles by the same cerebral aqueduct plug failed to stimulate either feeding or hyperglycemia. Therefore, it appears that lateral ICV injections of glycolytic inhibitors act on the caudal brainstem mechanism and not on forebrain substrates (the hypothalamic nuclei) as originally proposed (e.g., Himsworth, 1970).

Contrasts in the action of different metabolically active agents may provide evidence for separate interoceptive mechanisms. Although fourth ventricular injections of 5-TG elicit both feeding and hyperglycemia (Ritter *et al.*, 1981; Flynn & Grill, 1985) fourth ventricle phlorizin elicited feeding but failed to elicit sympathoadrenal hyperglycemia (Flynn & Grill, 1985). Alloxan, another drug affecting glucose metabolism, although in a way not yet understood, also stimulate feeding but not hyperglycemia when administered into the fourth ventricle (Ritter & Strang, 1982). Thus, the distinctive actions of the various drugs discussed appear to uncover different interoceptive processes and, moreover, speak to the existence of separate systems controlling feeding and the sympathoadrenal response to glucoprivation.

The separability of neural systems controlling compensatory behavioral and autonomic responses can be highlighted in other ways. First, the induced hyperglycemia appears to develop earlier than the feeding response to the metabolic challenge provided by 2-DG (Houpt & Epstein, 1973). Second, the neural system mediating feeding in response to systemic 2-DG is more susceptible to pretreatment with toxic fourth ventricle doses of alloxan than the system mediating hyperglycemia (Ritter, Murane, & Landenheim, 1982). Finally, compensatory behavioral and autonomic responses are dissociable by intravenous infusions of different metabolic fuels (Stricker *et al.*, 1977).

Autonomic and behavioral responses can be separately activated by different metabolic manipulations. These results suggest that the integrative substrates mediating these two types of compensatory responses are at least partially separable. The identification of the underlying neural circuits, however, is hampered by our ignorance of metabolic receptor mechanisms in general and, in particular, of those that trigger feeding. Indeed, it is not clear whether metabolically active agents act on neurons, glia, or other specialized cells (e.g., cells of the circumventricular organs). A priority for future research is to distinguish the metabolic mechanism(s) underlying the actions of these drugs by studying their effects on a variety of organ systems. Development of new metabolically active agents may provide more selective stimulation (or inhibition) of different interoceptive channels. Such developments will aid in localizing the interoceptors and their associated integrative networks.

PRIMACY OF CAUDAL BRAINSTEM FUNCTION: CONCLUSIONS AND QUALIFICATIONS

The data just described provide compelling evidence that the caudal brainstem is sufficient for production of coordinated consummatory behavior and for the integration of exteroceptive and interoceptive signals that regulate food intake. The primacy of the caudal brainstem in the control of feeding is supported, but not proven, by the fact that many responses mediated by the caudal brainstem, when in isolation from forebrain influence, resemble those obtained in the intact animal. This section considers other substrates that may contain interoceptors or perform regulatory integration that parallel or supplant caudal brainstem function in the intact animal. This section also considers the role of the caudal brainstem in relation to appetitive function not seen in the decerebrate.

INTEROCEPTOR LOCATIONS

As just discussed, some of the relevant interoceptors for feeding are located within the brain posterior to the hypothalamus. Others, such as those in the periphery, also contribute to the initiation and termination of feeding. For example, various experiments suggest that feeding is influenced by interoceptors in or near the stomach (e.g., Deutsch, 1983; Smith *et al.*, 1985), duodenum (e.g., McHugh & Moran, 1985), and liver (e.g., Russek, 1963; Tordoff & Friedman, 1986). Supporting the role of the peripheral interoceptor populations is the elimination of associated feeding effects following section of relevant peripheral nerves (i.e., branches of the vagus and/or splanchnic nerves: Smith *et al.*, 1985; Friedman & Granneman, 1983; Deutsch, 1983; Stricker & McCann, 1985). The principal integrative and output mechanisms of the caudal brainstem in such instances are left intact and presumably responsive to interoceptive influences derived from central sources and from peripheral receptors whose centripetal connections remain intact.

Still other ingestion-relevant interoceptors not residing in the caudal brainstem are found in the forebrain. Hormone interoceptors for regulatory hydromineral systems (e.g., steroid and peptide receptors) have been located in the hypothalamus and other forebrain regions (Coirini, Marusic, DeNicola, Rainbow, & McEwen, 1983). The forebrain location of critical interoceptors is supported by the failure of the chronic decerebrate rat to demonstrate compensatory consummatory responses to orally applied water or NaCl solutions while under osmotic challenge (Grill & Miselis, 1981) and to NaCl solutions when sodium deficient (Grill, Schulkin, & Flynn, 1986). Insofar as decerebrate rats emit discriminative consummatory responses to NaCl and water stimuli, the failure of such regulatory challenges to affect ingestive behavior may reflect a "disconnection" between forebrain hormonal receptors and the caudal brainstem mechanism that mediates the gustatory influence on the consummatory pattern. It is of interest to consider the decerebrate results in light of osmosensitive and sodium-sensitive cells of the liver. It is very likely that these hepatic interoceptors were affected during osmotic challenge or sodium depletion. Their centripetal connections (vagus) were intact, but they were unable to influence the behavior of the decerebrates. These receptors may constitute an afferent limb for autonomic compensatory responses to hydromineral challenge, but their role in behavioral control under these conditions must be discounted. It is possible, of course, that such hepatic receptors normally contribute to ingestive compensation in the intact rat. If so, the results obtained in decerebrate rat experiments would suggest that such hepatic information is processed in the forebrain.

When evidence favors the functional relevance of interoceptors located within the caudal brainstem, it is appropriate to ask whether these receptors are necessary or sufficient for homeostatic control in the intact animal. This question becomes particularly important when receptors located elsewhere are active under the same metabolic conditions. Thus, in addition to the caudal brainstem receptors sensitive to correlates of energy availability, other receptors that appear to be sensitive to the same conditions are located in both the periphery and forebrain. This prompts a consideration of the relative roles of the different interoceptor populations.

The sufficiency of caudal brainstem receptors to stimulate feeding and autonomic compensatory responses to metabolic challenge was demonstrated by the effectiveness of microinfusions of metabolic agents whose action was confined to the fourth ventricle (see above). Nevertheless, stimulation of peripheral receptors by similar treatments can trigger behavioral compensation. Moreover, peripheral receptors can operate independently of brain interoception; fructose, a metabolic fuel whose utilization is restricted to the periphery, eliminates the insulin-generated behavioral compensation despite the fact that fructose fails to block autonomic compensation (Stricker et al., 1977). We have, then, two receptor populations that are each capable of independently driving the system. In the intact animal, both groups may participate in the control of feeding responses; neither group can be established as superordinate at this time. By contrast, recall that the primacy of caudal brainstem interoception for the sympathoadrenal response has been demonstrated (see above).

The existence of metabolically sensitive interoreceptors in the hypothalamus has been proposed (Oomura, Ooyama, Yamamota, & Naka, 1967; Marrazzi, 1976). There is no convincing evidence, however, that such receptors contribute to either behavioral or autonomic compensation for metabolic challenge. Glucose antimetabolites introduced into the lateral ventricle and restricted to the forebrain by means of a cerebral aqueduct plug fail to elicit either type of compensatory response (see above). Furthermore, selective intrahypothalamic infusions of these agents, more effectively stimulating the putative interoceptors, also fail to elicit either feeding (Epstein, Nicolaidis, & Miselis, 1975; Panksepp, 1975) or hyperglycemia (see discussion by DiRocco & Grill, 1979).

Although the case for hypothalamic interoceptive involvement in compensatory responses to metabolic challenge is weak, it is likely that the hypothalamus plays an *integrative* role in the neural control of feeding behavior in the intact animal. The hypothalamus receives and processes oral and visceral information via ascending pathways as shown by anatomical (e.g., Norgren, 1976) and physiological experiments (e.g., Schmitt, 1973; Nicolaidis, 1969; Swanson & Sawchenko, 1983). Furthermore, the rat's physiological state can modulate baseline activity and responsiveness of individual hypothalamic neurons (Burton, Rolls, & Mora, 1976). Therefore, the hypothalamus is a site of integration of potential relevance to the regulation of compensatory responses. Hypothalamic output, in turn, can affect sensory, motor, autonomic motor, and internuncial systems residing in the caudal brainstem; arguments for the functional roles of descending pathways have been presented (Bereiter, Berthoud, & Jeanrenaud, 1981; Matsuo, Nobuaki, & Kusano, 1984; Murzi, Hernandez, & Baptista, 1986; Rogers & Hermann, 1986.)

The interaction between hypothalamic integrative and caudal brainstem integrative mechanisms in the intact animal may be conceptualized in various ways. For example, (1) regulatory effects on the behavior of the intact rat may be under the exclusive control of integrative mechanisms of the hypothalamus, whose descending outputs engage motor or premotor systems in the caudal brainstem. According to this scheme, integrative substrates that mediate regulatory responses in the chronic decerebrate are not relevant to the control of ingestion in the intact animal. This places the hypothalamus as the primary integrator in the network. (2) Integrative processing endemic to the caudal brainstem takes place but is only weakly reflected in the behavior of the intact animal; outputs of caudal brainstem integrators are subordinate to the descending outputs from the forebrain. (3) Hypothalamic and caudal brainstem outputs have equal footing with regard to activity in the final common path. Whether the two regions perform redundant or nonoverlapping integrative functions, their outputs should be combined on the basis of their relative strength in order to determine the system's output.

A fourth conceptualization of the interaction between forebrain and caudal brainstem processing places the caudal brainstem in a dominant position in the neural system controlling feeding in the intact animal. Ascending projections from the caudal brainstem may be the source of information about taste and peripheral or caudal brainstem interoception for specific hypothalamic integrative processes. In addition to providing afferent information, widespread caudal brainstem projections may more globally affect feeding-relevant neural processing in various hypothalamic and other forebrain regions. The potency of caudal brainstem output in the modulation of forebrain processing continues to be emphasized in contemporary thinking on physiological and behavioral activation and behavioral state control (Steriade, Ropert, Kitsikis, & Oakson, 1980; Hobson, McCarley, & Wyzinski, 1975; Scheibel & Scheibel, 1958). The plausibility of this type of model for feeding control in the intact animal is supported by the integrative capacity of the chronic decerebrate and by the existence of direct and indirect ascending pathways. These pathways appear to relay relevant information to the hypothalamus and other forebrain systems (see above) and may, indeed, set the tone of forebrain processing. Such brainstem-centered models for feeding in the intact animal have not yet been seriously considered by investigators in this field.

Given the current level of understanding of the neural basis of feeding, it is premature to favor one model over others listed above. Certainly, neuroanatomical evidence can be rallied to support virtually any conceivable scheme. Forebrain-centered and caudal-brainstem-centered models for regulatory integration should each receive attention on the basis of their ability to guide future work and to identify physiological or behavioral tests that can lead to rejection of incorrect conceptualizations.

Appetitive Behavior

As noted above, chronic decerebrate rats die without oral or gastric infusion of nutrients. These rats, no matter how depleted, do not approach food stimuli. The forebrain, therefore appears to be necessary for the normal expression of appetitive behavior. It seems reasonable to ascribe to the forebrain various processes that serve the appetitive behavior of the intact animal. For example, the identification of certain stimulus patterns as food often requires processing of olfactory information or "higher-order" visual analysis not available to the isolated caudal brainstem. Forebrain processes involving complex spatial relationships and stored information are likely to be important in the selection and execution of foraging strategies when the food stimulus does not directly stimulate sensory receptors.

Certain evidence suggests that an intact forebrain may be necessary for the animal to modify its behavior effectively as a function of previous experience with food. For example, chronic decerebrates have failed thus far to demonstrate convincingly taste aversion conditioning, an extremely robust phenomenon in intact animals (Grill & Norgren, 1978d; A. C. Spector & H. J. Grill, unpublished data). However, it is unclear whether this failure reflects a sensory or performance deficit or a deficit in the ability to associate a taste stimulus with particular visceral stimuli (see discussion by Spector, Breslin, & Grill, 1988). Further experiments must be conducted before any confidence can be attached to the suggestion that decerebrates lack all associative faculties of relevance to feeding. For example, it is important to investigate whether decerebrates can alter their consummatory responses to the oral delivery of a normally rejected taste like quinine when it has been repeatedly followed by the oral delivery of a normally ingested taste like sucrose. Intact rats receiving this sequence of intraoral infusions can be trained to alter their taste reactivity profile to quinine from aversive to ingestive (Breslin, Davidson, & Grill, 1990). It is also important to investigate the ability of decerebrates to display operant behavior reinforced by taste stimulation. Although it is unlikely that this preparation could be induced to perform a variety of traditional operants such as the bar press, it is possible that these rats will demonstrate instrumental learning for taste reward when other operants (e.g., head turning, nose poking, tail movement) are explored.

It is important to separate the various systems that contribute to selection and execution of appetitive behavior from those that arouse the animal. We think the latter lie closer to the heart of the regulatory process. In the absence of the forebrain, caudal brainstem mechanisms responsive to reductions in metabolic fuels can activate the animal in a general sense; chronic decerebrate rats

increase their locomotory and general activity level in response to food deprivation but do not direct their behavior toward food (Junquera & Racotta, 1980).
When coordinated search and approach to food are initiated by reductions in
metabolic fuels in the intact animal, the same caudal brainstem mechanisms may
"drive" the forebrain systems (see previous section, scheme 4) involved in directing behavior relative to the environment. Conversely, the caudal brainstem
mechanism responsible for the satiation of intraoral intake in the chronic decerebrate may allow the intact animal to disengage from the food stimulus and
initiate appetitive sequences pertinent to goal objects other than food.

COMMENTS

Throughout this chapter we have made the case for the fundamental importance of caudal brainstem structures in the production and regulation of consummatory behavior. At the same time we stress that there is no substrate within
the caudal brainstem that should be viewed as a "center." Caudal brainstem
structures communicate with the forebrain and the periphery as part of an
organized neural network for feeding control in the intact animal. Many contemporary students of the hypothalamus have incorporated study of other brain
regions into their research program, bringing new and powerful neuroanatomical, neurochemical, and metabolic methods to bear on the interactions
between regions under study. Many have chosen to limit their attention to forebrain structures. We propose that attention to caudal brainstem function will
facilitate the analyses of forebrain contributions to the overall feeding control
system. We illustrate with a discussion of selected neurochemical systems linked
to feeding function whose forebrain distributions are more extensive and of
higher density than their brainstem complements.

Administration of benzodiazepine-related compounds produces an increase
in ingestion that apparently acts by enhancing food palatability (Cooper & Estall,
1985; Berridge & Treit, 1986). Benzodiazepine binding sites are found in highest density in the forebrain, including cortex, amygdala, and hypothalamus. The
suggestion has been made, based in part on results of benzodiazepine microinjection studies (e.g., Nagy, Zambo, & Desci, 1979), that forebrain receptors are
important in the alteration of feeding behavior. Not discouraged by the lower-
density benzodiazepine binding in the caudal brainstem, Berridge (1988) sought
to assess the ability of receptors at that level to mediate the behavioral effects of
the drug using the chronic decerebrate rat. Decerebrates' ingestive consummatory responses to sucrose were enhanced by doses of chlordiazepoxide in similar
proportion to the effects obtained from the intact controls. We concur with
Berridge's interpretation:

> The conclusion from the present study should not be that benzodiazepine effects upon
> feeding are unrelated to forebrain receptors or processing circuitry, but rather that
> relevant receptors and circuits do exist in the caudal brainstem, . . . and that brainstem
> systems must therefore be incorporated into our understanding of the control of feed-
> ing by benzodiazepines.

The notion of endogenous CCK as a contributor to normal satiety is still
controversial, but the intake-suppressive effect of peripherally administered
CCK constitutes a behavioral probe into the physiology of CCK action and
interactions with the central mechanisms of feeding control. To develop the

CCK model, researchers have identified particular subsets among the extensive central and peripheral distributions of the various molecular forms of CCK and associated receptors (see Vanderhaeghen & Crawley, 1985) as critical for intake suppression. One focus is the paraventricular nucleus of the hypothalamus (PVN). Infusions of CCK into the PVN suppress feeding (Faris, Scallet, Olney, Della-Fera, & Baile, 1983), and lesions of the PVN block the reduction of food intake otherwise produced by i.p. CCK (Crawley & Kiss, 1985; see also Myers & McCaleb, 1981). The involvement of the PVN, however is not an absolute requirement for CCK action on feeding, as is evident from the potent supressiton of sucrose intake and ingestive taste reactivity that follows peripheral administration of CCK-8 in the chronic decerebrate rat (Grill & Smith, 1988; see above). A caudal brainstem emphasis is further supported by the demonstration that it is the sulfated form of CCK-8 that mediates its feeding action. Receptors for sulfated CCK are concentrated in the caudal brainstem, especially in the solitary complex and area postrema; outside of the sulfated CCK-8 receptors located in the posterior hypothalamus, very few are found in the forebrain (Moran, Robinson, Goldrich, & McHugh, 1986). In addition, the activation of sulfated CCK-8 receptors in the pyloris and in nerve terminals shown to be necessary for intake supression following peripheral administration of CCK-8 is reported to the caudal brainstem via the vagus (Smith *et al.*, 1985). These results taken together suggest that if the PVN, posterior hypothalamus, or other forebrain sites contribute to the inhibitory effect of exogenous CCK on feeding in the intact rat, they work in concert with brainstem mechanisms of CCK action.

Biogenic amine (BA) systems receive a great deal of attention from researchers interested in feeding. Convergent lines of experimentation have led this group to a focus on BA synapses in the circumscribed regions of the hypothalamus and certain other forebrain regions. First, depending on cannula placement, localized intracerebral microinfusion of a BA can reliably influence subsequent consumption. For example, intake is elevated by localized infusions of norepinephrine (NE) that specifically perfuse the PVN (e.g., Leibowitz, 1978). Conclusions derived from such experiments have received support from selective lesion studies demonstrating the abolition of the effects of BA microinfusions and of other treatments thought to depend on BA mechanisms localized to a given region (e.g., Myers & McCaleb, 1981; Sclafani & Kirchgessner, 1986). A complementary approach exploits recent advances in the regional *in vivo* measurements of endogenous BAs and related metabolites. Researchers have been able to identify localized changes in BA activity that vary over the short term in relation to the animal's physiological state and to the time frame of spontaneous meals and meals elicited by experimental treatments (e.g., Smythe, Grunstein, Bradshaw, Nicholson, & Compton, 1984; Hernandez & Hoebel, 1988; Myers, Peinado, Miñado, 1988; see Leibowitz & Myers, 1987). For example, NE levels in the medial hypothalamus (including the PVN) increase during deprivation-induced feeding (Kruissink, van der Gugten, & Slangen, 1986; Martin & Myers, 1975). Continued development of *in vivo* assays for BA activity will improve their spatial and temporal resolution as well as their sensitivity (see Leibowitz & Myers, 1987). A more comprehensive model of the feeding-relevant interactions among the biogenic amines and other neurochemical systems of the forebrain will be possible if these technical advances are applied in experiments that correlate neurochemical data with detailed measurements of sensory and behavioral variables.

A profile of forebrain BA involvement in feeding control, however cogent, will remain incomplete without due consideration of the cells from which BA in the forebrain originates. These cells, situated in nuclear groups within the caudal brainstem (Dahlstrom & Fuxe, 1964; Moore & Bloom, 1979), are themselves postsynaptic targets of BA axons arising from neighboring cells and from other brainstem nuclei. By virtue of their extensive projections to non-BA targets in the caudal brainstem, the BA nuclei appear to be in position to directly influence key neural components of the feeding system. As such, pre- and postsynaptic elements of the BA system endemic to the caudal brainstem may comprise a "primary" circuit for BA actions on feeding. We expect that this system eventually will receive as much attention from feeding scientists as have the forebrain hot spots.

An experiment that fails to uncover differences between decerebrate and intact rats can be quite dramatic. The addition of another complex intake-regulatory function to the already remarkable repertoire of the isolated caudal brainstem can be both unsettling and refreshing at the same time. Such an experiment can initiate a period of rapid progress in the study of a given feeding function because so much tissue, including favored substrates of the past, may be put aside as attention is directed to the development of a brainstem-based model of ingestion control. As information about the location, connectivity, and physiology of the rudimentary integrative substrates accumulates, forebrain mechanisms can be more effectively reconsidered. The weight of existing evidence makes it clear that a better understanding of the fundamental contribution of caudal brainstem mechanisms to the control and regulation of feeding is within our grasp using contemporary methods and should receive the highest priority.

Acknowledgments

We thank Beverly Oritz for her valuable assistance in the production of this manuscript. The critical comments of Alan Spector, Eva Kosar, Beverly Oritz, Shari Berk, and Meg Waraczynski are greatly appreciated. The research is supported by NIH grant DK 21397.

REFERENCES

Anand, B. K. & Brobeck, J. R., (1985). Hypothalamic control of food intake in rats and cats. *Yale Journal of Biology and Medicine, 24,* 123–140.

Bard, P. & Macht, M. B., (1958). The behavior of chronically decerebrate cats. In G. E. W. Wolstenholme, C. M. O'Connor, & A. Churchill (Eds.), *Ciba Foundation symposium on the neurological bases of behavior,* (pp. 55–71). London: Churchill.

Bereiter, D. A., Berthoud, H. R., & Jeanrenaud, (1981). Chorda tympani and vagus nerve convergence onto caudal brainstem neurons in the rat. *Brain Research Bulletin, 7,* 261–266.

Berridge, K. C. (1988). Brainstem systems mediate the enhancement of palatability by chlorodiazepoxide. *Brain Research, 447,* 262–268.

Berridge, K. C., & Treit, D. (1986). Chlordiazepoxide directly enhances positive ingestive reactions. *Phamacology Biochemistry and Behavior, 24,* 217–221.

Berthoud, H. R., & Mogenson, G. J. (1977). Ingestive behavior after intracerebral and intracerebroventricular infusions of glucose and 2-deoxy-D-glucose. *American Journal of Physiology, 233,* R127–R133.

Bradbury, M. (1979). *The concept of a blood–brain barrier.* New York: Wiley.

Breslin, P., Davidson, T., & Grill, H. J. (1990). Conditioned reversal of reactions to normally-avoided tastes. *Physiology and Behavior. 47,* 535–538.

Burton, M. J., Rolls, E. T. & Mora, F. (1976). Effects of hunger on the responses of neurons in the lateral hypothalamus to the sight and taste of food. *Experimental Neurology, 51*, 668–677.

Campfield, L. A. & Smith, F. J. (1986). Functional coupling between transient declines in blood glucose and feeding behavior: Temporal relationships. *Brain Research Bulletin, 17*, 427–433.

Cannon, W. B., McIver, M. A. & Bliss, S. W. (1924). A sympathetic and adrenal mechanism for mobilizing sugar in hypoglycemia. *American Journal of Physiology, 69*, 46–66.

Chandler, S. H., & Tal, M. (1986). The effects of brainstem transections on the neuronal networks responsible for rhythmical jaw muscle activity in the guinea pig. *Journal of Neurosciences, 6*, 1831–1842.

Chhina, G. S., Anand, B. K. & Rao, P. S. (1971). Effect of glucose on hypothalamic feeding centers in deafferented animals. *American Journal of Physiology, 221*, 662–667

Coimbra, C. C., Gross, J. I., & Migliorini, R. H. (1979). Intraventricular 2-deoxyglucose, insulin and free fatty acid mobilization. *American Journal of Physiology, 207*, E317–E329.

Coirini, H., Marusic, E. T., DeNicola, A. F., Rainbow, T. C., & McEwen, B. S. (1983). Identification of mineral corticoid binding sites in rat brain by competition studies and density gradient centrifugation. *Neuroendocrinology, 37*, 354–360.

Collier, G. H. (1985). Satiety: An ecological perspective. *Brain Research Bulletin, 14*, 693–700.

Coons, E. E., Levak, M. & Miller, N. E. (1965). Lateral hypothalamus: Learning of food-seeking response motivated by electrical stimulation. *Science, 150*, 1320–1321.

Cooper, S. J., & Estall, L. B. (1985). Behavioral pharmacology of food, water and salt intake in relation to drug actions at benzodiazepine receptors. *Neuroscience and Biobehavioral Reviews, 9*, 5–19.

Craig, W. (1918). Appetites and aversions as constituents of instincts. *Biology Bulletin, 34*, 91–107.

Crawley J. N., & Kiss, J. Z. (1985). Paraventricular nucleus lesions abolish the inhibition of feeding induced by systemic cholecystokinin. *Peptides, 6*, 927–935.

Dahlstrom, A., & Fuxe, K. (1964). Evidence for the existence of monoamine-containing neurons in the central nervous system. I. Demonstration of monoamines in the cell bodies of brainstem neurons. *Acta Physiologica Scandinavica, Supplementum, 232*, 1–55.

Davis, J. D., & Levine, M. W. (1977). A model for the control of ingestion. *Psychological Review, 84*, 379–412.

Desiraju, T., Benerjee, M. G., & Anand, B. K. (1968). Activity of single neurons in the hypothalamic feeding centers: Effect of 2-deoxy-D-glucose. *Physiology and Behavior, 3*, 757–760.

Deutsch, J. A. (1983). Dietary control and the stomach. *Progress in Neurobiology, 20*, 313–332.

DiRocco, R. J., & Grill, H. J. (1979). The forebrain is not essential for sympathoadrenal hyperglycemic response to glucoprivation. *Science, 204*, 1112–1114.

Doty, R. W., & Bosma, J. F. (1956). An electomyographic analysis of reflex deglutition. *Journal of Neurophysiology, 19*, 44–60.

Epstein, A. N., Nicolaidis, S., & Miselis, R. (1975). The glucoprivic control of food intake and the glucostatic theory of feeding behavior. In G. J. Mogenson & F. R. Calaresu (Eds.), *Neural integration of physiological mechanisms and behavior*. (pp. 148–168) Toronto: University of Toronto Press.

Faris, P. L., Scallet, A. C., Olney, J. W. Della-Fera, M. A. & Baile, C. A. (1983). Behavioral and immunohistochemical analysis of the function of cholecystokinin in the hypothalamic paraventricular nucleus. *Neuroscience Abstracts, 9*, 56.3.

Fluharty, S. J., & Grill, H. J. (1981). Taste reactivity of lateral hypothalamic lesioned rats: Effects of deprivation and tube feeding. *Neuroscience Abstracts, 7*, 28.

Flynn, F. W. (1987). Effects of bombesin administration on taste-elicited behaviors of intact and chronic decerebrate rats. *Neuroscience Abstracts, 13*, 881.

Flynn, F. W. & Grill, H. J. (1983). Insulin elicits ingestion in decerebrate rats. *Science, 221*, 188–190.

Flynn, F. W. & Grill, H. J. (1985). Fourth ventricular phlorizin dissociates feeding from hyperglycemia in rats. *Brain Research, 341*, 331–336.

Flynn, F. W., & Grill, H. J. (1988). Intraoral intake and taste reactivity responses elicited by sucrose and sodium chloride in chronic decerebrate rats. *Behavioral Neuroscience, 102*, 934–941.

Fonberg, E. (1974). Amygdala function within the alimentary system. *Acta Neurobiologiae Experimentalis, 34*, 435–466.

Friedman, M. I., & Granneman, J. (1983). Food intake and peripheral factors after recovery from insulin-induced hypoglycemia. *American Journal of Physiology, 244*, R374–R382.

Glick, Z., & Mayer, J. (1968). Hyperphagia caused by cerebral ventricular infusion of phlorizin. *Nature, 219*, 1374.

Gold, R. M. (1973). Hypothalamic obesity: The myth of the ventromedial nucleus. *Science, 182*, 488–490.

Grill, H. J. (1980). Production and regulation of ingestive consummatory behavior in the chronic decerebrate rat. *Brain Research Bulletin, 5(Suppl. 4)*, 79–87.

Grill, H. J., Berridge, K. C. & Ganster, D. (1984). Oral glucose is the prime elicitor of preabsorptive insulin secretion. *American Journal of Physiology, 246*, R88–R95.

Grill, H. J., & Kaplan, J. M. (in press). Sham feeding in chronic decerebrate and intact rats. *American Journal of Physiology.*

Grill, H. J., & Miselis, R. R. (1981). Lack of ingestive compensation to osmotic stimuli in chronic decerebrate rats. *American Journal of Physiology, 240*, 81–86.

Grill, H. J. & Norgren, R. (1978a). The taste reactivity test. I. Mimetic responses to gustatory stimuli in neurologically normal rats. *Brain Research, 143*, 263–279.

Grill, H. J., & Norgren, R. (1978b). The taste reactivity test. II. Mimetic responses to gustatory stimuli in chronic thalamic and chronic decerebrate rats. *Brain Research, 143*, 281–297.

Grill, H. J., & Norgren, R. (1978c). Neurological tests and behavioral deficits in chronic thalamic and chronic decerebrate rats. *Brain Research. 143*, 299–312.

Grill, H. J., & Norgren, R. (1978d). Chronically decerebrate rats demonstrate satiation but not bait-shyness. *Science, 201*, 267–269.

Grill, H. J., Schulkin, J., & Flynn, F. W. (1986). Sodium homeostasis in chronic decerebrate rats. *Behavioral Neuroscience, 100*, 536–543.

Grill, H. J., & Smith, G. P. (1988). Cholecystokinin decreases sucrose intake in chronic decerebrate rats. *American Journal of Physiology, 23*, R853–R856.

Grill, H. J., Spector, A. C., Schwartz, G. J., Kaplan, J. M., & Flynn, F. W. (1987). Evaluating taste effects on ingestive behavior. In N. Rowland & F. Toates (Eds.), *Methods and techniques to study feeding and drinking behavior*, (pp. 151–188). Amsterdam: Elsevier.

Grillner, S. (1985). Neurobiological bases of rhythmic motor acts in vertebrates. *Science, 228*, 143–149.

Hernandez, L., & Hoebel, B. G. (1988). Feeding and hypothalamic stimulation increase dopamine turnover in the accumbens. *Physiology and Behavior, 44*, 599–606.

Himsworth, R. L. (1970). Hypothalamic control of adrenalin secretion in response to insufficient glucose. *Journal of Physiology, 206*, 411–417.

Hobson, J. A., McCarley, R. W., & Wyzinski, P. W. (1975). Sleep cycle oscillation: Reciprocal discharge by two brain stem neuronal groups. *Science, 189*, 55–58.

Houpt, K. A. (1982). Gastrointestinal factors in hunger and satiety. *Neuroscience and Behavioral Reviews, 6*, 145–164.

Houpt, K. A., & Epstein, A. N. (1973). Ontogeny of controls of food intake in the rat: GI fill and glucoprivation. *American Journal of Physiology, 225*, 58–66.

Jean, A., & Car, A. (1979). Inputs to the swallowing medullary neurons from the peripheral afferent fibers and the swallowing cortical area. *Brain Research, 178*, 567–572.

Junquera, J. & Racotta, R. (1980). Manifestations of peripherally induced satiety in mesencephalic rats. *Brain Research Bulletin, (Suppl. 4)*, 75–77.

Kaplan, J. M., & Grill, H. J. (1987). Oropharyngeal correlates of meal progress in the rat. *Neuroscience Abstracts, 13*, 335.

Kaplan, J. M., & Grill, H. J. (1989). Swallowing during ongoing fluid ingestion in the rat. *Brain Research, 499*, 63–80.

Kawamura, Y. & Yamamoto, T. (1978). Studies on neural mechanisms of the gustatory–salivary reflex in rabbits. *Japanese Journal of Physiology, 285*, 35–47.

Kruissink, N., van der Gugten, J., & Slangen, J. L. (1986). Short-term feeding-related changes in mediodorsal hypothalamic catecholamine release. *Pharmacology, Biochemistry and Behavior, 24*, 575–579.

Leibowitz, S. F. (1978). Paraventricular nucleus: A primary site mediating adrenergic stimulation of feeding and drinking. *Pharmacology Biochemistry Behavior, 8*, 163–175.

Leibowitz, S. F. & Myers, R. D. (1987). The neurochemistry of ingestion: Chemical stimulation of the brain and *in vivo* measurement of transmitter release. In F. M. Toates & N. E. Rowland (Eds.), *Techniques in the behavioral and neural sciences, Vol. I: Feeding and drinking.* (pp. 271–316). New York: Elsevier.

LeMagnen J., & Tallon, S. (1966). La periodicite spontanee de la prise d'aliments ad libitum du rat blanc. *Journal de Physiologie, 58*, 323–349.

Lowe, A. A. (1981). The neural regulation of tongue movements. *Progress in Neurobiology, 15*, 295–344.

Lund, J. P., & Dellow, P. G. (1971). The influence of interactive stimuli on rhythmical masticatory movements in rabbits. *Archives of Oral Biology, 16*, 215–223.

Luschei, E. S., & Goldberg L. J. (1981). Neural mechanisms of mandibular control: Mastication and voluntary biting. *Handbook of physiology, Vol. II: Motor control,* (pp. 1237–1274). Bethesda: American Physiological Society.

Macht, M. B. (1951). Subcortical localization of certain "taste" responses in the cat. *Federation Proceedings, 10,* 88.

Marrazzi, M. A. (1976). Hypothalamic glucoreceptor response-biphasic nature of unit potential changes. In D. Novin, W. Wyrwicka, & G. A. Bray (Eds.), *Hunger: Basic mechanisms and clinical implications* (pp. 171–178). New York: Raven Press.

Martin, G. E., & Myers, R. D. (1975). Evoked release of [14C]norepinephrine from the rat hypothalamus during feeding. *American Journal of Physiology, 229,* 1547–1555.

Matsuo, R., Nobuaki, & Kusano, K. (1984). Lateral hypothalamic modulation of oral sensory afferent activity in nucleus tractus solitarius neurons of rats. *Journal of Neuroscience, 4,* 1201–1207.

McCleery, R. H. (1977). On satiation curves. *Animal Behavior, 25,* 1005–1015.

McHugh, P. R., & Moran, T. H. (1985). The stomach: A conception of its dynamic role in satiety. *Progress in Psychobiology and Physiological Psychology, 11,* 197–230.

Miller, A. J. (1982). Deglutition. *Physiological Reviews, 62,* 129–184.

Miller, F. R. & Sherrington, C. S. (1915). Some observations on the bucco-pharyngeal stage of reflex deglutition in the cat. *Quarterly Journal of Experimental Physiology, 9,* 147–186.

Miselis, R. R. & Epstein, A. N. (1975). Feeding induced by intracerebroventricular 2-deoxy-D-glucose in the rat. *American Journal of Physiology, 229,* 1438–1447.

Moore, R. Y., & Bloom, F. E. (1979). Central catecholamine neuron systems: Anatomy and physiology of the norepinephrine and epinephrine systems. *Annual Review of Neuroscience, 2,* 113–168.

Moran, T. H., Robinson, P. H., Goldrich, M. S., & McHugh, P. R. (1986). Two brain cholecystokinin receptors: Implications for behavioral actions. *Brain Research, 362,* 175–179.

Morgane, P. J. (1961). Alterations in feeding and drinking behavior of rats with lesions in globi pallidi. *American Journal of Physiology, 201,* 420–428.

Murzi, E., Hernandez, L. & Baptista, T. (1986). Lateral hypothalamic sites eliciting eating affect medullary taste neurons in rats. *Physiology and Behavior, 36,* 829–834.

Myers, R. D. & McCaleb, M. L. (1981). Peripheral and intrahypothalamic cholecystokinin act on the noradrenergic feeding circuit in the rat's diencephalon. *Neuroscience, 6,* 645–655.

Myers, R. D., Peinado, J. M., & Miñano, F. J. (1988). Monoamine transmitter activity in lateral hypothalamus during its perfusion with insulin or 2-DG in sated and fasted rat. *Physiology and Behavior, 44,* 633–643.

Nagy, J., Zambo, K., & Desci, L. (1979). Anti-anxiety action of diazepam after intra-amygdaloid application in the rat. *Neuropharmacology, 18,* 573–576.

Nicolaidis, S. (1969). Discriminatory responses of hypothalamic osmosensitive units to gustatory stimulation in cats. In C. Pfaffmann (Ed.), *Olfaction and taste* (pp. 569–573). New York: Rockefeller University Press.

Norgren, R. (1976). Taste pathways to hypothalamus and amygdala. *Journal of Comparative Neurology, 166,* 17–30.

Oomura, Y., Ono, T., Ooyama, H. & Wayner, M. J. (1969). Glucose and osmosensitive neurons of the rat hypothalamus. *Nature, 222,* 282–284.

Oomura, Y., Ooyama, H., Yamamoto, T. & Naka, F. (1967). Reciprocal relationship of the lateral and ventromedial hypothalamus in the regulation of food intake. *Physiology and Behavior, 2,* 97–115.

Panksepp, J. (1975). Central metabolic and humoral factors involved in the neural regulation of feeding. *Biochemistry and Behavior, 3, (Suppl. 1),* 107–119.

Parker, S. W. & Feldman, S. M. (1967). Effect of mesencephalic lesions on feeding behavior in rats. *Experimental Neurology, 17,* 1313–1326.

Ritter, R. C., Slusser, P. G. & Stone, S. (1981). Glucoreceptors controlling feeding and blood glucose: Location in the hindbrain. *Science, 213,* 451–453.

Ritter, S., Murane, J. M., & Landenheim, E. E. (1982). Glucoprivic feeding is impaired by lateral or fourth ventricular alloxan injection. *American Journal of Physiology, 243,* R312–R317.

Ritter, S. & Strang, M. (1982). Fourth ventricular alloxan injection causes feeding but not hyperglycemia in rats. *Brain Research, 249,* 198–201.

Rogers, R. C., & Hermann, G. E. (1986). Hypothalamic paraventricular nucleus stimulation-induced gastric acid secretion and bradycardia suppressed by oxytocin antagonist. *Peptides, 7,* 695–700.

Russek, M. (1963). An hypothesis on the participation of hepatic glucoreceptors in the control of food intake. *Nature, 197,* 79–80.

Scheibel, M. E., & Scheibel, A. E. (1958). Structural substrates for integrative patterns in the brain stem reticular core. In H. H. Jasper, L. D. Proctor, R. S. Knighton, W. S. Noshay, & R. T. Costello (Eds.), *Reticular formation of the brain* (pp. 31–35). Boston: Little, Brown.

Schmitt, M. (1973). Influences of hepatic portal receptors on hypothalamic feeding and satiety centers. *American Journal of Physiology, 225,* 1089–1095.

Schwartz, G. J. & Grill, H. J. (1984). Relationship between taste reactivity and intake in neurologically intact rats. *Chemical Senses, 9,* 249–272.

Sclafani, A., & Kirchgessner, A. (1986). The role of the medial hypothalamus in the control of food intake: An update. In R. C. Ritter, S. Ritter, & C. D. Barnes (Eds.), *Feeding behavior: Neural and humoral controls* (pp. 27–66). New York: Academic Press.

Smith, G. P., & Epstein, A. N. (1969). Increased feeding in response to decreased glucose utilization in the rat and monkey. *American Journal of Physiology, 217,* 1083–1087.

Smith, G. P., Gibbs, J. & Kulkosky, P. J. (1982). Relationships between brain–gut peptides and neurons in the control of food intake. In B. G. Hoebel & D. Novin (Eds.), *The neural basis of feeding and reward* (pp. 149–165). Brunswick: Haer Institute.

Smith, G. P., Jerome, C. & Norgren, R. (1985). Afferent axons in abdominal vagus mediate satiety effect of cholecystokinin in rats. *American Journal of Physiology, 249,* R638–R641.

Smythe, G. A., Grunstein, H. S., Bradshaw, J. E., Nicholson, V., & Compton, P. J. (1984). Relationship between brain noradrenergic activity and blood glucose. *Nature, 308,* 65–67.

Spector, A. C., Breslin, P., & Grill, H. J. (1988). Taste reactivity as a dependent measure of the rapid formation of conditioned taste aversion: A tool for the neural analysis of taste–visceral associations. *Behavioral Neuroscience, 102,* 942–952.

Steiner, J. E. (1973). The gustofacial response: observations on normal and anencephalic newborn infants. In J. F. Bosma (Ed.), *Fourth symposium on oral sensation and perception: Development in the fetus and infant* (pp. 125–167). Bethesda: U.S. Department of Health, Education, and Welfare.

Stellar, E., & Hill, J. H. (1952). The rat's rate of drinking as a function of water deprivation. *Journal of Comparative and Physiological Psychology, 45,* 96–102.

Steriade M., Ropert, N., Kitsikis, A., & Oakson, G. (1980). Ascending activating neuronal networks in midbrain reticular core and related rostral systems. In J. A. Hobson & M. B. Brazier (Eds.), *The reticular formation revisited* (pp. 125–167). New York: Raven Press.

Stricker, E. M. (1984). Biological bases of hunger and satiety: Therapeutic implications. *Nutrition Review, 42,* 333–340.

Stricker, E. M., & McCann, M. J. (1985). Visceral factors in the control of food intake. *Brain Research Bulletin, 14,* 687–692.

Stricker, E. M., Rowland, N., Saller, C., & Friedman, M. I. (1979). Homeostatis during hypoglycemia: Central control of adrenal secretion and peripheral control of feeding. *Science, 196,* 79–81.

Swanson, L. W. & Sawchenko, P. E. (1983). Hypothalamic integration: Organization of the paraventricular and supraoptic nuclei. *Annual Review of Neuroscience, 6,* 269–324.

Thexton, A. J. & Griffiths, C. (1979). Reflex oral activity in decerebrate rats of different age. *Brain Research, 175,* 1–9.

Tordoff, M. G., & Friedman, M. I. (1986). Hepatic portal glucose infusions decrease food intake and increase food preference. *American Journal of Physiology, 251,* R192–R196.

Travers, J. B. & Norgren, R. (1983). Afferent projections to the oral motor nuclei in the rat. *Journal of Comparative Neurology, 220,* 280–298.

Travers, J. B., Travers, S. P. & Norgren, R. (1987). Gustatory neural processing in the hindbrain. *Annual Review of Neuroscience, 10,* 595–632.

Vanderhaeghen, J. & Crawley, J. N. (1985). Neuronal cholecystokinin. *Annals of the New York Academy of Sciences, 448.*

Waldbillig, R. J. (1975). Attack, eating, drinking and gnawing elicited by electrical stimulation of rat mesencephalon and pons. *Journal of Comparative and Physiological Psychology, 89,* 200–212.

Weijnen, J. A. W. M., Wouters, J., & van Hest, J. M. H. H. (1984). Interaction between licking and swallowing in the drinking rat. *Brain Behavior and Evolution, 25,* 117–127.

Weingarten, H. P. (1984). Meal initiation controlled by learned cues: Basic behavioral properties. *Appetite, 5,* 147–158.

Weingarten, H. P. & Watson, S. D. (1982). Sham feeding as a procedure for assessing the influence of diet palatability on food intake. *Physiology and Behavior, 28,* 401–407.

Woods, J. W. (1964). Behavior of chronic decerebrate rats. *Journal of Neurophysiology, 27,* 635–644.

Yamamoto, T., Fujiwara, T., Matsuo, R. & Kawamura, Y. (1982). Hypoglossal motor nerve activity elicited by taste and thermal stimuli applied to the tongue of rats. *Brain Research, 238,* 89–104.

7

Food Intake
Gastric Factors

J. A. Deutsch

Introduction and Historical Overview

This chapter begins with a brief historical overview of research on the function of the stomach in hunger satiety, focusing on methodological issues that are still relevant today, such as the method of gastric loading and criteria of satiety. The remainder of the chapter is divided into sections each of which addresses itself to answering a particular question.

1. Is hunger satiety produced by events in the upper gastrointestinal tract?
2. Are satiety signals produced in the duodenum or in the stomach?
3. Are such signals generated by intragastric volume, pressure, or break-down products of nutrient?
4. What part in the control of ingestion do oropharyngeal signals play?
5. Are gastric satiety signals calibrated through a process of learning?
6. How do satiety signals travel to the central nervous system? Is the mode of transmission neural? What is the involvement of cholecystokinin?

One of the earliest experiments to focus research interest on the stomach was that of Hull, Livingstone, Rouse, and Barker (1951). These workers reported that a dog with an esophageal fistula would consume huge amounts of food in a single meal. This finding led to a very large number of experiments in which various substances were placed directly in the stomach (summarized by K. A. Houpt, 1982) in an attempt to discover what factors produced satiety. Satiety was measured as a decrease in food intake subsequent to the insertion of some substance in the stomach. On the basis of such experiments, many theories concerning the mechanisms of gastric satiety have been suggested, such as that of Grossman (1955) concerning stretch receptors and of Harper and Spivey (1958) concerning

J. A. Deutsch   Department of Psychology, University of California, San Diego, LaJolla, California 92093.

osmoreceptors. Nevertheless, the exact role of the stomach in satiety remained unsettled. Part of the difficulty in this regard may result from using a reduction in food intake as the operational definition of satiety.

In retrospect, it seems obvious that such a criterion is highly ambiguous. Indeed, Miller and Kessen (1952) pointed out that meal size reduction after intragastric loads could not result only from satiety but also from malaise or nausea. (There is, of course, the problem of distinguishing satiety from malaise or nausea, and this will be discussed below.) Moreover, Booth, Lovett, and McSherry (1972) have suggested that grossly hypertonic glucose consumed normally, even when ingested by mouth, could produce an aversive state of affairs.

How is such aversion detected or measured in animals, and especially the rat, on which most feeding research is done? The rat does not vomit, and until recently no direct unconditioned reaction to visceral malaise has been available as an index. Such a direct index of malaise has now been discovered (Stricker, McCann, Flanagan, & Verbalis, 1988), and it consists of an elevation of oxytocin secretion. [However, this biological marker is useful only when known alternative bases for oxytocin secretion (e.g., dehydration, suckling) are not present. Oxytocin is a marker that occurs in the rat. Arginine vasopressin (AVP) may be used as an equivalent marker in humans and nonhuman primates.] A less direct test of aversion, but one that needs very little specialized equipment, consists of setting up a conditioned taste aversion (Garcia & Ervin, 1968).

What is conditioned taste aversion in the context of gastric research? If inserting a substance into the stomach produces satiety, then we should expect such a manipulation to be rewarding to the animal. If, on the other hand, inserting such a substance produces some untoward effect such as malaise or nausea, then much a manipulation should be negatively reinforcing. Tests have been developed that enable us to infer whether a state of affairs is rewarding or aversive to the animal. Such tests are based on the observation that if a rat is exposed to a neutral stimulus such as a flavor and then made ill, it will then tend to avoid such a flavor stimulus when it is presented on a future occasion. Such a learning to avoid, or conditioned taste aversion, makes a very convenient test. First, the interval between the exposure to a neutral flavor and the subsequent administration of the agent that causes illness may be as long as many hours. Second, the effect is large, and acquisition can occur even with one pairing of the flavor and illness. Whether a conditioned taste aversion is always produced by a state that in humans would be introspectively reported as illness, malaise, or nausea is difficult to know, especially as there are some agents such as ethanol that produce conditioned taste aversion in rats. Now it is possible that rats are made sick by substances that do not have this effect in humans or that an aversion can be triggered in rats by substances that are potentially dangerous without the mediation of malaise. (A substance that causes ataxia and reduction in fear and vigilance is potentially life-threatening to a heavily preyed-upon rodent.) However, it is not necessary to decide such questions before using conditioned taste aversion as an informative test. If a previously neutral stimulus is avoided after it has been followed by administration of a substance, it is plausible to conclude that the administration of such a substance had been aversive rather than rewarding, even if we do not know the nature of such aversiveness. Such an interpretation of a conditioned taste aversion test is rendered even more persuasive by the fact that the pairing of a neutral stimulus, such as a flavor, with relief from a vitamin deficiency or an induction of satiety makes such a neutral stimulus more attractive (Holman, 1968; Puerto, Deutsch, Molina, & Roll, 1976).

Although malaise or nausea may decrease food intake, few experiments involving intragastric loads of nutrients contain controls for such a confounding factor. Consequently, we must now view with caution the evidence stemming from such experiments and reexamine the conclusion that the stomach (or upper gastrointestinal tract, for that matter) is involved in sensing satiety.

Consider the following example from our laboratory (Deutsch, Molina, & Puerto, 1976). We injected sesame oil into the stomach of a rat to determine whether a conditioned taste aversion would develop. Sesame oil is highly palatable to rats, which drink about 4 mL of the oil per day even when chow pellets and water are freely available. When 1 mL of sesame oil was injected after the rats had drunk a flavored liquid, a very strong conditioned taste aversion to the liquid quickly developed. Indeed, we even were able to use intragastric sesame oil to produce a conditioned taste aversion to normal drinking water despite the fact that it has been much more difficult to produce a conditioned aversion to a familiar taste than to an unfamiliar taste (Revusky & Bedarf, 1967; Revusky & Garcia, 1970; Nachman, 1970; Maier, Zahorik, & Albin, 1971). Specifically, rats were allowed to drink water for 15 minutes per day, a taste with which they had been familiar all their lives. Another 15 minutes later, 1 mL of sesame oil was placed in their stomach. They were then given access to water flavored with almond or banana extract for 1 hour, 90 minutes later to prevent progressive dehydration and to show that total water intake did not decrease as a result of the experimental treatment. The rats drank 9.6 mL of unflavored water and 3.3 mL of flavored water on the first day. On the 13th day they drank 2 mL of unflavored water and 7.6 mL of flavored water. This aversion to tap water generalized from the experimental boxes in which the rats were tested to tap water in their own home cages.

This initial experiment showed that a palatable nutrient placed directly in the stomach produced malaise of sufficient magnitude to cause a powerful conditioned taste aversion. However, the time course of such malaise was not well defined, and its onset could have been considerably delayed with respect to the time of the actual gastric injection. To explain the reduction in eating that is seen after intragastric injection, the development of malaise must be very prompt. Accordingly, we ran an experiment in which intragastric injection of oil was paired with the choice of one of two different solutions of flavored water. Each daily session was restricted to 15 minutes. One flavored solution was paired with intragastric injections of 0.1 mL of saline per 1 mL drunk, and the other was paired with 0.1 mL of sesame oil per 1 mL drunk. No changes in preference were seen during the first 4 days of testing. Thus, starting on the fifth day the intragastric injections were increased to 0.5 mL per 1 mL of flavored water drunk. The amount consumed on the fourth day was 4.2 mL (S.D. ± 0.97 mL) of the oil-paired flavor and 4.6 mL (S.D. ± 0.75 mL) of the saline-paired flavor, but on the fifth day the amounts drunk were 1.75 mL (S.D. ± 0.75 mL) and 4.25 mL (S.D. ± 1.5 mL), respectively. Intake of the oil-paired flavor was diminished further for the next 7 days until the termination of the experiment, while no diminution was seen in the intake of the saline-paired flavor.

From these results it can be concluded that intragastric injection of nutrient can produce malaise within 15 minutes of injection. Indeed, it probably develops very much more quickly, because the rat is able to discriminate which of the two flavors that it drinks is paired with the oil. If malaise took a long time to develop after injection, then there should be no preferential decrease in the intake of one flavor but a reduction in the intake of both flavors sampled instead. We have noted that other nutrients, such as casein hydrolysate, also produce taste aversions when injected intragastrically. Why such intragastric injections produce taste aversions is discussed in a later section.

The demonstration that the insertion of substances into the stomach leads to learned taste aversions has been recounted here in some detail because of its continuing relevance to research on satiety. It appears that research on this subject has produced unintelligible results merely because the criterion of satiety used, a reduction in intake or meal size, was simplistic and ambiguous. As will be seen below, much present research is flawed by the use of the same criterion.

Why might the insertion into the stomach of nutrients, or even palatable nutrients, produce aversive consequences? The answer seems to be twofold. First, such nutrients may not have been modified chemically by processes that normally occur in the mouth. Second, the act of pumping material into the stomach may produce mechanical abnormalities in the stomach. Let us first review the evidence that suggests that biochemical abnormalities arise when food is placed directly in the stomach. Pavlov (1910) noted that the major stimuli to gastric secretion are the expectation of food and the olfactory and visual stimulation provided by it. [This phenomenon, reviewed by Powley (1977), is called the cephalic-phase reflex.] When such preparatory stimulation does not occur, the normal flow of digestive juice is not evoked by the vagus. Such sensory stimulation even promotes the secretion of pancreatic enzymes (Sarles, Dani, Prezelin, Souville, & Figarella, 1968). The direct placement of food in the stomach eliminates such preparatory sensory stimulation and will therefore reduce the normal secretion of digestive juices and lessen the rate of digestive breakdown of nutrient in the stomach. Though a reduction in the rate of digestive breakdown might be expected to slow the process of gastric emptying, the opposite result seems to occur. Because there is less hydrochloric acid secreted when food is placed directly in the stomach, stomach contents empty more rapidly into the duodenum. The arrival of hydrochloric acid in the duodenum releases cholecystokinin (CCK) and secretin. Such gastrointestinal hormones in their turn decrease the rate of gastric emptying (Chey, Hitanaut, Hendricks, & Lorber, 1970; Chvasta, Weisbrodt, & Cooke, 1971). The reduced secretion of gastrin also would lead to an increased rate of gastric emptying (Cooke, Chvasta, & Weisbrodt, 1972), as would a reduction in pancreatic enzymes (Mallinson, 1968).

Other biochemical factors are present that would speed up the emptying of stomach contents into the duodenum when food is loaded directly into the stomach. The mouth secretes enzymes that, as Malhotra (1967) has pointed out, are of importance in the digestive process. The effects of these in the breakdown of long-chain carbohydrates is well known. Less well known is the effect of Ebner's gland, situated at the base of the tongue, on lipolysis (Plucinski, Hamosh, & Hamosh, 1979; Gonzalez & Deutsch, 1985).

Beside such biochemical factors that lead to premature emptying of stomach contents into the duodenum, there are also mechanical factors related to direct stomach loading that lead to the same result. When an animal spontaneously loads its own stomach by eating, every time swallowing occurs, the stomach increases its volume reflexly in order to accommodate the extra incoming load.

This has been called the receptive relaxation of the stomach (Cannon & Lieb, 1911). Thus, in spite of the increase in volume caused by the arrival of the extra load, the stomach maintains a steady intragastric pressure. As a result, except at the very outset of a normal meal, intragastric pressure remains very constant though gastric volume increases dramatically (Young & Deutsch, 1980). On the other hand, we have found that when nutrient or bulk is infused into the stomach of an animal at the same rate as it eats, intragastric pressure increases as volume goes up. Receptive relaxation does not occur because it depends on the act of swallowing. When nutrient or bulk is infused into the stomach but with the pylorus blocked, the rate at which pressure rises becomes much steeper. This means that when the pylorus is not blocked, material is forced down the pressure gradient into the duodenum, leading to a less steep increase of pressure in the stomach. So it seems that the mechanical factors act in the same direction as the chemical factors when the stomach is filled and the animal itself is not feeding. Nutrient is transferred prematurely into the duodenum.

Such a transfer has aversive consequences. Rapid gastric emptying into the duodenum causes feelings of nausea and discomfort such as occur clinically after substantial gastrectomy. It is known as the dumping syndrome (Roberts, 1967).

So far the explanation of why nutrient placed in the stomach is aversive rests on indirect evidence. However, such an explanation has more direct experimental evidence to support it. We carried out an experiment (Deutsch, Puerto, & Wang, 1977) to test if the premature arrival of poorly digested nutrient in the duodenum is aversive. A section of surgical rubber tubing was inserted inside the pyloric sphincter of rats to keep it permanently open. There were two groups of experimental rats. One was given a choice of solid food and water as against fresh milk. The second group was given a choice of solid food and water as against milk that was predigested by being taken from the stomach of donor rats. A third group, which underwent a sham operation without the insertion of tubing, was given a choice of solid food and water as against fresh milk. The tube inserted in the pylorus made no difference in the rat's choice of predigested milk. At the end of 5 days, the rats in the experimental group (given a choice of predigested milk) were drinking over 12 mL of milk on the average. The control rats given a choice of fresh milk drank a closely matched amount. On the other hand, the other group of experimental rats given fresh milk as a choice drank less than 3 mL each at the end of 5 days.

To control for the possibility that predigested milk is preferred to fresh milk, unoperated rats were given a choice for 5 days between predigested and normal milk under the conditions of the main experiment. The mean percentage volume of predigested milk drunk each day was 44.5% of the total volume of milk ingested. We can therefore reject the hypothesis that the results of the main experiment reflect a preference for predigested milk.

It seems then that the entry into the duodenum of milk that has received somewhat less than the normal amount of digestion can override the normally rewarding consequences of milk ingestion, presumably because of some aversive consequences. In a further experiment we measured whether nutrients, when placed directly in the stomach, receive less digestive modification than if they are normally ingested. We further determined whether such nutrients, placed directly in the stomach, pass more quickly into the duodenum than when they have been normally eaten. We examined the rate of lipolysis of fresh milk after it had been normally drunk and after it had been placed in the stomach (Molina, Thiel,

Deutsch, & Puerto, 1977). Lipase is mixed with a meal as it passes Ebner's gland at the base of the tongue. A reduced rate of lipolysis when the throat is bypassed is therefore to be expected. A group of rats was implanted with intragastric fistulae. This group was then divided into two. The first group was allowed to drink milk normally. The second group was intubated with the same volume of milk at the same rate as it had been drunk by the first group. Milk was then withdrawn from the stomach after 5, 20, or 45 minutes of delay. The withdrawn samples of milk were then immediately analyzed, and the proportion of free fatty acid in each sample determined. It was found that there is much less lypolysis of milk, even after 45 minutes, when it is injected into the stomach than when it is normally drunk.

A further experiment was performed to find out whether nutrient injected directly into the stomach passes more quickly into the duodenum than the same quantity of nutrient normally ingested. To do this rats were sacrificed, and stomach volume was measured in two conditions. In the first, rats were allowed to ingest a known volume of dextrin. In the second condition, the same volume was injected intragastrically at the same rate. The results showed that the gastric emptying of contents that had been normally drunk was much slower.

These experiments show that the digestive process is abnormal when the arrival of nutrient in the stomach bypasses the mouth. Such abnormalities could explain the aversive or nonrewarding nature of injected nutrient.

The Upper Gastrointestinal Tract and Satiety

As was stated at the outset, when food is not allowed to reach the stomach, gross overeating results (Hull *et al.*, 1951; Janowitz & Grossman, 1949; Share, Martiniuk, & Grossman, 1952; Mook, 1963). The food is diverted before it reaches the stomach by the use of an esophageal fistula. When food is aspirated from the stomach, abnormally high intakes during a meal are again observed (Davis & Campbell, 1973). When food is withdrawn from the stomach after a meal is finished (Snowdon, 1970), the rat begins to ingest nutrient again with a median latency of about 3 minutes (Davis & Campbell, 1973). Even when the amount withdrawn is quite small (about 3 mL), the amount ingested in compensation is quite close. Such compensation for the amount withdrawn is as good 40–50 minutes after the original meal as after shorter times.

These findings demonstrate that some events either in the stomach or past the stomach terminate a meal. Davis and Campbell (1973) suggested that such compensatory drinking is caused by a reduction of stomach distention. They also stated an alternative interpretation of their findings. This was that a lessening of arrival of nutrient in the duodenum produces a reduction of duodenal satiety signals and thus an enhancement of nutrient ingestion. In a more general way the experiments in which stomach content reduction produces an increase in ingestion give rise to two broad questions. (1) Where does such a reduction in nutrient act? (2) What aspect of such a reduction triggers compensatory intake? Let us first examine the evidence that satiety signals originate in the duodenum and thus that removal of (or reduction in) gastric contents produces its effects of increasing food intake by lessening duodenal contents.

It has been claimed that satiation occurs as a result of intraduodenal injection (Snowdon, 1975; Ehman, Albert, & Jamieson, 1971; Campbell & Davis,

1974; Vanderweele, Novin, Rezek, & Sanderson, 1974). However, satiation here was defined as a simple reduction in subsequent food intake, and no controls were run for alternative explanations of ~uch a reduction in intake such as malaise or nausea. To evaluate this evidence, we therefore ran such a control. A nutrient was used for intraduodenal injection that Snowdon had claimed produced optimal satiation, and it was injected at a rate and in a concentration that he had used in his experiment. Intraduodenal cannulae were implanted in rats according to the method described by him (Snowdon, 1975). The rats drank one of two flavors each day for 15 minutes. Immediately after drinking one of the flavors, the rats were intraduodenally injected with a 1 M (18%) glucose solution at the rate of 0.6 mL/min for 5 minutes (Snowdon, 1975). Immediately after drinking the other flavor (on alternate days), the rats were injected in the same way with physiological saline. There was a considerable reduction in the intake of the flavor paired with glucose (Deutsch et al., 1976).

An interesting point of relevance here has been made by McHugh and Moran (1986a,b). In an important experiment they found that in fasted, hungry monkeys there is little effect on feeding of small volumes of glucose that enter the intestine from the stomach. On the other hand, direct infusions into the intestine of glucose similar in volume have a powerful inhibitory effect on subsequent feeding. "An effect of feeding produced in any one compartment cannot alone be assumed to be illuminating in its physiological role since aspects of the delivery of the nutrients into that compartment may be important in evoking the behavioral effect" (McHugh & Moran, 1986a,b). These results throw in doubt the conclusions of other studies that posit an important role for intestinal factors in satiety (Gibbs, Maddison, & Rolls, 1981; Houpt, Anika, & Houpt, 1979; Reidelberger, Kalogeris, Leung & Mendel, 1983).

Liebling, Eisner, Gibbs, and Smith (1975) attempted to show that satiety was produced by duodenal stimulation by using a different method. A drainage plug was inserted into the stomach while a tube was placed in the duodenum. Rats continued to eat excessively while nutrient was allowed to drain out of the stomach through the drainage plug. However, when food was injected into the duodenum through the tube while the food that was being eaten was draining from the stomach, the rats would stop eating. This led Liebling et al. to conclude that satiety is produced by the entry of food into the duodenum. This view is supported by the observation that nutrient enters the duodenum rapidly when the animal normally eats (Balagura & Fibiger, 1968; Wiepkema, Alingh Prins, & Steffens, 1972).

However, the interpretation of the experiment may be different. Cessation of eating when nutrient is introduced into the duodenum could be a result of nausea rather than satiety. Liebling et al. (1975) dealt with this objection by running a control for the aversive properties of intraduodenal injection. Rats were allowed to drink for only 10 minutes a day and given a highly palatable solution of saccharin (0.25%) to drink. After drinking the saccharin, the first group of rats was intraduodenally injected with nutrient. A second group was injected with 0.15 M lithium chloride at 0.6% body weight diluted to a volume of 6 mL. The third group was injected with the same volume of physiological saline. When the test for conditioned taste aversion was run, all the rats were extremely thirsty, as they had only 10 minutes of access to water each of the 9 or 11 previous days. Further, during the test only the saccharin solution was presented during the 10 minutes of the test, so that the rat, to satisfy its thirst had no choice but to drink the saccharin. To show a conditioned taste aversion the rats would have to overcome an extreme thirst. The rats that had been injected with LiCl did drink less, whereas there was no difference between the intraduodenally and saline-injected groups. However, the test of conditioned taste aversion used is rather insensitive, because the rat's tendency to satisfy its thirst is pitted against its tendency to stay away from a flavor of water that had been associated with malaise. Essentially, the rat is given a choice between dying of poison or dying of thirst. A more satisfactory assay of conditioned taste aversion had been described in the literature (Rozin, 1969; Grote & Brown, 1971; Dragoin, McCleary, & McCleary, 1971). Here a neutral flavor of water is offered along with the flavor paired with the presumed poison in the test, and conditioned taste aversion is evaluated by looking at the proportion of the neutral flavor consumed. We ourselves (Deutsch & Hardy, 1977) have shown large and highly significant conditioned aversion using this latter more sensitive method where the method used by Liebling *et al.* (1975) failed to reveal a conditioned aversion.

THE STOMACH OR DUODENUM?

Another way to ascertain if satiety messages are generated by the duodenum is to confine nutrient to the stomach while the animal is eating. If satiety is produced by messages from the duodenum, eating should continue while no nutrient is present in the duodenum. But if the exit from the stomach is blocked while the animal eats, eating might stop because of abnormal volume or the buildup of excessive pressure. On the other hand, if we allow nutrient to drain from the stomach (as in the experiment of Liebling *et al.* above), the stomach cannot attain the volume of nutrient that normally correlates with satiety, and the stomach cannot retain partially digested food. These problems can, however, be solved if a valve is implanted in the stomach through which the stomach contents can escape when a certain pressure, namely the intragastric pressure at normal satiety, is exceeded. Such a valve is simple. Basically, we used a vertical tube of a length equal to the height of the column of fluid supported by the stomach at satiety when the pylorus is unobstructed. Such an overflow device provided an exit from the stomach that is hydraulically similar to the pylorus and the duodenum. With this device in place and the pylorus blocked, pressure as the animal eats can build up to the normal level without exceeding it, so that the stomach can become as full as during normal eating (without the pylorus obstructed). On the other hand, pressure can never exceed that which occurs during normal eating. [Further research revealed that the implantation of such a device was actually unnecessary because pressure in the stomach after an initial rise stays quite flat as the stomach expands (Young & Deutsch, 1980).]

Experiments have been conducted with rats fitted with an inflatable cuff around the pylorus and the overflow valve (Deutsch, Young, & Kalogeris, 1978). Eight rats 12 hours hungry were allowed to drink milk for 30 minutes. When the pyloric cuff was inflated, they drank a mean of 11.5 mL. With the pylorus unobstructed they drank a mean of 13.6 mL. A similar result was obtained with another six rats run under the same conditions. These drank a mean of 15.8 mL without the cuff inflated and a mean of 14.8 mL with the cuff inflated. (Similar results were obtained by Kraly & Smith, 1978.)

However, there is another way to show the involvement of the duodenum. As we saw above, Snowdon (1970) and Davis and Campbell (1973) demonstrated that compensatory drinking of nutrient occurred after nutrient was withdrawn from the stomach at the end of a meal. If we inflate a cuff around the pylorus at the end of a meal and then withdraw nutrient from the stomach, this should show us where the satiating stimuli act. If the signals of satiety originate from the duodenum, there should be compensatory drinking only when nutrient is removed from the stomach under the condition when the pylorus is unobstructed. If compensatory drinking occurs equally with cuff inflated and uninflated, this is difficult to reconcile with the hypothesis that duodenal signals produce satiety. If on the other hand the signals of satiety emanate from the stomach, there should be compensatory drinking that is the same in amount whether the pyloric cuff is open or closed.

The following experiment was run to investigate the effects of pyloric closure on compensatory ingestion. Rats ($n = 6$) 12 hours hungry were allowed to drink milk for 20 minutes without the pyloric cuff inflated. Then the milk was made unavailable for 5 minutes. Then the rats were permitted to drink again for 25 minutes under four conditions. In the first condition the pyloric cuff was inflated. In the second the pyloric cuff was inflated, and 10 mL of stomach contents was aspirated. In the third condition the pyloric cuff was left uninflated. Finally, in the fourth condition 10 mL of stomach contents was aspirated without the cuff having been inflated. The results were as follows. With the cuff inflated the rats drank 6.9 mL of milk more after 10 mL of gastric contents were aspirated than when no such removal took place ($P < 0.01$). With the cuff uninflated the rats drank 7.4 mL milk more after removal of 10 mL of gastric contents ($P = 0.01$). The difference in the amount drunk in compensation with cuff on and cuff off is only 0.5 mL and is not significant. If the cuff around the pylorus is inflated, and duodenal content determines intake, then the rat should drink the same amount independent of whether nutrient is left in the stomach or withdrawn from the stomach. But rats drink different amounts when stomach contents are made to vary even though duodenal content stays constant. This provides good evidence that satiety signals do not emanate from the duodenum but rather from the stomach.

Other experiments corroborate this conclusion. Gastric catheters and a pyloric cuff were implanted in seven rats. They were allowed to ingest nutrient in the test session under two conditions while 15 hours hungry. They were treated the same under both conditions for the first 10 minutes without the pylorus occluded. At the end of the first 10 minutes, the pylorus was occluded in the first condition for the next 30 minutes, and a mean of 4.1 mL was withdrawn from the stomach as soon as the pylorus was occluded. The mean amount consumed in the first 10 minutes was 4.6 mL, and then after gastric withdrawal and pyloric occlusion, 3.6 mL in the next 15 minutes and 0.4 mL in the last 15 minutes. In the second condition no nutrient was withdrawn from the stomach, but the pylorus was occluded after the first 10 minutes of drinking. The mean amount drunk in this second condition was 4.3 mL in the first 10 minutes, 0.6 mL in the next 15 minutes, and 0.3 mL in the last 15 minutes. Thus, withdrawal with cuff inflated led to a highly significant and accurate compensation by subsequent drinking (Deutsch & Gonzalez, 1980). In a further experiment (Deutsch & Gonzalez, 1981) gastric catheters and pyloric cuffs were surgically implanted in six rats. They were permitted to drink a 50 : 50 Mazola corn oil–water emulsion for 30 minutes. They drank a mean of 8.7 mL with cuffs inflated and a mean of 9.7 mL with cuffs uninflated.

It seems then that the signals that lead to the termination of a meal emanate from the stomach and not the duodenum, at least under the experimental conditions described above. However, we shall see that under different conditions, gastric signals are ignored and oropharyngeal cues dominate.

THE NATURE OF THE SATIETY SIGNAL

It seems that the stomach senses some effects of nutrient ingestion. A suggestion that has often been made is that it is the distention of the stomach that acts as a satiety signal. That distention of the stomach acts as a satiety signal is a view with many adherents such as Cannon and Washburn (1912) and Paintal (1973). This view has recently been championed by McHugh and Moran (1986a, 1986b). In a series of elegant experiments they demonstrated that gastric emptying occurred at a constant rate in terms of the caloric content of the stomach. (However, such a relationship is not universal. Fructose which provides the same amount of energy as glucose (Hunt & Spurrell, 1951; Moran & McHugh, 1981), empties much faster, and xylose, though it yields fewer calories, empties at the same rate as glucose.) This finding led them to hypothesize that gastric distention in terms of both its extent and its duration could provide a source of visceral signals that could influence food intake by reducing meal size and prolonging meal intervals in step with the gastric caloric content. They went on to demonstrate that the rate of gastric emptying was controlled by inhibitory effects on gastric emptying by calories in the intestine.

Initially, when the duodenum is empty, there is an initial rush of nutrient into the duodenum at the outset of a meal. This intraduodenal nutrient then generates an inhibitory influence (that prevents further emptying of the stomach) proportionate in duration to the calories present in the intraduodenal nutrient. It was also shown that this effect was not an osmolar effect (McHugh, 1979; McHugh, Moran, & Wirth, 1982). Although each discharge into the duodenum may be different in size, the averaged outcome in such a system is that gastric emptying occurs at a constant rate in terms of nutrient content, though this occurs in fits and starts over any short time segment. However, stomach distention is an ambiguous concept. It could refer simply to a gastric volume that is above a certain level, or it could refer to gastric pressure that is above a certain level, or some combination of both. We ran the following experiment to find out whether the rat eats until its stomach reaches a certain volume.

Pyloric cuffs and stomach catheters were surgically implanted in 16 rats. The rats were trained to drink milk from U-tubes. As the rats drank, the level of the fluid in the U-tube dropped, and this interrupted current flow across two electrodes in the other end of the U-tube. This switched on a first pump, which acted to restore the fluid level in the U-tube until current flowed through the two electrodes. In half the tests a second pump was switched on at the same time as the first, but it simultaneously pumped physiological saline through the gastric catheter of the rat at the same rate as the rat drank. If volume signals shut off ingestion, then the rats should have stopped eating very much earlier than usual when the excess volume was being placed in the stomach. The results were mixed. In some rats there was a reduction in the amount drunk when extra volume was pumped into the stomach, whereas other animals' intake were unaffected by this procedure (Deutsch, Gonzalez, & Young, 1980).

There was, however, a difference between these two groups of animals. Those that were responsive to the extra gastric volume were, on the whole, large drinkers. Those that did not reduce their intake when double the volume was placed in their stomach tended to be small drinkers. There was a strong correlation between the amount normally drunk and the decrease caused by extra gastric volume ($r = 0.77$, $P < 0.001$). A similar relationship emerged when the experiment was run with cuff inflated ($r = 0.675$, $P < 0.002$). The data suggest that stomach volume does play a part in limiting intake but only after about 20 mL has been consumed. However, at lower volumes something else operates to shut off intake.

This conclusion was checked by giving rats calorically denser nutrients so that all the rats would become small drinkers. This was done by feeding some rats ($N = 11$) Nutrico,®☆ a complete high-calorie diet, and other rats an emulsion of Mazola®™ corn oil (Deutsch & Gonzalez, 1980, 1981). Physiological saline at half the rate the rats drank was pumped into their stomachs. This 50% extra volume made no difference to the size of the meal. In the experiment in which the rats drank Nutrico® for 30 minutes, they drank a mean of 4.9 mL in the base-line condition (when they were drinking normally). In the experimental condition, when extra saline was pumped into the stomach, they drank a mean of 5.1 mL.

The results were very similar in the experiment in which the rats ($N = 6$) drank a 50 : 50 Mazola corn oil–water emulsion for 30 minutes. On one-half of the trials 50% saline was pumped into the stomach as the rats drank (Deutsch & Gonzalez, 1981). The rats drank a mean of 9.7 mL in the base-line condition and 8.9 mL when saline was simultaneously injected into the stomach. We have also measured pressure in the stomach directly (as apart from volume) by using an electronic pressure sensor while the rat ate to satiety in the normal manner to see if some change in pressure correlated with meal termination (Young & Deutsch, 1980). After a brief initial increase at the beginning of a meal, intragastric pressure stays surprisingly level, and we were unable to find any change in pressure that correlated with the end of a meal. Such a result is not unexpected, given the receptive relaxation of the stomach (Cannon & Lieb, 1911).

So far then the experimental evidence shows that stimuli mediated by the stomach are sufficient to terminate eating at the normal amount and that volume or distention or pressure does not operate to produce such termination until the volume in the stomach is very large. It seems then that some other nonmechanical factor acts in the stomach or through the stomach to terminate eating. This factor is related to the amount of nutrient in the stomach seemingly independent of its dilution, at least over a large range (see Table 1 for a summary of data).

It does not seem that intragastric nutrient is sensed directly. The simultaneous intragastric injection of fresh corn oil emulsion in the same volume as the corn oil emulsion that the rat drinks by mouth has little effect on the size of the meal. Rats ($N = 8$) drank a mean of 8.1 mL in 30 minutes when an equal volume of fresh 50 : 50 corn oil emulsion was pumped simultaneously into the stomach. They drank a mean of 8.6 mL when half this volume of saline was pumped into the stomach instead. As mentioned above, the gastric breakdown of fats depends on Ebner's glands, which are placed at the back of the tongue. Lipase is secreted by these glands as food passes over them during the act of ingestion. One would then expect much less lipolysis of a given volume of oil emulsion if a part of it did not pass over Ebner's glands. To render the lipolysis of the oil emulsion infused directly into the stomach more similar to that ingested naturally, we collected oil emulsion from the stomachs of donor rats by siphoning it out after the end of a meal of oil emulsion. We then found that the simultaneous intragastric injection of predigested oil does reduce intake. Rats (N = 12) drank a mean of 15.2 mL of oil emulsion in 30 minutes when saline in equal volumes was intragastrically injected. However, when 4.9 mL (mean quantity) of predigested oil emulsion is injected instead, the rats drank a mean of 10.8 mL. Whatever it is in the predigested emulsion that reduces intake appears to reside in the oily fraction of the emulsion. We took the predigested oil emulsion and separated it by a centrifuge into the oily and aqueous fractions. Each fraction was then taken and mixed with an equal volume of water. In the baseline condition another group of rats (N = 12) drank a mean of 14.1 mL of oil emulsion while saline in half the volume was concurrently injected into the stomach. When 5 mL of the aqueous fraction plus water was injected as the rats drank oil emulsion, they drank a mean of 12.9 mL. When a mean of 4.8 mL of the oil fraction plus water was infused into the stomach as the rats drank, they ingested a mean of 10.9 mL (Gonzalez & Deutsch, 1985). The fact that something in predigested oil reduces intake does not necessarily show that it also produces satiety. It might reduce intake by producing malaise. Accordingly, we tested to see if intragastric predigested oil produces conditioned taste aversion.

Ten rats were given two flavors of 50 : 50 corn oil–water emulsion. Five of the rats had one of the flavors paired with a simultaneous intragastric injection of 5 mL predigested emulsion. The other five rats had the other flavor paired with such an intragastric injection. On the second day the flavor not presented on the first day was given, and no intragastric injection was paired with it. On the third day the rats were given a choice between the two flavors of emulsion. The rats drank a mean of 4.6 mL of the flavor paired with the intragastric oil injection and a mean of 5.7 mL of the other flavor. The experiment was then repeated on the same rats a second time. This time they drank 5.3 mL of the

TABLE 1. BASE-LINE DRINKING COMPARED WITH DRINKING WITH GASTRIC SIPHON, EXTRA GASTRIC VOLUME, AND PYLORIC OCCLUSION[a]

	Nutrico	Oil emulsion
Base-line drinking (mL)	4.9	9.7
Drinking (mL) when 5 mL siphoned from stomach	8.8*	13.5*
Difference (mL)	3.9	3.8
Estimate of volume of nutrient siphoned from stomach (mL)	3.7	5.0
Drinking when 50% extra volume arrived in stomach (mL)	5.1	8.9
Drinking when pylorus occluded (mL)	—	8.7

[a]Asterisks indicate $P < 0.01$. Only removal of nutrient from the stomach makes a significant difference.

intragastric-oil-paired flavor and 6.1 mL of the other flavor. After a third repetition, they drank 4.2 mL of the flavor paired with the intragastric oil emulsion and 5.9 mL of the other flavor. There was no significant trend to decrease the choice of the flavor paired with the intragastric injection. A parallel experiment was performed with lithium chloride, a substance known to produce malaise. First we established that 3 mL of 1.5% LiCl solution produced approximately the same suppression of intake when it was intragastrically injected while the rats drank oil emulsion as 5 mL of predigested oil emulsion. After the first pairing with LiCl the rats chose a mean of 2.5 mL of the flavor paired with LiCl but 8.6 mL of the other flavor. Because there was clear evidence of conditioned taste aversion after this first pairing, the experiment was concluded (Gonzalez & Deutsch, 1985).

The evidence that we have been considering shows quite clearly that meal termination, or satiety, is caused by some gastric signal that arises when a certain quantity of material derived digestively from the nutrient is present, seemingly independent of its dilution. For instance, the compensation for the amount of nutrient lost from the stomach is made accurately either when we use a balanced mixture of the major macronutrients (Nutrico®) or a water–oil emulsion. Further, the amount drunk by mouth stays the same whether the stomach contents are diluted as the rat drinks by an intragastric infusion of saline. It seems that the rat regulates its intake by inspecting the amount of nutrient (or nutrient derivative) in the stomach.

Oropharyngeal Factors

However, such a conclusion is directly at variance with findings that show that it takes several days for animals to adjust their intake when they are shifted from one dilution of a diet to another (Janowitz & Hollander, 1956; Le Magnen, 1967; Share et al., 1952; Booth, 1972; Booth & Davis, 1973). In these experiments the animal initially eats the same amount as previously, irrespective of gastric nutrient content, until it learns through repeated experience that it needs to eat more (or less) to obtain the same amount of nourishment. We have ourselves confirmed such results under our conditions, so that the observed discrepancy cannot be caused by some methodological differences.

From one point of view, taking nutrient out of the stomach after it has been ingested and diluting nutrient before it is ingested appear the same. Both procedures alter the amount that must be ingested to produce the amount in the stomach that had previously signaled satiety. However, when food is siphoned out of the stomach, and so nutritionally "diluted," intake is increased within the same meal. But when food is so altered before it is eaten that more must be ingested to keep the stomach content the same as before this occurred, the amount ingested does not rise to meet the shortfall, at least for a few days.

There is a possible difference between the two situations. There may be a confounding variable in the case of dilution or condensation of nutrient. In none of the experiments where nutrient was made more or less dilute was there any stringent test of whether such a manipulation altered the taste as well as the density of the resulting food. It is possible that when taste remains the same and entirely familiar, gastric factors may control intake. However, control of intake may be entirely oropharyngeal if the taste of the nutrient is novel. Such a hypothesis is simple to test. Compensation for gastric loss should occur when rats drink a familiar nutrient. However, such a compensation for gastric loss should not occur when rats drink the same nutrient but flavored differently for the first time.

To test this hypothesis, 24 rats were implanted with a stomach tube and accustomed to drinking an unflavored 50 : 50 corn oil–water emulsion in a restraining cage (Deutsch, 1983). Then 6 mL was siphoned from their stomachs as they drank under two conditions. In the first, they drank the familiar oil–water emulsion. In the second condition the same emulsion was drunk but flavored with banana extract for the first time. (Unsiphoned controls showed that the encounter with the banana flavor did not change the amount normally drunk.) The mean when the familiar unflavored emulsion was drunk without siphoning was 13.7 mL, and the mean when the emulsion was flavored with banana for the first time was 13.3 mL. As in other experiments, rats compensated well for siphoning, drinking 17.8 mL of the unflavored emulsion. This contrasts with 14.2 mL drunk by the group that were siphoned when they were introduced to the banana-flavored emulsion for the first time. It seems then that when the flavor is new, oropharyngeal signals determine intake, and stomach signals are disregarded. However, as the banana flavor becomes familiar, compensation for siphoning gradually increases over 3 days. After 7 days of the new flavor, rats drink 17.9 mL. Though in a continuation of the experiment rats exposed to both the flavored and unflavored emulsion drank while being siphoned for 8 days, the amount drunk immediately returned to base line as soon as siphoning was discontinued. This shows again that it is the amount in the stomach that controls meal size once taste is familiar.

If a perceived change in taste results in a failure to compensate for a shortfall in nutrient reaching the stomach, then we might expect that there should be a compensation for dietary dilution, but only if such change were below a noticeable change in taste. To test this idea, two groups of 14 rats each were trained to drink a 5 : 5 corn oil–water emulsion. When intake had stabilized, both groups were divided into two subgroups, matched on the basis of intake. On the experimental day the first subgroup was given an oil emulsion consisting of 5 parts of oil to 9 parts of water, and the second subgroup was kept on the original emulsion. The first subgroup, which now drank the more dilute emulsion, drank a mean of 20.8 mL, whereas the second subgroup drank 14.1 mL (difference significant at $P < 0.002$). The other group (also divided into two subgroups) was given the same treatment as the first two subgroups except that the two different emulsions were newly flavored with 0.5% banana extract. This abolished the difference in intake seen in subgroups 1 and 2. The subgroup drinking the 5 : 5 emulsion drank a mean of 14 mL, whereas the subgroup drinking the 5 : 9 emulsion drank a mean of 14 mL. The new flavor by itself again made no difference to intake when dilution was not changed but seemed to prevent compensation for dilution. When diluted and flavored oil emulsion was repeatedly presented, the rats learned to compensate over a few days.

We also ran an experiment in which rats ($N = 8$) were habituated to drinking a 5 : 5 water–oil emulsion, which was then switched to an emulsion of 5 parts of oil to 15 parts of water. This change produced no compensation. The mean amount drunk for the 7 days before the change was 12 mL. When rats were first given the diluted emulsion (with no flavor added), the mean amount drunk was 11.2 mL.

As we saw above, a reduction of intake of oil emulsion occurs when about 5 mL of predigested oil is injected into the stomach. There is then a decrease of about 4.5 mL in the amount of oil emulsion ingested by mouth. If the presence of a novel flavor leads to a disregarding of gastric cues, then no compensation should occur when predigested oil is infused into the stomach when a novel flavor is being tasted. Three groups of six rats, implanted with stomach tubes, were habituated to drinking 50 : 50 corn oil–water emulsion. After such habituation the first group was injected with 5 mL saline when drinking for 30 minutes the emulsion flavored with 0.5% banana for the first time. This group drank a mean of 10.6 mL as compared with a mean of 10.4 mL as the average of the five previous days of the unflavored emulsion. The second group of rats was injected with 5 mL of predigested oil emulsion as it drank (again for 30 minutes) the emulsion flavored with banana (as above) for the first time. This second group drank a mean of 8.7 mL as compared to a mean of 9.4 mL over the previous 5 days of the unflavored emulsion. The third group was injected with a solution of 0.6% LiCl up to a maximum of 5 mL under the same conditions as the first two groups. This dose of LiCl had been found to decrease intake to a similar degree as an injection of 5 mL of predigested emulsion when the flavor of the emulsion drunk by mouth was left unaltered. This third group of rats drank a mean of 4.1 mL as compared to a mean of 9.6 mL for the previous 5 days. (Only a mean of 3.7 mL of LiCl was injected because of the low amount drunk by mouth.) This result is of interest because it can be developed into a test of whether a procedure results in satiety or malaise. Rats respond to cues of sickness even though they are tasting a novel flavor, but they do not then respond to gastric satiety signals.

Mook, Bryner, Rainey, and Wall (1980) report another case in which oropharyngeal, and not gastric, signals regulate intake. Alteration of the gastric consequences of saccharin intake does not alter the amount of saccharin drunk by mouth. Glucose intake is also regulated oropharyngeally when the glucose escapes through an esophageal fistula as the rat drinks. On the other hand, gross overdrinking occurs when the glucose drains through a gastric fistula. Such results are confirmed and extended by Sclafani and Nissenbaum (1985a, 1985b). Although saccharin and glucose stimulate gustatory receptors in the mouth, only glucose stimulates receptors in the stomach. Mei (1983) reports evidence for carbohydrate-sensitive receptors in the stomach. We may hypothesize that when gastric receptors are not stimulated, control of intake switches to oropharyngeal metering, as in the case where unfamiliar food is being ingested.

Learning and Gastric Signals

It seems then that gastric signals govern intake only when the food being ingested is familiar and not when it is unfamiliar. This suggests that these signals mediated by the stomach have to be calibrated by a process of learning and that there is a calibration specific to each food, as sensed by the mouth and the nose. The satiating value of a gastric message therefore depends on the messages received in temporal proximity by sensory organs in the head and on past experience with the metabolic consequences that follow such cephalic and gastric signals. If this is the case, then signals of satiety are not hard-wired. To test this conclusion we attempted to see if gastric signals could be arbitrarily paired with a food. As the food we used a 50 : 50 corn oil–water emulsion. As the gastric signals we used amino acids and free fatty acids. These are digestive products of proteins and fats, respectively, and as such they might be expected to stimulate chemoreceptors concerned with food intake regulation. Amino acids cannot be produced by the digestive breakdown of fats and would therefore be ideal for the study of arbitrary pairing of a food with gastric signals. Rats totally unfamiliar with oil or oil emulsion were surgically implanted with gastric tubes. On recovery from the operation they were introduced to the oil emulsion, and as they drank it, an amino acid in small quantity was pumped into their stomach. For instance, a 3% concentration of L-lysine was pumped into their stomach at 20% of the rate they drank the oil emulsion by mouth, this rate being determined by an electronically controlled pump. A control group was intragastrically infused with saline at the same 20% rate.

To find out whether the substance infused into the stomach had in fact come to be utilized as a satiety signal, we carried out the following test. After intake of the oil emulsion had stabilized, the experimental group (those rats receiving intragastric amino acid) was switched to an intragastric infusion of saline at the same rate. The control group was switched from saline to the intragastric infusion of the amino acid that the experimental groups had received. When this switch took place, the experimental group's intake went up, and there was no change in the intake of the control group (Deutsch & Tabuena, 1986). The results of a number of experiments using this paradigm are summarized in Table 2.

An increase in intake when the intragastric infusion of amino acid was omitted would be predicted from the hypothesis that it had become a gastric signal of satiety. The lack of change in intake in the control group into which the amino acid was suddenly introduced is consistent with the gastric signal hypothesis but could not be predicted from it before we obtained our particular result for the following reason. We had no reason in advance to reject the idea that all the various chemical stimuli act on a single receptor or that, if there are various receptors, their outputs converge and so the information from them becomes pooled. Satiety would then be signaled by variations in the magnitude of a signal arriving in the central nervous system. On this hypothesis the new chemical signal would mimic an overall increase in intensity of stimuli that had been present all along. However, the results clearly show that when novel substances are pumped into the stomach, no decrease in oral intake occurs. On the other hand, when the substances introduced into the stomach are the same as those already used as signals (such as oleic acid or predigested oil), oral intake decreases. This indicates that the amino acids used are recognized as different from the substances resulting from the digestive breakdown of oil emulsion. There is therefore no single satiety signal that ascends from the stomach to the central nervous system.

The results in Table 2 also raise a number of other points. The fact that a sudden introduction of amino acids into the stomach does not decrease intake makes it unlikely that a sudden release from nausea or malaise is a likely explanation of an increase in intake when such amino acids are first omitted. Nor is gastric reflux a plausible explanation of the results, because the rat ignores gastric signals when oral flavor changes and drinks the base-line amount. It is also unlikely that the increase in intake we see on omitting amino acids is caused by a nutritional or caloric dilution of stomach contents. In the case of L-tryptophan, its omission amounts to about 0.16% of the calories of the meal. Such a

TABLE 2. RESULTS OF INTRODUCING VARIOUS SUBSTANCES INTO THE STOMACH OR OMITTING THEM

	Substance									
	L-Lysine L-Threonine L-Tryptophan	L-Lysine L-Threonine	L-Lysine	L-Tryptophan	Oleic acid			Predigested oil emulsion	
Concentration at which infused	4% 4% 4%	4% 4%	3%	0.8%	100%			100%	
Rate at which infused (% of rate of drinking)	10	10	10	10	10 20		5	100 (5 mL infused)	(5 mL withdrawn)
Mean percentage change in intake when suddenly introduced		+0.3	−13.7	+13.1	−52 −58.3			−4.8 mL	
Mean percentage change in intake when suddenly omitted	+58.1	+26.7	+37.3	+25.5			+35		+4.1 mL
S.E.M. of percentages	11.2	8.1	10.9	6.0	5.4	4.4	6.8	9.9 8.6	15.4	0.9 mL	1.3 mL
n	9	6	6	6	6	7	7	11 11	10	11	6
P (from correlated t-test, 2-tailed)	<0.002	<0.05	n.s.	<0.002	n.s.	<0.002	n.s.	<0.002 <0.002	<0.05	<0.002	<0.05

small percentage decrease nutritionally, however, causes an increase of about 25% in intake.

SATIETY SIGNAL PATHWAYS

So far the evidence shows that signals from the stomach play an important role in regulating meal size. When nutrient is removed from the stomach, even when the pylorus is clamped shut by a cuff, compensatory eating rapidly occurs that accurately makes up for the quantity of nutrient removed. There are two types of signal terminating food intake within a meal that originate in the stomach. The first type of signal relays information about gastric volume and comes into operation only when the stomach is quite distended. The second type of signal concerns gastric nutrient content irrespective of volume or dilution. It is probably this signal that also functions as a positive reinforcer of eating, whereas the first (concerning excessive volume) is probably negatively reinforcing. To have an effect on eating, the signals described above must have an effect on the central nervous system, and we may therefore ask how such gastric signals travel to it.

The function of the vagus has been investigated by Kraly and Gibbs (1980), who performed a bilateral subdiaphragmatic vagotomy or a sham vagotomy in different groups of rats. A pyloric noose was implanted at the same time. Meal size was compared when nooses occluded the pylorus and when the pylorus was left open. Meal size was unaffected by the closure of the pylorus in both the vagotomized and sham-operated groups. This again shows that satiety signals equal in effectiveness to those generated normally are generated by the stomach alone. That they are not conveyed by the vagus nerve is not, however, established by the Kraly and Gibbs experiment. The rat can use oropharyngeal metering to control intake and could switch from gastric to oral control if the gastric cues were removed. This is why we conducted the test for gastric satiety cues by withdrawing material from the stomach and looking at compensatory intake rather than measuring original intake.

Let us first consider the evidence concerning volume signals. We found that subdiaphragmatic vagotomy in the rat abolished responsiveness to gastric distention. On the other hand, we found undiminished responsiveness to gastric nutrient content (Gonzalez & Deutsch, 1981). What this means is that vagotomized rats compensate for nutrient that is siphoned from their stomachs in a manner similar to the controls. On the other hand, when saline is infused into the stomach as they drink, the vagotomized rats do not reduce their oral intake under conditions where normal controls do. Because the rats in our experiment were maintained on a liquid diet, the stomachs of the vagotomized rats were of normal size. Kraly and Gibbs (1980) also found that vagotomized rats did not increase meal size, even when the meal was confined to the stomach by pyloric closure.

The greater splanchnic nerves innervate the upper gastrointestinal tract along with the vagi. This raises the possibility that these splanchnic nerves could convey nutrient sensing to the central nervous system. However, such a possibility seems contraindicated by experiments on celiacectomized or splanchnicotomized animals in which little change in normal food intake occurs (Opsahl, 1977; Rezek, Schneider, & Novin, 1975). The lack of a change in normal food intake in splanchnicotomized animals is not conclusive evidence against the involvement of the splanchnics. A rat that lacks gastric nutrient sensing might still show little or no effect of such an absence because it possesses an alternative, or back-up, system using oropharyngeal cues alone to control intake. This system is used when the hungry rat ingests saccharin (Mook *et al.*, 1980) or a meal with a novel flavor (Deutsch, 1983). However, when this alternative system is in control, changes in gastric nutrient content during a meal leave oral intake unchanged, even though normal intake (without alteration of gastric contents) appears unaffected. If this oropharyngeal system assumes control over eating when gastric nutrient signals no longer inhibit it, no overt change in normal eating would occur after splanchnicotomy.

In order to see if gastric nutrient sensing was removed by splanchnicotomy, we would have to observe the rats' reaction to change in gastric nutrient content. This can be done by siphoning out nutrient during a meal. We therefore measured compensatory intake after siphoning in newly splanchnicotomized rats. Newly splanchnicotomized rats, unlike sham-operated controls, do not compensate for gastric siphoning. On the other hand, newly splanchnicotomized controls, not siphoned, eat the usual amount (Deutsch & Ahn, 1986). However, these results must be viewed with caution. The experiment is technically difficult, and a second replication yielded negative results. It is possible that damage to the splanchnic nerve occurred during preparatory surgery. It is also possible that splanchnicotomy causes a switch from gastric to oropharyngeal control though the splanchnic nerve itself does not carry gastric satiety signals. Such a switch from gastric to oropharyngeal occurs when novel, instead of familiar, flavor is introduced even though the pathway from stomach to brain must remain surgically intact. Further experiments are necessary to clarify the problem.

Other hypotheses can be advanced to explain how signals travel from the upper gastrointestinal tract to the brain. These hypotheses postulate humoral rather than neural transmission.

It is possible that small amounts of nutrient are absorbed by the stomach and pass into the bloodstream to be monitored directly by some organ in the periphery or some central nervous structure. Evidence that some of the factors concerned with either the evocation or satiation of hunger are humoral has been obtained by Davis and colleagues (Davis, Gallagher, & Ladove, 1967; Davis, Gallagher, Ladove, & Turavsky, 1969). Blood is transfused between a pair of rats when one is hungry and the other is satiated. Two milliliters of blood is withdrawn through a chronic intravenous cannula from both rats simultaneously, and then the blood from one rat is injected into the other. This procedure is repeated until 26 mL of blood has been exchanged between the two rats. Immediately after the transfusion, each rat is given access to food.

In one experiment, one member of the pair was kept on an *ad libitum* diet, and the other was kept without food for 23.5 hours each day. The blood of each pair was transfused 30 minutes before the food-deprived animal was fed each day. When the food intake of the 23.5-hour deprived rat was measured after it had received blood from the satiated animal, it was found that its intake dropped to 50% of the amount it ingested when no transfusion was given. Thus, it seems as if the blood of a satiated rat contains a factor that produces satiety in a hungry rat. However, the reduction in food intake might be caused by some nonspecific factor connected with transfusion and not necessarily with satiety as normally produced. To control for this possibility, the donor rat, which had been on an unrestricted diet, was deprived of food 24 hours before its blood was injected into the recipient. (This recipient was on a 23.5-hour schedule.) In this case the food intake of the recipient was not reduced below the normal level. This showed that the effect of the blood transfusion on the recipient was a function of the satiety or hunger of the donor. Transfusion in itself had not caused a drop in food intake.

However, we still do not know what is being transferred. One possibility, as Davis believes, is that a satiety factor is being transferred from the donor to the recipient rat. The other possibility is that the factor that produces or evokes hunger in the recipient is being diluted by the blood of the donor where it is not present. Either way, the reduction of food intake that was observed would occur.

To investigate these effects further, Davis kept both members of the pair of rats without food 23.5 hours each day. The donor member of the pair was fed 30 minutes before transfusion, so that it was satiated. However, the recipient of this freshly satiated blood did not show a significant reduction in intake. One interpretation of this is that 30 minutes after feeding the humoral state that initiates eating has not been reversed by the processes of digestion. Therefore, when the blood of a freshly satiated donor is used, it does not dilute the humoral state that initiates eating in the recipient. If this interpretation is correct, it would imply that the signals that normally turn eating off (satiety signals) are not transferred from donor to recipient. This might be because such signals are not humoral. However, it seems that the factor that causes the initiation of eating is humoral.

Another version of the humoral hypothesis postulates that gastrointestinal hormones are secreted on the arrival of nutrient in the upper digestive tract and that these are then monitored either peripherally or centrally. More specifically, it has been proposed that the gastrointestinal hormones cholecystokinin (CCK) and bombesin (BBS) are satiety signals. Cholecystokinin is a gastrointestinal hormone secreted predominantly by the duodenum and jejunum. There are CCK receptors in the pancreas and gastric fundic glands, and CCK-specific binding has been localized by autoradiography in the circular smooth layer of the pyloric sphincter in its distalmost portion (Smith *et al.*, 1984). Cholecystokinin has also been found in portions of the gastrointestinal tract associated with neural and neuroendocrine cells. Such nerves are most frequent in the colon in the myenteric and submucosal ganglia, smooth muscle, and the mucosa.

Cholecystokinin has also been found in vagal afferent fibers, and CCK receptors have also been detected in the vagus nerve of rats. It has been shown (Smith, Jerome, & Norgren, 1985) that afferent axons in the abdominal vagus mediate the "satiety" effect of peripherally administered cholecystokinin in rats. Supporting this is the finding of Edwards, Ladenheim, and Ritter (1986) that lesions of the medial and commissural subnuclei of the nucleus of the solitary tract together with the area postrema produced a significant alteration of the effects of CCK on eating. This lesion includes most of the terminal field from the gastric branch of the vagus nerve. Corroborating evidence has been reported by Crawley, Kiss, and Mezey (1984). Cholecystokinin is secreted by the duodenum and jejunum in response to free fatty acid and amino acid stimulation. It has a demonstrated physiological function in causing the gallbladder to contract and the pancreas to secrete digestive enzymes. It also has effects on intestinal motility retarding gastric emptying by its effects on the pyloric sphincter (McHugh & Moran, 1986a,b). It has been found by a large number of investigators (for a good review, see Morley, Bartness, Gosnell, & Levine, 1985) that injections of these hormones produce a temporary decrease in food intake. It has also been argued that after the injections of these hormones animals show a set of behaviors characteristic of normal satiety (Antin, Gibbs, Holt, Young, & Smith, 1975). It was also originally argued that to qualify as a satiety hormone, a substance should not produce conditioned taste aversion, and evidence was adduced that CCK did not produce conditioned taste aversion.

The argument that CCK and BBS are satiety hormones because they produce a reduction in food intake is, taken by itself, quite inconclusive. Many manipulations produce food intake reduction. The hormones could produce malaise, sleepiness, fear, or hyperactivity that would also reduce intake.

The second line of evidence, if correct, would be more persuasive. Satiety is supposed to produce a characteristic behavior sequence, and CCK produces the same sequence (Antin et al., 1975). However, the problem with the behavioral sequence of satiety is that, in spite of its name, it has not been shown to be uniquely evoked by satiety. In the original study proposing this criterion of satiety, it was claimed that after eating, various behaviors occur, such as grooming and resting. When CCK is administered, such behaviors seem to occur also, but not when food intake is frustrated by the addition of quinine to the food. Furthermore, it was found that these behaviors do not occur when food is drained from the stomach as eating takes place. There are, however, methodological inadequacies in this study of Antin et al. (1975). Only five rats were used per group, only means are reported, and the observation was not performed double blind. Further, to be used as a criterion of satiety, it must be shown that the behavioral sequence is not also a characteristic of other states that could reduce food intake, such as nausea or sleepiness. Actual inspection of the data indicates that only the amount of resting is likely to be different between the groups, and resting is a behavior that can occur for a multitude of reasons.

To test the validity of the concept of behavioral sequence of satiety we performed a further study (J. A. Deutsch, B. O. Moore, & W. G. Young, unpublished data). We injected rats with saline, the C-terminal octapeptide of CCK ($40U \ kg^{-1}$), lithium chloride ($45 \ mg \ kg^{-1}$), and 2-deoxy-D-glucose ($500 \ mg \ kg^{-1}$). There were eight rats in each of the above groups. In addition, there was a group of four rats injected with 2-deoxy-D-glucose at $750 \ mg \ kg^{-1}$. An observer who did not know the purpose of the experiment or the drugs used scored the behavior of the rats. The criteria used were those used by Antin et al. (1975). The rats were placed after injection in a plexiglas enclosure without food and observed for 30 minutes. The only significant differences that were noted, compared to saline, were in the amount of resting, and these were both only found in the groups injected with 2-deoxy-D-glucose. It is to be noted that 2-deoxy-D-glucose in these amounts produces large increases in food intake (Smith & Epstein, 1970).

A second experiment was a variation of the first insofar as this time we allowed the rats access to food in the 30-minute observation period. Further, the 2-deoxy-D-glucose (2DG) groups were omitted. There were again eight (new) rats per group. The rats injected with saline rested 8.5% (S.E.M. ± 3.2) of the time, those with CCK rested 18% (S.E.M. ± 4.5) of the time, whereas those treated with LiCl rested 22.6% (S.E.M. ± 7.2) of the time. LiCl reliably produces nausea and malaise.

It is therefore clear that the so-called behavioral sequence of satiety as described by Antin *et al.* (1975) cannot be used to discriminate satiety from an increase in hunger or an attack of malaise. E. M. Stricker (personal communication) reports a similar experiment to that of Deutsch *et al.* (above) (using LiCl but not 2DG) that yielded a similar outcome, showing that the behavioral sequence of satiety was not peculiar to satiety. Indeed, Stricker states that the same effects were found even when a meal was interrupted by simply removing food from the cage, thus indicating that grooming followed by resting appears to be what caged rats do when eating is interrupted for whatever reason.

The third piece of evidence originally adduced to show that CCK was a satiety hormone was that it did not produce a conditioned taste aversion (Holt, Antin, Gibbs, Young, & Smith, 1974). However, the conditioned taste aversion test used to show that CCK was not aversive was quite insensitive. A large dose of CCK reduces food intake by about 50% at the most, and then for only about 15 minutes. If such a reduction were caused by malaise, the malaise would evidently have to be quite mild. On the other hand the test used by Holt *et al.* (1974) was a one-bottle test, run when the rats were exceedingly thirsty. Deutsch and Hardy (1977) showed that conditioned taste aversion to CCK developed after three pairings in a more sensitive two-bottle situation and when the rats were less thirsty during the test.

The response to this finding that CCK produces conditioned taste aversion has been threefold. The first argument states (Gibbs & Smith, 1980) that the conditioned taste paradigm

> can no longer be considered a critical test of the presence or absence of malaise. Since a variety of agents (including isotonic saline and chlorpromazine, an antinausea drug) which serve as effective unconditioned stimuli for the formation of conditioned taste aversions do not produce sickness, the conditioned taste aversion test cannot be used as evidence of sickness.

However, the vast majority of agents that produce conditioned taste aversion also plausibly induce sickness. So suppose that 1% of the agents that produce conditioned taste aversion (CTA) do not produce sickness. This would still mean that any new agent that produces CTA and whose sickness-producing status is uncertain still has a 99% chance of being a sickness-producing agent. But we can use a less general argument to greater effect. Let us look at the contention that physiological saline produces a CTA even though it does not produce sickness. As the only basis for their claim, Gibbs and Smith (1980) quote a study by Revusky, Smith, and Chalmer (1971). In this study 10 mL of saline was intracardially injected in under an hour into rats some of which weighed under 200 g. It should be remembered that normal blood volume in rats is only 4.3 mL/100 g. It is difficult to understand why Gibbs and Smith believe that no sickness was produced by saline in these circumstances. Their contention regarding chlorpromazine is also open to doubt. "Paradoxically, although the antipsychotics are used in the treatment of nausea and vomiting, they can also induce these disturbances. The mechanism is unknown but may result from anticholinergic effects of decreased mobility and relaxation of the stomach" (Sack, 1977).

A second, different line of argument has been used by Smith, Gibbs, and Kulkosky (1982). This consists of quoting studies that do not show a CTA caused by CCK. However, it should be noted that all the studies in their review consist of only one pairing of CCK. Deutsch and Hardy (1977) have already found that three pairings were necessary. [However, some of the experiments quoted by Smith, Gibbs, and Kulkosky (1982) do show an effect after only one pairing.]

There is therefore no reason to reject the idea that CCK causes CTA. It seems again evident that CCK does not produce a strong effect in terms of CTA. But then, CCK produces a relatively small effect on appetite and one consistent with the production of only a mild malaise. The third argument to deal with the result that CCK produces a CTA states that satiety is a form of malaise anyhow, so that a satiety agent could very well produce a CTA. A fourth related but somewhat different argument has also been used. Results showing avoidance of the flavor paired with the hormones are not caused by conditioned aversion but by conditioned satiety. Therefore, when the rat consumes the flavor paired with the hormone, it will drink less because such a flavor will be more satiating.

These last two arguments have been juxtaposed because they are both inconsistent with the same experimental evidence. When satiety is produced using intragastric nutrient injection (Holman, 1968; Puerto et al., 1976), the rats actually come to prefer the flavor that has been paired with the satiating agent. For instance, in the experiment of Puerto et al. (1976), gastric tubes were implanted in eight rats. One flavor was presented for 7 minutes on odd days and another for the same time on even days. One of the two flavors for half the group was always followed by an intragastric injection of 8–12 mL of predigested milk at 0.6 mL/min. The two flavors were then presented simultaneously on the ninth day. At the end of 7 minutes the rats had drunk a mean of 7.1 mL of the milk-paired flavor and 1.7 mL of the flavor that had not been nutrient paired ($P <$ 0.025). Such results are clearly inconsistent with the idea that satiety is aversive or that conditioned satiety causes a reduction of intake of the conditioned stimulus.

However, a hypothesis is not readily abandoned unless a more plausible hypothesis can take its place. Is there some hypothesis that could explain the reduction in eating consequent on a CCK injection? An alternative hypothesis holds that the gastrointestinal hormones, in the doses injected, produce sickness or malaise. Such malaise could cause a reduction in eating and could account in a straightforward way for the fact that the administration of such hormones, in doses that are used to demonstrate food intake reduction, also cause CTAs (Deutsch & Hardy, 1977; Deutsch & Parsons, 1981; Young, Deutsch, & Tom, 1982). Swerdlow, Van der Kooy, Koob, and Wenger (1983) demonstrated that hungry rats develop an aversion to a place where they are injected with CCK, in doses used by others to demonstrate "satiety." But rats satiated in the same place by being fed there approach it instead.

Another approach to this problem was taken by Moore and Deutsch (1985). There were two related considerations. If CCK-induced reductions in food intake occurred through the mechanism of normal satiety, CCK-induced satiety and normal satiety should respond in the same way to pharmacological challenge. Also, if CCK exerts its action on food intake by reducing malaise or sickness, an antiemetic should reduce CCK's effect on eating. We used a set dose of trimethobenzamide on its own and together with an ascending range of doses of CCK to observe its effect on food intake. Trimethobenzamide is an antiemetic that acts specifically on the chemoreceptor trigger zone and has been shown to be useful in alleviating nausea in humans. The dose of the drug that we used was shown by Coil et al. (1978) to attenuate LiCl-induced aversions. We found that the antiemetic did not increase normal consumption and, if anything, slightly reduced it. However, the antiemetic increased intake when CCK was administered by about 26% for each dose of CCK administered. Since the antiemetic increased intake by about 26% for each dose of CCK, normal intake (without exogenous CCK) should have increased by 26% also, as the antiemetic should have been effective against the endogenous CCK. Instead, the antiemetic-treated rats consumed 95% of their normal amount. Thus, CCK-induced feeding termination and normal satiety behave differently under the same pharmacological challenge. Further, because the antiemetic is antidotal to the effects of CCK (as it is to LiCl), and because the antiemetic is not antidotal to normal satiety, it seems very likely, first, that malaise is the factor that causes food intake reduction after CCK and, second, that malaise is not a factor producing normal satiety.

Collins, Walker, Forsyth, and Belbeck (1983), Shillabeer and Davidson (1984), Schneider, Gibbs, and Smith (1986) used a related approach in which proglumide, a CCK antagonist, was used to see if it had the same effect on "satiety" whether normally produced or induced by CCK administration. Whereas proglumide antagonized the reduction in intake induced by CCK administration, only Shillabeer and Davison (1984) report evidence that proglumide administration increases normal food intake. Schneider et al. (1986), in a thorough repetition study, were not able to show such an increase. This is important because, as Schneider et al. (1986) state, "the results of Shillabeer and Davison are the only evidence for the satiating effect of endogenous CCK."

More recently, there have been reports that CCK antagonists do produce an increased meal size (Dourish, Coughlan, Hawley, Clark, & Iversen, 1988; Dourish, Rycroft, & Iversen, 1989; Hewson, Leighton, Hill, & Hughes, 1988). However in the study of Hewson et al. (1988), while the CCK antagonist increased intake of a palatable diet, it did not increase feeding when the rats were made hungry and given their regular diet, although other differences in the two experiments could have been responsible for the discrepancy. Although the positive results may at first sight support the idea that CCK signals satiety, their correct explanation could easily lie elsewhere. First, the inconsistency in the various experiments suggests that the correct variables giving rise to this phenomenon have not yet been identified. This makes the claim that the phenomenon is due to a manipulation of satiety premature.

Second, other explanations of the increase in meal size after CCK antagonist administration have not been tested. For example, it has been suggested (B. Moore, personal communication) that the CCK antagonist produces an effect similar to that of vagotomy. Gonzalez and Deutsch (1981) showed that such surgery results in an animal that can tolerate degrees of distention aversive to intact rats, while not affecting the operates' ability to compensate for withdrawal of nutrient from the stomach. It is this latter mechanism that is related to normal satiety. The CCK antagonist might thus abolish the pain or discomfort due to excessive gastric volume. Such a suggestion is made plausible by the finding (Dourish, Hawley, & Iversen, 1988; Watkins, Kinscheck, & Mayer, 1984) that CCK antagonists potentiate analgesia. These agents could thus have an effect on the signaling of aversive stimuli from the periphery, such as result from pronounced gastric distention. In the very convincing demonstration of the CCK antagonist effect by Dourish, Rycroft, and Iversen (1989), the degree of food deprivation before the intake test would have necessitated the ingestion of nutrient volume in quite excessive quantities. In that part of the study of Hewson et al. (1988) that yielded positive results, rats (approximately 200 g) not treated with antagonist ate over 10 g in 30 min.

Another possible explanation of the effect of CCK antagonists on food intake lies in their action on the pancreas (Dourish, Rycroft, & Iversen, 1989). If, for example, these antagonists produced a fall in blood glucose, the observed increase in eating could be due not to decreased satiety but to increased hunger. In sum, the finding that CCK antagonists (in some studies) increase food intake suffers from ambiguities of interpretation in much the same way as does the finding that CCK reduces food intake.

So far we have used behavioral criteria to evaluate CCK's function. There have also been studies using the physiological consequences of CCK administration. As we saw above, it has been claimed that the administration of CCK produces the behavioral sequence of satiety. The premise that makes such a claim interesting is that if CCK really signals satiety, then the overt behavior after CCK administration should resemble in important respects behavior that follows normal satiety. The same premise applies to physiological as well as behavioral measures. Some studies have been carried out examining physiological variables. In the first study (Deutsch, Thiel, & Greenberg, 1978) duodenal motility was monitored during food intake after both CCK injection and saline injection. In the saline condition, there was a large and uniform increase in the amplitude of peristaltic waves during eating and satiety. On the other hand, such waves were almost entirely obliterated after an injection of CCK.

We criticized the results from experiments on behavioral satiety on the grounds that many other states besides satiety could generate such a sequence. However, such an argument does not exclude the possibility that such a behavioral sequence indicates satiety after CCK is administered. It simply renders such an indication much less probable. On the other hand, the criticism here has a different basis, and that is that CCK administration and CCK secreted during normal satiety actually have different physiological effects. This makes a much stronger argument against equating CCK-induced eating termination with normally induced satiety. The amount of CCK secreted during normal satiety does not produce an obliteration of peristaltic waves, as does the exogenous administration of CCK. It is difficult to escape the conclusion that the exogenously administered CCK needed to produce food intake reduction must be very much larger than the endogenous level normally secreted during eating.

The fact that the exogenous CCK produces a cessation of duodenal peristalsis fits well with the hypothesis that CCK administration in the doses used to demonstrate food intake reduction gives rise to malaise. Another attempt to estimate whether the amount of exogenous CCK that was necessary to produce a reduction in meal size was comparable to CCK produced endogenously was made by Pappas, Melendez, Strah, and Debas (1985) in the dog. Intraduodenal infusion of sodium oleate, which had a significant effect on CCK secretion as judged by gallbladder contraction and pancreatic secretion, had no significant effect on sham feeding. The D_{50} value of exogenous CCK-8 for gallbladder contraction is 22 pmol $kg^{-1} h^{-1}$, 120 pmol $kg^{-1} h^{-1}$ for pancreatic protein response. In contrast the D_{50} of CCK-8 to show an inhibiting effect on feeding is 340 pmol $kg^{-1} h^{-1}$, which exceeds the maximal dose for its gallbladder and pancreatic action. Again it seems that the effects of CCK on food intake occur only when an unnaturally high dose is used and do not occur when CCK is secreted physiologically.

These studies indicate that the amount of CCK injected to produce "satiety" is much higher than that secreted after a meal. Such a conclusion is corroborated by the measurement of CCK levels. "Most of the available radioimmunoassay techniques for CCK measure a postprandial rise in plasma immunoreactive CCK that is equivalent to that produced by an infusion of 5–15 pmol kg^{-1} h^{-1}" (Pappas *et al.*, 1985). In the rat, in which most investigations have been performed, the doses administered to induce satiety range from 200 to 12,000 pmol (based on a conversion rate of 3 IDU = 875 pmol). This estimate is shared by Yalow, Eng, and Straus (1983). "... The reported effects of peripheral administration of these peptides on regulation of satiety are achieved only at levels greatly in excess of the physiological range achieved in response to feeding. Thus, endogenous peripheral CCK alone would appear not to be the physiological satiety factor."

Related work in the domestic fowl (Savory, 1987) investigated the effects of BBS and CCK-8 peripherally administered in doses that produce "satiety." It was found that such doses caused complete inhibition of peristalsis for short periods that coincided with complete inhibition of feeding, and longer periods of reduced feeding when abnormal gastrointestinal motility occurred. Increases in heart rate coinciding with such periods also occurred. It was therefore suggested that peripherally administered BBS and CCK-8 may act on feeding by distracting animals with abdominal discomfort.

Some of the gastrointestinal effects of BBS have also been examined (Young *et al.*, 1982). Intragastric pressure following doses of BBS reported to cause "satiety" were measured. Intragastric pressure in the rat does not generally exceed 4 cm of water during eating. However, within 3–4 minutes of a BBS injection (16 mg/kg), intragastric pressure rises to a steady 15 cm of water and above with frequent brief spasms. Such effects are completely reversed by an administration of diazepam (5 mg/kg). If BBS decreases food intake as a consequence of the elevation of intragastric pressure and associated spasms, then such reduction of intake should be reversed on the administration of the same dose of diazepam as abolishes the gastric abnormalities. Although such a dose of diazepam on its own did not increase food intake, it did abolish the food intake reduction produced by BBS alone. This again suggests that the effect of BBS is not to produce satiety but to produce malaise. Two separate experiments (Deutsch & Parsons, 1981; Young *et al.*, 1982) favor this hypothesis. It was found that BBS produces CTA when it is paired with either flavored water or flavored nutrient. The CTA appears at both the doses used (16 mg/kg and 8 mg/kg) to demonstrate satiety.

The secretion of the neurohypophyseal hormone oxytocin has been used by Verbalis, McCann, McHale, and Stricker (1986) as an index of visceral illness in rats. They found that agents such as lithium chloride, apomorphine, and copper sulfate that cause conditioned taste aversions also cause secretion of oxytocin (OT) in a dose-dependent manner. They used this fact to ascertain whether CCK produced OT secretion that resembled satiety or visceral illness. "Administration of cholecystokinin (CCK) to rats caused a dose-dependent increase in plasma levels of OT. The OT secretion was comparable to that found in response to nausea-producing chemical agents that cause learned taste aversions" (Verbalis *et al.*, 1986). A much smaller rise in OT secretion occurs during normal satiety. The authors conclude that "thus, after administration of large doses of CCK, vagally mediated activation of central nausea pathways seem to be predominantly responsible for the subsequent decrease in food intake."

Stricker *et al.* (1988) have been able to explain the effect of CCK on food intake in a plausible manner. They noted in other experiments that gastric distention in rats produces only small increases in the secretion of OT. On the other hand, the effects of distention were considerably augmented by simultaneous administration of CCK and rapidly caused increases in the levels of OT comparable to those seen after large doses of lithium chloride, copper sulfate, or apomorphine. From these results, the above authors conclude that "CCK acts to inhibit food intake mainly by stimulating CCK receptors on the gastric vagus, thereby amplifying the effects of gastric distention and mimicking the effects of a distressingly large meal or a very pronounced distention." It might seem more direct to answer the question concerning the felt effects of CCK by administering it to people. It would seem to be a simple matter to ask human subjects whether they felt malaise or satiety after CCK. The best we can do using rats as subjects is to draw inferences about their sensory experiences. Experiments on human subjects have been conducted (Kissileff, Pi-Sunyer, Thornton, & Smith, 1981; Pi-Sunyer, Kissileff, Thornton, & Smith, 1982; Stacher, Steinringer, Schmierer, Schneider, & Winklehner, 1982; Shaw *et al.*, 1985), but these are beset with methodological problems, not the least of which are the uncalibrated descriptive habits of the human subjects. Malaise is often reported, but does it correlate sufficiently with food intake reduction to imply a causal relationship? The most convincing study so far (Miaskiewicz, Stricker, & Verbalis, 1989) shows that human subjects report nausea and show emesis when administered doses are comparable to those that reduce food intake in rats. At somewhat lower doses, the subjects do not report nausea or show emesis, though a disinclination to eat is accompanied by the sensation of severe gastric cramps. As interesting as these subjective reports are, the study also documents that administration of CCK (even in very low doses, such as 0.05 mg/kg) stimulates neurohypophyseal secretion of AVP in humans (as it does in monkeys). Nausea is a potent stimulus of AVP secretion in humans and monkeys (Rowe, Shelton, Helderman, Vistal, & Robertson, 1979; Verbalis, Richardson, & Stricker, 1987). On the other hand, the consumption of a large meal to satiety produced no elevation of AVP. This evidence again shows that the effects of exogenous CCK are not the same as those of naturally evoked satiety but that they resemble those of emetic agents (e.g., Moore & Deutsch, 1985).

It seems then that CCK is not a satiety signal. We have seen above that whatever it is that is a satiety signal proceeds from the stomach and not the duodenum. On the other hand, CCK is not generated by the stomach but the duodenum. Further section of the afferent vagus abolishes the effect of CCK. First, this shows that CCK is not used as a signal to carry information directly to the cental nervous system. Second, it has been shown that satiety information reaches the brain in spite of vagotomy. It follows that CCK, when it produces signals in the vagus, is not producing satiety signals. The only plausible evidence that CCK is a satiety agent is that it causes a brief and relatively small reduction in food intake. But such a reduction is not a unique property of satiety-inducing substances. Such a reduction is also consistent with production of a brief, relatively mild malaise. And it is this second possibility that is supported by the evidence. CCK administration produces CTA to neutral stimuli with which it has been paired. The argument that malaise and satiety are in many ways the same and therefore the conditioned aversion test is inconclusive is controverted by two pieces of evidence. Known satiety-producing stimuli in the same paradigm induce preference and not aversion. Second, a drug that reduces malaise does not lead to lessened satiety. At a more physiological level, it has been shown in numerous studies that the amount of CCK used to show "satiety" is very much larger than the amount that is secreted physiologically. Also proglumide, a CCK blocker, does not increase food intake as the hypothesis that CCK is a satiety agent would predict. Finally, CCK triggers a release of OT in rats and AVP in humans, characteristic of sickness-producing agents and different from that evoked by satiety consequent on normal eating.

CONCLUSION

An interesting picture has emerged from research on the function of the stomach in meal size regulation. Signals for the termination of a meal of familiar food emanate from the stomach and not the duodenum. Such signals are of two kinds. The first are generated by bulk but only begin to function at relatively high levels of volume, presumably acting to prevent damage. The second, chemical in nature, function at lower levels, and they are produced by the food itself or by the partial breakdown products of digestion. Such gastric chemical stimuli come to function as regulatory signals by a process of learning. Before such learning takes place, and when the food is still unfamiliar, regulation is not gastric but oropharyngeal. The gastric volume signals are conveyed to the central nervous system by the vagus, whereas the chemical signals may travel via the splanchnic nerve. Many problems still remain unresolved. We do not know how many different types of chemical signal emanate from the stomach that can be used to regulate intake. We must still find the receptors that generate such signals. Further, much work remains to be done on the relationship between oropharyngeal and gastric metering and on the way that regulatory signals reach the central nervous system. To end on a note of caution, it should be observed that most of the recent evidence concerning the physiological basis of meal size regulation has been obtained on the rat. The generality of such findings remains to be determined.

REFERENCES

Antin, J., Gibbs, J., Holt, J., Young, R. C., & Smith, G. P. (1975). Cholecystokinin elicits the complete behavioural sequence of satiety in rats. *Journal of Comparative and Physiological Psychology, 89,* 784–790.

Balagura, S., & Fibiger, H. C. (1968). Tube feeding: Intestinal factors in gastric loading. *Psychonomic Science, 10,* 373–374.

Booth, D. A. (1972). Conditioned satiety in the rat. *Journal of Comparative and Physiological Psychology, 81,* 457–471.

Booth, D. A., & Davis, J. D. (1973). Gastrointestinal factors in the acquisition of oral sensory control of satiation. *Physiology and Behavior, 11,* 23–29.

Booth, D. A., Lovett, D., & McSherry, M. (1972). Postingestive modulation of the sweetness preference gradient in the rat. *Journal of Comparative and Physiological Psychology, 78,* 485–512.

Campbell, C. S. & Davis, J. D. (1974). Licking rate of rats is reduced by intraduodenal and intraportal glucose infusion. *Physiology and Behavior, 12,* 357–365.

Cannon, W. G. & Lieb, C. W. (1911). The receptive relation of the stomach. *American Journal of Physiology, 27,* 267–273.

Cannon, W. B., & Washburn, A. L. (1912). An explanation of hunger. *American Journal of Physiology, 29,* 441–455.

Chey, W. Y., Hitanaut, S., Hendricks, J. & Lorber, S. H. (1970). Effect of secretin and cholecystokinin on gastric emptying and gastric secretion in man. *Gastroenterology, 58*, 820–827.

Chvasta, T., Weisbrodt, N. W., & Cooke, A. R. (1971). The effect of pentagastrin and secretin on gastric emptying in the dog. *Clinical Research, 19*, 656.

Coil, J. D., Hankins, W. G., Jenden, D. J. & Garcia, J. (1978). The attenuation of a specific cue-to-consequence association by antiemetic agents. *Psychopharmacology, 56*, 21–25.

Collins, S., Walker, D., Forsyth, P., & Belbeck, L. (1983). The effects of proglumide on cholecystokinin-, bombesin- and glucagon-induced satiety in the rat. *Life Sciences, 32*, 2223–2229.

Cooke, A. R., Chvasta, T. E., & Weisbrodt, N. W. (1972). Effect of pentagastrin on emptying and electrical and motor activity of the dog stomach. *American Journal of Physiology, 223*, 934–938.

Crawley, J. N., Kiss, J. Z., & Mezey, E. (1984). Bilateral midbrain transections block the behavioral effects of cholecystokinin on feeding and exploration in the rat. *Brain Research, 322*, 316–321.

Davis, J. D., & Campbell, C. S. (1973). Peripheral control of meal size in the rat: Effect of sham feeding on meal size and drinking rate. *Journal of Comparative and Physiological Psychology, 83*, 379–387.

Davis, J. D., Gallagher, R. L., & Ladove, R. (1967). Food intake controlled by a blood factor. *Science, 156*, 1247–1248.

Davis, J. D., Gallagher, R. J., Ladove, R. F. & Turavsky, A. J. (1969). Inhibition of food intake by a humoral factor. *Journal of Comparative and Physiological Psychology, 67*, 407–414.

Deutsch, J. A. (1983). Dietary control and the stomach. *Progress in Neurobiology, 20*, 313–332.

Deutsch, J. A., & Ahn, S. J. (1986). The splanchnic nerve carries satiety signals. *Behavioral and Neural Biology, 45*, 43–47.

Deutsch, J. A., & Gonzalez, M. F. (1980). Gastric nutrient content signals satiety. *Behavioral and Neural Biology, 30*, 113–116.

Deutsch, J. A. & Gonzalez, M. F. (1981). Gastric fat content and satiety. *Physiology and Behavior, 26*, 673–676.

Deutsch, J. A., Gonzalez, M. F. & Young, W. G. (1980). Two factors control meal size. *Brain Research Bulletin, 5, (suppl. 4)*, 55–57.

Deutsch, J. A. & Hardy, W. T. (1977). Cholecystokinin produces bait shyness in rats. *Nature, 266*, 196.

Deutsch, J. A., Molina, F., & Puerto, A. (1976). Conditioned taste aversion caused by palatable nontoxic nutrients. *Behavioral Biology, 16*, 161–174.

Deutsch, J. A., & Parsons, S. L. (1981). Bombesin produces taste aversion in rats. *Behavioral and Neural Biology, 31*, 110–113.

Deutsch, J. A., Puerto, A., & Wang, M.-L. (1977). The pyloric sphincter and differential food preference. *Behavioral Biology, 19*, 543–547.

Deutsch, J. A., & Tabuena, O. O. (1986). Learning of gastrointestinal satiety signals. *Behavioral and Neural Biology, 45*, 282–299.

Deutsch, J. A., Thiel, T. R., & Greenberg, L. H. (1978). Duodenal motility after cholecystokinin injection or satiety. *Behavioral Biology, 24*, 393–399.

Deutsch, J. A., Young, W. G., & Kalogeris, T. J. (1978). The stomach signals satiety. *Science, 201*, 165–167.

Dourish, C. T., Coughlan, J., Hawley, D., Clark, M. L., & Iversen, S. D. (1988). In R. Y. Wang and R. Shoenfeld (Eds.) Cholecystokinin antagonists, New York: Liss 1988.

Dourish, C. T., Hawley, D., & Iversen, S. D. (1988). Enhancement of morphine analgesia and prevention of morphine tolerance in the rat by the cholecystokinin antagonist L-364, 718. *European Journal of Pharmacology, 147*, 469–472.

Dourish, C. T., Rycroft, W., & Iversen, S. D. (1989). Postponement of satiety by blockade of brain cholecystokinin (CCK-B) receptors. *Science, 245*, 1509–1511.

Dragoin, W., McCleary, G. E., & McCleary, P. (1971). A comparison of two methods of measuring conditioning taste aversions. *Behavioral Research Methods and Instruments, 3*, 309–310.

Edwards, G. L., Ladenheim, E. E., & Ritter, R. C. (1986). Dorsomedial hindbrain participation in cholecystokinin-induced satiety. *American Journal of Physiology, 252(5)*, R971–977.

Ehman, G. K., Albert, D. J., & Jamieson, J. C. (1971). A comparison of two methods of measuring conditioning taste aversions. *Behavioral Research Methods and Instruments, 3*, 309–310.

Garcia, J., & Ervin, F. R. (1968). Gustatory–visceral and telereceptor–cutaneous conditioning: Adaptation in external and internal milieus. *Communications in Behavioral Biology, 1*, 389–415.

Gibbs, J., Fauser, D. J., Rowe, E. A., Rolls, B. J., Rolls, E. T., & Maddison, S. P., (1979). Bombesin suppresses feeding in rats. Nature (London), *282*, 208–210.

Gibbs, J., & Smith, G. P. (1980). Bombesin—satiety or malaise? *Nature, 285*, 592.

Gibbs, J., Maddison, S. P., & Rolls, E. T. (1981). The satiety role of the small intestine examined in sham feeding rhesus monkeys. *Journal of Comparative and Physiological Psychology, 95,* 1003–1005.

Gonzalez, M. F., & Deutsch, J. A. (1981). Vagotomy abolishes cues of satiety produced by gastric distention. *Science, 212,* 1283–1284.

Gonzalez, M. F., & Deutsch, J. A. (1985). Intragastric injections of partially digested triglyceride suppress feeding in the rat. *Physiology and Behaviour, 35,* 861–865.

Grossman, M. I. (1955). Integration of current views on regulation of hunger and appetite. *Annals of the New York Academy of Sciences, 63,* 76–91.

Grote, F. W., & Brown, R. T. (1971). Conditioned taste aversions: Two-stimulus tests are more sensitive than one-stimulus tests. *Behavioral Research Methods and Instruments, 3,* 311–312.

Harper, A. E., & Spivey, H. E. (1958). Relationship between food intake and osmotic effect of dietary carbohydrates. *American Journal of Physiology, 193,* 483–487.

Hewson, G., Leighton, G. E., Hill, R. G., & Hughes, J. (1988). The cholecystokinin receptor antagonist L364,718 increases food intake in the rat by attenuation of the action of endogenous cholecystokinin. *British Journal of Pharmacology, 93,* 79–84.

Holman, G. L. (1968). Intragastric reinforcement effect. *Journal of Comparative and Physiological Psychology, 69,* 432–441.

Holt, J., Antin, J., Gibbs, J., Young, R. C. & Smith, G. P. (1974). Cholecystokinin does not produce bait shyness in rats. *Physiology and Behavior, 12,* 497–498.

Houpt, K. A. (1982). Gastrointestinal factors in hunger and satiety. *Neuroscience and Biobehavioral Reviews, 6,* 145–164.

Houpt, T. R., Anika, S. N., & Houpt, K. A. (1979). Preabsorptive intestinal satiety controls food intake in pigs. *American Journal of Physiology, 236,* R328–R337.

Hull, C. L., Livingston, J. R., Rouse, R. O. & Barker, A. N. (1951). True, sham, and esophageal feeding as reinforcements. *Journal of Comparative and Physiological Psychology, 44,* 236–245.

Hunt, J. N., & Spurrell, W. R. (1951). The pattern of gastric emptying of the human stomach. *Journal of Physiology (London), 113,* 157–168.

Janowitz, H. D. & Grossman, M. I. (1949). Some factors affecting the food intake of normal dogs and of dogs with esophagostomy and gastric fistula. *American Journal of Physiology, 159,* 143–148.

Kissileff, H. R., Pi-Sunyer, F. X., Thornton, J., & Smith, G. P. (1981). Cholecystokinin-octapeptide (CCK-8) decreases food intake. *American Journal of Clinical Nutrition, 34,* 154–160.

Kraly, F. S. & Gibbs, J. (1980). Vagotomy fails to block the satiating effects of food in the stomach. *Physiology and Behavior, 24,* 1007–1010.

Kraly, F. S., & Smith, G. P. (1978). Combined pregastric and gastric stimulation by food is sufficient for normal meal size. *Physiology and Behavior, 21,* 405–408.

LeMagnen, J. (1967). Habits and food intake. In *Handbook of physiology, Section 6, Vol. 1, The alimentary canal.* Bethesda: American Physiological Society.

Liebling, D. S., Eisner, J. D., Gibbs, J. & Smith, G. P. (1985). Intestinal satiety in rats. *Journal of Comparative and Physiological Psychology, 89,* 955–965.

Maier, S. F., Zahorik, D. M., & Albin, R. W. (1971). Relative novelty of solid and liquid diet during thiamine deficiency determining development of thiamine-specific hunger. *Journal of Comparative and Physiological Psychology, 74,* 254–262.

Malhotra, S. L. (1967). Effect of saliva on gastric emptying. *American Journal of Physiology, 213,* 169–173.

Mallison, C. N. (1968). Effect of pancreatic insufficiency and intestinal lactase deficiency on the gastric emptying of starch and lactose. *Gut, 9,* 737.

McHugh, P. R. (1979). Aspects of the control of feeding: Application quantitation in psychobiology. *Johns Hopkins Medical Journal, 144,* 147–155.

McHugh, P. R. & Moran, T. H. (1986a). The inhibition of feeding produced by direct intraintestinal infusion of glucose: Is this satiety? *Brain Research Bulletin, 17,* 415–418.

McHugh, P. R., & Moran, T. H. (1986b). The stomach and satiety. In *The interaction of the chemical senses with nutrition,* (pp. 167–180). Orlando, FL: Academic Press.

McHugh, P. R., Moran, T. H., & Wirth, J. B. (1982). Postpyloric regulation of gastric emptying in rhesus monkeys. *American Journal of Physiology, 243,* R408–R415.

Mei, N. (1983). Sensory structures in the viscera. In H. Autrim (Eds.), *Progress in Sensory Physiology,* (ed. 4, pp. 1–42). Berlin: Springer Verlag.

Miaskiewicz, S. L., Stricker, E. M., & Verbalis, J. G. (1989). Neurohypophyseal secretion in response to cholecystokinin but not meal-induced gastric distention in humans. *Journal of Clinical Endocrinology and Metabolism, 68,* 837–843.

Miller, N. E., & Kessen, M. L. (1952). Reward effects of food via stomach fistula compared with those of food via mouth. *Journal of Comparative and Physiological Psychology, 45,* 555–564.

Molina, F., Thiel, T., Deutsch, J. A., & Puerto, A. (1977). Comparison between some digestive processes after eating and gastric loading in rats. *Pharmacology, Biochemistry and Behavior, 7,* 347–350.

Mook, D. G. (1963). Oral and postingestional determinants of the intake of various solution in rats with esophageal fistulas. Journal of Comparative and Physiological Psychology, 56, 645–659.

Mook, D. G., Bryner, C. A., Rainey, L. D. & Wall, C. L. (1980). Release of feeding by the sweet taste in rats: Oropharyngeal satiety. *Appetite, 1,* 299–315.

Moore, B. O., & Deutsch, J. A. (1985). An antiemetic is antidotal to the satiety effects of cholecystokinin. *Nature, 315,* 321–322.

Moran, T. H., & McHugh, P. R. (1981). Distinctions amongst three sugars in their effects on gastric emptying and satiety. *American Journal of Physiology, 241,* R25–R30.

Morley, J. E., Bartness, T. J., Gosnell, B. A., & Levine, A. S. (1985). Peptidergic regulation of feeding. *International Review of Neurobiology, 27,* 207–298.

Nachman, M. (1980). Learned taste and temperature aversions due to lithium chloride sickness after temporal delays. *Journal of Comparative and Physiological Psychology, 73,* 22–30.

Opsahl, C. A. (1977). Sympathetic nervous system involvement in the lateral hypothalamic lesion syndrome. *American Journal of Physiology, 232,* R128–R136.

Paintal, A. S. (1973). Vagal sensory receptors and their reflex effects. *Physiological Review, 53,* 159–227.

Pappas, T. N., Melendez, R. L., Strah, K. M., & Debas, H. T. (1985). Cholecystokinin is not a peripheral satiety signal in the dog. *American Journal of Physiology, 249(6),* G733–738.

Pavlov, I. P. (1910). *The work of the digestive glands.* London: Griffin.

Pi-Sunyer, F. X., Kissileff, H. R., Thornton, J. & Smith, G. P. (1982). C-terminal octapeptide of cholecystokinin decreases food intake in obese men. *Physiology and Behavior, 29,* 627–630.

Plucinski, T. M., Hamosh, M., & Hamosh, P. (1979). Fat digestion in rat: Role of lingual lipase. *American Journal of Physiology, 237(6),* E541–E547.

Powley, T. L. (1977). The ventromedial hypothalamic syndrome, satiety, and a cephalic phase hypothesis. *Psychological Review, 84,* 89–126.

Puerto, A., Deutsch, J. A., Molina, F., & Roll, P. L. (1976). Rapid discrimination of rewarding nutrient by the upper gastrointestinal tract. *Science, 192,* 485–487.

Reidelberger, R. D., Kalogeris, T. J., Leung, P. M. & Mendel, V. E. (1983). Postgastric satiety in the sham-feeding rat. *American Journal of Physiology, 244:* R872–R881.

Revusky, S. H., & Bedarf, E. W. (1967). Association of illness with prior ingestion of novel foods. *Science, 155,* 219–220.

Revusky, S. H., & Garcia, J. (1970). Learned associations over long delays. In G. H. Bower & J. T. Spence (Eds.),:*The psychology of learning and motivation: Advances in research and theory IV* (pp. 1–85). New York: Academic Press.

Revusky, S. H., Smith, S. M. H., Jr., & Chalmer, D. V. (1971). Flavor preference: Effects of ingestion-contingent intravenous saline or glucose. *Physiology and Behavior, 6,* 341–343.

Rezek, M., Schneider, K., & Novin, D. (1975). Regulation of food intake after vagotomy, coeliactomy and a combination of both procedures. *Physiology and Behavior, 15,* 517–522.

Roberts, K. (1967). The dumping syndrome. In J. M. Thompson, D. Berkowitz, & E. Polish (Eds.) *The stomach,* New York:

Rowe, J. W., Shelton, R. L., Helderman, J. H., Vistal, R. E., & Robertson, G. L. (1979). Influence of the emetic reflex on vasopressin release in man. *Kidney International, 16,* 729–735.

Rozin, P. (1969). Central or peripheral mediation of learning with long CS–US intervals in the feeding system. *Journal of Comparative and Physiological Psychology, 67,* 421–429.

Sack, R. L. (1977). Side effects of and adverse reactions to psychotropic mediations. In J. D. Barchas, (Eds.), *Psychopharmacology, from theory to practice,* (pp. 276–288). New York: Oxford University Press.

Sarles, H., Dani, R., Prezelin, G., Souville, C., & Figarella, C. (1968). Cephalic phase of pancreatic secretion in man. *Gut, 9,* 214–221.

Savory, C. J. (1987). An alternative explanation for apparent satiating properties of peripherally administered bombesin and cholecystokinin in domestic fowls. *Physiology and Behavior, 39(2),* 191–202.

Schneider, L. H., Gibbs, J., & Smith, G. P. (1986). Proglumide fails to increase food intake after an ingested preload. *Peptides, 7,* 135–140.

Sclafani, A., & Nissenbaum, J. W. (1985a). Is gastric sham feeding really sham feeding? *American Journal of Physiology, 248*, R387–R390.

Sclafani, A., & Nissenbaum, J. W. (1985b). On the role of mouth and gut in the control of saccharin and sugar intake: A reexamination of the sham feeding preparation. *Brain Research Bulletin, 14*, 569–576.

Share, I., Martiniuk, E., & Grossman, M. I. (1952). Effect of prolonged intragastric feeding on oral food intake in dogs. *American Journal of Physiology, 169*, 229–235.

Shaw, M. J., Hughes, J. J., Morley, J. E., Levine, A. S., Silvis, S. E., & Shafer, R. B. (1985). Cholecystokinin octapeptide action on gastric emptying and food intake in normal and vagotomized man. *Annals of the New York Academy of Sciences, 48*, 640–641.

Shillabeer, G., & Davison, J. S. (1984). The cholecystokinin antagonist, proglumide, increases food intake in the rat. *Regulatory Peptides, 8*, 171–176.

Smith, G. P., & Epstein, A. N. (1970). Increased feeding in response to decreased glucose utilization in the rat and monkey. *American Journal of Physiology, 217*, 1083–1087.

Smith, G. P., Gibbs, J., & Kulkosky, P. J. (1982). Relationships between brain–gut peptides and neurons in the control of food intake. In B. G. Hoebel & D. Novin (Eds.) *The neural basis of feeding and reward* (pp. 149–165). Brunswick, ME: Haer Institute.

Smith, G. P., Jerome, C. & Norgren, R. (1985). Afferent axons in abdominal vagus mediate satiety effect of cholecystokinin in rats. *American Journal of Physiology, 249(5)*, R638–641.

Smith, G. T., Moran, T. H., Coyle, J. T., Kuhar, M. J., O'Donahue, T. L., & McHugh, P. R. (1984). Anatomical localization of cholecystokinin receptors to the pyloric sphincter. *American Journal of Physiology, 246*, R127–R130.

Snowdon, C. T. (1970). Gastrointestinal sensory and motor control of food intake. *Journal of Comparative and Physiological Psychology, 71*, 68–76.

Snowdon, C. T. (1975). Production of satiety with small intraduodenal infusion in the rat. *Journal of Comparative and Physiological Psychology, 88*, 231–239.

Stacher, G., Steinringer, H., Schmierer, G., Schneider, C., & Winklehner, S. (1982). Cholecystokinin octapeptide decreases intake of solid food in man. *Peptides, 3*, 133–136.

Stricker, E. M., McCann, M. J., Flanagan, L. M., & Verbalis, J. G. (1988). Neurohypophyseal secretion and gastric function: Biological correlates of nausea. In H. Takagi, Y. Oomura, M. Ito, & M. Otsuka (Eds.) *Biowarning systems in the brain* (pp. 293–305). Tokyo: University of Tokyo Press.

Swerdlow, N. R., Van der Kooy, D., Koob, G. F., & Wenger, J. R. (1983). Cholecystokinin produces conditioned place-aversions, not place preferences, in food-deprived rats: Evidence against involvement in satiety. *Life Sciences, 32*, 2087–2093.

Vanderweele, D. A., Novin, D., Rezek, M., & Sanderson, J. D. (1974). Duodenal or hepatic-portal glucose perfusion: Evidence for duodenally based satiety. *Physiology and Behavior, 12*, 467–473.

Verbalis, J. G., McCann, M. J., McHale, C. M., & Stricker, E. M. (1986). Oxytocin secretion in response to cholecystokinin and food: Differentiation of nausea from satiety. *Science, 232*, 1417–1419.

Verbalis, J. G., Richardson, D. W., & Stricker, E. M. (1987). Vasopressin release in response to nausea-producing agents and cholecystokinin in monkeys. *American Journal of Physiology, 252*, R749–R753.

Watkins, L. R., Kinscheck, I. B., & Mayer, D. J. (1984). Potentiation of opiate analgesia and apparent reversal of morphine tolerance. *Science, 224*, 395–396.

Wiepkema, P. R., Alingh Prins, A. J., & Steffens, A. B. (1972). Gastrointestinal food transport in relation to meal occurrence in rats. *Physiology and Behavior, 9*, 759–763.

Yalow, R. S., Eng, J., & Straus, E. (1983). The role of CCK-like peptides in appetite regulation. *Advances in Metabolic Disorders, 10*, 435–456.

Young, W. G., & Deutsch, J. A. (1980). Intragastric pressure and receptive relaxation in the rat. *Physiology and Behavior, 25*, 973–975.

Young, W. G., Deutsch, J. A., & Tom, T. D. (1982). Diazepam reverses bombesin-induced gastric spasms and food intake reduction in the rat. *Behavioral Brain Research, 4*, 401–410.

8

Systemic Factors in the Control of Food Intake
Evidence for Patterns as Signals

L. Arthur Campfield and Francoise J. Smith

Molecules, Stores, and Patterns: What is the Form of Critical Signals?

The study of the control of food intake has addressed a variety of major and minor issues over the years. The following questions have been among the major issues: What factors are responsible for the initiation of feeding? What factors are responsible for the size of individual meals? What factors are responsible for the termination of meals? What factors are responsible for the duration of the intermeal interval? What factors match food intake to energetic demands? Each of these has been further subdivided into classes of factors (e.g., sensory, gastrointestinal, metabolites, hormonal, neural), site of origin or action, and metabolic states. The dominant theoretical approach to each of these questions has been the so-called "depletion/repletion" model. In this construct, a critical variable, usually associated with the magnitude of a specific "store," will cause ingestive behavior when it is reduced below an implicit threshold. Feeding behavior will cease when the level of this variable has returned to the "defended" level. This model was derived from a direct translation of the negative feedback control system to the physiological problem of the regulation of energy balance. Thus, changes in carbohydrate, fat, or/and protein body stores were postulated to lead to appropriate changes in food intake to return the regulated store or stores to the desired level. A large experimental literature has accumulated based on this theoretical approach.

L. Arthur Campfield and Francoise J. Smith Neurobiology and Obesity Research, Hoffmann-La Roche, Inc., Nutley, New Jersey 07110.

Although the body fat store was fairly easy to estimate and manipulate, carbohydrate and protein stores were not. The search for a representation of body energy stores in the plasma led to another major theoretical construct in the field, the regulation of blood levels of metabolic substrates. The most important example was the glucostatic theory of Mayer (1953, 1955). This theory held that the rate of glucose utilization in a privileged brain region controlled food intake. Blood glucose concentration and/or its anterial–venous difference was used to infer the rates of glucose utilization. Thus, hypoglycemia and its associated decrease in glucose utilization led to initiation of feeding, whereas postprandial hyperglycemia resulted in cessation of feeding and a period of satiety. Although the properties and special position of glucose in terms of brain metabolism were central points in the arguments for the glucostatic theory, Mayer's focus remained on glucose utilization and its relation to body carbohydrate stores. However, as the theory spread throughout the field, much of the experimental focus and debate became centered on blood glucose concentrations rather than on rates of glucose utilization in an unknown brain region.

This shift in focus led many in the field to consider the glucose molecule itself to be the signal for the control of food intake. This was consistent with the trend in the field to consider molecules that circulate in the blood (e.g. glucose, insulin, glucagon, CCK, gastrin, catecholamines) to be candidate control signals for the regulation of food intake. This concept emerged from endocrinology and metabolic physiology and gave the field a much longer list of candidate signals. Many of these signals could be accounted for or associated with the glucostatic, aminostatic, and lipostatic theories of food intake control. However, other signals required or emerged from the renewed interest in gastric and intestinal mechanisms in the digestion and absorption of nutrients as well as the control of food intake. The importance of gut factors in the process of satiation lead to important theoretical constructs advanced by Gibbs, Young, and Smith (1973) and Moran and McHugh (1982) that focused attention on molecules arising from the gut in the control of feeding. At about the same time, the hypothesis of Woods and Porte (1976) that brain insulin concentrations provided integrated information about body fat stores to the CNS also reinforced the search for molecules as important signals in food intake.

Two other developments that also contributed to the "molecule as signal" paradigm in this field were the discovery and elucidation of brain–gut peptides that had powerful and often dramatic effects on food intake and the striking advances achieved in receptor biochemistry and molecular biology. In recent years, the effects of brain and/or gut peptides, peptide fragments, mixtures of peptides, and peptide receptor antagonists on food intake have been studied. Also, studies of neuro- or regulatory peptides have awakened interest in the classical neurotransmitters, and the linkage between molecular signals and body stores, common to earlier work, is now often nonexistent. Thus, we study molecules *per se* and construct theories of feeding control based on molecules: the insulin, CCK, and serotonin hypotheses.

Recently, attention has been focused on behavioral sequences or patterns. Well-established orosensory motor patterns have been described by Grill and Norgren (1978). Thus, complex motor programs underlying feeding reside in the CNS, and signals related to body stores or specific molecules may "trigger" them under appropriate circumstances. An alternative theoretical construct has emerged from the consideration of sequences in which a complex temporal and/or spatial pattern fulfills the role of initiator of these motor programs. Thus, the temporal pattern of a specific molecule (e.g., transient decline in blood glucose), the pattern of several molecules (e.g., nutrient flux across the intestine), the temporal pattern of a specific molecule in a specific context of other patterns (e.g., insulin pre-, during, or postmeal) or the spatial pattern caused by the passage of a specific molecule through body compartments (e.g, glucose interacting with multiple glucose-receptive neural elements) may act as control signals with which the CNS organizes feeding behavior. Within this conceptual framework, it is not the store or molecule but rather the pattern that conveys critical information to the CNS. The demonstration of the utility of this conceptual framework, its ability to link feeding behavior tightly to physiology, and its potential to refocus our field on the message rather than the messenger form the focus of this chapter. This point of view also has the potential to synthesize much of our field because, rather than debating the merits of transient declines in blood glucose or hypothalamic norepinephrine or hindbrain CCK, we can ask how all of these components or elements of the pattern are integrated to elicit specific behaviors.

Another key element of the pattern-as-signal concept is the notion of representation of peripheral events and states within the CNS. The CNS may contain "representations" of metabolic and possibly behavioral states—absorptive and postabsorptive states, hunger, and satiety. Although often thought of in terms of spatial maps, transformations, or homunculi, representation in this context is considered to be a dynamic pattern of activity in a set of neurons that function as a central analogue of sensory, somatic, or visceral events. Thus, we seek these representations of peripheral body stores, metabolic state, and the external world and how these representations interact to control feeding behavior. What form would these representations take? Based on our knowledge of the nervous system, and by analogy to other brain systems involved in sensorimotor integration, these representations would probably be patterns of electrical and/or neurochemical activity of one or more neural networks controlling feeding. Thus, in this theoretical construct, patterns in the periphery carry information that is detected by central neural networks, which results in modified patterns of activity of these neural networks. Specific patterns of activity of these networks are postulated to correspond to the behavioral states associated with meal initiation and meal termination. Transition from the representation corresponding to satiety during intermeal intervals to the representation corresponding to meal initiation within the CNS will cause the onset of feeding. Meal termination will occur when the representation corresponding to satiety is restored by reversal of the meal initiation pattern activity by the complementary pattern corresponding to satiation.

If we could capture a visual image of the patterns of activity of the feeding network before, during, and after a meal, we would see a transition from the pattern characteristic of the intermeal interval to that of meal initiation and food ingestion followed by a transition back to the pattern corresponding to satiety. Thus, activity patterns corresponding to two active processes, meal initiation and satiation, will be superimposed on the hypothesized passive or low-activity steady state corresponding to satiety each time an animal eats a meal. In this theoretical construct, the components of a behavioral sequence become distinct patterns of activity in discrete neuronal networks and are linked to the underlying physiological and biochemical dynamics of these networks. Mechanistic decomposition of the behavioral sequences of feeding will require the identification and characterization of at least the major components of the key elements of that sequence: initiation, maintenance, and termination. Perturbation or shaping of feeding behavior would be then translated into modulation of these complex activity patterns and their integration by the brain.

In this chapter, attention is focused on such a pattern that has important information content in the postgastric control of feeding: the transient decline in blood glucose and its role in signaling meal initiation.

THE PATTERN OF BLOOD GLUCOSE DYNAMICS THAT SIGNALS MEAL INITIATION

Glucose uptake and utilization has been a central feature of many hypotheses for meal initiation since first suggested by Carlson (1916) and developed by Mayer (1953, 1955) into the glucostatic theory. Mayer proposed that decreased glucose utilization, which was detected by the brain at glucosensitive sites, represented the stimulus for meal initiation. Numerous experimental studies emphasize the role of decreased glucose utilization or decreased intracellular glucose concentrations rather than the absolute level of blood glucose as the stimulus for meal initiation. The observed induction of feeding by administration of pharmacological doses of insulin (Booth & Pitt, 1968; Epstein & Teitelbaum, 1967; Larue-Achagiotis & LeMagnen, 1979; MacKay, Callaway, & Barnes, 1940; Morgan & Morgan, 1940) or of nonmetabolizable glucose analogues (Likuski, Debons, & Cloutier, 1967; Smith & Epstein, 1969; Stricker & Rowland, 1978), the satiating effects of small glucose infusions or gastric loads (Novin, 1976; Novin, VanderWeele, & Rezek, 1975; Novin, Gonzalez, & Sanderson, 1976), and effects of central injections of glucose and 2-deoxyglucose (Booth, 1968; Miselis & Epstein, 1975) all strongly suggest a role for decreased glucose uptake and utilization, possibly modulated by insulin, at a target site or sites in the control of meal initiation. However, other experimental results appeared inconsistent with the glucostatic theory. When intravenous glucose infusions with or without insulin were administered prior to meals, no delay in meal initiation or reduction in meal size was observed (Bernstein & Grossman, 1956; Grinker, Cohn, & Hirsch, 1971; Nicolaidis & Rowland, 1976). Furthermore, the observations that large, prolonged decreases in blood glucose were required to induce feeding following insulin administration and that the onset of feeding often occurred when the blood glucose had returned to baseline have also been used as evidence against the glucostatic theory (Danguir & Nicolaidis, 1980; Smith, Gibbs, Strohmayer, & Stukes, 1972; Steffens, 1969).

More recently, a signal for the initiation of freely taken meals in rats with continuous access to familiar food has been identified: a brief fall and rise in blood glucose concentration. The identification of this signal was a direct result of a series of technological innovations leading to computer-based continuous monitoring of blood glucose in awake rats (Steffens, 1969; Nicolaidis, Rowland, Meile, Marfaing-Jallat, & Pesez, 1974; Louis-Sylvestre & LeMagnen, 1980; Campfield, Brandon, & Smith, 1985). The ability to monitor blood glucose continuously in freely behaving animals led to renewed consideration of the role of blood glucose in meal initiation. Transient declines in blood glucose were first described by Louis-Sylvestre and LeMagnen (1980), who showed that a fall in blood glucose was correlated with meal initiation in the rat. They observed that blood glucose concentration declined 6% to 8% 5.0 ± 0.3 minutes prior to meal onset in the dark and light phases. We have confirmed and extended these findings and now propose that transient declines in blood glucose of the correct magnitude and time course are signals for meal initiation that are detected by peripheral and central glucoreceptive elements and mapped into feeding behavior. The evidence we have accumulated in our laboratories over the past 5 years to support this assertion is reviewed.

DESCRIPTION OF THE SIGNAL: TRANSIENT DECLINES IN BLOOD GLUCOSE

Blood glucose concentration and meal pattern were continuously monitored in rats feeding freely using methods described previously in detail (Campfield *et al.*, 1985). Since all of the studies reviewed have utilized these same techniques, and the evidence supporting a role for glucose in meal initiation is critically dependent on these technological innovations, it is important that the reader have a clear understanding of these methods. Thus, they will be briefly described.

Adult female Wistar rats were housed in individual cages with free access to powdered rat chow and tap water. Powdered rat chow was placed in a food cup fitted with strain gauge weighing apparatus. Animals were kept in a temperature-controlled room with 12 : 12-hour light–dark schedule.

Rats were implanted with chronic cardiac cannulae. Following a 7-day recovery period characterized by resumption of consistent gain in body weight and normal meal pattern, experimental studies were conducted in awake rats in their home cages. Heparin (200 U) was injected intravenously and, 45 minutes later, blood withdrawal (25 μL/min) for continuous blood glucose monitoring was initiated. Blood was withdrawn from freely behaving rats through polyethylene tubing attached to the cardiac cannula and injected into the sample chamber of a glucose analyzer (YSI Model 23A). The analogue outputs from both the food cup and the glucose analyzer were sampled 8–10 times a minute, amplified, digitized, and interfaced to a microcomputer. Blood glucose monitoring was continued for up to 3 hours. Experiments have been conducted throughout the light–dark cycle.

When the blood glucose concentration and meal pattern were monitored continuously in free-feeding rats, a transient fall and rise in blood glucose was observed before each meal independent of the light–dark cycle. The average time course of blood glucose in nine early experiments is shown in Figure 1. The blood glucose concentration was expressed as percentage change from baseline; the zero time reference point was chosen as the minimum blood glucose, and the data points were averaged at 1-minute increments before and after the nadir. During an average decline, blood glucose concentration fell gradually to 11.6 ± 1.2% below baseline. Blood glucose concentration began to decline 12.1 ± 1.7 minutes before the onset of food intake and continued to decline to a minimum at 5.4 ± 1.5 minutes before meal onset. Note that these average times were measured from the beginning of the meal as opposed to the minimum glucose concentration, as shown in Figure 1. Correlations between meal size and percentage maximum decline in blood glucose concentration and the total duration of the decline were both nonsignificant. Thus, transient declines in blood glucose predict meal initiation but not meal size.

L. ARTHUR
CAMPFIELD AND
FRANCOISE J.
SMITH

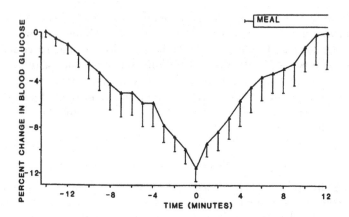

Figure 1. Average waveform of the transient decline in blood glucose. In this figure, the minimum glucose concentration was used as the time-zero reference. The data are expressed as percentage change from baseline. This figure provides a representation of the shape of the decline in blood glucose independent of the time of meal initiation. Data are mean ± S.E.M. for $N = 9$. (Reprinted from Campfield *et al.*, 1985, with permission.)

Random fluctuations in blood glucose occurred in these experiments that were not related to feeding. Analysis of the lower limits of transient declines in blood glucose that were associated with meal initiation indicated that blood glucose decreases with magnitudes greater than 6% and durations of more than 6 minutes invariably preceded food intake. When these criteria are applied to what by now has become a very large data set (approximately 225 experiments), we conclude that no such transient decline in blood glucose has been observed in the absence of food intake, nor has meal initiation been observed in the absence of such transient decline in blood glucose concentration when feeding in response to deprivation.

These findings are consistent with and confirm the original observations of a premeal decline in blood glucose by Louis-Sylvestre and LeMagnen (1980). They also are consistent with the decrease in *metabolisme de fond* before meal initiation reported by Nicolaidis and Even (1984). These results also permit the reevaluation of the reports that meal initiation following insulin administration occurred when blood glucose was near or at baseline levels (Danguir & Nicolaidis, 1980; Smith *et al.*, 1972; Steffens, 1969). Although these results have been used to argue against the glucostatic theory of Mayer (1953, 1955), the temporal relationship between blood glucose and meal initiation is remarkably similar following both intravenous insulin and spontaneous transient declines in blood glucose.

NATURE OF THE SIGNAL

Having documented the temporal relationship between transient declines in blood glucose and meal initiation in free-feeding conditions, we next examined the nature of this systemic signal for meal initiation, its magnitude or strength, and the time course of the functional coupling between blood glucose dynamics and feeding behavior. These questions were addressed by preventing access to food before and during declines in blood glucose, restoring access to food at various times following the beginning of the decline in blood glucose, and measuring the latency to feeding and the next decline in blood glucose (Campfield & Smith, 1986a).

In nine experiments conducted across the light–dark transition, the food cup was covered before and during the transient fall of blood glucose and was uncovered after blood glucose had returned to the baseline concentration. The time course of blood glucose concentration before an expected meal, expressed as percentage change from baseline, in these experiments is compared in Figure 2 to similar experiments in which rats had free access to food. The time course of blood glucose concentration in the food-deprived and free-feeding groups was very similar. Food-seeking behavior (i.e., orienting and/or moving to the food cup, sniffing, trying to remove the cover) occurred with a latency of 10.1 ± 1.6 minutes after the beginning of the fall in blood glucose and 3.4 ± 1.3 minutes after the nadir in glucose, values comparable to those seen in free-feeding animals. Food-seeking behavior was transient (less than 5 minutes) and ended as the glucose concentration increased toward baseline.

In four similar experiments, the cage and food cup were thoroughly cleaned prior to the experiment, and no food was placed in the food cup. The time course of blood glucose and the latency to food-seeking behavior observed were not different in the total absence of food. Thus, neither preventing access to food nor the absence of food affected the time course of the transient fall and rise of blood glucose before an expected meal and the latency to food-seeking behavior.

In these experiments, the food cup was uncovered 6 to 8 minutes after blood glucose returned to the baseline concentration, and the latency to meal initiation was measured. In all cases, food-seeking behavior ceased prior to un-

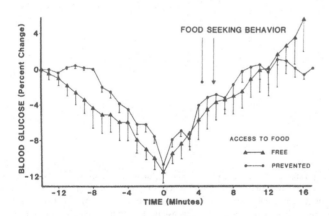

Figure 2. Average time course of the transient decline in blood glucose when access to food was free or prevented. Blood glucose concentrations are expressed as percentage change from the intermeal baseline concentrations in this figure. The minimum glucose concentration has been taken as the time-zero reference. Data were selected each minute from the reference point and averaged in each of nine experiments. Data are mean ± S.E.M. The onset of food-seeking behavior is indicated by the vertical arrows in free-access (filled triangles) and prevented-access (filled circles) groups. Mean baseline glucose concentrations were 104 ± 5 and 96 ± 3 mg/dL, respectively. Note the similar time course of blood glucose and onset of food-seeking behavior when food access was free and prevented. (Reprinted from Campfield & Smith, 1986a, with permission.)

covering the food cup. A composite of these experiments is shown in Figure 3. The data for the rats with deprived access to food from Figure 2 are repeated on the left, and the average time course of the second glucose transient and timing of meal initiation following restoration of access to food are shown on the right. In all these experiments, feeding was not observed immediately after the uncovering of the food cup but rather after a normal intermeal interval and a second transient fall in blood glucose and rise toward baseline. Meal initiation following the second decline in blood glucose occurred an average of 84 ± 8 minutes after the beginning of the first decline in blood glucose. This compares to the average intermeal for these rats at this phase of the photoperiod of 98 ± 7 minutes. The mean coefficient of variation of blood glucose concentration during the interdecline interval was 2.1 ± 0.5%. The size of the meal was 1.5 ± 0.3 g and was not different from meals eaten at this time by these rats in the free-access condition (2.0 ± 0.4 g).

In five other experiments, the food cup was uncovered 4–9 minutes before blood glucose returned to baseline. In these experiments, rats moved to the food cup and feeding began within 2 minutes following removal of the food cup cover. These experiments indicate that when access to food was restored as the blood glucose was rising toward baseline, meal initiation occurred as expected.

Combination of these studies and our studies reviewed above in the free-feeding condition (Campfield *et al.*, 1985) allowed calculation of the approximate temporal evolution of the functional coupling between blood glucose and

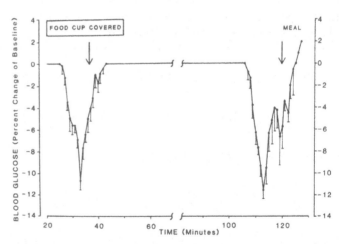

Figure 3. Composite representation of the blood glucose concentration in nine experiments in which access to food was prevented. Blood glucose concentrations are expressed as percentage change from intermeal baseline concentrations. Data are mean ± S.E.M. The experiment began with the food cup covered prior to and during a transient decline in blood glucose. Access to food was restored 6 to 8 minutes after the blood glucose concentration had returned to baseline. The left-hand portion of the figure repeats data presented in Figure 2. Note that time scale has been compressed in this figure. Coefficients of variation for blood glucose concentration in the predecline and interdecline averaged 2.1 ± 0.5%. The onset of food-seeking behavior (first decline) and meal initiation (second decline) are indicated by the vertical arrows. Mean baseline glucose concentrations prior to the first and second declines were 96 ± 3 and 97 ± 4 mg/dL, respectively. Note that following restoration of access to food, feeding did not occur until a delay comparable to an intermeal interval and a second transient decline in blood glucose. (Reprinted from Campfield & Smith, 1986a, with permission.)

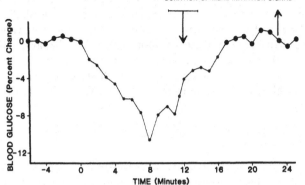

Figure 4. Composite time course of the functional coupling between transient decline in blood glucose and meal initiation. Average time course of blood glucose concentration in the prevented-access condition was expressed as percentage change from baseline concentrations and repeated from Figure 2. Data are mean values. The latest estimates of the onset and offset times for the meal initiation signal to be above threshold are indicated by downward and upward arrows, respectively. For details, see text. (Reprinted from Campfield & Smith, 1986a, with permission.)

meal initiation. The resulting composite time course is shown in Figure 4. The latency to meal initiation in the free-feeding condition (12.1 ± 1.7 minutes from the beginning of the decline in glucose) was taken to be the latest time that the glucose-dependent signal for meal initiation exceeded its threshold and coupled blood glucose to feeding behavior (downward arrow). Since meals were initiated with a short latency when access to food was restored while blood glucose was rising toward baseline, the meal initiation signal was considered to be above threshold throughout this period. The termination of the functional coupling was then taken as 6 minutes after blood glucose concentration returned to baseline because when food access was restored at this time or later, a normal inter-meal interval and a second transient decline in blood glucose preceded meal initiation (upward arrow). Therefore, the functional coupling between blood glucose and feeding behavior was of short duration (approximately 12 minutes) and persisted less than 6 minutes after the end of a blood glucose decline. The transient nature of this coupling required a second decline in blood glucose to initiate feeding when the rat was unable to eat within this narrow temporal window.

In 14 experiments, a novel food (orange slice, potato chip, or chocolate chip cookie) was presented 30 minutes after the food cup was uncovered. Access to powdered chow had been restored 8 minutes after the end of the transient decline in blood glucose. The novel food was eaten in seven experiments with an average latency of 2.5 ± 0.9 minutes following presentation, but without a prior decline in blood glucose. These studies demonstrate that novel foods with strong sensory qualities can be eaten without any prior changes in blood glucose concentration.

These studies demonstrate that the time course of the transient fall in blood glucose before expected meals and the latency to food-seeking behavior were not dependent on the presence of or access to familiar food. Rather, these results further emphasize the role of patterns of blood glucose dynamics, small transients in blood glucose concentration, as control signals, or the reflection of control signals, that may organize the onset of feeding. These results suggested that transients in blood glucose reflected an endogenous, metabolically dependent control signal that was functionally coupled to feeding behavior. This control signal resulted in a strong activation of food-seeking behavior, which led to meal initiation when food was available. Indeed, the strength of the functional coupling could account for the tight correlation between premeal transient declines in blood glucose and meal initiation previously observed in free-feeding rats (Louis-Sylvestre & LeMagnen, 1980; Campfield *et al.*, 1985).

EVIDENCE FOR CAUSALITY

Based on their initial studies, Louis-Sylvestre and LeMagnen (1980) concluded that blood glucose concentration was among the feedback signals in the regulation of feeding and body energy storage, and the observed decline in blood glucose concentration prior to meal onset was either the signal for meal initiation or a consequence of the true signal. In either case, a causal relationship between the decline in blood glucose concentration and feeding was proposed (LeMagnen, 1981). This proposal has proved to be controversial. Several investigators argued that the premeal decline in blood glucose concentration was correlated with rather than causally related to meal initiation (see commentaries to LeMagnen, 1981). Motivated by the controversy over the causality of the observed premeal decline in blood glucose, we have conducted a series of experiments designed to determine whether transient declines in blood glucose induce meal initiation.

Our first approach was to infuse glucose intravenously to partially block the premeal decline in blood glucose and assess the effect on subsequent feeding behavior. In seven experiments, 10% glucose (up to 0.2 mL) was infused over a 5-minute period beginning as soon as a fall in blood glucose before an expected meal was detected. Glucose infusions were administered each time a decline in glucose was detected. Isotonic saline also was administered in separate experiments to control for the nonspecific effects of infusions. The transient declines in blood glucose observed in these experiments are summarized in Figure 5.

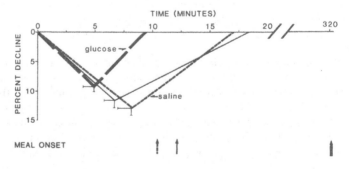

Figure 5. Effects of glucose clamp on premeal decline in blood glucose and latency to the anticipated meal. Comparison of the timing and magnitude of the maximum percentage decline and time of the return to baseline in experiments without infusions (solid line), with saline infusions (dotted line), and with glucose infusion (heavy dotted line). The arrows represent the time of onset of the next meal in each group. Note decreased decline duration and increased latency to meal initiation with glucose infusion. (Reprinted from Campfield *et al.*, 1985, with permission.)

Comparison of the results obtained indicates that although the initial rate of decline, the timing, and the magnitude of the glucose nadir were not significantly affected by the glucose infusions, the duration of the decline was decreased, and the latency to the anticipated meal was increased markedly (318 ± 94 minutes compared to 12.1 ± 1.7 minutes). It is important to note that 0.2 mL of 10% glucose contains at most 9% of the calories consumed in the smallest meal.

These studies suggest that it is the shape or pattern of the transient decline in blood glucose that affects meal initiation. Further support for this hypothesis was provided by the results of additional experiments in which glucose was infused during the rising phase at the end of the transient decline. In these experiments, the anticipated meal occurred with normal latency despite the glucose infusion (data not shown). These observations suggest that it is not glucose itself that uncouples the transient declines in blood glucose from meal initiation but rather a modification of the shape of the transient decline in blood glucose. Taken together these results offer an explanation for the previously reported failure of intravenous glucose infusions to delay the onset of feeding or reduce meal size (Danguir & Nicolaidis, 1980; Smith *et al.*, 1972; Steffens, 1969). Our studies demonstrate that in order to block meal initiation, a glucose infusion must occur during the early phase of the transient decline in blood glucose. Thus, meal initiation would not have been blocked if glucose had been infused either before the decline or towards its end (Campfield *et al.*, 1985).

Additional experiments were performed in which glycine, β-hydroxybutyrate, or fructose was infused intravenously instead of glucose. The effects of these infusions on the parameters of the transient decline and the latency to meal initiation are shown in Table 1. It can be seen that only glucose infusions blocked meal initiation. This result suggests that changes in blood glucose rather than any other nutrient provide the signal for meal initiation.

Additional supportive evidence was obtained from experiments in which fructose was infused intravenously in free-feeding rats to produce transient declines in blood glucose observed prior to meal initiation (Smith, Driscoll, & Campfield, 1988). Random sequences of fructose (doses range 0.05–0.2 mL of 10%) separated by at least 30 minutes were infused intravenously within 2 minutes during intermeal intervals. During the early dark phase, fructose was followed by slight decreases or increases in blood glucose (−4% to +10% at 6 minutes). In the light phase, however, three types of dose-dependent declines in blood glucose were observed as shown in Figure 6. The first pattern was a fall to a suppressed level (−6%) that was maintained for at least 30 minutes. The second pattern was a transient fall (−10% at 8 minutes) and return to baseline at 28 minutes. However, the third pattern mimicked the transient decline in blood glucose observed prior to meal initiation: a transient fall (−9% at 8 minutes) and return to baseline at 17 minutes. In these four experiments, meal initiation occurred with a latency within the normal range for spontaneous meals following transient declines in blood glucose. No feeding was observed following the other blood glucose response patterns induced by fructose infusion. The observation of feeding during a period of low probability for spontaneous meals only following a fall and rise in blood glucose induced by fructose that mimicked the transient decline in blood glucose provides further evidence that transients in blood glucose dynamics are a causal signal for meal initiation.

In order to evaluate the role of insulin in meal initiation, another series of

TABLE 1. PARAMETERS OF TRANSIENT DECLINES IN BLOOD GLUCOSE AND SUBSEQUENT MEALS[a]

Infusion administered	Transient decline in blood glucose				Meal	
	Maximum decline (%)	Time of nadir (min)	Total duration (min)	Percentage blockade (%)	Latency to next meal (min)	Meal size (g)
None (N = 9)	−11.6 ± 1.2	6.7 ± 0.9	18.5 ± 2.4	0	12.1 ± 1.7	1.3 ± 0.2
Saline (N = 6)	−12.8 ± 1.2	8.2 ± 1.2	17 ± 4	0	10.2 ± 3.4	1.2 ± 0.3
Glucose (10%) (N = 7)	−8.9 ± 0.9	5.0 ± 1.0	9.2 ± 1.8	46.5 ± 13.4*	318 ± 94*	1.4 ± 0.1
Glycine (4.2%) (N = 4)	−9.9 ± 0.5	9.1 ± 3.1	24 ± 5.3	0	14.6 ± 3.3	0.9 ± 0.3
β-Hydroxybutyrate (7%) (N = 3)	−8.7 ± 1.2	14.6 ± 4.7	28 ± 8.7	0	23.3 ± 5.2	1.2 ± 0.4
Fructose (10%) (N = 3)	−8.8 ± 3.4	8.2 ± 1.6	18.2 ± 1.2	0	13.3 ± 1.4	0.8 ± 0.1

[a]N = number of experiments; *$P < 0.5$. Data are mean ± S.E.M. Concentrations were chosen to maintain isotonicity with 10% glucose.

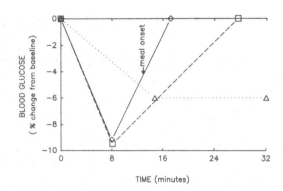

Figure 6. Response patterns in blood glucose following fructose during the light phase. Comparison of the timing and magnitude of the percentage decline and time of the return to baseline in experiments in which fructose was administered during an intermeal interval. Note three distinct response patterns, which were independent of the dose. The number of experiments was four in each group.

studies characterized the time course of plasma insulin concentrations before, during, and after meal initiation. In these studies, blood was continuously withdrawn from awake male and female Wistar rats at the rate of 25 μL/min and pooled over 4-minute intervals. Sampling began at least 40 minutes before an anticipated meal and was continued up to 100 minutes. Plasma insulin was stable during intermeal intervals and then briefly increased by about 50% before returning to basal levels. The peak of the insulin spike occurred 26 (males) to 13 (female) minutes before meal initiation and preceded the transient decline in blood glucose. In contrast, plasma insulin remained constant (coefficient of variation = 11%) throughout the sampling period in experiments in which feeding did not occur. Plasma insulin reached its lowest point before the meal in most experiments. These data suggest that a spike of insulin in the plasma, possibly vagally mediated, may play a role in the origin of the transient declines in blood glucose that precede meal initiation (Campfield & Smith, 1986c; Campfield, Smith, Driscoll & Spirt, 1988).

The observation of a transient spike of insulin before the transient decline in blood glucose suggested another way to determine whether that change was responsible for meal initiation. If a transient decline in blood glucose could be induced that mimicked the spontaneous premeal glucose transient, then meal initiation should result. Perhaps a brief release of insulin from the pancreas could be used to induce such a transient decline. Our previous studies with a peripherally acting acetylcholine analogue, carbamyl-β-methylcholine chloride (bethanechol), had shown that it caused a brief spike of insulin release that was very similar to the insulin spike observed before feeding (Kajinuma, Kaneto, Kuzuya, & Nakao, 1968; Campfield et al., 1988; Campfield & Smith, 1986c; Smith and Campfield, 1986a). Therefore, a series of 15 experiments were conducted in which various doses of acetylcholine analogue were infused intravenously in 16 female Wistar rats during the long intermeal interval in the light phase. Random sequences of up to four doses were infused at least 20 minutes apart until a fall in blood glucose, if any, was observed. In nine experiments, meal initiation occurred when analogue infusion was followed by a transient decline in blood glucose with a shape similar to that observed before meals. In six

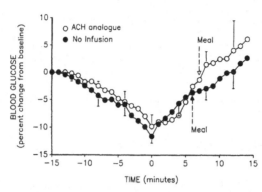

Figure 7. Transient declines in blood glucose and meal initiation induced by acetylcholine analogue infusion during the light phase. Blood glucose concentrations are expressed as percentage change from the intermeal baseline concentrations. The minimum glucose concentration has been taken as the time-zero reference in the ACh-induced (open symbols) and spontaneous (closed symbols) transient declines. Data are mean ± S.E.M., and N = 9.

other experiments, blood glucose changed by less than 5% after a smaller average dose of the analogue, and no feeding occurred.

Analysis of these data revealed a clear state dependence for meal initiation. The frequency of induction of a transient decline that mimicked the spontaneous premeal glucose transient averaged 28% in the early light phase (a period of low probability for meal initiation) and rose to 79% in the late light and light–dark transition (a period of high probability of feeding).

These studies demonstrate the experimental induction of meal initiation during the long intermeal intervals of the light phase in nondeprived rats, but only following a brief fall and rise in blood glucose that mimicked the spontaneous blood glucose transients prior to meal initiation (Figure 7). The latency of meal initiation following infusion of acetylcholine analogue corresponds to the temporal relationship between the spike in plasma insulin and meal initiation in free-feeding rats. These studies provide strong support for a causal relationship between transient declines in blood glucose and meal initiation under free-feeding conditions (Campfield & Smith, 1987b).

Combining the experimental induction of meal initiation using two different probes, acetylcholine analogue and fructose, probably acting through different mechanisms (fructose is not a potent stimulus for insulin secretion), underscores the importance not of the glucose molecule nor blood glucose concentration but instead of the temporal pattern of blood glucose dynamics in the control of meal initiation.

DETECTION OF THE SIGNAL

In an attempt to identify a role for the parasympathetic nervous system in the origin of the decline in blood glucose and/or its detection by the hepatic glucose-receptive elements described by Niijima (1979, 1980, 1983), rats receiving transection of the subdiaphragmatic vagus nerve or its hepatic branch were studied (Campfield & Smith, 1986b). Completeness of vagotomy was verified by a combination of behavioral and physiological measures (Campfield, Smith, & LeMagnen, 1983). Transient declines in blood glucose qualitatively similar to those observed in intact rats occurred prior to meals in both groups with successful vagotomies. However, the major finding of these studies was the failure of transient declines in blood glucose to reliably predict meal initiation in 45% of the trials in both vagotomized groups. This is the only experimental situation we have studied in which glucose declines within normal limits fail to predict meal initiation. This finding may be analogous to the observation that the variance of mean blood pressure markedly increases following denervation of the carotid sinus. The resulting decrement in the precision of the regulation of mean blood pressure has been attributed to the absence of the peripheral detectors. In our studies, an increase in the variance of intermeal blood glucose concentrations, particularly in the subdiaphragmatic vagotomized group, was observed. These studies in vagotomized rats suggest that detection of the transient declines by peripheral glucoreceptive elements innervated by the vagus nerve is important in the robust and reliable coupling between blood glucose dynamics and meal initiation observed in free-feeding intact rats.

Although these studies indicate that central and nonvagally dependent peripheral glucoreceptive elements unaffected by subdiaphragmatic or hepatic vagotomy are able to detect a transient decline in blood glucose and participate in the neural network that maps or transforms that signal into meal initiation, they also demonstrate that the resultant coupling between blood glucose dynamics and meal initiation was much less reliable in these rats. Thus, these results suggest that vagally dependent peripheral receptors, possibly through a lower threshold, provide the faithful coupling observed between blood glucose and meal initiation in intact rats. In these experiments, we observed an increased frequency of "smaller" transient declines in both vagotomized groups that just met or fell below the criteria (magnitude, slope, duration, and/or area below baseline) for transient declines that mapped successfully into meal initiation. This observation is consistent with the notion that the peripheral glucoreceptors have the equivalent of a lower threshold for detection of the systemic signal compared to the central detection elements. The importance of central integration of peripheral signals in the control of appetite is further demonstrated in these experiments (Campfield & Smith, 1986a).

In summary, the studies reviewed above suggest that transient declines in blood glucose with the necessary shape and magnitude are reliable signals for feeding that are detected by peripheral and central nervous system glucose-receptive elements and are mapped into meal initiation.

The mechanism underlying the observed functional coupling between transient declines in blood glucose and meal initiation remain elusive. The hepatic glucoreceptive afferent described by Niijima (1979, 1980, 1983) and the central glucoreceptive neurons described by Oomura (1983; Oomura, Shimizu, Miyahara, & Hattori, 1982) provide a possible afferent pathway for the detection, measurement, and/or relay of control information contained in the transient premeal decline in blood glucose to the caudal brainstem (Grill & Norgren, 1978; Norgren, 1983) and the lateral hypothalamic neuronal systems (Oomura, 1983; Oomura et al., 1982; Rolls, 1981) that are thought to initiate and sequence the motor programs for feeding (Norgen, 1983; Ono, Hishino, Sasaki, Fukoda, & Muramoto, 1980; Rolls, 1981). One possibility consistent with our studies is that the afferent elements recognize the pattern of blood glucose dynamics, perhaps by integrating the area of the glucose excursion below baseline, and food seeking and feeding are initiated once the critical pattern or component of the pattern is detected.

Possible mechanisms for the origin of the transient decrease in systemic blood glucose include decreased glucose absorption from the intestine (LeMagnen, 1981), decreased glucose production from the liver, increased central and/or peripheral glucose utilization, or a combination of transient alterations in both glucose production and utilization. A role for plasma insulin is suggested by studies described above, and other hormonal or neural signals involved in glucoregulation also may play a role. The identification of the mechanism or mechanisms and the nature of the efferent signals responsible for these modifications in glucose homeostasis must await further experimentation. Whether the transient decline in blood glucose was a reflection of a change in brain utilization of glucose or an event of peripheral origin or both also cannot be determined by these studies and must await further research.

Transient Declines in Blood Glucose in VMH-Lesioned Rats

Following the characterization of the role of transient declines in blood glucose in signaling meal initiation, we were interested in extending these studies to rats with altered feeding patterns. The syndrome of hyperphagia, hyperinsulinemia, altered autonomic function, and obesity following bilateral electrolytic lesion of the ventromedial hypothalamus (VMH lesion) provides a model with coexisting altered patterns of feeding (Bray & York, 1979; Brooks & Lambert, 1946; Heatherington & Ranson, 1942; Powley, Opsahl, Cox, & Weingarten, 1980), metabolic flux across intestine (Booth, Toates & Platt, 1976; Booth, 1978), pancreatic insulin secretion (Berthoud & Jeanrenaud, 1979; Rohner-Jeanrenaud & Jeanrenaud, 1980; King, Carpenter, Stamoutsos, Frohman, & Grossman, 1978; Inoue, Campfield, & Bray, 1977; Campfield & Smith, 1983b; Campfield, Smith and Larue-Achagiotis, 1986; Smith & Campfield, 1986a), autonomic neural activity and metabolic state (Bray & York, 1979; Powley *et al.*, 1980). Several lines of convergent evidence support the hypothesis that many of the characteristic alterations of the VMH syndrome, including feeding behavior, are secondary to altered autonomic neural activity (Bray & York, 1979; Inoue & Bray, 1979, 1980; Powley & Opsahl, 1974; Powley *et al.*, 1980). This hypothesis leads to the perspective that the VMH-lesioned rat has altered patterns of autonomic activity, which in turn causes altered patterns of intestinal absorption and pancreatic hormone secretion.

Blood glucose and meal pattern were monitored continuously in rats that had received successful bilateral electrolytic VMH lesions. Chronic cardiac cannulas were implanted in intact and VMH-lesioned (30 days post-operative) female Wistar rats 1 week prior to the experiments. Transient declines in blood glucose were observed prior to meal initiation in VMH-lesioned rats; the parameters of these declines in blood glucose were within the ranges previously ob-

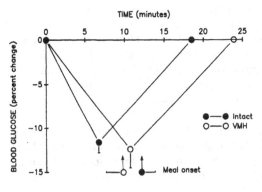

Figure 8. Transient declines in blood glucose in VMH-lesioned rats. Comparison of the timing and magnitude of the maximum percentage decline and time of the return to baseline in experiments in 30-day VMH-lesioned obese rats ($N = 12$) and intact rats ($N = 9$). The onset of feeding in each group is indicated by the symbol and S.E.M. at the bottom of the figure. Note that meal initiation occurs near the nadir of the decline in VMH-lesioned rats.

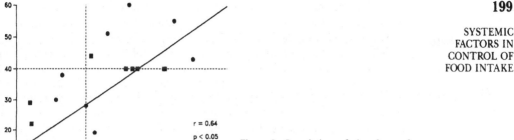

Figure 9. Correlation of the slope of blood glucose during the initial 10 minutes of feeding and subsequent intermeal interval in a group of VMH-lesioned (circle, $N = 12$) and intact rats (square, $N = 9$).

served in normal rats (see Figure 8). However, the striking difference in the VMH-lesioned group was the significantly earlier onset of feeding; meal initiation occurred at the nadir of the transient decline rather than during the rising phase of the decline. When the individual blood glucose and meal initiation records were analyzed, it was discovered that meal initiation occurred as the blood glucose begin to decline in six experiments and as the blood glucose rose toward baseline in six other experiments. Both temporal patterns were observed in different trials in the same animal. It was also noted that the intermeal interval was much shorter in the experiments in which meal initiation occurred earlier during the transient decline in blood glucose. This analysis suggested an association between the slope of the blood glucose trajectory during feeding and the intermeal interval (Smith & Campfield, 1986b).

To test this suggestion, the slope of blood glucose concentrations during the initial 10 minutes of feeding and subsequent intermeal intervals were computed from the digitized records of experiments in VMH-lesioned and intact rats. A significant positive correlation was found between the slope of the blood glucose trajectory and intermeal interval (Figure 9). Two groups emerged from this analysis: for intermeal intervals >40 minutes, the average slope was 1.6 ± 0.3%/min; for intermeal intervals <40 minutes, the average slope was −0.7 ± 0.2%/min. Negative slopes were observed during 50% of the meals in VMH-lesioned rats but only in approximately 20% of meals in intact rats. Thus, if blood glucose concentration is constant or falls during the initial 10 minutes of feeding, subsequent feeding will occur with a short latency (17.8 ± 4.8 minutes), whereas if blood glucose rises during feeding, the expected longer (98 ± 6.8 minutes) intermeal interval will be observed. These studies suggest that the central nervous system may be monitoring not only excursions of blood glucose below baseline prior to feeding to signal meal initiation but also the blood glucose trajectory during feeding as a determinant of the intermeal interval. These results support the importance of the pattern of blood glucose dynamics and the peripheral and/or central detection and processing of glucose-related inputs by the central nervous system in the control of ingestive behavior (Campfield & Smith, 1987a).

The study of transient declines in blood glucose prior to meals in VMH-lesioned rats led us to the observation that the pattern of blood glucose dynamics during feeding may also play an important role in the control of feeding behavior. This leads quite naturally to a consideration of the patterns of metabolic substrates and hormones.

The time course of absorption of metabolic substrates in rats eating standard chow has been defined. Blood glucose increases within 8–10 minutes following the beginning of a meal, and blood amino acids and fatty acids rise slightly after blood glucose. All plasma substrates increase to peak values and then gradually are cleared from the circulation and return to levels characteristic of the postabsorptive state (for review, see LeMagnen, 1985). In similar studies conducted in human subjects, meals are followed by a similar although delayed time course of absorption. Thus, ingestion of a meal consists of the following patterns in plasma fuel substrates. Prior to the meal, plasma substrates are at low, constant levels characteristic of the postabsorptive state. During the initial, pre-absorption phase of ingestion, no change in substrate concentrations in blood occur while ingestion continues. Later, first plasma glucose and then amino acid, fatty acid, and triglyceride concentrations begin to increase, followed by the termination of ingestion. After a rise to peak levels the concentrations of blood substrates return to low postabsorptive levels (Gerich, Charles, & Grodsky, 1976; Unger, Dobbs, & Orci, 1978).

Superimposed on the pattern of metabolic substrates are changes in plasma pancreatic and intestinal hormone concentrations. The postabsorptive state is also characterized by relatively constant hormone concentrations: at low levels for insulin and GI hormones and at high levels for such counterregulatory hormones as glucagon, epinephrine, norepinephrine, and cortisol.

In the rat, the earliest documented change in hormone concentrations is the transient spike in plasma insulin that occurs before the transient decline in blood glucose (26 to 13 minutes before meal initiation) described above (Campfield and Smith, 1986c; Campfield et al., 1988). This increase appears to be vagally mediated, and it may contribute to the observed reliable coupling between the premeal pattern of blood glucose and meal initiation.

A second change in hormone levels related to ingestion consists of the well-studied cephalic-phase transient increases in insulin and glucagon. These brief pulses of hormone release, which also are vagally mediated (Berthoud & Jeanrenaud, 1979; Powley et al., 1980; Kral, Powley, & Brooks, 1983; Woods & Porte, 1974; Louis-Sylvestre, 1978), can be elicited by the sight, smell, or the anticipation of food presentation and occur before any ingestion of food.

The actual ingestion of food causes a major change in intestinal and pancreatic hormone release. Nutrients in the intestine are powerful stimuli for the release of GI hormones, in particular CCK and GIP. When metabolic substrates are being absorbed across the intestine, the fourth wave of pancreatic hormone secretion occurs. Plasma insulin increases in response to increasing glucose and amino acid concentrations, while glucagon, stimulated by amino acids but inhibited by glucose, remains essentially constant or slightly increased (Steffens, 1969; Unger et al., 1978). Thus, insulin secretion from pancreatic β cells is stimulated three or four times as we move from the anticipation of a meal to satiety. This observation is an example of why the insulin molecule *per se* cannot convey an unambiguous message to the CNS. Instead, increases in plasma insulin concentration may have different meanings depending on the context of antecedent events. Thus, it is the pattern of plasma insulin concentrations that may convey vital information to the CNS related to metabolic state.

Insulin secretion from the endocrine pancreas is controlled by a variety of factors including metabolic, neural, hormonal, and ionic inputs in addition to neurotransmitters, prostaglandins, cyclic nucleotides, paracrine effects, and blood flow (for review see Gerich *et al.*, 1976; Smith, Woods, & Porte, 1979; Unger *et al.*, 1978; Wollheim & Sharp, 1981). Several laboratories, including ours, have focused on the interaction of metabolic and autonomic neural inputs in the control of insulin secretion (for review see Campfield *et al.*, 1986). Evidence has accumulated that the autonomic nervous system can alter pancreatic hormone response by three possible mechanisms: (1) changes in descending autonomic motor outflow to the pancreas (for review see Woods & Porte, 1974; Campfield *et al.*, 1983; Girardier, Seydoux, & Campfield, 1976; Kita, Niijima, Oomura, Ishizuka, Aou, Yamabe, & Yoshimatsu, 1980); (2) changes in the pancreatic islet responsiveness as a result of changes in autonomic neural input (Campfield & Smith, 1983a, 1983b; Campfield *et al.*, 1986; Kennedy & Parker, 1963; Smith & Campfield, 1986a); (3) alterations in the sensory afferents to the central nervous system (for review see Kral *et al.*, 1983; Campfield *et al.*, 1983, 1986).

Ventromedial hypothalamic lesions result in an obese state that is characterized by increased β-cell responsiveness to glucose. Following bilateral electrolytic lesion, these animals have an immediate sustained increase in the plasma concentration of insulin, and the β cell is hyperresponsive to both basal and stimulatory glucose concentrations (Berthoud & Jeanrenaud, 1979; Bray & York, 1979; Inoue *et al.*, 1977; Inoue & Bray, 1977; Inoue, Mullen, & Bray, 1983; King *et al.*, 1978; Rohner-Jeanrenaud & Jeanrenaud, 1980; Campfield & Smith, 1983b; Campfield *et al.*, 1986; Smith & Campfield, 1986a; Yoshimatsu, Niijima, Oomura, Yamabe, & Katafuchi, 1984). Experimental studies in our laboratory have demonstrated that increased glucose responsiveness after VMH lesion was persistent and not progressive (Campfield *et al.*, 1986).

The studies reviewed above demonstrate that postprandial insulin secretion is increased in rats with VMH lesions, and they suggest that an increased insulin secretory response to other stimuli is possible. Thus, it may be inferred that spikes in insulin that precede meal initiation may be exaggerated in terms of magnitude or/and duration. Alterations in the patterns of plasma insulin may play a role in the earlier meal onset with respect to the beginning and nadir of the transient decline in blood glucose. Enhanced spikes in premeal plasma insulin may also mediate, in part, the decreased slope of the change in blood glucose concentration during the first 10 minutes of feeding. It is at least a possibility that increased glucose responsiveness of the pancreatic β cell in rats with VMH lesions may play a role in the three alterations we have observed in the processes of meal initiation in these rats: (1) earlier meal onset with respect to the transient decline in blood glucose; (2) decreased or negative slope of blood glucose during the first 10 minutes of feeding; and (3) a smaller intermeal interval. These alternatives would tend to lead to more frequent, larger meals, which are characteristic of rats with VMH lesions. The validation of this hypothesis must await further experimental work.

CONCLUSIONS

An example of a pattern with information content critical to the regulation of feeding behavior has been considered in some detail. The evidence supporting transient declines in blood glucose as signals for meal initiation has been reviewed. The associated patterns of plasma insulin concentration and their potential role have also been described. The central message of this example is not that glucose and insulin are important signals in feeding but rather that the temporal patterns of blood glucose and plasma insulin concentrations have important information embedded within them, and they convey this information to the critical neural elements controlling feeding behavior within the brain.

Experimental perturbation of the pattern of blood glucose dynamics results in modification of feeding behavior. For example, infusion of glucose intravenously to blunt transient declines in blood glucose delayed meal initiation (Campfield et al., 1985), whereas inducing this pattern in blood glucose concentration with an acetylcholine analogue led to meal initiation (Campfield & Smith, 1987b). In addition, alterations in this pattern and its temporal relationship to feeding behavior were observed in VMH-lesioned rats as described above. The results of these and other perturbation studies together with the descriptive studies provide evidence for the importance of patterns of blood glucose and possibly plasma insulin as signals in the control of feeding behavior. The pattern of blood glucose dynamics under discussion provides at least partial mechanisms for two of the major issues in feeding control identified in the introduction: factors responsible for initiation of feeding and the duration of the intermeal interval. Prior to the focus on temporal patterns of blood glucose concentration, which could be followed as a result of technological innovations allowing its continuous monitoring, blood glucose had been considered but rejected as an explanation of either physiological process. Focus on patterns has allowed a reconsideration of these two processes in terms of the dynamics of blood glucose rather than its concentration and has provided a new context for studying these processes.

Is the "pattern as signal" conceptualization advocated here simply a complementary theoretical construct to the other, more established concepts in the control of feeding mentioned in the introduction, or is it a competing, alternative notion with more explanatory power? The ability of this construct to account adequately for or explain meal initiation and/or the intermeal interval or other aspects of feeding behavior remains to be determined by future research. These patterns must be subjected to further and more rigorous experimental tests, and additional patterns have to be identified and tested. However, the initial success of this approach suggests to the authors that its application to the numerous disparate and competing findings in feeding research may provide insight and possibly facilitate the development of a more complete understanding.

This optimism is based on three points: utility, power to integrate, and focus on the message rather than the messenger. First, this construct has refocused two and possibly three regulatory issues in our field. The role of blood glucose in meal initiation and the role of hyperinsulinemia in the VMH-lesioned obese syndrome were, at best, controversial. However, when attention was focused on the patterns of blood glucose dynamics and of meal-associated insulin secretion, as has been done in this chapter, signals for meal initiation and perhaps the intermeal interval were revealed, and a basis for the commonly observed alterations in feeding in VMH-lesioned rats was proposed. Second, this hypothesis is based on the integration of multiple patterns by the neuronal networks controlling feeding. Thus, the goal of the search is not the pattern but rather the patterns that together form the central representations of meal initiation, maintenance, and termination. The integrative focus of this conceptualization has much in common with and is a descendent of earlier integrative approaches of combined energy flux in the plasma (Booth, Toates, & Platt, 1976; Booth, 1978) and neurons that integrate specific molecular signals (Nicolaidis, 1974). However, this concept has an even broader focus on the totality of central representations of peripheral events related to feeding. An analogy may be that the brain is responding, conditioned on its current state, to the symphony rather than the voice of individual sections of the orchestra. Third, the ability of the "pattern-assignal" construct to integrate the dynamics of blood glucose before, during, and after meals into one or more important messages related to feeding behavior suggests that a similar focus on the patterns of insulin and CCK rather than a focus on the location or concentration of these hormones may also yield important insights.

Whether this optimism turns out to be justified must await further research. However, the process of experimental investigation and testing of these and other patterns with embedded information content, which will require the determination and perturbation of the temporal profiles of other candidate signals, should by itself lead to new insights and more complete understanding of the control of feeding behavior. If this concept continues to be useful in thinking about feeding behavior and its physiological bases, then a small change in blood glucose will have indeed had a large effect on behavior.

Acknowledgment

The participation of P. Brandon, K. Kamp, D. Driscoll, and N. Spirt in many of these studies is gratefully acknowledged. The authors would like to thank the Pharmacology Word Processing Unit (M. Johnson, P. Lotito, S. Maloney, N. Manion, and L. Zielenski) for preparing this manuscript for publication.

REFERENCES

Bernstein, L. M., & Grossman, M. I. (1956). An experimental test of the glucostatic theory of the regulation or food intake. *Journal of Clinical Investigation, 35,* 627–633.

Berthoud, H. R., & Jeanrenaud, B. (1979). Acute hyperinsulinemia and its reversal by vagotomy after lesions of the ventromedial hypothalamus in anesthetized rats. *Endocrinology, 105,* 146–151.

Booth, D. A. (1968). Effects of intrahypothalamic glucose injection on eating and drinking elicited by insulin. *Journal of Comparative and Physiological Psychology, 65,* 13–16.

Booth, D. A. (1978). Prediction of feeding behavior from energy flows in the rat. In D. A. Booth (Ed.), *Hunger models* (pp. 227–278). London: Academic Press.

Booth, D. A., & Pitt, M. E. (1968). The role of glucose in insulin-induced feeding and drinking. *Physiology and Behavior, 3,* 447–453.

Booth, D. A., Toates, F. M., & Platt, S. V. (1976). Control system for hunger and its implication in animals and man. In D. Novin, W. Wyrwicka, & G. A. Bray (Eds.), *Hunger: Basic mechanisms and its clinical implications* (pp. 127–143). New York: Raven Press.

Bray, G. A., & York, D. A. (1979). Hypothalamic and genetic obesity in experimental models: An autonomic and endocrine hypothesis. *Physiological Review, 59,* 719–809.

Brooks, C. M., & Lambert, E. F. (1946). A study of the effects of limitation of food intake and the method of feeding on the rate of weight gain during hypothalamic obesity in the albino rat. *American Journal of Physiology 147,* 695–707.

Campfield, L. A., Brandon, P., & Smith, F. (1985). On-line continuous measurement of blood glucose and meal pattern in free-feeding rats: The role of glucose in meal initiation. *Brain Research Bulletin, 14,* 605–616.

Campfield, L. A., & Smith, F. J. (1983a). Neural control of insulin secretion: Interaction of norepinephrine and acetylcholine. *American Journal of Physiology, 244,* R629–R634.

Campfield, L. A., & Smith, F. J. (1983b). Alteration of islet neurotransmitter sensitivity following ventromedial hypothalamic lesion. *American Journal of Physiology, 244,* R635–R640.

Campfield, L. A., & Smith, F. J. (1986a). Functional coupling between transient declines in blood glucose and feeding behavior: Temporal relationships. *Brain Research Bulletin, 17,* 427–433.

Campfield, L. A., & Smith, F. J. (1986b). Transient declines in blood glucose and meal initiation: Evidence for a functional role for peripheral glucoreceptors. In *Proceedings 9th International Conference of Physiology of Food and Fluid Intake,* Seattle, WA.

Campfield, L. A., & Smith, F. J. (1986c). Blood glucose and meal initiation: A role for insulin? *Society for Neuroscience Abstracts, 12,* 109.

Campfield, L. A., & Smith, F. J. (1987a). Glucose dynamics during feeding predict frequency of ingestion. *Federation Proceedings, 144,* 901.

Campfield, L. A., & Smith, F. J. (1987b). Meal initiation following induction of transient declines in blood glucose: Evidence for causality. *Society for Neuroscience Abstracts, 13,* 466.

Campfield, L. A., Smith, F. J., Driscoll, D. W., & Spirt, N. (1988). Vagotomy blocks insulin spike that precedes meal initiation. *Society for Neuroscience Abstracts, 14,* 1197.

Campfield, L. A., Smith, F. J., & Larue-Achagiotis, C. (1986). Temporal evolution of altered islet neurotransmitter sensitivity after VMH lesion *American Journal of Physiology, 251,* R63–R69.

Campfield, L. A., Smith, F. J., & LeMagnen, J. (1983). Altered endocrine pancreatic function following vagotomy: Possible behavioral and metabolic bases for assessing completeness of vagotomy. *Journal of the Autonomic Nervous System, 9,* 283–300.

Carlson, A. J. (1916). *The control of hunger in health and disease.* Chicago: University of Chicago Press.

Danguir, J., & Nicolaidis, S. (1980). Circadian sleep and feeding patterns in the rat: Possible dependence on lipogenesis and lipolysis. *American Journal of Physiology, 238,* E223–E230.

Epstein, A. N., & Teitelbaum, P. (1967). Specific loss of the hypoglycemic control of feeding in recovered lateral rats. *American Journal of Physiology, 213,* 1159–1167.

Gerich, J. E., Charles, M. S., & Grodsky, G. M. (1976). Regulation of pancreatic insulin and glucagon secretion. *Annual Reviews of Physiology, 38,* 353–388.

Gibbs, J., Young, R. C., & Smith, G. P. (1973). Cholecystokinin decreases food intake in rats. *Journal of Comparative and Physiological Psychology, 84,* 488–495.

Girardier, L., Seydoux, J., & Campfield, L. A. (1976). Control of A and B cells *in vivo* by sympathetic nervous input and selective hyper- or hypoglycemia in dog pancreas. *Journal de Physiologie (Paris), 72,* 801–814.

Grill, H. J., & Norgren, R. (1978). The taste reactivity test. II. Mimetic responses to gustatory stimuli in chronic thalamic and chronic decerebrate rats. *Brain Research, 143,* 281–297.

Grinker, J., Cohn, C. K., & Hirsch, J. (1971). The effects of intravenous administration of glucose, saline, and mannitol on meal regulation in normal-weight human subjects. *Behavioral Biology, 6,* 203–208.

Hetherington, A. W., & Ranson, S. W. (1942). The spontaneous activity and food intake of rats with hypothalamic lesions. *American Journal of Physiology, 136,* 609–617.

Inoue, S., & Bray, G. A. (1977). The effects of subdiaphragmatic vagotomy in rats with ventromedial hypothalamic obesity. *Endocrinology, 100,* 108–114.

Inoue, S., & Bray, G. A. (1979). An autonomic hypothesis for hypothalamic obesity. *Life Sciences, 25,* 561–566.

Inoue, S., & Bray, G. A. (1980). Role of the autonomic nervous system in the development of ventromedial hypothalamic obesity. *Brain Research Bulletin, 5(Suppl. 4),* 119–125.

Inoue, S., Campfield, L. A., & Bray, G. A. (1977). Comparison of metabolic alterations in hypothalamic and high fat diet-induced obesity. *American Journal of Physiology, 233,* R162–R168.

Inoue, S., Mullen, Y. S., & Bray, G. A. (1983). Hyperinsulinemia in rats with hypothalamic obesity: Effects of autonomic drugs and glucose. *American Journal of Physiology, 245,* R372–R378.

Kajinuma, H., Kaneto, A., Kuzuya, K., & Nakao, K. (1968). Effects of methacholine on insulin secretion in man. *Journal of Clinical Endocrinology and Metabolism, 28,* 1384–1387.

Kennedy, G. C., & Parker, R. A. (1963). The islets of langerhans in rats with hypothalamic obesity. *Lancet, 2,* 981–982.

King, B. M., Carpenter, R. G., Stamoutsos, B. A., Frohman, L. A., & Grossman, S. P. (1978). Hyperphagia and obesity following ventromedial hypothalamic lesion in rats with subdiaphragmatic vagotomy. *Physiology and Behavior, 20,* 643–651.

Kita, H., Niijima, A., Oomura, Y., Ishizuka, S., Aou, S., Yamabe, K., & Yoshimatsu, H. (1980). Pancreatic nerve response induced by hypothalamic stimulation in rats. *Brain Research Bulletin, 5(Suppl. 4),* 163–168.

Kral, J. G., Powley, T. L., & Brooks, C. McC. (Eds.). (1983). *Vagal nerve function: Behavioral and methodological considerations.* Elsevier: New York.

Larue-Achagiotis, C., & LeMagnen, J. (1979). The different effects of continuous night and day-time insulin infusion on the meal pattern of rats: Comparison with the meal pattern of hyperphagic hypothalamic rats. *Physiology and Behavior, 22,* 435–440.

LeMagnen, J. (1981). The metabolic basis of the dual periodicity of feeding in rats. *Behavioral and Brain Sciences, 4,* 561–607.

LeMagnen, J. (1985). *Hunger.* London: Cambridge University Press.

Likuski, H. J., Debons, A. F., & Cloutier, R. J. (1967). Inhibition of gold thioglucose-induced hypothalamic obesity by glucose analogues. *American Journal of Physiology, 212,* 669–676.

Louis-Sylvestre, J. (1978). Feeding and metabolic patterns in rats with truncular vagotomy or with transplanted beta cells. *American Journal of Physiology, 235,* E119–E125.

Louis-Sylvestre, J., & LeMagnen, J. (1980). A fall in blood glucose level precedes meal onset in free-feeding rats. *Neuroscience and Biobehavioral Reviews, 4,* 13–15.

MacKay, E. M., Callaway, J. W., & Barnes, R. F. (1940). Hyperalimentation in normal animals produced by protamine insulin. *Journal of Nutrition, 20,* 59–66.

Mayer, J. (1953). Glucostatic mechanisms of regulation of food intake. *New England Journal of Medicine, 249,* 13–16.

Mayer, J. (1955). Regulation of energy intake and the body weight, the glucostatic theory and the lipostatic hypothesis. *Annals of the New York Academy of Sciences, 63,* 15–43.

Miselis, R. R., & Epstein, A. N. (1975). Feeding induced by intracerebroventricular 2-deoxy-d-glucose in the rat. *American Journal of Physiology, 229,* 1438–1447.

Moran, T. H., & McHugh, P. R. (1982) Cholecystokinin suppresses food intake by inhibiting gastric emptying. *American Journal of Physiology, 242,* R491–R497.

Morgan, C. T., & Morgan, J. D. (1940). Studies in hunger: the relationship of gastric denervation and dietary sugar to the effect of insulin upon food intake in the rat. *Journal of Genetics Psychology, 57,* 153–167.

Nicolaidis, S. (1974). Short term and long term regulation of energy balance. *Proceeding of the XXVI International Union of Physiological Sciences (IUPS)* (pp. 122–123). New Delhi, India.

Nicolaidis, S., & Even, P. (1984). Mesure du metabolisme de fond en relation avec la prise alimentaire: Hypothese ischymetrique. *Comptes Rendus des Seances de l'Academie des Sciences CD, 298,* 295–300.

Nicolaidis, S., & Rowland, N. (1976). Metering of intravenous versus oral nutrients and regulation of energy balance. *American Journal of Physiology, 231,* 661–668.

Nicolaidis, S., Rowland, N., Meile, M.-J., Marfaing-Jallat, P., & Pesez, A. (1974). A flexible technique for long term infusions in unrestrained rats. *Pharmacology Biochemistry and Behavior, 2,* 131–136.

Niijima, A. (1979). Control of liver function and neuroendocrine regulation of blood glucose levels. In C. M. Brooks, K. Koizumi, & A. Sato (Eds.), *Integrative functions of the autonomic nervous system* (pp. 68–83). Amsterdam: Elsevier/North-Holland.

Niijima, A. (1980). Glucose sensitive afferent nerve fibers in the liver and regulation of blood glucose. *Brain Research Bulletin 5* (Suppl. 4), 1975–1979.

Niijima, A. (1983). Glucose sensitive afferent nerve fibers in the liver and their role in food intake and blood glucose regulation. *Journal of the Autonomic Nervous System, 9,* 207–220.

Norgren, R. (1983). Afferent interactions of cranial nerves involved in ingestion. *Journal of the Autonomic Nervous System, 9,* 67–77.

Novin, D. (1976). Visceral mechanisms in the control of food intake. In D. Novin, W. Wyrwicka, & G. Bray (Eds.), *Hunger: Basic mechanisms and clinical implications* (pp. 357–369). New York: Raven Press.

Novin, D., Gonzalez, M. F., & Sanderson, J. D. (1976). Paradoxical increased feeding following glucose infusions in recovered lateral rats. *American Journal of Physiology, 230,* 1084–1089.

Novin, D., VanderWeele, D. A., & Rezek, M. (1975). Infusions of 2-deoxy-d-glucose into the hepatic portal system causes eating: Evidence for peripheral glucoreceptors. *Science, 181,* 852–860.

Ono, T., Nishino, H., Sasaki, K., Fukoda, M., & Muramoto, K. (1980). Role of the lateral hypothalamus and amygdala in feeding behavior. *Brain Research Bulletin 5(Suppl. 4),* 143–149.

Oomura, Y. (1983). Glucose as a regulator of neuronal activity. *Advances in Metabolic Diseases, 10,* 31–65.

Oomura, Y., Shimizu, N., Miyahara, S., & Hattori, K. (1982). Chemosensitive neurons in the hypothalamus: Do they relate to feeding behavior. In B. G. Hoebel & D. Novin (Eds.), *The neural basis of feeding and reward* (pp. 551–566). Brunswick, ME: Haer Institute for Electrophysiological Research.

Powley, T. L., & Opsahl, C. A. (1974). Ventromedial hypothalamic obesity abolished by subdiaphragmatic vagotomy. *American Journal of Physiology, 226,* 25–33.

Powley, T. L., Opsahl, C. A., Cox, J. E., & Weingarten, H. P. (1980). The role of the hypothalamus in energy homeostasis. In P. J. Morgane & J. Panksepp (Eds.), *Handbook of the hypothalamus. Behavioral studies of the hypothalamus* (vol. 3A, pp. 211–298). New York: Marcel Dekker.

Rohner-Jeanrenaud, F., & Jeanrenaud, B. (1980). Consequences of ventro-medial hypothalamic lesions upon insulin and glucagon secretion by subsequently isolated perfused pancreases in the rat. *Journal of Clinical Investigation, 65,* 902–910.

Rolls, E. T. (1981). Central nervous mechanisms related to feeding and appetite. *British Medical Bulletin, 37,* 131–134.

Smith, F. J., & Campfield, L. A. (1986a). Pancreatic adaptation in VMH obesity: *In vivo* compensatory responses to altered neural input. *American Journal of Physiology, 251,* R70–R76, 1986a.

Smith, F. J., & Campfield, L. A. (1986b). Perturbations in glucose homeostasis initiate feeding in rats: Altered temporal relationships in VMH obesity. *Society for Neuroscience Abstracts, 12,* 109.

Smith, F. J., Driscoll, D. W., & Campfield, L. A. (1988). Short term effects of fructose on blood glucose dynamics and meal initiation. *Physiology and Behavior, 44,* 625–631.

Smith, G. P., & Epstein, A. N. (1969). Increased feeding in response to decreased glucose utilization in the rat and monkey. *American Journal of Physiology, 217,* 1083–1087.

Smith, G. P., Gibbs, J., Strohmayer, A. J., & Stukes, P. E. (1972). Threshold doses of 2-deoxy-D-glucose for hyperglycemia and feeding in rats and monkeys. *American Journal of Physiology, 222,* 77–81.

Smith, P. H., Woods, S. C., & Porte, D., Jr. (1979). Control of the endocrine pancreas by the autonomic nervous system and related neural factors. In C. M. Brooks, K. Koizumi, & A. Sato (Eds.), *Integrative functions of the autonomic nervous system* (pp. 84–97). Amsterdam: Elsevier/North-Holland.

Steffens, A. B. (1969). The influence of insulin injections and infusions on eating and blood glucose in the rat. *Physiology and Behavior, 4,* 823–828.

Stricker, E. M., & Rowland, N. (1978). Hepatic versus cerebral origin of stimulus for feeding induced by 2 deoxy-d-glucose in rats. *Journal of Comparative and Physiological Psychology, 92,* 126–132.

Unger, R. H., Dobbs, R. E., & Orci, L. (1978). Insulin, glucagon and somatostatin in the regulation of metabolism. *Annual Review of Physiology, 40,* 307–343.

Wollheim, C. B., & Sharp, G. W. G. (1981). Regulation of insulin release by calcium. *Physiological Reviews, 61,* 914–973.

Woods, S. C., & Porte, D., Jr. (1974). Neural control of the endocrine pancreas. *Physiological Review, 54,* 596–619.

Woods, S. C., & Porte, D., Jr. (1976). Insulin and the set-point regulation of body weight. In D. Novin, W. Wyrwicka, & G. Bray (Eds.), *Hunger: Basic mechanisms and clinical implications* (pp. 273–280). New York: Raven Press.

Yoshimatsu, H., Niijima, A., Oomura, Y., Yamabe, K., & Katafuchi, T. (1984). Effects of hypothalamic lesion on pancreatic autonomic nerve activity in the rat. *Brain Research, 303,* 147–152.

9

Human Obesity
A Problem in Body Energy Economics

THEODORE B. VANITALLIE AND HARRY R. KISSILEFF

INTRODUCTION

The principal purpose of this chapter is to provide a biological perspective from which to view human obesity, particularly its severe form. It is our contention that, to be better understood, the phenomenon of obesity should be considered in the context of the body's energy economy—a system that has three major components: intake, expenditure, and storage. In this chapter we consider the physiological mechanisms by which the three components are coordinated to accomplish the biological goals of optimizing energy storage when food is abundant and of conserving energy stores during privation.

Our overall hypothesis about the obesity phenomenon embodies the notion that the human species has been predisposed by its evolutionary history to store fat in times of plenty, thereby enabling survival during periods of famine. Thus, the hypothesis would predict a strong genetic–environmental interaction in which those individuals who are especially predisposed to storing and sparing triglyceride would be more likely than others to become obese when food is abundant and attractive. This part of the hypothesis is not new; the same idea was eloquently enunciated by J. V. Neel in 1962, when he coined the expression "thrifty genotype." However, we have built on Neel's formulation in several respects; for example, we postulate that one of the major brakes that limited severity of fatness in the past, namely, the requirement to perform physical work in order to obtain food, has been largely eliminated by our mechanized society. We point out as well that, from an econometric standpoint, the body's energy storehouse should be kept as full as possible in time of plenty. Third, in our

THEODORE B. VANITALLIE   Department of Medicine, Columbia University College of Physicians and Surgeons at St. Luke's-Roosevelt Hospital Center, New York, New York 10025.   HARRY R. KISSILEFF   Departments of Medicine and Psychiatry, Columbia University College of Physicians and Surgeons at St. Luke's-Roosevelt Hospital Center, New York, New York 10025.

THEODORE B.
VaNITALLIE AND
HARRY R.
KISSILEFF

review, we attempt to identify some of the sites at which the thrifty genotype might be expressed.

One of the remarkable features of the human adipose organ is the extraordinary range (20-fold or more) of its capacity to store triglyceride. In order to explain the large variability that occurs in the size of the fat depot, we have first examined the epidemiologic evidence that both genetic and environmental factors contribute to the naturally occurring variability observed in body fatness. Next, we have considered the physiology of the energy economy, with particular attention paid to the findings of experimental studies on the major determinants of energy intake, output, and storage.

One cannot discuss depot fat without considering the adipose tissue in which it is stored. Thus, we have examined the contribution of the adipocytes themselves to the dynamics of fat storage and release. Even a small genetically determined imbalance between triglyceride storage and release could lead to obesity.

Having reviewed this physiology, we then consider the possibility that the fat stores are regulated by a system embodying a set point. We find that the evidence in support of the set-point theory of body weight regulation is inconclusive and that the phenomena the theory attempts to elucidate are explained at least as well by alternative formulations that more directly address the mechanisms involved. It is our position that the size of the depot is not determined by a mechanism that compares the current level of the fat stores with some "desired" level. We believe that the size of the body's inventory of stored fat is necessarily flexible and determined by events taking place in the overall body energy economy.

The view that the fat depot is regulated at a value determined by a set point is also incompatible with the metabolic role of the adipose tissue, which is to subserve the continual need of the body's cells for fuel. Therefore, we have introduced an econometric model designed to provide a picture of the energy transfer system that we believe to be more consistent with the available physiological evidence.

Another issue taken up in our review is the presumed effect of the size of the energy storage depot on energy intake. This consideration has included attention to the various routes (neural, humoral, metabolic) by which the adipose tissue's fat stores could influence food intake. In addition, we have discussed various real-life situations in which body fat content is altered. Such alterations may be programmed into some animal species in order to accommodate certain biologically imperative activities like hibernation, migration, and lactation. They may also result from metabolic changes within the body occurring secondarily to illness or use of certain pharmacologically active agents, or arising from volitional changes in eating or exercise behaviors. Finally, we have applied ideas arising from this review to the thorny problem of helping formerly obese individuals maintain their reduced weights. Current experimental evidence suggests that one approach to this goal could be to develop medications capable of overcoming the metabolic characteristics that predispose certain individuals to excessive fat storage.

DETERMINANTS OF BODY FAT CONTENT AND THEIR INTERACTION

There is evidence that the body fat content is affected (*inter alia*) by sex, age, genetic endowment, level of physical activity, and dietary environment. One way

of visualizing how genetic and environmental factors influence fatness is by
adaptation of a formula employed by Bouchard (1985) to account for the variations that occur in any of a number of biological variables. Thus,

$$V_{BF} = V_G + V_E + V_{GXE} + e$$

V_{BF} is variance in body fat content, G refers to genetic factors, E refers to
environmental factors, GXE to an interaction between genetic and environmental factors, and e is an error term. This formula embodies the notion that environmental and genetic factors may act independently of one another or may
interact to affect body fat content (Poehlman et al., 1986).

GENETIC FACTORS

It has become increasingly evident that genetic factors play an important
role in explaining variability in body fat content. Examples of the different lines
of evidence in support of this role include the following: (1) the existence of a
number of animal models of genetic obesity [i.e., the ob/ob obese–hyperglycemic
mouse and the fa/fa (Zucker) rat] (Sclafani, 1984); (2) adoption studies indicating
that the body mass indices [determined by dividing body weight in kilograms by
the square of the height in meters (kg/m²)] of adopted children correlate with the
body mass indices (BMIs) of their biological parents and not those of their
adoptive parents (Stunkard, Foch, & Hrubec, 1986; Stunkard, Sörensen, et al.,
1986; Price, Cadoret, Stunkard, & Troughton, 1987)]; (3) studies that indicate
that, in body fat topography and percentage body fat, members of monozygotic
twin pairs resemble each other more closely than do members of dizygotic twin
pairs (Bouchard, 1985); (4) the fact that members of certain races are at higher
risk of becoming obese than others, even if allowance is made for differences in
socioeconomic status and level of education. Notable examples are the relatively
high prevalences of obesity in the Pima Indians of the southwestern United
States and in middle-aged black women (Bogardus et al., 1986; VanItallie &
Woteki, 1987); and (5) the finding that when both parents are obese, their
children have a very high risk of becoming obese (Garn, 1985).

ENVIRONMENTAL FACTORS

Epidemiologic studies have provided useful clues concerning the role of the
environment as a determinant of body fat content. Among such studies have
been (1) observations comparing the BMIs of populations exposed to different
environments (Gordon, Kagan, & Rhoads, 1975), (2) comparisons of BMIs of
individuals who have emigrated with those of their siblings who remained at
home, and (3) observations of the changing frequency distributions of relative
weights exhibited by populations whose environment has undergone significant
secular changes.

According to recent data reported by the Health and Welfare Ministry of
Japan (1986), Japanese men and women have BMIs substantially lower than
those of American men and women of comparable stature and age. Although
some of this "leanness" may be attributable to genetic factors, it also seems likely
that the Japanese life style plays a significant role, particularly the low-fat, calorically dilute diet consumed by most Japanese. This supposition is strengthened
by the fact that Japanese men who emigrate to Hawaii and California gain weight

THEODORE B.
VANITALLIE AND
HARRY R.
KISSILEFF

and have been reported to be some 15 lb heavier than men of comparable age and stature in Japan (Kagan *et al.*, 1974).

Finally, Americans themselves appear to have become heavier during the last century (VanItallie, 1979), although the various population samples on which this observation is based (Civil War soldiers, World War I and II draftees, insured policy holders, and other groups such as those examined during three National Health Surveys between 1960 and 1980) are not equally representative of the population at large or necessarily comparable in genetic makeup. Moreover, the methods used to measure height and weight presumably differed across surveys.

DIET. In view of the secular increases in relative weight that seem to have occurred in the United States since the middle of the last century, it is instructive to consider the changes in the U.S. diet that have taken place since the turn of the century (Guthrie, Habicht, Johnson, & VanItallie, 1986). The most striking change has been in the proportion of dietary calories contributed by fat and sugar (Figure 1). In 1909, the U.S. food supply, on a per-capita per-day basis, provided 3510 kcal and 126 g of fat (32% of total calories). In 1982 the food supply provided 3380 kcal and 162 g of fat (43% of total calories). Although the food supply is only an indirect measure of actual consumption, it is pertinent to inquire whether there was any causal relationship between the 34% increase in total fat "availability" that was recorded during the 20th century and the rise in mean body weight in U.S. adults that presumably occurred during the same general period of time.

The per-capita level of carbohydrate provided by the U.S. food supply was highest during 1909–1913 [of 490 g per person per day, 154 g (31%) was simple carbohydrate]. In 1982, the carbohydrate provided by the U.S. food supply was

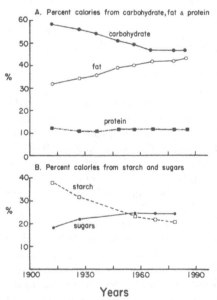

Figure 1. United States food supply (1909–1985). Values shown represent foods available for consumption. Because some fat is discarded during food preparation, including cooking, actual intake of fat by American adults as estimated by the U.S. Department of Agriculture's Continuing Survey of Food Intake by Individuals (CSFII) averaged about 37% of total calories in 1985. Source: U.S. Department of Agriculture, Human Nutrition Information Service.

388 g per person per day, of which 204 g (53%) was simple carbohydrate ("sugar"). Throughout the 20th century, about 76% of the carbohydrate in the food supply has come from grain products, sugars, and other caloric sweeteners. However, since 1901–1913, the proportion of carbohydrate from grain products has declined from 56% to 37%, while that from sugars and other caloric sweeteners has increased from 22% to 38%. The rise in consumption of carbohydrate from sugars and caloric sweeteners in the early part of the century is attributed to an increased use of table sugar (sucrose), and the rise since 1970 appears related to increased use of corn sweeteners.

Crude fiber provided by the U.S. food supply decreased from 6.9 g per person per day in 1909–1913 to 4.1 g per day in 1982. Comparable figures on *dietary* fiber are not available.

Overall, during the 20th century, the U.S. diet increased substantially in total fat and sugar content, with corresponding decreases occurring in complex carbohydrate and crude fiber. Possible implications of such changes for body weight are discussed in a later section.

OTHER POSSIBLE ENVIRONMENTAL DETERMINANTS. Two other changes in the U.S. environment during the 20th century may be pertinent to the growing problem of obesity in this country: one has to do with the extent to which machines have replaced muscular work since 1900 (VanItallie, 1979); the other is concerned with the change that has occurred in the food environment—increased food advertising, the rise of the supermarket, the proliferation of fast-food restaurants, food and candy vending machines, and other devices designed to make food more available, attractive, and tempting to the consumer. Although it is impossible to quantify the impact of these 20th century "newcomers" on eating behavior in the United States, information about the following "indices" (shown for 1987 in Table 1) is suggestive: (1) dollars spent for food advertising in the United States; (2) number of food/candy/soft drink vending machines in the United States; (3) number of supermarkets; (4) number of fast-food restaurants. Although in no way conclusive, these data are consistent with

TABLE 1. SOME STATISTICS ON CERTAIN "NEWCOMERS" TO THE 20TH CENTURY U.S. FOOD ENVIRONMENT (DATA FOR 1987)

Food advertising[a]
  $3.42 billion
Vending machines[b]
  Number: 3.7 million
  Units of foods/drinks/snacks dispensed:
    35 billion
Supermarkets[c]
  Number: 30,400
  Sales: $228.5 billion
Chain (mostly "fast food") restaurants[d]
  Number: 107,498
  Sales: $149 billion

[a]*Advertising Age*, September 28, 1988.
[b]*Vending Times*, census of the industry issue, 1988.
[c]*Progressive Grocer*, 55th annual report, April 1988.
[d]*Nation's Restaurant News*, August 8, 1988.

the hypothesis that changes in the food environment have contributed to the increasing prevalence of obesity observed in the U.S. population since the turn of the century.

Constitutional Determinants

GENDER. On the average, fat comprises a higher proportion of total body weight in women than in men (Durnin & Wormersley, 1974). The increased fat deposition first becomes conspicuous during puberty and is sustained throughout life (Zack, Harlan, Leaverton, & Cormoni-Huntley, 1979; Garn & Haskell, 1960). Men and women differ with respect to fat topography (Vague, 1947; VanItallie, 1988): women tend to deposit excess fat in hips and thighs ("femoral–gluteal," "gynoid," or "lower body" obesity), whereas men tend to deposit excess fat in the abdomen and upper trunk ("abdominal," "android," or "upper body" obesity). The propensity to store fat in hips and thighs may relate to woman's special reproductive role: evidence has been developed indicating that fat stored in these depots may constitute a dedicated reserve whose metabolic availability is enhanced during pregnancy and lactation (Rebuffé-Scrive et al., 1985).

AGE. Cross-sectional surveys have indicated that, on the average, adult men and women gain weight as they age, with men tending to reach a peak between the ages of 45 and 55 and women peaking a decade or more later (VanItallie, 1985). It is believed that, for the most part, the average increment in body weight during adult life (ca. 17 lb for men 5'10" tall and ca. 22 lb for women 5'6" tall) largely reflects increased storage of fat, since linear growth ordinarily ceases between the ages of 18 and 24 (Abraham, Johnson, & Najjar, 1979). This assumption is confirmed by many cross-sectional anthropometric studies indicating that, at several anatomic sites, the thickness of the subcutaneous fat layer increases with aging (Abraham, Carroll, Najjar, & Fulwood, 1983). Most conclusively, a variety of studies of body composition entailing the use of techniques such as hydrodensitometry, measurement of total body ^{40}K, and measurement of total body electrical conductivity all show that, on the average, adult men and women become fatter as they age.

It used to be thought that at least part of the age-related increase in fatness could be attributed to the gradual decline in resting metabolic rate (about 2% per decade) exhibited by adults during the aging process (Durnin & Passmore, 1967). This argument proved to be circular, however, because of the observation that lean body mass also tends to decrease with age (Forbes & Reina, 1970). Since resting metabolic rate is to a considerable extent a function of the lean body mass, it may be more accurate to ascribe a substantial portion of the drop in lean body (and muscle) mass and the increase in fatness associated with aging to the decreases in physical activity and muscular work that usually occur as people get older.

Physiology of the Body's Energy Economy

Energy Balance and Energy Stores: Interrelationships

Mammals have three significant energy storage systems—the gastrointestinal tract, the fat depot, and liver and muscle glycogen. Liver glycogen constitutes a reserve that has the special function of supporting homeostasis of blood glucose. Muscle glycogen, for the most part, is available to support local muscle contractions. When the gastrointestinal tract is empty, the body must rely on liver glycogen and depot fat as reserve energy sources. Since there is less than a day's supply of glycogen in the liver, it is evident that the ability of an animal to survive when food is not available depends on the presence of fat in the adipose tissue (Glore, Layman, & Bechtel, 1988). Because fat stores can be drawn on as needed, the amount that would be appropriate to set aside must depend on a variety of considerations. Initially, it is helpful to examine the factors that can lead to a change (Δ) in total body fat content, and, as a first approximation, this consideration can be addressed by means of the following formula:

$$\Delta \text{ body fat content} = \text{net energy balance} \times \text{time}$$

If net energy balance is negative, body fat content obviously will decrease. The longer this condition prevails, the greater the fat loss. Of course, the opposite result occurs when net energy balance is positive.

The problem of understanding the relation of energy balance to fat storage becomes remarkably complicated as soon as one begins to examine more closely the terms of the conventional energy balance equation; namely, energy balance = energy intake − energy expenditure. For example, it can be misleading to view the dependence of changes in fat content on energy balance as simply a matter of "calories in" versus "calories out." This is because "calories" arise from the catabolism of energy-supplying nutrients such as protein, carbohydrate, and fat. As is discussed in more detail below, isocaloric quantities of these macronutrients are not necessarily equivalent as calorie sources in the human body even though they generate similar amounts of heat when they undergo combustion in a bomb calorimeter. Similarly, "calories out" seems to be a simple concept until one becomes aware of the multiplicity of biochemical and physiological mechanisms by which energy is expended. The problem is further complicated by the fact that the efficiency with which the healthy body uses fuel freshly derived from ingested food is highly variable, being affected by such factors as (1) the composition of the diet, (2) the amount ingested relative to maintenance energy needs, (3) the status of the energy stores, and (4) the amount and nature of muscular work habitually performed. In view of the complexity of the problem of energy balance and the difficulties encountered in explaining spontaneous variations in human fatness, it is necessary to take a closer look at both the components of energy expenditure and the factors that affect energy intake.

Energy Intake

"Equicaloric" quantities of fat, protein, carbohydrate, and ethanol are not necessarily interchangeable as energy sources when they are metabolized in the mammalian body. Although this fact has been long understood by scientists concerned with animal nutrition (Blaxter, 1971, 1973), its significance for the human energy economy has been neglected until recently. Workers engaged in animal husbandry have long-known that equicaloric diets of differing composition vary in their ability to fatten beef cattle (feed efficiency), but it was not until the classic overfeeding experiments conducted by Sims and co-workers (1968, 1973) on human volunteers that clinical nutritionists became aware of the fact that, excess calorie for excess calorie, a nonobese individual can be more readily fattened by a high-fat than a high-carbohydrate diet. Subsequent overfeeding studies have confirmed and extended the results reported by Sims et al. (Horton & Danforth, 1985).

Why is fattening achieved more easily by a high-fat than a low-fat, high-carbohydrate diet?

A partial answer to this question is provided by the results of studies in laboratory animals, certain farm animals, and human subjects indicating that excess calories from dietary fat are converted to adipose tissue triglyceride with greater efficiency than are excess calories from dietary carbohydrate (see Table 2). And, apart from these experimental findings, there are solid theoretical reasons to predict that the thermic effect of feeding excess carbohydrate (leading to inefficient storage of dietary energy) will be considerably greater than that of feeding excess fat (Flatt, 1985). Yet, although studies in man tend to support this principle, the magnitude of the effect on energy expenditure of overfeeding fat versus that of overfeeding carbohydrate can differ widely from experiment to experiment. For example, Schutz, Acheson, and Jéquier (1985) overfed carbohydrate for 7 days to three young men, each of whom spent a total of 10 days in a respiration chamber. In these subjects, carbohydrate overfeeding (in amounts ranging from 1550 to 2770 excess kcal/day) induced a progressive stimulation of 24-hour energy expenditure, amounting to 33% of the excess energy intake by the seventh day of overfeeding.

In a more recent study, Lean, James, and Garthwaite (1989) found (at best) modest effects on energy expenditure when they overfed either fat or carbohydrate for only 24 hours to two groups of age-matched women (five normal weight and five formerly obese). Both hypercaloric diets contained 50% more energy than required for energy balance; in one case, the excess calories were all from carbohydrate, in the other they were mostly from fat. Lean et al. found that excess dietary fat did not increase 24-hour energy expenditure (measured in a whole-body calorimeter) in either the normal-weight or the postobese group. In contrast, 50% carbohydrate overfeeding for 24 hours increased mean metabolic rate by 2.2% in the postobese group and not at all in the normal-weight subjects. This small increase in the postobese subjects was much less than would be predicted for the obligatory costs of storing the excess energy (Flatt, 1985). Also, these results were somewhat at odds with earlier studies conducted by Lean and

TABLE 2. ENERGY RETAINED BY ADULT ANIMALS WHEN NUTRIENTS ARE GIVEN TO PROMOTE ENERGY RETENTION[a]

		Energy retained (kcal/100 kcal nutrient)	Increment of heat (kcal/100 kcal nutrient)
Carbohydrate	Glucose	72	28
	Starch	75	25
Protein	Casein	66	34
	Fish protein	64	36
Fat	Arachis oil	82	18
	C_{18} fatty acids	83	17

[a]Adapted from Blaxter (1973).

James (1987) demonstrating an overall lower thermic effect from a high-fat, low-carbohydrate diet fed over 24 hours than from an isocaloric low-fat, high-carbohydrate diet.

Schutz *et al.* (1985) identified four factors that might be invoked to explain the progressive increase in oxygen consumption that they observed during prolonged carbohydrate overfeeding: (1) the energy cost of converting excess carbohydrate to glycogen and to fat (an energetically inefficient process producing a corresponding increase in obligatory thermogenesis); (2) incremental facultative thermogenesis resulting from stimulation of sympathetic nervous activity (urinary norepinephrine was significantly elevated during carbohydrate overfeeding); (3) a small effect of the increase in body weight (by about 8%) resulting from the increased energy cost of moving; (4) deposition of lean body tissue (inferred from measurements of nitrogen balance) increased the size of the lean body mass, thereby contributing to a slightly higher basal metabolic rate.

Thus, studies of the thermogenic effects of carbohydrate or fat, when one of these nutrients is added to a weight-maintenance diet, have yielded varying results, depending (*inter alia*) on the age, sex, nutritional status, and previous weight history of the subjects, the size and precise nature of the nutrient load (Dusmet, Schutz, Bessara, & Jéquier, 1985), the duration of overfeeding, and the method used to measure thermogenic effect. More consistent results could be anticipated if such experiments were done under the same standardized conditions by several laboratories.

A great many studies have been conducted in rodents to determine the efficiency with which dietary calorigenic nutrients are used to manufacture and store triglyceride. One way of obtaining such information is to feed rats or hamsters diets that vary in nutrient composition (i.e., high-fat, low-carbohydrate or high-carbohydrate, low-fat diets). A record is then kept of the amount that is eaten, the rate of weight gain, and, at the end of the study, the composition of the animal's carcass. One measure of efficiency of use of the dietary calories is taken to be the ratio of calories ingested to those incorporated into the body.

In a typical experiment (average values cited), Oscai, Brown, and Miller (1984) showed that rats that consumed 36,113 kcal of a high-fat diet weighed 880 g at the end of the experiment, whereas rats consuming 36,125 kcal of rat chow (low in fat) over the same time period reached a weight of 666 g. The carcasses of the rats on the high-fat diet contained 452 g of fat (51% of body weight), whereas the carcasses of the chow-fed rats contained 205 g of fat (30% of body weight).

Although epidemiologic data involving comparison of population groups show a positive relationship between per capita fat intake of adults and mean body mass index (West, 1978), long-term studies have yet to be done in man demonstrating that the enhanced availability of foods high in fat content will predictably increase body fat content in individuals who normally subsist on a high-carbohydrate, low-fat diet.

In 1987, Lissner, Levitsky, Strupp, Kalkwarf, and Roe reported the results of short-term experiments in young adults in whom they determined the effect on spontaneous energy intake of manipulating the fat content of the diet. Three levels of dietary fat (as percentage of total available calories) were provided: 15–20%, 30–35%, and 45–50%. The dietary fat content was increased by adding fat-rich items to foods that were normally low in fat (e.g., cream added to strawberries, mayonnaise to tuna sandwich, vegetable oil to muffins). Each of the three diets was provided to the subjects for 2 weeks with the order of the experimental periods randomized. When offered the low-fat diet, the subjects consumed an average of 2087 kcal/day and slowly lost weight; when offered the high-fat diet during another experimental period, the same subjects consumed an average of 2714 kcal/day and slowly gained weight. When offered the diet that provided 30–35% fat, the subjects consumed an average of 2352 kcal/day, an intake associated with virtually no change in body weight.

Although provocative, these results are difficult to interpret. For example, it is not clear whether the subjects consumed more calories on the high-fat diet because the foods tasted better, because the caloric density of the foods was higher, or for some other reason. According to the authors, the subjects gave the low-fat diet a good palatability rating; yet, there is a large body of evidence indicating that the taste and texture of many foods are enhanced by the addition of fat. Moreover, it is possible various foods differed in their satiating efficiency, thereby affecting in differential fashion intakes of the various foods (Kissileff, 1984). In any event, the observations reported by Lissner et al. are consistent with other studies showing that many people fail to exhibit commensurate caloric compensation when the energy density of the diet is systematically varied, either by changing its fat content or its carbohydrate content (Porikos & VanItallie, 1984).

Thus, such evidence as is available suggests that, over the short term, some people will spontaneously reduce calorie intake when shifted from a high-fat to a low-fat diet. It is still not known whether individuals who remain on a low-fat diet will continue to eat fewer calories and equilibrate at a reduced weight or whether they will gradually return to their original weight. Current evidence suggests that for at least several months, humans accustomed to a high-fat diet will spontaneously decrease energy intake in response to a low-fat diet. Thus, for some individuals, diets diminished in fat content may be useful in weight control in two respects: (1) by favoring a decreased intake of calories and (2) by reducing the efficiency with which calories are stored as fat.

Energy Expenditure

In the past, energy expenditure for 24 hours (or longer) was calculated by the factorial method, based on a physical activity diary. The diary was used to specify different activities and their duration, with the cost of each type of activity—sitting, standing, walking, etc.—either estimated from tabular data available in the literature or measured by indirect calorimetry. An example of the factorial approach is provided in Table 3 in which the 24-hour energy expenditure of a 70-kg reference man is calculated (Committee on Dietary Allowances, 1980).

Although the factorial approach may be useful for estimating the approximate daily energy expenditures of individuals whose daily agenda is known, it appears unable to explain why individuals whose overall physical activity appears to be similar may differ markedly in body fat content.

During the past two decades, students of energy balance physiology have begun to focus increasingly on a wider array of components of energy expenditure. The four components that have attracted greatest attention are (1) resting metabolic rate, (2) thermic effect of food, (3) spontaneous physical activity ("fidgeting"), and (4) voluntary physical activity. It has been of particular interest

TABLE 3. AN EXAMPLE OF USE OF THE FACTORIAL METHOD FOR CALCULATING 24-HOUR ENERGY EXPENDITURE[a]

Activity category[b]	Man, 70 kg		
	Time (h)	Rate (kcal/min)	Total (kcal)
Sleeping, reclining	8	1.0–1.2	540
Very light	12	up to 2.5	1300
Seated and standing activities, auto and truck driving, laboratory work, typing, playing musical instruments, sewing, ironing			
Light	3	2.5–4.9	600
Walking on level, 2.5–3 mph, tailoring, pressing, garage work,electrical trades, carpentry, restaurant trades, washing clothes, shopping with light load, golf, sailing			
Moderate	1	5.0–7.4	300
Walking 3.5—4 mph, weeding and hoeing, loading and stacking bales, scrubbing floors, shopping with heavy load, cycling, skiing, tennis, dancing			
Heavy	0	7.5–12.0	
Walking with load uphill, tree felling, work with pick and shovel, basketball, swimming, climbing, football			
Total	24		2740

[a]Format adapted from Committee on Dietary Allowances (1980).
[b]Data from Durnin and Passmore (1967).

to measure the proportional contribution of each of these components to total 24-hour energy expenditure. Such measurements can now be made by means of a respiratory chamber in which a volunteer subject remains in reasonable comfort for 24 hours or longer (Jéquier & Shutz, 1983). It is possible to equip such a chamber with a radar device that records spontaneous movements while the subject is "at rest." Oxygen consumption and CO_2 production and their changes are monitored continuously. Using a respiratory chamber of this type, Ravussin, Lillioja, Anderson, Christin, and Bogardus (1986) determined in subjects at rest the percentages of 24-hour energy expenditure contributed by resting metabolic rate, the thermic response to food, and spontaneous physical activity. The results in 118 Southwestern American Indians aged 18 to 65 years are summarized in Figure 2.

As shown in the figure an average of 78% of the total energy expended per 24 hours was contributed by the resting metabolic rate (RMR). Differences in lean body mass accounted for 81% of the variance in daily energy expenditure among people. More than half of the remaining variance was explained by family membership. In this same series thermic effect of food accounted for an average of 7%, and spontaneous physical activity accounted for an average of 15% of total energy expenditure per 24 hours. It was noteworthy that the contribution of spontaneous activity to total daily energy expenditure ranged from 6% to 30%, with much of the variance explained by family membership.

Spontaneous activity is not considered in activity diaries, and hence the contribution of this activity (which may exhibit a ninefold variation) to total daily energy expenditure has been neglected in the past. Moreover, a number of

THEODORE B.
VanITALLIE AND
HARRY R.
KISSILEFF

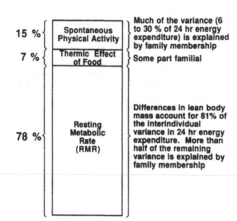

15 %	Spontaneous Physical Activity
7 %	Thermic Effect of Food
78 %	Resting Metabolic Rate (RMR)

Much of the variance (6 to 30 % of 24 hr energy expenditure) is explained by family membership

Some part familial

Differences in lean body mass account for 81% of the interindividual variance in 24 hr energy expenditure. More than half of the remaining variance is explained by family membership

Figure 2. Components of human energy expenditure as determined over a 24-hour period in a respiratory chamber. Note the influence of family membership on the variance associated with each component. (From Ravussin *et al.*, 1986.)

investigators have reported that the resting metabolic rate correlates well with the size of the lean body mass, particularly the body cell mass moiety (Webb, 1981; Weigle, Sande, Iverius, Monsen, & Brunzell, 1988). Despite the high positive correlation that obtains between RMR and lean body mass (LBM), RMR may vary fairly widely among individuals even after it has been normalized for LBM (mL O_2 consumption/h per kg LBM). For example, there is evidence that, compared to young men and women whose RMR/kg LBM is normal or above normal, age-matched individuals whose RMR/kg LBM is below normal are at significantly higher risk of gaining excess weight and becoming obese during a subsequent 4-year observation period (Ravussin *et al.*, 1988). It has also been shown that formerly obese women who had restricted calorie intake and lost their excess fat exhibited RMR values (normalized for LBM) about 15% lower than those of never-obese women of comparable body composition (Geissler, Miller, & Shah, 1987).

THERMIC RESPONSE TO FOOD

The size of the thermic response to (or thermic effect of) food appears to be influenced by metabolic factors within the individual (at least in part genetically determined) and by the nature of the ingested food. Thus, individuals who differ in their genetic makeup (i.e., with and without a strong family history of obesity) may also differ in the thermic response they exhibit following ingestion of a standard meal. Jéquier and Schutz (1985) have reported that a defect in diet-induced thermogenesis can be demonstrated in about one-third of an unselected group of obese women. Once demonstrated, this defect does not disappear following therapeutic weight reduction.

As Jéquier and Schutz and others (Trayhurn & James, 1981) have pointed out, the thermogenic response to food intake can be divided into two components—"obligatory" and "facultative." The obligatory component depends on the energy costs associated with nutrient ingestion, factors such as activity of gastrointestinal smooth muscle, the processes of digestion and absorption, and the metabolic work involved in processing and disposing of diet-derived substrates. The facultative component is related to the stimulating effect of food ingestion on various energy-dissipating processes, including activation of the sympathetic nervous system.

Because of many studies that have shown that brown adipose tissue (BAT) is an important energy-dissipating mechanism in rodents (Himms-Hagen, 1984), and that BAT thermogenesis is under sympathetic control (Rothwell & Stock, 1981b; Stock & Rothwell, 1985), investigators have searched assiduously for indications that BAT plays a significant physiological role in the adult human body. To date, no convincing evidence has been reported that BAT plays such a role in the energy economy of adult man (Welle, 1985).

Although a number of investigators have reported that some obese individuals exhibit a reduced thermogenic response to food ingestion, the basis of such "defective" responses is not well understood. Danforth, Daniels, Katzeff, Ravussin, and Garrow (1981) and Katzeff and Daniels (1985) have demonstrated that the thermogenic response of Pima Indians to graded doses of norepinephrine is similar to that of lean Caucasians; in addition, a number of workers have reported studies indicating that blunting of the thermogenic response to glucose of both obese patients and rodents can be related to insulin resistance. Thus, streptozotocin-induced diabetic rats (Rothwell & Stock, 1981a) and insulin-resistant patients (Golay et al., 1982; Ravussin et al., 1983) exhibit reduced thermogenic responses to glucose loads.

Even though BAT does not appear to be functional in adult man, its postulated role in certain rodents is extremely interesting. Thus, the concept has evolved that, in mice and rats (for example), BAT functions as an energy buffer in the regulation of energy balance (Himms-Hagen, 1984, 1985). This role is well illustrated in two animal models. (1) In the cold-acclimated rat, BAT is hypertrophied and hyperplastic, giving the animal an enhanced capacity for cold-induced nonshivering thermogenesis. Although the thermogenic activity resides in BAT, it is mediated by the sympathetic nervous system, which, in turn, is under hypothalamic control. In the cold-acclimated rat, which exhibits hyperphagia and a high metabolic rate, as much as 40% of its food energy can be used for thermogenesis in BAT. (2) Cafeteria-fed obese rats are both hyperphagic and hypermetabolic. The hypermetabolism, in turn, is related to increased sympathetic nervous system activity and increased thermogenesis in BAT. Indeed, like the cold-acclimated rat, the cafeteria-obese rat develops a larger BAT mass and an increased capacity to increase its metabolic rate in response to norepinephrine.

In contrast to the metabolic responses exhibited by overfed animals, fasting and food-restricted rats become hypometabolic, showing a reduced amount of BAT, a reduction in sympathetic activity, and a suppression in BAT thermogenesis. Also, the food-deprived rat exhibits a diminished ability to increase its metabolic rate in response to administered norepinephrine.

It is obvious that in the underfed rat, metabolic efficiency is enhanced via an adaptive reduction in BAT thermogenesis. Conversely, in the overfed rat, metabolic efficiency is reduced via an adaptive intensification in BAT thermogenesis. These changes in efficiency help to conserve energy when food is in short supply; moreover, they act to reduce excessive fluctuations in body fat content.

The energy-buffering mechanism that appears to operate in the rat has many of the attributes of a negative feedback control system. Thus, a change in the output of the system, which could be the organism's "awareness" of an insufficiency or excess of diet-derived fuel, induces an opposite change in the input, namely, the efficiency with which the fuel is used. Here, it would seem that the goal of the system is to ensure that, despite variations in the availability of dietary energy, the energy reserves are protected. Thus, when food is in short supply, dietary energy is used with frugality, helping to preserve fat stores. When food is abundant, the wasteful use of dietary energy could help to protect against an undesirable buildup of the fat depot. At the same time these apparently goal-directed processes could also have as their function the preservation of the LBM, for which the fat mass could act as a protective metabolic buffer. This notion is considered further in connection with the body weight set-point issue.

Although BAT thermogenesis underlies the energy-buffering mechanism in the rat, it is entirely conceivable that an analogous mechanism can operate in man in the absence of BAT. Such a concept is considered later.

METABOLIC ROLE OF STORED TRIGLYCERIDE

In man and many other species, the triglycerides stored in adipocytes serves as the body's major energy reserve. Thus, the adipose tissue is able to provide readily available fuel (as free fatty acids and glycerol) to certain working cells, particularly when other energy sources are in reduced supply. For several hours after a meal, the absorbed products of digestion such as glucose and other hexoses, "fat" (as chylomicrons, partial glycerides, or short- or medium-chain fatty acids), and amino acids are delivered to the circulating energy pool and become available to meet energy needs. In the postabsorptive phase, when these diet-derived substrates have been largely cleared from the circulation, the body must rely on liver glycogen and glucogenic amino acids released from protein stores to maintain the blood sugar at levels sufficient to permit the brain to function normally. At the same time, free fatty acids (FFA) derived from adipose tissue triglyceride take on the burden of serving as the principal fuel source for most of the body's nonencephalic metabolic functions.

The foregoing facts indicate the critical role played by adipose tissue triglyceride in the operation of the body's energy economy. Later, in this analysis, we illustrate this role by comparing it to the part played by an inventory of manufactured products in a business.

SPONTANEOUS VARIABILITY IN MAN OF BODY FAT CONTENT

One of the most striking features of the adipose tissue is the enormous variability in the quantity of fat that can be stored in this depot. A well-trained marathon runner who weighs 135 lb (60.8 kg) may possess a fat depot of 10.8 lb (4.9 kg), whereas a morbidly obese man or woman weighing 400 lb (180 kg) may have fat content of 215 lb (96.8 kg). This amounts to a 20-fold difference in absolute fat content. In order to understand such variability, it is helpful to examine the morphology of the adipose tissue.

At any given time, the quantity of fat stored in adipose tissue is a function of fat cell size and fat cell number (Björntorp & Sjöstrom, 1971). Thus,

Adipose tissue triglyceride content = mean fat cell size × fat cell number

The relationship between mean fat cell size and number is shown graphically in Figure 3. As the figure indicates, nonobese adults often possess somewhere

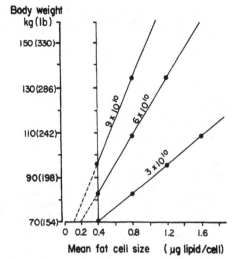

Figure 3. Relationship of fat cell number and mean fat cell size to body weight. Note that individuals with hyperplastic adipose tissue can only achieve "optimal" weight by reducing fat cell size below normal.

between 30 and 40 billion fat cells whose average size may range from 0.4 to 0.6 μg lipid per cell. (For example, 30 billion cells × 0.5 μg triglyceride/cell = 15 kg triglyceride, constituting approximately 21% of the weight of a 70-kg individual.)

Uncertain Relationship between Body Weight and Body Fat Content

Total body fat content can be estimated from relative weight (for example, weight relative to some standard of optimal weight for height), various height–weight indices (the most commonly used being the BMI), various combinations of skinfold thicknesses and other anthropometric measurements, hydrodensitometry, and technically sophisticated methods such as measurement of total body electrical conductivity, bioelectrical impedance, total body water, and total body potassium. When the results of these methods are compared, it is clear that relative weight and BMI are not necessarily good predictors of fatness. For example, in the first National Health and Nutrition Examination Survey (NHANES I) conducted during 1971–1974 by the National Center for Health Statistics (NCHS), both the body mass index and the quantity of subcutaneous fat (estimated from the sum of the triceps and subscapular skinfold thickness) were determined in a representative sample of the U.S. population. When the resulting data were cross-classified, using NCHS criteria for overweight (based on BMI) and obesity (based on skinfold thickness), the authors identified five categories of individuals: (1) underweight, not obese; (2) normal weight, not obese; (3) normal weight, obese; (4) overweight, not obese; (5) overweight, obese (Abraham *et al.*, 1983). The proportions of the NHANES I population corresponding to these designations are shown in Table 4.

TABLE 4. CROSS-CLASSIFICATION OF U.S. ADULT POPULATION[a] ACCORDING TO THE
DISTRIBUTION OF BODY MASS INDEX[b] AND OF TRICEPS PLUS SUBSCAPULAR SKINFOLD
MEASUREMENTS[c]

Categories[d]	Prevalence (%)	
	Men	Women
Underweight, not obese	18.1	17.8
Normal weight, not obese	52.4	46.7
Normal weight, obese	6.7	6.0
Overweight, not obese	10.2	7.9
Overweight, obese	12.6	21.6

[a]The reference population on which the percentages are based was a representative sample of civilian noninstitutionalized U.S. men and nonpregnant women 20–29 years old who were examined during the first National Health and Nutrition Examination Survey (NHANES I), 1971–1974.
[b]Body mass index is W/H^P, where P is 2 for men and 1.5 for women.
[c]Data are from Abraham et al. (1983).
[d]Underweight, normal weight, and overweight are defined with respect to the 15th and 85th percentiles of the distribution of the body mass index. Nonobese and obese are separated with respect to the 85th percentile of the distribution of the triceps plus subscapular skinfold thickness.

THE "SET-POINT" THEORY OF BODY WEIGHT REGULATION

It is impossible to consider the physiology of the body's energy stores without sooner or later confronting the concept that the depot fat is regulated at a specified level or "set point." A corollary of this notion is that obesity ". . . represents a condition of regulation at an elevated set point. . ." (Keesey, 1986).

Keesey (1986, 1989) has been one of the most eloquent and persuasive proponents of the "set-point" model of energy regulation as contrasted to the more traditional behavioral perspective. However, in our view, one need not take an "either/or" position on the regulation of body fat or the origins of obesity; as we have pointed out, the size of the fat depot at any given time is the resultant of many influences, which may include both learned "maladaptive eating patterns" and genetic effects.

In a comprehensive review of the set-point theory, Keesey (1986) emphasized (1) the relative constancy of body weight in adult humans and other adult mammals and (2) the spontaneous reversion to the "normally maintained" weight level that occurs in rats whose weight has been artificially reduced by dietary restriction or raised by force feeding. Further, Keesey called attention to some of the mechanisms that act to restore an artificially altered weight to normal: (1) anorexia and weight loss in rats whose weights were previously elevated by lateral hypothalamic stimulation; (2) partial dissipation of excess dietary energy as heat in healthy volunteer subjects made obese by experimental overfeeding; (3) an apparent increase in efficiency of energy utilization in weight-reduced rats, permitting spontaneous restoration to control weights when the rats are given only the amounts of food they normally ate prior to caloric restriction; and (4) a corresponding decrease in efficiency of energy utilization in overfed humans and rats, acting to limit weight gain.

Keesey holds that the energy-buffering mechanisms previously discussed constitute evidence for a set-point type of regulation. In taking note of the fact that two individuals of the same age, sex, and height may maintain widely differing body weights for prolonged periods of time, he asks, "What determines or sets the body weight each individual maintains and defends?"

As Keesey stresses, studies of animals in which the lateral hypothalamus (LH) has been systematically damaged appear to provide the best evidence in support of the set-point model. As is well known, LH-lesioned animals initially exhibit aphagia or anorexia and lose weight to a lower level, which they then defend. But if the body weight of a rat is sufficiently reduced by dietary restriction prior to LH lesioning, the postlesion aphagia and anorexia are eliminated or greatly reduced. It appears that when they stabilize at a reduced weight, LH-lesioned rats maintain a RMR that is appropriate for their new size; in contrast, nonlesioned weight-reduced rats are said to exhibit decreases in RMR significantly lower than would be expected on the basis of their reduced metabolic mass. Finally, if LH-lesioned rats are stimulated by a cafeteria-type diet to achieve body weights corresponding to the level of nonlesioned control rats fed a conventional laboratory diet, their RMRs may be some 30% higher than those of the nonlesioned animals. Indeed, immediately after placement of LH lesions and before they have lost weight, LH animals exhibit an acutely elevated RMR.

Keesey acknowledges that body-weight set points can be altered in a manner that may be analogous to the rise in body-temperature "set point" that is postulated to occur during the course of a febrile illness. As examples, he cites the weight changes seen in hibernating rodents. Ground squirrels, for example, show a prolonged phase of hyperphagia and weight gain before hibernation. Mrosovsky and Fisher (1970) have described this phenomenon as an example of a "sliding set point." Keesey also suggests that a high-fat diet can elevate the body-weight set point, whereas an increase in physical activity can lower it. Conceivably, the putative set point in humans is also subject to alteration by starting or ceasing the habit of cigarette smoking, by aging, or by certain changes in endocrine function.

INADEQUACIES OF THE SET-POINT MODEL

Despite the fluency of the arguments marshaled in support of the set-point model, we question whether this model provides the most parsimonious explanation of the various phenomena that have been invoked as evidence for the existence of a set point for body weight.

It is therefore appropriate to consider the problems raised by the set-point hypothesis for body weight regulation. The first problem is the nature of the detected output. It is unlikely that body weight *per se* is monitored. Since the components involved in body weight change are rarely measured, it is often assumed that a change in body weight is a change in body fat content. This may not be the case. When people (and animals) gain weight, there is usually an increase in the lean body mass as well as in the size of the fat depot. Conversely, when human beings lose weight, a substantial proportion of the loss may be in the form of lean body mass (VanItallie & Yang, 1977). Thus, when an animal that is experimentally underfed or overfed appears to make homeostatic adjustments in metabolism or food intake, it is still not clear whether some portion of these adaptations might not represent responses to a change in lean body mass rather than to a change in body fat content. Indeed, many of the metabolic responses that are now uncritically attributed to depletion or excess of the body fat stores may in fact be partly or wholly caused by concomitant changes occurring in the fat-free portion of the body. In future studies, it will be essential to specify the changes in body composition that occur as a result of dietary (or other) manipulations that affect energy stores. Then it should become possible to gain a clearer picture of which component of weight change, if any, the animal detects.

In assessing the set-point model, it is also important to have an integrated perspective on the role of the energy stores in metabolism. As mentioned earlier, the adipose tissue serves as an energy reservoir subordinated to the ongoing energy needs of the body. Thus, although the body must always have a supply of energy in order to survive, it does not require a reserve of a specific size for survival. To be sure, there are special cases. Some birds need to store a supply of fat that is sufficient to fuel their migratory flights (Odum, 1960). Ground squirrels need enough extra fat to sustain them during hibernation (Mrosovsky, 1985; Mrosovsky & Faust, 1985). But, by its very nature, it is adaptive for the adipose storage depot to be flexible; hence, fat cells have the capacity to shrink to a very small volume, to increase to a large size, and, in some instances, to proliferate (Faust, Johnson, Stern, & Hirsch, 1978). Because the adipose tissue's triglyceride store serves to provide a constantly changing flow of fuel to a body whose needs for energy can vary over time, there would seem to be no biological reason for this depot to be maintained at some "set" level. Thus, we do not believe that energy intake is precisely adjusted to changes in energy expenditure in order to maintain the depot fat at some predetermined quantity.

This is not to say that the body does not have mechanisms to restore a depleted fuel supply or to protect itself from surfeit. The energy-buffering systems discussed earlier clearly help to fulfill these objectives. However, it seems that the biological function of the body's fuel reserves is quite different from that of blood glucose or of body temperature. For example, maintenance of blood glucose above a certain level is essential for normal brain metabolism and, ultimately, life. For the sake of argument, one might concede that a system containing a set point can explain with some plausibility the precision with which the blood glucose is regulated in the face of the frequent changes that occur in the availability of dietary carbohydrate and in the rate of glucose utilization by the body cell mass. Factors that keep the blood glucose level within certain limits include the liver's supply and rate of hydrolysis of glycogen, the pattern of hormone secretions—insulin, glucagon, growth hormone, and catecholamines—and the activity and balance of the autonomic nervous system. However, it would be difficult to defend the concept that both the blood glucose level and the liver glycogen content are simultaneously maintained at relatively constant values by the operation of systems that include set points. Because the liver supplies glucose needed to prevent the occurrence of life-threatening hypoglycemia, simultaneous homeostatic control of blood glucose and liver glycogen is impossible.

Similarly, the adipose tissue subserves the body's constant need for fuel, and, quite clearly, the critical need for energy must take precedence over a putative need to maintain the fuel supply at any given "set" level.

Counterregulatory Mechanisms

Yet, as mentioned earlier, the body cannot long survive if its reserves of liver glycogen or fat become unduly depleted. It is logical to expect that as the body's reserve of triglyceride diminishes, mechanisms will come into play that favor its replenishment. This requirement can be fulfilled by ingestion of food in excess of that required for proximate energy needs and, when food is in short supply, by cutting back on energy expenditure, whether in the form of physical activity or resting metabolic rate, or by improving the efficiency of energy use.

Proponents of the set-point model predict that the reduction of body fat content below an individual's "set" level will trigger various mechanisms (described earlier) designed to replenish this energy reserve. The fact that an increase in physical activity (as might occur during training for a long-distance race) can substantially decrease body fatness without actuating restorative metabolic mechanisms is explained by these same proponents as being one of the many situations in which the set point is altered.

AN INVENTORY CONTROL MODEL

At this juncture, it is appropriate to consider a different view of the regulation of the body's energy reserves. This view regards the size of the fat stores as being integrated into the body's overall energy economy and therefore determined by the varying needs of that economy.

In other words, the size of the fat depot is seen to be the result of an equilibrium of positive and negative energy fluxes within the body—fluxes that are determined by such factors as the composition of the diet, the pattern of food availability, the energy that must be expended in the course of obtaining food, the size of the meal (governed in part by the size of the stomach), the energy cost to the organism of sustaining its metabolic processes (the resting metabolic rate), and the physical work it performs, including the energy cost of moving the body from one place to another.

In this regard, we have found it instructive to compare the human energy economy to the economy of a manufacturing business (VanItallie & Kissileff, 1985). In a manufacturing business, increasing the production run of a manufactured item reduces the per-unit cost of production. Obviously, it would be wasteful to set up an assembly line in an automobile factory to produce only a few cars. On the other hand, if sales of cars are lagging, the inventory of manufactured cars will rise, and with that rise the carrying cost of the inventory will also increase. At some point, the carrying cost of the expanded inventory will become sufficiently high to offset the advantage gained by sustaining production at its previous level. In this case, production will have to be cut back, or some way must be found to reduce the inventory.

In the case of man (for example), one can regard the amount of food taken at the time of a meal as being comparable to a production run. If there is an energy cost involved in obtaining a certain amount of food (by hunting or foraging), then it becomes economical to increase the amount of food consumed per meal so as to reduce the unit (per calorie) cost involved in procuring fuel for the body. The ability to make such an increase is limited by the size of the stomach. Thus, if it costs a certain number of calories of muscular work to obtain a meal, it is obvious that increasingly larger meals will increase the benefit : cost ratio of the food procurement process. However, there are conditions under which such a favorable benefit : cost ratio could lead to an excessively high inventory of stored energy (fat) associated with a progressive deterioration of the benefit : cost ratio. In this instance, the increased weight and size of the fat depot would increase to an uneconomical level the energy cost of sustaining the body's resting metabolism and its movements. At some point then, the increased cost of sustaining and moving the body ("carrying cost") would limit the size of the fat depot, which would then tend to stabilize at an equilibrium between maintenance energy needs plus the energy cost of food procurement and the energy content of the food procured.

The Concept of "Carrying Cost"

One of the factors acting to favor stability of the body's inventory of energy is the change in carrying cost that occurs as the size of that inventory changes. According to Jéquier and Schutz (1983), for every kilogram that is added to man's body weight the daily energy expenditure increases by about 20 kcal. Conversely, for every kilogram lost, daily energy expenditure decreases by approximately 20 kcal. As mentioned earlier, these changes come about because the resting metabolic rate tracks changes in body size; in addition, a change in weight produces a corresponding change in the energy cost of moving the body.

When energy balance becomes positive, this change, if sustained, produces an enlargement of the energy depot. Enlargement of the depot, in turn, progressively increases energy expenditure, which has the effect of gradually bringing energy intake and output once more into equilibrium (Figure 4). It is noteworthy that this equilibrating effect can occur without any change in food intake or physical activity pattern (VanItallie, Kissileff, & Yang, 1986).

By means of the foregoing comments, we have pointed out that when the body's energy economy changes, the size of the triglyceride depot readily increases or decreases, and also that this lability has survival value for the organism in certain situations and environments. The operating levels of the various biochemical and physiological systems involved in the control of body fat presumably have been shaped by selection pressures over the course of evolution. In order to be precise about such levels, it is necessary to specify the conditions under which they are measured.

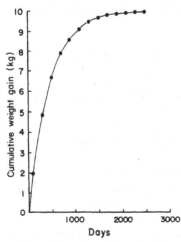

Figure 4. Self-limitation of weight gain following an indefinitely sustained increase in energy intake of 200 kcal/day without any change in the pattern of physical activity. The curve illustrates how the rate of weight gain progressively diminishes, reaching an asymptote after a cumulative gain approximating 10 kg. The equation used to generate this curve was derived from the following two assumptions. (1) The individual consumed an extra 200 kcal per day, which was converted to fat with 100% efficiency. This addition would result in a 22.222 g [200 kcal/(9 kcal/g)] weight gain. (2) The individual would lose each day an amount equal to 20 kcal for each kilogram of extra accumulated weight (Jéquier & Schutz, 1983). Conversion of this energy into fat loss at 100% efficiency would result in a fractional weight loss equal to 0.00222 [20 kcal/(1000 g × 9 kcal/g)]. The actual loss on a given day would be the algebraic sum of the fractional loss and the cumulative weight gain from the previous day. The net gain each day would therefore be computed by subtracting the actual loss from the daily increment.

Triglyceride is stored in adipocytes, and the amount of triglyceride that can be accommodated in adipose tissue is limited by the ability of the fat cell to enlarge and by the number of fat cells. Moreover, adipocytes themselves tend to maintain a particular size because their rate of basal lipolysis increases as they enlarge and decreases as they diminish in size (Björntorp, 1972; Bray & Campfield, 1975). In addition, as fat cells increase in size they may undergo other metabolic changes (e.g., reduced insulin sensitivity) that seem to make them increasingly resistant to further enlargement. Such behavior of individual fat cells obviously contributes to the relative constancy of body weight observed in most adult mammals (Flatt, 1972, 1988).

If body fat content is regarded as being, in part, the product of an equilibrium of energy fluxes, it follows that a chronic displacement of that equilibrium (as would be caused by therapeutic caloric restriction in an obese individual) would decrease "body weight." If the reduction in calorie intake is not too severe, the individual who was in negative caloric balance will eventually return to a state of energy equilibrium, albeit at a lower level of energy exchange. The new equilibrium achieved would reflect both a reduced calorie intake and a reduced energy expenditure. Everyone who has treated severely obese patients would agree that, after they have lost weight, such patients have a strong tendency to regain it; however, to attribute such recidivism to the operation of a set point is to neglect the simpler explanation that people have a varying storage capacity for fat that appears to depend in part on fat cell number. Thus, the return to an elevated weight may depend in large part on the fact that fat cells emptied of much of their triglyceride content are metabolically disposed to reaccumulate fat.

Although an expanded storage capacity for fat based on adipocyte hypercellularity is believed to make it more difficult for a weight-reduced individual to avoid returning to the obese state (Krotkiewski et al., 1977), many formerly corpulent people have been able to remain slender by virtue of maintaining appropriate changes in their eating and exercise behaviors. These changes have altered the dynamics of body's energy economy, but this is not the same as saying that a change has occurred in a set point for body weight.

Some investigators have suggested that children with a strong family history of obesity exhibit accelerated proliferation of their adipose tissue between infancy and adolescence (Brook, 1972; Salans, Cushman, & Weismann, 1973). If the existence of such a genetically determined phenomenon can be confirmed, the presence of adipose tissue hyperplasia would be sufficient to explain their increased storage of fat; it is therefore not necessary to invoke the operation of an elevated set point to explain this form of childhood obesity.

Evidence is mounting that many preobese or formerly obese persons may have a reduced rate of energy expenditure that somehow increases their risk of accumulating or reaccumulating excess fat. It has been suggested that, in Pima Indians (Ravussin et al., 1988) and infants born to overweight mothers (Roberts, Savage, Coward, Chew, & Lucas, 1988), this reduced rate is genetically determined. In preobese or postobese persons, the several components of energy expenditure we mentioned earlier have been proposed to explain the individual's disposition to become obese: a lower resting energy expenditure (Ravussin et al., 1988), a lower nonresting energy expenditure (Weigle et al., 1988), a

THEODORE B.
VanITALLIE AND
HARRY R.
KISSILEFF

reduced physical activity level (Roberts *et al.*, 1988), a low 24-hour metabolic rate (Geissler *et al.*, 1987), and a reduced thermic effect of food (Jéquier & Schutz, 1985). In addition, certain pharmacologically active agents (fluvoxamine or nicotine) that increase energy expenditure can reduce body weight, and agents (imipramine and amitriptyline) that decrease expenditure can increase body weight (Hofstetter, Schutz, Jéquier, & Wahren, 1986; Jacobs & Gottenborg, 1981; Fernstrom, Epstein, Spiker, & Kupfer, 1985; Fernstrom & Kupfer, 1988). Thus, all other factors being held constant, a reduced rate of energy expenditure will result in a greater fat accumulation, whereas an increased rate of expenditure will decrease body fat content.

One explanation for these phenomena would be to attribute the expenditure-related alterations in body weight to a changing set point (i.e., drugs like imipramine and fluvoxamine respectively raise and lower the set point for body weight; Stunkard, 1982). Another interpretation (the one we favor) is that body fat content is determined, at least in part, by the effect of these drugs on the rate of energy expenditure and hence on the balance between rate of storage and rate of mobilization of adipose tissue triglyceride. If there is a sustained increase in rate of mobilization, the size of the fat cell (and the fat depot) will decrease until a new equilibrium is reached between storage and mobilization.

The Triglyceride Depot Viewed as an Energy Reservoir

In summary, in this part of our consideration of the possible determinants of the body fat content, we have argued that neither the variation in the size of the fat depot that occurs among individuals nor the relative constancy observed within an individual is best explained by a set-point model. Although such a model has certain appealing features, the very notion of a set point in biology implies the "function of automatically maintaining an important physiological variable within a narrow range in the face of disturbances that may act on the system" (Brobeck, 1965). In our view, the body's triglyceride depot is more like an unattended reservoir whose state of fullness is subordinated to the overriding requirement to supply water to a community of users. When rainfall is adequate, the amount of water in the reservoir is usually determined by its storage capacity, including the height of the dam, not by a set point. When the reservoir becomes depleted, the community takes (or should take) measures to conserve water, but not in order to maintain a certain water content in the reservoir. Similarly, when the food supply is adequate and energy expenditure is relatively low, the size of the fat depot tends to be determined by the storage capacity of the adipose tissue; however, during periods of food scarcity, or when energy expenditure is high (Woo, 1985), the depot may shrink to a level considerably below capacity. If one were to give this notion a teleological flavor, the triglyceride in the adipose depot is there to respond to the body's need for energy, not to be maintained at some preset level.

Within a relatively homogeneous population that shares the same environment, interindividual variations in fatness are attributable in considerable part to genetic influences (Price, 1987). In addition, when the food supply is abundant, the sedentary individual's storage capacity for fat is likely to be filled. If an overweight individual loses fat because of transitory adherence to a calorically restricted diet, there is every reason to expect that all of the fat loss will be regained once the dieter has returned to his prediet eating habits. If each individual has a characteristic storage capacity for fat, then it should not be surprising that, after undergoing therapeutic weight loss, he or she is likely to return to or near the same weight that obtained prior to the period of caloric restriction. Such "recidivism" does not have to be explained by the operation of a system that embodies a set point; we believe that it can be adequately understood on the basis of the equilibratory mechanisms discussed above.

Eating Behavior in Relation to the Size of the Energy Stores

There is abundant evidence suggesting that eating behavior is influenced by the state of the energy stores (Kissileff & VanItallie, 1982). Thus, previously nonobese animals and humans whose triglyceride reserves have been depleted by means of dietary caloric restriction tend to overeat and regain their lost weight after they are given unlimited access to food. However, as we pointed out earlier, this picture is complicated by the fact that both fat and lean body mass are lost during weight reduction induced by caloric restriction.

For example, during the first 77 days of experimental semistarvation, 60% of the weight lost by the 32 previously healthy and nonobese volunteer subjects studied by Keys *et al.* in Minnesota during World War II was lean body mass (Keys, Brozek, Henschel, Mickelsen, & Taylor, 1950). One wonders what proportion of the poststarvation hyperphagia exhibited by these subjects was generated by depletion of fat stores and what proportion by depletion of the lean body mass.

This issue is underscored by the fact that after 3 months of semistarvation the Minnesota subjects had a body-fat content that was reduced to an average of 8.7% of body weight (from an original 14%). This degree of leanness is also approached by some elite marathon runners; however, marathoners are generally healthy and energetic, whereas the food-deprived Minnesota subjects were chronically weak, fatigued, cold, apathetic, and depressed. It is noteworthy, however, that the Minnesota subjects lost about 19% of their body cell mass ("active tissue") during the first 3 months of semistarvation. In contrast, long-distance runners lose fat during training, but there is every reason to believe that their lean body mass remains intact or even enlarges somewhat as physical strength and endurance increase (Wilmore, 1983). Marathon runners are free to consume as much food as they want; yet, they obviously do not consume sufficient extra calories to replenish their "depleted" triglyceride reserves.

Unfortunately, it is difficult to manipulate the body's content of adipose tissue triglyceride independently of the lean body mass (LBM). Until experiments are performed in which the body's fat content is systematically depleted (or increased) while LBM remains unchanged (and this includes studies of parabiotic pairs), it will be difficult to interpret changes in ingestive behavior that follow a period of under- or overfeeding.

With that caveat in mind, one can proceed to examine the possible ways in which the size of the triglyceride storage depot might influence eating behavior. Questions about this overall issue include the following:

1. Is body fat actually regulated? Brobeck (1965) has insisted that for true regulation to occur, some function of the regulated variable must be monitored.

2. By what mechanism(s) can the brain monitor the status of the fat stores?
3. Does the size of the triglyceride storage depot affect eating behavior by some indirect mechanism that does not involve active monitoring by the brain of the body's energy reserves?

Because the literature bearing on these questions is so voluminous, we can consider only a few highlights.

PUTATIVE HUMORAL SIGNALS FROM ADIPOSE TISSUE

First, it is still unclear whether the brain is able to monitor the status of the fat stores. If such monitoring occurs, there are several ways in which the appropriate information might be transmitted to the brain. One possibility is that some agent is released into the circulation by fat cells that reflects adipose tissue status and is detected indirectly or directly by the central nervous system. A number of substances have been suggested as possibly playing this role, including free fatty acids, glycerol, and (most recently) adipsin. To date, none of these putative messengers has gained substantial acceptance among investigators in the field of ingestive behavior.

There is preliminary evidence that lack of adipsin, a protein that is nearly identical with complement D, may contribute to the development of obesity in certain types of genetically obese rodents (Flier, Cook, Usher, & Spiegelman, 1987). Adipsin is manufactured by fat cells and then secreted into the circulation (Cook, Min, Johnson, Chaplinsky, Flier, Hunt, & Spiegelmen, 1987; Spiegelman & Green, 1980). Conceivably, this protein could act as a negative feedback signal to the brain, turning off appetite when fat cells get too large. Absence of adipsin would create an open-loop situation resulting in chronic overeating and excessive fat storage. Enthusiasm for adipsin as a possible marker of genetic obesity has been somewhat dampened by recently reported studies of adipsin mRNA in adipose tissue of genetically obese (Zucker) rats, where no differences between obese animals and their lean littermates were detected, either in young pups (16 days old) or after hyperinsulinemia had developed (Lavau, Dugail, Quignard-Boulange, Bazin, & Guerre-Mills, 1989).

PUTATIVE NEURAL SIGNALS

Another possibility is that information about the status of the adipose depot is transmitted to the brain via neural pathways. Although it is true that some adipocytes are directly innervated (Hausberger, 1934; Wertheimer & Shapiro, 1948), it is hard to imagine what characteristic of the cell could be detected by a nerve fiber that would accurately reflect the cell's content of triglyceride. Moreover, the innervated cells would have to be representative of fat cells overall. This circumstance seems unlikely in view of the fact that fat cell size can vary significantly from site to site (i.e., abdominal subcutaneous fat versus intraabdominal fat versus gluteal subcutaneous fat).

It has also been suggested by Nicolaidis (1974, 1981) that there might be lipid-bearing cells in the CNS that epitomize the peripheral adipose tissue and whose changes in fat content are monitored by a neuronal system concerned with the control of food intake.

The notion that metabolic events in specialized brain cells may parallel events taking place in peripheral (i.e., liver) cells has been taken up by Martin and colleagues (see below) and also (in a sense) by Friedman (Chapter 19), who suggests a hepatic control of food intake with some sort of communication between the liver and neuronal systems in the brain that manage ingestive behavior.

INSULIN AS AN ADIPOSITY SIGNAL

Woods *et al.* (1985) have proposed that the concentration of insulin in plasma provides the brain with information about the size of the fat depot. This theory (Porte & Woods, 1981) has some appeal because the blood insulin level is maintained at higher than normal values in people whose fat cells are enlarged (hypertrophic obesity). According to Woods *et al.* insulin from plasma is transported into the central nervous system via a specific insulin transport system situated in brain capillary endothelial cells. The concept that plasma insulin has a relatively rapid access to the brain has replaced the original hypothesis of Woods *et al.* that insulin enters the brain slowly by way of the cerebrospinal fluid. Despite the attractiveness of insulin as an adiposity signal, it must be kept in mind that individuals with hyperplastic obesity (whose adipose tissue contains an increased number of fat cells of normal size) can possess a greatly expanded triglyceride depot without exhibiting any increase in fasting or meal-stimulated plasma insulin levels. Nevertheless, the insulin theory has received some support from the fact that when insulin is injected into the CSF of primates, food intake diminishes and weight is lost. Also, it has been reported that, in rats, feeding was stimulated by the administration of insulin antibodies into an area of the hypothalamus concerned with food intake control (Strubbe & Mein, 1977).

EFFECT OF NUTRITIONAL STATUS ON REGIONAL BRAIN METABOLISM

Apart from the issue of whether the size of the fat stores can be monitored by the brain, there is growing evidence that nutritional status—whether the animal is in the fed state or has been food deprived—affects the metabolic behavior of central nervous system neurons concerned with food intake control (Jhanwar-uniyal & Leibowitz, 1986; Chafetz, Parks, Diaz, & Leibowitz, 1986; Leibowitz, 1988). For example, nutritional deprivation may have the effect of down-regulating α_2-noradrenergic receptor activity in the paraventricular nucleus. This reduction in α_2 activity is said to be followed by a selective increase in carbohydrate intake. This increase appears to be limited to the early phase of the dark period.

γ-Aminobutyric acid (GABA) is one of a group of neurotransmitters that can influence food ingestion. GABA may serve a specific intermediary function in the link between glucose metabolism and feeding behavior. In this regard, Beverly and Martin (1989) have shown that, in many respects, changes in hypothalamic metabolism parallel those in the liver cell. For example, in the hypothalamus, glucose metabolism via the pentose phosphate and GABA shunt pathways is affected by nutritional state. In animals that have been food deprived and depleted of fat stores, the activity of the GABA shunt is decreased in the lateral hypothalamus and increased in the medial hypothalamus. This effect appears to be mediated via glutamate decarboxylase, the rate-limiting enzyme of the GABA shunt. The activity of this enzyme is increased in the medial hypothalamus of food-restricted rats. On the basis of these findings, Martin and colleagues have proposed that information about energy status is communicated in the hypothalamus through alteration of activity of specific metabolic pathways that influence neurotransmitter status.

In summary, evidence continues to accumulate that changes in overall nutritional status can affect in various ways the metabolism of neuronal systems in the brain concerned with food intake control. When one is attempting to cope with new data on changes in regional brain metabolism as they appear to relate to changes in appetitive behavior and food intake control, it may be helpful to keep at least three principles in mind. First, in directing feeding the brain must integrate a multiplicity of signals, inhibitory and excitatory, arising from a variety of loci; second, scientific information that is generated about interrelationships between nutritional and metabolic status and brain-mediated ingestive behavior has to be interpreted in the light of an understanding, based on systematic observation, of how animals and people actually eat; third, efforts have to continue to be made to determine how information transmitted to the brain about proximate nutritional status (empty stomach, empty small intestine, liver glycogen depletion, falling blood glucose level, etc.) is integrated (if, indeed, it is) with information about the fat stores. Does information about the status of the fat depot modulate the brain's response to other signals of depletion and repletion? Alternatively, does the state of the triglyceride depot modulate other metabolic signals of depletion and repletion so that the brain receives only indirect information about the fat stores?

Studies of animal and human eating behavior strongly suggest that, if the status of the fat depot influences the brain's control of body fat, it must do so very subtly. Animals and people with "dietary obesity" do not show a disposition to undereat that is the mirror image of the disposition to overeat exhibited by food-deprived, nutritionally depleted animals. Yet, dietary obese rats are far less "hyperphagic" than normal-weight rats during sham feeding (VanderWeele & VanItallie, 1983). A number of studies have shown that if dietary obesity in the rat is sustained for a sufficient period of time, the animal is likely to defend its increased weight, even if the obesity-producing diet is replaced by conventional laboratory pellets (Rowe & Rolls, 1979; Rolls, Rowe, & Turner, 1980). In this latter situation, the animal's adipose tissue may have proliferated (Faust *et al.*, 1978); thus, one is led to suspect that the presence of additional fat cells is somehow sustaining the new level of fat storage.

ALTERING THE BODY'S INVENTORY OF STORED ENERGY

In certain animals, the size of the energy stores may change predictably in association with a pattern of instinctual appetitive behaviors that serve to safeguard the individual and/or perpetuate the species. As mentioned earlier, animals may "overeat" and store extra fat in preparation for such activities as migration, hibernation, mating, pregnancy, lactation, or nesting. In these situations, it is also quite possible that fat storage is promoted by metabolic changes within the organism that favor a shift toward lipogenesis and away from lipolysis. During this accretion period, fat stores may be less readily mobilized, making it necessary for the animal to eat more frequently in order to meet the energy needs of the nonadipose cell mass.

In contrast, for a time during the mating season, certain male sea lions do not eat, devoting themselves exclusively to defense of the harem against competing males. As a result they lose weight rapidly. Similarly, while penguins are incubating their eggs they are apparently anorexic and resist being tempted away from the nest by proffered food (Mrosovsky & Sherry, 1980). Such biologically mandated periods of change in ingestive behavior and in the size of the fat depot have been explained by certain investigators as being indicative of a "sliding set point" for body weight (Mrosovsky & Fisher, 1970). However, the enlarged fat depot that results from premigratory hyperphagia is not "defended" by the migrating bird; it is, of course, used to fuel the migratory flight. In other words, the extra triglyceride is there to be used, not to be maintained. Thus, although the animal may alter its ingestive behavior and, consequently, its energy stores in anticipation of certain predictable biological needs, it does not seem necessary to explain this kind of adaptive regulation by postulating a system that embodies a changing set point for "body weight."

(It is difficult, however, to devise an experiment to disprove the set-point hypothesis. This is because advocates of the hypothesis can always attribute any alterations in weight and fatness that result from a sustained shift in diet composition, a change in physical activity level, or the use of certain medications to the operation of a set point that had been reset by the manipulation in question. But this kind of explanation seems to beg the question, distracting attention from the task of gaining a more detailed understanding of the actual biochemical and physiological mechanisms that underlie and account for the observed changes in appetitive behavior.)

Effects of Changes in Energy Expenditure on Depot Fat

In man the size of the fat depot may change involuntarily as a result of shifts in the body's energy economy. One example is the weight gain that follows cessation of cigarette smoking (Jacobs & Gottenborg, 1981). People also may increase their fat content while taking certain types of medication; for example, the tricyclic antidepressant amitriptyline (Fernstrom et al., 1985). Cigarette smoking appears to increase resting energy expenditure, and cessation of the habit is followed by a significant decrease in the resting metabolic rate (Dalloss & James, 1984; Hofstetter et al., 1986). The precise mechanism of this effect is not clearly understood, but an effect of nicotine on the activity of the sympathetic nervous system may be partly responsible. Whatever the case, it appears that when the resting metabolic rate is lowered by cessation of smoking or by the use of certain antidepressant medications, the size of the fat depot increases. In other words, neither the ex-smoker nor the patient on antidepressant medication spontaneously reduces caloric intake enough to compensate for the reduced energy expenditure. Although it is not known whether nicotine or amitriptyline affects the mobilizability of the fat stores, it seems clear that a change in energy expenditure, whether it results from a change in physical activity or in metabolic rate, can alter the size of the fat depot, which then tends to stabilize at a new level.

The mechanism by which a change in energy expenditure alters the body's content of stored fat is not clearly understood. One possible explanation has been advanced recently by Flatt (1988). He points out that "in metabolic terms the steady state of weight maintenance depends on the oxidation of a fuel mix

THEODORE B.
VANITALLIE AND
HARRY R.
KISSILEFF

whose average composition matches that of the diet." To ascertain whether or not such a match obtains, one must first determine the relative contributions made by glucose and FFA to the fuel mixture being oxidized within the body. The proportion contributed to the mix by each of these two substrates is disclosed by the nonprotein respiratory quotient (RQ). Next, in order to determine the dietary fuel mix, one measures the ratio of CO_2 produced to O_2 consumed during oxidation of an aliquot sample of the diet being consumed (again corrected for protein content). This ratio is called the nonprotein food quotient or FQ. When the average RQ exceeds the FQ, the mixture of fuels being burned contains less fat than the mixture of foods being consumed. According to Flatt, since the regulation of food intake will be geared to prevent glycogen depletion, a sustained period with an average RQ greater than FQ implies accumulation of fat. In contrast, if the RQ is less than the FQ, the mixture of fuels oxidized contains more fat than that supplied by the diet, a state of affairs that reflects net withdrawal of fat from adipose tissue.

As Flatt emphasizes, sustained exercise leads to a gradual decline in the RQ. Thus, by promoting fat oxidation exercise may reduce the average RQ below the level of the FQ, thereby sparing liver glycogen and permitting glycogen balance to continue to be maintained during energy deficit. In contrast, a reduction in energy expenditure (as would occur after cessation of cigarette smoking) would raise RQ above FQ and result in increased fat storage.

In man, cognitive factors are capable of overriding physiologically based urges to eat. Thus, some individuals ("hunger strikers") have subjected themselves to prolonged starvation in order to call attention to a problem about which they feel deeply. Young women with a phobia about becoming fat voluntarily restrict food intake and can develop all the signs of severe semistarvation (Strober, 1986). Nevertheless, the body's strong inherent urge to forestall nutritional depletion is well illustrated by the behavior of patients with bulimia (Fairburn & Cooper, 1982). In order to control calorie intake, they self-induce vomiting after meals; however, after a time the cumulative effects of this deprivation appear to overcome their efforts to restrain food intake, with the emergence of intermittent bouts of gorging (Wardle, 1987).

The Problem of Weight Regain in Formerly Obese Individuals

Although it is evident that obese individuals can restrict calorie intake and lose substantial quantities of weight, the very high incidence of weight regain among the formerly obese is a continuing cause of frustration and concern. As mentioned earlier in this review, proponents of the "set-point hypothesis" explain this high rate of relapse by suggesting that obesity represents a condition of regulation at an elevated set point. According to this hypothesis, obese individuals who have reduced their weight by adhering to a calorically restricted diet are below their set point for body weight. Being below set-point weight implies that the body reacts to this challenge by making active adjustments in its energy economy designed to return weight to its set-point level. Such adjustments might include an exaggerated reduction in rate of energy expenditure combined with an increased appetite and an enhanced efficiency of utilization of dietary energy.

Another way of explaining the problem of frequent relapse in reduced obese individuals is by reference to the behavior of the adipose tissue after weight reduction. It is quite possible that, compared to the adipose tissue of spontaneously nonobese individuals, the adipose tissue of at least some obese persons is less disposed, on a cell-for-cell basis, to yield its stored fat in response to the usual lipolytic stimuli. Perhaps the lipolytic stimuli themselves are weaker in such obese persons. In either case, after a sufficiently prolonged period of energy deficit, when mean fat cell size has become smaller, the special disposition of the adipocytes to retain triglyceride would become more manifest. This state of affairs would result in a higher than normal nonprotein RQ, reflecting a reduced rate of fat oxidation and an intensified rate of carbohydrate combustion. If the desire to eat and the rate of gastric emptying are coupled to the carbohydrate supply (Granneman & Stricker, 1984), the relative inhibition of lipolysis in fat-reduced obese individuals could well initiate a process resulting in an enhanced desire for food, a desire that would be increasingly difficult to control.

Also, by definition, compared to the "never obese," many formerly obese individuals appear to have an expanded storage capacity for triglyceride. For them, the pronouncement that "nature abhors a vacuum" might well be applicable inasmuch as fat cells that have been triglyceride depleted have a greater propensity to store fat than cells already well stocked with triglyceride.

Some preliminary experimental evidence supports the "lipophilia" theory described above (Froidevaux, Schutz, Christin, Jéquier, 1988); however, a great deal of confirmatory research remains to be done. But at least the theory is stated in a form that permits the design of experiments to test its validity.

Finally, one may ask whether the information examined in this review can provide useful clues for the more effective treatment of obesity. Because the major problem faced by obese dieters is weight maintenance after weight loss, our focus is on this consideration. However, we have pointed out that the way in which an obese individual loses weight may affect his or her physiological and psychological response to weight reduction. If weight is reduced by drastically restricting calorie intake, a significant proportion of the lost weight is likely to be lean body mass (VanItallie, 1989). In contrast, if weight is reduced by a sustained increase in physical activity, little if any lean body mass will be lost (Wilmore, 1983). The metabolic responses to these two modes of weight reduction are different; one might expect an individual with a reduced lean body mass to exhibit corresponding reductions in resting metabolic rate and thermic response to food. Such physiological changes do not appear to occur in persons who lose fat by increasing physical activity.

Because dietary fat is used more efficiently than carbohydrate for triglyceride storage, diets prescribed for weight maintenance are low in fat and high in carbohydrate content. As Flatt (1988) puts it, ". . . maintenance of an RQ lower than the FQ is a necessary and sufficient condition for weight loss. . . . The RQ/FQ concept helps one realize that it must be easier to maintain the average RQ below the diet's FQ when consuming a diet with a high FQ (i.e., a diet high in carbohydrates, and low in fats)."

Apart from the metabolic advantage to the dieter of consuming a high-carbohydrate, low-fat diet is the fact that the RQ/FQ ratio varies directly with the energy intake/energy expenditure ratio (EI/EE). Thus, when energy expenditure exceeds energy intake, the RQ/FQ ratio is also decreased.

If one combines sustained physical exercise with a high-carbohydrate low-fat diet, the RQ/FQ ratio will be more diminished, decreasing further the likelihood of weight regain.

The ability of weight-conscious individuals to adhere to a low-fat diet may well be facilitated by the availability of acceptable nonfat or fat-reduced products that have the organoleptic qualities of fat. Also, it may be possible to identify or fabricate specific food items with unique satiating ability. An efficiently satiating food will reduce immediately subsequent food intake by an appreciably greater amount than the energy the "satiating" food contains (Kissileff, 1984).

Although a low-fat diet and increased physical exercise are useful in helping formerly obese persons maintain reduced weights, the ability of a weight-reduced individual to avoid relapse also depends on constitutional factors such as the 24-hour energy expenditure (Is it significantly lower than that of never-obese individuals of the same sex matched for height, age, and body composition?), the storage capacity for fat (Is the adipose tissue hyperplastic?), and the metabolic behavior of the adipocytes (Do they have an unusual propensity to store fat or to resist lipolytic stimuli?). If such factors are found to frustrate weight maintenance efforts in a high proportion of formerly obese individuals, the problem of maintaining weight may prove to be much more difficult than originally anticipated. Indeed, successful treatment ultimately may require the development of new medications designed to modify or overcome the metabolic characteristics that predispose individuals to excessive fat storage.

REFERENCES

Abraham, S., Carroll, M. D., Najjar, M. F., & Fulwood, R. (1983). *Obese and overweight adults in the United States.* U.S. Dept. of Health and Human Services Publication No. (PHS) 83-1680, Vital and Health Statistics, Series 11, No. 230. Hyattsville, MD: National Center for Health Statistics.

Abraham, S., Johnson, C. L., & Najjar, M. F. (1979). *Weight by height and age for adults 15–74 years. United States, 1971–74.* DHEW Publication No. (PHS) 79-1656. Hyattsville, MD: National Center for Health Statistics.

Beverly, J. R., & Martin R. J. (1989). Increased GABA shunt activity in VMN of three hyperphagic rat models. *American Journal of Physiology, 256,* R1225–R1231.

Björntorp, P. (1972). Disturbances in the regulation of food intake. Obesity: Anatomic and physiologic–biochemical observations. *Advances in Psychosomatic Medicine, 7,* 116–127.

Björntorp, P., & Sjöstrom, L. (1971). Number and size of adipose tissue fat cells in relation to metabolism in human obesity. *Metabolism, 20,* 703–713.

Blaxter, K. L. (1971). Methods of measuring the energy metabolism of animals and interpretation of results obtained. *Federation Proceedings, 30,* 1436–1443.

Blaxter, K. L. (1973). Energy utilization and obesity in domesticated animals. In G. A. Bray (Ed.), *Obesity in perspective.* DHEW Publication No. (NIH) 75-708 (pp. 127–135). Hyattsville, MD: Department of Health, Education and Welfare.

Bogardus, C., Lillioja, S., Ravussin, E., Abbott, W., Zawadzki, J. K., Young, A., Knowler, W. C., Jacobowitz, R., & Moli, P. P. (1986). Familial dependence of the resting metabolic rate. *New England Journal of Medicine, 315,* 96–100.

Bouchard, C. (1985). Inheritance of fat distribution and adipose tissue metabolism. In J. Vague, P. Björntorp, B. Guy-Grand, M. Rebuffé-Scrive, & P. Vague (Eds.), *Metabolic complications of human obesities* (pp. 87–97). Amsterdam: Elsevier.

Bray, G. A., & Campfield, L. A. (1975). Metabolic factors in the control of energy stores. *Metabolism, 24(1),* 99–117.

Brobeck, J. R. (1965). Exchange, control and regulation. In W. S. Yamamoto & J. R. Brobeck (Eds.), *Physiological controls and regulations* (pp. 1–13). Philadelphia: W. B. Saunders.

Brook, C. G. D. (1972). Evidence for a sensitive period in adipose-cell replication in man. *Lancet, 2,* 624–627.

Chafetz, M. D., Parks, K., Diaz, S., & Leibowitz, S. F. (1986). Relationships between medial hypothalamic α_2-receptor binding, norepinephrine, and circulating glucose. *Brain Research, 384,* 404–408.

Committee on Dietary Allowances. (1980). Recommended dietary allowances. Washington, DC: Food and Nutrition Board, National Academy of Sciences.

Cook, K. S., Min, H. Y., Johnson, D., Chaplinsky, R. J., Flier, J. S., Hunt, C. R., & Spiegelman, B. M. (1987). Adipsin: A circulating serine protease homolog secreted by adipose tissue and sciatic nerve. *Science, 237,* 402–405.

Dalloss, H. M., & James, W. P. T. (1984). The role of smoking in the regulation of energy balance. *International Journal of Obesity, 8,* 365–375.

Danforth, E., Jr., Daniels, R. J., Katzeff, H. L., Ravussin, E., & Garrow, J. S. (1981). Thermogenesis in Pima Indians. *Clinical Research, 29,* 663A.

Durnin, J. V. G. A., & Passmore, R. (1967). *Energy work and leisure.* London: Heinemann Educational Books.

Durnin, J. V. G. A., & Wormersley, J. (1974). Body fat assessed from total body density and its estimation from skinfold thickness: Measurements of 481 men and women aged from 16 to 72 years. *British Journal of Nutrition, 32,* 77–92.

Dusmet, M., Schutz, Y., Bessara, T., & Jéquier, E. (1985). Metabolic response to high protein overfeeding. *International Journal of Obesity, 9 (Suppl 2),* A26.

Fairburn, C. G., & Cooper, P. J. (1982). Self-induced vomiting and bulimia nervosa: An undetected problem. *British Medical Journal, 284,* 1153–1155.

Faust, I. M., Johnson, P. R., Stern, J. S., & Hirsch, J. (1978). Diet-induced adipocyte number increase in adult rats: A new model of obesity. *American Journal of Physiology, 235,* E279–E286.

Fernstrom, M. H., Epstein, L. H., Spiker, D. G., & Kupfer, D. J. (1985). Resting metabolic rate is reduced in patients treated with antidepressants. *Biological Psychiatry, 20,* 688–692.

Fernstrom, M. H., & Kupfer, D. J. (1988). Antidepressant-induced weight gain: A comparison study of four medications. *Psychiatry Research, 26,* 265–271.

Flatt, J. P. (1972). Role of the increased adipose tissue mass in the apparent insulin insensitivity of obesity. *American Journal of Clinical Nutrition, 25,* 1189–1192.

Flatt, J. P. (1985). Energetics of intermediary metabolism. In J. S. Garrow & D. Halliday (Eds.), *Substrate and energy metabolism* (pp. 58–69). London: John Libbey.

Flatt, J. P. (1988). Importance of nutrient balance in body weight regulation. *Diabetes/Metabolism Review, 4,* 571–581.

Flier, J. S., Cook, K. S., Usher, P., & Spiegelman, B. M. (1987). Severely impaired adipsin expression in genetic and acquired obesity. *Science, 237,* 405–408.

Forbes, G. B., & Reina, J. C. (1970). Adult lean body mass declines with age. *Metabolism, 19,* 653–663.

Friedman, M. I., Tordoff, M. G., & Ramírez, I. (1986). Integrated metabolic control of food intake. *Brain Research Bulletin, 17,* 855–859.

Froidevaux, I., Schutz, Y., Christin, L., & Jéquier, E. (1988). Twenty-four hour energy expenditure after weight loss in obese subjects. *Experientia, 44,* A31.

Garn, S. M. (1985). Continuities and changes in fatness from infancy to adulthood. *Current Problems in Pediatrics, 15,* 1–47.

Garn, S. M., & Haskell, J. A. (1960). Fat thickness and developmental status in childhood and adolescence. *AMA Journal of Diseases of Children, 99,* 746–751.

Geissler, C. A., Miller, D. S., & Shah, M. (1987). The daily metabolic rate of the post-obese and the lean. *American Journal of Clinical Nutrition, 45,* 914–920.

Glore, S. R., Layman, D. K., & Bechtel, P. J. (1988). Adaptations of adult lean and obese Zucker rats during fat restriction. *Nutrition Research, 8,* 1403–1412.

Golay, A., Schutz, Y., Meyer, H. U., Thiebaud, D., Curchod, B., Maeder, E., Felber, J.-P., & Jéquier, E. (1982). Glucose induced thermogenesis in nondiabetic and diabetic obese subjects. *Diabetes, 31,* 1023–1028.

Gordon, T., Kagan, A., & Rhoads, G. (1975). Relative weights in different populations. *American Journal of Clinical Nutrition, 28,* 304–305.

Granneman, J., & Stricker, E. M. (1984). Food intake and gastric emptying in rats with streptozotocin-induced diabetes. *American Journal of Physiology, 247,* R1054–R1061.

Guthrie, H. A., Habicht, J.-P., Johnson, S. R., & VanItallie, T. B. (1986). *Nutrition monitoring in the United States. A progress report from the Joint Nutrition Monitoring Evaluation Committee.* U.S. Dept. of Health and Human Services Publication No. (PHS) 86-1255. Hyattsville, MD: U.S. Department of Health and Human Services.

Hausberger, F. X. (1934). Uber das Inervation der Fettorgane. *Zeitschrift fur Mikroskopische und Anatomische Forschung, 36,* 231–236.

Health and Welfare Ministry of Japan. (1986). Data published in *The Daily Yomiuri,* August 28, 1986.

Himms-Hagen, J. (1984). Brown adipose tissue thermogenesis as an energy buffer: implications for obesity. *New England Journal of Medicine, 311,* 1549–1558.

Himms-Hagen, J. (1985). Defective brown adipose tissue thermogenesis in obese mice. *International Journal of Obesity, 9 (Suppl. 2),* 17–24.

Hofstetter, A., Schutz, Y., Jéquier, E., & Wahren, J. (1986). Increased 24-hour energy expenditure in cigarette smokers. *New England Journal of Medicine, 314,* 79–82.

Horton, E. S., & Danforth, E., Jr. (Eds.). (1985). *Regulation of energy expenditure.* London: John Libbey.

Jacobs, D. R., Jr., & Gottenborg, S. (1981). Smoking and weight: The Minnesota Lipid Research Clinic. *American Journal of Public Health, 71,* 391–396.

Jéquier, E., & Schutz, Y. (1983). Long-term measurements of energy expenditure in humans using a respiration chamber. *American Journal of Clinical Nutrition, 38,* 989–998.

Jéquier, E., & Schutz, Y. (1985). New evidence for a thermogenic defect in human obesity. *International Journal of Obesity, 9* (Suppl. 2), 1–7.

Jhanwar-uniyal, M., & Leibowitz, S. F. (1986). Impact of food deprivation on α_1- and α_2-noradrenergic receptors in the paraventricular nucleus and other hypothalamic areas. *Brain Research Bulletin, 17,* 889–896.

Kagan, A., Harris, B. R., Winkelstein, W., Jr., Johnston, K. G., Kato, H., Syme, S. L., Rhoads, G. G., Gay, M. L., Nichaman, M. Z., Hamilton, H. B., & Tillotson, J. (1974). Epidemiologic studies of coronary heart disease and stroke in Japanese men living in Japan, Hawaii and California: Demographic, physical, dietary and biochemical characteristics. *Journal of Chronic Diseases, 27,* 345–364.

Katzeff, H. L., & Daniels, R. (1985). The sympathetic nervous system in human obesity. *International Journal of Obesity, 9* (Suppl. 2), 131–137.

Keesey, R. E. (1986). A set-point theory of obesity. In K. D. Brownell & J. P. Foreyt (Eds.), *Handbook of eating disorders* (pp. 63–87). New York: Basic Books.

Keesey, R. E. (1989). Physiological regulation of body weight and the issue of obesity. *Medical Clinics of North America, 73,* 15–27.

Keys, A., Brozek, J., Henschel, A., Mickelsen, O., & Taylor, H. L. (1950). *The biology of human starvation.* Minneapolis: University of Minnesota Press.

Kissileff, H. R. (1984). Satiating efficiency and a strategy for conducting food loading experiments. *Neuroscience and Biobehavioral Reviews, 8,* 129–135.

Kissileff, H. R., & VanItallie, T. B. (1982). Physiology of the control of food intake. *Annual Review of Nutrition, 2,* 371–418.

Krotkiewski, M., Sjöstrom, L., Björntorp, P., Carlgren, G., Garelick, G., & Smith, U. (1977). Adipose tissue cellularity in relation to prognosis for weight reduction. *International Journal of Obesity, 1,* 395–416.

Lavau, M., Dugail, I., Quignard-Boulange, A., Bazin, R., & Guerre-Millo, M. (1989). Adipsin in different obesity models. In P. Björntorp & S. Rössner (Eds.), *Obesity in Europe 88* (pp. 167–172). London: John Libbey.

Lean, M. E. J., & James, W. P. T. (1987). Metabolic effects of isoenergetic nutrient exchange over 24 hours in relation to obesity in women. *International Journal of Obesity, 12,* 15–27.

Lean, M. E. J., James, W. P. T., & Garthwaite, P. H. (1989). Obesity without overeating? Reduced diet-induced thermogenesis in post-obese women, dependent on carbohydrate and not fat intake. In P. Björntorp & S. Rössner (Eds.), *Obesity in Europe 88. Proceedings of the 1st European congress on obesity* (pp. 281–286). London: John Libbey.

Leibowitz, S. F. (1988). Hypothalamic paraventricular nucleus: Interaction between α_2-noradrenergic system and circulating hormones and nutrients in relation to energy balance. *Neuroscience Biobehavioral Review, 12,* 101–109.

Lissner, L., Levitsky, D. A., Strupp, B. J., Kalkwarf, H. J., & Roe, D. A. (1987). Dietary fat and the regulation of energy intake in human subjects. *American Journal of Clinical Nutrition, 46,* 886–892.

Mrosovsky, N. (1985). Cyclical obesity in hibernators: The search for the adjustable regulator. In J. Hirsch & T. B. VanItallie (Eds.), *Recent advances in obesity research: IV* (pp. 45–56). London: John Libbey.

Mrosovsky, N., & Faust, I. M. (1985). Cycles of body fat in hibernators. *International Journal of Obesity, 9,* 93–98.

Mrosovsky, N., & Fisher, K. C. (1970). Sliding set points for body weight in ground squirrels during the hibernation season. *Canadian Journal of Zoology, 48,* 241–247.

Mrosovsky, N., & Sherry, D. F. (1980). Animal anorexias. *Science, 207,* 837–842.

Neel, J. V. (1962). Diabetes mellitus: A "thrifty" genotype rendered detrimental by "progress"? *American Journal of Human Genetics, 14,* 353–362.

Nicolaidis, S. (1974). A possible molecular basis of regulation of energy balance. In *Psychology of food and fluid intake, 26th International Congress of Physiological Sciences* (p. 97). Jerusalem: Satellite Symposia.

Nicolaidis, S. (1981). Lateral hypothalamic control of metabolic factors related to feeding. *Diabetologia, 20,* 426–434.

Odum, E. P. (1960). Premigratory hyperphagia in birds. *American Journal of Clinical Nutrition, 8*, 621–629.

Oscai, L. B., Brown, M. M., & Miller, W. C. (1984). Effect of dietary fat on food intake, growth and body composition in rats. *Growth, 48*, 415–424.

Poehlman, E.T., Tremblay, A., Despres, J.-P., Fontaine, E., Perusse, L., Theriault, G., & Bouchard, C. (1986). Genotype-controlled changes in body composition and fat morphology following over-feeding in twins. *American Journal of Clinical Nutrition, 43*, 723–731.

Porikos, K. P., & VanItallie, T. B. (1984). Efficacy of low-calorie sweeteners in reducing food intake: Studies with aspartame. In L. D. Stegink & L. J. Filer, Jr. (Eds.), *Aspartame: Physiology and Biochemistry* (pp. 273–286). New York: Marcel Dekker.

Porte, D., Jr., & Woods, S. C. (1981). Regulation of food intake and body weight by insulin. *Diabetologia, 20*, 274–280.

Price, R. A. (1987). Genetics of human obesity. *Annals of Behavioral Medicine, 9*, 9–14.

Price, R. A., Cadoret, R. J., Stunkard, A. J., & Troughton, E. (1987). Genetic contributions to human fatness: An adoption study. *American Journal of Psychiatry, 144*, 1003–1008.

Ravussin, E., Bogardus, C., Schwartz, R. S., Robbins, D. C., Wolfe, R., Horton, E. S., Danforth, E., Jr., & Sims, E. A. H. (1983). Thermic effect of infused glucose in man: Decreased response associated with insulin resistance in obesity and non-insulin-dependent diabetes mellitus. *Journal of Clinical Investigation, 72*, 893–902.

Ravussin, E., Lillioja, S., Anderson, T. E., Christin, L., & Bogardus, C. (1986). Determinants of 24-hour energy expenditure in man: methods and results using a respiratory chamber. *Journal of Clinical Investigation, 78*, 1568–1578.

Ravussin, E., Lillioja, S., Knowlde, W. C., Christin, L., Freymond, D., Abbott, W. G. H., Boyce, V., Howard, B. V., & Bogardus, C. (1988). Reduced rate of energy expenditure as a risk factor for body-weight gain. *New England Journal of Medicine, 318*, 467–472.

Rebuffé-Scrive, M., Enk, L., Crona, N., Lönnroth, P., Abrahamsson, L., Smith, U., & Björntorp, P. (1985). Fat cell metabolism in different regions in women. Effect of menstrual cycle, pregnancy, and lactation. *Journal of Clinical Investigation, 75*, 1973–1976.

Roberts, S. B., Savage, J., Coward, W. A., Chew, B., & Lucas, A. (1988). Energy expenditure and intake in infants born to lean and overweight mothers. *New England Journal of Medicine, 318*, 461–466.

Rolls, B. J., Rowe, E. A., & Turner, R. C. (1980). Persistent obesity in rats following a period of consumption of a mixed high energy diet. *Journal of Physiology (London), 298*, 415–427.

Rothwell, N. J., & Stock, M. J. (1981a). A role for insulin in the diet-induced thermogenesis of cafeteria-fed rats. *Metabolism, 30*, 673–678.

Rothwell, N. J., & Stock, M. J. (1981b). Regulation of energy balance. *Annual Review of Nutrition, 1*, 235–256.

Rowe, E. A., & Rolls, B. J. (1979). Exercise and the development and persistence of dietary obesity in male and female rats. *Physiology and Behavior, 23*, 241–247.

Salans, L. B., Cushman, S. W., & Weismann, R. E. (1973). Studies of human adipose tissue. Adipose cell size and number in nonobese and obese patients. *Journal of Clinical Investigation, 52*, 929–941.

Schutz, Y., Acheson, R. J., & Jéquier, E. (1985). Twenty-four-hour energy expenditure and thermogenesis: Response to progressive carbohydrate overfeeding in man. *International Journal of Obesity, 9 (Suppl. 2)*, 111–114.

Sclafani, A. (1984). Animal models of obesity: Classification and characterization. *International Journal of Obesity, 8*, 491–508.

Sims, E. A. H., Danforth, E., Horton, E. S., Bray, G., Glennon, J., & Salans, L. B. (1973). Endocrine and metabolic effects of experimental obesity in man. *Recent Progress in Hormone Research, 29*, 457–496.

Sims, E. A. H., Goldman, R. F., Gluck, C. M., Horton, E. S., Kelleher, P. C., & Rowe, D. W. (1968). Experimental obesity in man. *Transactions of the Association of American Physicians, 81*, 153–170.

Spiegelman, B. M., & Green, H. (1980). Control of specific protein biosynthesis during the adipose conversion of 3T3 cells. *Journal of Biology and Chemistry, 255*, 8811–8818.

Stock, M. J., & Rothwell, N. J. (1985). Factors influencing brown fat and the capacity for diet-induced thermogenesis. *International Journal of Obesity, 9 (Suppl. 2)*, 9–15.

Strober, M. (1986). Anorexia nervosa: History and psychological concepts. In K. D. Brownell & J. P. Foreyt (Eds.), *Handbook of eating disorders. Physiology, psychology, and treatment of obesity, anorexia, and bulimia* (pp. 231–246). New York: Basic Books.

Strubbe, J. H., & Mein, C. G. (1977). Increased feeding in response to bilateral injection of insulin antibodies in the VMH. *Physiology and Behavior, 19*, 309–313.

Stunkard, A. J. (1982). Anorectic agents lower a body weight set point. *Life Sciences, 30,* 2043–2055.

Stunkard, A. J., Foch, T. T., & Hrubec, Z. (1986). A twin study of human obesity. *Journal of American Medical Association, 256,* 51–54.

Stunkard, A. J., Sörensen, T. I. A., Hanis, C., Teasdale, T. W., Chakraborty, R., Schull, W. J., & Schulsinger, F. (1986). An adoption study of human obesity. *New England Journal of Medicine, 314,* 193–198.

Trayhurn, P., & James, W. P. T. (1981). Thermogenesis: Dietary and non-shivering aspects. In L. A. Cioffi, W. P. T. James, & T. B. VanItallie (Eds.), *The body weight regulatory system: Normal and disturbed mechanisms* (pp. 97–105). New York: Raven Press.

Vague, J. (1947). La différenciation sexuelle, facteur déterminant des formes de l'obésité. *La Presse Médicale, 30,* 339–340.

VanderWeele, D. A., & VanItallie, T. B. (1983). Sham feeding is inhibited by dietary-induced obesity in rats. *Physiology and Behavior, 31,* 533–537.

VanItallie, T. B. (1979). Obesity: The American disease. *Food Technology, 1979,* 43–47.

VanItallie, T. B. (1985). Health implications of overweight and obesity in the United States. *Annals of Internal Medicine, 103,* 983–988.

VanItallie, T. B. (1988). Topography of body fat: Relationship of risk of cardiovascular and other diseases. In T. G. Lohman, A. F. Roche, & R. Martorell (Eds.), *Anthropometric standardization reference manual* (pp. 143–149). Champaign, IL: Human Kinetics Books.

VanItallie, T. B. (1989). The dietary treatment of severe obesity. In J. E. Hamner 3rd (Ed.), *The 1988 Distinguished Visiting Professorship Lectures* (pp. 33–44). Memphis TE: Braun-Brumfield.

VanItallie, T. B., & Kissileff, H. R. (1985). Physiology of energy intake: An inventory control model. *American Journal of Clinical Nutrition, 42,* 914–923.

VanItallie, T. B., Kissileff, H. R., & Yang, M. U. (1986). The human energy economy: Adaptation to energy imbalance. In E. Ferrari & F. Brambilla (Eds.), *Disorders of eating behaviour: A psycho-neuroendocrine approach* (pp. 99–104). Oxford: Pergamon Press.

VanItallie, T. B., & Woteki, C. E. (1987). Who gets fat? In A. E. Bender, L. J. Brookes (Eds.), *Body weight control: The physiology, clinical treatment and prevention of obesity* (pp. 39–52). Edinburgh: Churchill Livingstone.

VanItallie, T. B., & Yang, M-V. (1977). Current concepts in nutrition: Diet and weight loss. *New England Journal of Medicine, 297,* 1158–1161.

Wardle, J. (1987). Compulsive eating and dietary restraint. *British Journal of Clinical Psychology, 26,* 47–55.

Webb, P. (1981). Energy expenditure and fat-free mass in men and women. *American Journal of Clinical Nutrition, 34,* 1816–1826.

Weigle, D. S., Sande, K. J., Iverius, P.-H., Monsen, E. R., & Brunzell, J. D. (1988). Weight loss leads to a marked decrease in nonresting energy expenditure in ambulatory human subjects. *Metabolism, 37,* 930–936.

Welle, S. (1985). Evidence that the sympathetic nervous system does not regulate dietary thermogenesis in humans. *International Journal of Obesity, 9* (Suppl. 2), 115–121.

Wertheimer, E., & Shapiro, B. (1948). Physiology of adipose tissue. *Physiology Reviews, 28,* 451–464.

West, K. M. (1978). *Epidemiology of diabetes and its vascular lesions.* New York: Elsevier.

Wilmore, J. H. (1983). Body composition in sport and exercise: Directions for future research. *Medical Science Sports and Exercise, 15(1),* 21–31.

Woo, R. (1985). The effect of increasing physical activity on voluntary food intake and energy balance. *International Journal of Obesity, 9* (Suppl. 2), 155–160.

Woods, S. C., Porte, D., Bobbioni, E., Ionescu, E., Sauter, J.-F., Rohner-Jeanrenaud, F., & Jeanrenaud, B. (1985). Insulin: Its relationship to the central nervous system and to the control of food intake and body weight. *American Journal of Clinical Nutrition, 42,* 1063–1071.

Zack, P. M., Harlan, W. R., Leaverton, P. E., & Cormoni-Huntley, J. (1979). A longitudinal study of body fatness in childhood and adolescence. *Journal of Pediatrics, 95,* 126–130.

PART III
Food Selection

10

Gustatory Control of Food Selection

Thomas R. Scott

Introduction

The chemical senses are the most primitive of specialized sensory systems, with an evolutionary history of some 500 million years. Accordingly, they are involved in mediating the functions most basic to the survival of the individual and species: feeding and reproduction. Taste has become more specialized for the former; olfaction, the latter. The sense of taste largely manages dietary selection, not only by its analysis of the quality and intensity of potential foods but also through communication with the visceral senses. The receptors for taste are situated at the threshold of the body, at the junction between the external and internal milieux. Here selection is completed and digestion begins. It is fitting, then, that the sensory code for taste is influenced not only by the chemical composition of potential foods but also by the experiences of the organism and its physiological state at the time of sampling.

Animals must ingest the chemicals necessary for energy production and the maintenance of biochemical processes and actively reject those that are toxic or from which the body will derive no benefit. Thus, taste stimuli are rarely treated neutrally; they carry a hedonic tag. Taste differs from the nonchemical senses, which, during wakefulness, are constantly and often dispassionately analyzing the external world. The taste system is presented with potential foods discretely and, in each case, is called on to determine their acceptability. To discharge this responsibility, the taste system incorporates or maintains intimate contact with the neural substrates of hedonic appeal. Its message must, at the least, provide a dimension on which the physiological value of a potential food can be determined for subsequent hedonic coloring by higher-order neurons.

Physiological needs are in constant flux. In just hours, the definition of which chemicals are acceptable may be subject to modification as the dangers of

Thomas R. Scott   Department of Psychology, University of Delaware, Newark, Delaware 19716.

malnutrition weigh against those of toxicity. After only minutes of feeding the balance may reverse again. If taste is to provide the information on which the digital decision to swallow or reject is based, its signals should be modifiable to reflect those changes.

This chapter summarizes recent evidence from rats and monkeys that suggests that the sense of taste is responsive both to the external chemical environment and to the internal milieu of the organism. With its outward face, taste evaluates chemical stimuli along a physiological dimension bounded by nutrients and toxins at the hedonically positive and negative extremes. Its analysis converges in the central nervous system with those from olfactory and somesthetic receptors to create an appreciation of flavor. With its inward face, taste remains sensitive to physiological condition and to the effects of gustatory experience. Its susceptibility to modification in response to nutritional need may underlie the notion of body wisdom.

ORGANIZATION OF THE TASTE SYSTEM

The relationship between the sense of taste and metabolic state is established at its most fundamental level by this finding: the neural code for taste quality is based on a dimension of physiological welfare.

No issue is more basic to the characterization of a sensory system than a definition of the dimension(s) on which its perceptions are based. The study of auditory perception may proceed in a logical fashion because it is known that pitch derives largely from the frequency of the incident pressure waves. Similarly with the relationship between color perception and stimulus wavelength, form perception and locus of retinal stimulation, pressure sensation and skin deformation, and temperature perception and the thermal qualities of the impinging stimulus. It is clear from a study of gustatory receptor mechanisms that taste perceptions also are a product of the physical characteristics of relevant stimuli. However, these are largely independent for each of the basic tastes. Hydrogen ion concentration may partly explain sourness—it relates not at all to saltiness. There is no apparent single physical dimension on which the study of the sense of taste may be organized. This realization has led to the suggestion that taste is not an integrated sensory system but a series of independent modalities sensitive to unrelated features of the chemical environment.

The introduction of multidimensional scaling techniques into taste research permitted a direct approach to this issue (Erickson, Doetsch, & Marshall, 1965; Schiffman & Erickson, 1971). With no *a priori* knowledge of the system's organization, a wide range of sapid stimuli could be applied to the tongue, and the profile of either neural or behavioral activity elicited by each chemical determined. Similarity measures among the profiles, as provided by correlation coefficients or other statistics, then could be used in a multidimensional scaling program to generate a spatial representation of relative stimulus similarity. For example, the activity evoked from perhaps 50 single neurons by the application of saline to the subject's tongue would constitute the response profile for NaCl. This profile can be compared with that representing each of the other stimuli by computing the Pearson product–moment correlation between them. The list of correlation coefficients so obtained offers a measure of the similarity of NaCl to each of the other stimuli employed. Performing the same calculation between all possible pairs of profiles provides a full matrix of coefficients that describes the relative taste qualities of all stimuli. Similarly with behavioral data, where neuronal spikes may be replaced by number of licks from a spout containing each stimulus, or psychophysical data, where a stimulus profile may be generated by the subject's level of agreement with each of a series of adjectives (adjectives replace neurons; level of agreement replaces spikes). In each case, the resulting matrix of relative similarity is then represented spatially by multidimensional scaling techniques. The axes of the space in which the chemicals are located must represent those stimulus characteristics that underlie gustatory discriminability, for the amount of each characteristic a stimulus possesses is what determines its taste quality and hence its position in the space. Therefore, the dimension along which taste quality perception is organized may be determined by finding the optimal match between a stimulus characteristic and each axis of the multidimensional space. The importance of any characteristic in determining taste quality is proportional to the total data variance accounted for by the axis with which it is matched.

This approach has now been used to interpret both psychophysical (Schiffman & Erickson, 1971) and electrophysiological data (Scott & Mark, 1987). The common result of these studies—the latter of which is elaborated below—is that a major dimension along which taste stimuli may be organized relates not to any one physical feature of the stimulus molecule but rather to a physiological characteristic: its effect on the welfare of the organism.

In Figure 1 a multidimensional space is presented in which 16 chemicals are placed according to the relative similarities of the response profiles they evoked from 42 taste cells in the nucleus tractus solitarius of the rat. The two dimensions of Figure 1 (top) account for 95% of the data variance, a preponderance of which, 91%, pertains to dimension 1 alone. Thus, the stimulus characteristic that corresponds to this dimension is the major factor in permitting these chemicals to be neurally distinguished by the taste system. Stimulus placement on this dimension correlates 0.83 ($P < 0.001$) with stimulus toxicity, as indexed here by the rat's oral LD_{50}. Stimulus toxicity, then, provides an excellent basis for predicting relative taste quality across a wide range of chemicals. This dimension is shown in isolation in Figure 1 (bottom). Dimension 2, accounting for 4% of the variance, corresponds to differences in the effectiveness of various solutions in driving the system. The mean number of spikes evoked across all 42 neurons during the 5-second response period correlates 0.76 with placement on this dimension ($P < 0.001$), implying that it is a measure of total response magnitude.

Although the relationship between LD_{50} and stimulus position on the dominant first dimension is highly significant, it does not explain the full discriminative capacity of the rat. The organic acids all generate patterns that correlate about +0.90 with that of strychnine, which is 1000 times more toxic and is easily discriminated from the acids. Thus, the rat has access to more information than is assayed by this analysis. That additional input may derive from the temporal characteristics of the evoked response. In the analysis above, only the total spikes that accumulated from each stimulus application during 5 seconds of evoked activity were considered. Yet the time course of this accumulation not only carries reliable information regarding taste quality (DiLorenzo & Schwartzbaum, 1982; Nagai & Ueda, 1981) but also may be sufficient to activate appropriate reflexive responses to chemicals in behaving rats. Therefore, temporal profiles, each composed of 50 100-millisecond bins collapsed across neurons, were gener-

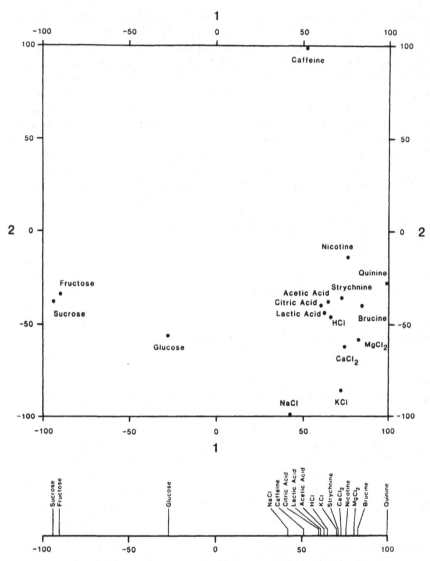

Figure 1. Top: A two-dimensional space representing relative similarities among taste qualities as determined by activity profiles across neurons. Dimension 1 accounts for 91% of the data variance, and the position of a stimulus along its length correlates 0.83 ($P < 0.001$) with stimulus toxicity (rat oral LD_{50}). Dimension 2 accounts for 4% of the variance, and stimulus position along it correlates 0.76 ($P < 0.001$) with total evoked activity across all neurons. Five percent of the variance is unaccounted for by these two dimensions. Bottom: Dimension 1 shown in isolation. (Reprinted from Scott & Mark, 1987, with permission.)

ated for every stimulus, and the correlation matrix and multidimensional representation were calculated as before. The result is shown in Figure 2 (top). Dimension 1 is again dominant, accounting for 89% of the variance, and placement on it correlates +0.85 with oral LD_{50} ($P < 0.001$). It is represented in isolation in Figure 2 (bottom). Dimension 2, accounting for 6% of the variance, is undefined.

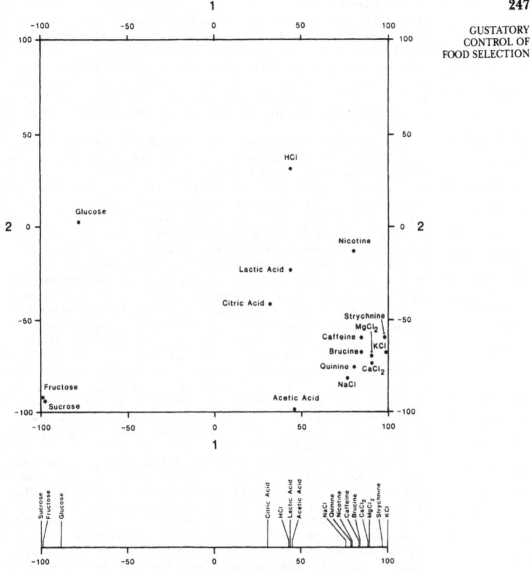

Figure 2. Top: A two-dimensional space representing relative similarities among taste qualities as determined by activity profiles across time. Dimension 1 accounts for 89% of the data variance, and the position of a stimulus along its length correlates 0.85 ($P < 0.001$) with stimulus toxicity. Dimension 2 accounts for 6% of the variance and is undefined. 5% of the variance is unaccounted for by the two dimensions. Bottom: Dimension 1 shown in isolation. (Reprinted from Scott & Mark, 1987, with permission.)

The temporal analysis provides the separation of strychnine from the organic acids, all of whose temporal discharge patterns correlate only in the low 40s with that of the alkaloid. It also introduces coefficients that would be perplexing (the temporal pattern of salty NaCl correlates nearly +0.90 with those of bitter $MgCl_2$ and $CaCl_2$) if the earlier analysis (in which these coefficients reach only the mid-50s) were not available. Certain stimulus pairs, then, are clearly discriminable based on the response distribution either across neurons or across time. Other pairs—quinine and sucrose—are readily discriminable by either means. When neither factor provides separation, however, as with $MgCl_2$ and $CaCl_2$, the rat cannot easily make a behavioral discrimination. The conclusion is that taste quality information in the hindbrain of the rat is carried in a spatiotemporal code, both the spatial and temporal aspects of which are organized predominantly along dimensions that relate to the rat's welfare, measured here by stimulus toxicity.

The finding of a dimension of physiological welfare underlying the organization of taste quality is not simply a recreation of a sweet–bitter dichotomy. It is a more fundamental physiological dimension on which the psychological perceptions of sweetness and bitterness may be based. Chemicals in the environment promote or disrupt physiological functions in animals, providing nutrition or causing illness or death. Selection among foragers then favors the taste system that activates appropriate hedonic tone (attraction to nutrients, revulsion by toxins) to match the physiological consequences of ingestion. It is hardly surprising that the taste system evolved to protect its host from ingesting toxins through the unpleasant sensation that humans label bitterness, or that it motivates the organism to consume carbohydrates through positive experiences called sweetness. Those ancient creatures who preferred the taste of toxins would have contributed little to the subsequent gene pool. What is less expected is that the perceptions we derive from this system are based not on the physical characteristics of the stimuli, as they are in all nonchemical senses, but on the physiological consequences of ingestion. By analogy, the visual system may be instrumental in permitting us to avoid predators and detect prey, but predator and prey are not coded as such in the optic nerve, except perhaps in quite primitive animals. This distinction reinforces the unique quality of taste as the final arbiter of which substances physically enter the body and reaffirms its role in recognizing both the external and internal environments.

There is ample behavioral evidence that the hedonic value of a taste stimulus is monitored in the hindbrain. Acceptance–rejection reflexes are readily identifiable across a wide phylogenetic and ontogenetic range (Steiner, 1979; Steiner & Glaser, 1985) and remain essentially constant with only the caudal brainstem intact (Grill & Norgren, 1978a, 1978b; Steiner, 1973). Steiner concluded that facial expressions are adaptive both in dealing with the chemical—swallowing if appetitive, clearing the mouth if aversive—and in communicating a hedonic dimension to other members of the species. The "hedonic monitor" is innate, is located in the brainstem (Pfaffmann, Norgren, & Grill, 1977), and is neurally intact in humans by the seventh gestational month (Steiner, 1979). Thus, the neural organization on which the hedonics of feeding may be based in the rat is available in the hindbrain gustatory code, and the reflexive behavior commensurate with the ultimate hedonic appreciation is organized here.

PLASTICITY OF TASTE

The preceding section argues that the taste system is organized to perform a general differentiation of toxins from nutrients, that this is accomplished at a lower-order neural level, and that the analysis incorporates or directly influences

powerful hedonic experiences. While providing a broad and effective system for
maintaining the biochemical welfare of the species, this organization would not
recognize the idiosyncratic allergies or needs of the individual or be sensitive to
changes in those needs over time. To serve these requirements, the taste code
must be plastic.

PLASTICITY BASED ON EXPERIENCE

An animal's experience has a pronounced and lasting effect on its behavioral
reactions to taste stimuli. The gustatory experiences of suckling rats establish
taste preferences that persist into adulthood (Capretta & Rawls, 1974). Preferences also develop through association of a taste with positive reinforcement, in
particular with a visceral reinforcement such as occurs with the administration of
a nutrient of which the animal has been deprived (Revusky, Smith, & Chalmers,
1971). Gustatory preferences can be established in humans and other animals
even by mere familiarity through constant exposure (Domjan, 1976). However,
the most reliable effect of experience on subsequent preference obtains from the
establishment of a conditioned taste aversion (CTA). This is an especially efficient form of conditioning in which an intense aversion may be developed
through a single pairing of a novel taste (the conditioned stimulus, or CS) with
gastrointestinal malaise (the unconditioned stimulus, or US) (Garcia, Kimmeldorf, & Koelling, 1955). The aversion to a conditioned taste solution is so
readily established, so potent, and so resistant to extinction that the CTA protocol has itself become a standard tool for studying physiological processes and
taste-related behavior (Smotherman & Levine, 1978).

The neural substrates of conditioned taste aversions have been investigated
in scores of experiments, most of which have involved ablating selected structures and testing the ability of subjects to retain former CTAs or to develop
subsequent aversions. Although these studies have implicated the cortex, amygdala, hippocampus, hypothalamus, thalamus, olfactory bulb, and area postrema
as having some involvement in aversion learning, only amygdaloid and hypothalamic participation seems unequivocal. Rarely have recordings been made
from neurons of conditioned animals to determine the effects of a CTA on taste-
evoked activity. Aleksanyan, Buresova, and Bures (1976) reported that the preponderance of hypothalamic activity evoked by saccharin in rats shifted from the
lateral to the ventromedial nucleus with formation of a saccharin CTA. Di-
Lorenzo (1985) recorded the responses evoked by a series of taste stimuli in the
pontine parabrachial nucleus of rats, then paired the taste of NaCl with gastrointestinal malaise and repeated the recordings. The response to NaCl increased
significantly and selectively in a subset of gustatory neurons.

Chang and Scott (1984) recorded single-unit gustatory-evoked activity from
the nucleus tractus solitarius of three groups of rats: unconditioned (exposed
only to the taste of the saccharin CS with no induced nausea), pseudoconditioned
(experienced only the US, nausea, with no gustatory referent), and conditioned
(taste of saccharin CS paired with gastrointestinal malaise). Comparisons were
performed among the groups' responses to an array of 12 stimuli, including the
saccharin CS, a more concentrated saccharin solution, and saccharides, salts,
acids, and an alkaloid, through which alterations in the entire gustatory code
resulting from this taste-learning experience could be evaluated. Chang and
Scott reported that the CS evoked a significantly larger response from condi-

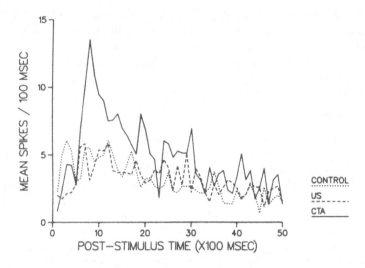

Figure 3. Poststimulus time histograms of the mean responses among sweet-sensitive neurons to 0.0025 M Na saccharin in the three groups of rats. The enhanced activity during the first 3 seconds of evoked response in conditioned subjects may represent the increased salience of the saccharin taste for these animals. (Reprinted from Chang & Scott, 1984, with permission.)

tioned animals than from those of either control group and that the effect was limited to the 30% of neurons that showed a sweet-sensitive profile of responsiveness. Temporal analyses of the activity evoked from this subgroup of saccharin-sensitive neurons revealed that nearly the entire increase in discharge rate was attributable to a burst of activity that diverged from control group levels 600 milliseconds following stimulus onset, peaked at 900 milliseconds, and returned to control levels by 3000 milliseconds (Figure 3). Thus, the major consequence of the conditioning procedure was to increase responsiveness to the saccharin CS through a well-defined peak of activity. The same enhanced response and temporal pattern was evoked to a lesser extent by other sweet stimuli—fructose, glucose, sucrose—providing a likely neural counterpart to generalization of the aversion.

Since a range of taste stimuli was used in this study, the effect of an increased response to the CS and related chemicals could be evaluated in terms of taste quality. The pattern of activity evoked by saccharin was altered by the conditioning procedure. How did the new pattern relate to those of other chemicals? Correlation matrices were calculated, and multidimensional spaces were constructed (as described earlier) in three dimensions from the response profiles of unconditioned and conditioned rats (Figure 4). In the former group (Figure 4A), representing a normally functioning taste system, the stimulus arrangement is similar to that seen by others (Doetsch & Erickson, 1970). The basic distinction between sweet and nonsweet chemicals is apparent, as is the precise arrangement of stimuli within the nonsweet group: the four Na–Li salts should be virtually indistinguishable by the neural patterns they elicit. The complex sweet–salty–bitter taste of concentrated (0.2 M) sodium saccharin is represented appropriately between the sweet and nonsweet clusters. The consequence of the conditioning procedure is to disrupt this clear organization (Figure 4B). The relative

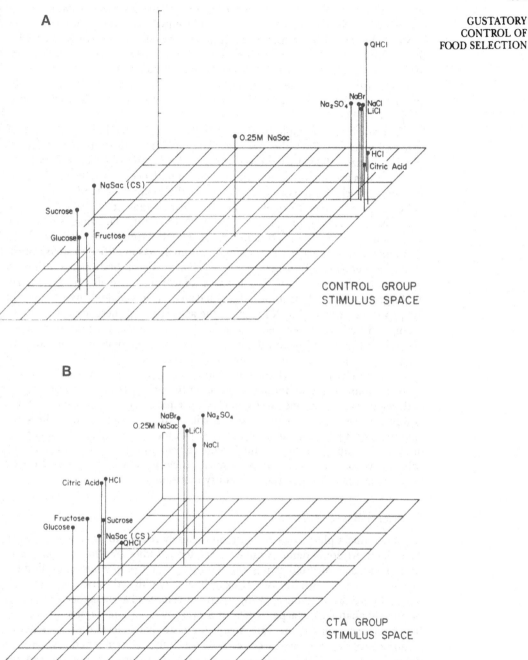

Figure 4. Three-dimensional spaces representing relative similarities among stimuli as determined from neural responses in the control group (A) and the CTA group (B). Sodium saccharin (NaSac, CS) is fixed at the same coordinates in each space. (Reprinted from Chang & Scott, 1984, with permission.)

similarity among nonsweet qualities is reduced; the sharp distinction between sweet and nonsweet is blurred. Moreover, the relative increase in similarity between sweet and nonsweet chemicals is differential, with the greatest increase between the saccharin CS and bitter quinine. The rearrangement is so decisive that quinine is nearly as close to the sweet CS as it is to the acids, a finding that would be quite aberrant in a normally functioning taste system. This may offer a neural concomitant to the increasingly similar behavioral reaction evoked by quinine and sweet chemicals to which an aversion has been conditioned (Grill & Norgren, 1978a).

There are several implications of these findings for the interrelationship of taste and ingestion. First, they reinforce the reports by Aleksanyan *et al.* (1976) and DiLorenzo (1985) that the responses of brainstem and indeed hindbrain taste neurons are subject to modification by experience. Secondly, conditioned aversions, which in humans affect primarily the hedonic evaluation of the CS rather than its perceived quality, caused a rearrangement of the neural taste space that was based on responses from the NTS of rats. This suggests that the hindbrain neural code in rats combines sensory discrimination with a hedonic component. Finally, activity in these intact animals was modified in ways appropriate for mediation of the behavioral aversion, yet decerebrate rats—with NTS and its vagal afferents intact—are incapable of learning or retaining a CTA (Grill & Norgren, 1978a). This implies hypothalamic or amygdaloid involvement in the processes expressed in NTS, given the association of these areas with motivation and hedonics, their required integrity for aversion learning to proceed (Kemble & Nagel, 1973), and their close, reciprocal anatomic relationship with NTS. A direct test of this implication, however, refuted it. Mark and Scott (1988) treated decerebrate rats with the same conditioning protocol as in the Chang and Scott study and, based on facial reactions, confirmed the failure of these animals to develop aversions to the saccharin CS. Subsequent recordings from the NTS, however, disclosed a peak of evoked activity to the CS with virtually the same amplitude and time course as was reported in intact rats. Thus, the source of modification must be located caudal to the colliculi (the plane of decerebration). Selective vagotomies and lesions of nuclei such as area postrema and dorsal motor nuclei may be required to identify that source.

Plasticity Based on Need

An animal's physiological condition is closely related to its choice of foods. The "body wisdom" demonstrated in cafeteria studies by Richter appears to result from taste-directed changes in food selection. Compensatory feeding behavior has been shown in cases of experimentally induced deficiencies of thiamine (Seward & Greathouse, 1973), threonine (Halstead & Gallagher, 1962), and histidine (Sanahuja & Harper, 1962). It is presumed that the physiological benefits of dietary repletion are paired with the taste that preceded those benefits, creating a conditioned taste preference by which the associated hedonic value of the taste is enhanced. Since physiological needs are in constant flux, the hedonic value of a taste experience must be quite labile.

Long-Term Needs: The Case of Sodium. A constant and perhaps innate preference exists for sodium. Evolving in sodium-deficient environments, most mammals seek out and consume salt wherever it is to be found and, when

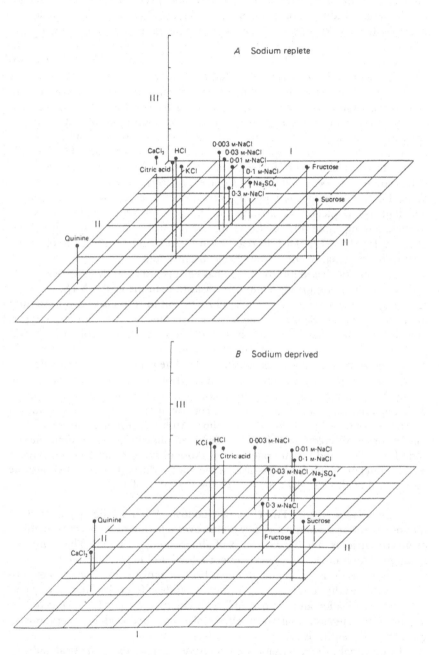

Figure 5. Three-dimensional spaces representing relative similarities among stimuli as determined from neural responses in sodium-replete (A) and sodium-deprived (B) rats. (Reprinted from Jacobs *et al.*, 1988, with permission.)

plentiful, in excess of need. Both rats and humans select Na salts in their diets even when sodium replete (Denton, 1976). This preference becomes exaggerated under conditions of salt deficiency. Humans depleted by pathological states (Wilkins & Richter, 1940) or by experimental manipulations (McCance, 1936) often show a pronounced craving for salt. Rodents subjected to uncontrolled urinary Na loss following adrenalectomy (Richter, 1936), to dietary sodium restrictions (Fregley, Harper, & Radford, 1965), or to acute loss of plasma volume (Stricker & Jalowiec, 1970) show sharp increases in salt consumption. This compensatory response to the physiological need for salt results from a change in the hedonic value of tasted sodium. Concentrations of NaCl that had been evaluated negatively and rejected under conditions of sodium repletion evoke a positive hedonic response and acceptance in deprivation. The hedonic change has been thought to follow from a decreased gustatory sensitivity to salt that accompanies sodium depletion. Contreras (1977) analyzed the responses of single fibers in the chorda tympani nerve to NaCl in replete and in sodium-deficient rats. Depletion was accompanied by a specific reduction in salt responsiveness among those 40% of fibers that were most sodium responsive (Contreras & Frank, 1979). This decreased sensitivity could result in the observed shift in the acceptance curve to higher concentrations.

This intensity-based interpretation has been recast recently by Jacobs, Mark, and Scott (1988). Recording from central taste neurons in the NTS, these researchers confirmed a moderate overall reduction in responsiveness to sodium in salt-deprived rats. Separate analyses of activity among different neuron types, however, revealed that the responsiveness of the salt-sensitive group of cells was profoundly depressed and that this effect was partially offset by a remarkable increase in activity among sweet-sensitive cells. The net effect was to transfer the burden of coding sodium from salt- to sweet-sensitive neurons. This implies not so much a change in perceived intensity as in perceived quality in sodium-deficient rats: salt should now taste "sweet" or, if sweetness is only a human construct, perhaps "good" to such an animal. Multidimensional spaces based on the responses of replete and of sodium-deprived rats confirm shifts in the neural code for sodium and lithium salts toward those of sucrose and fructose (Figure 5). This interpretation explains the eagerness with which deprived rats consume sodium, an avidity usually reserved for the ingestion of sugars.

SHORT-TERM NEEDS: THE CASE OF GLUCOSE. Whereas the appreciation of micronutrient or sodium deficiency occurs over a period of days, the availability of macronutrients, notably glucose, is of almost hourly concern. There must be accommodation of the decision to accept a food and in the hedonic evaluation that drives this decision, as energy needs become more acute. Common experience reinforces the results of psychophysical studies: foods become more palatable with deprivation and less palatable with satiety (Cabanac, 1971). The following evidence suggests that these effects also are mediated by alterations in gustatory afferent activity.

In mammals, the term satiety encompasses a complex of physical and biochemical mechanisms that may operate through independent neural channels. It is clear that alterations in gustatory activity are involved in this process. Glenn and Erickson (1976) recorded multiunit activity from the NTS of freely fed rats as they induced gastric distension and observed a pattern of differential modification in taste responsiveness. Distension severely depressed activity evoked by sucrose, followed in descending order by that to NaCl, HCl, and finally quinine, the responses to which were unmodified. Relief from distension reversed the effect over a 45-minute period. When rats were deprived of food for 48–72 hours, however, the influence of distension on taste was lost, suggesting that the modulating processes may be sensitive to the overall nutritive state of the animal.

Giza and Scott have investigated the effects of other satiety factors on taste-evoked activity in the NTS. Multiunit responses to taste stimuli were recorded before and after intravenous loads of 0.5 g/kg glucose or a vehicle (Giza & Scott, 1983). The glucose infusion had a selectively suppressive effect on taste activity. Elevated blood glucose was associated with a significant reduction in gustatory responsiveness to glucose, with a maximum effect occurring 8 minutes following the intravenous load. Recovery took place over 60 minutes as blood glucose approached normal levels. Responsiveness to NaCl and HCl was suppressed to a lesser degree and for a briefer period, and quinine-evoked responses were unaffected. Similarly, an intravenous administration of 0.5 U/kg insulin resulted in a transient suppression of taste activity evoked by glucose and fructose (Giza & Scott, 1987b). Therefore, hyperglycemia and modest hyperinsulinemia, both of which result in depression of food intake, are associated with reductions in the afferent activity evoked by hedonically positive tastes. These findings imply that the pleasure that sustains feeding is reduced, making termination of the meal more likely.

If taste activity in NTS is influenced by the rat's nutritional state, then intensity judgments should change with satiety. Psychophysical studies of human subjects, although not fully consistent among themselves, generally do not support this position. Humans typically report that the hedonic value of appetitive tastes declines with satiety but that intensity judgments are affected to a lesser extent or not at all (Rolls, Rolls, Rowe, & Sweeney, 1981). There are at least three levels of ambiguity that cloud the interpretation of these conflicting results: first, species differences—rats versus humans; secondly, neural level—electrophysiological data are from the hindbrain, whereas psychophysical responses presumably reflect a cortical influence; finally, the possible effects of barbiturate anesthetics on the rats. A resolution of the implied conflict, then, requires both an analysis of the rat's intensity perceptions and of the human's taste-evoked activity in the hindbrain.

To examine intensity perception in the rat, conditioned taste aversions were formed to 1.0 M glucose, and the degree of behavioral generalization from this to a range of other concentrations was measured (Scott & Giza, 1987). Hyperglycemic rats reacted to glucose concentrations as if they were 50% less intense than did conditioned animals with no glucose load (Giza & Scott, 1987a). Thus, the neural suppression in the hindbrain that results from an intravenous glucose load appears to be manifested in the perception of reduced intensity.

Although it is reassuring that the rat's behavior conforms to the implications of its neural responses, the original conflict with psychophysical results remains unresolved. To complete the puzzle, information is needed on the influence of satiety on taste-evoked activity at various synaptic levels of the human. The closest available approximation to these data may be supplied by subhuman primates. First, the response characteristics of taste neurons in the NTS of cynomolgus monkeys were defined (Scott, Yaxley, Sienkiewicz, & Rolls, 1986). Then the activity of small clusters of these cells was monitored as mildly food-deprived monkeys were fed to satiety with glucose (Yaxley, Rolls, Sienkiewicz, &

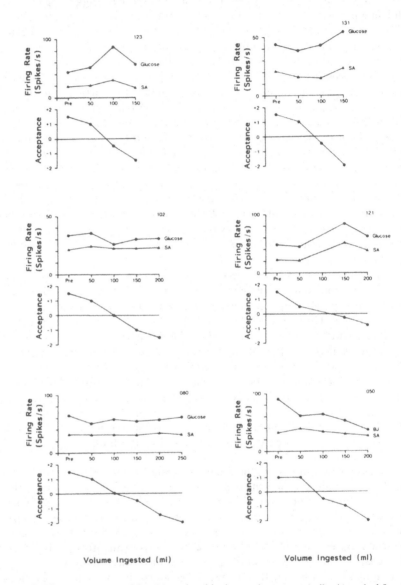

Figure 6. The spontaneous activity (SA) and multiunit neural responses (spikes/s) evoked from the NTS by the taste solutions on which the monkey was fed to satiety. Each graph represents the results of a separate experiment during which the monkey consumed the satiating solution in 50-mL aliquots, as labeled on the abscissa. Represented below the neural response data for each experiment is a behavioral measure of acceptance of the satiating solution on a scale of +2.0 (avid acceptance) to −2.0 (active rejection). The satiating solution is labeled for each graph. BJ, blackcurrant juice. (Reprinted from Yaxley et al., 1985, with permission.)

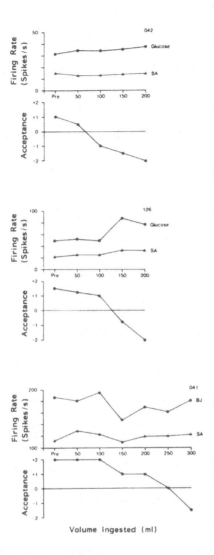

Figure 6. (*continued*)

Scott, 1985). Satiety was measured behaviorally as the monkeys progressed from avid acceptance to active rejection of glucose, typically after consuming 200–300 mL (Figure 6, bottom of each frame). Despite the effects of gastric distension and elevated blood glucose and insulin levels this procedure was designed to cause, the responsiveness of NTS neurons to the taste of a range of solutions, including glucose, was unmodified (Figure 6, top of each frame). These results are in marked contrast to those reported in anesthetized rats, where similar physiological manipulations caused a reduction in responsiveness to sugars of up to 50% (Giza & Scott, 1983; Glenn & Erickson, 1976).

The same approach has been extended to single neurons in cortical taste areas of the frontal operculum (Rolls, Scott, Sienkiewicz, & Yaxley, 1988) and

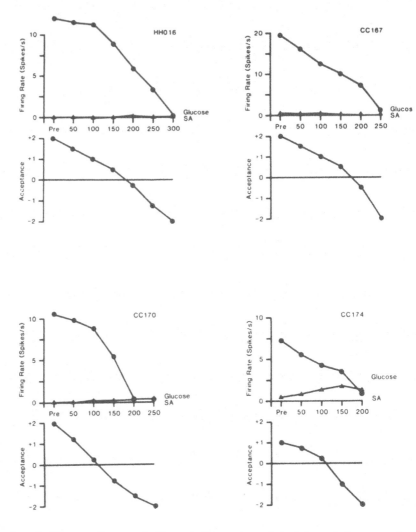

Figure 7. The same format as in Figure 6. Responses are derived from single neurons in the orbitofrontal cortex of the monkey. At this level of processing, discharge rate is related to level of satiety rather than the purely sensory aspects of a stimulus.

anterior insula (Yaxley, Rolls, & Sienkiewicz, 1989) with similar results. Therefore, it appears that the decreased acceptance and reduced hedonic value associated with satiety do not result from a decrement in gustatory responsiveness at any level up to and including primary gustatory cortex. Rather, activity here is related to sensory quality independent of physiological state.

The situation changed when neurons of the monkey's orbitofrontal cortex (OFC) were studied (Rolls, Yaxley, Sienkiewicz, & Scott, 1985). Taste-responsive cells showed vigorous activity to preferred solutions when the monkey was deprived. As satiety increased—and acceptance turned to rejection—the responsiveness declined to near spontaneous rate (Figure 7). There was no change in

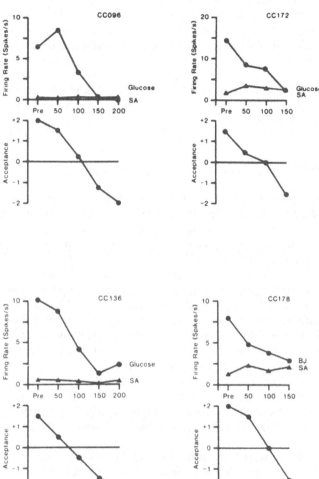

Figure 7. (*continued*)

activity elicited by other stimuli, however, even those to whose qualities the taste of the satiating stimulus would readily generalize. Thus, effects that are apparent in the rat's hindbrain are not manifested in the macaque until the afferent signal reaches an advanced stage of cortical processing in OFC. There the association between a stimulus and its reward value could occur, depending on the need state of the monkey. Through alterations in OFC activity, the animal could modify its behavior according to the availability of environmental resources in relation to its own needs. Separate populations of neurons have also been identified in the monkey's lateral hypothalamic area that responded to the taste of preferred foods if the animal was hungry (Burton, Rolls, & Mora, 1976). As in OFC, the induction of satiety suppressed this activity. The electrophysiological evidence, then, supports the position that gustatory incentives for initiation and maintenance of feeding may be modulated by momentary physiological needs. In primates, however, this influence is evident only after several stages of synaptic processing, including cortical relays at which the quality-intensity evaluation is held independent of hedonic appreciation. It would not be surprising if the macaque joined humans in their ability to evaluate the sensory aspects of food separately from its appeal. A resolution to the conflict between rat electrophysiological and human psychophysical data, then, lies not in whether hedonic evaluations are part of the gustatory neural code but in the neural level at which the interaction occurs.

CONCLUSION

I propose that the sense of taste is like a Janus head placed at the gateway to the city. One face is turned outward to its environment, to warn of and resist the incursion of chemical perils while recognizing and encouraging the receipt of required goods. The other looks inward to monitor the effects of admitted wares on the city's activity and to remain current with its needs.

The outward face of taste signals the quality and intensity of chemicals through a spatiotemporal code, both the spatial and temporal aspects of which are organized on a dimension of physiological welfare. This analysis offers a first approximation of the appropriateness of consuming a chemical. The capacity to perform it is genetically endowed and, one supposes, derives from evolutionary pressures to avoid chemicals that are toxic and to consume those that provide nutrients. We are, after all, the progeny of those who, among other things, selected wisely from among the chemicals in their environment. This analytical process is inherent to the structure and function of the taste system and so applies across all members of a species and at all times. It forms the first layer of what may be seen as a three-tiered system that underlies the concept of body wisdom.

The second layer, which, with the third, requires both faces of taste, permits fine adjustments to the inherited code as well as tailoring to the physiology of the individual. A deficient enzyme may render a normally nutritious substance indigestible such that its ingestion causes nausea, vomiting, and diarrhea. This experience alters gustatory responsiveness in a manner appropriate to accommodate the idiosyncracy. Thus, a chemical that generates a response pattern similar to those of preferred tastes in the naive rat may assume negative characteristics after its ingestion is associated with malaise. Conditioning occurs on a time scale that relates to visceral rather than operant processes. In tandem with neophobic reactions, it serves to limit the harm of ingesting a toxin to that caused by a single, mild exposure. Whereas the effects of conditioning apply only to the individual, the mechanism that permits this extraordinarily powerful association between taste and physiological consequences is inherent to the structure of the gustatory–visceral complex.

The third tier involves short-term fluctuations in sensitivity that promote or inhibit feeding and encourage consumption of a nutritionally replete diet. Positive hedonics are suppressed by satiety factors such as those resulting from gastric distension and the postabsorptive delivery of metabolically utilizable calories. Thus, a ready endogenous energy supply reduces the reinforcing value of food and so promotes cessation of feeding. Conversely, the need for sodium arouses enhanced responsiveness to salt among taste neurons whose activity is associated with positive hedonics, thereby encouraging salt consumption.

The taste system evaluates substances at the interface between discrimination and digestion. It combines, in its Janus head, the qualities of rapid stimulus identification and spatiotemporal coding associated with the exteroceptive senses with the slower recognition and still mysterious codes of the visceral senses. The result is a system that reduces the diverse and frequently hostile environment to a chemical subset that effectively satisfies the complex and changing requirements of the individual's internal milieu.

Acknowledgments

During preparation of this chapter, the author was supported by research grants from the National Institutes of Health, the National Science Foundation, and the Campbell Soup Company. Dr. Barbara K. Giza helped organize the references, and Ms. Judy A. Fingerle typed the manuscript.

REFERENCES

Aleksanyan, A. A., Buresova, O., & Bures, J. (1976). Modification of unit responses to gustatory stimuli by conditioned taste aversion in rats. *Physiology and Behavior, 17,* 173–179.

Burton, M., Rolls, E., & Mora, F. (1976). Effects of hunger on the response of neurons in the lateral hypothalamus to the sight and taste of food. *Experimental Neurology, 51,* 668–677.

Cabanac, M. (1971). Physiological role of pleasure. *Science, 173,* 1103–1107.

Capretta, P. J., & Rawls, L. H. (1974). Establishment of a flavor preference in rats: Importance of nursing and weaning experience. *Journal of Comparative and Physiological Psychology, 86,* 670–673.

Chang, F-C. T., & Scott, T. R. (1984). Conditioned taste aversions modify neural responses in the rat nucleus tractus solitarius. *Journal of Neuroscience, 4,* 1850–1862.

Contreras, R. (1977). Changes in gustatory nerve discharges with sodium deficiency: A single unit analysis. *Brain Research, 121,* 373–378.

Contreras, R., & Frank, M. (1979). Sodium deprivation alters neural responses to gustatory stimuli. *Journal of General Physiology, 73,* 569–594.

Denton, D. A. (1976). Hypertension: A malady of civilization? In M. P. Sambhi (Ed.), *Systemic effects of antihypertensive agents* (pp. 577–583). New York: Stratton Intercontinental Medical Books.

DiLorenzo, P. M. (1985). Responses to NaCl of parabrachial units that were conditioned with intravenous LiCl. *Chemical Senses, 10,* 438.

DiLorenzo, P. M., & Schwartzbaum, J. S. (1982). Coding of gustatory information in the pontine parabrachial nuclei of the rabbit: Temporal patterns of neural response. *Brain Research, 251,* 245–257.

Doetsch, G. S., & Erickson, R. P. (1970). Synaptic processing of taste-quality information in the nucleus tractus solitarius of the rat. *Journal of Neurophysiology, 33,* 490–507.

Domjan, M. (1976). Determinants of the enhancement of flavored-water intake by prior exposure. *Journal of Experimental Psychology: Animal Behavior Processes, 2,* 17–27.

Erickson, R. P., Doetsch, G. S., & Marshall, D. A. (1965). The gustatory neural response function. *Journal of General Physiology, 49,* 247–263.

Fregley, M. J., Harper, J. M., & Radford, E. P., Jr. (1965). Regulation of sodium chloride intake by rats. *American Journal of Physiology, 209,* 287–292.

Garcia, J., Kimmeldorf, D. J., & Koelling, R. A. (1955). Conditional aversion to saccharin resulting from exposure to gamma radiation. *Science, 122,* 157–158.

Giza, B. K., & Scott, T. R. (1983). Blood glucose selectively affects taste-evoked activity in the rat nucleus tractus solitarius. *Physiology and Behavior, 31,* 643–650.

Giza, B. K., & Scott, T. R. (1987a). Blood glucose level affects perceived sweetness intensity in rats. *Physiology and Behavior, 41,* 459–464.

Giza, B. K., & Scott, T. R. (1987b). Intravenous insulin infusions in rats decrease gustatory-evoked responses to sugars. *American Journal of Physiology, 252,* R994–R1002.

Glenn, J. F., & Erickson, R. P. (1976). Gastric modulation of gustatory afferent activity. *Physiology and Behavior, 16,* 561–568.

Grill, H. J. & Norgren, R. (1978a). The taste reactivity test: I. Mimetic responses to gustatory stimuli in neurologically normal rats. *Brain Research, 143,* 263–279.

Grill, H. J., & Norgren, R. (1978b). The taste reactivity test. II. Mimetic responses to gustatory stimuli in chronic thalamic and chronic decerebrate rats. *Brain Research, 143,* 281–297.

Halstead, W. C., & Gallagher, B. B. (1962). Autoregulation of amino acid intake in the albino rat. *Journal of Comparative and Physiological Psychology, 55,* 107–111.

Jacobs, K. M., Mark, G. P., & Scott, T. R. (1988). Taste responses in the nucleus tractus solitarius of sodium-deprived rats. *Journal of Physiology (London), 406,* 393–410.

Kemble, E. D., & Nagel, J. A. (1973). Failure to form a learned taste aversion in rats with amygdaloid lesions. *Bulletin of the Psychonomic Society, 2,* 155–156.

Mark, G. P., & Scott, T. R. (1988). Conditioned taste aversions affect gustatory activity in the NTS of chronic decerebrate rats. *Neuroscience Abstracts, 14,* 1185.

McCance, R. A. (1936). Experimental sodium chloride deficiency in man. *Proceedings of the Royal Society, London, Series B, 119,* 245–268.

Nagai, T., & Ueda, K. (1981). Stochastic properties of gustatory impulse discharges in rat chorda tympani fibers. *Journal of Neurophysiology, 45,* 574–592.

Pfaffmann, C., Norgren, R., & Grill, H. J. (1977). Sensory affect and motivation. *Annals of the New York Academy of Sciences, 290,* 18–34.

Revusky, S. H., Smith, M. H., & Chalmers, D. V. (1971). Flavor preferences: Effects of ingestion contingent intravenous saline or glucose. *Physiology and Behavior, 6,* 341–343.

Richter, C. P. (1936). Increased salt appetite in adrenalectomized rats. *American Journal of Physiology, 115,* 155–161.

Rolls, B. J., Rolls, E. T., Rowe, E. A., & Sweeney, K. (1981). Sensory specific satiety in man. *Physiology and Behavior, 27,* 137–142.

Rolls, E. T., Scott, T. R., Sienkiewicz, Z. J., & Yaxley, S. (1988). The responsiveness of neurons in the frontal opercular gustatory cortex of the monkey is independent of hunger. *Journal of Physiology, 56,* 876–890.

Rolls, E. T., Yaxley, S., Sienkiewicz, Z. J., & Scott, T. R. (1985). Gustatory responses of single neurons in the orbitofrontal cortex of the macaque monkey. *Chemical Senses, 10,* 443.

Sanahuja, J. C., & Harper, A. E. (1962). Effect of amino acid imbalance on food intake and preference. *American Journal of Physiology, 202,* 165–170.

Schiffman, S. S., & Erickson, R. P. (1971). A psychophysical model for gustatory quality. *Physiology and Behavior, 7,* 617–633.

Scott, T. R., & Giza, B. K. (1987). A measure of taste intensity discrimination in the rat through conditioned taste aversions. *Physiology and Behavior, 41,* 315–320.

Scott, T. R., & Mark, G. P. (1987). The taste system encodes stimulus toxicity. *Brain Research, 414,* 197–203.

Scott, T. R., Yaxley, S., Sienkiewicz, Z. J., & Rolls, E. T. (1986). Gustatory responses in the nucleus tractus solitarius of the alert cynomolgus monkey. *Journal of Neurophysiology, 55,* 182–200.

Seward, J. P., & Greathouse, S. R. (1973). Appetitive and aversive conditioning in thiamine-deficient rats. *Journal of Comparative Physiology, 83,* 157–167.

Smotherman, W. P., & Levine, S. (1978). ACTH and ACTH$_{4-10}$ modification of neophobia and taste aversion responses in the rat. *Journal of Comparative and Physiological Psychology, 92,* 22–33.

Steiner, J. E. (1973). The gustofacial response: Observation on normal and anencephalic newborn infants. In J. F. Bosma (Ed.), *Symposium on oral sensation and perception. IV* (pp. 254–278). Bethesda: NIH-DHEW.

Steiner, J. E. (1979). Human facial expressions in response to taste and smell stimulation. In H. W. Reese & L. Lipsett (Eds.), *Advances in child development* (vol. 13, pp. 257–295). New York: Academic Press.

Steiner, J. E., & Glaser, D. (1985). Orofacial motor behavior-patterns induced by gustatory stimuli in apes. *Chemical Senses, 10,* 452.

Stricker, E. M., & Jalowiec, J. E. (1970). Restoration of intravascular fluid volume following acute hypovolemia in rats. *American Journal of Physiology, 218,* 191–196.

Wilkins, L., & Richter, C. P. (1940). A great craving for salt by a child with corticoadrenal insufficiency. *Journal of the American Medical Association, 114,* 866–868.

Yaxley, S., Rolls, E. T., & Sienkiewicz, Z. J. (1989). The responsiveness of neurons in the insular gustatory cortex of the macaque monkey is independent of hunger. *Physiology and Behavior, 42,* 223–229.

Yaxley, S., Rolls, E. T., Sienkiewicz, Z. J., & Scott, T. R. (1985). Satiety does not affect gustatory activity in the nucleus of the solitary tract of the alert monkey. *Brain Research, 347,* 85–93.

11

Comparative Studies of Feeding

F. Reed Hainsworth and Larry L. Wolf

Introduction

The Design of Feeding

Comparative studies seek to explain similarities and differences among species. All animals have the similar problem of obtaining food to survive and reproduce, but there is tremendous diversity in types of food and mechanisms for obtaining and using it. Feeding has been classified by type of food consumed (e.g., carnivore, herbivore, frugivore, insectivore, detritivore, nectarivore, omnivore) and the ingestive mechanism used (e.g., biting, filtering, licking, chewing, gulping, sucking, fermenting), but classification does little to elucidate function. What shapes and explains this diversity?

Evolution through natural selection should produce animals that are efficient in their feeding. Efficiency usually is defined in economic terms of costs and benefits. The ability to survive and reproduce should be maximum when costs and benefits have particular values. Comparative studies are concerned with how feeding has been molded by natural selection to solve the cost–benefit problems of survival and reproduction for animals that occur in a variety of environments (Hainsworth & Wolf, 1979; Stephens & Krebs, 1986).

The effort to understand feeding design relies heavily on engineering concepts. Engineers design efficient machines to solve particular problems. Comparative physiologists work in reverse to try to understand the cost–benefit problems that have led to the design of animals through natural selection. Ideas have been borrowed from two major areas of engineering: control theory (Ashby, 1956; Riggs, 1963) and optimization theory (Alexander, 1982a). The former involves the familiar concepts of regulation, feedback, and control (Brobeck,

F. Reed Hainsworth and Larry L. Wolf   Department of Biology, Syracuse University, Syracuse, New York 13244.

1965; Yamamoto, 1965). The latter is a branch of applied mathematics concerned with design of efficient machines.

Use of engineering models has potential pitfalls. When concepts are borrowed from control theory, it is tempting to assume specific details about design without the required evidence (Booth, 1980; Booth, Toates, & Platt, 1976; Davis, 1980; Wirtshafter & Davis, 1977). Similarly, use of optimization theory in biology has been criticized for assuming that animals are perfect machines when there is considerable evidence to the contrary (Gould, 1986; Gould & Lewontin, 1979).

On the other hand, engineering models have been extremely useful for investigating functional design in several areas of physiology. Negative feedback control features have been used to study control of respiration (Hitzig & Jackson, 1978), regulation of body temperature (Hardy, 1965; Hardy & Stitt, 1976; Heller & Colliver, 1974), contraction of muscle (McMahon, 1984), and hemoglobin function (Prosser, 1986), among others. Optimization theory has been used to study design of animal locomotion (Alexander, 1982b) and population sex ratios (Charnov, 1982), among other phenomena. In each case the success of the approaches has depended on availability of evidence to test predictions arising from particular models of functional design (Maynard Smith, 1978).

If feeding is to be understood in terms of cost–benefit models, then it is important to study organisms in which costs and benefits can be quantified. It also is important to study a diversity of organisms and conditions that influence survival. Features of feeding design ultimately must explain feeding under natural conditions and over the lifetimes of animals. These criteria are stringent, and most studies of feeding address only a small component of food intake or use (Hainsworth & Wolf, 1979).

PREVIEW

In this chapter we examine three major areas of the comparative study of feeding. The first involves broad comparisons of animals of vastly different sizes to ask how major differences in costs and benefits for energy might produce different patterns of feeding and energy use. We identify small endothermic animals as those facing extreme energy regulation problems. Also, we emphasize the need to include both costs and benefits when considering energy use patterns by examining some problems from traditional approaches that consider only costs.

The second area concerns optimal foraging. We briefly explore economic decisions among feeding alternatives, the importance of time scale for feeding choices, and the types of choices made under natural conditions that could influence where, what, when, and how much to eat.

The third area concerns nectar feeders as a model system to study the design of feeding based on ease of quantifying energy costs and benefits, inherently severe regulatory problems from small body size for endothermic nectar feeders, and the ease of testing predictions that these features allow. We examine evidence for rules thought to govern design for search, detection, ingestion, assimilation, and storage of energy in nectar feeders. We find that many studies of optimal feeding design do not involve an appropriate role for hunger (defined as an energy deficit; see Figure 5), and we propose a way to recognize its importance.

VARIATION IN ENERGY COSTS AND BENEFITS WITH BODY SIZE

The traditional physiological approach of measuring energy expenditures without considering consequences for feeding produces problems in interpreting measurements. To explore these problems we examine how energy demands and supplies vary with body size. We hope to foster more critical thinking about energy as a resource by integrating cost and benefit functions for energy.

Larger animals have more body tissue cells and must expend more energy per time. Measurements extending over many endothermic species (mice to elephants; hummingbirds to ostriches) show a power function relationship between minimum energy use per time (y) and body mass (kg): $y = a(kg)^B$, with the exponent B equal to 0.75 (Peters, 1983; Schmidt-Nielsen, 1984). Measurements restricted to individual species often yield an exponent of 0.67, however, which would be expected if surface area/volume influenced energy expenditures (Feldman & McMahon, 1983; Heusner, 1982). A slope less than 1.0 indicates that increases in body size produce smaller increases in energy demand and feeding. Thus, when energy supplies are available, more mass can be supported by larger animals than by smaller.

How do energy supplies change as size changes? This is a more difficult question to address because food quality and quantity are highly variable. One approach is to examine how digestive storage organs vary with body size. These contain the meals consumed by animals and may reflect energy available between feeding episodes. For a variety of animals, measurements indicate that gut volume varies with mass raised to an exponent of 1.0 (Belovsky, 1986; Calder, 1984; Demment & Van Soest, 1985; Hainsworth & Wolf, 1972a; Martin, Chivers, MacLarnon, & Hladik, 1985). An exponent of 1.0 is consistent with linear scaling of volumes with mass in other regulatory systems; for example, lung volume in birds (Lasiewski & Calder, 1971) and mammals (Stahl, 1967) increases linearly with mass (exponents 0.94 to 1.06), blood volume increases linearly with body mass (exponent 1.02; Stahl, 1967), and heart size increases linearly with body mass for mammals (exponent 0.98; Prothero, 1979) and birds (exponent 0.91; Hartman, 1955).

ADVANTAGES AND DISADVANTAGES AS BODY SIZE VARIES

The different exponents relating metabolic demands ($B = 0.75$) and digestive supplies ($B = 1.0$) of energy to body size in endotherms produce advantages and disadvantages as size varies (Figure 1). The supply and demand lines converge as size decreases and diverge as size increases. If food quality is similar (which is not always the case; see below), then after consuming a meal proportional to body size, small animals could meet their metabolic requirements for less time than large animals. This is consistent with the common observation that small endotherms feed more frequently than large endotherms (Calder, 1984). Hummingbirds, for example, feed 3–12 times/h (Wolf & Hainsworth, 1977). A lower limit to endotherm body size may occur when continuous feeding would be insufficient even with high-quality food. At this extreme the only way to survive would be to lower energy demands, such as with torpor (Hainsworth & Wolf, 1978).

F. REED
HAINSWORTH
AND LARRY L.
WOLF

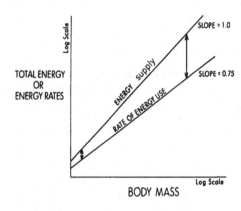

Figure 1. Different slopes for energy supply and rate of energy use versus body mass produce advantages and disadvantages as body size varies. The small difference between supply and rate of use at small mass produces high-frequency feeding, whereas animals of large mass would feed less frequently and may have a greater ability to store energy.

As body size increases, meals could represent larger energy reserves relative to demands, and hence energy demands could be met for longer periods between meals (still assuming food quality is similar). Also, there may be a greater ability to store energy from what has been consumed. Energy storage depends on the fraction of meals not used for maintenance between meals (Hainsworth & Wolf, 1979; LeMagnen, 1976; Wolf & Hainsworth, 1977), and this may increase as size increases. Calder (1984), for example, estimates an exponent on body mass for fat content of 1.19.

An increased ability of large animals to consume and store energy relative to their demands may explain some size-dependent phenomena influencing use of energy. One is Bergmann's rule, that body mass is correlated inversely with environmental temperature. A colder environment requires higher rates of energy use for temperature regulation, and an advantage would be obtained from the increased ability to consume and store energy associated with larger size. Another is hibernation with torpor. Mammals that become torpid seasonally, such as ground squirrels and marmots, are of intermediate size. No large mammals become torpid seasonally [although bears become inactive in winter, they do not drop their body temperatures more than a few degrees (Folk, 1967)], whereas small mammals become torpid daily. If the ability to consume and store energy relative to rate of use increases with body size, then a large mammal may not be faced with the problem of starvation, and torpor may not be necessary.

Our interpretation differs from more traditional explanations for these phenomena that are based on advantages or disadvantages associated only with costs. Rates of energy use often are expressed as weight-specific metabolic rates that take the form: $(kg)^{0.75}/(kg)^{1.0} = (kg)^{-0.25}$, so weight-specific rates decrease with increasing body size (the presumed advantage associated with being bigger). The disadvantage of being small usually is attributed to constraints associated with very high metabolic costs per gram (Schmidt-Nielsen, 1972, 1984; Tracy, 1977). The problem with this view is failure to consider benefits together with total costs. Whether an animal survives depends on both its total costs and the supplies to meet them (Garland, 1983). There is no simple cost advantage to being bigger, because total costs always increase with size. Advantages and disadvantages can be identified only when total costs are compared with supplies (Figure 1). When total costs change with $(kg)^{0.75}$ and supplies change with $(kg)^{1.0}$, then supplies/costs changes with $(kg)^{0.25}$, or costs/supplies changes with

$(kg)^{-0.25}$. As body size varies, the capacity to store energy changes. This produces a disadvantage at small body size and an advantage at large body size.

Body Size and Variation in Food Quality

We have assumed for simplicity that food quality does not vary. In such a case the upper limit to body size would be set by availability of food. But there is a negative relationship between the abundance and quality of some foods. Some plants are abundant, and fiber content increases with available biomass (Demment & Van Soest, 1985). Efficient digestion of fiber requires ruminant fermentation with selective retention of food, and efficient fermentation takes time. Demment & Van Soest (1985) measured retention times of 35–45 hours for effective energy yield from plant foods in a fermentation chamber. Based on metabolic costs per gut capacity, which decrease with increasing body size, small animals would have insufficient time for efficient fermentation, whereas large animals could retain food for sufficient periods. Here the advantage of large size permits use of a low-quality but abundant food rather than reduced meal frequency or increased energy storage ability per meal.

Body Size and Variation in Foraging Costs

Comparisons based on resting metabolic rates neglect the obvious requirement that animals must be active to feed. Although the energy costs for movement have been studied in detail for a large number of animals (Alexander, 1978; Schmidt-Nielsen, 1972, 1984), such costs depend on movement speed, and it often is not clear how speed should be adjusted for feeding to be efficient (Pyke, 1981a). Efficient feeding may depend on the rate of capture, so the difference between rate of food gain and rate of energy expended is maximum while an animal feeds (Norberg, 1981). This is an optimality criterion that we consider in detail below, but it is only one of several possible criteria.

Hummingbirds provide an opportunity to examine costs for feeding together with gains. A major cost is from hovering. It is easy to measure rates of energy use for hovering hummingbirds, because they remain stationary, and thus many measurements have been reported (Bartholomew & Lighton, 1986; Berger & Hart, 1972; Epting, 1980; Lasiewski, 1963; Wolf & Hainsworth, 1971). This behavior also is interesting because it is the most costly form of flight (Hainsworth & Wolf, 1972b, 1974; Lighthill, 1977; Rayner, 1979). Why do feeding hummingbirds hover instead of perch? Use of hovering by very small endotherms appears to exacerbate their already severe problems in energy regulation.

The weight-specific rate of oxygen consumption for hovering hummingbirds is 43 mL O_2/g per h regardless of the size of the species studied (Bartholomew & Lighton, 1986; Hainsworth & Wolf, 1972b). The metabolic costs associated with increasing size are less than expected based on geometric similarity, because hummingbird wings increase in length disproportionately as size increases (Alexander, 1982b; Epting & Casey, 1973; Greenewalt, 1962; Hainsworth & Wolf, 1974). Nevertheless, the relationship between total hovering costs and body mass makes hovering more expensive relative to perching for larger hummingbirds (Hainsworth & Wolf, 1972b). Most relatively small nectar-feeding birds (such as hummingbirds) hover to feed, whereas most relatively large nec-

F. REED
HAINSWORTH
AND LARRY L.
WOLF

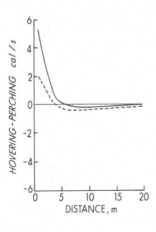

Figure 2. Calculated differences between rates of net energy gain for an 8-g hummingbird for hovering versus perching as a function of round-trip distance. Values assume a flight speed of 2 m/s at 85% of hovering costs, an ambient temperature of 20°C, a body temperature of 41°C, and a total intake of 15 cal. The solid line assumes an intake rate of 15 cal/s while hovering and 13.6 cal/s while sitting. The dashed line assumes these rates are halved. (From Wolf & Hainsworth, 1983.)

tar-feeding birds (such as sunbirds and honeyeaters) perch. If the rate of net energy gain while feeding [(energy consumed − energy expended)/time] is calculated for both modes of feeding, hovering could yield higher rates of gain than perching when body size is small and feeding is more rapid with hovering (Pyke, 1981b; Wolf & Hainsworth, 1983). Figure 2 illustrates the impact of distance from a perch to food on rates of net energy gain for the two foraging behaviors, assuming a constant rate of movement (Wolf & Hainsworth, 1983).

There has been an attempt to study hovering versus perching (Miller, 1985). Two species of small hummingbirds (3.0–3.5 g) almost always perched when given the opportunity (which they often do not have at the flowers they visit). Although this suggests that the ideas discussed above are wrong, it should be noted that in this study (Miller, 1985) distances to the feeder were large, and according to Figure 2 perching would be expected (Hainsworth, 1986). It would be necessary to vary distance systematically to determine whether hovering occurred with short distances, perching with long, and whether the behaviors can be switched back and forth depending on the economics of the feeding situation.

Although we can identify certain types of animals and behaviors as representing severe energy regulatory problems based on size-related variables, studies of feeding design ultimately must account for numerous other factors influencing the economics of food consumption and use. Optimal foraging theory offers an approach to possible solutions of a variety of complex feeding problems. This approach is considered in the next section.

Optimal Foraging

Natural Selection and Feeding

Feeding is an intricate set of external and internal processes triggered by and influencing the internal state of an animal. It probably has a genetic basis that may include influences on size and morphology of the consumer (e.g., Price, 1987; Price & Grant, 1984; Wolf, Hainsworth, & Stiles, 1972), locations of feeding (e.g., Rausher, 1984; Via, 1986), and the type of feeding behavior used (e.g., Arnold, 1981; Sokolowski, 1985). Because feeding is required for survival and reproduction and is part of an organism's phenotype, it should be molded by natural selection.

A feeding animal faces what can be considered a series of times when alternative actions might be taken (reviewed in Stephens & Krebs, 1986). The alternatives could be to eat one item or another (Emlen, 1966; Werner & Hall, 1974), to move in one direction or another (Gill & Wolf, 1977; Heinrich, 1979a; Pyke, 1978a, 1978b, 1984), or perhaps to produce a digestive enzyme at one rate or another. Each alternative actually performed depends on both the genetic and environmental influences on the organism, and the survival and reproductive consequences may vary among individuals using different alternatives.

Natural selection becomes an optimizing force, continually moving the genetic constitution of the population toward the optimal alternatives. The evolutionary viewpoint looks at natural selection as a design process, with success of alternative designs determined by differential reproduction of individuals using each. Studies of feeding as an optimality problem try to analyze why a particular design is better than other plausible designs.

Feeding cannot be divorced from the remainder of the phenotype, and natural selection will act on potential feeding alternatives in the context of other activities influencing survival and reproduction. Predators, competitors, mates, temperatures, and so on can influence which alternative will lead to the most descendants. In addition, the current morphological and physiological capabilities of the animal (the outcome of earlier evolution) influence the range of possible alternatives and their selective values. This evolutionary history is important in understanding the value of alternative actions and the responses to possible experimental manipulations. A manipulation for which an animal has no historical background makes interpretations difficult. For example, nonnutritional sugars elicit feeding in blowflies (Dethier, 1976), a seemingly maladaptive response, perhaps attributable to a lack of such sugars in the past natural environment.

Alternative feeding activities may vary in both their benefits and costs. Ultimately, these benefits and costs should be measured in units of relative genetic representation in future generations (Schoener, 1971; Sih, 1982). Translating feeding actions directly into these units is difficult, however, and other success measures more directly related to feeding usually are used (Schoener, 1971). For the following example, we use arbitrary benefit and cost units.

Figure 3 considers as an example a rather specific aspect of feeding, what

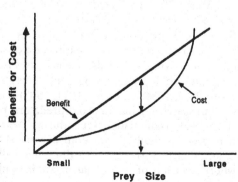

Figure 3. A general benefit and cost function for variation in prey size. The greatest difference between benefits and costs would occur at an intermediate value (arrow).

size food to eat. Benefits are assumed to increase linearly with size of potential prey. The cost curve reflects necessary time, energy, injury probabilities, and so on associated with capturing, handling, eating, and processing prey of different sizes. Small prey presumably require less total effort than large prey, but some costs, such as those for searching, may be fixed and may exceed potential benefits of very small prey. Very large prey may be exceedingly difficult to capture, subdue, or ingest, in which case costs go up faster than benefits; hence, a positively accelerating cost curve. Some alternatives have higher costs than benefits, and in most cases they should be avoided. An intermediate prey size maximizes net benefits, or if benefit and cost are considered as rates, then rate of net benefit (energy intake) is maximized at an intermediate prey size.

PROBLEMS AND CRITERIA

The first question in an optimality approach to feeding is to define the problem and possible alternative solutions. Problems could range from internal processes such as assimilation rates and efficiencies or digestive passage rates to external ones such as what, when, or where to eat and how long to eat in one place before moving. Although these problems are interrelated to varying degrees, initially we simplify matters by narrowing the scope to a single problem.

The next question is to identify the criterion for the optimization process. We noted earlier that relative number of breeding offspring is the criterion for evolution by natural selection, but translating the results of feeding directly into offspring is difficult, especially in the short term. Theorists have resorted to other criteria more directly related to feeding but presumed to influence offspring production through either survival or reproduction. A commonly used criterion is maximizing rate of net energy intake during foraging. This assumes that animals forage for energy and not nutrients and that accumulating energy at the fastest possible rate during foraging produces the most offspring. Indeterminant growers, especially those with reproduction and survival positively related to size, might be expected to fit this criterion best. For example, egg production of female fish often is a direct function of size, and smaller fish often have higher mortality rates (Bagenal, 1967).

Some studies of the harvest of nectar by various animals have used maximizing rate of net energy gain during foraging as the criterion producing predictions that closely fit observations (Pyke, 1984). For example, bumblebees (*Bombus*) sometimes leave nectar in flowers they visit. Using the net rate criterion and the observation that rate of intake decreased as a flower was emptied, Hodges and Wolf (1981) predicted very closely the amount left in flowers. The bees emptied flowers in an area with low volumes per flower and left about 1.5 µL per flower in an area with higher volumes per flower.

Other criteria also are plausible, especially if we view feeding in the context of other activities. Animals in some cases may be selected to minimize feeding time consistent with meeting some energy requirement for survival (Hixon, 1980; Pyke, 1984; Schoener, 1971, 1983). Time released from feeding might be used to search for mates, avoid predators, build a nest, and so on. Maximizing short-term accumulation rates during feeding and minimizing foraging time may produce similar predictions in some situations but different predictions in others. The two criteria can be discriminated using some nectar-feeding animals, where selection of foods that maximize rates of energy gain during foraging yields less total energy intake.

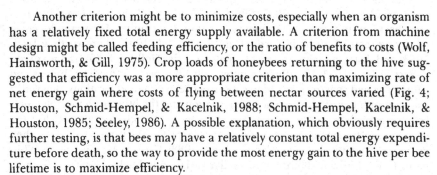

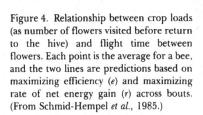

Figure 4. Relationship between crop loads (as number of flowers visited before return to the hive) and flight time between flowers. Each point is the average for a bee, and the two lines are predictions based on maximizing efficiency (*e*) and maximizing rate of net energy gain (*r*) across bouts. (From Schmid-Hempel *et al.*, 1985.)

Another criterion might be to minimize costs, especially when an organism has a relatively fixed total energy supply available. A criterion from machine design might be called feeding efficiency, or the ratio of benefits to costs (Wolf, Hainsworth, & Gill, 1975). Crop loads of honeybees returning to the hive suggested that efficiency was a more appropriate criterion than maximizing rate of net energy gain where costs of flying between nectar sources varied (Fig. 4; Houston, Schmid-Hempel, & Kacelnik, 1988; Schmid-Hempel, Kacelnik, & Houston, 1985; Seeley, 1986). A possible explanation, which obviously requires further testing, is that bees may have a relatively constant total energy expenditure before death, so the way to provide the most energy gain to the hive per bee lifetime is to maximize efficiency.

Time scale is an important consideration for many criteria. When time units are part of a criterion (rate or time criteria), then the time for maximization or minimization must be carefully considered. Is rate of net energy gain maximized over seconds, minutes, hours, days, weeks, a season, or a lifetime? The scale will be dictated partly by the biology of the animals. Determinant growers cannot continue to maximize over a long period without become obese. Hummingbirds may sometimes maximize rates of energy gain during feeding bouts, but not over the entire feeding day. Honeybees may maximize over a season to produce the most queens and drones at the end of a season (Seeley, 1985). Time minimizers may do so, for example, during hot times of the day or all day. We will explore the question of time scales in detail for nectar feeders in our discussion of a model system where there is considerable evidence that what is maximized is net energy gain per meal rather than rate of energy gain during a meal.

Once an appropriate criterion is identified, it is necessary to relate the benefits and costs of feeding to the criterion to predict the optimal alternatives under specified sets of environmental conditions. In the example of honeybees, Schmid-Hempel *et al.* (1985) calculated efficiencies and rates of net energy gain of carrying various crop loads at different flower densities (Figure 4). The predictions then were compared to what the bees actually did. Poor fits would force a reevaluation of the model in terms of both criteria and other possible constraints. Good fits do not eliminate alternative models but lend support to the current model.

Many constraints other than those used to calculate direct benefits and costs could influence feeding. So far we have neglected nutrients, but they must be considered as well. For example, moose diets and foraging locations in Isle Royale National Park, U.S.A., are predicted best with a model considering the requirement for sodium as well as for energy (Belovsky, 1978). Aquatic and terrestrial vegetation differ in energy and sodium contents, and moose appear to eat an optimal mix to meet sodium requirements and provide the best possible rate of energy gain. Other possible constraints include competitors (Milinski, 1982), defense of a territory (Ydenberg, 1987), mating, and predator avoidance (Godin & Smith, 1988; Lima, 1985; Milinski & Heller, 1978; Werner, Gilliam, Hall, & Mittelbach, 1983).

DYNAMIC PROGRAMMING AND ACCOUNTING FOR HUNGER

Most optimal foraging theories and experiments implicitly assume that the energy state of the animal does not vary. However, a few studies have considered hunger as a variable internal state or tried to control for state changes (Hainsworth, Tardif, & Wolf, 1981; Krebs, Erichsen, Webber, & Charnov, 1977; Molles & Pietruszka, 1987; Perry, 1987; Sih, 1982). Energy deficits have been shown to feed back on important components of feeding, including, for example, crop-emptying rates of hummingbirds (Hainsworth et al., 1981; Hainsworth & Wolf, 1983) and distances mantids react by stalking a fly (Holling, 1966), and we might expect hunger levels to influence potentially all aspects of feeding. Juncos and white-crowned sparrows forced to choose between two food locations, one with a constant number of seeds at each presentation and the other with variable numbers of seeds but the same long-term average as the other location, select the constant location when in positive energy balance but the variable location when in negative energy balance (Caraco, 1981, 1983; Caraco, Martindale, & Whitham, 1980). Individual birds show similar variation in feeding location. In these experiments hunger (measured as energy deficits) influenced choice of foraging situation.

The changing internal state of an animal as it feeds, coupled with long-term optimality criteria, means that feeding might vary depending on how a choice in the current state can best meet a long-term goal. Theorists now are beginning to use dynamic programming as a way to model these more complex situations (Houston & McNamara, 1988; Mangel & Clark, 1986). Empiricists will need to investigate more directly the influence of state changes on feeding activities.

INFORMATION

Choices in optimal foraging are made among possible future alternatives. The costs and benefits of these future options must be predictable if an animal is to choose reliably among them. Predictability requires some information about the future options, information generally derived from past and current activities (Green, 1980, 1984; Iwasa, Higashi, & Yamamura, 1981). In diet theory, information about absolute abundances of the best diet item, handling time, and benefits from alternatives were necessary for the early general models (Krebs et al., 1977; Werner & Hall, 1974). Early models of how long an animal remained feeding in a particular patch of food assumed perfect information about quality of alternative patches and the costs of traveling between patches (Charnov, 1976; Cibula & Zimmerman, 1987; Cowie & Krebs, 1979). In controlled laboratory experiments with extensive pretrial experience, the animals may have learned what the alternatives were and the net benefits of each (Krebs, Kacelnik, & Taylor, 1978).

In nature conditions continually change, partly as a result of activities of the feeding animals. In many cases, estimates of future conditions can be derived from recent experience (e.g., Green, 1980, 1984; Kileen, 1982a, 1982b; Lima, 1983), but the precision of estimates depends on the temporal and spatial scales of variation. Variations on a small temporal or spatial scale means that experiences occurring several choices in the past probably are not useful predictors, whereas large-scale variation means more historical variation can be integrated into predictions of alternatives in a subsequent choice (Cibula & Zimmerman, 1987; Ollason, 1980, 1987). Perhaps some measure of change of hunger or energy reserves can be used to integrate the environmental information.

Continually changing conditions may make it worthwhile to sample the environment to monitor changes. In some cases the information will be acquired as a result of optimal feeding behavior; in other circumstances it may be necessary to use "suboptimal" options. Bumblebees feeding at flowers vertically arranged on a plant often leave before probing all flowers because the expected net reward from the next flower is less than expected from a flower on a nearby plant. But if the predictability of net reward in the next flower on the plant changes through time, it may pay to sample a flower that would not otherwise be visited (Hodges, 1981, 1985). On average, sampling will lower rates of net accumulation, meaning the animal is trading off maximizing its current rate of reserve accumulation for information acquisition, another optimality problem in how to sample (Shettleworth, Krebs, Stephens, & Gibbon, 1988).

Animals also may be able to store information for recall at particular times. Some nectar-feeding birds apparently can remember to visit preferentially plants that were visited least recently, allowing for maximum nectar accumulation between visits (Gill & Wolf, 1977; Kamil, 1978). The information used is not known, but it may involve using different sections of a foraging range in sequence (Davies & Houston, 1981; Gill & Wolf, 1977) and probably depends on the number of locations of food that must be remembered (Wolf & Hainsworth, 1983). The reliability of this information depends also on whether other feeders visit the plants; the more competitors, the less likely the predicted reward will be realized. We might expect this kind of memory-based information to be used mostly where only a single or a few feeders are present, either because of low consumer densities or resource defense by an individual.

Nectar Feeders as Model Systems

Nectar feeders provide a convenient system to test ideas about feeding. Their food is relatively easy to quantify and reproduce, and it is easy to observe feeding behavior because most potential food locations have evolved to be conspicuous (Hainsworth & Wolf, 1979; Heinrich, 1979a, 1979b; Wolf & Hainsworth, 1983).

Although there is interesting variation (Baker & Baker, 1975; Heyneman, 1983), many nectars are relatively simple sugar solutions (Hainsworth, 1973; Hainsworth & Wolf, 1972a, 1976; Heinrich, 1979b), so protein requirements of nectar feeders may dictate consumption of other foods such as insects and pollen. Insect catching also may provide efficient energy intake for hummingbirds (Hainsworth, 1977), but time budgets indicate that nectar is the main energy source for hummingbirds and sunbirds (Gass, Angehr, & Cents, 1976; Gill, 1978; Montgomerie, 1979; Wolf & Hainsworth, 1971; Wolf et al., 1975). Bumblebees and honeybees often forage for nectar and pollen from separate plants, so their feeding can be partitioned (Heinrich, 1979b). Energy per milliliter generally is higher for nectars consumed by insects, whereas nectar production rates usually are higher for bird-pollinated plants (Baker, 1975; Pyke & Waser, 1981).

Foraging increases costs while providing access to food. Costs of foraging in hummingbirds and sunbirds are estimated from time budgets using laboratory measurements of oxygen consumption rates for perching and flight (Bartholomew & Lighton, 1986; Gass et al., 1976; Gill & Wolf, 1975; Hainsworth & Wolf, 1972b; Montgomerie, 1979; Wolf & Hainsworth, 1971). The foraging behavior of hummingbirds is influenced by their size and their strategy of hovering to feed. Bumblebees often thermoregulate while foraging, with costs varying with environmental temperature and foraging mode (Heinrich, 1979b). Unfortunately, little is known of natural feeding in adult blowflies, but laboratory studies provide information on feeding controls (Dethier, 1976). Studies of feeding in butterflies concern thermoregulation, food detection, and fluid ingestion (Anderson, 1932; Kingsolver, 1983; Kingsolver & Daniel, 1979; May, 1985; Pivnick & McNeil, 1985; Watt, Hoch, & Mills, 1974).

We divide feeding into a cycle of consecutive stages: hunger, search, detection, pursuit, capture, ingestion, digestion, assimilation, and hunger (Hainsworth, 1981; Wolf & Hainsworth, 1983). Pursuit and capture are trivial components for most nectar feeders, as plants provide the nectar. The stages of search to ingestion mainly involve interaction with external environments, whereas the stages of digestion to hunger depend on internal processing. Hunger initiates search after a period depending on amount previously assimilated, requirements for energy storage, and current rates of expenditure. We define hunger for energy as the difference between an optimal energy state that maximizes performance and the current energy state (Figure 5). Both components of the difference can vary, so hunger will vary in intensity in different situations. Energy states maximizing performance vary over long periods, whereas current energy state varies over short periods.

Hunger can take on both positive and negative (satiety) values (Figure 5). When meals are used, meal initiation and termination should occur at positive and negative values, respectively. The rate of change from negative to positive (influenced by consumed food quality and quantity, rate of expenditure and requirements for energy storage) sets intermeal intervals, whereas the rate of change from positive to negative is the rate of net energy gain during a meal.

Major concerns in the following discussion are the extent to which and how hunger variation influences other components of feeding. Optimal foraging studies usually consider hunger to be a consequence of variation in food abundance; i.e., animals are hungry because food is less available. The focus then is on how rate of net energy gain during foraging can be maximized for a particular level of food availability (Charnov, 1976; Richards, 1983). Hunger seldom is considered to vary during foraging or to be under the control of feeding animals, yet hunger should vary for any animal that consumes a meal (Figure 5). We shall find an important role for hunger variation in each feeding component.

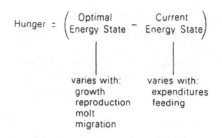

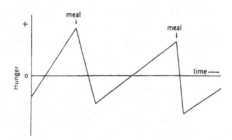

Figure 5. A definition for hunger based on energy with major variables influencing hunger for energy. Variation in hunger should occur within and between meals.

Also, we review evidence that rate of net energy gain during foraging is not always maximized for nectar feeders, and we propose hunger minimization as a more general and basic criterion for feeding design.

SEARCH AND DETECTION

FORAGING PATHS. Problems locating food are severe when food is cryptic and dispersed. Blowflies are attracted to some foods by odor, but individual sugar drops (such as excreted by aphids on leaves or stems) may not be detected until contacted by tarsal chemoreceptors (Dethier, 1976). When a hungry fly contacts only one sugar drop, it performs an intricate dance by turning in the vicinity of stimulation with an intensity dependent on hunger level and sugar concentration (Dethier, 1957, 1976; Nelson, 1977). This is a classic example of "area-restricted search" (Tinbergen, Impekoven, & Frank, 1967), where a straight path takes a forager through areas of poor quality rapidly but turning permits effective patch exploitation. Similar behavior has been described for birds (Smith, 1974a, 1974b; Zach & Falls, 1976a, 1976b), bumblebees (Pyke, 1978b), and honeybees (Schmid-Hempel, 1984), and it has been modeled as a mechanism to maximize rate of net energy gain during feeding (Pyke, 1978a).

Foraging paths can vary. For example, Soltz (1986) found no tendency for a variety of bee species to turn in response to patch quality when visiting *Vicia* flowers. Variation will occur when hunger influences search intensity as in blowflies (Dethier, 1957, 1976; Nelson, 1977). An intensive search in a restricted area will occur early in a foraging bout when energy reserves are low. As feeding progresses, search intensity will decrease as hunger wanes. It often is not clear how hunger variation influences search movements under natural conditions. However, rates of recruitment of additional foraging honeybees to artificial food sources vary depending on honey stores in a hive. A hive with large stores yields recruits at a lower rate than one with low stores (T. D. Seeley, personal communication).

Although not concerned with nectar feeders, study of patch use by the aquatic insect *Notonecta hoffmanni* clearly showed a role for hunger, because only starved individuals showed high energy intake rates (Sih, 1982).

TERRITORIALITY. Search can be confined and foraging effectiveness potentially increased within territories. Territories apparently are adjusted relative to energy requirements of hummingbirds (Gass, 1979; Gass et al., 1976; Hixon, Carpenter, & Patton, 1983; Kodric-Brown & Brown, 1978) and sunbirds (Gill & Wolf, 1975). However, not all species are territorial, as some resources are not economically defensible (Carpenter & McMillen, 1976; Gill & Wolf, 1975; Wolf, 1978), and most female hummingbirds are nonterritorial (Stiles, 1973; Wolf, 1969; Wolf & Wolf, 1971).

Search efficiency could be enhanced if a forager visited defended flowers at sufficiently long intervals to allow for nectar renewal. Information on plant quality could be obtained before a visit if foraging occurred in a spatial sequence and/or from remembering locations (Gill & Wolf, 1977; Kamil, 1978). In laboratory experiments involving visits to pairs of artificial flowers, adult hummingbirds more readily learned to visit a different location for a small reward than to return to the same location (Cole, Hainsworth, Kamil, Mercier, & Wolf, 1982). For bananaquits this differential learning occurs in adults but not in naive juveniles raised in the lab (Wunderle & Martinez, 1987).

Large flower numbers within many territories make it difficult to pattern visits to individual flowers. Some plants provide information on rewards through systematic changes in nectar volumes from bottom to top of a flowering stalk. Bumblebees appear to leave these plants based on diminishing reward rates (Best & Bierzychudek, 1982; Hodges, 1981, 1985; Pyke, 1981c). Hawkmoths (*Hemaris fuciformis*) visiting red catchfly flowers arranged in two vertical rows tend to move from bottom to top within a row. Shifting rows when an empty flower is visited limits visits to multiple empty flowers (Dreisig, 1985). Other plant species provide less specific information. Hummingbirds foraging at *Ipomopsis aggregata* obtain no useful information on volumes in nearby flowers from nectar volumes in visited flowers (Wolf & Hainsworth, 1986). Only a fraction of flowers on an individual plant are visited, with departure related to quality of flowers and the probability of a revisit to an empty flower. In this situation food is cryptic, so mechanisms for search in a restricted area may improve foraging once a bout is started (Wolf & Hainsworth, 1986).

Hunger intensity may influence territoriality and search within territories, but we know of no studies in which energy reserves have been varied systematically. Evidence is from the more easily studied adjustments to short-term changes in energy supplies (Ewald & Bransfield, 1987; Gass, 1979; Gass et al., 1976; Kodric-Brown & Brown, 1978; Powers, 1987).

FOOD CHOICE. Studies of search paths and territoriality do not discriminate between the criteria of minimizing hunger and maximizing rates of net energy gain, because maximizing rates of net energy gain during feeding with these mechanisms can also minimize hunger. Optimal foraging studies of food choice also assess selection based on maximizing net benefits per time feeding (Stephens & Krebs, 1986). For many animals this criterion also would minimize hunger, but there is considerable evidence that food choice by nectar-feeding animals does not maximize rates of net energy gain but does minimize hunger by providing more net energy per meal.

Blowflies show typical preference–aversion functions for sucrose and glucose when volume consumption is measured daily. Average consumption increases as a function of concentration to a peak and then declines at higher concentrations (Dethier & Rhoades, 1954). Total sugar consumption increases over the concentration range, and flies with a simultaneous choice always consume more sugar from the greater sucrose concentration (up to 3.0 versus 1.0 M; Dethier & Rhoades, 1954). This choice could minimize hunger by providing more net energy per meal. Rate of sugar intake during single feedings is maximum at a sucrose concentration of 1.0 M, so blowflies evidently do not select the food that maximizes energy gain rate during feeding (Dethier, Evans, & Rhoades, 1956; Dethier & Rhoades, 1954). It is important to note, however, that preference experiments measured intakes over several meals, and some of each food was consumed. This could either be related to "mistakes" or to a systematic change in preference within a meal. The latter could be consistent with maximizing rate of net energy gain over an entire meal if feeding influences subsequent choices within a meal (Stephens, Lynch, Sorensen, & Gordon, 1986). The dependence of tarsal chemoreceptor responses on sugar concentration (Dethier, 1976) suggests that such a systematic change in preference within a meal is unlikely and that preferences should minimize hunger by maximizing net meal energy.

Tarsal thresholds for proboscis extension depend on energy state, because starved blowflies fed a single meal have lower thresholds than flies fed *ad libitum* having the same crop volume (Edgecomb, Murdock, Smith, & Stephen, 1987). Labellar lobe spreading prior to ingestion also depends on energy state (Pollack, 1977). The stretch receptors associated with the median abdominal nerve (Gelperin, 1971) and the recurrent nerve (Gelperin, 1967), which monitor short-term changes in internal food availability, provide proportional inhibitory input to the CNS, which is integrated with excitatory input from peripheral chemoreceptors to control feeding thresholds and meal sizes (Bowdon & Dethier, 1986; Dethier & Gelperin, 1967). The effects of energy reserves are thought to involve additional factors, as tarsal thresholds for proboscis extension were higher in flies fed *ad libitum* even after transection of the recurrent nerve (Edgecombe *et al.*, 1987). An additional factor modulating feeding also may explain the continued increase in central excitatory state with deprivation after the crop is empty (Dethier, Solomon, & Turner, 1965).

With these controls, variation in energy state during feeding will produce different sequential food choices. Because feeding changes both short- and long-term internal energy supplies, diet choice will continuously change as feeding progresses. This effect may explain the lack of a sharp threshold for selection of low-quality food items as the proportion of low-quality items increases in a sequentially selected diet (Stephens & Krebs, 1986). Sharp thresholds are predicted on the assumption that hunger does not vary during a series of food choices. Decreases in hunger from feeding could continually shift thresholds and produce a gradual change in sequential diet choice (McNamara & Houston, 1987).

In simultaneous tests hummingbirds prefer solutions with natural combinations of sucrose, glucose, and fructose but not fructose or glucose alone (Hainsworth & Wolf, 1976). Because natural foods are acceptable in composition, the most important variables influencing rates of energy intake are those affecting energy value and time for extraction. Extraction time increases with corolla length, nectar volume and nectar concentration (Hainsworth, 1973; Hainsworth & Wolf, 1976, 1979; Montgomerie, 1984; Wolf & Hainsworth, 1983). Corolla length is particularly important when the tongue must be extended far beyond the tip of the bill, and hummingbirds prefer feeders with shorter artificial corollas when corolla lengths are long if nectar concentration and volume are held constant (Hainsworth & Wolf, 1976).

Natural nectar concentrations of hummingbird-pollinated plants vary from 0.25 to 2.2 M sucrose [8–75% (wt/vol), mean for 222 species = 23%; Baker, 1975; Hainsworth, 1973, 1974; Heyneman, 1983; Pyke & Waser, 1981; Stiles, 1975; Wolf, Stiles, & Hainsworth, 1976]. There are interesting differences between habitats; many lowland tropical species of plants have relatively high nectar concentrations, whereas many highland species have lower concentrations (Wolf et al., 1976). However, within a habitat many concentrations are similar, and most plants differ in corolla lengths and nectar volumes per flower. Exceptions are some plants with high concentrations and small corollas pollinated by very small hummingbirds (Salvia and Tropaeolum) (Hainsworth, 1974; Hainsworth & Wolf, 1972c). Nectar concentrations for these plant species may be higher because of competition with small, bee-pollinated plant species that the small hummingbirds might visit (Hainsworth & Wolf, 1976).

In simultaneous preference experiments, hummingbirds selected detectably higher sucrose concentrations up to 37% (Hainsworth & Wolf, 1976). Only at 55% sucrose did they select a lower concentration in the experiments of Tamm and Gass (1986). Another study indicated preference for concentrated fluids up to at least 78% sucrose (Pyke & Waser, 1981).

Because corolla length and sugar concentration have opposite effects on energy gain rates, controlled laboratory experiments can determine whether hummingbirds prefer feeders with a lower concentration but higher rate of net energy gain because of a short corolla or feeders with a higher concentration but a lower rate of net energy gain because of a long corolla. The results indicated that they select the higher concentration despite lower rates of net energy gain (Hainsworth & Wolf, 1976; Montgomerie, Eadie, & Harder, 1984). Also, one hummingbird species in the field was found to be territorial at a plant with a higher nectar concentration but a lower rate of net energy gain than available alternatives (Wolf et al., 1976). Montgomerie et al. (1984) showed that hummingbirds select foods that maximize net energy per volume consumed, and this maximizes net meal energy. As with blowflies, food selection by hummingbirds does not maximize rates of net energy gain during feeding, but it does minimize hunger.

Honeybees increase recruitment and dance intensity with concentration up to at least 2.0 M sucrose (Frisch, 1967), also suggesting preference for concentrated foods that may have lower rates of energy gain. Honeybees may integrate energy costs and energy gains during foraging (Waddington, 1985); obviously, costs could influence choices from effects on net energy gains.

INGESTION

Engineering models have some of their more complex and explicit applications in studies of design of the apparatus used to transport sugar water. Three structures have been examined in detail: the proboscis of butterflies and the tongues of hummingbirds and bumblebees. Here also emphasis has been on determinants of short-term rates of net energy gain. There is little to suggest that short-term rates are important for nectar feeders, but because short-term factors predominate in studies of their feeding, it is important to examine the data in some detail.

BUTTERFLIES. Ingestion by butterflies is modeled as fluid flow through a tube using Bernoulli's equation, measurements of butterfly proboscis dimensions, and assumptions for laminar flow, the nature of the fluid, and the pressure head developed for flow (Heyneman, 1983; Kingsolver & Daniel, 1979; May, 1985; Pivnick & McNeil, 1985). Because viscosity increases with concentration, flow decreases with a given pressure head, work for flow increases, and net energy gain per time is maximum at an intermediate concentration.

Assuming a constant pressure head for flow, Kingsolver and Daniel (1979) calculated a concentration of 20–30% sucrose for maximum rate of net energy gain through a butterfly proboscis and claimed a match between this and natural nectar concentrations. However, Pivnick and McNeil (1985) and May (1985) noted that the butterflies they studied often fed on nectars of much higher concentration (40–65% sucrose). Like butterflies, humans sucking fluid rapidly through a pipette were found to have a maximum rate of net energy ingestion at 35–40% sucrose (May, 1985; Pivnick & McNeil, 1985). The model was modified to account for a variable pressure head. The pressure difference increases with concentration, and the maximum rate of energy ingestion occurs at the somewhat higher concentration (May, 1985; Pivnick & McNeil, 1985).

Should butterflies prefer 35–40% sucrose in a simultaneous preference test if their feeding apparatus is designed to maximize rate of net energy ingestion at this concentration? This question assumes that feeding preference is determined only by short-term energy ingestion rates. As mentioned above, other nectar feeders sacrifice short-term rates of net energy gain to minimize hunger by selecting energy-rich foods (Dethier & Rhoades, 1954; Hainsworth & Wolf, 1976). Despite lower ingestion rates, total sugar consumption increases with sucrose concentration for meals consumed by the butterfly *Thymelicus lineola*, and natural nectar concentrations can be as high as 65% sucrose (Pivnick & McNeil, 1985). Recent measurements of sugar preferences of butterflies within meals indicate they also select concentrated fluids with low ingestion rates that minimize hunger (Hainsworth, 1989). Such findings argue strongly that the proboscis and food choice are not designed to maximize rates of energy gain during feeding.

HUMMINGBIRDS. The open grooves on the tongues of hummingbirds and sunbirds cannot sustain suction pressure (Ewald & Williams, 1982; Hainsworth, 1973; Schlamowitz, Hainsworth, & Wolf, 1976; Weymouth, Lasiewski, & Berger, 1964). Fluid transport occurs from surface tension capillarity coupled with licking (Ewald & Williams, 1982; Kingsolver & Daniel, 1983). Models developed for time-dependent (multiple licks) and length-dependent (single lick) flow yield different optimal concentration predictions for maximum rate of net energy ingestion (35–40% versus 20–25%, respectively; Kingsolver & Daniel, 1983). Small volumes per flower should be emptied with a single lick, and concentrations should be lower, but several plants with small volumes per flower (*Salvia*, *Tropaeolum*) have high concentrations (32–64%; Hainsworth, 1974; Hainsworth

& Wolf, 1972c). Hummingbirds visiting artificial feeders that required multiple licks maximized rates of energy gain with 50% sucrose (Tamm & Gass, 1986), which is a higher concentration than predicted by either model and much higher than found in flowers with large nectar volumes (Bolten & Feinsinger, 1978). Also, hummingbird tongues, like the blowfly's and butterfly's proboscis, are not designed to provide rapid ingestion of the preferred, concentrated fluids that minimize hunger by maximizing net meal energy (Hainsworth & Wolf, 1976; Montgomerie *et al.*, 1984).

BUMBLEBEES. Bumblebees lap nectar with a hairy tongue, so rate of ingestion depends on lap rate and volume per lap. Sucrose concentrations below 40% have no effect on volume ingestion rates, but volume per lap falls at 50–65% sucrose because of higher viscosity (Harder, 1986). Rates of net energy gain during ingestion depend on volume per flower and concentration, with maxima generally at 50–65% sucrose (Harder, 1986). Because honeybees prefer energy-rich foods (Frisch, 1967), hunger minimization also may override short-term ingestion rate factors influencing their feeding.

"OPTIMAL" NECTAR CONCENTRATIONS? A corollary to studies of ingestion is the idea that nectar may have evolved to be efficiently consumed. Plants usually are pollinated by nectar feeders, and it may benefit the plants to evolve characteristics that make feeding efficient for pollinators; otherwise, they may go elsewhere. This idea has motivated search for "optimal" nectar concentrations; for example, the claim of a match between natural nectar concentrations and maximum rates of energy ingestion (Heyneman, 1983; Kingsolver & Daniel, 1979). The preferences of nectar feeders belie this. Because preference is for more concentrated food, the low natural nectar concentrations must be for reasons other than the preferences of pollinators.

MEAL SIZES. Most models of optimal meal size also involve maximizing rate of net energy gain during foraging, where animals returning to a central place, such as a hive or perch, are predicted to increase meal sizes with increased energy costs (Orians & Pearson, 1979; Stephens & Krebs, 1986). Like the short-term models for ingestion discussed above, these models neglect the consequences of ingestion on meal patterns. Although meal energy may be consumed quickly for some foods, it does not necessarily always follow that the meal will provide energy for long periods. This is particularly true for meals of nectar.

Studies of hummingbird meal sizes have examined rates of net energy gain across meals and three alternative long-term criteria: minimizing time feeding, maximizing net energy gain per bout, and maximizing efficiency (net gain per cost) across meals (DeBenedictis, Gill, Hainsworth, Pyke, & Wolf, 1978). The cost of carrying a meal led to predictions of less than maximum possible meal sizes for maximizing both efficiency and rate of net energy gain from bout to bout, whereas feeding time minimization and maximizing net gain per bout always predicted full crops each meal. Hummingbird meal sizes are variable and often less than maximum capacity (Wolf & Hainsworth, 1977), so they have been viewed as a way to maximize rate of net energy storage across meals (Hainsworth & Wolf, 1983; Schmid-Hempel *et al.*, 1985). Some caution is required, however, because observed meal sizes could not distinguish between criteria of maximizing efficiency and rate of energy storage (DeBenedictis *et al.*, 1978).

In one of two hummingbird species studied after systematic deprivation, meal sizes varied directly with variation in daily rates of energy storage, so meal sizes may have contributed to maximization of long-term rate of energy storage only when energy reserves were low or hunger intensity was high (Hainsworth *et al.*, 1981).

Rate of net energy gain maximization has recently been shown not to apply to meal sizes in honeybees, which instead take meals that maximize efficiency (gain per cost) from bout to bout (Schmidt-Hempel, 1987; Schmid-Hempel *et al.*, 1985; Schmidt-Hempel & Schmidt-Hempel, 1987; Seeley, 1986). This was attributed to deterioration of individual flight performance with time, so more energy is accumulated over time when efficiency is maximized (Houston *et al.*, 1988). If hummingbirds are influenced in performance as they accumulate energy (such as from increased mass leading to higher flight costs), then perhaps they also may consume meals that maximize efficiency. Also, a more efficient animal operating at a maximum rate of net energy gain across meals would accumulate more energy per volume of food consumed than a less efficient one (DeBenedictis *et al.*, 1978).

Digestion and Assimilation

Optimal foraging designs involving a maximum rate of net energy gain during feeding are concerned only with energy acquisition (usually search to ingestion). However, because acquired food is not used until it is assimilated, perhaps design of digestive processes might be governed by the same optimality criterion. This view has been formalized in models that consider food within a digestive tract as a patch passing through the animal (Cochran, 1987; Sibly, 1981; Sibly & Calow, 1986; Speakman, 1987; Taghon, Self, & Jumars, 1978). When cumulative gains from assimilation follow a diminishing rate of return, the patch should remain within the digestive system for an optimal length of time to yield a maximum rate of net energy gain between meals.

A more specific engineering approach considers digestive systems as chemical reactors (Penry & Jumars, 1986, 1987). There are three basic categories of reactors: batch, steady-state flow, and semibatch (unsteady flow). Some animals eating discrete meals could use batch reaction (some echinoderms and annelids), whereas steady-state flow reaction may involve plug-flow (geese, corophiid amphipods, and deposit-feeding polychaetes) or continuous-flow, stirred-tank reactors in series with plug-flow reactors (ruminants) (Penry & Jumars, 1986, 1987).

A basic set of three operating variables characterizes any reactor: (1) concentration of nutrient ingested, (2) volume ingested and/or gut volume, and (3) holding or throughput time. Few studies measure these in such a way that digestion can be compared with chemical reactor design (Penry & Jumars, 1986), but some evidence on gut volumes and food quality suggest that adaptations involve these features. For example, Milton (1981) compared howler and spider monkey diets and digestive volumes and found howler monkeys ferment leaf material slowly with a large-volume hindgut, whereas spider monkeys process a large volume of fruit per time through a smaller hindgut.

Experimental tests of optimal digestion theories have involved animals that eat low-quality foods and have continuous flow of food through the intestines; i.e., they do not eat discrete meals. Tests have involved measuring gut volumes of herbivorous birds fed different quality foods (Kenward & Sibly, 1977) and studying body size effects on diet selection and digestive processing in ruminants (Baker & Hobbs, 1987) and particle size selection by deposit feeders (Taghon, 1982). Results are in general agreement with predictions, but the animals and their foods represent cases in which digestion is dominated by essentially continuous intestinal assimilation.

Recent modifications of the theories to accommodate features of meal use suggest that assimilation efficiency should vary depending on time and costs between meals (Cochran, 1987; Penry & Jumars, 1987; Sibly & Calow, 1986). For example, with low time and energy cost between meals, voiding food from the digestive tract at low assimilation could raise the rate of net energy gain across meals. Although this could occur in some animals, there is considerable evidence that animals eating meals of high-quality food do not vary assimilation efficiency, and they have rates of net energy gain between meals that vary with demands for energy. A variety of studies show negative feedback control of stomach or crop emptying, so energy is assimilated at relatively low rates that match the animals' short-term demands for energy between meals (Abisgold & Simpson, 1987; Booth, 1978; Gelperin & Dethier, 1967; Hunt, 1980; McHugh & Moran, 1979). Higher short- or long-term energy demands increase rate of meal processing in proportion to the demands. Gastric emptying rates of glucose in laboratory rats increase with a diabetes-induced decrease in glucose utilization (Granneman & Stricker, 1984) and with insulin-induced hypoglycemia (McCann & Stricker, 1986); gastric emptying of a liquid diet in rats increases with a short-term increase in demand for heat production from cold exposure (Kraly & Blass, 1976); long-term rates of energy storage between meals vary in proportion to prior-night energy expenditures in hummingbirds (Hainsworth, 1980) and sparrows (Kendeigh, Kontogiannis, Malzac, & Roth, 1969); crop-emptying rate in blow-flies increases with a deficit produced by deprivation (Edgecomb *et al.*, 1987); and crop-emptying rates, energy storage rates between meals, and feeding frequencies in hummingbirds increase with energy deficits (Hainsworth *et al.*, 1981). Rate of net energy gain between meals is not maximized for animals eating high-quality foods unless short- or long-term demands for energy are extreme.

Hummingbirds assimilate essentially all of the sugars passing through their digestive tract (Hainsworth, 1974; Karasov, Phan, Diamond, & Carpenter, 1986). Glucose absorption sites have normal binding constants but are present at extremely high densities (Karasov *et al.*, 1986), which represents an adaptation to small body size. Crop-emptying times have been measured for restrained birds using X-ray photography, refilling (Hainsworth & Wolf, 1972a), or double-isotope dilution methods (Karasov *et al.*, 1986). A 100-μL meal of 18–20% sucrose is emptied in about 15–30 minutes, but crop-emptying rates probably vary with volumes and concentrations as has been described in detail for blowflies (Edgecomb *et al.*, 1987; Gelperin, 1966; Thomson & Holling, 1977). For example, meal frequencies of hummingbirds vary with sugar concentration in a manner suggesting more rapid emptying at lower concentrations (Wolf & Hainsworth, 1977).

Based on the few direct measurements of crop-emptying rates in hummingbirds and the high weight-specific rate of sugar absorption (the presumed rate-limiting step), it has been suggested that their digestive function is geared always to occur as fast as possible to produce a maximum rate of net energy gain (Diamond, Karasov, Phan, & Carpenter, 1986; Karasov *et al.*, 1986). But meal patterns suggest that hummingbirds have variable control over digestive processing rates through rate of crop emptying (Hainsworth *et al.*, 1981; Wolf & Hainsworth, 1977). When hummingbirds were systematically deprived of food to reduce energy reserves, their crop-emptying rates measured from meal patterns increased from day to day, and daily rates of energy storage also increased (Hainsworth *et al.*, 1981). Hummingbirds only appear to process food at a maximum rate because they are small endotherms. Their digestive processing actually is adjustable and contributes to hunger minimization rather than rate of net energy gain maximization from meal to meal.

HUNGER

We have emphasized energy state for hunger because energy is an important and easily measured currency. However, there can be interactions between nutrients. Female blowflies, for example, show periods of preference for protein associated with egg development (Belzer, 1978), but the preference will not occur unless a fly first ingests sugar and carbohydrate reserves are above a threshold amount (Rachman, 1980).

We have found it necessary to consider hunger variation to interpret every component of feeding. Its importance is most notable where the criterion of rate of net energy gain maximization during feeding or between meals fails to explain function. Selection of energy-rich foods by hummingbirds, butterflies, and blowflies does not maximize rate of net energy gain while the animals are feeding, but it improves performance by reducing hunger for long periods between meals. Stomach and crop emptying of high-quality foods does not produce maximum rates of net energy gain between meals unless energy deficits are extreme. As a more general design criterion, we suggest that animals feed and use food to *minimize hunger*.

This criterion is similar to minimizing the probability of starvation, as proposed with theories involving dynamic programming (Houston & McNamara, 1985; McNamara & Houston, 1986), but hunger should involve more than just survival (Figure 5). Also, models for minimizing starvation probability have emphasized consequences of variation in food characteristics rather than also focusing on physiological mechanisms, such as meals, that can influence energy storage rates and starvation probability.

For animals using meals, the variables contributing to hunger minimization through foraging can be summarized with an equation for time between meals, where hunger is minimized when time between meals is maximized:

$$T = [(cM - fh) - a]/(m + s), \tag{1}$$

where T is time between meals, c is food energy concentration (cal/mL), M is meal size (mL), f is rate of energy expenditure during feeding (cal/time), h is feeding time, a is excreted energy, m is rate of energy expenditure for maintenance between meals (cal/time), and s is rate of energy storage between meals (cal/time) (Hainsworth, 1990). The factor $(cM - fh)$ recognizes that food choice by hummingbirds maximizes *net* energy per volume consumed (Montgomerie *et al.*, 1984), whereas s represents the effects of energy deficits on digestive processing and feeding frequency (see above; Hainsworth *et al.*, 1981). Note that because hunger depends on variable "optimal" energy states (Figure 5), s should vary depending on these states.

Except with extreme corolla lengths that increased fh, T in hummingbirds was maximized with maximum c, and rate of net energy gain for consuming a meal $[(cM - fh)/h]$ was less than maximum. For blowflies and butterflies fh is so low that $(cM - fh)$ is maximum with maximum c, and $[(cM - fh)/h]$ is less than maximum. For other consumers, however, maximizing $[(cM - fh)/h]$ also could maximize $(cM - fh)$ and T, when other types of food of high energy value do not take a disproportionate amount of time to consume. When items of the same quality are consumed in succession with diminishing returns (patch exploitation), maximizing $[(cM - fh)/h]$ maximizes $(cM - fh)$ by minimizing fh for a meal. Thus, when it occurs, maximizing rate of net energy gain during foraging is a mechanism to minimize hunger by maximizing net meal energy.

These arguments and equation 1 place no constraints on the time for eating a meal. If an energy-rich food takes a long time to consume, and if time for a meal is limited, then more net energy could be obtained by eating a lower-quality food that can be consumed faster. There should be a critical meal time set by the time needed to eat the same net energy from a high-quality food as from a complete meal of a low-quality food. When time for a meal is constrained to be less than this critical time, the low-quality food should be eaten. It would be interesting to test this hypothesis and to determine the precision with which animals monitor net energy gains.

Animals maximize the intensity of feeding when energy deficits are extreme, and a decrease in energy reserves should produce this behavior to minimize hunger (through s in equation 1). Animals should become less hungry as they feed and energy reserves increase. This will influence area-restricted search (Dethier, 1957; Nelson, 1977) and sequential food choice (Bence & Murdoch, 1986; Molles & Pietruszka, 1987; Perry, 1987), so the behaviors change during foraging.

Hunger variation can interact with other processes (such as mating, predation, etc.), so hunger is not always minimized as rapidly as possible. Whenever variation in energy state influences the rate of change of energy state from feeding (through s in equation 1) in a negative proportional manner, hunger is minimized rapidly only at low values of energy state. This demonstrates a trade-off between foraging and other functions. The trade-offs are structured so foraging has a lower priority when energy reserves are relatively high but a higher priority when energy reserves are relatively low (Caraco, 1981; Hainsworth & Wolf, 1983; Millinski & Heller, 1978; Sibly & McFarland, 1976).

For many organisms meal-to-meal or daily energy storage rates (s in equation 1) are difficult to link to feeding because of the relatively long intervals between meals and the large number of variables influencing long-term net energy gains (diet composition, amount consumed, assimilation efficiencies, maintenance costs, etc.). Although some information can be obtained by examining long-term patterns of weight changes (Kendeigh et al., 1969; Mrosovsky & Barnes, 1974), more detail comes from study of animals that feed rapidly on simple foods, i.e., the hummingbird–nectar system.

Energy storage rates in hummingbirds vary from day to day but not within days. When their meals are measured and flight and maintenance costs are subtracted from energy gains throughout days, the birds are observed to store energy at a constant rate within days even after they have been challenged with a period of food deprivation that substantially reduces energy reserves (Hainsworth, 1980; Hainsworth et al., 1981; Tooze & Gass, 1985). There usually is a brief increase in rate of energy storage in the period immediately after deprivation, but this and subsequent feeding are insufficient to restore deficits, and the birds end the day with lower levels of energy reserves than those feeding *ad libitum* (Hainsworth et al., 1981; Tooze & Gass, 1985). When energy reserves are not sufficient for overnight temperature regulation, the birds enter torpor and drastically reduce the rate at which remaining energy reserves are used (Hainsworth, Collins, & Wolf, 1977; Tooze & Gass, 1985). The next day the energy storage rate is set at a higher value in proportion to the extent of energy reserve depletion (Hainsworth et al., 1981). Crop-emptying rates (as intermeal intervals/meal energy) and meal sizes vary from day to day and are related to variable rates of energy storage between days (Hainsworth et al., 1981).

This pattern demonstrates that regulation of energy reserves by feeding is not a short-term phenomenon even in an animal with rapid turnover of energy. Hummingbirds may set storage rates for a day at dawn depending on energy reserves, but the level of stored reserves will be subject to factors that influenced access to food the previous day (competition, inclement weather; Gass & Lertzman, 1980) and expenditure rates overnight (environmental temperature, use of torpor).

The rate of energy storage in hummingbirds appears to depend on the extent to which energy reserves are depleted at the start of a day, but how is the extent determined? In engineering jargon we wish to know the "set point" so we can assess deviations from it. Based on the variable long-term demands of hummingbirds for energy for migration, reproduction, and molt, the "set point" is likely to be variable, so hunger and behaviors to minimize hunger are likely to vary at different times. There is very little information on this issue, and long-term experiments are needed. A recent report that a rufous hummingbird with considerable energy reserves entered torpor the night before leaving on migration (Carpenter & Hixon, 1988) is consistent with a variable "set point" feature for energy regulation through adjustment of expenditures (also see Schuchmann, Kruger, & Prinzinger, 1983, for data on torpor use in hummingbirds). Perhaps feeding adjustments occur as well.

Hummingbirds are not unique in these long-term patterns of feeding and energy regulation. Hamsters are noted for failing to respond to a short-term food deprivation challenge (Borer, Rowland, Mirow, Borer, & Kelch, 1979; Rowland, 1982; Silverman & Zucker, 1976). Recent experiments indicate that they may utilize a food hoard when deprived of other food, and the size of the hoard may be adjusted based on energy demands (Lea & Tarpy, 1986). Food may be stored externally or internally by some animals, and the extent of the storage may depend on predictability of long-term access to food (Andersson & Krebs, 1978).

A great deal of information exists on meal patterns for various species studied under *ad libitum* conditions in the laboratory (e.g., Hirsch, 1973; Kanarek, 1975; LeMagnen & Devos, 1970; Petersen, 1976; Sanderson & Vanderweele, 1975; Schuchmann & Abersfelder, 1986; Simpson & Ludlow, 1986; Wolf & Hainsworth, 1977; Zeigler, Green, & Lehrer, 1971). For most, meals are consumed and then expended, with a fraction stored, rather than representing a way to replenish expenditures since the last meal (Hainsworth & Wolf, 1979; Wolf & Hainsworth, 1977). Little attention has been paid to the fraction stored (influencing s in equation 1) and how it may vary, yet long-term patterns of energy storage will have a major impact on feeding through effects on hunger.

Each meal is a short-term increment on a cumulative energy storage curve; the amount expended to the next meal sets the slope of the cumulative gain curve. Knowing how the energy storage function varies should contribute substantially to understanding feeding and the effects of hunger.

Conclusions

The reader may conclude that there is little value to studies of optimal foraging design to date because some have involved inappropriate criteria or have not considered feeding as a way to control hunger. We do not wish to draw this conclusion. However, we do wish to appeal for a broader approach to comparative studies of feeding, one that tries to unify ecological and evolutionary approaches with physiological studies of feeding.

Optimal foraging theories provide their greatest utility through easily testable hypotheses. They foster explicit formulation of ideas and predictions. Predictions about feeding should be continually refined and tested. Falsification of hypotheses represents progress, and it is easier to falsify a hypothesis that is explicit in its predictions. We certainly have found this to be the case in testing the idea that food choice is designed for maximizing rate of net energy gain during foraging in nectar feeders.

The evolutionary approach to feeding also forces investigators to view feeding as a set of activities that influence survival and reproduction in particular environments. This natural context may help to interpret results that vary between species, individuals, and even within individuals.

Optimal foraging studies are not meant to replace other ways to understand feeding, and a diversity of approaches should aid progress by asking questions in different ways or from different points of view. Optimal foraging studies initially tend to concentrate on "why" questions about feeding and gradually move toward "how" questions to understand the details of mechanisms. The reverse also occurs. Physiologists move from details about "how" feeding is organized to ask "why" it may be organized that way and not some other way. Despite this, few investigators using either approach seem to be aware of the other. We believe neither approach should be abandoned, but they both should be expanded. Ecological and evolutionary studies of feeding obviously need to examine consequences beyond short-term factors influencing intake of food, and physiologists need to address interesting ecological and evolutionary questions. In general, more effort seems needed to integrate the approaches and to recognize that we all are interested in the same things.

References

Abisgold, J. D., & Simpson, S. J. (1987). The physiology of compensation by locusts for changes in dietary protein. *Journal of Experimental Biology, 129*, 329–346.

Alexander, R. McN. (1978). Mechanics and scaling of terrestrial locomotion. In T. J. Pedley (Ed.), *Scale effects in animal locomotion* (pp. 93–110). New York: Academic Press.

Alexander, R. McN. (1982a). *Optima for animals*. London: Edward Arnold.

Alexander, R. McN. (1982b). *Locomotion of animals*. New York: Chapman Hall.

Anderson, A. L. (1932). The sensitivity of the legs of common butterflies to sugars. *Journal of Experimental Zoology, 63*, 235–259.

Andersson, M., & Krebs, J. (1978). On the evolution of hoarding behaviour. *Animal Behaviour, 26,* 707–711.

Arnold, S. (1981). The microevolution of feeding behavior. In A. C. Kamil & T. D. Sargent (Eds.), *Foraging behavior, ecological, ethological, and psychological approaches* (pp. 409–453). New York: Garland STPM Press.

Ashby, W. R. (1956). *An introduction to cybernetics.* New York: John Wiley & Sons.

Bagenal, T. B. (1967). A short review of fish fecundity. In S. D. Gerking (Ed.), *The biological basis of fish production* (pp. 89–111). New York: John Wiley & Sons.

Baker, D. L., & Hobbs, N. T. (1987). Strategies of digestion: Digestive efficiency and retention time of forage diets in montane ungulates. *Canadian Journal of Zoology, 65,* 1978–1984.

Baker, H. G. (1975). Sugar concentrations in nectars from hummingbird flowers. *Biotropica, 7,* 137–141.

Baker, H. G., & Baker, I. (1975). Studies of nectar-constitution and pollinator-plant coevolution. In L. E. Gilbert & P. H. Raven (Eds.), *Coevolution of animals and plants* (pp. 100–140). Austin: University of Texas Press.

Bartholomew, G. A., & Lighton, J. R. B. (1986). Oxygen consumption during hover-feeding in free-ranging Anna hummingbirds. *Journal of Experimental Biology, 123,* 191–199.

Belovsky, G. E. (1978). Diet optimization in a generalist herbivore: The moose. *Theoretical Population Biology, 14,* 105–134.

Belovsky, G. E. (1986). Optimal foraging and community structure: Implications for a guild of generalist grassland herbivores. *Oecologia, 70,* 35–52.

Belzer, W. (1978). Patterns of selective protein ingestion by the blowfly *Phormia regina. Physiological Entomology, 3,* 169–175.

Bence, J. R., & Murdoch, W. W. (1986). Prey size selection by the mosquitofish: Relation to optimal diet theory. *Ecology, 67,* 324–336.

Berger, M., & Hart, J. S. (1972). Die Atmung biem Kolibri *Amazilia fimbriata* wahrend des Schwirr-fluges bei Verschiedenen Umgebungstemperaturen. *Journal of Comparative Physiology, 81,* 363–380.

Best, L. S., & Bierzychudek, P. (1982). Pollinator foraging on foxglove (*Digitaria purpurea)*: A test of a new model. *Evolution, 36,* 70–79.

Bolten, A. B., & Feinsinger, P. (1978). Why do hummingbird flowers secrete dilute nectars? *Biotropica, 10,* 307–310.

Booth, D. A. (1978). Prediction of feeding behaviour from energy flows in the rat. In D. A. Booth (Ed.), *Hunger models* (pp. 227–278). London: Academic Press.

Booth, D. A. (1980). Conditioned reactions in motivation. In F. M. Toates & T. R. Halliday (Eds.), *Analysis of motivational processes* (pp. 77–102). New York: Academic Press.

Booth, D. A., Toates, F. M., & Platt, S. V. (1976). Control system for hunger and its implications in animals and man. In D. Novin, W. Wyrwicka, & G. Bray (Eds.), *Hunger: Basic mechanisms and clinical implications* (pp. 127–143). New York: Raven Press.

Borer, K. T., Rowland, N., Mirow, A., Borer, R. C., & Kelch, R. P. (1979). Physiological and behavioral responses to starvation in the golden hamster. *American Journal of Physiology, 236,* E105–E112.

Bowdon, E., & Dethier, V. G. (1986). Coordination of a dual inhibitory system regulating feeding behaviour in the blowfly. *Journal of Comparative Physiology, 158,* 713–722.

Brobeck, J. R. (1965). Exchange, control, and regulation. In W. S. Yamamoto & J. R. Brobeck (Eds.), *Physiological controls and regulations* (pp. 1–13). Philadelphia: W. B. Saunders.

Calder, W. A. (1984). *Size, function and life history.* Cambridge: Harvard University Press.

Caraco, T. (1981). Energy budgets, risk and foraging preferences in dark-eyed juncos (*Junco hyemalis). Behavioral Ecology and Sociobiology, 8,* 213–217.

Caraco, T. (1983). White-crowned sparrows (*Zonotrichia leucophrys*): foraging preferences in a risky environment. *Behavioral Ecology and Sociobiology, 12,* 63–69.

Caraco, T., Martindale, S., & Whitham, T. S. (1980). An empirical demonstration of risk sensitive foraging preferences. *Animal Behaviour, 28,* 820–830.

Carpenter, F. L., & Hixon, M. A. (1988). A new function for torpor: Fat conservation in a wild migrant hummingbird. *Condor, 90,* 373–378.

Carpenter, F. L., & McMillen, R. E. (1976). Threshold model of feeding territoriality and test with a Hawaiian honeycreeper. *Science, 194,* 639–642.

Charnov, E. L. (1976). Optimal foraging: Attack strategy of a mantid. *American Naturalist, 110,* 141–151.

Charnov, E. L. (1982). *The theory of sex allocation.* Princeton: Princeton University Press.

Cibula, D. A., & Zimmerman, M. (1987). Bumblebee foraging behavior: Changes in departure decisions as a function of experimental nectar manipulations. *American Midland Naturalist, 117,* 386–394.

Cochran, P. A. (1987). Optimal digestion in a batch-reactor gut—the analogy to partial prey consumption. *Oikos, 50,* 268–269.

Cole, S., Hainsworth, F. R., Kamil, A. C., Mercier, T., & Wolf, L. L. (1982). Spatial learning as an adaptation in hummingbirds. *Science, 217,* 655–657.

Cowie, R. J., & Krebs, J. R. (1979). Optimal foraging in patchy environments. In R. M. Anderson, B. D. Turner, & L. R. Taylor (Eds.), *The British ecological symposium, Vol. 20, Population dynamics* (pp. 183–205). Oxford: Blackwell Scientific Publications.

Davies, N. B., & Houston, A. I. (1981). Owners and satellites: The economics of territory defense in the pied wagtail, *Motacilla alba. Journal of Animal Ecology, 50,* 157–180.

Davis, J. D. (1980). Homeostasis, feedback and motivation. In F. M. Toates & T. R. Halliday (Eds.), *Analysis of motivational processes* (pp. 23–38). New York: Academic Press.

DeBenedictis, P., Gill, F. B., Hainsworth, F. R., Pyke, G. H., & Wolf, L. L. (1978). Optimal meal size in hummingbirds. *American Naturalist 112,* 301–316.

Demment, M. W., & Van Soest, P. J. (1985). A nutritional explanation for body-size patterns of ruminant and nonruminant herbivores. *American Naturalist, 125,* 641–672.

Dethier, V. G. (1957). Communication by insects: Physiology of dancing. *Science, 125,* 331–336.

Dethier, V. G. (1976). *The hungry fly.* Cambridge: Harvard University Press.

Dethier, V. G., Evans, D. R., & Rhoades, M. V. (1956). Some factors controlling the ingestion of carbohydrates by the blowfly. *Biological Bulletin, 111,* 204–222.

Dethier, V. G., & Gelperin, A. (1967). Hyperphagia in the blowfly. *Journal of Experimental Biology, 47,* 191–200.

Dethier, V. G., & Rhoades, M. V. (1954). Sugar preference–aversion functions for the blowfly. *Journal of Experimental Zoology, 126,* 177–204.

Dethier, V. G., Solomon, R. L., & Turner, L. H. (1965). Sensory input and central excitation and inhibition in the blowfly. *Journal of Comparative and Physiological Psychology, 60,* 303–313.

Diamond, J. M., Karasov, W. H., Phan, D., & Carpenter, F. L. (1986). Digestive physiology is a determinant of foraging bout frequency in hummingbirds. *Nature, 320,* 62–63.

Dreisig, H. (1985). Movement patterns of a clear-wing hawkmoth, *Hemaris fuciformis,* foraging at red catchfly, *Viscaria vulgaris. Oecologia, 67,* 360–366.

Edgecomb, R. S., Murdock, L. L., Smith, A. B., & Stephen, M. D. (1987). Regulation of tarsal taste threshold in the blowfly, *Phormia regina. Journal of Experimental Biology, 127,* 79–94.

Emlen, J. M. (1966). The role of time and energy in food preferences. *American Naturalist, 100,* 611–617.

Epting, R. J. (1980). Functional dependence of the power for hovering on wing disc loading in hummingbirds. *Physiological Zoology, 53,* 347–357.

Epting, R. J., & Casey, T. M. (1973). Power output and wing disc loading in hovering hummingbirds. *American Naturalist, 107,* 761–765.

Ewald, P. W., & Bransfield, R. J. (1987). Territory quality and territorial behavior in two sympatric species of hummingbirds. *Behavioral Ecology and Sociobiology, 20,* 285–293.

Ewald, P. W., & Williams, W. A. (1982). Function of the bill and tongue in nectar uptake by hummingbirds. *Auk, 99,* 573–576.

Feldman, H. A., & McMahon, T. (1983). The ¾ mass exponent for energy metabolism is not a statistical artifact. *Respiratory Physiology, 52,* 149–163.

Folk, G. E., Jr. (1967). Physiological observations of subarctic bears under winter den conditions. In K. C. Fisher, A. R. Dawe, C. P. Lyman, E. Schonbaum, & F. E. South, Jr. (Eds.), *Mammalian hibernation III* (pp. 75–85). New York: Elsevier.

Frisch, K. (1967). *The dance language and orientation of bees.* Cambridge: Harvard University Press.

Garland, T., Jr. (1983). Scaling the ecological cost of transport to body mass in terrestrial mammals. *American Naturalist, 121,* 571–587.

Gass, C. L. (1979). Territory regulation, tenure, and migration in rufous hummingbirds. *Canadian Journal of Zoology, 57,* 914–923.

Gass, C. L., Angehr, G., & Centa, J. (1976). Regulation of food supply by feeding territoriality in the rufous hummingbird. *Canadian Journal of Zoology, 54,* 2046–2054.

Gass, C. L., & Lertzman, K. P. (1980). Capricious mountain weather: A driving variable in hummingbird territorial dynamics. *Canadian Journal of Zoology, 58,* 1964–1968.

Gelperin, A. (1966). Control of crop emptying in the blowfly. *Journal of Insect Physiology, 12,* 331–345.

Gelperin, A. (1967). Stretch receptors in the foregut of the blowfly. *Science, 157,* 208–210.

Gelperin, A. (1971). Abdominal sensory neurons providing negative feedback to the feeding behavior of the blowfly. *Zeitschrift fur vergleichende Physiologie, 72,* 17–31.

Gelperin, A., & Dethier, V. G. (1967). Long-term regulation of sugar intake by the blowfly. *Physiological Zoology, 40,* 218–228.

Gill, F. B. (1978). Proximate costs of competition for nectar. *American Zoologist, 18,* 753–763.

Gill, F. B., & Wolf, L. L. (1975). Economics of territoriality in the golden-winged sunbird. *Ecology, 56,* 333–345.

Gill, F. B., & Wolf, L. L. (1977). Non-random foraging by sunbirds in a patchy environment. *Ecology, 58,* 1284–1296.

Godin, J.-G. J., & Smith, S. A. (1988). A fitness cost of foraging in the guppy. *Nature, 333,* 69–71.

Gould, S. J. (1986). Evolution and the triumph of homology, or why history matters. *American Scientist, 74,* 60–69.

Gould, S. J., & Lewontin, R. C. (1979). The spandrels of San Marco and the Panglossian paradigm: A critique of the adaptationist programme. *Proceedings of the Royal Society of London [B], 205,* 581–598.

Granneman, J. G., & Stricker, E. M. (1984). Food intake and gastric emptying in rats with streptozotocin-induced diabetes. *American Journal of Physiology, 247,* R1054–R1061.

Green, R. F. (1980). Batyesian birds: A simple example of Oaten's stochastic model of optimal foraging. *Theoretical Population Biology, 18,* 244–256.

Green, R. F. (1984). Stopping rules for optimal foragers. *American Naturalist, 123,* 30–40.

Greenewalt, C. H. (1962). Dimensional relationships for flying animals. *Smithsonian Miscellaneous Collections, 144(2),* 1–46.

Hainsworth, F. R. (1973). On the tongue of a hummingbird: Its role in the rate and energetics of feeding. *Comparative Biochemistry and Physiology, 46,* 65–78.

Hainsworth, F. R. (1974). Food quality and foraging efficiency: The efficiency of sugar assimilation by hummingbirds. *Journal of Comparative Physiology, 88,* 425–431.

Hainsworth, F. R. (1977). Foraging efficiency and parental care in *Colibri coruscans. Condor, 79,* 69–75.

Hainsworth, F. R. (1978). Feeding: Models of costs and benefits in energy regulation. *American Zoologist, 18,* 701–714.

Hainsworth, F. R. (1980). Patterns of energy use in birds. In R. Nohring (Ed.), *Acta XVII Congressus Internationalis Ornithologici* (pp. 287–291). Berlin: Deutsche Ornithologen-Gesellschaft.

Hainsworth, F. R. (1981). Energy regulation in hummingbirds. *American Scientist, 69,* 420–429.

Hainsworth, F. R. (1986). Why hummingbirds hover: A commentary. *Auk, 103,* 832–833.

Hainsworth, F. R. (1989). "Fast food" *vs.* "haute cuisine": Painted ladies, *Vanessa cardui(L.),* select food to maximize net meal energy. *Functional Ecology, 3,* 701–707.

Hainsworth, F. R. (1990). Criteria for efficient energy use: Using exceptions to prove rules. *Journal für Ornithologie, 131,* 311–317.

Hainsworth, F. R., Collins, B. G., & Wolf, L. L. (1977). The function of torpor in hummingbirds. *Physiological Zoology, 50,* 215–222.

Hainsworth, F. R., Tardiff, M. F., & Wolf, L. L. (1981). Proportional control for daily energy regulation in hummingbirds. *Physiological Zoology, 54,* 452–462.

Hainsworth, F. R., & Wolf, L. L. (1972a). Crop volume, nectar concentration and hummingbird energetics. *Comparative Biochemistry and Physiology, 42A,* 359–366.

Hainsworth, F. R., & Wolf, L. L. (1972b). Power for hovering flight in relation to body size in hummingbirds. *American Naturalist, 106,* 589–596.

Hainsworth, F. R., & Wolf, L. L. (1972c). Energetics of nectar extraction in a small, high altitude, tropical hummingbird. *Journal of Comparative Physiology, 80,* 377–387.

Hainsworth, F. R., & Wolf, L. L. (1974). Wing disc loading: Implications and importance for hummingbird energetics. *American Naturalist, 108,* 229–233.

Hainsworth, F. R., & Wolf, L. L. (1976). Nectar characteristics and food selection by hummingbirds. *Oecologia, 25,* 101–113.

Hainsworth, F. R., & Wolf, L. L. (1978). The economics of temperature regulation and torpor in nonmammalian organisms. In L. C. H. Wang & J. W. Hudson (Eds.), *Strategies in cold: Natural torpidity and thermogenesis* (pp. 147–184). New York: Academic Press.

Hainsworth, F. R., & Wolf, L. L. (1979). Feeding: An ecological approach. *Advances in the Study of Behavior, 9,* 53–96.

Hainsworth, F. R., & Wolf, L. L. (1983). Models and evidence for feeding control of energy. *American Zoologist, 23,* 261–272.

Harder, L. D. (1986). Effects of nectar concentration and flower depth on flower handling efficiency of bumble bees. *Oecologia, 69,* 309–315.

Hardy, J. D. (1965). The "set-point" concept in physiological temperature regulation. In W. S. Yamamoto & J. R. Brobeck (Eds.), *Physiological controls and regulations* (pp. 98–116). Philadelphia: W. B. Saunders.

Hardy, J. D., & Stitt, J. T. (1976). Interaction of hypothalamic and skin temperature in cold thermogenesis in the rabbit. *Israel Journal of Medical Sciences, 12,* 1052–1055.

Hartman, F. A. (1955). Heart weight in birds. *Condor, 57,* 221–238.

Heinrich, B. (1979a). Resource heterogeneity and patterns of movement in foraging bumblebees. *Oecologia, 140,* 235–245.

Heinrich, B. (1979b). *Bumblebee economics.* Cambridge: Harvard University Press.

Heller, H. C., & Colliver, G. W. (1974). CNS regulation of body temperature during hibernation. *American Journal of Physiology, 227,* 583–589.

Heusner, A. A. (1982). Energy metabolism and body size. I. Is the 0.75 mass exponent of Kleiber's equation a statistical artifact? *Respiratory Physiology, 48,* 1–12.

Heyneman, A. J. (1983). Optimal sugar concentrations of floral nectars—dependence on sugar intake efficiency and foraging costs. *Oecologia, 60,* 198–213.

Hirsch, E. (1973). Some determinants of intake and patterns of feeding in the guinea pig. *Physiology and Behavior, 11,* 687–704.

Hitzig, B. M., & Jackson, D. C. (1978). Central chemical control of ventilation in the unanesthetized turtle. *American Journal of Physiology, 235,* R257–R264.

Hixon, M. (1980). Food production and competitor density as the determinants of feeding territory size. *American Naturalist, 115,* 510–530.

Hixon, M. A., Carpenter, F. L., & Patton, D. C. (1983). Territory area, flower density, and time budgeting in hummingbirds: An experimental and theoretical analysis. *American Naturalist, 122,* 366–391.

Hodges, C. M. (1981). *Optimal foraging in bumblebees; patterns of time allocation among feeding sites and tests of foraging theory.* Ph.D. dissertation, Syracuse University, Syracuse, NY.

Hodges, C. M. (1985). Bumblebee foraging: Energetic consequences of using a threshold departure rule. *Ecology, 66,* 188–197.

Hodges, C. M., & Wolf, L. L. (1981). Optimal foraging in bumblebees: Why is nectar left behind in flowers? *Behavioral Ecology and Sociobiology, 9,* 41–44.

Holling, C. S. (1966). The functional response of invertebrate predators to prey density. *Memoirs of the Entomology Society of Canada, 48,* 1–86.

Houston, A. I., & McNamara, J. M. (1985). The choice of two prey types that minimizes the probability of starvation. *Behavioral Ecology and Sociobiology, 17,* 135–141.

Houston, A. I., & McNamara, J. M. (1988). A framework for the functional analysis of behavior. *Behavioral and Brain Sciences, 11,* 117–130.

Houston, A., Schmid-Hempel, P., & Kacelnik, A. (1988). Foraging strategy, worker mortality, and the growth of the colony in social insects. *American Naturalist, 131,* 107–114.

Hunt, J. N. (1980). A possible relation between the regulation of gastric emptying and food intake. *American Journal of Physiology, 239,* G1–G4.

Iwasa, Y., Higashi, M., & Yamamura, N. (1981). Prey distribution as a factor determining the choice of optimal foraging strategy. *American Naturalist, 117,* 710–723.

Kamil, A. C. (1978). Systematic foraging for nectar by amakihi, *Loxops virens. Journal of Comparative and Physiological Psychology, 92,* 388–396.

Kanarek, R. B. (1975). Availability and caloric density of the diet as determinants of meal patterns in cats. *Physiology and Behavior, 15,* 611–618.

Karasov, W. H., Phan, D., Diamond, J. M., & Carpenter, F. L. (1986). Food passage and intestinal nutrient absorption in hummingbirds. *Auk, 103,* 453–464.

Kendeigh, S. C., Kontogiannis, J. E., Malzac, A., & Roth, R. R. (1969). Environmental regulation of food intake by birds. *Comparative Biochemistry and Physiology, 31,* 941–957.

Kenward, R. E., & Sibly, R. M. (1977). A woodpigeon (*Columbia palumbus*) feeding preference explained by a digestive bottleneck. *Journal of Applied Ecology, 14,* 815–826.

Kileen, P. R. (1982a). Incentive theory. In D. J. Bernstein (Ed.), *Nebraska symposium on motivation, 1981: Response structure and organization* (pp. 169–216). Lincoln: University of Nebraska Press.

Kileen, P. R. (1982b). Incentive theory: II. Models for choice. *Journal of the Experimental Analysis of Behavior, 38,* 217–233.

Kingsolver, J. G. (1983). Ecological significance of flight activity in *Colias* butterflies: Implications for reproductive strategy and population structure. *Ecology, 64,* 546–551.

Kingsolver, J. G., & Daniel, T. L. (1979). On the mechanics and energetics of nectar feeding in butterflies. *Journal of Theoretical Biology, 76,* 167–179.

Kingsolver, J. G., & Daniel, T. L. (1983). Mechanical determinants of nectar feeding strategy in hummingbirds: Energetics, tongue morphology, and licking behavior. *Oecologia, 60,* 214–226.

Kodric-Brown, A., & Brown, J. H. (1978). Influence of economics, interspecific competition, and sexual dimorphism on territoriality of migrant rufous hummingbirds. *Ecology, 59,* 285–296.

Kraly, F. S., & Blass, E. M. (1976). Increased feeding in rats in a low ambient temperature. In D. Novin, W. Wyrwicka, & G. Bray (Eds.), *Hunger: Basic mechanisms and clinical implications* (pp. 77–87). New York: Raven Press.

Krebs, J. R., Erichsen, J. T., Webber, M. I., & Charnov, E. L. (1977). Optimal prey selection by the great tit (*Parus major*). *Animal Behaviour, 25,* 30–38.

Krebs, J. R., Kacelnik, A., & Taylor, P. (1978). Test of optimal sampling by foraging great tits. *Nature, 275,* 27–31.

Lasiewski, R. C. (1963). Oxygen consumption of torpid, resting, active, and flying hummingbirds. *Physiological Zoology, 36,* 122–140.

Lasiewski, R. C., & Calder, W. A., Jr. (1971). A preliminary allometric analysis of respiratory variables in resting birds. *Respiratory Physiology, 11,* 152–166.

Lea, S. E. G., & Tarpy, R. M. (1986). Hamsters' demand for food to eat and hoard as a function of deprivation and cost. *Animal Behaviour, 34,* 1759–1768.

LeMagnen, J. (1976). Interactions of glucostatic and lipostatic mechanisms in the regulatory control of feeding. In D. Novin, W. Wyrwicka, & G. Bray (Eds.), *Hunger: Basic mechanisms and clinical implications* (pp. 89–102). New York: Raven Press.

LeMagnen, J., & Devos, M. (1970). Metabolic correlates of the meal onset in the free food intake of rats. *Physiology and Behavior, 5,* 805–814.

Lighthill, J. (1977). Introduction to the scaling of aerial locomotion. In T. J. Pedley (Ed.), *Scale effects in animal locomotion* (pp. 365–404). New York: Academic Press.

Lima, S. L. (1983). Downy woodpecker foraging behavior: Efficient sampling in simple stochastic environments. *Ecology, 65,* 166–174.

Lima, S. L. (1985). Maximizing feeding efficiency and minimizing time exposed to predators: A trade-off in the black-capped chickadee. *Oecologia, 66,* 60–67.

Mangel, M., & Clark, C. W. (1986). Towards a unified foraging theory. *Ecology, 67,* 1127–1138.

Martin, R. D., Chivers, D. J., MacLarnon, A. M., & Hladik, C. M. (1985). Gastrointestinal allometry in primates and other mammals. In W. L. Jungers (Ed.), *Size and scaling in primate biology* (pp. 61–89). New York: Plenum Press.

May, P. G. (1985). Nectar uptake rates and optimal nectar concentrations of two butterfly species. *Oecologia, 66,* 381–386.

Maynard Smith, J. (1978). Optimization theory in evolution. *Annual Review of Ecology and Systematics, 9,* 31–56.

McCann, M. J., & Stricker, E. M. (1986). Gastric emptying of glucose loads in rats: Effects of insulin-induced hypoglycemia. *American Journal of Physiology, 251,* R609–R613.

McHugh, P. R., & Moran, T. H. (1979). Calories and gastric emptying: A regulatory capacity with implications for feeding. *American Journal of Physiology, 236,* R254–R260.

McMahon, T. A. (1984). *Muscles, reflexes, and locomotion.* Princeton: Princeton University Press.

McNamara, J. M., & Houston, A. I. (1986). The common currency for behavioral decisions. *American Naturalist, 127,* 358–378.

McNamara, J. M., & Houston, A. I. (1987). Partial preferences and foraging. *Animal Behaviour, 35,* 1084–1099.

Milinski, M. (1982). Optimal foraging: The influence of intraspecific competition on diet selection. *Behavioral Ecology and Sociobiology, 11,* 109–115.

Milinski, M., & Heller, R. (1978). Influence of a predator on the optimal foraging behaviour of sticklebacks (*Gasterosteus aculeatus* L.). *Nature, 275,* 642–644.

Miller, R. S. (1985). Why hummingbirds hover. *Auk, 102,* 722–726.

Milton, K. (1981). Food choice and digestive strategies of two sympatric primate species. *American Naturalist, 117,* 496–505.

Molles, M. C., Jr., & Pietruszka, R. D. (1987). Prey selection by a stonefly: The influence of hunger and prey size. *Oecologia, 72,* 473–478.

Montgomerie, R. D. (1979). *The energetics of foraging and competition in some Mexican hummingbirds.* Ph.D. thesis, McGill University, Montreal, Quebec.

Montgomerie, R. D. (1984). Nectar extraction by hummingbirds: Response to different floral characters. *Oecologia, 63,* 229–236.

Montgomerie, R. D., Eadie, J. M., & Harder, L. D. (1984). What do foraging hummingbirds maximize? *Oecologia, 63,* 357–363.

Mrosovsky, N., & Barnes, D. S. (1974). Anorexia, food deprivation and hibernation. *Physiology and Behavior, 12*, 265–270.

Nelson, M. C. (1977). The blowfly's dance: Role in the regulation of food intake. *Journal of Insect Physiology, 23*, 603–612.

Norberg, R. A. (1981). Why foraging birds in trees should climb and hop upwards rather than downwards. *Ibis, 123*, 281–288.

Ollason, J. G. (1980). Learning to forage—optimally? *Theoretical Population Biology, 18*, 44–56.

Ollason, J. G. (1987). Learning to forage in a regenerating patchy environment. *Theoretical Population Biology, 31*, 13–32.

Oriens, G. H., & Pearson, N. E. (1979). On the theory of central place foraging. In D. J. Horn, R. D. Mitchell, & G. R. Stairs (Eds.), *Analysis of ecological systems* (pp. 154–177). Columbus: Ohio State University Press.

Penry, D. L., & Jumars, P. A. (1986). Chemical reactor analysis and optimal digestion. *Bioscience, 36*, 310–315.

Penry, D. L., & Jumars, P. A. (1987). Modeling animal guts as chemical reactors. *American Naturalist, 129*, 69–96.

Perry, D. M. (1987). Optimal diet theory: Behavior of a starved predatory snail. *Oecologia, 72*, 360–365.

Peters, R. H. (1983). *The ecological implications of body size.* Cambridge: Cambridge University Press.

Petersen, S. (1976). The temporal pattern of feeding over the oestrous cycle of the mouse. *Animal Behaviour, 24*, 939–955.

Pivnick, K. A., & McNeil, J. N. (1985). Effects of nectar concentration on butterfly feeding: Measured feeding rates for *Thymelicus lineola* (Lepidoptera: Hesperiidae) and a general feeding model for adult Lepidoptera. *Oecologia, 66*, 226–237.

Pollack, G. S. (1977). Labellar lobe spreading in the blowfly: Regulation by taste and satiety. *Journal of Comparative Physiology, 121*, 115–134.

Powers, D. R. (1987). Effects of variation in food quality on the breeding territoriality of the male Anna hummingbird (*Calypte anna*). *Condor, 89*, 103–111.

Price, T. (1987). Diet variation in a population of Darwin's finches. *Ecology, 68*, 1015–1028.

Price, T. D., & Grant, P. R. (1984). Life history traits and natural selection for small body size in a population of Darwin's finches. *Evolution, 38*, 483–494.

Prosser, C. L. (1986). *Adaptational biology: Molecules to organisms.* New York: John Wiley & Sons.

Prothero, J. (1979). Heart weight as a function of body weight in mammals. *Growth, 43*, 139–150.

Pyke, G. H. (1978a). Optimal foraging: Movement patterns of bumblebees between inflorescences. *Theoretical Population Biology, 13*, 72–98.

Pyke, G. H. (1978b). Are animals efficient harvesters? *Animal Behaviour, 26*, 241–250.

Pyke, G. H. (1981a). Optimal travel speeds of animals. *American Naturalist, 118*, 475–487.

Pyke, G. H. (1981b). Why hummingbirds hover and honeyeaters perch. *Animal Behaviour, 29*, 861–867.

Pyke, G. H. (1981c). Optimal foraging in bumblebees and coevolution with their plants. *Oecologia, 36*, 281–293.

Pyke, G. H. (1984). Optimal foraging theory: A critical review. *Annual Review of Ecology and Systematics, 15*, 523–575.

Pyke, G. H., & Waser, N. M. (1981). The production of dilute nectars by hummingbird and honeyeater flowers. *Biotropica, 13*, 260–270.

Rachman, N. J. (1980). Physiology of feeding preference patterns of female black blowflies (*Phormia regina*) I. The role of carbohydrate reserves. *Journal of Comparative Physiology, 139*, 59–66.

Rausher, M. D. (1984). Tradeoffs in performance on different hosts: Evidence from within and between site variation in the beetle, *Deloyala guttata. Evolution, 38*, 582–595.

Rayner, J. M. V. (1979). A vortex theory of animal flight. Part 1. The vortex wake of a hovering animal. *Journal of Fluid Mechanics, 91*, 705–738.

Richards, L. J. (1983). Hunger and the optimal diet. *American Naturalist, 122*, 326–334.

Riggs, D. S. (1963). *The mathematical approach to physiological problems.* Baltimore: Williams & Wilkins.

Rowland, N. (1982). Failure by deprived hamsters to increase food intake: Some behavioral and physiological determinants. *Journal of Comparative and Physiological Psychology, 96*, 591–603.

Sanderson, J. D., & Vanderweele, D. A. (1975). Analysis of feeding patterns in normal and vagotomized rabbits. *Physiology and Behavior, 15*, 357–364.

Schlamowitz, R. F., Hainsworth, F. R., & Wolf, L. L. (1976). On the tongues of sunbirds. *Condor, 78*, 104–107.

Schmid-Hempel, P. (1984). The importance of handling time for the flight directionality in bees. *Behavioral Ecology and Sociobiology, 15,* 303–309.

Schmid-Hempel, P. (1987). Efficient nectar-collecting by honeybees I. Economic models. *Journal of Animal Ecology, 56,* 209–218.

Schmid-Hempel, P., Kacelnik, A., & Houston, A. I. (1985). Honeybees maximize efficiency by not filling their crop. *Behavioral Ecology and Sociobiology, 17,* 61–66.

Schmid-Hempel, P., & Schmid-Hempel, R. (1987). Efficient nectar-collecting by honeybees II. Response to factors determining nectar availability. *Journal of Animal Ecology, 56,* 219–227.

Schmidt-Nielsen, K. (1972). *How animals work.* Cambridge: Cambridge University Press.

Schmidt-Nielsen, K. (1984). *Scaling: Why is animal size so important?* Cambridge: Cambridge University Press.

Schoener, T. W. (1971). Theory of feeding strategies. *Annual Review of Ecology and Systematics, 2,* 369–404.

Schoener, T. W. (1983). Some simple models of optimal feeding—territory size: A reconciliation. *American Naturalist, 121,* 608–629.

Schuchmann, K.-L., & Abersfelder, F. (1986). Energieregulation und ZeitKoordination der Nahrungsaufnahme einer andinen Kolibriart, *Aglaeactis cupripennis. Journal für Ornithologie, 127,* 205–215.

Schuchmann, K.-L., Kruger, K., & Prinzinger, R. (1983). Torpor in hummingbirds. *Bonn Zoologische Beitrage, 34,* 273–277.

Seeley, T. D. (1985). *Honeybee ecology, A study of adaptation in social life.* Princeton: Princeton University Press.

Seeley, T. D. (1986). Social foraging by honeybees: How colonies allocate foragers among patches of flowers. *Behavioral Ecology and Sociobiology, 19,* 343–354.

Shettleworth, S. J., Krebs, J. R., Stephens, D. W., & Gibbons, J. (1988). Tracking a fluctuating environment: A study of sampling. *Animal Behaviour, 36,* 87–105.

Sibly, R. M. (1981). Strategies of digestion and defecation. In C. R. Townsend & P. Calow (Eds.), *Physiological ecology: An evolutionary approach to resource use* (pp. 109–139). Oxford: Blackwell Scientific Publications.

Sibly, R. M., & Calow, P. (1986). *Physiological ecology of animals, An evolutionary approach.* Oxford: Blackwell Scientific Publications.

Sibly, R., & McFarland, D. (1976). On the fitness of behavior sequences. *American Naturalist, 110,* 601–617.

Sih, A. (1982). Optimal patch use: Variation in selective pressure for efficient foraging. *American Naturalist, 120,* 666–685.

Silverman, H. J., & Zucker, I. (1976). Absence of post-fast food compensation in the golden hamster (*Mesocricetus auratus*). *Physiology and Behavior, 17,* 271–285.

Smith, J. N. M. (1974a). The food searching behavior of two European thrushes. I. Description and analysis of search paths. *Behaviour, 48,* 276–302.

Smith, J. N. M. (1974b). The food searching behaviour of two European thrushes. II. The adaptiveness of the search patterns. *Behaviour, 49,* 1–61.

Sokolowski, M. B. (1985). Genetics and ecology of *Drosophila melanogaster* larval foraging and pupation behavior. *Journal of Insect Physiology, 31,* 857–864.

Soltz, R. L. (1986). Foraging path selection in bumblebees: Hindsight or foresight? *Behaviour, 99,* 1–21.

Speakman, J. R. (1987). Apparent absorption efficiencies for redshank (*Tringa totanus* L.) and oyster-catcher (*Haematopus ostralegus* L.): Implications for the predictions of optimal foraging models. *American Naturalist, 130,* 677–691.

Stahl, W. R. (1967). Scaling of respiratory variables in mammals. *Journal of Applied Physiology, 22,* 453–460.

Stephens, D. W., & Krebs, J. R. (1986). *Foraging theory.* Princeton: Princeton University Press.

Stephens, D. W., Lynch, J. F., Sorensen, A. E., & Gordon, D. (1986). Preference and profitability: Theory and experiment. *American Naturalist, 127,* 533–553.

Stiles, F. G. (1973). Food supply and the annual cycle of the Anna hummingbird. *University of California Publication in Zoology, 97,* 1–116.

Stiles, F. G. (1975). Ecology, flowering phenology, and hummingbird pollination of some Costa Rican *Heliconia* species. *Ecology, 56,* 285–301.

Taghon, G. L. (1982). Optimal foraging by deposit-feeding invertebrates: Roles of particle size and organic coating. *Oecologia, 52,* 295–304.

Taghon, G. L., Self, R. F. L., & Jumars, P. A. (1978). Predicting particle selection by deposit feeders: A model and its implications. *Limnology and Oceanography, 23,* 752–759.

Tamm, S., & Gass, C. L. (1986). Energy intake rates and nectar concentration preferences by hummingbirds. *Oecologia, 70*, 20–23.

Thomson, A. J., & Holling, C. S. (1977). A model of carbohydrate nutrition in the blowfly *Phormia regina* (Diptera, Calliphoridae). *Canadian Entomology, 109*, 1181–1198.

Tinbergen, N., Impekoven, M., & Frank, D. (1967). An experiment on spacing-out as a defense against predation. *Behaviour, 28*, 307–321.

Tooze, Z. J., & Gass, C. L. (1985). Responses of rufous hummingbirds to midday fasts. *Canadian Journal of Zoology, 63*, 2249–2253.

Tracy, C. R. (1977). Minimum size of mammalian homeotherms: Role of the thermal environment. *Science, 198*, 1034–1035.

Via, S. (1986). Genetic covariance between oviposition preferences and larval performances in an insect herbivore. *Evolution, 40*, 778–785.

Waddington, K. D. (1985). Cost-intake information used in foraging. *Journal of Insect Physiology, 31*, 891–897.

Watt, W. B., Hoch, P. C., & Mills, S. G. (1974). Nectar resource use by *colias* butterflies: Chemical and visual aspects. *Oecologia, 14*, 353–374.

Werner, E. E., Gilliam, J. F., Hall, D. J., & Mittelbach, G. G. (1983). An experimental test of the effects of predation risk on habitat use in fish. *Ecology, 64*, 1540–1548.

Werner, E. E., & Hall, D. J. (1974). Optimal foraging and the size selection of prey by the bluegill sunfish (*Lepomis macrochirus*). *Ecology, 55*, 1042–1052.

Weymouth, R., Lasiewski, R., & Berger, A. (1964). The tongue apparatus in hummingbirds. *Acta Anatomica, 58*, 252–270.

Wirtshafter, D., & Davis, J. D. (1977). Set-points, settling points, and the control of body weight. *Physiology and Behavior, 19*, 75–78.

Wolf, L. L. (1969). Female territoriality in a tropical hummingbird. *Auk, 86*, 490–504.

Wolf, L. L. (1978). Aggressive social organization in nectarivorous birds. *American Zoologist, 18*, 765–778.

Wolf, L. L., & Hainsworth, F. R. (1971). Time and energy budgets of territorial hummingbirds. *Ecology, 52*, 980–988.

Wolf, L. L., & Hainsworth, F. R. (1977). Temporal patterning of feeding by hummingbirds. *Animal Behaviour, 25*, 976–989.

Wolf, L. L., & Hainsworth, F. R. (1983). Economics of foraging strategies in sunbirds and hummingbirds. In W. P. Aspey & S. I. Lustick (Eds.), *Behavioral energetics: The cost of survival in vertebrates* (pp. 223–264). Columbus: Ohio University Press.

Wolf, L. L., & Hainsworth, F. R. (1986). Information and hummingbird foraging at individual inflorescences of *Ipomopsis aggregata*. *Oikos, 46*, 15–22.

Wolf, L. L., Hainsworth, F. R., & Gill, F. B. (1975). Foraging efficiencies and time budgets of nectar feeding birds. *Ecology, 56*, 117–128.

Wolf, L. L., Hainsworth, F. R., & Stiles, F. G. (1972). Energetics of foraging: Rate and efficiency of nectar extraction by hummingbirds. *Science, 176*, 1351–1352.

Wolf, L. L., Stiles, F. G., & Hainsworth, F. R. (1976). Ecological organization of a tropical, highland hummingbird community. *Journal of Animal Ecology, 45*, 349–379.

Wolf, L. L., & Wolf, J. S. (1971). Nesting of the purple-throated carib hummingbird. *Ibis, 113*, 306–315.

Wunderle, J. M., & Martinez, J. S. (1987). Spatial learning in the nectarivorous bananaquit: juveniles versus adults. *Animal Behaviour, 35*, 652–658.

Yamamoto, W. S. (1965). Homeostasis, continuity, and feedback. In W. S. Yamamoto & J. R. Brobeck (Eds.), *Physiological controls and regulations* (pp. 14–31). Philadelphia: W. B. Saunders.

Ydenberg, R. C. (1987). Foraging vs. territorial vigilance: The selection of feeding sites by male great tits (*Parus major* L.). *Ethology, 74*, 33–38.

Zach, R., & Falls, J. B. (1976a). Ovenbird (Aves: Parulidae) hunting behavior in a patchy environment: An experimental study. *Canadian Journal of Zoology, 54*, 1863–1879.

Zach, R., & Falls, J. B. (1976b). Foraging behavior, learning, and exploration by captive ovenbirds (Aves: Parulidae). *Canadian Journal of Zoology, 54*, 1880–1893.

Zeigler, H. P., Green, H. L., & Lehrer, R. (1971). Patterns of feeding behavior in the pigeon. *Journal of Comparative and Physiological Psychology, 76*, 468–477.

12

Food Selection

Paul N. Rozin and Jay Schulkin

Early Work on Food Selection

Curt Richter, the great pioneer in the study of food selection, died this year at the age of 94. This indefatigable, infinitely curious, and ingenious student of nature stands as the great giant of this field. In a series of classic studies in the 1930s and 1940s, Richter (1943) demonstrated adaptive food choice in rats over a wide range of nutrients and conditions. Specifically, he showed that rats changed their food selection in such a way as to optimize nutrition when challenged by a variety of nutrient deficiencies or conditions such as pregnancy that make special nutritional demands. This work was of major significance in establishing the idea of behavioral homeostasis, placing Richter in a direct line from Claude Bernard and Walter Cannon. Richter also began an analysis of the behavioral basis for the adaptive selection, using salt hunger as the example. We will have many occasions to refer to Curt Richter in this chapter. On the sad occasion of his death, we dedicate this chapter to him.

Food selection was studied under controlled conditions, but in a scattered way before the work of Curt Richter. Evvard (1915) reported that swine grew well when faced with a nine-food-item cafeteria. Osborne and Mendel (1918) offered rats a choice between diets with balanced or unbalanced amino acid distributions in the protein fraction. Most of the rats preferred the complete-protein diet. Green (1925) showed that bone consumption in South African cattle was an adaptive response to phosphorus deficiency, since phosphorus replacement terminated the behavior. Already apparent in these early studies was what was to become the central question of the field: Is diet selection innate or acquired? Green (1925) suggested that his cattle learned that bone ingestion

PAUL N. ROZIN Department of Psychology, University of Pennsylvania, Philadelphia, Pennsylvania 19104. JAY SCHULKIN Department of Anatomy, University of Pennsylvania, Philadelphia, Pennsylvania 19104.

produced positive effects. Harris, Clay, Hargreaves, and Ward (1933) reported that rats often failed to choose the B vitamins they needed when faced with a large number of choices but that they could be educated, by selective exposure, to make adaptive choices. This line of work suggested learning as the mechanism for adaptive food selection by omnivores. P. T. Young (1948, 1961) also argued for learning mechanisms, emphasizing the importance of environmental variables in food choice and showed how habits could interfere with adaptive selection (Young & Chaplin, 1945). On the other hand, Dove (1935), based primarily on studies of self-selection by chicks, presumed that some individual animals innately choose optimal diets. Katz (1937), in an excellent review of the field, also emphasized the role of innate factors. The champion of this view was Curt Richter (1943), who held that specific hungers resulted from deficiency-triggered innate recognition of needed nutrients. Richter supported this view primarily with his results on salt appetite.

The facts were and are that neither innate nor learned mechanisms are sufficient to explain food selection. The innate model may work well for specialist animals that eat only a narrow range of foods, but it would be unreasonable to assume that generalists, who eat a very wide range of foods, would have separate innate systems to respond to each of the 30–50 required nutrients. If they did, it would be hard to explain the frequent failures in self-selection studies (e.g., Kon, 1931; Scott & Quint, 1946b; Harris et al., 1933). On the other hand, learning models faced serious obstacles. To learn rapidly about the consequences of ingestion requires an ability to associate a flavor with events that occurred many minutes or hours after, a feat that seemed beyond the known learning capacity of animals. Finally, to understand humans, one clearly had to deal with more than innate preferences or simple associative learning; many authors (e.g., Remington, 1936; Renner, 1944; Mead, 1943) emphasized the powerful importance of transmitted knowledge in determining human food choices. In short, no theory or combination of theories seemed adequate to the task of explaining food choice. We consider and reconsider this issue as we explore different arenas of progress in research on food selection over the past few decades.

THE CENTRALITY OF FOOD SELECTION IN BIOLOGY AND CULTURE

The quest for food is a very frequent and necessary activity and probably places more demands on structure and behavior than any other activity. For many animal species, no waking activity occupies more time or attention than food selection, whether it is the "passive" grazing of an animal that lives on or in its food or the active search of the predator.

Food selection is a major force in evolution, as indicated by the prominence of food habits in the description of particular species and in the identification or naming of higher-level taxonomic groups (e.g., the carnivores or insectivores, among mammals). Surely, the prominence of food selection holds for humans as well. On most accounts, food selection, including an omnivorous or carnivorous heritage, figures prominently in human evolution. The growth of the brain and the advent of bipedal locomotion, allowing for better visualization of prey in high grass and for the handling of weapons, are often attributed to selection pressures based on food acquisition. Domestication of plant and animal food sources is a major feature of the evolution of cultures. And as with other animals, food seeking, selection, preparation, and ingestion constitute *a* if not *the* major waking activity of humans.

Neglect of the Study of Food Selection

Although the subject of food selection received considerable attention in Morgan and Stellar's (1950) classic physiological psychology text, it receives little if any attention in contemporary texts on physiological psychology or introductory psychology. In all of these texts, hunger and thirst are well represented. We believe this neglect stems from a number of sources (see Rozin, 1977).

First, whereas thirst, hunger, and salt appetite fall neatly within the homeostatic tradition, which provides a compelling and attractive model for research, acquisition of other nutrients does not.

Second, it seemed more likely that learning was involved in food selection than in hunger or thirst. There were no available mechanisms of learning that could be invoked to deal with the long delays between ingestion and its consequences. Psychologists during the middle of this century had a great interest in learning, but food selection did not seem to be an example of what they were studying. Furthermore, the abiological cast of American behavioristic psychology caused it to deemphasize the biological importance of food selection as a domain in which learning could be studied.

Third, although hunger is associated with obesity, food selection is not closely related to a salient pathology.

Fourth, the powerful role of cultural factors in human food choice may have discouraged psychological research. Most psychologists interested in food adopt a natural science paradigm and shun the complexities of cultural research.

Fifth, Curt Richter, at the Johns Hopkins Medical School, was across town from the Ph.D.-granting biology and psychology departments at Hopkins and so sponsored very few Ph.D. theses and generated few followers. Moreover, Paul Thomas Young, the second major promoter of food selection research during the middle of this century, emphasized the role of affect in food choice. Unfortunately, the behavioristic temper of psychology in the mid-20th century made psychologists unsympathetic to a "mental" concept such as affect, even though Young objectified it.

Thus, according to our analysis, multiple factors were at work in turning the attention of the field away from the study of food selection. In recent years, however, there has been a resurgence of interest in food selection, as indicated by studies of animal foraging, learning about poisons, affect, and palatability in food choice, social factors in food preference, and psychological and anthropological studies of food choice in humans.

The Problem of Food Selection

The food selection "cycle" consists, in its most general form, of (1) arousal of an interest in food (or a specific food), (2) search for appropriate foods, (3) recognition of such foods, (4) a "decision" to consume such foods, (5) capture, (6) processing, and (7) ingestion. The issue of arousal became specialized to the study of a few major nutritional categories: food calories in investigations of hunger, water in thirst, and sodium in salt hunger. Search, capture, and processing fell primarily into the domain of natural history until the recent spurt of work on foraging. Within psychology, since Richter's classic work, the focus of experimental work has been on the selection of readily available foods, and the main question has been ontogenetic: to what extent is the choice innate or learned?

There are two subproblems that arise at the point of food selection. One deals with selection among different foods with different nutrient composition. This encompasses most of the work and is personified in Richter's classic work. A second concerns selection between different exemplars of the same general category of food (e.g., optimal size of prey) and is treated briefly in our discussion of the foraging literature below and in more detail in Chapter 11.

Any consideration of food selection must take into account some fundamental aspects of nutrition and physiology, which we discuss in this section.

Storage and Internal versus Behavioral Solutions

There are many thousands of different molecules that are necessary components of any complex organism. Most are synthesized within the organism, including all proteins. From the point of view of food selection, the issue centers on molecules that cannot be synthesized at a rate sufficient to maintain life or to meet special needs, such as gestation. For such chemicals, ingestive behavior is required.

Storage in the body may minimize day-to-day needs for specific nutrients. At one extreme is oxygen. Storage is minimal, such that behavior is required for nutrient acquisition on less than a minute-by-minute basis. For water or essential amino acids, there is minimal storage, so that deficit effects appear within hours to a few days. For some of the B vitamins, storage prevents deficiency signs for weeks, and for still others such as vitamin A, months of deprivation are normally necessary to produce frank symptoms.

Regulation of Nutrient Levels

Some nutrients are regulated; that is, body levels are maintained within a rather narrow range. For example, body fluid osmolality, closely related to body levels of both sodium and water, is tightly regulated and varies little. For other nutrients, such as many vitamins, there is simply a lower limit; excesses are stored, metabolized, or excreted, but there is no behavioral satiety effect. The design demands on a nonregulatory system are minimal; indeed, there need be no nutrient-specific internal state that signals deficiency, and a state of general malaise (see below) can motivate a change in food selection.

The Loose Mapping between Nutrient and Sensory Categories

There are fundamental nutrient categories, such as protein, carbohydrate, fat, water, and sodium. Molecules that are members of such categories have a common metabolic fate or function, but they need not share any specific structural properties. Nutrient categories are not equally susceptible to identification by receptor systems. It might not even be possible to design a single receptor that would detect the density of food calories in a potential food. On the other hand, a sodium receptor would be easy to design and is widespread in animals. If it is important to have an energy (or a carbohydrate) detector, it may be necessary to construct a disjunctive system that senses a set of alternative structures, all of which feed into the same category. For example, hexoses and starches seem to have little in common on sensory grounds.

It is an interesting problem for selection to determine how an animal "seeking" carbohydrate identifies it. There have been few studies directed at the learning of such difficult nutritional categories, though there is abundant evidence that animals seeking quick energy may increase their intake of low- or high-molecular-weight carbohydrates. A series of studies in the last decade by Sclafani and his colleagues (reviewed in Sclafani, 1987) speaks directly to this issue. Evidence from a number of sources suggests that rats have a polysaccharide receptor that is separate from the sugar receptor and leads to a qualitatively different sensation; for example, taste aversions to the polysaccharide polycose do not generalize to sucrose. The existence of such a receptor could result from the impossibility of building a single receptor system to detect these very different molecules, or it could have the adaptive value of allowing distinctions between small-molecule, rapidly absorbed carbohydrates and the more slowly absorbed polysaccharides (Sclafani, 1987). This disjunctive system allows for the identification of carbohydrates. It also permits distinctions between subcategories; glucose, but neither other monosaccharides nor the glucose-based disaccharide maltose, induces an oral cephalic-phase insulin response in the rat (Grill, Berridge, & Ganster, 1984). Here we have an example of a sensory system making a much finer distinction than seems called for on nutritional grounds.

STRATEGIES OF FOOD SELECTION

There are two fundamental components to food selection (Rozin, 1976a). One is in an internal state detector, which arouses appropriate food search based in some way on nutritional state. A second is a food detector or recognition device, which identifies appropriate objects in the environment.

THE SPECIALISTS

The specialists employ the simplest strategy of food selection. In the simplest of cases, only one food species or one well-defined food type is consumed. As a result, there is need for only one state detector and one food detector. The simplicity of the system probably makes it more likely that both the state and food detectors will be specified in the genes. The "univore" solution, involving consumption of just one species, is quite rare, but slight variants of the univore solution that encompass single categories of foods with distinct unifying sensory and nutritional properties are quite common. Such categories often have taxonomic status. Koalas, which consume only a few species of eucalyptus leaves, are close to pure univores. More generally, animals consuming insects exclusively, such as some frog species, have a basic specialist design: the food detector senses small, dark moving objects, and it can be activated by a single state detector responding to energy deficit.

A variant on the specialist theme is a set of specialist state and food detector pairings in the same animal. This solution serves when there are a few nutritionally distinct and perceptually well-defined food categories. For example, Dethier's (1976) stunning analysis of food selection in the blowfly, *Phormia regina*, reveals separate state and food detector systems for energy, water, and protein.

The carnivores, which as a group face the most difficult problems of prey location and capture, can survive with simple state and food detectors. Since their food tends to have a similar chemical composition to themselves, the food is likely to be nutritionally complete and nontoxic. Dietary imbalances are unlikely, and a single nutrient deprivation signal may suffice. For some carnivores, it also is possible that the features of prey can be described simply (e.g., moving objects of a certain size), so that food recognition may be a rather simple matter.

The Generalists

The major design problems in food selection systems occur for animals eating a wide range of foods, the generalists. Many of these foods, as exclusive diets, would be either toxic or inadequate sources of nutrients. These generalists can be subdivided into two categories: generalist herbivores and omnivores. For these animals, food identification is a real problem, because there may be no simple way of describing the edible category in such a way as to omit all inedibles. Learning mechanisms seem to predominate in such cases. With respect to internal state, generalists face the problem of multiple nutrient deficiencies and must be able to have some information about the nature of a deficit they face as well as the appropriate food to correct it. Given the prevalence of toxins in plants (e.g., Freeland & Janzen, 1974), the generalist also must be able to identify and either completely avoid such plants or, for cases of mild toxicity, limit intake.

It is the generalist model, well represented by three of the most successful animals in the world, the cockroach (e.g., *Periplaneta americana*), the rat (e.g., *Rattus norvegicus*), and humans, that has most engaged the interest of psychologists. Psychologists have long had an interest in learning, and learning seems to play a fundamental role in generalist food selection.

The Specialist in the Generalist

The specialist's innately specified systems and the generalist strategy, presumably relying more on learning, are extremes. Combinations are possible. For example, the quintessential generalist, the rat, has some specialized systems to handle specific food selection problems. Salt appetite is the prime example (see below and Chapters 15 and 18), with innately specified state and food (sodium) detectors. For thirst there also are innate state detectors as well as oral (if not visual) recognition of water. For energy deficits, there appears to be an innately specified state detector corresponding to hunger. The proclivity for sweets may constitute an innately determined part of the system detecting food energy. Finally, recent evidence suggests an unlearned specific appetite for protein (Deutsch, Moore, & Heinrichs, 1989).

All mammals are specialists (univores) early in life, with milk as their sole food. There are many adaptations to facilitate the switch from this diet to the broader adult diet (Rozin & Pelchat, 1988).

In this section, we consider three prototypical cases: salt hunger as an example of an innate specialist mechanism (in this case, a specialist within the generalist) and poison avoidance and thiamine-specific hunger as examples of a general learning strategy that is characteristic of generalists.

THE SPECIALIST MODEL: THE INNATE HUNGER FOR SALT

When our evolutionary ancestors emerged from the sea, and hence lost a continually accessible source of sodium, selection pressures emerged that promoted physiological mechanisms to conserve sodium and behavioral mechanisms for seeking and identifying sodium in the environment. There is particular urgency for this because sodium is essential for the regulation of extracellular fluid balance and other basic functions and because the body has limited sodium stores (see Denton, 1982).

There is abundant evidence from both naturalistic and laboratory studies indicating a preference for salty tastes by sodium-deficient animals. Many animals in the wild are known to migrate great distances to gain access to salt licks. This is particularly true of herbivorous mammals far from the sea, whose sodium sources in food are diluted by rain (Denton, 1982).

Richter (1936) reported that adrenalectomized rats, which would normally die of sodium depletion within a matter of days, would survive indefinitely when given access to sodium salts. Richter (1939, 1956) suggested that this specific appetite was innate. He believed that sodium deficiency induced changes both in sodium receptors, making them more sensitive to sodium, and in the central processing and evaluation of input from taste neurons signaling the presence of sodium. He presumed that a state detector sensed sodium deficiency and influenced both the sensitivity and preference of an innate sodium detection system. Much of the subsequent research on salt hunger has been devoted to defining how sodium deficiency is detected (see Chapter 15.)

Richter's analysis was basically correct. However, although there are physiological changes in the gustatory system resulting from sodium deficiency (Contreras, 1977; Jacobs, Mark, & Scott, 1988), there is no behavioral evidence for a change in detection threshold (e.g., Bare, 1949; Koh & Teitelbaum, 1961). In other words, it is the preference threshold for sodium that drops with deficiency, not the absolute threshold (see Chapter 10).

Salt hunger is the paradigmatic example of an innate specific hunger (Richter, 1936, 1956; Wolf, 1969; Denton, 1982). Rats and sheep, the two principal species studied, show enhanced ingestion of sodium salts the first time they are rendered sodium deficient (Epstein & Stellar, 1955; Denton & Sabine, 1963; Nachman, 1962) and do so within seconds on first exposure to sodium salts (Nachman, 1962; Wolf, 1969). Furthermore, sodium-deficient rats show the same avid and immediate response to LiCl, which tastes almost the same as NaCl but does not reduce sodium depletion and may have toxic effects (Nachman, 1963; Schulkin, 1982). These observations argue against a sodium preference based on the aftereffects of sodium ingestion.

There is a distinct hedonic component to salt appetite. In sodium deficiency in the rat, the set of sodium-responsive neurons in the nucleus tractus solitarius shifts to include many neurons that respond primarily to sugar (implying a positive hedonic response) and not to salt in the sodium-replete animal (Jacobs *et al.*, 1988). Rats change their orofacial response to hypertonic NaCl to indicate a more positive expression when rendered sodium deficient, and do so on the first exposure to sodium salts after their first bout of sodium deficiency (Berridge, Flynn, Schulkin, & Grill, 1984). The hedonic response thus appears to be unlearned and is part of the expression of the innate hunger.

The preference change that occurs following sodium depletion is highly specific to sodium salts (Richter, 1936; Wolf, 1969; Denton, 1982), although animals will on occasion also prefer other salts such as LiCl or occasionally KCl. Both NaCl and LiCl have, for humans, the paradigmatic "salty" taste. The sodium-deficient animal may be searching for this salty taste (Schulkin, 1982).

There is some evidence for salt hunger in humans. Wilkins and Richter (1940) describe the case of a child with adrenal insufficiency who showed a powerful salt preference and a preoccupation with things salty. Salt preference and salt liking of adults increase when they are salt deprived (Beauchamp, Bertino, & Engelman, 1983).

Although there is an innate salt hunger, this is not to say that learning plays no role in sodium identification. While thirsty, animals can learn the location of sodium sources, how to acquire sodium, and with what it is associated. When subsequently sodium deprived, they display this knowledge (Krieckhaus & Wolf, 1968; Krieckhaus, 1970).

LEARNING ABOUT FOODS: POISON AVOIDANCE

Although learning mechanisms had been suggested to account for most specific hungers from the earliest researchers onwards, no articulated theory that was in accord with known principles of learning was forthcoming, and the learning approach to specific hungers stagnated. It was rescued from an unexpected direction: research on poisoning.

In all the consideration by psychologists of food selection prior to the late 1960s, the problem was always framed in terms of selection of proper nutrients, as opposed to avoidance of imbalances or toxins. A major problem for generalists in nature is the avoidance of abundant plant toxins, which presumably evolved as animal deterrents (e.g., Janzen, 1977; Freeland & Janzen, 1974). Animals consuming a wide variety of plants need some mechanism for avoiding or limiting ingestion of such toxins. Even a strategy of spreading the risk (Freeland & Janzen, 1974), that is, not consuming too much of anything and counting on internal mechanisms to detoxify low levels of any specific toxin, would fail with potent poisons. Presumably, the widespread avoidance of bitter tastes (Garcia & Hankins, 1975) is one adaptation to this situation. Rats and other animals can learn to avoid such toxins. It was this learning ability, manifested in response to man-made poisons, that came to the attention of psychologists.

One of the pioneers in this area was the ubiquitous Curt Richter, who took on the problem of rodent control and poisoning in conjunction with World War II defense issues. Richter (1953) noted that the rats surviving a first round of poisoning seemed resistant to subsequent poisoning attempts and suggested that learning was involved. Since these poisons were man-made, there could hardly be an innate aversion to them, and their initial success argued against such an explanation. Richter placed low levels of poison in foods, such that rats would become ill but not die from ingesting the amounts offered. He noted that most rats subsequently avoided the food in question after one poisoning episode, even though the poison in question did not produce clear negative symptoms until 15 to 20 minutes post-ingestion. He favored a learning interpretation ". . . . . . under conditions in which because of the quick action of the poison they could associate the ill effects with the eating of the food or with some characteristic of the poison." Richter, the champion of innate mechanisms, saw their limits in this case. Other work on poison avoidance by rats in the same period indicated the success of rats at avoiding poisons (e.g., Chitty & Southern, 1954; Barnett, 1956, 1963; Rzoska, 1953).

In parallel to the poison work, a number of investigators, most prominently John Garcia and his colleagues (e.g., Garcia, Kimeldorf, & Hunt, 1961) and J. C. Smith and his colleagues (reviewed in Smith, 1971), had shown that rats came to avoid a taste (often a sugar solution) when its ingestion was followed by x radiation. It was well known that the effects of x radiation were not apparent for hours or more post-radiation. Like the poison avoidance, this seemed to violate a basic principle of learning, which required a close temporal contingency between events to be associated. It was this line of work, and the apparent contradiction between the likely learning interpretation and the impossibility of learning, that prompted Garcia and his colleagues to perform the two critical experiments that changed the face of food selection research and had a major influence on learning theory as well.

LONG-DELAY LEARNING. The critical problem was the apparent long delay between ingestion and its consequences, but it always was possible to suppose that there were some early effects of poison or x radiation that could support learning. Garcia's genius was to eliminate that interpretation by not administering the negative event (toxin or x ray) until a substantial amount of time elapsed after the termination of ingestion. Garcia, Ervin, and Koelling (1966) reported that rats learned, in one trial, to avoid a taste when its ingestion was followed 30 minutes later by injection of the nausea-producing drug apomorphine. This study identified the critical learning mechanism that was necessary to explain a whole host of self-selection and poison avoidance behaviors. A long-delay learning mechanism was just what would be called for in a system in which an action (ingestion) produced inherent delays because of the intermediation of the digestive system.

BELONGINGNESS. There were still other problems in accounting for successful poison or x-ray avoidance. How did the rat know *what* to avoid? It might avoid the environment in which it was poisoned, any object present in it, etc. Some selection principle was necessary, and it was discovered by Garcia and Koelling (1966) in a second classic experiment. They showed that rats poisoned after drinking a solution with a distinctive taste and associated with distinctive light and noise stimuli avoided the taste but not the light and noise. In contrast, rats shocked after drinking the same solution with associated light and sound subsequently avoided the light and sound but not the taste. This principle of belongingness coupled with the long-delay principle gave a satisfactory account for most aspects of poison avoidance. Taste, as the receptor system that guards the entry into the body, would obviously be primed for connection with some types of "malaise." In general, however, the proper description of belongingness is that for any particular species, those modalities or submodalities that are particularly associated with food identification would be primed to be associated with the consequences of ingestion (Rozin & Kalat, 1971).

POTENTIATION. Animals normally identify foods distally (e.g., by sight or smell), and most decisions about whether to eat or not probably are made at this level. In many cases, experiences following ingestion form the basis for changes in the significance of distal cues. Such distal cues should change in value with respect to food only insofar as they are linked to eating and its effects. Garcia, Rusiniak, and their colleagues (Garcia & Rusiniak, 1980), in an extension of the belongingness principle, have shown that a major determinant of whether an odor becomes negative following poisoning is the presence of a mediating taste. The mediating taste, of course, is an indicator of the fact that something was eaten. This phenomenon has been called potentiation.

There are a number of other adaptations, besides long-delay learning, belongingness, and potentiation, that facilitate learning about foods and their negative consequences.

THE IMPORTANCE OF THE NOVEL–FAMILIAR DIMENSION. From well before the classic studies by Garcia, work on rat poisoning emphasized a basic distinction between foods already sampled and new foods. The suspiciousness of wild rats toward new foods (neophobia) had been described a number of times (e.g., Barnett, 1956, 1963; Richter, 1953). It is an adaptive response in a world in which both man-made and natural poisons abound. This neophobia becomes accentuated after poisoning experiences, and under extreme conditions, some wild rats starve to death rather than try new foods (Richter, 1953). The novel–familiar dimension is of central importance in learning about poisons as well. Revusky and Bedarf (1967) showed that when a laboratory rat is poisoned after consuming both a novel and a familiar food, an aversion develops to the novel but not the familiar food.

SAMPLING. Perhaps as a result of neophobia, when rats finally sample a new food, they tend to take only a small amount (Rzoska, 1953). This strategy allows for a "test" of toxicity while avoiding lethal consequences in most cases. Sampling is discussed again in the section on thiamine-specific hunger.

SOCIAL INFLUENCES. Rats and many other animal species are highly social. It would be natural to assume that some social learning about poisoning takes place. Surprisingly, although the social route of information transfer has turned out to be very important in food selection (Galef, 1988; Chapter 13), it does not seem to operate by direct transmission of information about harmful foods; rather, rats come to avoid poisons by social information that directs them to foods that are safe (Galef, 1988). It is possible that other species communicate learned aversions. Glander (1975) reports that howler monkeys successfully avoid trees with toxic leaves and suggests that this may occur as a result of the learning by a few pioneers. The mechanism of such learning has yet to be elucidated.

GENERALITY. Conditioned taste aversions have been demonstrated a number of times in humans, initially by questionnaire studies (Garb & Stunkard,

1974) and later under experimental conditions (e.g., Bernstein, 1978). There has **307**
been further investigation of the nature of the aversion and the role of affect in
this process, and these are discussed below in the section on affect. FOOD SELECTION

SUMMARY. A set of learning abilities and choice strategies permit the rat to
successfully avoid most toxic foods. The demonstration of these abilities had a
major influence on the study of animal learning (e.g., Garcia, Hankins, & Rusiniak, 1974; Seligman, 1970; Rozin & Kalat, 1971; Rozin & Schull, 1988; Shettleworth, 1972; Domjan, 1983). It also laid the groundwork for a learning explanation of acquired preferences.

ACQUIRED PREFERENCES: THE PARADIGMATIC CASE
OF THIAMINE-SPECIFIC HUNGER

THE BASIC PHENOMENON. Thiamine-specific hunger, enhanced ingestion of
thiamine or B vitamins by thiamine-deficient animals, was the first specific hunger to be explained by learning mechanisms (reviewed in Rozin, 1967b, 1976a).
In a classic study, Harris *et al.* (1933) adopted the standard specific hunger
design: animals are maintained on a deficient diet until signs of deficiency appear and then are offered a choice of the deficient diet and one or more other
diets, one of which contains the missing nutrient. Control animals experience the
same diets and choices but are given supplements so that they do not develop a
deficiency. Harris *et al.* reported that rats with multiple B-vitamin deficiencies
reliably chose foods rich in B vitamins. The selection of the vitamin-B-rich diet
from a choice of three foods by deficient rats typically occurred on the first day
and lasted for at least 1 week. This phenomenon was confirmed for thiamine
deficiency by Scott and his colleagues, who offered rats a choice between a
deficient diet and that same diet supplemented with thiamine (Scott & Quint,
1946a; Scott & Verney, 1949).

EARLY THEORIES. Harris *et al.* (1933) and later Scott and Verney (1947,
1949) proposed that thiamine- or B-vitamin-deficient rats learn to prefer
thiamine-rich foods on the basis of prompt beneficial effects of the vitamin.
There was much evidence for this hypothesis. (1) There was no evidence in the
work of these authors for an immediate preference for the thiamine-rich food.
(2) When offered a B-vitamin-rich source as one of six to ten choices, many rats
failed to discover the vitamin (Harris *et al.*, 1933). (3) When the vitamin-fortified
and deficient choices were distinctively labeled in a two-choice situation, rats
would initially favor the vitamin-rich food. However, when flavors were switched
so that the flavor originally marking the vitamin-rich food was now in the deficient foods, rats would prefer the distinctive flavor that was associated with the
vitamin and not the vitamin source itself. This suggests that the rats were detecting an arbitrary flavor associated with recovery and not the vitamin itself (Harris
et al., 1933; Scott & Verney, 1947).

Richter, Holt, and Barelare (1937) reported an immediate and avid response to thiamine in solution by thiamine-deficient rats in a "cafeteria" situation
in which many choices were available. The immediacy of the preference led
them to propose an innate avidity for thiamine induced by thiamine deficiency.

The main argument against the learning view was that it made learning
demands on the animal that were well beyond the rat's acknowledged learning
capacity. Perhaps for this reason, the problem lay unexplored for 15 years after

Scott. Additional problems for a learning position were raised by Rozin and his colleagues, who reported a failure to get thiamine-specific hunger when the vitamin was presented in water as opposed to food, although rapid recovery ensued upon ingestion (Rozin, Wells, & Mayer, 1964), and a preference for thiamine-rich foods by rats recovered from deficiency by injection days before being offered the choice (Rozin, 1965). In both of these demonstrations, the beneficial effects of thiamine seemed not to be involved in the specific hunger.

THE EXPLANATION OF THIAMINE-SPECIFIC HUNGER

Learned Aversions. The solution to this problem proceeded from this point. First, it was shown that thiamine-deficient animals prefer any new food offered in opposition to the deficient diet (Rodgers & Rozin, 1966). The novelty preference might account for the apparent immediacy of thiamine preference in deficient animals; in most studies, the vitamin-enriched food also was the new food. Second, the novelty effect was explained as a specific aversion to the deficient diet; the deficient rat's behavior to the deficient diet when it is the only available food is like the behavior of rats to innately unpalatable foods (Rozin, 1967a). Furthermore, rats recovered from thiamine deficiency on a vitamin-rich diet refused to eat the deficient diet when offered. This explains the preference for thiamine-rich foods by recovered rats as a retained aversion to the deficient diet.

These findings suggested that a major feature of thiamine-specific hunger is learning that the deficient diet is undesirable (Rozin, 1967a, 1967b). This shifted the focus of learning to the prechoice period and suggested a strong parallel between specific hungers and poison avoidance. The two critical findings of Garcia, which were published after the research on aversions was completed, provided exactly the mechanisms needed to account for the aversion to deficient diet.

Neophobia. Richter (1953), Rzoska (1953), and Barnett (1956) all reported neophobia in wild rats that was enhanced following poisoning experiences. Yet the above results suggested an increased interest in new foods by rats "poisoned" on a thiamine-deficient diet. This apparent contradiction was easily settled. The choice offered in these experiments was between the old, deficient diet and a new diet. Shyness toward the new diet may have been masked by the strong aversion to the deficient diet. When wild or domestic thiamine-deficient rats were offered a choice among the old deficient diet, a new diet, and a familiar safe diet that they had eaten before deficiency, they showed a strong preference for the old safe diet; what had been described as a "neophilia" was really a "paleophobia" (Rozin, 1968a).

Learned Preferences. Thiamine-specific hunger cannot be explained entirely as an aversion. Thiamine-deficient rats learn to favor thiamine-rich diets (Garcia, Ervin, York, & Koelling, 1967). This diet is not treated simply as a safe familiar diet. It is preferred to other familiar safe diets that are not associated with recovery from deficiency (Zahorik, Maier, & Pies, 1974; Revusky, 1967, previously showed this effect in hungry rats for a food associated with satiety).

Selection Principles. One of the two basic problems in learning about foods is the discovery of which food or other object or event is to be associated with aversive or positive events. The belongingness principle focuses on a taste or food as the target. The novelty principle narrows the field to new foods. Two other principles have been invoked.

One is sampling. It has been proposed that the rat's natural pattern of food sampling tends to isolate the effects of individual foods so that the effects of such ingestion can be attributed to particular foods (Rozin, 1969). The idea of adaptive sampling can be decomposed into two separate claims. One is that all rats—deficient, recovered, previously poisoned, and naive—tend to eat one food at a time. The evidence is mixed on this point. Rats often sample multiple new and familiar foods when first exposed to them. However, subsequent meals tend to be primarily of one particular food source (Rozin, 1969; Beck, Hitchcock, & Galef, 1988). A second claim is that prior poisoning or deficiency exaggerates this sampling behavior. There were suggestive data on this point (Rozin, 1969), but a more thorough recent study suggests that there is no change in feeding pattern associated with a poisoning experience (Beck *et al.*, 1988).

Recent research indicates a powerful importance for a second selection principal, social guidance in generating food preferences (Galef, 1988). Thiamine hunger has not been used in this context, but these factors would surely be at work in this case (see discussion of social factors below).

Other Behavioral Adaptations to Thiamine Deficiency. Vitamin-B_1-deficient rats, as well as rats deficient in some other substances, show an increase in ingestion of their own feces (Richter & Rice, 1945). This is an adaptive response, since the flora of the hindgut synthesize a number of vitamins that can be utilized only if ingested; indeed, Barnes (1962) has suggested that the hindgut is functionally a processing center that acts to generate nutrients for reingestion. There also are adaptive changes in macronutrient selection in thiamine-deficient rats. Fats spare thiamine. Richter and Hawkes (1941) showed enhanced fat intake and decreased carbohydrate intake in vitamin-B-deficient rats, and Scott, Verney, and Morissey (1950) demonstrated this phenomenon in thiamine-deficient rats.

OTHER SPECIFIC HUNGERS. We have mapped out two distinct solutions to food selection problems in the rat: an innate salt hunger and a general-purpose learning mechanism for thiamine hunger. What remains is to ask about all of the other nutrients.

Water hunger, that is, thirst, is like salt hunger. This is appropriate, since water and sodium are basic and interrelated components of extracellular fluid. There is evidence for an innate internal state detector; for example, the first time chickens are water deprived they drink an amount appropriate to their water deficit (Stricker & Sterritt, 1967). There is also evidence for an innately determined water detection system (a water taste, e.g., Bartoshuk, 1968), which is tied to the internal detector. However, chickens appear to have to learn that the visual stimulus of water corresponds to the taste of water (Morgan, 1894). Pigeons seem to be able to learn to recognize water visually in a wide variety of contexts, from water drops to lakes (Herrnstein, Loveland, & Cable, 1976).

There is no other well-established specialist (innate) system in rats. However, there is evidence that a variety of mineral deficiencies lead to enhanced preference for NaCl, perhaps because the sensory quality of saltiness is correlated with the occurrence of minerals other than sodium (Schulkin, 1986). "Salt licks" usually contain many minerals in addition to sodium (Jones & Hanson, 1985). Laboratory experiments indicate that animals with deficiencies in minerals other than sodium show enhanced sodium salt ingestion (Lewis, 1968; Adam & Dawborn, 1972; Schulkin, 1986). The innate hunger for sodium may serve as a guide for food selection in all mineral deficiencies; the cooccurrence of sodium and other minerals in nature makes the search for a salty taste an adaptive solution (Schulkin, 1986). This may be a case in which a specialist mechanism is generalized to a broader selection context.

There is evidence of specific hungers for a number of minerals, including potassium, calcium, iron, phosphorus, and zinc (e.g., Richter & Eckert, 1937; Richter, 1943; Scott et al., 1950; Milner & Zucker, 1965), all of which may be based in part on an innate bias to ingest salty tasting substances and on general learning mechanisms. Although rats tend initially to consume NaCl when potassium deficient, in the absence of a sodium source, they develop a preference for potassium salts (Adam & Dawborn, 1972).

Specific hungers for all nutrients other than sodium may be examples of the general-purpose learning mechanism described for thiamine. The hallmark of the operation of such a system is the presence of anorexia in deficient animals, a sign of a learned aversion. Anorexia appears in many vitamin deficiencies. There is evidence for the operation of the general learning mechanism for other B vitamins, such as pyridoxine, pantothenate, and riboflavin (Scott & Quint, 1946a; Rozin & Rodgers, 1967; Rodgers, 1967). Specific hungers for vitamins A and D have been difficult to demonstrate. Neither of these vitamin deficiencies has anorexia as a prominent symptom. They also have slow onsets of deficiency and slow recovery.

Specific hungers are elicited not only by diets deficient in nutrients but by special metabolic demands. For example, Richter and Barelare (1938) showed that during pregnancy and lactation, sodium and calcium intakes are elevated (see also Denton, 1982). Naturalistic observations have shown that a variety of herbivorous and omnivorous female mammals and birds migrate to salt licks during the reproductive season (Denton, 1982). These preferences probably have a hormonal basis. Even virgin female rats show a higher base line of sodium and calcium intake than males, but this effect is not present in females ovariectomized before 12 days of age (Krecek, Novikova, & Stribral, 1972).

There are different ways of responding to deficiency, as we saw in the case of thiamine hunger: deficient diet avoidance, thiamine preference, carbohydrate avoidance, and coprophagy. The case of vitamin D deficiency is instructive. Although there is no vitamin-D-specific hunger, the calcium deficiency that is the major symptom of vitamin D deficiency forms the basis for a learned increase in calcium intake. Thus, the rat can "treat" vitamin D deficiency without ingestion of the vitamin (Brommage & DeLuca, 1984).

Given the essential need for protein and its low level in many plant foods, one would expect to see a clear specific hunger for protein. There is recent evidence for an immediate (hence, presumably innate) taste-mediated protein preference in protein-deficient animals (Deutsch, Moore, & Heinrichs, 1989). All of the other evidence favors a learning model. Of course, both mechanisms could be operative.

Individual animals in early studies with cafeterias often failed to thrive because of low protein intake (e.g., Scott & Quint, 1946b; Lat, 1967). Much of this failure can be attributed to the unpalatability of casein, the usual protein source. (Interestingly, Deutsch's protein-deficient rats show immediate protein recognition and preference for a variety of proteins, but not casein.) Nonetheless, a variety of studies, dating back to Osborne and Mendel (1918), indicate adaptive choice of adequate-protein diets in most animals. Evidence for a protein-specific hunger (reviewed by Anderson, 1979; Li & Anderson, 1983) comes from a number of sources. (1) Protein-deficient rats usually choose sources with adequate protein. (2) Rats respond to protein dilution by increasing protein intake, suggesting a regulation of protein intake (Rozin, 1968b; Musten, Peace, & Anderson, 1974). Furthermore, rats consume relatively more protein when the protein is of low quality, thus guaranteeing sufficient intake of the limiting amino acid (Ashley & Anderson, 1975). (3) The increased protein response is specific to protein need. When energy but not protein needs are increased, rats selectively increase carbohydrate or fat intake (Donhoffer, 1960). (4) Rats learn to associate an arbitrary flavor with protein repletion. A flavor paired with protein repletion will be preferred subsequently when the rat is in a protein-depleted state. However, this preference disappears when protein is intubated before the preference test (Gibson & Booth, 1986).

Rats also are able to avoid amino acid imbalanced diets and to select diets containing limiting essential amino acids (Harper, 1967; Booth & Simson, 1971; Anderson, 1979).

THE WISDOM OF THE BODY: NEED-FREE SELECTION

DEFICIT CORRECTION OR DEFICIT AVOIDANCE

Can some generalists self-select a balanced or even optimal diet in the absence of any deficiency and without any prior experience? If they could, it would argue strongly for the role of innate factors, since in the need-free state there would be no positive consequences of nutrient ingestion that might support learning. It would further constrain innate models, since models that claim that prewired circuits detect a deficiency and activate a particular food or taste "search image" also require a deficiency state as a prerequisite. In fact, we have no model for how naive need-free selection might be optimized.

The bulwark of support for adaptive, need-free self-selection comes from two sources. Davis (1928, 1939) offered the choice of a variety of wholesome foods to newly weaned human infants over a period of some months. Infants indicated their preference for particular foods in the display of about 15 in front of them by pointing, and the nurse obliged by offering a spoon of the desired food. Davis reports that children ate a balanced diet, were extremely healthy, and grew very well on this regimen. However, one cannot conclude too much about the infant's abilities to self-select in the need-free state. All the foods used were of high nutritional value; no spices or flavorings (e.g., sugar) were added. It is very likely that random selection among the dairy, meat, vegetable, grain, and fruit choices available would support excellent growth. Furthermore (see Rozin, 1976a, for more detailed analysis), the two most favored foods were milk and fruit, the two sweetest items among the choices. Similarly, Richter (Richter, Holt, & Barelare, 1938; Richter, 1943) found that rats flourished on a cafeteria choice including pure nutrients such as exemplars of the three basic macronutrient categories and specific mineral and vitamin sources. Again, most rats in these studies grew very well on this regimen.

There are four problems with the results of such cafeteria studies (see also

Chapter 13). First, there are many examples of failure to regulate (see Chapter 13; Kon, 1931; Scott, 1946; see Lat, 1967, for a review), Often, this failure comes down to inadequate selection of sufficient protein. Second, in many cases, random selection would produce an adequate diet. Third, palatability is known to be a major factor in food choice; the experimenter's selection of foods or nutrients as choices could strongly determine the results. As Galef and Beck (Chapter 13) note, need-free self-selection appears to be adaptive in benign environments. Fourth, in cases where there is adequate self-selection, one must be sure that incipient deficiencies did not appear, since such deficiencies could engage a set of innate or learned mechanisms that fall within the deficit-correction (as opposed to deficit-avoidance) model.

We conclude that there has not been a definitive demonstration of adaptive food selection under conditions where this might reasonably fail, in need-free animals.

THE FORAGING/OPTIMIZATION APPROACH

One of the major advances in the study of food selection in the last few decades has been the study of optimal foraging. This approach arises out of a concern to study the adaptiveness of behavior in a quantitative and rigorous way. It assumes that through natural selection, animals have evolved an ability to make optimal choices of the sort that they face in their natural environment. Hence, an analysis of costs and benefits associated with a choice of behaviors will yield testable predictions about behavior. When the behavior does not meet the prediction, it usually is assumed that the cost–benefit analysis was in error. The virtue of this approach is that it promotes careful, quantitative analysis of the organism, the environment, and the interaction between them. Because feeding occurs frequently, is easily measured, and has measurable adaptive consequences (growth or survival), it has provided the most useful framework within which to test optimality models (for reviews of this area, see Krebs & Davies, 1984; Collier, 1983; Chapter 11).

The choices at issue in the foraging work include choice of qualitatively different foods, but more often they encompass other aspects of food selection: differences in food choices may not be related to nutrients but to the amount of work necessary to extract nutrient, time to extract it, or density of food items in different locations. There is more emphasis on the search stage in the food selection cycle, with consideration of problems such as when to start searching, where to search, when to abandon a search area, and the pattern of search. The basic principle invoked is that an organism should continue in its current consumption pattern so long as the alternatives have a poorer cost–benefit ratio. For example, rats prefer shelled to unshelled sunflower seeds, which is "justified" on cost–benefit grounds because of the additional time and energy costs of extracting the nut from the shell. However, when the ratio of unshelled to shelled nuts in the mixed food source exceeds about five to one, rats begin to consume unshelled nuts, presumably because the search costs for shelled nuts have become higher (Collier, 1983). In another case, shore crabs (*Carcinus maenus*) are consistently faced with the choice of whether to process (break the shell) and consume a particular mussel they encounter. Larger mussels yield more energy, but breaking the shell is more difficult. Calculations of the optimal size mussel, in terms of cost and benefit, predict the favored size of the crabs (Elner & Hughes, 1978).

Until recently, the focus of optimal foraging work has been on analysis of costs and benefits and demonstration of the fit between predictions and behavior. The concern has been with steady-state performances such that the animal is assumed to already know the relevant information. There now is an increasing amount of research on the mechanism of choice, including concern with the mechanism of optimization [e.g., detection of highest rates of reinforcement through continuous preference shifts toward this alternative (Herrnstein & Vaughan, 1980)] or mode of memorizing the location of prey and prey sites already visited (Shettleworth & Krebs, 1982). These studies also touch on the ontogenetic issue: what is the source of information and the processing of it? That is, what has been specified in the genes? Nonetheless, in contrast to the literature on choice of nutrients, most of the foraging literature emphasizes function and immediate rather than ontogenetic causes of behavior.

THE PROBLEM OF CONTEXT

Both cafeteria and foraging studies raise the critical issue of context. In designing a self-selection study, any investigator is faced with a conflict between creating a natural environment and refining choices so that basic nutritional variables will be isolated. At one extreme are pure nutrient choices (e.g., sugar, casein, olive oil, sodium chloride) as employed by Richter. In between are mixed-diet choices, with each diet particularly high in a specific ingredient (e.g., high protein, high carbohydrate), as in the studies by Anderson (1979). At the other extreme, employed primarily in foraging studies, are natural foods. There is no simple rule to be invoked; it is a trade-off between fidelity to nature and analyzability of the results. However, if the endpoint is understanding choices in the face of real foods, then purified ingredients are useful tools only insofar as they preserve critical features of natural foods. In the case of humans, for example, efforts to understand sweet preference in terms of reaction to different concentrations of sugars in water may have limited generality; the preferred concentration of sugar in a food depends on the particular food or beverage. Pangborn (1980) made this point clearly with respect to sugar preference in humans, and Cowart and Beauchamp (1986) have done so for sodium preference in humans. In the limits, the point is clear for humans; though a person may like sweetness very much, he or she is unlikely to pour sugar on a steak. Furthermore, preference may depend on the relationship of components. The optimal sweetness of a dairy product, as judged by human adults, is a function of the fat content (Drewnowski & Greenwood, 1983).

Preference data in animals also depends on the form of foods. Thiamine-deficient rats show a clear preference for thiamine in solid foods but not in liquids (Rozin et al., 1964). Rats show clear preferences for sodium salts only when they are presented in a liquid medium (Beauchamp & Bertino, 1985).

Context extends beyond food characteristics. In a sheltered laboratory environment, where predation and search for shelter may occupy no time, extra costs associated with obtaining a preferred food may be tolerable. But the costs of such a preference may increase when time pressure becomes a part of the situation. The optimal foraging research goes much further than the standard laboratory self-selection study to take context into account, but it is rare even among such studies to vary nonfood alternatives. It is sobering to realize that a basic parameter such as number of meals eaten by a rat in a laboratory setting is altered in a major way by providing rats with a comfortable place to sleep (Nicolaidis, Danguir, & Mather, 1979).

The Hedonic Component in Food Choice

The major proponent of the importance of hedonic factors in food choice has been Paul Thomas Young. In a vigorous and substantial program of research in the 1940s and 1950s, Young emphasized that food choice was, in large part, determined by stimulus properties of foods (Young, 1948, 1961). Young introduced the concept of palatability, defined as "... the acceptability of foodstuffs as determined by the characteristics of the food stimulus" (Young, 1948, p. 301). He elaborated this position, holding that "the term palatability refers to the immediate affective reaction (liking or disliking) of an organism which occurs when a food comes in contact with the head receptors" (Young, 1948, p. 310). Young argued, both theoretically and through an extensive set of empirical studies, that palatability was a basic aspect of food choice and that it could not be adequately determined by the standard procedure of measuring relative intake over long periods of time. Different levels of intake of the same nutrient at different concentrations often indicated palatability of the different concentrations rather than any regulation of amount ingested. In order to investigate palatability, Young measured responses in the first seconds of contact with a pair of foods, which minimized postingestional effects. He also emphasized that clear food preferences were observable in need-free animals.

Young's position was not adequtely appreciated at the time, probably because of the questionable "mental" status of hedonic processes, even though his notion of palatability was defined behaviorally. In fact, Young was right on the mark; a psychology of food choice, like other aspects of motivation, cannot avoid the powerful role of affect. Liking is a dominant aspect of human experience with food; indeed, liking the flavor of a food is the best predictor of human food choice or intake in the absence of economic and availability constraints (Schutz & Judge, 1984). There is now a return of interest in affective processes, which has its roots at a time well before the work of P. T. Young.

Charles Darwin presented evidence that hedonic events are part of the fabric of behavior. His book *The Expression of the Emotions in Man and Animal* (1872) is replete with examples of alimentary reactions to food sources, sometimes with pleasure, other times with distaste or disgust. Facial expression was one of the bodily responses that Darwin focused on as an index of emotion or affect. In recent years, these expressions have been studied extensively, described in objective terms, and assigned to different basic emotions (e.g., Ekman & Friesen, 1975). The analysis of facial responses to tastants in human infants has revealed distinctive negative responses to bitter and sour and positive, accepting responses to sweet stimuli (Steiner, 1979; Rosenstein & Oster, 1988).

The power and salience of affective responses is hard to deny in humans. Because of the easy availability of verbal report, human research on hedonic aspects of choice has employed liking rating scales rather than analysis of facial expressions. There is a certain directness in using verbal report as a measure of affect, as in the research of Cabanac (1971), who emphasized the role of pleasure (measured on a hedonic scale) in the regulation of energy intake. Under ideal conditions, hedonic responses measured on rating scales are interpreted in relation to other relevant measures (Blundell, 1983). However, because of the absence of verbal report from animals, and inadequate assimilation of Young's methodological approaches to the measurement of palatability, the animal literature paid little attention to affect. Individual researchers (Pfaffmann, 1960; Stellar, 1974) emphasized its importance, but research did not proceed.

An objective analogue to human facial expression was unavailable until the recent research of Grill and Norgren (1978a). They showed that rats display characteristic facial and other gestures to bitter and sweet stimuli. These responses seem analogous to those of humans and, like the human responses (Steiner, 1979), are organized in the brainstem, as indicated by their presence in decerebrates (Grill & Norgren, 1978b).

Berridge and Grill (1984; see also Nowlis, 1977) have suggested that hedonic responses do not lie on a continuum; rather, analysis of facial expressions of rats to a variety of stimuli indicates independent positive and negative hedonic systems. These systems may compete for control of the ultimate behavior, ingestion, but appear behaviorally in expressive gestures.

The Grill–Norgren technique has opened the door for the study of affective processes in animals. For example, with respect to food choice, it has been shown that the sodium-deficient rat shows expressions of greater positivity to sodium tastes than the nondeficient rat (Berridge et al., 1984) and that the lithium-chloride-poisoned rat shows a quininelike expression when sampling the previously poisoned food (Parker, 1982; Pelchat, Grill, Rozin & Jacobs, 1983) (see section on acquisition of preferences, below, for more details).

Recent evidence from Scott and his colleagues (e.g., Scott & Mark, 1986) suggests experience-based changes in the profile of response of gustatory neurons in the brainstem that correspond to the hedonic changes that have been reported. Multidimensional scaling analyses of response patterns to a variety of tastants suggest that following a taste aversion procedure, the target taste profile looks more like the bitter profile. Furthermore, the salt profile looks more like the sugar profile when sodium deficiency is induced. These results suggest that low-level sensory coding changes may be at least partly responsible for hedonic changes (see Chapter 10).

The reemergence of palatability and affect as important aspects of food choice has prompted a fuller analysis of the concept of preference.

The Structure of Food Preferences and Avoidances in Humans and Other Animals

Preferences and avoidances are measured in terms of relative intake of different food choices or as responses to direct verbal questions. Liking, or affect, is reflected in some of these preferences but not all, in the sense that a dieter might like ice cream better than cottage cheese but prefer to eat cottage cheese because it contains fewer calories. This suggests that, especially for humans, some analysis must be done of subtypes of preferences. Analysis of questionnaires and interviews with humans about the basis for their food preferences suggests that there are three underlying motives for food acceptance or rejection. These are (1) sensory–affective motivation (affectively based, good or bad taste or smell), (2) anticipated consequences of ingestion (e.g., harmful if swallowed), and (3) ideational motivation (limited to humans, and based on knowledge of the origin or nature of a particular food) (Rozin & Fallon, 1980; see Table 1).

For the case of food rejections, four subcategories can be generated from these three motivations (Table 1): *distastes*, rejected primarily because of bad sensory properties, *dangers*, rejected primarily because of anticipated harmful consequences, *inappropriates*, rejected primarily because of negative response to the idea or nature of the item (e.g., paper), and *disgusts*, also rejected on ideational grounds, but with a strong distaste component as well (e.g., worms).

The difference between danger and distaste illustrates a fundamental principle in food choice. Dislike means dislike of the taste; danger refers to fear of consequences. A person who rejects shrimp as food on the grounds of distaste is in a very different state from one who rejects shrimp as dangerous. A food can become classified as dangerous because a person hears that it is harmful or because a person directly experiences its negative effects. The interesting psychological question is what makes a food distasteful? We know of one critical factor. Humans who experience nausea after eating a food tend to come to dislike the food (an acquired distaste). This happens much less frequently after negative gastrointestinal events other than nausea, and even less frequently after negative nongastrointestinal events such as skin eruptions or respiratory problems. After negative events other than nausea, the associated food tends to become classified as dangerous. The human data suggest a special role for nausea in generating distastes (Pelchat & Rozin, 1982). This result has been confirmed in animals, using the analysis of rat facial expressions pioneered by Grill and Norgren (1978a; see above). Foods paired with lithium chloride, which presumably causes nausea in rats as it does in humans (Verbalis, McHale, Gardiner, & Stricker, 1986), come to evoke expressions such as gaping, indicating negative affect and aversion. However, pairing of food with intubation of lactose, which leads to lower gastrointestinal pain, or with electric shock leads to avoidance, but without negative expressive gestures (Parker, 1982; Pelchat *et al.*, 1983). This parallels closely the distinction between distaste and danger in humans.

The Acquisition of Preferences

In the domain of preferences, the basic distinction (Table 1) is between foods liked for their sensory properties ("good tastes") and foods eaten for their positive consequences ("beneficials"). Preference acquisition is usually slower than aversion acquisition, and there is no powerful manipulation corresponding to nausea that reliably produces good tastes. Rather, there is evidence for weak contributions from a number of sources, with social factors emerging as the most powerful influence. We divide explanations for acquisition of liking into three categories: exposure, Pavlovian processes, and social factors. We then consider, briefly, special mechanisms that may be involved in the reversal of innate aversions (see Beauchamp, 1981; Birch, 1987; Booth, 1982; Rozin & Vollmecke, 1986, for more extensive reviews).

Nonsocial Factors

In general, exposure tends to increase liking, and Zajonc (1968) has proposed that mere exposure is a sufficient explanation for this. There is evidence

TABLE 1. Psychological Categorization of Acceptance and Rejection[a]

Dimensions	Rejections				Acceptances			
	Distaste	Danger	Inappropriate	Disgust	Good taste	Beneficial	Appropriate	Transvalued
Sensory affective	−			−	+			+
Anticipated consequences		−				+		
Ideational		?	−	−		?	+	+
Contaminant		−		−				+
Examples	Beer Chili Spinach	Allergy foods Carcinogens	Grass Sand	Feces Insects	Saccharine Favorite foods	Medicines Healthy foods	Ritual foods	Leavings of heroes, loved ones, or deities

[a]From Fallon and Rozin (1983).

for "mere exposure" in the domain of food (e.g., Pliner, 1982). Two problems emerge. First, is "mere" exposure sufficient, or is exposure providing the opportunity for other processes (e.g., Pavlovian conditioning) to take place? Second, exposure, especially when intense and very frequent, often leads to a decrement in liking. This phenomenon, originally described in animals by Katz (1937; see also LeMagnen, 1956), has been clearly demonstrated and analyzed in humans by Rolls and her colleagues (Rolls, Hetherington, Burley, & van Duijvenvoorde, 1986). The relationship between sensory-specific satiety and exposure has yet to be elaborated, but it seems that the effects of sensory-specific satiety are more time limited.

Pavlovian conditioning may be the principal process in the acquisition of liking, as it is in the conditioned taste aversion phenomenon (Martin & Levey, 1978; Rozin & Zellner, 1985). Two categories of US have been identified. One is positive postingestive consequences, such as rapid satiety or termination of drug withdrawal symptoms. When such events follow a flavor, there may be enhanced liking for the flavor in humans (e.g., Booth, Mather, & Fuller, 1982) and animals (Zellner, Berridge, Grill, & Ternes, 1985; Bolles, Hayward, & Crandall, 1981; Booth, Stoloff, & Nicholls, 1974; Tordoff & Friedman, 1986). The most striking example of enhanced preferences of this type came from recent work by Sclafani and his colleagues using polycose, a partially hydrolized starch, as an unconditioned stimulus (Sclafani & Nissenbaum, 1988). A second category of US is positive tastes (such as sweetness). After pairing of a neutral taste with an already positive taste, humans like the neutral taste better (Zellner, Rozin, Aron, & Kulish, 1983), and animals show a corresponding preference (Holman, 1975). However, neither of these Pavlovian USs seems very potent. Social events are much more powerful. Their operation may well be cast in Pavlovian terms, but we consider them under a separate heading.

SOCIAL FACTORS

Work by Galef and his collaborators over the last decade (reviewed in Galef, 1988; Chapter 13) has added a new dimension to our understanding of preference development in the rat by showing the importance of "passive" social factors. Both wild and domestic rats develop preferences for foods eaten by conspecifics. More particularly, young rats feed in the presence of older rats, and this experience causes an enhanced preference, as does exposure to food properties by nursing animals exposed to mother's milk. Most strikingly, exposure to a conspecific ("demonstrator") that has recently eaten a specific food induces a substantial preference for that food. This holds even for unpalatable foods, as, for example, foods containing modest amounts of chili pepper (Galef, 1989). Exposure to a demonstrator who has eaten diet A also serves to weaken markedly the subsequent aversion a rat will develop when consumption of diet A is followed by poisoning. These social effects require little exposure and are robust. They seem to be mediated by the presence of the taste/odor of the food on the demonstrator animal in association with carbon disulfide, a chemical present in the breath of rats (Galef, Mason, Preti, & Bean, 1988). This is a remarkably simple mechanism for these rich social effects. Surprisingly, there is no evidence for socially learned aversions in rats, nor for any preference-inducing social mechanism other than passive social exposure.

Research by Birch on children (e.g., Birch, 1980; Birch, Zimmerman, & Hind, 1980; reviewed in Birch, 1987; see also Rozin, 1988), greatly extending the original findings of Duncker (1938), indicates that perception that a food is valued by peers or respected by others tends to enhance liking for that food. Whereas Galef's studies of animals may be accounted for by passive exposure to the presence of conspecifics, the studies on humans seem to depend much more on evaluative responses emanating from the conspecific. Also, although there are little or no data suggesting active instruction in food choice by animal parents, this behavior is very common in humans.

Liking for Innately Unpalatable Substances

Humans, almost uniquely among generalists, develop strong likings for innately unpalatable foods. Common examples are irritants (e.g., chili pepper, black pepper, horseradish) and bitter substances (e.g., coffee, alcohol). With regard to the acquisition of a liking of chili pepper (reviewed in Rozin, 1990a), there is evidence for the operation of all the factors described above. In the traditional setting, there is mere exposure. Chili, as a meal constituent, is followed by satiety, and the production of saliva that it induces enhances the flavor of the foods consumed. Perhaps most critically, it is consumed by young children in an environment in which older siblings and parents consume and enjoy it. In addition, it is possible that there are special mechanisms that are involved in reversing aversions, that is, mechanisms that depend on the initial aversion. Two examples are the induction of an endogenous opiate response following multiple self-induced exposures to a painful/negative stimulus and the enjoyment of the fact that the body signals rejection but the mind knows there is no danger (benign masochism) (Rozin & Schiller, 1980).

Culture, Biology, and Human Food Choice

The reversal of innate aversions under the influence of cultural forces raises the question of the role of culture in human food choice. The most informative fact that one can learn about a human, in terms of predicting food preference, is his or her culture or ethnic group. It is obvious that culture-specific food likes are acquired. The question is whether this cultural information is itself a distilled form of nutritional wisdom gained by individuals in the past by the acquisition processes we have already described. There is, of course, one other process: humans may both manipulate their environment and think about causes and effects. Hence, they may reason their way to a conclusion that, for example, scurvy can be prevented by consuming fresh fruits, a conclusion that took hundreds of years to mature. Such nutritional wisdom may be passed along and may influence likes by influencing exposure.

Every culture has a body of rules, preferences, and likings concerning foods that constitute their cuisine. This tradition includes constraints or predispositions with respect to the basic staple foods consumed, specific modes of preparation, and specific flavorings (E. Rozin, 1982). The question is, what is the origin and significance of such traditions? Marvin Harris (e.g., 1985, 1987) has argued most forcefully that the major part of human culinary institutions can be accounted for as nutritional adaptations, often to consuming sufficient protein.

Solomon Katz (e.g., 1982) has demonstrated the adaptive value of culinary practices in a number of cases; for example, traditional maize preparation techniques (the making of tortillas) in Mexico and Central America greatly enhance the nutritional value of the corn. Others (Remington, 1936; Mead, 1943; de Garine, 1972; Rozin, 1990b) propose that much of cuisine must be understood in a larger social context, where food serves more than nutritional functions. For example, in India, food is a moral substance and serves as a social marker in maintaining caste differences (Marriott, 1968; Appadurai, 1981). Who prepared a food is a powerful determinant of its acceptability.

It surely is the case that there is truth in both positions, and research is now directed more to explication of specific cases and uncovering the roots and functions of traditions (see Rozin, 1982, for a review). Since there are rather few innately determined behavioral biases in generalists, one cannot expect a major influence on cuisine from this quarter. However, where such biases exist, most clearly in the preference for sweets, we see a massive effect on human culinary institutions. The widespread use of sugar, development of sugar technology, and artificial sweetener technology are motivated by the innate liking for sweet (Mintz, 1985; Rozin, 1982). More common, as in the examples developed by Katz (1982), is the use of human reasoning power or natural selection to increase the use of those culinary techniques that promote survival. Thus, for example, the development of milk fermentation technology, leading to yogurt and cheese, is a compensation for the inability of humans to digest lactose in raw milk (Simoons, 1982). Interestingly, in this case, the domestication of animals led to the availability of milk as an adult food, which in turn established a selection pressure for the ability of adults to digest milk. These cultural events have led to a change in the genetics of certain groups of Northern European origin. so that such people are now lactose tolerant as adults (Simoons, 1982). The point is that there is a complex interaction of biology, psychology, and culture in human food preferences, with each "domain" influencing the other (Rozin, 1982). The stage has been set for analysis of these interactions, and a few cases (e.g., sugar, milk) have already been elaborated.

Human Food Choice: Attitudes to Animal Foods

The human response to animal foods illustrates many of the points we have made and serves to both summarize the state of the art and set some priorities for the future. Meat is both the most preferred and most tabooed food among humans. The attraction to meat can easily be explained on biological grounds. For a long period in their history, humans had a diet in which meat played a substantial part. Nutritionally, it is an ideal food. It would not be surprising if there was an innate preference for meat (though probably not present at birth) mediated by a preference for either fatty textures or protein.

The fact that ingestion of meat is allowed preferentially for senior males in many traditional cultures suggests that it is a highly prized food. Yet, there is a strong aversion to consuming meat, as witnessed for example by the large number of meat taboos and a long history of vegetarian movements dating back to ancient Greece. Most significantly, the strongest category of food rejections in humans is disgust, and this category consists almost entirely of animal products (Angyal, 1941; Rozin & Fallon, 1987).

The roots of meat aversion are many, and all are of cultural origin. There is the longstanding belief that meat is unhealthy and that it is not a natural food for humans (Dombrowski, 1984). Note that in the *Old Testament,* man is a vegetarian until God allows the ingestion of animals after the Flood. Then there are the moral views opposing the killing of animals for food and the ecological opposition to consumption of animal foods on the grounds that it is wasteful. Finally, there is the psychological concern that ingestion of animals will make the eater more animallike. This is a special case of the principle of "you are what you eat," which is widely held in traditional cultures (Nemeroff & Rozin, 1989). Contemporary attitudes to meat, in any culture, may include an influence of all of these factors and must be understood within the context of related cultural attitudes. An integrated approach to meat as a human food requires inputs from many disciplines in the natural and social sciences. This is a project for the future.

Summary: Approaches to Food Choice

The broad multidisciplinary approach to food choice in both animals and humans has two aspects. In one view, it means consideration of behavioral and physiological mechanisms and interest in different time frames: accounting for phenomena in terms of immediate causes, explaining development and phylogeny, and examining function. In this regard, Dethier's (1976) classic work on the blowfly is a model of breadth. In another view, the multidisciplinary approach, including humans, requires the different types of explanation just referred to and also encompasses contributions from various disciplines in the natural and social sciences. It is best exemplified, in the current literature, in Denton's (1982) monumental monograph *The Hunger for Salt,* in the volume on *Sweetness* edited by Dobbing (1988), and in more general terms, in the edited volume *Psychobiology of Human Food Selection* (Barker, 1982). So far, we have only scratched the surface or, more appropriately, taken a little bite out of a very big pie.

Acknowledgments

Preparation of this chapter was supported by funds awarded to Paul Rozin from the University of Pennsylvania Research Fund and the Whitehall Foundation and funds from an NIMH Research Career Development Award to Jay Schulkin.

References

Adam, W. R., & Dawborn, J. K. (1972). Effect of potassium depletion on mineral appetite in the rat. *Journal of Comparative and Physiological Psychology, 418,* 51–58.

Anderson, G. H. (1979). Control of protein and energy intake: Role of plasma amino acids and brain neurotransmitters. *Canadian Journal of Physiology and Pharmacology, 57,* 1043–1057.

Angyal, A. (1941). Disgust and related aversions. *Journal of Abnormal and Social Psychology, 36,* 393–412.

Appadurai, A. (1981). Gastro-politics in Hindu South Asia. *American Ethnologist, 8,* 494–511.

Ashley, D. V. M., & Anderson, G. H. (1975). Food intake regulation in the weanling rat: The effect of the most limiting essential amino acids of gluten, casein and zein on the self selection of protein and energy. *Journal of Nutrition, 105,* 1405–1411.

Bare, J. K. (1949). The specific hunger for sodium chloride in normal and adrenalectomized white rats. *Journal of Comparative and Physiological Psychology, 42,* 242–253.

Barker, L. M. (Ed.). (1982). *The psychobiology of human food selection*. Bridgeport, CT: AVI.

Barnes, R. H. (1962). Nutritional implications of coprophagy. *Nutrition Reviews, 20,* 289–291.

Barnett, S. A. (1956). Behaviour components in the feeding of wild and laboratory rats. *Behaviour, 9,* 24–43.

Barnett, S. A. (1963). *A study in behaviour. Principles of ethology and behavioral physiology, displayed mainly in the rat.* London: Methuen.

Bartoshuk, L. (1968). Water taste in man. *Perception and Psychophysics, 3,* 69–72.

Beauchamp, G. K. (1981). Ontogenesis of taste preference. In D. Walcher & N. Kretchmer (Eds.), *Food, nutrition and evolution* (pp. 49–57). New York: Masson.

Beauchamp, G. K., & Bertino, M. (1985). Rats (*Rattus norvegicus*) do not prefer salted solid food. *Journal of Comparative Psychology, 99,* 240–247.

Beauchamp, G. K., Bertino, M., & Engelman, K. (1983). Modification of salt taste. *Annals of Internal Medicine, 98,* 763–769.

Beck, M., Hitchcock, C. L., & Galef, B. G., Jr. (1988). Diet sampling by wild Norway rats offered several unfamiliar foods. *Animal Learning and Behavior, 16,* 224–230.

Bernstein, I. L. (1978). Learned taste aversions in children receiving chemotherapy. *Science, 200,* 1302–1303.

Berridge, K. C., Flynn, F. W., Schulkin, J., & Grill, H. J. (1984). Sodium depletion enhances salt palatability in rats. *Behavioral Neuroscience, 98,* 652–660.

Berridge, K. C., & Grill, H. J. (1984). Isohedonic tastes support a two-dimensional hypothesis of palatability. *Appetite, 5,* 221–231.

Birch, L. L. (1980). Effect of peer model's food choices and eating behaviors on pre-schoolers food preferences. *Child Development, 51,* 489–496.

Birch, L. L. (1987). Children's food preferences: Developmental patterns and environmental influences. In G. Whitehurst & R. Vasta (Eds.), *Annals of Child Development.* (Vol. 4, pp. 171–208) Greenwich, CT: JAI.

Birch, L. L., Zimmerman, S. I., & Hind, H. (1980). The influence of social–affective context on the formation of children's food preferences. *Child Development, 51,* 856–861.

Blundell, J. E. (1983). Problems and processes underlying the control of food selection and nutrient intake. In R. J. Wurtman & J. J. Wurtman (Eds.), *Nutrition and the brain.* (Vol. 6, pp. 163–231). New York: Raven Press.

Bolles, R. C., Hayward, L., & Crandall, C. (1981). Conditioned taste preferences based on caloric density. *Journal of Experimental Psychology: Animal Behavior Processes, 7,* 59–69.

Booth, D. A. (1982). Normal control of omnivore intake by taste and smell. In J. Steiner & J. Ganchrow (Eds.), *The determination of behavior by chemical stimuli. ECRO symposium* (pp. 233–243). London: Information Retrieval.

Booth, D. A., Mather, P., & Fuller, J. (1982). Starch content of ordinary foods associatively conditions human appetite and satiation, indexed by intake and eating pleasantness of starch-paired flavors. *Appetite, 3,* 163–184.

Booth, D. A., & Simson, P. C. (1971). Food preferences acquired by association with variations in amino acid nutrition. *Quarterly Journal of Experimental Psychology, 23,* 135–145.

Booth, D. A., Stoloff, R., & Nicholls, J. (1974). Dietary flavor acceptance in infant rats established by association with effects of nutrient composition. *Physiological Psychology, 2,* 313–319.

Brommage, R., & DeLuca, H. F. (1984). Self-selection of a high calcium diet by vitamin D-deficient lactating rats increases food consumption and milk production. *Journal of Nutrition, 114,* 1377–1385.

Cabanac, M. (1971). Physiological role of pleasure. *Science, 173,* 1103–1107.

Chitty, D., & Southern, H. N. (1954). *Control of rats and mice.* London: Oxford University Press.

Collier, G. (1983). Life in a closed economy: The ecology of learning and motivation. In M. D. Zeiler & P. Harzem (Eds.), *Advances in the analysis of behavior: Vol. 3. Biological factors in learning* (pp. 223–274). Chichester: John Wiley & Sons.

Contreras, R. J. (1977). Changes in gustatory nerve discharges with sodium deficiency: A single unit analysis. *Brain Research, 121,* 373–378.

Cowart, B. J., & Beauchamp, G. K. (1986). The importance of sensory context in young children's acceptance of salty tastes. *Child Development, 57,* 1034–1039.

Darwin, C. R. (1872). *The expression of emotions in man and animals.* London: John Murray. Reprinted Chicago: University of Chicago Press, 1965.

Davis, C. (1928). Self-selection of diets by newly weaned infants: An experimental study. *American Journal of Diseases of Children, 36,* 651–679.

Davis, C. (1939). Results of the self-selection of diets by young children. *Canadian Medical Association Journal, 41,* 257–261.

De Garine, I. (1972). The socio-cultural aspects of nutrition. *Ecology of Food and Nutrition, 1,* 143–163.

Denton, D. (1982). *The hunger for salt.* Berlin: Springer-Verlag.

Denton, D. A., & Sabine, J. R. (1963). The behaviour of Na deficient sheep. *Behaviour, 16,* 364–376.

Dethier, V. G. (1976). *The hungry fly.* Cambridge, MA: Harvard University Press.

Deutsch, J. A., Moore, B. O., & Heinrichs, S. C. (1989). Unlearned specific appetite protein. *Physiology and Behavior, 46,* 619–624.

Dobbing, J. (Ed.) (1988). *Sweetness.* London: Springer-Verlag.

Dombrowski, D. A. (1984). *The philosophy of vegetarianism.* Amherst, MA: University of Massachusetts Press.

Domjan, M. (1983). Biological constraints on instrumental and classical conditioning: Implications for general process learning theory. In G. H. Bower (Ed.), *Psychology of learning and motivation* (Vol. 17, pp. 215–277) New York: Academic Press.

Donhoffer, S. (1960). Spontaneous selection of food. *Triangle, 4,* 233–239.

Dove, W. F. (1935). A study of individuality in the nutritive instincts and of the causes and effects of variations in the selection of foods. *American Naturalist, 69,* 469–544.

Drewnowski, A., & Greenwood, M. R. C. (1983). Cream and sugar: Human preferences for high-fat foods. *Physiology and Behavior, 30,* 629–633.

Duncker, K. (1938). Experimental modifications of children's food preferences through social suggestion. *Journal of Abnormal and Social Psychology, 33,* 489–507.

Ekman, P., & Friesen, W. V. (1975). *Unmasking the face.* Englewood Cliffs, NJ: Prentice-Hall.

Elner, R. W., & Hughes, R. N. (1978). Energy maximization in the diet of the shore crab, *Carcinus maenus. Journal of Animal Ecology, 47,* 103–116.

Epstein, A. N., & Stellar, E. (1955). The control of salt preference in the adrenalectomized rat. *Journal of Comparative and Physiological Psychology, 48,* 167–172.

Evvard, J. M. (1915). Is the appetite of swine a reliable indication of physiological need. *Proceedings of the Iowa Academy of Science, 22,* 375–402.

Fallon, A. E., & Rozin, P. (1983). The psychological bases of food rejections by humans. *Ecology of Food and Nutrition, 13,* 15–21.

Freeland, W. J., & Janzen, D. H. (1974). Strategies in herbivory by mammals: The role of plant secondary compounds. *American Naturalist, 108,* 269–289.

Galef, B. G., Jr. (1988). Communication of information concerning distant diets in a social sentral-place foraging species: *Rattus norvegicus.* In T. Zentall & B. G. Galef, Jr. (Eds.), *Social learning: A comparative approach* (pp. 119–140). Hillsdale, NJ: Lawrence Erlbaum.

Galef, B. G., Jr. (1989). Enduring social enhancement of rats' preferences for the palatable and the piquant. *Appetite, 13,* 81–92.

Galef, B. G., Jr., Mason, J. R., Preti, G., & Bean, N. J. (1988). Carbon disulfide: A semiochemical mediating socially-induced diet choice in rats. *Physiology and Behavior, 42,* 119–124.

Garb, J., & Stunkard, A. J. (1974). Taste aversions in man. *American Journal of Psychiatry, 131,* 1204–1207.

Garcia, J., Ervin, F. R., & Koelling, R. A. (1966). Learning with prolonged delay of reinforcement. *Psychonomic Science, 5,* 121–122.

Garcia, J., Ervin, F. R., Yorke, C. H., & Koelling, R. A. (1967). Conditioning with delayed vitamin injection. *Science, 155,* 716–718.

Garcia, J., & Hankins, W. G. (1975). The evolution of bitter and the acquisition of toxipohobia. In D. Denton & J. P. Coghlan (Eds.), *Fifth international symposium on olfaction and taste* (pp. 1–12). New York: Academic Press.

Garcia, J., Hankins, W. G., & Rusiniak, K. W. (1974). Behavioral regulation of the milieu interne in man and rat. *Science, 185,* 824–831.

Garcia, J., Kimeldorf, D. J., & Hunt, E. L. (1961). The use of ionizing radiation as a motivating stimulus. *Psychological Review, 68,* 383–395.

Garcia, J., & Koelling, R. A. (1966). The relation of cue to consequence in avoidance learning. *Psychonomic Science, 5,* 123–124.

Garcia, J., & Rusiniak, K. W. (1980). What the nose learns from the mouth. In D. Muller-Schwarze & R. M. Silverstein (Eds.), *Chemical signals* (pp. 141–156). New York: Plenum Press.

Gibson, E. L., & Booth, D. A. (1986). Acquired protein appetite in rats: Dependence on a protein specific need state. *Experientia, 42,* 1003–1004.

Glander, K. E. (1975). Habitat description and resource utilization: A preliminary report on mantled howler monkey ecology. In R. H. Tuttle (Eds.), *Socioecology and psychology* (pp. 37–57). The Hague: Mouton.

Green, H. H. (1925). Perverted appetites. *Physiology Review, 5,* 336–348.

Grill, H. J., Berridge, K. C., & Ganster, D. J. (1984). Oral glucose is the prime elicitor of preabsorptive insulin secretion. *American Journal of Physiology, 246,* R88–R95.

Grill, H. J., & Norgren, R. (1978a). The taste reactivity test. I. Oro-facial responses to gustatory stimuli in neurologically normal rats. *Brain Research, 143,* 263–279.

Grill, H. J., & Norgren, R. (1978b). The taste reactivity test. II. Mimetic responses to gustatory stimuli in chronic thalamicand chronic decerebrate rats. *Brain Research, 143,* 281–297.

Harper, A. E. (1967). Effect of dietary protein content and amino acid pattern on food intake and preference. In C. F. Code (Eds.), *Handbook of physiology. Section 6. Alimentary canal. Vol. 1: Control of food and water intake* (pp. 399–410). Washington, DC: American Physiological Society.

Harris, L. J., Clay, J. Hargreaves, F., & Ward, A. (1933). Appetite and choice of diet: The ability of the vitamin B deficient rat to discriminate between diets containing and lacking the vitamin. *Proceedings of the Royal Society of London (Series B), 113,* 161–190.

Harris, M. (1985). *Good to eat: Riddles of food and culture.* New York: Simon & Schuster.

Harris, M. (1987). Foodways: Historical overview and theoretical prolegomenon. In M. Harris & E. B. Ross (Eds.), *Food and evolution: Toward a theory of human food habits* (pp. 57–90). Philadelphia: Temple University Press.

Herrnstein, R. J., Loveland, D. H., & Cable, C. (1976). Natural concepts in pigeons. *Journal of Experimental Psychology: Animal Behavior Processes, 2,* 285–302.

Herrnstein, R. J., & Vaughan, W. (1980). Melioration and behavioral allocation. In J. E. R. Staddon (Eds.), *Limits to action: The allocation of individual behavior* (pp. 143–176). New York: Academic Press.

Holman, E. (1975). Immediate and delayed reinforcers for flavor preferences in rats. *Learning and Motivation, 6,* 91–100.

Jacobs, K. M., Mark, G. P., & Scott, T. R. (1988). Taste responses in the nucleus tractus solitarius of sodium-deprived rats. *Journal of Physiology, 406,* 393–410.

Janzen, D. H. (1977). Why fruit rots, seeds mold and meat spoils. *American Naturalist, 111,* 691–713.

Jones, R. L., & Hanson, H. C. (1985). *Mineral licks: Geography and biogeochemistry of North American ungulates.* Iowa City: Iowa State University Press.

Katz, D. (1937). *Animals and men: Studies in comparative psychology.* London: Longmans.

Katz, S. H. (1982). Food, behavior, and biocultural evolution. In L. M. Barker (Ed.), *The psychobiology of human food selection* (pp. 171–188). Westport, CT: AVI.

Koh, S. D., & Teitelbaum, P. (1961). Absolute behavioral taste thresholds in the rat. *Journal of Comparative and Physiological Psychology, 54,* 223–229.

Kon, S. K. (1931). LVIII. The self-selection of food constituents by the rat. *Biochemical Journal, 25,* 473–481.

Krebs, J. R., & Davies, N. B. (Eds.). (1984). *Behavioral ecology* (2nd ed.). Oxford: Blackwell Scientific Publications.

Krecek, J., Novakova, V., & Stribal, K. (1972). Sex differences in the taste preference for a salt solution in the rat. *Physiology and Behavior, 8,* 183–188.

Krieckhaus, E. E. (1970). "Innate recognition" aids rats in sodium regulation. *Journal of Comparative and Physiological Psychology, 73,* 117–122.

Krieckhaus, E. E., & Wolf, G. (1968). Acquisition of sodium by rats: Interaction of innate mechanisms and latent learning. *Journal of Comparative and Physiological Psychology, 65,* 197–201.

Lat, J. (1967). Self-selection of dietary components. In C. F. Code (Ed.), *Handbook of physiology. Section 6. Alimentary canal. Vol. 1: Control of food and water intake* (pp. 367–386). Washington, DC: American Physiological Society.

LeMagnen, J. (1956). Hyperphagie provoquee chez le rat blanc par alteration du mecanisme de satiete peripherique. *Comptes Rendus de la Societe de Biologie, 150,* 32.

Lewis, M. (1968). Discrimination between drives for sodium chloride and calcium. *Journal of Comparative and Physiological Psychology, 65,* 208–212.

Li, E. T. S., & Anderson, G. H. (1983). Amino acids in the regulation of food intake. *Nutrition Abstracts and Reviews, 53,* 169–181.

Marriott, M. (1968). Caste ranking and food transactions: A matrix analysis. In M. Singer & B. S. Cohn (Eds.), *Structure and change in Indian society* (pp. 133–171). Chicago: Aldine.

Martin, I., & Levey, A. B. (1978). Evaluative conditioning. *Advances in Behavior Research and Therapy, 1,* 57–102.

Mead, M. (1943). The problem of changing food habits. *Bulletin of the National Research Council, 108,* 20–31.

Milner, P., & Zucker, P. (1965). Specific hunger for potassium in the rat. *Psychonomic Science, 2,* 17–18.

Mintz, S. W. (1985). *Sweetness and power.* New York: Viking Press.

Morgan, C. L. (1894). *An introduction to comparative psychology.* London: Walter Scott.

Morgan, C. T., & Stellar, E. (1950). *Physiological psychology.* New York: McGraw-Hill.

Musten, B., Peace, D., & Anderson, G. H. (1974). Food intake regulation in the weanling rat: Self-selection of protein and energy. *Journal of Nutrition, 104,* 563–572.

Nachman, M. (1962a). Taste preferences for sodium salts by adrenalectomized rats. *Journal of Comparative and Physiological Psychology, 55,* 1124–1129.

Nachman, M. (1963). Taste preferences for lithium chloride by adrenalectomized rats. *American Journal of Physiology, 205,* 219–221.

Nemeroff, C., & Rozin, P. (1989). An unacknowledged belief in "you are what you eat" among college students in the United States: An application of the demand-free "impressions" technique. *Ethos. The Journal of Psychological Anthropology, 17,* 50–69.

Nicolaidis, S., Danguir, J., & Mather, P. (1979). A new approach of sleep and feeding behaviors in the laboratory rat. *Physiology & Behavior, 23,* 717–722.

Nowlis, G. H. (1977). From reflex to representation: Taste-elicited tongue movements in the human newborn. In J. M. Weiffenbach (Eds.), *Taste and development. The genesis of sweet preference* (pp. 190–203). Bethesda, MD: U.S. Department of Health, Education & Welfare, Publication No. (NIH) 77-1068.

Osborne, T. B., & Mendel, L. B. (1918). The choice between adequate and inadequate diets, as made by rats. *Journal of Biological Chemistry, 35,* 19–27.

Pangborn, R. M. (1980). A critical analysis of sensory responses to sweetness. In P. Koivistoinen & L. Hyvonen (Eds.), *Carbohydrate sweeteners in foods and nutrition* (pp. 87–110). London: Academic Press.

Parker, L. A. (1982). Nonconsummatory and consummatory behavioral CRs elicited by lithium- and amphetamine-paired flavors. *Learning and Motivation, 13,* 281–303.

Pelchat, M. L., Grill, H. J., Rozin, P., & Jacobs, J. (1983). Quality of acquired responses to tastes by *Rattus norvegicus* depends on type of associated discomfort. *Journal of Comparative Psychology, 97,* 140–153.

Pelchat, M. L., & Rozin, P. (1982). The special role of nausea in the acquisition of food dislikes by humans. *Appetite, 3,* 341–351.

Pfaffmann, C. (1960). The pleasures of sensation. *Psychological Review, 67,* 253–268.

Pliner, P. (1982). The effects of mere exposure on liking for edible substances. *Appetite, 3,* 283–290.

Remington, R. E. (1936). The social origins of dietary habits. *Scientific Monthly,* 193–204.

Renner, H. D. (1944). *The origin of food habits.* London: Faber & Faber.

Revusky, S. H. (1967). Hunger level during food consumption: Effects on subsequent preference. *Psychonomic Science, 7,* 109–110.

Revusky, S. H., & Bedarf, E. W. (1967). Association of illness with prior ingestion of novel foods. *Science, 155,* 219–220.

Richter, C. P. (1936). Increased salt appetite in adrenalectomized rats. *American Journal of Physiology, 115,* 155–161.

Richter, C. P. (1939). Salt taste thresholds of normal and adrenalectomized rats. *Endocrinology, 24,* 367–371.

Richter, C. P. (1943). Total self regulatory functions in animals and human beings. *Harvey Lecture Series, 38,* 63–103.

Richter, C. P. (1953). Experimentally produced reactions to food poisoning in wild and domesticated rats. *Annals of the New York Academy of Sciences, 56,* 225–239.

Richter, C. P. (1956). Salt appetite of mammals: Its dependence on instinct and metabolism. *L'instinct dans le comportement des animaux et de l'homme* (pp. 577–629). Paris: Masson.

Richter, C. P., & Barelare, B., Jr. (1938). Nutritional requirements of pregnant and lactating rats studied by the self-selection method. *Endocrinology, 23,* 15–24.

Richter, C. P., & Eckert, J. F. (1937). Increased calcium appetite of parathyroidectomized rats. *Endocrinology, 21,* 50–54.

Richter, C. P., & Hawkes, C. D. (1941). The dependence of the carbohydrate, fat and protein appetite of rats on the various components of the vitamin B complex. *American Journal of Physiology, 131,* 639–649.

Richter, C. P., Holt, L. E., & Barelare, B., Jr. (1937). Vitamin B craving in rats. *Science, 86,* 354–355.

Richter, C. P., Holt, L. E., Jr., & Barelare, B., Jr. (1938). Nutritional requirements for normal growth and reproduction in rats studied by the self-selection method. *American Journal of Physiology, 122*, 734–744.

Richter, C. P., & Rice, K. K. (1945). Self-selection studies on coprophagy as a source of vitamin B complex. *American Journal of Physiology, 143*, 344–354.

Rodgers, W. L. (1967). Specificity of specific hungers. *Journal of Comparative and Physiological Psychology, 64*, 49–58.

Rodgers, W., & Rozin, P. (1966). Novel food preferences in thiamine deficient rats. *Journal of Comparative & Physiological Psychology, 61*, 1–4.

Rolls, B. J., Hetherington, M., Burley, V. J., & van Duijvenvoorde, P. M. (1986). Changing hedonic responses to foods during and after a meal. In M. A. Kare & J. G. Brand (Eds.), *Interaction of the chemical senses with nutrition* (pp. 247–268). New York: Academic Press.

Rosenstein, D., & Oster, H. (1988). Differential facial responses to four basic tastes in newborns. *Child Development, 59*, 1555–1568.

Rozin, E. (1982). The structure of cuisine. In L. M. Barker (Ed.), *The psychobiology of human food selection* (pp. 189–203). Westport, CT: AVI.

Rozin, P. (1965). Specific hunger for thiamine: Recovery from deficiency and thiamine preference. *Journal of Comparative and Physiological Psychology, 59*, 98–101.

Rozin, P. (1967a). Thiamine specific hunger. In C. F. Code (Ed.), *Handbook of physiology. Section 6. Alimentary canal. Vol. 1: Control of food and water intake* (pp. 411–432). Washington, DC: American Physiological Society.

Rozin, P. (1967b). Specific aversions as a component of specific hungers. *Journal of Comparative and Physiological Psychology, 64*, 237–242.

Rozin, P. (1968a). Specific aversions and neophobia as a consequence of vitamin deficiency and/or poisoning in half-wild and domestic rats. *Journal of Comparative and Physiological Psychology, 64*, 237–242.

Rozin, P. (1968b). Are carbohydrate and protein intakes separately regulated? *Journal of Comparative and Physiological Psychology, 659*, 23–29.

Rozin, P. (1969). Adaptive food sampling patterns in vitamin deficient rats. *Journal of Comparative and Physiological Psychology, 69*, 126–132.

Rozin, P. (1976a). The selection of foods by rats, humans, and other animals. In J. Rosenblatt, R. A. Hinde, C. Beer, & E. Shaw (Eds.), *Advances in the study of behavior* (Vol. 6, pp. 21–76). New York: Academic Press.

Rozin P. (1976b). Psychobiological and cultural determinants of food choice. In T. Silverstone (Ed.), *Appetite and food intake* (pp. 285–312). Berlin: Dahlem Konferenzen.

Rozin, P. (1977). The significance of learning mechanisms in food selection: Some biology, psychology and sociology of science. In L. M. Barker, M. Best, & M. Domjan (Eds.), *Learning mechanisms in food selection* (pp. 557–589). Waco, TX: Baylor University Press.

Rozin, P. (1982). Human food selection: The interaction of biology, culture and individual experience. In L. M. Barker (Ed.), *The psychobiology of human food selection* (pp. 225–254). Westport, CT: AVI.

Rozin, P. (1988). Social learning about foods by humans. In T. Zentall & B. G. Galef, Jr. (Eds.), *Social learning: A comparative approach* (pp. 165–187). Hillsdale, NJ: Lawrence Erlbaum.

Rozin, P. (1990a). Getting to like the burn of chili pepper. Biological, psychological and cultural perspectives. In B. G. Green, J. R. Mason, & M. R. Kare (Eds.), *Chemical senses*, Vol. 2: *Irritation* (pp. 231–269). Potomac, MD: Lawrence Erlbaum.

Rozin, P. (1990b). Social and moral aspects of food and eating. In I. Rock (Ed.), *The legacy of Solomon E. Asch: Essays in cognition and social psychology* (pp. 97–110). Potomac, MD: Lawrence Erlbaum.

Rozin, P., & Fallon, A. E. (1980). Psychological categorization of foods and nonfoods: A preliminary taxonomy of food rejections. *Appetite, 1*, 193–201.

Rozin, P., & Fallon, A. E. (1987). A perspective on disgust. *Psychological Review, 94*, 23–41.

Rozin, P., & Kalat, J. W. (1971). Specific hungers and poison avoidance as adaptive specializations of learning. *Psychological Review, 78*, 459–486.

Rozin, P., & Pelchat, M. (1988). Memories of mammaries: Adaptations to weaning from milk in mammals. In A. N. Epstein & A. Morrison (Eds.) *Advances in psychobiology* (Vol. 13, pp. 1–29). New York: Academic Press.

Rozin, P., & Rodgers, W. (1967). Novel diet preferences in vitamin deficient rats and rats recovered from vitamin deficiency. *Journal of Comparative and Physiological Psychology, 63*, 421–428.

Rozin, P., & Schiller, D. (1980). The nature and acquisition of a preference for chili pepper by humans. *Motivation and Emotion, 4*, 77–101.

Rozin, P., & Schull, J. (1988). The adaptive-evolutionary point of view in experimental psychology. In R. C. Atkinson, R. J. Herrnstein, G. Lindzey, & R. D. Luce (Eds.), *Handbook of experimental psychology* (pp. 503–546). New York: Wiley-Interscience.

Rozin, P., & Vollmecke, T. A. (1986). Food likes and dislikes. *Annual Review of Nutrition, 6*, 433–456.

Rozin, P., Wells, C., & Mayer, J. (1964). Specific hunger for thiamine: Vitamin in water versus vitamin in food. *Journal of Comparative and Physiological Psychology, 57*, 78–84.

Rozin, P., & Zellner, D. A. (1985). The role of Pavlovian conditioning in the acquisition of food likes and dislikes. *Annals of the New York Academy of Sciences, 443*, 189–202.

Rzoska, J. (1953). Bait shyness, a study in rat behavior. *British Journal of Animal Behaviour, 1*, 128–135.

Schulkin, J. (1982). Behavior of sodium deficient rats: The search for a salt taste. *Journal of Comparative and Physiological Psychology, 96*, 628–634.

Schulkin, J. (1986). The evolution and expression of salt appetite. In G. deCaro, A. N. Epstein, & M. Massi (Eds.), *The physiology of thirst and sodium appetite* (pp. 491–496). New York: Plenum Press.

Schutz, H. G., & Judge, D. S. (1984). Consumer perceptions of food quality. In J. V. McLoughlin & B. M. McKenna (Eds.), *Research in food science and nutrition. Vol. 4: Food science and human welfare* (pp. 229–242). Dublin: Boole Press.

Sclafani, A. (1987). Carbohydrate taste, appetite and obesity: An overview. *Neuroscience and Biobehavioral Reviews, 11*, 131–153.

Sclafani, A., & Nissenbaum, J. W. (1988). Robust conditioned flavor preference produced by intragastric starch infusions in rats. *American Journal of Physiology, 255*, R672–R675.

Scott, E. M. (1946). Self selection of diet. I. Selection of purified components. *Journal of Nutrition, 31*, 397–405.

Scott, E. M., & Quint, E. (1946a). Self selection of diet. III. Appetite for B vitamins. *Journal of Nutrition, 32*, 285–291.

Scott, E. M., & Quint, E. (1946b). Self selection of diet. IV. Appetite for protein. *Journal of Nutrition, 32*, 293–301.

Scott, E. M., & Verney, E. L. (1947). Self selection of diet. VI. The nature of appetites for B vitamins. *Journal of Nutrition, 34*, 471–480.

Scott, E. M., & Verney, E. L. (1949). Self selection of diet. IX. The appetite for thiamine. *Journal of Nutrition, 37*, 81–91.

Scott, E. M., Verney, E. L., & Morissey, P. D. (1950). Self selection of diet. XII. Effects of B vitamin deficiencies on selection of food components. *Journal of Nutrition, 41*, 373–382.

Scott, T. R., & Mark, G. P. (1986). Feeding and taste, *Progress in Neurobiology, 27*, 293–317.

Seligman, M. E. P. (1970). On the generality of the laws of learning. *Psychological Review, 77*, 406–418.

Shettleworth, S. J. (1972). (1972). Constraints on learning. In D. S. Lehrman, R. A. Hinde, & E. Shaw (Eds.), *Advances in the study of behavior* (vol. 4, pp. 1–68). New York: Academic Press.

Shettleworth, S. J., & Krebs, J. R. (1982). How marsh tits find their hoards. The roles of site preference and spatial memory. *Journal of Experimental Psychology: Animal Behavior Processes, 8*, 354–375.

Simoons, F. J. (1982). Geography and genetics as factors in the psychobiology of human food selection. In L. M. Barker (Eds.), *The psychobiology of human food selection* (pp. 205–224). Westport, CT: AVI.

Smith, J. C. (1971). Radiation: Its detection and its effect on taste preferences. In E. Stellar & J. M. Sprague (Eds.), *Progress in physiological psychology* (pp. 53–118). New York: Academic Press.

Steiner, J. E. (1979). Human facial expressions in response to taste and smell stimulation. In H. W. Reese & L. P. Lipsitt (Eds.), *Advances in child development and behavior* (vol. 13, pp. 257–295). New York: Academic Press.

Stellar, E. (1974). Brain mechanisms in hunger and other hedonic experiences. *Proceedings of the American Philosophical Society, 118*, 276–282.

Stricker, E. M., & Sterritt, G. M. (1967). Osmoregulation in the newly hatched domestic chick. *Physiology and Behavior, 2*, 117–119.

Tordoff, M. G., & Friedman, M. I. (1986). Hepatic portal glucose infusions decrease food intake and increase food preference. *American Journal of Physiology, 251* (Regulatory Integrative Comparative Physiology, 20), R192–R196.

Verbalis, J. G., McHale, C. M., Gardiner, T. W., & Stricker, E. M. (1986). Oxytocin and vasopressin secretion in response to stimuli producing learned taste aversions in rats. *Behavioral Neuroscience, 100*, 466–475.

Wilkins, L., & Richter, C. P. (1940). A great craving for salt by a child with cortico-adrenal insufficiency. *Journal of the American Medical Association, 114*, 866–868.

Wolf, G. (1969). Innate mechanisms for regulation of sodium intake. In C. Pfaffmann (Ed.), *Olfaction and taste* (pp. 548–553). New York: Rockefeller University Press.

Young, P. T. (1948). Appetite, palatability and feeding habit: A critical review. *Psychological Bulletin, 45*, 289–320.

Young, P. T. (1961). *Motivation and emotion. A survey of the determinants of human and animal activity.* New York: John Wiley & Sons.

Young, P. T., & Chaplin, J. P. (1945). Studies of food preference, appetite and dietary habit: III. Palatability and appetite in relation to bodily need. *Comparative Psychology Monographs, 18*, 1–45.

Zahorik, D. M., Maier, S. F., & Pies, R. W. (1974). Preferences for tastes paired with recovery from thiamine deficiency in rats: Appetitive conditioning or learned safety. *Journal of Comparative and Physiological Psychology, 87*, 1083–1091.

Zajonc, R. B. (1968). Attitudinal effects of mere exposure. *Journal of Personality and Social Psychology, 9* (part 2), 1–27.

Zellner, D. A., Berridge, K. C., Grill, H. J., & Ternes, J. W. (1985). Rats learn to like the taste of morphine. *Behavioral Neuroscience, 99*, 290–300.

Zellner, D. A., Rozin, P., Aron, M., & Kulish, C. (1983). Conditioned enhancement of human's liking for flavors by pairing with sweetness. *Learning and Motivation, 14*, 338–350.

13

Diet Selection and Poison Avoidance by Mammals Individually and in Social Groups

Bennett G. Galef, Jr. and Matthew Beck

Introduction

There is great diversity in the feeding behavior of mammals. Some are essentially monophagous, feeding on a single food or class of foods; others are eclectic in their ingestive behavior, composing exceptionally varied diets.

True ingestive specialists, such as vampire bats (*Desmodus rotundus*), sustained entirely by mammalian blood, acquire nutritionally balanced diets simply by identifying and then eating their staple food. Omnivores cannot compose an adequate diet so easily. Be they men, rats, gulls, or roaches, omnivores must ingest a suitable mix of different foods if they are to thrive outside the laboratory.

Lát (1967) has proposed that the need for omnivores to ingest selectively a number of different foods arises from the fact that, for an omnivore, there is no single natural food that contains adequate amounts of all the constituents required for optimal living. The same fact, that most free-living omnivores live in areas in which no single food provides an adequate diet, can be viewed quite differently. Development of an ability to compose an adequate diet by selectively eating a number of different foods allows omnivores to occupy areas in which no single satisfactory food is available and in which no monophage could sur-

BENNETT G. GALEF, JR. AND MATTHEW BECK   Department of Psychology, McMaster University, Hamilton, Ontario L8S 4K1, Canada.

vive. Omnivory thus both broadens the range of environments in which mammals can live and requires greater sophistication in selective ingestion than does monophagy.

Because ingestive specialists can meet their nutritional requirements by identifying and ingesting a single class of foods, they have little need to contact or ingest potential toxins. Omnivores, sampling broadly among foods in an attempt to locate sources of needed nutrients, should be exposed to any toxic agents in an environment more frequently than are specialists. In consequence, life as an omnivore could be facilitated not only by the ability to choose a balanced diet but also by the ability to identify and reject toxic substances (Rozin, 1976b). Each omnivore must steer a narrow course between the Scylla of dietary insufficiency and the Charybdis of ingestion-induced toxicosis, neither selecting such a restricted range of foods that it fails to include all required nutrients nor sampling so widely that it ingests harmful quantities of toxins.

Selecting an adequate diet and avoiding ingestion of toxins, the two main behavioral challenges faced by omnivores as a consequence of their omnivory, usually have been treated as separate problems. Investigators working on diet selection and those studying poison avoidance have developed different research paradigms, contributed to relatively independent literatures, and employed different theoretical perspectives.

In a landmark series of papers, Rozin (1965, 1967a, 1967b, 1968, 1976a, 1976b; Rozin & Kalat, 1971) provided an integrative framework that, quite unlike Richter's (1943) earlier model, permitted discussion of diet selection and poison avoidance as a unitary problem. Richter (1943) had hypothesized that each of the dozens of nutrients required by omnivores is regulated by an innate mechanism that detects a specific deficiency state, detects a needed nutrient in foods, and motivates its ingestion. Rozin proposed that although intake of three critical dietary elements (e.g., water, sodium, and calories) was controlled in the manner Richter had suggested, intake of most nutrients was not so controlled. On Rozin's model, sensitivity to the novelty or familiarity of a food, together with a capacity to respond behaviorally to the physiological consequences of eating a food, underlies both selection of an adequate mix of unregulated nutrients and avoidance of repeated ingestion of toxins. For example, an omnivore maintained on an inadequate food develops an aversive deficiency state that is similar to toxicosis. This aversive deficiency state produces a learned aversion to the taste of the inadequate diet. Conversely, eating an isolated meal of an unfamiliar food that produces recovery from illness leads to a learned preference for the flavor of the unfamiliar food.

If diet choice is to be optimized, consequences of ingestion must be associated with food-related stimuli despite long delays between eating a food and experience of consequences. Normally, association of stimuli with their consequences does not occur when they are separated by more than a few seconds. Rozin argued from data provided by Garcia and his co-workers (Garcia & Koelling, 1966, 1967; Garcia, Kimeldorf, & Hunt, 1961) that the learning of dietary aversions and preferences rests on two adaptive specializations of a more general Pavlovian learning process: first, attribution of changes in internal state to the consequences of ingestion ("belongingness"), and, second, a capacity to tolerate long delays between ingestion and the onset of the consequences of ingestion ("long-delay learning"). On this view, belongingness (Revusky, 1971) and long-delay learning have evolved expressly to facilitate solution of the problems of diet selection and poison avoidance faced by omnivores in natural habitat.

Rozin's model of food choice explains the results of many laboratory experiments in which rats succeeded in avoiding repeated ingestion of toxic foods or in selecting from among several diets the sole diet permitting redress of a deficiency state. Although we accept Rozin's general view of the causes of rats' successes both in selecting adequate diets and in avoiding poisons, we believe that it is incomplete in one important respect.

Rozin's model was designed to explain how omnivores select adequate diets and avoid eating poisons; it neither pays much attention to nor offers any explanation of their failures. We argue below that consideration of failures, both of diet selection in the laboratory and of poison avoidance in the field, may be of considerable importance in understanding the capacity of omnivores to select foods successfully in either setting.

It is important to keep in mind, especially when discussing Norway rats or other mammalian species capable of rapid rates of population growth (Pianka, 1970), that individuals often fail to respond adequately when challenged by the environment. Failure to find shelter, failure to avoid predators, failure to find safe and adequate food, failure to resist disease, failure of some sort results in the death of more than 90% of rats before they reproduce (Brooks, 1973). Such failures can be as instructive as less frequent successes in understanding the feeding-related behavioral capacities of omnivores. Focus on failure may also provide a useful cautionary message in this age of Panglossian explanations of apparently ubiquitous behavioral optimality. Animals fail when their behavioral capacities are insufficient to respond to environmental demands, and failure is common.

DIET SELECTION

CAN RATS SELF-SELECT ADEQUATE DIETS?

It has long been known that rats choosing from a cafeteria of purified nutrients can select a diet adequate for normal rates of growth. In a classic study, Richter, Holt, and Barelare (1938) found that rats offered 11 containers, each containing a different, relatively purified nutrient (casein, dextrose, olive oil, baker's yeast, water, salt mix, etc.) gained weight more rapidly and on fewer calories than did a control group of rats maintained on a standard laboratory chow. Young (1944) confirmed this result in groups of 12 rats, each group having access to the 11-food cafeteria used by Richter *et al.* (1938). Clearly, under some circumstances, rats can self-select an adequate diet from among a variety of potential foods.

Richter had expected his subjects to be successful in the cafeteria feeding situation. "The survival of animals and humans in the wild state in which the diet had to be selected from a variety of beneficial, useless and even harmful substances *is proof* [emphasis added] of this ability . . . to make dietary selections which are conducive to normal growth and development" (Richter *et al.*, 1938, p. 734). In Richter's view, the laboratory data illustrated a capacity for diet selection that could be deduced from the survival of omnivores in the natural world. Consequently, when omnivores failed to self-select an adequate diet in some experimental situation, the failure was interpreted by Richter as the result of an artifact of one sort or another: the use of complex natural foods (Richter *et al.*, 1938), inherited defects of the sensory system in domesticated subjects, age-related exhaustion of regulatory functions, or, in humans, perverse cultural influences (Richter, 1943).

Yet, review of the literature on dietary self-selection by rats in cafeteria feeding situations reveals that failure, as indicated by markedly slowed growth, is at least as common as success (Kon, 1931; Scott, 1946; Pilgrim & Patton, 1947; see Epstein, 1967; Lát, 1967, for reviews). Such failures to self-select, together with the theoretical considerations discussed below, suggest that Richter's deduction from the fact of the persistence of omnivores in nature, of a capacity for dietary self-selection in cafeteria feeding situations, was overdrawn.

Members of any species, even those as cosmopolitan in distribution as Norway rats or "primitive" *H. sapiens*, are not found everywhere within their respective species' ranges. By definition, individuals can survive only in those portions of the environment that both provide all resources necessary for life and lack insurmountable threats. An area would be devoid of rats if it contained either lethal substances that rats were unable to learn to avoid or a necessary nutrient only in a form that rats couldn't learn to eat. Existence of an omnivorous species in nature tells us little about the range of environments in which species members have the ability to self-select nutritionally adequate, safe diets. Persistence of omnivores outside the laboratory shows only that there exist portions of the environment where the behavioral capacities of species members are sufficient to permit development of a dietary repertoire conducive to self-maintenance, growth, and reproduction. In particular, existence of an omnivorous species in nature does not suggest that in the laboratory omnivores should be able to select nutritionally adequate diets from among a cafeteria of purified dietary components.

CAUSES OF FAILURE IN SELF-SELECTION

In discussing selection of foods containing nutrients other than salt, neither Rozin (1976b) nor Richter paid much attention to the relative palatability of foods. The flavor of a food is not necessarily a reliable indicator of its nutritive content. Consequently, selection of foods on the basis of their flavors cannot explain successful diet selection. Palatability may, however, be an important determinant of failure of self-selection (Richter, 1943). As Epstein (1967, p. 201) concluded in his review of the literature on dietary self-selection by rats, "Rats tend to eat what they like, and what they like is determined largely by the palatabilities of the foods they are offered." In laboratory cafeteria feeding studies, growth rate has been determined largely by protein intake (Lát, 1967). As Scott (1946, p. 403) indicated, "[individual] rats either do or do not like casein; if they like it, they eat an average of 3 grams a day and grow well; if they do not, they eat less than 0.1 gram per day, lose weight, and die within a short period."

Such laboratory findings can be extrapolated, with caution, to extralaboratory environments. How successful animals are in selecting an adequate diet will depend on their intake of the relatively less palatable available foods that contain necessary nutrients. If an animal eats enough of a relatively unpalatable food to allow evaluation of the consequences of eating that food, then the problem of dietary self-selection can, at least potentially, be solved. If an animal cannot overcome its initial aversion to the taste of some food so as to eat sufficient amounts of the food to permit its evaluation, then a capacity to learn about consequences of ingestion will be of no use.

The question "Can omnivores self-select adequate diets?" is unanswerable in that form. In benign ingestive environments, such as the one constructed by Richter *et al.* (1938), the relative palatabilities of the foods provided may lead directly to selection of an adequate diet by all individuals. In the Richter *et al.* (1938) study, the presence both of multiple sources of protein and of a relatively unpalatable carbohydrate (Epstein, 1967) probably rendered the problem immediately soluble (Scott & Quint, 1946). In less benign environments, spontaneous sampling, coupled with response to positive postingestional consequences, suffices to permit most subjects to overcome their initial avoidance of the least-palatable needed food or foods. In yet more challenging situations, no individual's spontaneous behavior may be sufficient to overcome the flavor-induced avoidance of one or more needed foods.

Failure to solve a diet-selection problem can result from the unpalatability of foods containing one or more necessary nutrients limiting ingestion of valuable foods to the point that discovery of their beneficial properties is unlikely. Results of several classic studies of diet selection in which rats were taught to solve cafeteria feeding problems they did not solve spontaneously suggest that rats often fail to solve problems that are, in principle, soluble. For example, when six of 12 rats selecting among four purified diet components (casein, sugar, salt mix, and fat) steadily lost weight because of low protein intake, McDonald, Stern, and Hahn (1963) fed them only casein for 4 weeks and then returned them to the choice situation. All subsequently ate casein and showed excellent weight gain. Similarly Harris, Clay, Hargreaves, and Ward (1933) found that vitamin-B-deficient rats, unable to select a single vitamin-B-rich food from among ten foods, could solve the problem after 2 or 3 days of access only to that food rich in vitamin B. The initial problem was not insoluble. Rather, the rats' behavioral proclivities were not sufficient to produce an ingestive pattern leading to a solution.

Contradictory outcomes both within and between studies of dietary self-selection are consistent with the view that both the particular foods offered to subjects and individual differences in dietary preference are significant determinants of success. One should not ask whether rats can select adequate diets but rather inquire as to the characteristics of foods that affect the probability of adaptive dietary selection by individuals with differing dietary preferences. In benign ingestive environments, like that of Richter *et al.* (1938), the palatability spectra of many animals may be sufficient to produce a relatively immediate solution. In less benign environments, spontaneous sampling and sensitivity to the consequences of ingestion can sometimes produce solutions. In harsh environments, animals are unable spontaneously to create situations leading to identification of needed nutrients.

The results of Davis' (1928) pioneering studies of diet selection by human infants, like the results of Richter's studies of dietary self-selection by rats in cafeteria feeding situations, have frequently been interpreted as providing evidence that naive omnivores can instinctively select a well-balanced diet from a cafeteria (Story & Brown, 1987). Davis herself, however, was well aware that human infants could self-select an adequate diet only when nutritious unsweetened foods were offered as choices. As Davis (1939, p. 261) stated, "Self-selection can have no, or but doubtful value, if the diet must be selected from inferior foods." In humans, as in rats, the capacity to self-select adequate diets depends on the diets offered for choice. The provision of an assortment of nutritious, roughly equipalatable foods rather than a capacity to self-select needed foods probably underlay much of the success of Davis' subjects (Rozin, 1976b).

Are Cafeteria Feeding Studies Adequate Analogues of Diet Selection in Natural Habitat? A Developmental Perspective

For the first few weeks or months of life, young mammals are ingestive specialists, sustained by a single nutritionally adequate diet, mother's milk. At some point in development, each individual's caloric needs outstrip either the energy transduction capacities of its dam (Babicky, Ostradalova, Parizek, Kolar, & Bibr, 1970) or her willingness to invest further resources (Trivers, 1974). It is at this point in development, at weaning, that each young omnivore must undertake the transition to omnivory, the potentially arduous task of developing an adequate diet of solid foods. Although adults may be challenged occasionally by a failure of one or another resource on which they have come to depend, every weanling must deal with withdrawal of its major source of sustenance. Thus, one would expect weanlings to solve laboratory analogues of diet selection problems that occur in nature.

It is, therefore, surprising that weanling rats presented with a self-selection task in the laboratory, even one that adults solve easily, often fail dismally in maintaining their body weight. Tribe (1954, 1955) presented 15 100-g female rats with seven foods (corn starch, glucose, margarine, yeast, casein, salt mix, and cod liver oil). Thirteen solved the problem and grew, but only two of ten rat pups survived placement "at weaning" in the same situation. Scott, Smith, and Verney (1948) offered 12- to 15-week-old rats a choice of but four foods [casein, vegetable oil, sucrose, and salt mix (vitamins were fed by pill)]. Thirteen of 20 12- to 15-week-olds gained weight, and all survived, but only nine of 31 weanlings survived in the same situation. Kon (1931) offered 28-day-old rats three dietary elements (casein, sucrose, and salt mix) supplemented with vitamins by hand. Two of his four weanling subjects died, and one gained no weight for 7 weeks. Such failures of weanling rats to solve dietary selection problems far simpler than one might expect them to face in natural circumstances suggest that the laboratory analogue of dietary selection, the cafeteria feeding experiment, does not capture some important aspect of dietary self-selection as it occurs outside the laboratory.

Social Solutions to Self-Selection Problems

If the ability to self-select an adequate diet is typical of weaning omnivores, then the not infrequent failure of weanling rats to survive diet selection studies in the laboratory poses a problem. Either the genotypes of domestic strains of Norway rat are deficient, the environment in which rats are maintained in the laboratory is so deviant as to preclude the development of a species-typical behavioral capacity, or the cafeteria feeding situation is not an appropriate analogue of the problem faced by weanling rats in natural circumstances.

One unchanging feature of the environment in which all mammals that survive to weaning age develop, absent in all reported laboratory studies of dietary self-selection, is the presence of a dam who, by her very existence and reproductive success, has demonstrated the adequacy of her dietary selections. Data from our laboratory suggest that the presence of an adult that has solved the problem of diet selection can be a major factor in the success of juveniles in composing an adequate diet.

In a recent experiment (Beck & Galef, 1989), we presented individual weanling rats with a cafeteria of four distinctively flavored diets. Three of these diets contained inadequate levels of protein (5%), and one had ample protein (20%) for the support of normal growth. We found (see Figure 1), as had Kon (1931), Scott *et al.* (1948), and Tribe (1954, 1955), that our weanling subjects did very poorly in such a situation. None was able to develop a preference for the protein-adequate diet. Each pup appeared well on its way to a premature demise if we had not terminated the experiment. Weanling rats faced with the same diet selection problem while in the presence of adults previously trained to ingest the protein-rich alternative grew rapidly in the experimental situation. Clearly, information acquired from a knowledgeable adult permitted weanlings to select an adequate diet in what we assume is the normal, species-typical fashion of rats.

Our analysis of the behavioral processes involved in social enhancement of dietary self-selection by weanling rats in a cafeteria situation is still in its early stages (Beck & Galef, 1989). However, enough is known about mechanisms of social influence on diet choice in young rats more generally to propose two alternative pictures of how the results shown in Figure 1 might be produced. First, juvenile rats prefer to eat at locations where adults are feeding rather than at locations that adults are ignoring (Galef & Clark, 1971). The simple presence of an adult rat at a feeding site attracts young to that site and causes them to begin feeding there (Galef, 1971, 1981). Adult rats also deposit residual olfactory cues around and in feeding sites they use, and these residual cues attract weanlings and cause them to eat where adults have eaten (Galef, 1986a; Galef & Beck, 1985; Galef & Heiber, 1983). Such social influences on diet selection are in a sense indirect. The presence or activity of adults at a feeding site attracts young and increases the probability that juveniles will eat the food to be found at the socially enhanced site (Galef, 1977). There is, however, no direct specification of what food to eat.

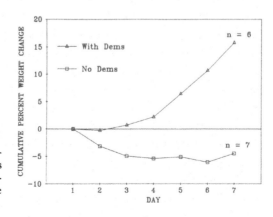

Figure 1. Mean cumulative percentage weight change seen in juvenile rats choosing among four diets in the presence and absence of knowledgeable demonstrators.

There is, in addition, considerable evidence of social effects acting directly on diet selection by naive rats. If one rat (a "demonstrator") eats a diet and then interacts with a naive rat (an "observer"), the observer rat subsequently exhibits a markedly enhanced preference for the diet its demonstrator ate (Galef & Wigmore, 1983; Posadas-Andrews & Roper, 1983).

In sum, a naive rat pup can exhibit socially enhanced intake of a diet for one of two reasons: either an adult induced the pup to begin feeding at a location where a particular food is found, or an adult directly induced an enhanced preference in the pup for a food that the adult was eating. Either direct or indirect induction of eating of the protein-rich diet could be responsible for the social facilitation of protein ingestion shown in Figure 1.

The fragility of weanling rats and their frequent failure to self-select adequate diets when alone in laboratory cafeteria feeding studies has imposed a research strategy on laboratory investigators of dietary self-selection by rats. Investigators extend the young rat's period of total dependence on others for diet selection by weaning pups from mother's milk to a nutritionally adequate chow compounded by nutritionists. Weaning from milk to chow maintains animals in a state of naiveté with respect to problems of nutrient selection. Adults maintained on chow from weaning and faced for the first time with a need to self-select nutrients have served as model systems for studying a process that normally occurs at the time of weaning from mother's milk.

Naive adult rats, like naive weanlings, often have difficulty in solving even relatively simple diet-selection problems. McDonald *et al.* (1963) offered 12 individually housed 90-day-old rats a four-choice cafeteria (sugar, salt mix, casein, and vegetable oil), and six lost weight. Similarly, Schutz and Pilgrim (1954) found that 21 of 43 young adult rats were losing weight 3 weeks after being offered a four-choice cafeteria similar to those used by Scott (1946) and Pilgrim and Patton (1947). Rozin (1968) offered ten thiamine-deficient rats a choice among three novel diets only one of which contained thiamine, and four of his ten subjects failed to select the thiamine-rich food.

Adult rats, like weanlings, will show facilitation of diet selection when in the presence of knowledgeable adult conspecifics. As can be seen in Figure 2, when individually housed adult rats were offered a choice of four novel diets, only one

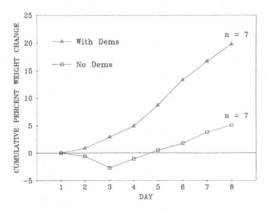

Figure 2. Mean cumulative percentage weight change seen in adult rats choosing among four diets in the presence and absence of knowledgeable demonstrators.

of which contained adequate protein, their weight gain was poor. Adult rats housed with a conspecific trained to eat the protein-rich diet gained weight rapidly.

From the point of view of a naive individual, an environment containing conspecifics that have already solved the problem of selecting an adequate diet is benign in comparison with the same environment lacking social sources of information. Rats that cannot select an adequate diet when alone in a cafeteria feeding situation do well in the presence of a successful other. The presence of knowledgeable conspecifics in demanding environments can expand the range of situations in which naive individuals are able to choose the foods they need to survive.

Poison Avoidance

As noted in the introduction to the present chapter, a specialist feeder may limit its encounters with most poisonous substances simply by eating a restricted range of foods. For a koala, the simple rule "eat it if it tastes and smells like eucalyptus, don't eat it if it doesn't" would go a long way both toward solving the problem of diet selection and toward reducing the probability of eating most environmental toxins. Omnivores, on the other hand, sampling widely among foods to locate needed nutrients, greatly increase their probability of eating poisons.

The congenital tendency of many mammals to reject bitter-tasting foods has been discussed as an adaptive response to flavors characteristic of many naturally occurring toxins (Garcia & Hankins, 1975; Young, 1968). Although some toxins could be avoided simply by rejecting bitter-tasting foods, the importance of palatability in identification and rejection of poisonous foods by free-living animals has been, so far as we can determine, very poorly explored (see, for review, Fowler, 1983). It is easy to imagine circumstances in which simple palatability-based avoidance of toxins would be sufficient to protect individuals against feeding on deleterious substances. On the other hand, it seems likely that for an omnivorous species to colonize areas containing diverse toxins, an ability to learn to recognize and avoid ingestion of not-unpalatable dangerous foods would be advantageous. Rozin (1976a) and others have suggested that the tendency of rats and other omnivores to take small initial meals of unfamiliar foods (Barnett, 1958; Rzoska, 1953) is a potentially useful tactic in minimizing ingestion of possible poisons. Similarly, belongingness and long-delay learning are capacities of possible use both in identifying toxic foods and in avoiding repeated ingestion of them.

As in diet selection, environmental characteristics will determine the behavioral sophistication needed for successful poison avoidance. In benign environments, a simple tendency to eat what tastes good might suffice. In more complex or demanding situations, tentative sampling of unfamiliar foods and sensitivity to their postingestional consequences could be useful. The question "How do animals avoid ingesting lethal quantities of toxic substances?" like the question "Can omnivores select nutritionally adequate diets?" is unanswerable in that form. It depends on the animal, the toxins, and the environment in which the toxins and the animal are located.

BENNETT G.
GALEF, Jr., AND
MATTHEW BECK

In study of taste-aversion learning, attention has focused on the ability of rats to learn to avoid repeated ingestion of good-tasting baits with long-delayed aversive postingestional consequences, what one of us has previously called "cryptic poison baits" (Domjan & Galef, 1983). Although such baits are used by exterminators to control rodent pests, there is some reason to question whether they are in any way typical of naturally evolved toxins (Domjan & Galef, 1983). Prey species that evolve means to manufacture or sequester toxins do so, at least in part, to deter their potential predators or consumers. Evolution of palatable toxins with delayed effects seems to us less likely than evolution of vile-tasting, fast-acting poisons. Immediately perceived unpalatability and rapid induction of illness or pain should be more reliable deterrents to potential predators than palatability coupled with long-delayed negative aftereffects. Rats would not have evolved an ability to cope with "cryptic" toxic foods unless the natural habitat provided exposure to cryptic toxic foods.

Review of the literature suggests that, because each rat has a variety of behavioral tactics to avoid ingesting lethal quantities of poison bait, killing free-living rats by poisoning them is a formidable task. It is somewhat surprising to find that, although total extermination of a population of rats by poisoning is difficult, killing 80% to 90% of a target population is not hard (Meehan, 1984). Even if one is quite crude in one's approach, the majority of rats usually succumb to an introduced poison. Chitty (1954), for example, found that poison baits, introduced without prebaiting into censused colonies of wild rats, typically eliminated 75% of colony members (median success in poisoning 37 colonies). Further, there was no evidence that survivors in Chitty's study (or in others) had learned to avoid eating the poison bait: perhaps some individuals found the bait unpalatable, were exceptionally neophobic, were unusually resistant to the poison, or failed to encounter the bait. The ability to learn to avoid a cryptic toxin may be even less well developed than Chitty's (1954) data suggest.

When exposed to cryptic toxins, both rats and mice (Tevis, 1956) die in large numbers. There is no need to explain how rodents survive human attempts to poison them. In general, they don't. By increasing the probability that rats will consume a lethal quantity of poison prior to the onset of symptoms of toxicosis (by using palatable baits, by prebaiting, by using highly toxic rodenticides with delayed onset of symptoms, etc.), one can create an environment in which individual rats have a very low probability of survival. Of course, if one's goal is total extermination of a rodent population, survival of even one pregnant female means failure.

ARE LABORATORY STUDIES OF TASTE-AVERSION LEARNING ADEQUATE ANALOGUES OF POISON AVOIDANCE OUTSIDE THE LABORATORY?

During the 20 years since discovery of the exceptional properties of taste–toxicosis conditioning in rats (Garcia & Koelling, 1966; Garcia, Ervin, & Koelling, 1966), there has been extensive discussion of the relationship between the characteristics of taste-aversion learning revealed by laboratory studies and the requirements for successful poison avoidance in the natural habitat (e.g., Rozin & Kalat, 1971; Shettleworth, 1984; Zahorik & Houpt, 1981). Treatment of taste–toxicosis conditioning as an adaptively specialized learning process shaped by selective pressures exerted on free-living rats by environmental toxins has been of heuristic value, both orienting research on taste–toxicosis conditioning in fruitful directions and promoting integration of field and laboratory studies of animal learning. As was the case in the development of models of diet selection, in discussions of poison avoidance attention has been focused on the success of omnivores. Once again, we shall focus on failure because such focus calls attention to often-ignored issues.

ADAPTIVE PATTERNS OF DIETARY SAMPLING. Laboratory-maintained rats, sustained throughout life on a single food and exposed to a single novel food prior to toxicosis induction, form an apparently adaptive aversion to the sole novel element in their diets. However, in the more complex feeding environments presumably found outside the laboratory, the decision as to what food to avoid after toxicosis is more difficult. If rats faced with a choice among novel foods ate discrete meals, each composed of a single food, and waited long enough between such meals to evaluate the postingestional consequences of each (e.g., Zahorik & Houpt, 1981), then one could with some confidence extrapolate from the simplified laboratory situation to the more complex outside world. If, to the contrary, rats tend to sample several unfamiliar foods in rapid succession, then the problem of toxin identification after onset of toxicosis is insoluble without further sampling of suspect foods.

It is commonly believed that rats faced with a number of new foods eat discrete meals of each unfamiliar, potentially dangerous food. Such discrete sampling would permit ready identification of a toxic food should illness occur after a meal of unfamiliar food. However, the few studies actually describing the behavior of rats the first time they encounter a number of novel foods do not support the assertion that rats sample discretely among them. Rozin's (1969) data, most frequently cited as demonstrating discrete sampling of novel foods by rats, in fact demonstrate the opposite. Five of the six subjects whose behavior Rozin described in detail ate either two or all three of the novel foods presented during the first 30 minutes of exposure to them.

Barnett (1956) reported that laboratory-born descendants of wild Norway rats, like Rozin's domesticated rats, tended to sample each of four novel foods available to them during their first feeding bout. It is difficult to see how a rat sampling several different novel foods upon first encounter with them could subsequently identify a toxic item if one were present (see also Beck, Hitchcock & Galef, 1988).

POISON-AVOIDANCE LEARNING OR DEVELOPMENT OF "FOOD PHOBIAS." Taste–toxicosis association learning over long delays has been extensively discussed as an adaptive specialization of a more general Pavlovian conditioning process. Implicit in this view is the assumption that, on average, it is beneficial for organisms outside the laboratory to learn an aversion to a novel food ingested several hours before an experience of nausea. However, learning of an aversion after a single pairing of novel taste and toxicosis carries potential costs as well as potential benefits. Chance ingestion of a food, particularly an unfamiliar one, in the hours prior to the onset of a bout of gastrointestinal distress induced by factors (bacterial, viral, or organic) unrelated to ingestion would result in learning of a maladaptive "food phobia" rather than learning of an adaptive aversion to a toxic agent. Indeed, if occurrence of gastrointestinal upset unrelated to ingestion is randomly distributed in time, the greater the tolerance for delays between experience of a taste and onset of illness in aversion learning, the greater the probability of learning maladaptive food phobias.

In the laboratory, rats are protected from many naturally occurring sources of malaise. Consequently, if a rat is fed an unfamiliar food and becomes ill, it appears obviously adaptive for the rat to learn an aversion to the unfamiliar food. In environments where occurrence of nausea is sometimes related to eating of a novel food and sometimes not, the utility of the acquisition of aversions to foods may be a less straightforward matter. Arguments to the effect that because rats can learn aversions over long delays the capacity to do so must be adaptive are only assertions without evidence that such learning actually enhances survival in natural environments.

We are ignorant of the frequency with which rats in natural habitat eat unfamiliar toxic foods and suffer toxicosis. We also do not know how often rats outside the laboratory eat unfamiliar safe foods and subsequently experience gastrointestinal upset. Hence, the usefulness, on average, of the capacity to learn aversions in a single trial over long delays between feeding and onset of illness remains speculative.

Outside the laboratory, the learning of "food phobias" as the result of fortuitous temporal associations of ingestion and illness is not a hypothetical event. Logue, Ophir, and Strauss (1981) asked 517 college undergraduates to describe their learned aversions. Respondents reported 23 aversions to eggs, 18 to shellfish, seven to organ meats, six each to Chinese food and mushrooms, three to hamburgers, etc., and none to true toxins. Similarly, M. L. Pelchat (personal communication, 1986), in recollecting her subjects' responses to questionnaires on learned flavor aversions (Pelchat & Rozin, 1982), could not recall a single instance of a learned aversion to a toxic substance other than alcohol. Even if the particular sample of egg or hot dog eaten by those who formed aversions to those foods was tainted and the cause of subsequent nausea, the learned aversion resulted in a long-lasting "food phobia" rather than adaptive avoidance of a true toxin. Although it is possible that some of these aversions to normally safe foods were, in fact, adaptive, perhaps the result of idiosyncratic allergic reactions to foods that most people can eat without ill effect, it seems reasonable to suggest that many were not adaptive. In humans there are costs of one-trial taste–toxicosis conditioning over long delays.

It might be proposed that the tendency to form aversions to relatively novel rather than to familiar foods, demonstrated many times in the laboratory (e.g., Kalat & Rozin, 1973; Revusky & Bedarf, 1967; Siegel, 1974), would provide protection against learning fortuitous, maladaptive aversions to safe foods. However, self-reports of behavior outside the laboratory are not consistent with such a view. Hot dogs, eggs, and hamburgers are unlikely to have been novel to those who formed aversions to them. Foods "known" to be safe are, at least occasionally, excluded from future ingestion by humans as the result of eating of a single tainted sample or happenstantial temporal pairing of ingestion with nausea unrelated to ingestion.

Although extrapolation from man to rat is as fraught with potential for error as is extrapolation in the opposite direction, the literature concerning taste-aversion learning in humans suggests that the learning of aversions after a single pairing of taste and toxicosis is not necessarily adaptive. Rats living in areas containing few toxic potential ingesta and providing frequent exposure to nausea-inducing bacteria, viruses, or parasites could gain little if anything from the capacity to form aversions in a single trial over long delays. Those rats occupying niches containing many toxic foods and providing few other sources of malaise could gain a great deal from the same capacity.

The point of the preceding arguments is not that taste-aversion learning is maladaptive or that individuals are unable to learn selectively to avoid naturally occurring toxins outside the laboratory. Obviously, under many conditions, animals can learn to avoid toxins, and taste–toxicosis conditioning is responsible for development of adaptive patterns of poison avoidance, Rather, we would conclude that the abilities of individual organisms to learn to identify and avoid toxins are far from perfect. Consequently, it is possible that individuals might benefit from learning processes other than taste–toxicosis conditioning in developing aversions to toxic foods.

SOCIAL SOLUTIONS TO SOME PROBLEMS IN POISON AVOIDANCE. In the literature, models of poison avoidance rest on five laboratory or field observations: (1) hesitancy of rats (particularly wild rats) to ingest unfamiliar foods (Barnett, 1958; Rzoska, 1953), (2) a tendency of rats to sample among several unfamiliar foods in such a way as to permit independent evaluation of each (Rozin, 1969; Zahorik & Houpt, 1981), (3) a tendency to attribute aversions to unfamiliar rather than familiar foods (Kalat & Rozin, 1973; Revusky & Bedarf, 1967), (4) attribution of illness to foods rather than other aspects of the environment (Garcia & Koelling, 1966), and (5) a capacity to form flavor–illness associations in spite of long delays between experience of an unfamiliar flavor and onset of illness (Garcia et al., 1966). Understanding of the processes of poison avoidance, like understanding of the process of diet selection, has been focused on the ability of individuals to identify appropriate substances for rejection. As we have already noted, outside the laboratory, rats live as members of social groups during part, if not all, of their lives. Many remain in their respective colonies from birth to death (Telle, 1966). Even those rats that disperse to become founders of new populations are obliged to spend their infancy and adolescence with their dams. Information acquired by any one colony member concerning the safety or toxicity of potential foods would be of use to all.

Review of the five capacities listed above, believed to be important in permitting individual rats to learn to avoid repeated ingestion of toxins, reveals that three of the five involve recognition of a potential food as novel. Weanling rats emerge from the nest into a totally unfamiliar environment without information permitting focus of concern on one or a few unfamiliar foods. Yet weanlings are in greatest need of strategies for developing safe, nutritionally adequate diets. The problem of identifying toxic foods appears more severe for naive weanlings than for knowledgeable adults that have developed inventories of known, safe foods to eat.

There is a simple strategy that would permit weanling rats to thrive in unfamiliar habitat. A new recruit to a population, whether a weanling or recent immigrant, could "assume" that living conspecifics had not eaten lethal doses of any poison food present in their shared environment. Naive individuals could also "assume" that, with high probability, senior colony members had already learned to avoid eating any noxious but nonlethal substances present in the environment. Thus, a naive individual could solve the problem of poison avoidance, as it could solve the problem of diet selection, by eating whichever foods adults of its group were eating. The result would be an avoidance of poisons dependent on socially acquired rather than individually acquired information. The same processes that could facilitate selection of nutritionally adequate diets could simultaneously preclude ingestion of toxins. Indeed, in stable environments, where the same foods were available for generations, the simple rule, eat what others are eating, would suffice for both nutrient selection and poison avoidance.

Information garnered from conspecifics could be used in a number of different ways to facilitate avoidance of poisons. First, as indicated above, an individual could avoid eating toxic foods simply by eating those foods others of its social group were eating. The results of a number of studies indicate that socially induced preference for one food results in avoidance by weanlings of potentially dangerous alternatives present in the environment. In wild rats, such socially induced preferences have resulted in total avoidance by weanlings of both feeding sites and foods that adults of their colony have learned to avoid (Galef & Clark, 1971).

Not only can socially acquired information play an indirect role in poison avoidance by directing ingestive behavior toward safe foods, socially transmitted information about foods can act to influence directly the course of poison-avoidance learning. As mentioned above, results of experiments in several laboratories have shown that interaction of a naive rat (an observer) with a conspecific previously fed some food (a demonstrator) results in substantial enhancement of the observer's preference for its demonstrator's diet (Galef & Wigmore, 1983; Posadas-Andrews & Roper, 1983; Strupp & Levitsky, 1984). Such socially mediated exposure to foods that others have eaten can act, as does actual ingestion of a diet (Revusky & Bedarf, 1967; Rozin & Kalat, 1971; Siegel, 1974), to attenuate subsequent learning of an aversion to socially experienced diets.

Consider a situation in which some members of a rat colony are exploiting a palatable, safe, nutritious food that one member of the colony has not yet eaten. The naive individual encounters the novel food for the first time, ingests it, and an hour or two later becomes ill, perhaps because the particular sample of novel food it ate was spoiled, perhaps because of a viral, bacterial, or parasitic infection unrelated to ingestion of the novel food. On the individual learning model, the naive individual, eating an unfamiliar food and becoming ill, should develop a profound aversion to the unfamiliar food. The naive individual would develop a "food phobia," foreclosing for some time exploitation of a potentially valuable source of nutrition. If, to the contrary, the naive individual could make use of socially acquired information, indicating that others are eating the unfamiliar food, the naive individual might ignore its own experience of toxicosis and continue to eat the food its fellows are eating.

In a recent experiment, Galef (1989) permitted naive, 23-hour food-deprived rats (observers) to interact for 20 minutes with either one (1-dem group) or two (2-dem group) conspecific demonstrators that had recently eaten a highly palatable diet (Normal Protein Test Diet, Teklad, Madison, WI, referred to below as Diet NPT). The hungry observers then were fed Diet NPT for 15 minutes and, immediately thereafter, were injected with LiCl solution to induce an aversion to Diet NPT. After a 24-hour period of recovery from illness, each observer was given a 22-hour choice between Diet NPT (the averted diet) and an unfamiliar, relatively unpalatable diet. As can be seen in Figure 3, many observ-

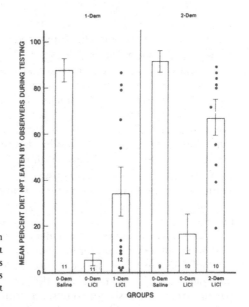

Figure 3. Mean amount of Diet NPT eaten by observers as a percentage of total amount consumed by observers during testing. Flags on histograms represent 1 S.E.M. Numbers within histograms are *n*s per group. See text for explanation of groups.

ers in both 1-dem and 2-dem groups failed to exhibit an aversion to Diet NPT. Prior interaction of an observer with a demonstrator or demonstrators that had eaten Diet NPT attenuated subsequent learning of an aversion to Diet NPT. Socially mediated dietary information had the capacity to prevent formation of a "food phobia."

Subjects in one control group (labeled group 0-dem, LiCl in Figure 3) were treated exactly as were observers in the experimental group described above but received no exposure to demonstrators fed Diet NPT. As one would expect, subjects in this control group exhibited profound aversions to Diet NPT. Subjects in a second control group (0-dem, Sal in Figure 3) were treated as were subjects in the first control group but were injected with isotonic saline solution rather than LiCl solution following ingestion of Diet NPT. These subjects exhibited a marked preference for Diet NPT.

Individual data points in Figure 3 reveal that behavior of observers in the experimental groups was highly variable. Some observers were unaffected by socially mediated exposure to Diet NPT and exhibited strong aversion to it. Aversion learning by other observers was completely blocked as a result of prior exposure to conspecifics that had eaten Diet NPT. Although the causes of such variability need to be determined, the data clearly indicate that socially mediated exposure to a diet can profoundly attenuate subsequent learning of an aversion to that diet. Rats may be substantially less likely to learn an aversion to a diet others of their social group have eaten than to totally unfamiliar foods.

This failure of rats to learn aversions to foods that conspecifics have eaten has been the subject of a number of recent studies in our laboratory. Two findings are of particular relevance to the present discussion.

First, approximately 50% of rats that ate a novel food (Diet NPT), were injected with LiCl, and therefore had learned an aversion to Diet NPT abandoned that aversion after a 15-minute period of interaction with each of two conspecific demonstrators that had eaten Diet NPT prior to interaction with averted subjects (Galef, 1985, 1986c).

Second, observer rats that interacted with a demonstrator recently fed one of two foods (either Diet N_1 or N_2), both unfamiliar to the observer, ate both Diets N_1 and N_2, and then suffered toxicosis, preferentially formed aversions to that novel food their respective demonstrators had not eaten. That is, those observers whose respective demonstrators had eaten Diet N_1 formed an aversion to Diet N_2; those observers whose demonstrators had eaten Diet N_2 formed an aversion to Diet N_1 (Galef, 1986b, 1987). Within this paradigm, substantial attenuation of learning of aversions by observers to the diets eaten by their respective demonstrators was seen even if interaction with a demonstrator fed Diet N_1 or N_2 occurred 8 days prior to observer sampling of Diets N_1 and N_2, toxicosis induction, and testing for aversion learning (Galef, 1987).

We know that the messages passing from demonstrators to observers, affecting the latters' aversion learning, can be olfactory (Galef & Wigmore, 1983; Posadas-Andrews & Roper, 1983). Because simply eating a diet does not enhance a subject's subsequent preference for that diet, whereas exposure to the same diet on a demonstrator does enhance an observer's subsequent preference for that diet (Galef, Kennett, & Stein, 1985), we suspect that each olfactory message passing from demonstrator to observer consists of two parts: first, an olfactory signal identifying the diet eaten by a demonstrator, and, second, a pheromone emitted by demonstrators that acts in concert with the diet-identifying olfactory signal to alter an observer's subsequent choice of food (Galef et al., 1985; Galef & Stein, 1985). The diet-identifying component of a message is simply the smell of a food a demonstrator has eaten diffusing both from food particles clinging to a demonstrator's pelage and from the demonstrator's digestive tract (Galef et al., 1985).

The contextual component of a message appears to be composed of volatile sulfur compounds, particularly carbon disulfide (CS_2), a chemical we have found on the breath of rats (Bean, Galef, & Mason, 1988; Galef, Mason, Preti, & Bean, 1988). Behavioral data support the contention that CS_2 may be an important component of messages that enhance diet preference in rats. Rats prefer foods moistened with CS_2 solution to those moistened with water (B. G. Galef, Jr., unpublished data). Rats preexposed to a diet moistened with a few drops of a dilute aqueous CS_2 solution, but not rats preexposed to the same diet moistened with an equal quantity of water, subsequently exhibit attenuation of aversion learning to the preexposed diet (Galef et al., 1988).

Socially mediated exposure to diets provides potential solutions, albeit imperfect ones, to both acquisition of "food phobias" and failure to discretely sample several unfamiliar foods. Interaction with conspecifics that have eaten an unfamiliar food can attenuate aversion learning to that unfamiliar food should chance illness strike during the hours after it was eaten. Encounters with conspecifics that have eaten a food have the potential to reverse a previously learned aversion to the food.

One can picture a naive rat as utilizing information extending far beyond its own experience of the consequences of ingestion of particular foods in deciding which foods to eat and which to avoid. A naive individual can exploit information acquired from its fellows both to select needed nutrients and to reduce the probability of excluding beneficial diets from its own feeding repertoire.

It is tempting to interpret the results of studies of social mediation of taste-aversion learning as suggesting that the ability of rats to avoid eating lethal doses of toxin is even more perfect than previously suspected. However, use of socially mediated information concerning foods, like individual taste–toxicosis association learning, can lead to error. Consider a rat that has eaten a lethal quantity of a toxic bait with effects delayed by several hours. During the interval between eating the lethal dose of toxin and death, the doomed individual could lead its fellows both to eat the toxic bait and to fail to form an aversion to the poison bait if they ate sublethal but illness-provoking first meals of it. There are rarely potential benefits without potential costs.

CONCLUSION

Because omnivores often live in environments where they must both select a mix of foods to formulate a balanced diet and learn to avoid repeated ingestion of any toxins they encounter, it has long been implicitly assumed that individual omnivores are adept at both diet selection and poison avoidance. Consequently, studies demonstrating the success of individual omnivores, either in selecting balanced diets or avoiding toxins, have received far more attention than studies demonstrating failure; the former studies appear to reflect real-world competence, whereas the latter appear to reflect laboratory artifacts.

The main argument of the present chapter is that failure of individual omnivores to self-select adequate diets or correctly identify toxins tells us as much about their behavioral capacities as does their success in diet selection or poison avoidance. The behavioral tactics available to individual omnivores for dealing with dietary insufficiencies or environmental poisons are adequate to cope with some environments but not with others. There is no simple answer to the question, "can omnivores avoid poisons or select balanced diets?" It depends on the omnivore, and it depends on the environment in which that omnivore is found. In benign environments, the sensory–affective systems of an animal may be sufficient to lead it to eat a balanced, safe mix of ingesta. In more demanding environments, the capacity of individuals to evaluate foods may be inadequate to the task of selecting a safe, balanced diet. Laboratory analogues both of benign and of demanding environments are informative.

The secondary argument of the present chapter is that, as social animals, omnivores can expand the range of environments in which they successfully avoid toxins and select balanced diets. An area inhabited by rats that have solved the nutritional problems posed by the foods found there may be far less challenging to naive newcomers than the same area would be in the absence of knowledgeable inhabitants. It is, for example, easy to imagine an area that contains so many palatable protein-deficient foods that only one in 100 naive rats introduced into that area would compose a protein-adequate diet before expiring. Yet one would not be at all surprised to find that all young born to a female that had located a reliable protein source in our imaginary area would thrive. The continued presence of groups of omnivores in an environment can not be used to infer that a naive individual would have the ability to survive there alone.

From the perspective of the present chapter, omnivores appear both better and worse at dietary selection and poison avoidance than they do from other perspectives. The individual omnivore is seen as less sophisticated in its abilities to find needed nutrients and avoid poisons; the social omnivore is seen as able to maintain populations in areas where the survival of a naive individual would be unlikely.

Evolution by natural selection is a process that, over generations, stabilizes unlikely events in a species' gene pool. Social learning can play a similar role at the behavioral level on a far shorter time scale. If naive individuals incorporate into their own behavioral repertoires the unlikely acquired behaviors of successful colleagues, they can thrive in environments that otherwise would be closed to them. The data suggest that social learning can augment individual learning both about nutritious, safe foods and about toxins and can thus expand the range of environments in which populations of omnivores can succeed.

W. C. Allee, who studied ways sociality could enhance fitness and the presence of conspecifics could render environmental challenges less severe, suggested that "many animals change or 'condition' an unfavorable medium so that others following or associating with them can survive better and thrive when they could not do so in a raw, unconditioned medium" (Allee, 1958, p. 210). The present review leads to a similar conclusion. The behavior of successful individuals in an area provides information to naive individuals that can lead them to succeed in selecting adequate diets and avoiding toxins where they otherwise would fail.

Acknowledgments

Preparation of this chapter was supported by grants from the Natural Sciences and Engineering Research Council of Canada and the McMaster University Research Board. We thank Harvey Weingarten, Mertice Clark, Paul Rozin, Evelyn Satinoff, and Edward Stricker for helpful comments on earlier drafts.

REFERENCES

Allee, W. C. (1958). *The social life of animals* (rev. ed.). Boston: Beacon Press.

Babicky, A., Ostradalova, I., Parizek, J., Kolar, J., & Bibr, B. (1970). Use of radioisotope techniques for determining the weaning period in experimental animals. *Physiologica Bohemoslovica, 19,* 457–467.

Barnett, S. A. (1956). Behaviour components in the feeding of wild and laboratory rats. *Behaviour, 9,* 24–42.

Barnett, S. A. (1958). Experiments on "neophobia" in wild and laboratory rats. *British Journal of Psychology, 49,* 195–201.

Bean, N. J., Galef, B. G., Jr., & Mason, R. J. (1988). At biologically significant concentrations, carbon disulfide both attracts mice and increases their consumption of bait. *Journal of Wildlife Management, 13,* 95–97.

Beck, M., & Galef, B. G., Jr. (1989). Social influences on the selection of a protein-sufficient diet by Norway rats (*Ratus norvegicus*). *Journal of Comparative Psychology, 103,* 132–139.

Beck, M., Hitchcock, C., & Galef, B. G., Jr. (1988). Diet sampling by wild Norway rats (*Ratus norvegicus*) offerered several unfamiliar foods. *Animal Learning & Behavior, 16,* 224–230.

Brooks, J. E. (1973). A review of commensal rodents and their control. *C. R. C. Critical Reviews in Environmental Control, 3,* 405–453.

Chitty, D. (1954). The study of the brown rat and its control by poison. In D. Chitty (Ed.), *Control of rats and mice,* Vol. 1 (pp. 160–305). Oxford: Clarendon Press.

Davis, C. M. (1928). Self-selection of diet by newly weaned infants: An experimental study. *American Journal of Diseases of Children, 36,* 651–679.

Davis, C. M. (1939). Results of the self-selection of diets by young children. *Canadian Medical Association Journal, 41,* 257–261.

Domjan, M., & Galef, B. G., Jr. (1983). Biological constraints on instrumental and classical conditioning: Retrospect and prospect. *Animal Learning and Behavior, 11,* 151–161.

Epstein, A. N. (1967). Oropharyngeal factors in feeding and drinking. In C. F. Code (Ed.), *Handbook of physiology, Vol. 1: Alimentary Canal* (pp. 197–218). Washington, DC: American Physiological Society.

Fowler, M. E. (1983). Plant poisoning in free-living wild animals: A review. *Journal of Wildlife Diseases, 19,* 34–43.

Galef, B. G., Jr. (1971). Social effects in the weaning of domestic rat pups. *Journal of Comparative and Physiological Psychology, 75,* 358–362.

Galef, B. G., Jr. (1977). Mechanisms for the transmission of acquired patterns of feeding from adult to juvenile rats. In L. M. Barker, M. R. Best, & M. Domjan (Eds.), *Learning mechanisms in food selection* (pp. 123–148). Waco, TX: Baylor University Press.

Galef, B. G., Jr. (1981). The development of olfactory control of feeding site selection in rat pups. *Journal of Comparative and Physiological Psychology, 95,* 615–622.

Galef, B. G., Jr. (1985). Socially induced diet preference can partially reverse a LiCl induced diet aversion. *Animal Learning and Behavior, 13,* 415–418.

Galef, B. G., Jr. (1986a). Olfactory communication among rats: Information concerning distant diets. In D. Duvall, D. Muller-Schwarze, & R. M. Silversted (Eds.), *Chemical signals in vertebrates, Vol. IV: Ecology, evolution, and comparative biology* (pp. 481–505). New York: Plenum Press.

Galef, B. G., Jr. (1986b). Social identification of toxic diets by Norway rats (*R. norvegicus*). *Journal of Comparative Psychology, 100,* 331–334.

Galef, B. G., Jr. (1986c). Social interaction substantially modifies learned aversions, sodium appetite, and both palatability and handling-time induced dietary preference in rats (*R. norvegicus*). *Journal of Comparative Psychology, 100,* 432–439.

Galef, B. G., Jr. (1987). Social influences on the identification of toxic foods by Norway rats. *Animal Learning and Behavior, 15,* 327–332.

Galef, B. G., Jr. (1989). Socially-mediated attenuation of taste-aversion learning in Norway rats: Preventing development of food phobias. *Animal Learning and Behavior, 17,* 468–474.

Galef, B. G., Jr., & Beck, M. (1985). Aversive and attractive marking of toxic and safe foods by Norway rats. *Behavioral and Neural Biology, 43,* 298–310.

Galef, B. G., Jr., & Clark, M. M. (1971). Social factors in the poison avoidance and feeding behavior of wild and domesticated rat pups. *Journal of Comparative and Physiological Psychology, 75.* 341–357.

Galef, B. G., Jr., & Heiber, L. (1976). The role of residual olfactory cues in the determination of feeding site selection and exploration patterns of domestic rats. *Journal of Comparative and Physiological Psychology, 90,* 727–739.

Galef, B. G., Jr., Kennett, D. J., & Stein, M. (1985). Demonstrator influence on observer diet preference: Effects of simple exposure and the presence of a demonstrator. *Animal Learning and Behavior, 13,* 25–30.

Galef, B. G., Jr., Mason, J. R., Preti, G., & Bean, N. J. (1988). Carbon disulfide can mediate socially-induced attenuation of taste-aversion learning in rats. *Physiology and Behavior, 42,* 119–124.

Galef, B. G., Jr., & Stein, M. (1985). Demonstrator influence on observer diet preference: Analysis of critical social interactions and olfactory signals. *Animal Learning and Behavior, 13,* 31–38.

Galef, B. G., Jr., & Wigmore, S. W. (1983). Transfer of information concerning distant food in rats: A laboratory investigation of the "information centre" hypothesis. *Animal Behaviour, 31,* 748–758.

Garcia, J., Ervin, F. R., & Koelling, R. A. (1966). Learning with prolonged delay of reinforcement. *Psychonomic Science, 5,* 121–122.

Garcia, J., & Hankins, W. G. (1975). The evolution of bitter and the acquisition of toxiphobia. In D. A. Denton & J. P. Coghan (Eds.), *Olfaction and taste V* (pp. 39–46). Academic Press: New York.

Garcia, J., Kimeldorf, D. J., & Hunt, E. L. (1961). The use of ionizing radiation as a motivating stimulus. *Psychological Review, 68,* 383–395.

Garcia, J., & Koelling, R. A. (1966). Relation of cue to consequence in avoidance learning. *Psychonomic Science, 4,* 123–124.

Garcia, J., & Koelling, R. A. (1967). A comparison of aversions induced by X-rays, toxins and drugs in the rat. *Radiation Research Supplement, 1,* 439–50.

Harris, L., Clay, J., Hargreaves, F., & Ward, A. (1933). The ability of vitamin B deficient rats to discriminate between diets containing and lacking the vitamin. *Proceedings of the Royal Society* (Section B), *113,* 161–190.

Kalat, J. W., & Rozin, P. (1973). "Learned safety" as a mechanism in long-delay taste-aversion learning in the rat. *Journal of Comparative and Physiological Psychology, 83*, 198–207.

Kon, S. K. (1931). LVIII. The self-selection of food constituents by the rat. *Biochemical Journal, 25*, 473–481.

Lát, J. (1967). Self-selection of dietary components. In C. F. Code (Ed.), *Handbook of physiology, Vol. 1: Alimentary canal* (pp. 367–386). Washington DC: American Physiological Society.

Logue, A. W., Ophir, I., & Strauss, K. E. (1981). The acquisition of taste aversions in humans. *Behavior Research and Therapy, 19*, 319–333.

McDonald, D. G., Stern, J. A., & Hahn, W. W. (1963). Effects of differential housing and stress on diet selection, water intake and body weight in the rat. *Journal of Applied Physiology, 18*, 937–942.

Meehan, A. P. (1984). *Rats and mice: Their biology and control.* Tonbridge, Kent: Brown, Knight & Truscott.

Pelchat, M. L., & Rozin, P. (1982). The special role of nausea in the acquisition of food dislikes by humans. *Appetite, 3*, 341–351.

Pianka, E. R. (1970). On r- and k-selection. *American Naturalist, 104*, 592–597.

Pilgrim, F. J., & Patton, R. A. (1947). Patterns of self-selection of purified dietary components by the rat. *Journal of Comparative and Physiological Psychology, 40*, 343–348.

Posadas-Andrews, A., & Roper, T. J. (1983). Social transmission of food preferences in adult rats. *Animal Behaviour, 31*, 265–271.

Revusky, S. H. (1971). The role of interference in association over a delay. In W. Honig & H. James (Eds.), *Animal memory* (pp. 155–213). New York: Academic Press.

Revusky, S. H., & Bedarf, E. W. (1967). Association of illness with prior ingestion of novel foods. *Science, 155*, 219–220.

Richter, C. P. (1943). Total self regulatory functions in animals and human beings. *Harvey Lecture Series, 38*, 63–103.

Richter, C. P., Holt, L., & Barelare, B. (1938). Nutritional requirements for normal growth and reproduction in rats studied by the self-selection method. *American Journal of Physiology, 122*, 734–744.

Rozin, P. (1965). Specific hunger for thiamine: Recovery from deficiency and thiamine preference. *Journal of Comparative and Physiological Psychology, 59*, 98–101.

Rozin, P. (1967a). Specific aversions as a component of specific hungers. *Journal of Comparative and Physiological Psychology, 64*, 237–242.

Rozin, P. (1967b). Thiamine specific hunger. In C. F. Code (Ed.), *Handbook of physiology, Vol. 1: Alimentary Canal* (pp. 411–431). Washington, DC: American Physiological Society.

Rozin, P. (1968). Specific aversions and neophobia resulting from vitamin deficiency or poisoning in half-wild and domestic rats. *Journal of Comparative and Physiological Psychology, 66*, 126–132.

Rozin, P. (1969). Adaptive food sampling patterns in vitamin deficient rats. *Journal of Comparative and Physiological Psychology, 69*, 126–132.

Rozin, P. (1976a). The significance of learning mechanisms in food selection: Some biology, psychology, and sociology of science. In L. M. Barker, M. R. Best, & M. Domjan (Eds.), *Learning mechanisms in food selection.* Waco, TX: Baylor University Press.

Rozin, P. (1976b). The selection of foods by rats, humans and other animals. In J. S. Rosenblatt, R. A. Hinde, E. Shaw, & C. Beer (Eds.), *Advances in the study of behavior (Vol. 6* pp. 21–26). New York: Academic Press.

Rozin, P., & Kalat, J. W. (1971). Specific hungers and poison avoidance as adaptive specializations of learning. *Psychological Review, 78*, 459–486.

Rzoska, J. (1953). Bait shyness, a study in rat behaviour. *British Journal of Animal Behaviour, 1*, 128–135.

Schutz, H. G., & Pilgrim, F. J. (1954). Changes in the self-selection pattern for purified dietary components by rats after starvation. *Journal of Comparative and Physiological Psychology, 47*, 444–449.

Scott, E. M. (1946). Self-selection of diet. I. Selection of purified components. *Journal of Nutrition, 31*, 397–406.

Scott, E. M., & Quint, E. (1946). Self-selection of diet. IV. Appetite for protein. *Journal of Nutrition, 32*, 293–301.

Scott, E. M., Smith, S., & Verney, E. (1948). Self-selection of diet. VII. The effect of age and pregnancy on selection. *Journal of Nutrition, 35*, 281–286.

Seligman, M. E. P. (1970). On the generality of the laws of learning. *Psychological Review, 77*, 406–418.

Shettleworth, S. J. (1984). Learning and behavioural ecology. In J. R. Krebs & N. B. Davies (Eds.), *Behavioural ecology,* 2nd ed. Sunderland, MA: Sinauer.

Siegel, S. (1974). Flavor pre-exposure and "learned safety." *Journal of Comparative and Physiological Psychology, 87,* 1073–1082.

Story, M., & Brown, J. E. (1987). Do young children instinctively know what to eat? The studies of Clara Davis revisited. *New England Journal of Medicine, 316,* 103–106.

Strupp, B. J., & Levitsky, D. A. (1984). Social transmission of food preference in adult hooded rats (*R. norvegicus*). *Journal of Comparative Psychology, 98,* 257–266.

Telle, H. J. (1966). Bietrag zur Kenntris der Verhaltensiweise von Ratten, vergleichend dargestellt bei *Rattus norvegicus* und Rattus rattus. Zeitschrift für Angewandte Zoologie, 53, 129–196.

Tevis, L., Jr. (1956). Behavior of a population of forest-mice when subjected to poison. *Journal of Mammalogy, 37,* 358–370.

Tribe, D. (1954). The self-selection of purified food constituents by the rat during growth, pregnancy and lactation. *Journal of Physiology, 124,* 64.

Tribe, D. (1955). Choice of diets by rats: The choice of purified food constituents during growth, pregnancy, and lactation. *British Journal of Nutrition, 9,* 103–109.

Trivers, R. L. (1974). Parent–offspring conflict. *American Zoologist, 14,* 249–264.

Young, P. (1944). Studies of food preference, appetite, and dietary habit. II. Group self-selection maintenance as a method in the study of food preferences. *Journal of Comparative Psychology, 37,* 371–391.

Young, P. T. (1968). Evaluation and preference in behavioral development. *Psychological Review, 75,* 222–241.

Zahorik, D. M., & Houpt, K. A. (1981). Species differences in feeding strategies, food hazards, and ability to learn food aversions. In A. C. Kamil & T. D. Sargent (Eds.), *Foraging behavior* (pp. 289–310). New York: Garland STPM Press.

PART IV
Thirst, Sodium Appetite, and Fluid Homeostasis

14

Thirst and Water Balance

DAVID J. RAMSAY AND TERRY N. THRASHER

INTRODUCTION

Plasma osmolality is one of the most tightly regulated homeostatic variables in mammals. Indeed, there is good evidence to show that osmolality in the extracellular compartment is carefully controlled in a wide variety of animal species (Ramsay, Thrasher, & Bie, 1988). Apart from the direct effects of osmolality on metabolic and membrane transport processes, there are a number of reasons why extracellular osmolality regulation is so critical. Sodium salts comprise in excess of 90% of osmotically active solutes in the extracellular fluid. Because cell membranes normally have very low permeabilities to sodium, any change in sodium concentration in the extracellular fluid must result in water movement between intra- and extracellular compartments.

For example, a reduction in extracellular fluid sodium concentration and thus its osmolality will result in cell swelling (Grantham, 1977). Apart from obvious problems that might arise from consequent changes in cytoarchitecture, cells contained in tissues and organs surrounded by structures of low elasticity will be subjected to compression. Brain tissue is particularly subject to damage, as the rigid cranium allows for no expansion. Thus, cellular hydration that accompanies dilution of extracellular osmolality has particularly deleterious effects on brain cells, and the resulting cerebral edema presents a medical emergency when it arises in humans (Arieff, Llach, & Massry, 1976).

A second consequence of cellular uptake of water during situations of low extracellular fluid osmolality is reduction in extracellular fluid volume. Because only one-third of the body water is usually distributed in the extracellular fluid compartment, a small fall in its osmolality is reflected in a large fall in extracellular fluid volume as water moves along osmotic gradients into cells. The

DAVID J. RAMSAY AND TERRY N. THRASHER   Department of Physiology, University of California, San Francisco, California 94143.

resulting fall in plasma volume—and rise in hematocrit—can have important effects on cardiovascular function and the maintenance of arterial blood pressure and tissue perfusion. Indeed, the characteristics of severe sodium depletion include low venous return and arterial pressure and failure to maintain adequate peripheral perfusion as well as cellular overhydration (Rampton & Ramsay, 1974). Although there are cellular mechanisms that can adjust cell osmotic content to situations of raised or lowered extracellular osmolality, and, not surprisingly, these are particularly well developed in brain cells, acutely cells behave as osmometers (Yannet, 1940; Ellory & Hall, 1988).

Regulation of water balance plays a critical role in the regulation of extracellular osmolality and volume in mammals. From the previous discussion it is not surprising that there are important inputs from both extracellular osmolality and the circulation that influence water balance. Thus, both a raised extracellular fluid osmolality, or cellular dehydration, and a reduced extracellular fluid volume, or extracellular dehydration, stimulate water conservation. The mechanisms that restore water balance following deficits are the development of thirst, which results in water-seeking behavior and the intake of water, and the secretion of vasopressin, which reduces renal water loss (Ramsay & Thrasher, 1986). Both mechanisms are necessary for successful restoration of water balance following a deficit. Stimulation of water intake unaccompanied by renal water retention would be unproductive. Conversely, the production of a concentrated urine without water intake would merely limit the severity of the developing deficit and not correct it. Activation of water intake and vasopressin secretion occur together, and there are many similarities between the mechanisms involved.

OSMOREGULATION

Although it is beyond the scope of this chapter to review in detail the historical development of theories of thirst, the experiments and thinking that led to the suggestion that extracellular fluid concentration was important in the control of thirst can be traced to Mayer (1900). He demonstrated an increase in plasma concentration during dehydration in dogs and suggested there was a center in the brainstem that was involved in stimulating thirst. These experiments, together with clinical observations, were important landmarks in the development of theories of thirst mechanisms and have been expertly reviewed (Fitzsimons, 1972, 1979; see Chapter 2).

For many centuries, the sensation of thirst had been linked with dryness of the mouth. In 1919, Cannon hypothesized that dehydration led to reduction of salivary gland secretion and drying of the mucous membranes of the mouth and pharynx, and the local sensations in the mouth were the cause of the sensation of thirst. It is undoubtedly true that dehydration leads to progressive reduction in salivary flow (Adolph, 1947). However, there are many examples of dissociation between water intake and dry mouth. For example, Montgomery (1931) showed that salivarectomized dogs do not increase basal water intake. Again, dogs with esophageal fistulas drink continuously—sham drinking—despite a wet mouth (Bellows, 1939). Thus, although dryness of the mouth may play a part in the subjective perception of thirst, it is not an essential component.

In 1937, Gilman showed that intravenous infusions of hypertonic sodium

chloride into dogs caused drinking, whereas similarly hypertonic solutions of
urea did not. Since urea does not cause osmotic withdrawal of water from cells,
he postulated that this was the important difference between the solutions. Localization of osmotically sensitive areas, however, awaited Verney's (1947) experiments on antidiuretic hormone. He showed that infusions of hypertonic saline
into exteriorized carotid arteries in water-loaded dogs caused the release of
antidiuretic hormone, whereas solutes that did not cause cellular dehydration
did not. The concept of osmoreceptors was born. Soon afterwards Wolf (1950)
proposed that these central osmoreceptors were also responsible for thirst.

The era of investigation of the localization of brain regions responsible for
drinking was begun by Andersson (1953). He demonstrated that injections of
hypertonic saline into the hypothalamus of goats caused copious drinking. The
localization of the osmoreceptor area responsible for antidiuretic hormone release to the anterior hypothalamus by Jewell and Verney (1957) strengthened
the view that these hypothalamic regions controlled both drinking and renal
water excretion. Lesion and local injection experiments in rats and rabbits confirmed localization of the receptors to the lateral preoptic area (Blass & Epstein,
1971; Peck & Blass, 1975; Peck & Novin, 1971).

CHARACTERISTICS OF OSMOREGULATION IN ANIMALS

In whole animals there is little doubt that administration of hypertonic salt
loads by various routes leads to stimulation of drinking. Following such administration in rats, drinking begins with short latency and is a robust phenomenon.
The amount of water consumed is appropriate to the quantity of salt given. This
was demonstrated clearly by Fitzsimons (1961a), who showed that when hypertonic sodium chloride is given to nephrectomized rats, where the complication of
renal excretion of the salt load is eliminated, sufficient water to correct the
hypertonicity is ingested. The only caution in these whole-animal experiments,
which have largely used the rat model, is that it is wise to avoid the intraperitoneal route for administration of hypertonic solutes because of the stress of
this procedure (Coburn & Stricker, 1978).

A similar approach to the study of vasopressin secretion has been pioneered
by Robertson (Robertson, Athar, & Shelton, 1977; Robertson, 1985). The development of accurate radioimmunoassays for vasopressin has allowed detailed
relationships between plasma osmolality and vasopressin to be described in a
number of species including man. A linear relationship between these two variables is usually present, and the intercept through the osmolality axis can be used
as an indication of the threshold, or set point, of the system (Figure 1). The use
of an analogue rating scale allows thirst and plasma osmolality to be related in a
similar way, and although a linear relationship is obtained, the threshold tends to
be higher than that for vasopressin (Robertson, 1984, 1985; Vokes & Robertson,
1988).

A problem with systemic administration of hypertonic solutions is that withdrawal of water from cells causes expansion of extracellular fluid volume, a
potential inhibitor of thirst. We therefore decided to use the preparation developed by Verney (1947) of dogs prepared with exteriorized carotid loops to
obviate this problem (Wood, Rolls, & Ramsay, 1977). In these experiments hypertonic solutions were infused bilaterally in the carotid arteries, and drinking was
measured. This allows the composition of the blood perfusing the forebrain to

DAVID J. RAMSAY
AND TERRY N.
THRASHER

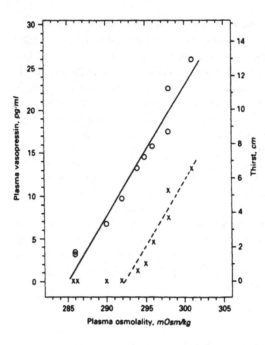

Figure 1. The relationship between plasma vasopressin (open circles) and index of thirst (crosses) and plasma osmolality in a healthy volunteer during infusion of 5% saline. (From Robertson, 1984, with permission.)

be altered without changing systemic plasma osmolality. When dogs are infused with hypertonic NaCl in this way for 10 minutes and drinking is measured in the last 5 minutes of the infusion, dose-related stimulation of drinking is seen (Figure 2). In contrast, intravenous infusions of the solutions at similar rates did not cause drinking. Intravenous infusions of sufficient hypertonic saline to raise plasma osmolality by 10 mosm/kg did cause drinking, but this was completely inhibited by simultaneous bilateral carotid infusion of sufficient water to remove the central osmotic stimulus. From these experiments we concluded that it was necessary to stimulate forebrain osmoreceptors for drinking to occur.

With similar techniques it is possible to study central osmotic regulation of vasopressin uncomplicated by systemic changes (Wade, Bie, Keil, & Ramsay,

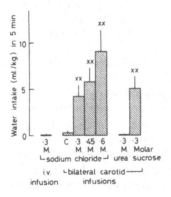

Figure 2. Water intake during infusions of hypertonic solutions in fluid-replete dogs (means ± S.E.M.). All carotid infusions were bilateral at a total rate of $0.6 \ ml \cdot kg^{-1} \ min^{-1}$ for 10 minutes, and water intake was measured for the last 5 minutes. Control infusion (C) was 0.15 M NcCl. (From Wood *et al.*, 1977, with permission.)

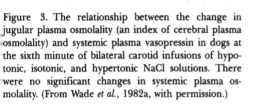

Figure 3. The relationship between the change in jugular plasma osmolality (an index of cerebral plasma osmolality) and systemic plasma vasopressin in dogs at the sixth minute of bilateral carotid infusions of hypotonic, isotonic, and hypertonic NaCl solutions. There were no significant changes in systemic plasma osmolality. (From Wade *et al.*, 1982a, with permission.)

1982a, 1982b). A linear relationship between plasma vasopressin and osmolality was obtained using carotid infusions of hypertonic saline in conscious dogs (Figure 3). However, when osmolality was lowered by carotid infusions of water, it was not possible to suppress vasopressin. In contrast, water loading or intravenous infusion of 5% glucose did suppress plasma vasopressin to undetectable levels. From these experiments we concluded that stimuli other than removal of the central osmotic stimulus alone, such as volume expansion, were necessary to suppress vasopressin secretion below values found in normally hydrated animals.

It is also of interest to depict these thirst and vasopressin relationships in dogs with plasma osmolality on the same graph (Figure 4). Under these circumstances, the elevation in plasma osmolality necessary to stimulate drinking and to increase plasma vasopressin above normal circulating levels are similar. In contrast, the data in Figure 1 indicate that the threshold for thirst is set some 10–15 mosm/kg higher than that for vasopressin. This may be related to the method of graphical analysis. It is certainly possible to extrapolate the vasopressin–osmolality relationship to the abscissa in Figures 3 and 4 and obtain a threshold for

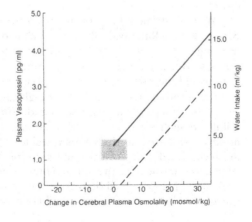

Figure 4. The relationship between plasma vasopressin concentration and water intake and change in cerebral plasma osmolality in dogs. Osmolality was altered using bilateral carotid infusions of hypotonic and hypertonic NaCl. The shaded area represents 95% frequency distribution for plasma in water-replete dogs determined from 154 measurements.

vasopressin that is lower than that for drinking. However, this does not happen physiologically, as animals do not spontaneously overdrink to suppress vasopressin secretion to zero. Normally hydrated animals secrete sufficient vasopressin to maintain urine more concentrated than plasma but do not drink continuously. If plasma osmolality is increased, for example by eating, this will stimulate increases in plasma vasopressin and cause drinking to correct the situation. Thirst and vasopressin mechanisms respond with equal sensitivity and similar thresholds to raised plasma osmolality above their normal operating points. In the case of plasma vasopressin, this normal operating point is 1–2 pg/mL, whereas with drinking, it is zero mL.

OSMORECEPTORS OR SODIUM RECEPTORS?

Andersson and his colleagues have proposed from the results of experiments in goats that the receptors responsible for thirst and vasopressin secretion evoked by cellular dehydration are close to the walls of the third cerebral ventricle and are sensitive to sodium concentration rather than to osmolality (Andersson, 1978). This proposal was based on the finding that intravenous infusions of NaCl or sucrose stimulated drinking and antidiuresis, where only intracerebroventricular administration of hypertonic NaCl was effective. Thus, equally hyperosmotic solutions of NaCl and sucrose should dehydrate the brain equally and raise CSF sodium but would have opposing effects on plasma sodium concentration. Furthermore, because central sucrose infusions would dilute CSF sodium concentrations by withdrawing water from brain cells, the conclusion was made that receptors sensitive to CSF sodium concentration must be important in cellular dehydration.

In a further elaboration of this theory, Olsson, Larsson, and Liljekvist (1970) and Olsson, Fyhrquist, Larsson, and Eriksson (1978) showed that intracerebroventricular infusions of isotonic or hypertonic glycerol blocked drinking in dehydrated goats. These central infusions decreased CSF sodium concentration. From these experiments it was deduced that receptors sensitive to CSF sodium were important and that peripheral osmoreceptors, which should have been stimulated by the dehydration, were not. However, some 20 minutes following the end of the central glycerol infusions, CSF sodium was still reduced by 25 meq/L, and yet the goats began to drink. It is difficult to reconcile these results with the theory of dominant CSF sodium receptors controlling thirst and antidiuresis.

We have studied this problem in a series of experiments in dogs (Thrasher, Brown, Keil, & Ramsay, 1980). It is possible to raise extracellular fluid osmolality by using a variety of solutes. Equiosmolar infusions of different solutes, however, can have differential effects on withdrawal of water from cells. Solutes such as sodium and sucrose are excluded from cells and cause general cellular shrinkage and dehydration. Solutes such as urea, glucose, and glycerol, however, pass with relative ease across cell membranes and do not act as effective solutes causing cellular dehydration. The blood–brain and blood–CSF barriers are different, however, and are selectively impermeable to these solutes. Clinicians have used intravenous infusions of urea and glycerol for many years to reduce cerebral edema and intraocular pressures. Thus, infusions of hypertonic NaCl, sucrose, urea, and glucose will all dehydrate the brain, whereas receptors on the blood side of the blood–brain barrier should only be dehydrated by NaCl and sucrose.

In these experiments, we found that intravenous infusions of equiosmolal hypertonic NaCl and sucrose had equivalent effects on drinking and on vasopressin secretion with a latency of about 15 minutes. In marked contrast, infusions of equally hyperosmotic glucose and urea had no effect on drinking and plasma vasopressin levels, even when continued for 45 minutes. However, the predicted increases in CSF sodium and osmolality occurred with all four solutes (Figure 5). It is difficult to support a role for CSF sodium or osmoreceptors, as systemic infusions of urea and glucose raised CSF concentrations and thus the environment of cells behind the blood–brain barrier. On the other hand, peripheral sucrose and sodium chloride infusions were equally effective in stimulating drinking and vasopressin despite opposite effects on plasma sodium. It is difficult to escape the conclusion that osmoreceptors are involved in the stimulation of drinking and vasopressin secretion and that they are sensitive to changes in plasma concentration, as Gilman (1937) first proposed.

This general conclusion was independently reached in studies on the control of drinking in sheep by McKinley, Denton, and Weisinger (1978). There is also evidence in rats that changes in plasma, rather than CSF, composition are correlated with drinking when various solutes are infused, showing the dominant role of peripheral osmoreception (Epstein, 1978).

The results of intracerebroventricular infusions of various solutes are less clear. In dogs, infusions of artificial CSF containing either 0.2 osmol/kg H_2O of NaCl or sucrose via the third ventricle were equally effective in stimulating drinking and vasopressin secretion (Thrasher, Jones, Keil, Brown, & Ramsay, 1980). Urea and glucose infusions were ineffective. The latency to drinking varied from 12 to 22 minutes, and the increase in CSF osmolality in the region of the third ventricle was approximately 50 mosm/kg H_2O. The effect of the centrally delivered stimulus, therefore, was weak, compared with the sensitivity of

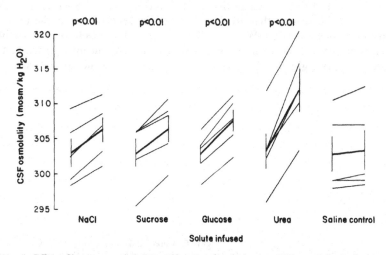

Figure 5. Effect of intravenous infusions of hypertonic solutions on CSF osmolality in five conscious dogs. The dark lines show the mean ± S.E.M. The initial point is the CSF osmolality before the infusion, and the final point is the CSF osmolality at the point when drinking occurs (NaCl and sucrose) or after 45 minutes of infusion (glucose, urea control) when no drinking occurred. (From Thrasher, Brown, *et al.*, 1980, with permission.)

drinking and vasopressin release following systemic infusions of solutes. In sheep, McKinley *et al.* (1978) and McKinley, Denton, Leksell, Tarjan, and Weisinger (1980) reported that central infusions of sodium salts were more effective than those of sucrose. They concluded that although peripheral osmoreceptors seemed to be more important, there was evidence for a central receptor sensitive to CSF sodium concentration. There is evidence in sheep, but not in other, nonruminant species, that CSF sodium concentration may have a particularly important role in the control of sodium appetite (Weisinger *et al.*, 1982). However, the control of water balance depends on osmoreceptors that sense plasma composition.

LOCATION OF OSMORECEPTORS

Although the arguments in the previous section clearly show that osmoreceptors that regulate water balance are sensitive to plasma composition, they are located in the brain. The experiments of Jewell and Verney (1957) localized the forebrain and hypothalamic regions as being important. A number of experiments that destroyed large amounts of brain tissue in the anterior wall of the third ventricle and lamina terminalis are known to cause adipsia, diuresis, and terminal dehydration unless interventions are made to maintain fluid balance (Andersson, 1978; Andersson & McCann, 1955). The experiments of Buggy and Johnson (1977) focused attention on the anterior ventral region of the brain bordering the third ventricle (AV3V). Lesions of this area that included the organum vasculosum laminae terminalis (OVLT) ventrally and extended to the anterior commissure dorsally caused profound disturbances of the regulation of water balance. The animals were adipsic, and despite increasing plasma osmolality, rats with these lesions still failed to drink and did not increase secretion rates of vasopressin (Brody & Johnson, 1980).

The infusion experiments in the previous section localized the osmoreceptors to a region of the brain sensitive to plasma composition. Such regions are circumventricular organs that are characterized by fenestrated capillaries and lack a blood–brain barrier (Weindl, 1973). Thus, cells in these organs should behave as peripheral tissue cells in their responses to systemic infusions of NaCl, sucrose, urea, and glucose. The AV3V lesions in rats and the larger lesions in

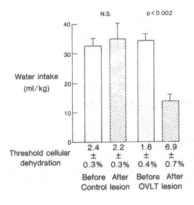

Figure 6. Water intake in response to a 30-minute intravenous infusion of hypertonic NaCl before and after lesions of the OVLT in dogs. The calculated cellular dehydration thresholds for drinking are shown below the bars. (From Thrasher, Keil, *et al.*, 1982a, with permission.)

dogs and goats include the OVLT, one of these circumventricular organs. We took advantage of the anatomy of the dog to make discrete electrolytic lesions of the OVLT without causing significant damage to surrounding tissue to test the hypothesis that the osmoreceptors were located in this structure (Thrasher, Keil, & Ramsay, 1982a; Thrasher & Keil, 1987).

The dogs were infused with hypertonic NaCl 2–3 weeks following the lesion when plasma osmolality was normal. At this time the dogs appeared normal and showed no obvious abnormality. However, the relationship between raised plasma osmolality and drinking was disrupted by the OVLT lesion compared to controls. When Wolf's (1950) equations, which calculate the degree of cellular shrinkage from the increment in plasma osmolality, were employed, approximately a fourfold increase in threshold cellular dehydration was found to be necessary to initiate drinking (Figure 6). Moreover, the volume of water consumed was much smaller than in dogs with control lesions. An increase of 24.1 ± 2.1 mosm/kg in plasma osmolality was required to stimulate drinking.

The data in Figure 7 further illustrate the profound disruption in the relationship between plasma osmolality and water intake. Not only was the threshold to vasopressin release increased, but there was no longer a significant correlation between plasma osmolality and water intake. A similar disruption of the relationship between plasma osmolality and vasopressin release occurs. McKinley and his colleagues have also shown interruption of osmotic control of drinking in sheep with lesions of the lamina terminalis that include the OVLT (McKinley *et al.*, 1982). An intriguing feature of animals with lesions that include the OVLT is that the capacity to drink or secrete vasopressin to physiological osmotic challenges is permanently impaired, yet they maintain a normal fluid balance as judged by daily intake–output measurements and plasma osmolality (Thrasher & Keil, 1987). This is discussed in more detail in a later section.

The other forebrain circumventricular organ that is a candidate as an osmoreceptor region is the subfornical organ (SFO). In dogs, lesions of this region do not cause deficits in drinking to acute osmotic stimuli in the amount consumed, latency to drink, or time of drinking (Thrasher, Simpson, & Ramsay, 1982). Some deficits, however, have been described in other species, although there is general agreement that the OVLT plays the dominant role (Hosutt, Rowland, & Stricker, 1981). It is likely that the SFO and OVLT are part of a similar forebrain circumventricular organ system concerned with water balance homeostasis. There is certainly evidence of rich interconnectivity between them. A degree of interspecies heterogeneity would be predicted and seems to be

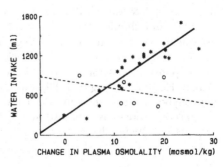

Figure 7. The relationship between plasma osmolality and water intake in dogs during intravenous infusion of hypertonic NaCl before (solid line) and at least 2 weeks following (broken line) lesions of the OVLT.

present. However, the major osmoregulatory function does localize in the OVLT.

THE SUPRAOPTIC NUCLEUS AS THE SITE OF THE OSMORECEPTORS

The supraoptic nucleus has been championed as the location of the osmoreceptors by Leng and his colleagues (Leng, Mason, & Dyer, 1982). The essential factor in favor of this suggestion is the known sensitivity of magnocellular neurons in the supraoptic nucleus to osmotic stimulation (Mason, 1980). In hypothalamic slice preparations, intracellular recording techniques have demonstrated that neurons in the supraoptic nucleus are depolarized when the osmolality of the medium is increased (Mason, 1980, 1982; Abe & Ogata, 1982). Furthermore, Leng et al. (1982) have argued that the supraoptic nucleus is highly vascularized, which would bring these osmosensitive neurons in close proximity with the blood, thus allowing them to "sense" blood osmolality. The osmosensitivity of the magnocellular neurons responsible for vasopressin secretion is not in dispute, although their sensitivity to osmotic challenges compared with increases in plasma osmolality known to increase vasopressin secretion calls their physiological role into question. However, the cells are on the brain side of the blood–brain barrier, and there is no evidence to suggest an imperfect barrier in this region. Thus, infusions of urea and glucose that dehydrate the brain should stimulate vasopression secretion if physiologically important osmoreceptors are in the supraoptic nucleus. As has been discussed, this does not happen.

We have shown that discrete lesions of the OVLT in dogs lead to loss of osmosensitivity with increases in plasma osmolality up to 24 mosm/kg (Thrasher, Keil, et al., 1982a). There is similar evidence in sheep and rats with large lesions of the anterior wall of the third ventricle that include the OVLT (McKinley et al., 1982; Brody & Johnson, 1980). Comparison of the osmotic sensitivity of hypothalamoneurohypophyseal explant preparations in tissue culture shows a markedly reduced secretion of vasopressin when such explants lack an OVLT (Sladek & Johnson, 1983). It does seem, therefore, that the OVLT is necessary for the expression of physiological osmosensitivity. Moreover, hypertonic injections made in the AV3V region excite magnocellular neurons in the SON (Honda, Negora, Higuchi, & Tadokoro, 1987), and there are cells in this area that are osmosensitive (Sayer, Hubbard, & Sirret, 1984).

Leng (Leng, Dyball, & Mason, 1985) has argued that the presence of the OVLT may be important to maintain the electrical sensitivity of supraoptic neurons. Removal of connections that provide excitatory input, as would be expected in a lesion experiment, certainly could reduce the likelihood that sufficient summation of EPSPs would occur to cause generation of action potentials (Leng, Dyball, & Russell, 1988). However, such electrical "depression" is of short duration and should certainly recover during a 2- to 3-week period before osmotic testing was carried out in dogs. There is evidence in rats that electrical activity of SON neurons recovers 2 weeks following destruction of the AV3V, yet the response to systemic osmotic challenges was largely inhibited (Chaudhry, Dyball, Honda, & Wright, 1989). Moreover, the capacity of the hypothalamoneurohypophyseal system in OVLT-lesioned dogs to secrete vasopressin with unaltered sensitivity is demonstrated by the normal vasopressin (and drinking) responses to hemorrhage, an extracellular dehydration stimulus (Thrasher &

Keil, 1987). Similarly, responses to hypovolemia in sheep with lamina terminalis lesions are not depressed and may be enhanced.

Recently Chaudhry *et al.* (1989), using extracellular recording techniques, examined the behavior of units in the AV3V region of rats identified antidromically by stimulation of the SON. Presumably such units project monosynaptically from the AV3V to the SON. The surprising finding was that such units were not responsive to osmotic stimuli. Unfortunately, the precise location of these units was not identified, so it is not clear if they were within the OVLT. Again, there is electrophysiological evidence for the presence of osmosensitive units within the OVLT of rats. If these projected to the SON via an intermediate synapse, then they would not have been included in the analysis. Further electrophysiological identification of receptors and pathways would certainly clarify this confused situation.

However, the overall facts seem clear. Osmosensitive receptors that control drinking and vasopressin secretion are located in the forebrain on the blood side of the blood–brain barrier, and the likely site is the OVLT. Without the OVLT, animals are incapable of responding by drinking or secreting vasopressin to osmotic stimuli over the normal physiological range, although responses to hypovolemia are intact. Whereas the exact interrelationships between the OVLT and SON require clarification, it is difficult to escape the conclusion that without precise information from the OVLT, the system fails to operate, and responses to physiological osmotic challenges are absent.

VOLUME REGULATION

The relationships between depletion of body fluids, sensation of thirst, and water intake have been appreciated since the previous century in reports in the clinical literature (Fitzsimons, 1979). However, it has proved more difficult to develop precise relationships between extracellular dehydration and drinking than with cellular dehydration. A contributing factor to this is the relative difficulty in the accurate assessment of plasma and/or extracellular fluid volume compared with estimations of plasma sodium or osmolality. Also, the threshold changes for stimulation of drinking and vasopressin secretion tend to be greater than for cellular dehydration, and the responses to moderate volume loss much less robust. A third problem is the compartmentalization of the extracellular fluid and the localization of receptors within specific loci of this highly heterogeneous compartment.

The technique that has allowed the development of precise relationships between extracellular fluid depletion and drinking has been the use of hyperoncotic colloid solutions (Fitzsimons, 1961b; Stricker, 1981). If a subcutaneous or peritoneal injection of polyethylene glycol (PEG) is made in rats, because of the disturbance of Starling forces across capillary endothelial membranes, an essentially protein-free plasma exudate is withdrawn from the circulation. This results in shrinkage of the plasma volume with no significant change in the concentration of crystalloids. An added advantage is that the extent of contraction of plasma, and blood volume, can be measured from changes in plasma protein concentration and hematocrit, at least acutely.

When rats are given subcutaneous injections of PEG, water drinking begins in approximately 1 hour and is related to the dose injected. The drinking is maintained over many hours, and in the absence of food (i.e., osmoles) is accompanied by dilution of extracellular fluid osmolality, a factor that tends to limit the intake of water (Stricker, 1969; Stricker & Jaloweic, 1970). If both water and saline solutions are offered throughout the testing period, water is ingested preferentially initially, but after 8–10 hours rats develop a salt appetite, which helps to restore plasma osmolality. There are many other techniques that have been used successfully to cause extracellular dehydration and stimulation of drinking and vasopressin secretion in a number of species, including hemorrhage (Fitzsimons, 1961b; Ramsay & Thrasher, 1986; Thrasher & Keil, 1987), sodium depletion (McCance, 1936) and mechanical restriction of venous return (Ramsay, Rolls, & Wood, 1975; Thrasher, Keil, & Ramsay, 1982b).

LOCATION OF RECEPTORS

Evidence that led to the present understanding that receptors that sense blood volume are located in the thorax came from studies of renal function. A number of maneuvers have been employed to alter the distribution of blood volume. For example, negative-pressure ventilation, which increases venous return, has been shown to be associated with diuresis, and positive-pressure ventilation with antidiuresis (Gauer & Henry, 1963). Again, isosmotic expansion of extracellular fluid volume in humans led to water diuresis if administered while the subject was supine but not if the subject was standing. Thus, the component of blood volume that seems to provide the stimulus is venous return, of obvious direct importance to the regulation of cardiac output and circulatory homeostasis.

There is certainly good evidence for the presence of stretch receptors in the right and left atria, particularly in the atriovenous junctions, that are responsive to increases in atrial volume and pressure (Linden & Kappagoda, 1982). Atrial stretch leads to activation of these receptors and increased vagal afferent activity, which has a major projection via the nucleus of the tractus solitarius to magnocellular neurones of the SON. Electrophysiological evidence shows inhibition of activity in SON during atrial stretch, and this can be correlated with reductions in plasma vasopressin (Menninger, 1979). Again, bilateral cold block of the vagosympathetic nerves in dogs leads to increased plasma vasopressin (Bishop, Thames, & Schmid, 1984). Stimulation of high-pressure baroreceptors in the carotid sinus region also causes inhibition of electrical activity in SON and a reduction in plasma vasopressin (Yamashita, 1977; Share & Levy, 1966). Thus, both low-pressure cardiopulmonary and high-pressure arterial baroreceptors can influence vasopressin secretion and thus renal water excretion.

There is good evidence from a number of sources that information from these baroreceptors can also influence water intake. Kozlowski and his co-workers showed that hemorrhage of 8–36% of the initial blood volume caused a decrease in the osmotic thirst threshold in dogs; that is, hemorrhage reduced the amount of cellular dehydration necessary to elicit drinking. Left cervical vagosympathectomy abolished this effect of hypovolemia on the osmotic thirst threshold (Kozlowski & Szczepanska-Sadowska, 1975). Cold or procaine block of the vagus nerve had similar effects. Conversely, Kaufman (1984) has shown that right atrial stretch in rats achieved by balloon inflation inhibits the drinking following PEG administration, an effect dependent on neural rather than humoral factors. Of great interest are observations that show that drinking in dogs following administration of dipsogens that also raise blood pressure is enhanced following chronic sinoaortic and cardiac denervation (Klingbeil, Quillen, Brooks, & Reid, 1986). Application of these techniques will allow the controversial area of the effects of volume on osmoregulatory thirst to be revisited.

Renin–Angiotensin System

The recognition that the renin–angiotensin system was an important stimulus to drinking came from the work of Fitzsimons (1964, 1969). He demonstrated that ligation of the inferior vena cava in the abdomen above the kidneys in the rat led to copious drinking. The drinking did not occur following bilateral nephrectomy. Drinking could also be elicited by partial occlusion of the abdominal aorta above the renal arteries. From these observations, it became apparent that the renin–angiotensin system could be associated with such drinking. Fitzsimons went on to show that intravenous infusions of angiotensin II restored the ability of nephrectomized rats to drink following vena cava ligation. Fitzsimons and Simons (1969) demonstrated that intravenous infusions of angiotensin II stimulated water intake in water-replete rats. The exquisite sensitivity of the dipsogenic effects of angiotensin II was shown by the vigorous drinking responses to small doses of the peptide when it was injected directly into various regions of the brain parenchyma (Epstein, Fitzsimons, & Rolls, 1970).

Since the demonstration that angiotensin II can cause drinking, many attempts have been made to evaluate its role in the control of water intake. There is no doubt that reductions in blood volume and blood pressure cause stimulation of renin secretion. Unloading of either the cardiopulmonary or the arterial baroreceptors leads to reflex increases in renal sympathetic outflow and renin secretion. Furthermore, reductions in renal perfusion pressure directly stimulate renin release via renal baroreceptor and macula densa mechanisms. Angiotensin II has a number of effects that counteract hypovolemia and hypotension in addition to stimulating water intake. It causes vasoconstriction, both directly and via the central nervous system, promotes adrenal secretion of aldosterone and thus sodium retention, and stimulates vasopressin and ACTH secretion (Ramsay, 1979). Additionally, in combination with raised plasma aldosterone, it stimulates sodium appetite in susceptible species (Epstein, 1986). In many respects, angiotensin II is ideally suited to be the "peptide of hypovolemia."

The mechanism of interaction of circulating angiotensin II with the brain was investigated by Simpson and Routtenberg (1973). Angiotensin II is a polar peptide and does not cross the blood–brain barrier. Simpson and Routtenberg showed that lesions of the SFO in rats completely obliterated the drinking following systemic administration of angiotensin II, with no effect on cellular dehydration thirst. This was the first of many demonstrations that the subfornical organ was a critical site for circulating angiotensin II to interact with the brain to cause drinking (Ramsay, 1979). Angiotensin II receptors have been demonstrated in the SFO, and there are high concentrations of converting enzyme there. Thus, circulating angiotensin I can be converted locally into angiotensin II, and high concentrations of the peptide are achieved in close proximity to its receptors. It is of great interest that local microinjection of angiotensin II into the SFO elicits vasopressin secretion and that this region may also be involved in the regulation of renal water excretion.

In dogs, it is clear that forebrain circumventricular organs are involved in the drinking responses to angiotensin II. Bilateral intracarotid infusions of low doses of angiotensin II stimulate water intake, whereas similar intravertebral or intravenous infusions do not (Reid, Brooks, Rudolph, & Keil, 1982). We have shown that SFO lesions in dogs eliminated drinking following systemic infusions of angiotensin II that cause drinking in normal dogs. These lesions were without effect on drinking to hypertonic saline (Thrasher, Simpson, et al., 1982).

There is good evidence that AV3V lesions in rats also eliminate drinking responses to systemic infusions of angiotensin II. Indeed, these lesions eliminate a number of centrally mediated responses to blood-borne angiotensin, including increases in blood pressure in both acute and chronic situations (Brody & Johnson, 1980). The problem in interpretation of these results is that the AV3V lesion is extensive and destroys a number of nuclei and pathways. Both SFO and OVLT have reciprocal connections to the nucleus medianus, SON, and PVN (Miselis, 1981). Lesions of the nucleus medianus have profound effects on water balance, causing adipsia and inappropriate diuresis, yet this structure is within the blood–brain barrier (Gardner, Mangiapane, & Simpson, 1981; Gardner, Verbalis, & Stricker, 1985; Thrasher, 1989). Presumably lesions of this nucleus interrupt pathways, destroy important synapses and cell bodies, or both. Presently it is difficult to establish the precise mechanisms that account for the multiple deficits of the AV3V lesion.

In dogs, as has been described, it is possible to ablate the OVLT with minimal damage to the surrounding tissue. Following OVLT lesions, systemic infusions of angiotensin II fail to elicit drinking and vasopressin secretion (Thrasher & Keil, 1987). Thus, lesions of this circumventricular organ cause loss of angiotensin II and osmotic effects on drinking and vasopressin secretion. It is unlikely that the cause is nonspecific depression, since hypovolemia from hemorrhage elicits normal drinking and vasopressin responses. There is much evidence in the literature that shows interaction between angiotensin II and osmotic stimulation of drinking. Indeed, hydration prevents angiotensin II stimulation of vasopressin secretion in dogs (Shimizu, Share, & Claybaugh, 1973; Claybaugh, 1976). Loss of the osmotic input as a result of OVLT lesions may prevent angiotensin II from stimulating vasopressin and drinking. Alternatively, the reciprocal circuitry between the OVLT and SON may be critical in the expression of angiotensin and osmotic responses (Chaudhry et al., 1989).

There is ample evidence that intracerebroventricular injections of very low doses of angiotensin II elicit drinking and vasopressin secretion (Phillips, 1987). Indeed, the full range of brain-mediated angiotensin actions can be provoked, including increased blood pressure and ACTH secretion. The AV3V, rather than the SFO, is important in these central responses, although the presence of angiotensinergic synapses in the forebrain complicates this simple interpretation. It is clear that components of the renin–angiotensin system are present in the central nervous system, and in situ hybridization techniques have shown the expression of the renin and angiotensinogen genes in the brain (Phillips, 1987). Thus, many have assumed that intracranial injections of angiotensin II mimic activation of the brain renin–angiotensin system. The physiological relevance of many of these studies remains to be clarified. There is a disappointing lack of precise data that link angiotensin II generation at specific brain loci to activation of identified angiotensinergic synapses by physiological stimuli. It is difficult to assess the importance of brain generation of angiotensin II without these critical data.

Participation of Circulating Angiotensin II in Drinking and Vasopressin Secretion

There is general agreement that systemic infusions of angiotensin II can cause drinking and vasopressin secretion in a number of species. Responses are enhanced if associated increases in blood pressure are prevented or its perception blunted by baroreceptor denervation. However, evaluation of the role of angiotensin II in drinking and vasopressin responses to extracellular dehydration is more difficult. The major difficulty is that angiotensin II is important in blood pressure maintenance in hypovolemia, and interventions that block the dipsogenic actions of angiotensin II also compromise the regulation of arterial pressure.

The original experiments on caval ligation drinking by Fitzsimons showed that bilateral nephrectomy inhibited the water intake and implicated angiotensin II as the mediator (Fitzsimons, 1961a). Stricker (1978) has argued that other factors, such as low blood pressure following nephrectomy, may so debilitate the animals that drinking is prevented. There is also controversy concerning correlation between plasma angiotensin II (or plasma renin activity) levels in various models of extracellular dehydration and drinking. Stricker (1978) has presented data that show the correlation to be very poor, whereas others (e.g., Mann, Johnson, Rascher, Genest, & Ganten, 1981; Johnson, Mann, Rascher, Johnson, & Ganten, 1981) have argued that plasma angiotensin II is dipsogenic at physiologically attainable concentrations.

Some insight may be offered by consideration of data from experiments with the β-adrenergic agonist isoproterenol. Subcutaneous injection of isoproterenol causes drinking and vasopressin release in a number of species (Houpt & Epstein, 1972; Fitzsimons & Szczepanska-Sadowska, 1974; Ramsay, Reid, Keil, & Ganong, 1978) but also causes arterial hypotension. Isoproterenol stimulates renin secretion, and because nephrectomy blocked the drinking and it was inhibited by angiotensin II receptor blockade, it was claimed that angiotensin mediated the effect (Rolls & Ramsay, 1975). However, others have shown that drinking persisted following nephrectomy (Fitzsimons and Szczepanska-Sadowska, 1974).

A solution to the dilemma may be proved by consideration of the dual actions of isoproterenol. This agent causes blood pressure to fall as well as renin release to be stimulated. As has already been discussed, a fall in blood pressure itself can cause drinking. In 1978, we showed that subcutaneous injection of a low dose of isoproterenol (6 μg/kg) stimulated renin secretion, drinking, and vasopressin secretion in dogs with a 14 mm Hg reduction in blood pressure. Central administration of saralasin completely blocked the drinking and vasopressin secretion to isoproterenol. A larger dose of isoproterenol (20 μg/kg) caused a larger fall in blood pressure and stimulation of drinking, vasopressin, and renin secretion. Blockade of the renin–angiotensin system with saralasin in this experiment did not block the drinking or vasopressin responses. Presumably with the greater reduction in blood pressure, the baroreceptor reflex component alone was sufficient to stimulate a water conservation mechanism (Ramsay,

1978). Similar conclusions were reached by Rettig, Ganten, and Johnson (1981) in rats.

Thus, during extracellular dehydration, at least two stimuli are normally acting together to cause drinking and antidiuresis: angiotensin II and baroreflexes. If either stimulus is withdrawn by some experimental manipulation, then the remaining input will be effective only if it is sufficiently intense to reach some critical threshold. There may also be other factors that contribute to the ability of the animal to drink, depending on specific experimental situations (Fitzsimons, 1972).

Another example of this controversy is found in the responses to caval ligation drinking in rats. We have reported that central administration of saralasin in doses shown to block the effect of high plasma concentration of angiotensin II fail to inhibit drinking following caval ligation in rats (Lee, Thrasher, & Ramsay, 1981). Thus, angiotensin II participation was not essential in the drinking in these experiments. Fitzsimons and Elfont (1982), however, using administration of captopril, showed suppression of drinking in cavally ligated rats and claimed that the involvement of angiotensin was essential. Apart from a discussion of differences in experimental design, the two results are not necessarily contradictory. Because both the lack of stimulation of cardiopulmonary and arterial baroreceptors and the increase in circulatory angiotensin II can cause drinking and vasopressin secretion, removal of the effects of angiotensin II need not necessarily block the effect of caval ligation. Indeed, the presence of drinking and vasopressin stimulation when there is proven angiotensin receptor blockade shows that the participation of angiotensin II is not essential to the expression of responses to extracellular dehydration. Additionally, the persistence of drinking in rats given polyethylene glycol following nephrectomy emphasizes that water intake may still be stimulated by hypovolemia in the absence of circulating angiotensin II (Fitzsimons, 1961b).

In dogs, proven central blockade of the renin–angiotensin system during caval constriction, which does not compromise the ability of the animal to maintain its blood pressure, does not prevent drinking responses and secretion of vasopressin (Thrasher, Keil, *et al.*, 1982b). Again, participation of angiotensin II does not seem to be essential. A compelling experimental result is the response of the OVLT-lesioned dog to hemorrhage (Thrasher & Keil, 1987). As has been described, such animals do not respond to angiotensin II or to cellular dehydration. The remaining input to drinking and vasopressin secretion is derived from the baroreflex. The fact that OVLT-lesioned dogs respond by drinking and secreting vasopressin following hemorrhage in a similar fashion to normal dogs clearly demonstrates that angiotensin II is not a necessary participant in extracellular dehydration water balance mechanisms.

CONNECTIVITY

The putative role of the SFO and OVLT as receptor sites that mediate drinking and vasopressin secretion in response to increases in circulating angiotensin II and osmolality has been inferred from deficits caused by lesions of these circumventricular organs. However, clear interpretation of these functional studies depends, in large part, on determining whether the forebrain structure possesses the necessary connections to influence the secretion of vasopressin and initiation of drinking. Neuroanatomical tracing studies published over the last 10 years have revealed that the necessary connections do exist between the SFO and OVLT and magnocellular neurons in the SON and PVN to influence vasopressin secretion and lateral preoptic and hypothalamic areas involved in the initiation of drinking behavior. The major studies have been conducted in rats, but preliminary studies in other species appear to confirm that the connections observed in rats are generally applicable to other mammalian species.

The first major investigation of connections between the SFO and hypothalamic areas implicated in the regulation of water balance was conducted by Miselis and colleagues (Miselis, Shapiro, & Hand, 1979). These studies were based on anterograde transport of tritiated amino acids into the SFO or adjacent structures to determine efferent projections from this CVO. The key findings were clear evidence of terminal fields in both the SON and PVN, which imply a direct influence on activity of magnocellular neuronal release of vasopressin. Also, terminal fields were identified in both the lateral preoptic area and the lateral hypothalamus, areas known to be intimately involved in the initiation of ingestive behavior. In addition, dense projections to the AV3V region, including the nucleus medianus (NM), the OVLT, and the network of periventricular neurons surrounding the anterior wall of the third ventricle were observed. The projections to the AV3V region also involve structures that have been postulated to play vital roles in the regulation of body fluid balance (Brody & Johnson, 1980). Thus, Miselis and colleagues provided a clear neuroanatomical basis for the SFO to influence hypothalamic structures involved in the regulation of water balance (Miselis, 1981).

Lind, van Hoesen, and Johnson (1982) confirmed the efferent projections of the SFO based on anterograde transport of horseradish peroxidase (HRP) injected into the SFO. Furthermore, retrograde transport of HRP revealed projections to the SFO from cell bodies clustered in the NM and diffuse projections from the medial septum, dorsal preoptic region, and throughout the medial preoptic and anterior hypothalamic areas (Figure 8).

Camacho and Phillips (1980) traced the afferent and efferent connections of the OVLT using injections of HRP into the OVLT via a transbuccal approach. They observed dense projections to the NM and further dorsally to the septal area and to the SFO. Also dense projections were observed to the lateral preoptic area and SON. Afferents to the OVLT were observed projecting from cell bodies in the NM and SFO dorsally and from medial preoptic area, lateral hypothalamus, SON, and PVN. This study together with the reports of Miselis and Lind *et al.* indicated that the efferent projections of both the SFO and OVLT can directly influence magnocellular neurons in the SON and lateral preoptic, lateral hypothalamic neurons involved in the initiation of drinking behavior. In addition, these studies show dense projections from both forebrain CVOs to the NM.

The connectivity of the NM was investigated by Saper and Levisohn (1983). Injection of HRP into the ventral NM revealed projections from neurons in the SFO, OVLT, periventricular preoptic region, pericellular portions of the PVN, medial and lateral preoptic areas, the parabrachial nucleus, and the nucleus of the solitary tract (NTS). The multiplicity of inputs from both forebrain circumventricular organs as well as efferents from medullary areas receiving baroreceptor afferents strongly suggests that the NM is a pivotal region for integration of signals concerning the status of body fluid balance.

DAVID J. RAMSAY
AND TERRY N.
THRASHER

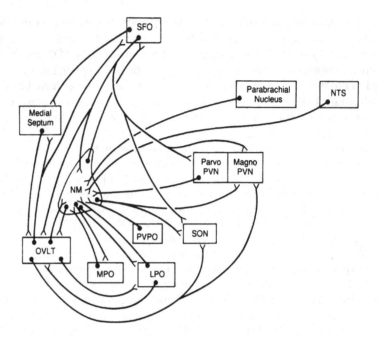

Figure 8. Schematic of neuroanatomical interconnections between forebrain circumventricular organs and the nucleus medianus in the control of water balance. SFO, subfornical organ; NM, nucleus medianus; OVLT, organum vasculosum laminae terminalis, medial preoptic area; LPO, lateral preoptic area; PVPO, periventricular preoptic area; SON, supraoptic nucleus; PVN, paraventricular nucleus; NTS, nucleus tractus solitarius.

Preliminary studies conducted in this laboratory have attempted to verify the neuroanatomical connections described above in the dog. Injections of HRP into the SON revealed retrogradely labeled neurons from both the OVLT and NM but minimal input from the SFO. In order to verify these results, HRP was also injected into the NM, just ventral to the anterior commissure. The major projections to the NM were from cell bodies in the OVLT and less dense but uniform labeling of neurons in the medial preoptic area. Very few cell bodies were apparent in the SFO. Labeled fibers were apparent in the SON and diffusely distributed throughout the medial and lateral preoptic areas. These preliminary results in dogs confirm the general pattern of connections elucidated in rats (Figure 8).

A number of electrophysiological studies have been conducted that confirm the connectivity of forebrain circumventricular organs to magnocellular circuits determined from the anatomic studies described above. Sgro, Ferguson, and Renaud (1984), Ferguson, Day, and Renaud (1984a, 1984b), and Ferguson (1988) have reported that stimulation of the SFO influenced the activity of magnocellular SON and PVN neurons identified by antidromic stimulation of the median eminence. Gutman, Ciriello, and Mogenson (1986) recorded from neurons in the SFO while stimulating the SON, PVN, or NM. They observed both antidromic and orthodromic activation of SFO units, confirming reciprocal connections between the SFO and the SON, PVN, and NM, respectively.

The electrophysiological connections of the OVLT have not been reported, but Honda *et al.* (1987) have observed that localized injection of hypertonic NaCl into the AV3V region excites units in the PVN. Furthermore, the effectiveness of the injection was directly related to the proximity of the cannula to the OVLT. Recently, Chaudhry *et al.* (1989) have reported that stimulation of units in the SON leads to both orthrodromic and antidromic activation of units in the AV3V region. However, as discussed previously, they did not indicate whether the units recorded from were in the OVLT or ventral NM. Nevertheless, the study by Chaudhry *et al.* (1989) confirms the existence of reciprocal connections between the AV3V area and magnocellular units in the SON.

RESPONSES TO WATER DEPRIVATION

Under normal conditions in their natural habitats, or in the laboratory, animals tend to drink intermittently. Although patterns of behavior differ widely among animals, water intake tends to be periprandial. For example, if dogs are fed once every 24 hours, 90% of their drinking is associated with feeding, most of it being postprandial. In their normal environments, hunting carnivorous animals may undergo even longer periods between feeding and thus drinking. The relationship between feeding and drinking patterns is not unexpected. Apart from behavioral effects, the main variable in the requirement for water is the intake of osmoles and thus feeding. Eating to provide substrates for energy necessitates the intake of osmoles such as sodium and potassium. Under steady-state conditions, metabolic requirements for these salts are minimal, and the excess is excreted in the urine, thus obligating water excretion. The ability to concentrate the urine is crucial to minimize water loss but cannot prevent it. Thus, between feeds and associated water intake, animals gradually become dehydrated.

The normal cycle in animals, therefore, is the gradual accumulation of water deficits, followed by their periodic correction (Ramsay *et al.*, 1988). When water intake occurs, the appropriate volume of water is consumed, and, except for a few special cases, overcorrection does not occur. This phenomenon of satiety is discussed in a later section. A simple measurement that shows that overcorrection does not occur is that random sampling of urine in animals with access to food and water always shows urine osmolality to be greater than that of plasma. Thus, vasopressin is being secreted to conserve water in response to developing deficits. The normal cycle of water balance, therefore, is the gradual appearance of a water deficit, followed by water intake to restore water balance.

This cycle may be simulated in the laboratory by depriving animals of water and studying their behavior. For example, dogs deprived of water but not food for 24 hours lose approximately 4% of their body weight (Ramsay, Rolls, & Wood, 1977). As Adolph (1967) has described in a number of species, when dehydrated animals are allowed access to water, appropriate amounts are consumed to repair the deficit. In dogs, drinking occurs very quickly. In our 24-hour water-deprived dogs it was completed in 2.5 minutes. In these experiments, plasma osmolality was increased by 12 mosm/kg H_2O during the dehydration period. A remarkable feature of these experiments was that drinking ceased a full 5 minutes before plasma osmolality began to be diluted. Indeed, plasma

DAVID J. RAMSAY
AND TERRY N.
THRASHER

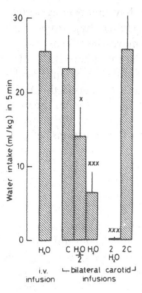

Figure 9. Water intake in eight water-deprived dogs. Bilateral carotid infusions of water were given to return cerebral plasma osmolality to values in euhydrated dogs (H_2O) or by half ($H_2O/2$) or double ($2 \cdot H_2O$) that value. Drinking was measured in the last 5 minutes of the 10-minute infusion period. (From Ramsay et al., 1977, with permission.)

osmolality took 50 minutes to return to normal. The mechanisms that allow the correct amounts of water be consumed are discussed later.

The rapid drinking behavior of the dog allows the stimulus responsible for the drinking to be manipulated. In these experiments we used 24-hour water-deprived dogs with exteriorized carotid loops (Ramsay et al., 1977). Bilateral carotid infusion of water at $0.3 \text{ mL} \cdot \text{kg}^{-1} \cdot \text{min}^{-1}$ removed the raised osmolality stimulus reaching the forebrain without affecting the systemic plasma significantly. This maneuver reduced drinking in the 24-hour water-deprived dogs by 72% (Figure 9). The dose-related effect of the phenomenon was shown by infusion of water into the carotids at half the rate, which reduced subsequent drinking by 35%. It is of great interest that in a later series of experiments, removal of the central osmotic stimulus in 24-hour water-deprived dogs reduced plasma vasopressin levels from 5.3 ± 0.6 pg/mL to 2.1 ± 0.8 pg/mL, a reduction of $70 \pm 5\%$ of the stimulated vasopressin levels (Wade, Keil, & Ramsay, 1983).

It is also possible to examine the contribution of extracellular dehydration to water intake following water deprivation. In these experiments in 24-hour water-deprived dogs, restoration of the volume deficit using systemic infusions of 0.15 M NaCl without change in osmolality reduced water intake by 27%. Combined intracarotid infusions of water to remove the central osmotic stimulus and 0.15 M NaCl to relieve the hypovolemia eliminated the water drinking in these water-deprived dogs (Figure 10). Reexpansion of the extracellular compartment in 24-hour water-deprived dogs in similar experiments reduced plasma vasopressin by $33 \pm 4\%$.

In dogs, therefore, it seems that 70% of the water intake and vasopressin responses to water deprivation can be accounted for by raised cerebral plasma osmolality and 30% by reduced extracellular fluid volume. The precise interaction between these two stimuli varies among species, but where it has been

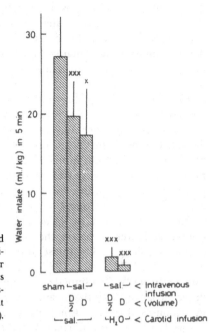

Figure 10. Water intake in eight water-deprived dogs. Intravenous infusions of saline were given either to completely restore the volume deficit (D) or to halve the deficit (D/2). Bilateral carotid infusions of water were sufficient to remove the central osmotic stimulus caused by water deprivation without affecting systemic plasma osmolality (Figure 9). (From Ramsay *et al.*, 1977, with permission.)

studied, both cellular and extracellular dehydration mechanisms contribute to water balance restoration following water deprivation.

These and similar responses to deficits have been termed regulatory drinking. Fitzsimons (1972) has stressed that much of the water intake in animals is not directly caused by deficit and should be termed nonregulatory. This is undoubtedly true, and many behavioral cues associated with eating behavior may be involved. However, it could be argued that although water intake may result from a complex of behavioral cues, the underlying controls are physiological. A simple example is seen when food is enriched with sodium. Water intake increases over a few days, and water balance and plasma osmolality are restored in the face of the increased intake of osmoles. Thus, the underpinning of determinants of fluid intake are physiological.

CHRONIC OVLT LESIONS

Much information about the importance of the OVLT in water balance has been gathered from studies of lesioned animals. As has been discussed, the original observations concerning the acute and chronic effects of the AV3V region were made by Johnson and Buggy (1978). Single midline lesions were made in rats and resulted in destruction of the nucleus medianus, OVLT, and the surrounding periventricular tissue. From the section on connectivity, it is clear that pathways from the OVLT, SFO, and nucleus medianus to SON and PVN will also be destroyed.

The primary effect of this lesion is the abolition of spontaneous water intake in the animals, combined with an inappropriately high urine output. Indeed, as the animals become dehydrated because of failure to drink and continued loss of urine, there is no increase in vasopressin secretion, and the process of dehydration continues. In fact, many of these animals would die if fluid therapy was not introduced. With careful management, some rats will return to spontaneous drinking and survive without further interventions. However, the impairment of drinking and vasopressin responses to acute cellular dehydration and to systemic infusions of angiotensin II is not restored in these chronically lesioned animals.

In dogs, Thrasher and Keil (1987) have described the chronic effects of discrete lesions of the OVLT. Immediately following the lesion, there is a period of polyuria for 24–36 hours unaccompanied by an increase in plasma vasopressin. There are increases in plasma osmolality, protein concentration, and plasma renin activity, all indices of negative water balance and contraction of the extracellular fluid volume. Over a period of about 2 weeks, all values returned to normal and remained so for as long as the dogs were studied. However, at all times tested after the lesion, there were no drinking or vasopressin responses to osmotic challenges. The question of how these dogs maintained water balance may be raised. Thrasher (1989) has pointed out that although the mean plasma osmolality of OVLT-lesioned dogs returned to normal, there was a large increase in the variance about the mean. Perturbation of body fluid homeostasis resulted in responses of plasma osmolality with greater oscillations. Responses to hemorrhage, however, were normal. Whether baroreflex regulation of drinking and vasopressin alone can compensate for loss of osmotic and angiotensin II inputs is not clear, and it is possible there are other as yet nonidentified inputs that participate in the regulation of water balance.

Lesions in sheep that are confined to the midline region of the anterior wall of the optic recess have similar effects (McKinley et al., 1982). These lesions contain the OVLT but are not confined to it. In rats, it is less clear if lesions of the OVLT alone produce a permanent loss of osmoregulatory ability.

Thrasher (1989) has reported that lesions of the nucleus medianus, confined mainly to its ventral portion, have more drastic effects on fluid balance. During 48 hours following the lesion, there was a pronounced polyuria, and hemoconcentration with a reduction in blood volume of $36 \pm 2\%$; plasma vasopressin concentration remained at basal levels. The dogs remained adipsic but would consume a liquid diet. The animals appeared behaviorally normal despite the loss of ability to maintain water balance. Because the nucleus medianus contains synapses and fibers of passage for connections among SFO, OVLT, PVN, and SON, it is tempting to argue that it subserves a vital integrative function for control of water balance. In sheep, larger lesions that include the nucleus medianus and OVLT result in a similar permanent loss of spontaneous drinking and control of vasopressin secretion, even when these animals are maintained for many months. The careful analysis of deficits in drinking behavior by Gardiner also shows the importance of the nucleus medianus (Gardiner et al., 1981; Gardiner & Stricker, 1985a, 1985b).

As Andersson and McCann (1955) first postulated, the anterior wall of the third ventricle contains structures vital to water balance control. In dogs, there seems to be more functional specificity between brain structures in this region than in other species so far studied. The SFO contains angiotensin II receptors and is responsible for mediating blood-borne effects of angiotensin II to cause drinking and vasopressin secretion. The OVLT is the site of osmoreceptors, and cellular dehydration stimulates drinking and vasopressin secretion via this structure. The nucleus medianus may play a role in integration of this information, or electrolytic lesions in this region may destroy important fibers of passage. Its precise role requires further clarification.

ROLE OF SODIUM BALANCE IN THE CONTROL OF PLASMA OSMOLALITY

Thus far the discussion of the control of plasma osmolality has been restricted to the regulation of water balance. Indeed, it has often been claimed that plasma osmolality is determined by water balance, whereas extracellular fluid volume is regulated by sodium balance. This separation of water and sodium balance is not helpful and may lead to erroneous conclusions. As has been discussed in the introduction to this chapter, changes in extracellular fluid osmolality always cause movement of water across cell membranes and thus changes in volume.

During dehydration, extracellular fluid volume is reduced, and it is well accepted that volume reduction leads to sodium retention via stimulation of the renin–angiotensin–aldosterone system. It was predicted and generally thought that dehydration was accompanied by sodium retention. In a little noticed report, Luke (1973) found that rats exhibited a natriuresis during water deprivation. Subsequently, this phenomenon of natriuresis during dehydration has been reported in a number of species (McKinley, Denton, Nelson, & Weisinger, 1983; Merrill, Skelton, & Cowley, 1986; Thrasher, Wade, Keil, & Ramsay, 1984; Zucker, Gleason, & Schneider, 1982). The phenomenon is easy to demonstrate.

In our experiments in dogs (Thrasher *et al.*, 1984), 24 hours of water, but not food, deprivation causes a marked natriuresis. On the following day, when free access to water is resumed, antinatriuresis occurs, and the "lost" sodium is regained. In these experiments, a number of animals reduced their food intake. It is clear, however, that reduction in food intake was not the cause of natriuresis. In our population of 19 dogs, a subgroup of seven showed no reduction in food intake, and the intensity of natriuresis was as great as in the remaining animals where there was reduction in food intake. The negative sodium balance during the 24 hours of water deprivation was 1.9 ± 0.2 meq/kg. The increase in plasma sodium during this period of dehydration was from 146 to 152 meq/L. Without the natriuresis, it can be calculated that the plasma sodium would have increased to 163 meq/L. Thus, loss of sodium ameliorates the rise in plasma osmolality that accompanies water deprivation. In this respect, it is of considerable interest that dogs subjected to longer periods of dehydration show little elevation in plasma osmolality after the first 24 hours. This is accomplished by a combination of continuing natriuresis and reduction in food, and thus osmolar, intake (Merrill *et al.*, 1986).

The relationship between natriuresis and food intake was examined in a further series of experiments (Metzler, Thrasher, Keil, & Ramsay, 1986). The cumulative sodium balance in 13 dogs for three consecutive days is shown in Figure 11. Dogs were fed at the beginning of each 24-hour period and deprived of water the second day. The rate of sodium excretion did not differ between control and dehydration days for the first 6 hours. Excretion of sodium takes place at similar rates during this postprandial period irrespective of the presence

DAVID J. RAMSAY
AND TERRY N.
THRASHER

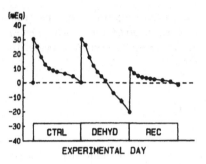

Figure 11. Cummulative sodium balance in dogs on three consecutive days with access to water (CTRL), during 24-hour water deprivation (DEHYD), and when water was returned to the home kennel (REC). The dogs were fed at the beginning of each 24-hour period. (From Metzler *et al.*, 1986, with permission.)

of water. However, on normal days, with access to water, the rate of sodium excretion slows after 8 hours until sodium balance is achieved. In contrast, however, the rapid postprandial urinary loss continues throughout the 24-hour period when water is withheld. It seems likely that the mechanisms responsible for the dehydration natriuresis are dissociated from the postprandial natriuresis. On the third day, when access to water is resumed, there is almost complete retention of the sodium taken in with the food.

The mechanisms responsible for the natriuresis are not clear. In our experiments, plasma aldosterone levels are not reduced during water deprivation. In contrast, Merrill *et al.* (1986) have reported reductions in plasma aldosterone during dehydration in sodium-restricted dogs. In these animals, plasma aldosterone is elevated because of stimulation of the renin–angiotensin system, and the raised plasma osmolality caused by dehydration is known to inhibit angiotensin-induced aldosterone secretion (Schneider, Taylor, Radke, & Davis, 1984). There are a number of other reports of dehydration natriuresis occurring without a fall in aldosterone levels (Lucke, Erbet, Herting, & Dohler, 1980; McKenna & Haines, 1981; Zucker *et al.*, 1982). Moreover, Luke (1973) has shown that injections of mineralocorticoid do not block dehydration natriuresis in rats.

The natriuresis cannot be accounted for by other factors known to be involved in the control of sodium excretion such as glomerular filtration rate, vasopressin, and angiotensin II (Metzler *et al.*, 1986). Moreover, atrial peptides and other putative natriuretic hormones described are stimulated by volume expansion, not contraction, and do not increase during dehydration. However, lesions of the lamina terminalis (McKinley, Denton, Park, & Weisinger, 1983; Thrasher *et al.*, 1983) or bilateral anterior hypothalamic knife cuts posterior to the OVLT (Bealer, Crofton, & Share, 1983) do block the natriuresis. The link between the neural deficit caused by the lesions or knife cuts and the dehydration natriuresis is not clear but does not appear to involve the pituitary or the renal nerves (Park, Congin, Denton, & McKinley, 1985).

As has been discussed previously, there is evidence that sodium-sensitive structures in the brain in the region of the hypothalamus and lamina terminalis are involved in salt and water balance. The concept of a sodium receptor within the blood–brain barrier that controls drinking and vasopressin secretion is not supported by evidence in dogs and rats. However, in sheep and goats there may be a role for an intracranial sodium sensor in the control of water balance and in salt appetite and sodium excretion (McKinley *et al.*, 1987). In all species so far examined, however, the OVLT and/or lamina terminalis is critical in the capacity of an animal to show a dehydration natriuresis. The mechanism involved, however, remains obscure.

The importance of dehydration natriuresis in the ability of an animal to maintain its plasma osmolality in the face of water deprivation can be demonstrated when the natriuresis is prevented. Dogs with discrete lesions of the OVLT do not show the natriuresis. In our studies, plasma osmolality increased by 8.8 ± 1.5 mosm/kg in dogs following 24-hour deprivation and by 15.2 ± 1.6 mosm/kg in the same dogs when the procedure was repeated after ablation of the OVLT (Thrasher *et al.*, 1983, Ramsay & Thrasher, 1986). In sheep, the elevation in plasma osmolality during dehydration following lamina terminalis lesions is even greater (McKinley *et al.*, 1982). The importance of natriuresis in the defense of plasma osmolality when water intake is inadequate is shown by these observations. Plasma osmolality control is not the province of water balance alone.

It is also of interest to point out that although large increases in plasma vasopressin occurs when normal dogs were water deprived, urine volume is not reduced. Presumably this is because of the increases in urinary sodium and osmolar clearance that accompany dehydration, thus obligating water loss. The effect of vasopressin, therefore, on reducing urinary water loss is minimal, being effectively prevented by the dehydration natriuresis. As O'Connor and Potts (1969) pointed out, water balance in dogs seems to be primarily regulated by water intake. Cowley, Skelton, and Merrill (1986) came to a similar conclusion in a series of chronic experiments in which plasma vasopressin was maintained constant by infusion. Thus, quantitative studies on thirst mechanisms assume an even greater significance in the consideration of water balance.

SATIETY

In 1939, Adolph reported that dehydrated dogs rapidly and accurately consume sufficient water to repair their deficits. Indeed, as mentioned previously, most dogs complete drinking within 2–3 minutes. Because dilution of the systemic plasma as a result of absorption of the ingested water does not begin for about 10 minutes, removal of cellular and extracellular dehydration thirst stimuli cannot account for the satiety. In this species, gastric distension does not seem to be an important factor in satiety, as gastric loading with water does not reduce subsequent water intake in dehydrated dogs (Adolph, 1950). Since dogs with esophageal fistulas drink in proportion to their deficits and show temporary satiation, oropharyngeal metering has been proposed as a mechanism for some time (Bellows, 1939; Adolph, 1950).

Not all animals restore fluid deficits following dehydration as rapidly as dogs (Adolph, 1967). Although all dehydrated animals begin drinking with a short latency, the rate of water intake may decline. For example, rats typically consume about one-half their deficits in 5 minutes and then drink much more slowly in the next hour to restore water balance to normal. Complete restoration of water balance may not occur until the next meal. Dogs, on the other hand, typically consume the total deficit during the first 5 minutes and rarely return to drink during the next hour. Most other mammals, including humans, fall between these two extremes (Rolls, *et al.*, 1980).

Thus, in rapid drinkers, the mechanisms that stop drinking must be different from those that cause it. The satiety, however, is only temporary. Permanent

DAVID J. RAMSAY
AND TERRY N.
THRASHER

satiety in any animal must entail correction of both cellular and extracellular fluid deficits. Slow drinkers, such as rats, can rely on correction of these body fluid deficits to reduce the thirst drive. The critical role of dilution of plasma osmolality in preventing further drinking in response to a number of dipsogenic stimuli in rats is well established. Again, the phenomenon of continuous water drinking (sham drinking) in dehydrated rats with esophageal or gastric fistulas shows that termination of drinking in this species must involve entry fistulae into the gut and/or its subsequent absorption (Blass, Jobaris, & Hall, 1976; Blass & Hall, 1976).

The study of drinking behavior in dehydrated dogs provides an excellent experimental model to separate temporary and permanent satiety mechanisms. In 1981, we reported the results of experiments in 24-hour water-deprived dogs during a number of procedures (Thrasher, Nistal-Herrera, Keil, & Ramsay, 1981). The dogs were prepared with gastric fistulas. When the fistula was closed with a plug, the dog's were essentially normal, and ingested fluid was absorbed from the gastrointestinal tract. With the fistula open, ingested fluid drained out and did not enter the duodenum. Moreover, fluid could be introduced into the stomach, thereby bypassing the mouth and esophagus. In these experiments, dehydrated dogs were offered either water or artificial extracellular fluid with the fistula either open or closed. In other experiments, fluid was introduced directly into the stomach. The amount of fluid consumed in 6 minutes was measured, and water was then offered 1 hour later to test satiety. In all experiments, the dogs had completed their drinking within the initial 6-minute period.

The data from these experiments are shown in Figure 12. The most striking result is that equivalent volumes of water or artificial extracellular fluid were consumed in the initial 6-minute period whether the fistula was open or closed. Temporary satiety, therefore, did not depend on water being the fluid consumed and did not depend on gastric or postgastric factors. Thus, although it has been suggested that putative receptors in the intestine and liver might be involved in satiety, such mechanisms are not essential in dogs. Furthermore, despite the fact that labeled water has been found in hepatic portal blood within 3 minutes of drinking in dogs, presenting a potential signal to hepatic receptors, it is difficult

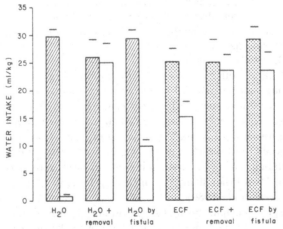

Figure 12. Fluid consumption after 24-hour water deprivation on presentation of water (hatched bars) or artificial ECF (stippled bars) or 60 minutes later (open bars). Removal: fluid was consumed and allowed to drain out via an open gastric fistula. Fistula: water or artificial ECF equal to the volume consumed by the dehydrated dogs when drinking spontaneously was introduced directly into the stomach via the gastric fistula. (From Thrasher *et al.*, 1981, with permission.)

to argue for their obligatory participation in temporary satiety in the light of these results. The original suggestion of Bellows (1939) that oropharyngeal metering provides the inhibitory input to water intake is supported by these observations.

It is also clear from these observations that only intake and the subsequent absorption of water provide permanent satiety as judged by intake of water when it is presented 1 hour later (Figure 11). When dehydrated dogs were first presented with water with the gastric fistula open, similar volumes were consumed. It is of interest to note that although the temporary satiety effects from oropharyngeal input had been quenched at the end of 1 hour, accurate oropharyngeal metering related to the deficit remained as similar volumes of water were drunk (Figure 12). The same argument can be applied to the situation when artificial extracellular fluid was consumed with the fistula open. The potency of oropharyngeal inhibitory input is also shown by the significant drinking 1 hour following intragastric administration of the appropriate volume of water to repair the deficit. Both cellular and extracellular deficits are corrected by that time, so theoretically there are no body fluid stimuli remaining. These important effects of oropharyngeal inputs have also been demonstrated by elegant experiments in which self-administration of intragastric or intravenous water in rats fails to satiate (Epstein, 1960; Nicolaidis & Rowland, 1974). Again, if the oropharynx is bypassed, important satiety inputs are missing. Thus, although oropharyngeal inputs are responsible for temporary satiety, they must also play an important part in the complex of events necessary to completely satisfy thirst drives.

Oropharyngeal inputs associated with drinking also affect renal conservation of water. The results in Figure 13 show that following ingestion of water, it

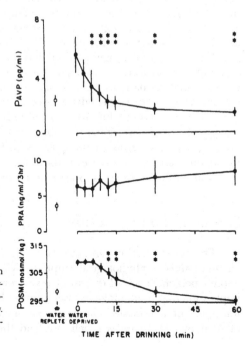

Figure 13. Effect of voluntary rehydration with water at time 0 in 24-hour water-deprived dogs on plasma osmolality, vasopressin concentration, and renin activity. (From Thrasher et al., 1981, with permission.)

takes 9–12 minutes for plasma osmolality to fall. Yet by 3 minutes plasma vasopressin is significantly reduced. Indeed, plasma vasopressin is reduced to pre-dehydration levels before plasma osmolality falls. This rapid initial fall in vasopressin occurs even if ingested water is removed via the gastric fistula or if artificial extracellular fluid rather than water is consumed. There is no initial fall in plasma vasopressin if water is introduced directly into the stomach. The rapid inhibition of vasopressin secretion is similar to temporary satiety in that it lasts only 20 minutes unless the ingested water is absorbed and results in correction of the original deficits. This rapid reduction in plasma vasopressin occurs in a number of species, including man (Blair-West, Gibson, Wood, & Brook, 1985; Geelen et al., 1984).

Nicolaidis (1969) described decreases in urine flow within 1 minute in hydrated rats following placement of hypertonic NaCl in the mouth or stomach. These anticipatory effects are likely to be of neural rather than humoral origin because of their rapid time course. The rapid fall in plasma vasopressin indicates that secretion of the peptide must have ceased early in the onset of drinking. Similar results have been reported in water-deprived monkeys, sheep, and humans. Electrophysiological evidence in monkeys shows that drinking caused immediate cessation of firing in supraoptic neurons (Vincent, Arnauld, & Biolac, 1972; Arnauld & duPont, 1982). Such experiments provide evidence for this phenomenon of rapid inhibition of vasopressin secretion.

The act of drinking is associated with an increase in heart rate and blood pressure (Hoffman, Phillips, Wilson, & Schmid, 1977; Thrasher, Keil, & Ramsay, 1987). It could be argued that the rise in blood pressure accounts for the reduction in vasopressin and temporary satiety. However, the increase in blood pressure with eating in dogs is similar to that with drinking, yet there is only inhibition of vasopressin and satiety following drinking. Moreover, similar increases in blood pressure can be achieved with bolus injections of phenylephrine, and this procedure fails to inhibit vasopressin (Thrasher et al., 1987). It is difficult to escape the conclusion that the oropharyngeal inputs associated with the act of drinking are responsible for inhibition of vasopressin secretion.

The nature of the ingested fluid may influence both satiety and vasopressin secretion in dehydrated animals. In monkeys, Arnauld and DuPont (1982) have reported that drinking hypertonic saline or glucose produces rapid increases in plasma vasopressin. In dehydrated dogs, we have found that drinking hypertonic saline or sucrose solutions does not result in profound reductions in plasma vasopressin and may enhance subsequent water drinking. Thus, although oropharyngeal metering of the quantity of fluid consumed appears to be the dominant feature, the nature of the fluid consumed may play a part in both satiety and inhibition of vasopressin secretion.

Summary and Conclusion

This review on thirst and the control of water balance has been written with an unashamedly physiological emphasis. We took the editor at his word and wrote a personal essay of the issues as we view them. We do not wish to leave the impression that fluid intake in association with eating or other behavioral cues is unimportant, and indeed, these cues play a major role in the pattern and timing of drinking. Drinking is a behavior and thus must take its place within the complex behavioral matrices specific to each species. However, underlying physiological systems are present to ensure that sufficient water intake occurs to allow homeostatic control of the volume and composition of the extracellular fluid. Water intake is intermittent, whereas the neurohumoral systems that regulate salt and water excretion operate continuously. When water intake occurs, satiety mechanisms allow physiologically appropriate volumes to be ingested. These mechanisms, together with integrated stimuli that regulate the intensity of the thirst stimulus, work to prevent overcorrection of fluid deficits. The physiological approach of studying water intake following the production of deficits mirrors the process that normally occurs in animals in hour-to-hour and day-to-day regulation of fluid balance. Without such physiological inputs, it is difficult to see how behavioral mechanisms could achieve such rigorously defended set points. The transduction mechanisms between physiological drive and species-specific expression of drinking behavior remain elusive.

REFERENCES

Abe, H., & Ogata, N. (1982). Ionic mechanisms for the osmotically induced depolorization in neurones of the guinea pig suprooptic nucleus *in vitro*. *Journal of Physiology (London)*, *327*, 157–171.

Adolph, E. F. (1939). Measurements of water drinking in dogs. *American Journal of Physiology*, *125*, 75–86.

Adolph, E. F. (1947). *Physiology of man in the desert*. New York: Interscience.

Adolph, E. F. (1950). Thirst and its inhibition in the stomach. *American Journal of Physiology*, *161*, 374–386.

Adolph, E. F. (1967). Regulation of water intake in relation to body water content. In C. F. Code (Ed.), *Handbook of physiology: Alimentary canal. Food and water intake* (section 6, vol. 1, pp. 163–171). Washington, DC: American Physiological Society.

Andersson, B. (1953). The effect of injections of hypertonic NaCl solutions into different parts of the hypothalamus of goats. *Acta Physiologica Scandinavica*, *28*, 188–201.

Andersson, B. (1978). Regulation of water intake. *Physiological Reviews*, *58*, 582–603.

Andersson, B., & McCann, S. M. (1955). The effect of hypothalamic lesions on the water intake of the dog. *Acta Physiologica Scandinavica*, *35*, 312–320.

Arieff, A. I., Llach, F., & Massry, S. G. (1976). Neurological manifestations and morbidity of hyponatremia: Correlation with brain water and electrolytes. *Medicine*, *55*, 121–129.

Arnauld, E., & duPont, J. (1982). Vasopressin release and firing of supraoptic neurones during drinking in the dehydrated monkey. *Pfluegers Archiv*, *394*, 195–201.

Bealer, S. L., Crofton, J. T., & Share, L. (1983). Hypothalamic knife cuts alter fluid regulation, vasopressin secretion and natriuresis during water deprivation. *Neuroendocrinology*, *36*, 364–370.

Bellows, R. T. (1939). Time factors in water drinking in dogs. *American Journal of Physiology*, *125*, 87–97.

Bishop, V. S., Thames, M. D., & Schmid, P. G. (1984). Effects of bilateral cold block on vasopressin in conscious dogs. *American Journal of Physiology*, *246*, R566–R569.

Blair-West, J. R., Gibson, A. P., Wood, R. L., & Brook, A. H. (1985). Acute reduction of plasma vasopressin by rehydration in sheep. *American Journal of Physiology*, *248*, R68–R71.

Blass, E. M., & Epstein, A. N. (1971). A lateral preoptic osmosensitive zone for thirst in the rat. *Journal of Comparative and Physiological Psychology*, *76*, 378–394.

Blass, E. M., & Hall, W. G. (1976). Drinking termination: Interactions among hydrational, orogastric and behavioral controls in rats. *Psychological Reviews*, *83*, 356–374.

Blass, E. M., Jobaris, R., & Hall, W. G. (1976). Oropharyngeal control of drinking in rats. *Journal of Comparative and Physiological Psychology*, *90*, 909–916.

Brody, M. J., & Johnson, A. K. (1980). Role of the anteroventral third ventricle region in fluid and electrolyte balance, arterial pressure regulation and hypertension. In L. Martin & W. F. Ganong (Eds.), *Frontiers in neuroendocrinology* (pp. 249–292). New York: Raven Press.

Buggy, J., & Johnson, A. K. (1977). Preoptic hypothalamic periventricular lesions: Thirst deficits and hypernatremia. *American Journal of Physiology*, *233*, R44–R52.

DAVID J. RAMSAY
AND TERRY N.
THRASHER

Camacho, A., & Phillips, M. I. (1980). Horseradish peroxidase study in rat of the neural connections of the organum vasculosum of the lamina terminalis. *Neuroscience Letters, 25*, 201–204.

Cannon, W. B. (1919). The physiological basis of thirst. *Proceedings of the Royal Society, London, 90*, 283–301.

Chaudhry, M. A., Dyball, R. E. J., Honda, K., & Wright, N. C. (1989). The role of interconnection between supraoptic nucleus and anterior third ventricular region in osmoregulation in the rat. *Journal of Physiology, (London), 410*, 123–135.

Claybough, J. R. (1976). The effect of dehydration on stimulation of ADH release by heterologous renin infusions on conscious dogs. *American Journal of Physiology, 231*, 655–660.

Coburn, P. C., & Stricker, E. M. (1978). Osmoregulatory thirst in rats after lateral preoptic lesions. *Journal of Comparative and Physiological Psychology, 92*, 350–361.

Cowley, A. W., Skelton, M. M., & Merrill, D. C. (1986). Osmoregulation during high salt intake: Relative importance of drinking and vasopressin secretion. *American Journal of Physiology, 251*, R878–R886.

Ellory, J. C., & Hall, A. C. (1988). Human red cell volume regulation in hypotonic media. *Comparative Biochemistry and Physiology, 90A*, 533–537.

Epstein, A. N. (1960). Water intake without the act of drinking. *Science, 131*, 497–498.

Epstein, A. N. (1978). Consensus, controversies and curiosities. *Federation Proceedings, 37*, 2711–2716.

Epstein, A. N. (1986). Hormonal synergy as the cause of salt appetite. In G. deCaro, A. N. Epstein, & M. Massi (Eds.) *The physiology of thirst and sodium appetite* (pp. 395–404). New York: Plenum Press.

Epstein, A. N., Fitzsimons, J. P., & Rolls, B. J. (1970). Drinking induced by injection of angiotensin into the brain of the rat. *Journal of Physiology (London), 210*, 457–474.

Ferguson, A. V. (1988). Systemic angiotensin acts at the subfornical organ to control the activity of paraventricular nucleus neurons with identified projections to the median eminence. *Neuroendocrinology, 47*, 489–497.

Ferguson, A. V., Day, T. A., & Renaud, L. P. (1984a). Subfornical organ stimulation excites paraventricular neurones projecting to dorsal medulla. *American Journal of Physiology, 247*, R1088–R1092.

Ferguson, A. V., Day, T. A., & Renaud, L. P. (1984b). Subfornical organ efferents influence the excitability of neurohypophyseal and tuberoinfundibular pariventricular nucleus neurones in the rat. *Neuroendocrinology, 39*, 423–428.

Fitzsimons, J. T. (1961a). Drinking by nephrectomized rats injected with various substances. *Journal of Physiology (London), 155*, 563–579.

Fitzsimons, J. T. (1961b). Drinking by rats depleted of body fluid without increase in osmotic pressure. *Journal of Physiology (London), 159*, 297–309.

Fitzsimons, J. T. (1964). Drinking caused by constriction of the inferior vena cava in the rat. *Nature, 204*, 479–480.

Fitzsimons, J. T. (1969). The role of the renal thirst factor in drinking induced by extracellular stimuli. *Journal of Physiology (London), 201*, 349–368.

Fitzsimons, J. T. (1972). Thirst. *Physiological Review, 52*, 468–561.

Fitzsimons, J. T. (1979). *The physiology of thirst and sodium appetite.* Cambridge: Cambridge University Press.

Fitzsimons, J. T., & Elfont, R. W. (1982). Angiotensin does contribute to drinking induced by caval ligation in the rat. *American Journal of Physiology, 243*, R558–R563.

Fitzsimons, J. T., & Simons, B. J. (1969). The effects on drinking in the rat of intravenous infusion of angiotensin, given alone or in combination with other stimuli of thirst. *Journal of Physiology (London), 203*, 45–57.

Fitzsimons, J. T., & Szczepanska-Sadowska, E. (1974). Drinking and antidiuresis elicited by isoprenaline in the dog. *Journal of Physiology (London), 239*, 251–260.

Gardiner, T. W., Mangiapane, M. L., & Simpson, J. B. (1981). Lesion of nucleus medianus but not organum vasculasum produce adipsia and thirst deficits in mice. *Neuroscience, 7*, 168–175.

Gardiner, T. W., & Stricker, E. M. (1985a). Hyperdipsia in rats after electrolytic lesions of nucleus medianus. *American Journal of Physiology, 248*, R214–R223.

Gardiner, T. W., & Stricker, E. M. (1985b). Impaired drinking responses of rats with lesions of nucleus medianus: Circadian dependence. *American Journal of Physiology, 248*, R224–R230.

Gardiner, T. W., Verbalis, J. G., & Stricker, E. M. (1985). Impaired secretion of vasopressin and oxytocin in rats after lesions of nucleus medianus. *American Journal of Physiology, 249*, R681–R688.

Gauer, O. H., & Henry, J. P. (1963). Circulatory basis of fluid volume control. *Physiological Reviews, 43*, 423–481.

Geelen, G., Keil, L. C., Kravik, S. E., Wade, C. E., Thrasher, T. N., Barnes, P. R., Pyka, G., Nesvig, C., & Greenleaf, J. E. (1984). Inhibition of plasma vasopressin after drinking in dehydrated humans. *American Journal of Physiology, 247,* R968–R971.

Gilman, A. (1937). The relation between blood osmotic pressure, fluid distribution and voluntary water intake. *American Journal of Physiology, 120,* 323–328.

Grantham, J. J. (1977). Pathophysiology of hypo-osmolar conditions: A cellular perspective. In T. E. Androli, J. J. Grantham, & F. C. Rector, Jr. (Eds.), *Disturbances in body fluid osmolality* (pp. 217–226). Bethesda: American Physiological Society.

Gutman, M. B., Ciriello, J., & Mogenson, G. J. (1986). Electrophysiological identification of forebrain connection of the subfornical organ. *Brain Research, 382,* 119–128.

Hoffman, W. E., Phillips, M. I., Wilson, E., & Schmid, P. G. (1977). A pressor response associated with drinking in rats. *Proceedings of the Society for Experimental Biology and Medicine, 154,* 121–125.

Honda, K., Negoro, H., Higuchi, T., & Tadokoro, Y. (1987). Activation of neurosecretory cells by osmotic stimulation of anteroventral third ventricle. *American Journal of Physiology, 252,* R1039–R1045.

Hosutt, J., Rowland, N., & Stricker, E. M. (1981). Impaired drinking responses of rats with lesions of the subfornical organ. *Journal of Comparative and Physiological Psychology, 95,* 104–113.

Houpt, K. A., & Epstein, A. W. (1972). The complete dependence of beta-adrenergic drinking on the renal dipsogen. *Physiology and Behavior, 7,* 897–905.

Jewell, P. A., & Verney, E. B. (1957). An experimental attempt to determine the site of the neurohypophysial osmoreceptors in the dog. *Philosophical Transactions of the Royal Society of London, Series B, 240,* 197–324.

Johnson, A. K. & Buggy, J. (1978). Periventricular preoptic-hypothalamus is vital for thirst and normal water economy. *American Journal of Physiology, 234,* 122–125.

Johnson, A. K., Mann, J. F. E., Racsher, W., Johnson, J. K., & Ganten, D. (1981). Plasma angiotensin II concentrations and experimentally induced thirst. *American Journal of Physiology, 240,* R229–R236.

Kaufman, S. (1984). Role of right atrial receptors in the control of drinking in the rat. *Journal of Physiology (London), 349,* 389–397.

Klingbeil, C. K., Quillen, E. W., Brooks, V. L., & Reid, T. A. (1986). Effect of baroreceptor denervation in the stimulation of drinking by angiotensin II. *Federation Proceedings, 45,* 904.

Kozlowski, S., & Szczepanska-Sadowska, E. (1975). Mechanisms of hypovolaemic thirst and interactions between hypovolaemia, hyperosmolality and the antiuretic system. In G. Peters, J. T. Fitzsimons, & L. Peters-Haefeli (Eds.), *Control mechanisms of drinking* (pp. 25–35). Berlin, Heidelberg: Springer-Verlag.

Lee, M. C., Thrasher, T. N., & Ramsay, D. J. (1981). Is angiotensin essential in drinking induced by water deprivation and caval ligation? *American Journal of Physiology, 240,* R75–R80.

Leng, G., Dyball, R. E. J., & Mason, W. T. (1985). Electrophysiology of osmoreceptors. In R. W. Schrier (Ed.), *Vasopressin* (pp. 333–342). New York: Raven Press.

Leng, G., Dyball, R. E. J., & Russell, J. A. (1988). Neurophysiology of body fluid homeostasis. *Comparative Biochemistry and Physiology, 90A,* 781–788.

Leng, G., Mason, W. T., & Dyer, R. G. (1982). The supraoptic nucleus as an osmoreceptor. *Neuroendocrinology, 34,* 75–82.

Lind, R. W., vanHoesen, G. W., & Johnson, A. K. (1982). An HRP study of the connections of the subfornical organ of the rat. *Journal of Comparative Neurology, 210,* 265–277.

Linden, R. J., & Kappagoda, C. T. (1982). *Atrial receptors.* Cambridge: Cambridge University Press.

Lucke, C., Erbet, H., Herting, T., & Dohler, K. D. (1980). Secretion of arginine vasopressin, aldosterone and corticosterone and plasma renin activity in water deprived rats. *Contributions to Nephrology, 19,* 63–70.

Luke, R. G. (1973). Natriuresis and chloruresis during hydropenia in the rat. *American Journal of Physiology, 224,* 13–20.

Mann, J. F. E., Johnson, A. K., Rascher, W., Genest, J., & Ganten, D. (1981). Thirst in the rat after ligation of the inferior vena cava: Role of angiotensin II. *Pharmacology, Biochemistry and Behavior, 15,* 337–344.

Mason, W. T. (1980). Supraoptic neurones of rat hypothalamus are osmosensitive. *Nature, 241,* 154–157.

Mason, W. T. (1982). Electrical properties of neurones recorded from rat supraoptic nucleus *in vitro. Proceedings of the Royal Society of London, Series B, 217,* 141–161.

Mayer, A. (1900). Variation de la tension osmotique de sang chez les animaux privés de liquides. *Comtes Rendus des Séances de la Société de Biologie et de ses Filiales, 52,* 153–155.

McCance, R. A. (1936). Experimental sodium chloride deficiency in man. *Proceedings of the Royal Society, London, Series B, 119*, 245–268.

McKenna, T. M., & Haines, H. (1981). Sodium metabolism during acclimation to water restriction by wild mice, *Mus musculus*. *American Journal of Physiology, 240*, R319–R325.

McKinley, M. J., Denton, D. A., Coghlan, J. P., Harvey, R. B., McDougall, J. G., Rundgren, M., Scoggins, B. A., & Weisinger, R. S. (1987). Cerebral osmoregulation of renal sodium excretion—a response analogous to thirst and vasopressin release. *Canadian Journal of Physiology and Pharmacology, 65*, 1724–1729.

McKinley, M. J., Denton, D. A., Leksell, L. C., Mouw, D. R., Scoggins, B. A., Smith, M. H., Weisinger, R. S., & Wright, R. D. (1982). Osmoregulatory thirst in sheep is disrupted by ablation of the anterior wall of the optic recess. *Brain Research, 236*, 210–215.

McKinley, M. J., Denton, D. A., Leksell, L. G., Tarjan, E., & Weisinger, R. S. (1980). Evidence for cerebral sodium sensors involved in water drinking in dogs. *Physiology and Behavior, 25*, 501–515.

McKinley, M. J., Denton, D. A., Nelson, J. F., & Weisinger, R. S. (1983). Dehydration induces sodium depletion in rats, rabbits and sheep. *American Journal of Physiology, 245*, R287–R292.

McKinley, M. J., Denton, D. A., Park, R. C., & Weisinger, R. S. (1983). Cerebral involvement in dehydration induced natriuresis. *Brain Research, 263*, 340–343.

McKinley, M. J., Denton, D. A., & Weisinger, B. S. (1978). Sensors for antidiuresis and thirst—osmoreceptors or CSF-sodium detectors? *Brain Research, 141*, 89–103.

Menninger, R. P. (1979). Response of supraoptic neurosecretory cells to changes in left atrial distension. *American Journal of Physiology, 236*, R261–R267.

Merrill, D. C., Skelton, M. M., & Cowley, A. J., Jr. (1986). Humoral control of water and electrolyte excretion during water restriction. *Kidney International, 29*, 1152–1161.

Metzler, C. H., Thrasher, T. N., Keil, L. C., & Ramsay, D. J. (1986). Endocrine mechanisms regulating sodium excretion during water deprivation in dogs. *American Journal of Physiology, 251*, R560–R568.

Miselis, R. R. (1981). The efferent projection of the subfornical organ of the rat: A circumventricular organ within a neural network subserving water balance. *Brain Research, 230*, 1–23.

Miselis, R. R., Shapiro, R. E., & Hand, P. J. (1979). Subfornical organ efferents to neural systems for control of body water. *Science, 205*, 1022–1025.

Montgomery, M. F. (1931). The role of the salivary glands in the thirst mechanism. *American Journal of Physiology, 96*, 221–227.

Nicolaidis, S. (1969). Early systemic responses to orogastric stimulation in the regulation of food and water balance: Functional and electrophysiological data. *Annals of the New York Academy of Sciences, 151*, 1176–1203.

Nicolaidis, S., & Rowland, N. (1974). Long-term self-intravenous "drinking" in the rat. *Journal of Comparative and Physiological Psychology, 87*, 1–15.

O'Connor, W. J., & Potts, D. J. (1969). The external water exchanges of normal laboratory dogs. *Quarterly Journal of Experimental Physiology, 54*, 244–265.

Olsson, K., Fyhrquist, F., Larsson, B., & Eriksson, L. (1978). Inhibition of vasopressin release during developing hypernatremia and plasma hyperosmolality: An effect of intracerebroventricular glycerol. *Acta Physiologica Scandinavica, 102*, 399–409.

Olsson, K., Larsson, B., & Liljekvist, E. (1970). Intracerebroventricular glycerol: A potent inhibitor of ADH-release and thirst. *Acta Physiologica Scandinavica, 98*, 470–477.

Park, R. G., Congin, M., Denton, D. A., & McKinley, M. J. (1985). Natriuresis induced by arginine vasopressin in sheep. *American Journal of Physiology, 249*, F799–F805.

Peck, J. W., & Blass, E. M. (1975). Localization of thirst and antidiuretic osmoreceptors by intracranial injections in rats. *American Journal of Physiology, 228*, 1501–1509.

Peck, J. W., & Novin, D. (1971). Evidence that osmoreceptors mediating drinking in rabbits are in the lateral preoptic area. *Journal of Comparative and Physiological Psychology, 74*, 134–147.

Phillips, M. I. (1987). Functions of angiotensin in the central nervous system. *Annual Reviews of Physiology, 49*, 413–435.

Rampton, D. S., & Ramsay, D. J. (1974). The effects of the production of sodium depletion by peritonial dialysis with 5% glucose of the volume and composition of extracellular fluid of dogs. *Journal of Physiology (London), 237*, 535–553.

Ramsay, D. J. (1978). Beta-adrenergic thirst and its relation to the renin–angiotensin system. *Federation Proceedings, 37*, 2689–2693.

Ramsay, D. J. (1979). The brain renin/angiotensin system. A re-evaluation. *Endocrinology, 104*, 672–676.

Ramsay, D. J. (1982). Effects of circulating angiotensin II on the brain. In W. F. Ganong & L. Martini (Eds.), *Frontiers in neuroendocrinology* (pp. 263–282). New York: Raven Press.

Ramsay, D. J., Reid, I. A., Keil, L. C., & Ganong, W. F. (1978). Evidence that the effects of iso-proterenol on water intake and vasopressin secretion are mediated by angiotensin. *Endocrinology*, *103*, 54–58.

Ramsay, D. J., Rolls, B. J., & Wood, R. J. (1975). The relationship between elevated water intake and edema associated with congestive cardiac failure in the dog. *Journal of Physiology (London,) 244*, 303–315.

Ramsay, D. J., Rolls, B. J., & Wood, R. J. (1977). Thirst following water deprivation in dogs. *American Journal of Physiology*, *232*, R93–R100.

Ramsay, D. J., & Thrasher, T. N. (1986). Hyperosmotic and hypovolemic thirst. In G. deCaro, A. N. Epstein, & M. Massi (Eds.), *The physiology of thirst and sodium appetite* (pp. 83–96). New York: Plenum Press.

Ramsay, D. J., Thrasher, T. N., & Bie, P. (1988). Endocrine components of body fluid homeostasis. *Comparative Biochemistry and Physiology*, *90A*, 777–780.

Reid, I. A., Brooks, V. L., Rudolph, C. D., & Keil, L. C. (1982). Analysis of actions of angiotensin on the central nervous system. *American Journal of Physiology*, *243*, R82–R91.

Rettig, R., Ganten, D., & Johnson, A. K. (1981). Isoproterenol-induced thirst: Renal and extrarenal mechanisms. *American Journal of Physiology*, *241*, R152–R158.

Robertson, G. L. (1984). Abnormalities of thirst regulation. *Kidney International, 25*, 460–469.

Robertson, G. L. (1985). Osmoregulation of thirst and vasopressin secretion: Functional properties and their relationship to water balance. In R. Schrier (Ed.), *Vasopressin* (pp. 202–213). New York: Raven Press.

Robertson, G. L., Athar, S., & Shelton, R. L. (1977). Osmotic control of vasopressin function. In T. E. Andocoli, J. J. Grantham, & F. C. Rector, Jr. (Eds.), *Disturbances in body fluid osmolality* (pp. 125–148). Bethesda: American Physiological Society.

Rolls, B. J., & Ramsay, D. J. (1975). The elevation of endogenous angiotensin and thirst in the dog. In G. Peters, J. T. Fitzsimons, & L. Peters-Hoepeli (Eds.), *Control mechanisms of drinking* (pp. 74–78). Berlin, Heidelberg: Springer-Verlag.

Rolls, B. J., Wood, R. J., Rolls, E. J., Lind, H., Lind, R. W., & Ledingham, J. C. G. (1980). Thirst following water deprivation in humans. *American Journal of Physiology*, *239*, R476–R482.

Saper, C. B., & Levisohn, D. (1983). Afferent connections of the median preoptic nucleus in the rat: Anatomical evidence for a cardiovascular integrative mechanism in the AV3V. *Brain Research, 288*, 21–31.

Sayer, R. J., Hubbard, J. C., & Sirret, N. E. (1984). Rat organum vasculosum laminae *in vitro*: Response to neurotransmitters. *American Journal of Physiology*, *247*, R374–R379.

Schneider, E. G., Taylor, R. E., Radke, K. J., & Davis, P. G. (1984). Effect of sodium concentration on aldosterone secretion by isolated, perfused canine adrenal glands. *Endocrinology, 115*, 2195–2204.

Sgro, S., Ferguson, A. V., & Renaud, L. P. (1984). Subfornical organ–supraoptic nucleus connections: An electrophysiological study in the rat. *Brain Research, 303*, 7–13.

Share, L., and Levy, M. N. (1966). Carotid sinus pulse pressure, a determinant of plasma antidiuretic hormone concentration. *American Journal of Physiology, 211*, 721–724.

Shimizu, K., Share, L., & Claybough, J. R. (1973). Potentiation of angiotensin II of the vasopressin response to an increasing plasma osmolality. *Endocrinology, 93*, 42–50.

Simpson, J. B., & Routeenberg, A. (1973). Subfornical organ: Site of drinking and elicitation by angiotensin II. *Science, 181*, 1172–1175.

Sladek, C. D., & Johnson, A. K. (1983). Effect of anteroventral third ventricle lesions on vasopressin release by organ-cultured hypothalamoneurohypophyseal explants. *Neuroendocrinology, 37*, 78–84.

Stricker, E. M. (1969). Osmoregulation and volume regulation in rats: Inhibition of hypovolemia thirst by water. *American Journal of Physiology, 217*, 98–105.

Stricker, E. M. (1978). The renin–angiotensin system and thirst—some unanswered questions. *Federation Proceedings, 37*, 2704–2710.

Stricker, E. M. (1981). Thirst and sodium appetite after colloid treatment in rats. *Journal of Comparative and Physiological Psychology, 95*, 1–25.

Stricker, E. M., & Jaloweic, J. E. (1970). Restoration of intravascular fluid volume following acute hypovolemia in rats. *American Journal of Physiology, 218*, 191–196.

Thrasher, T. N. (1989). Role of forebrain circumventricular organs in body fluid balance. *Acta Physiologica Scandinavica, 136*, 141–150.

Thrasher, T. N., Brown, C. J., Keil, L. C., & Ramsay, D. J. (1980). Thirst and vasopressin release in the dog: An osmoreceptor or sodium receptor mechanism? *American Journal of Physiology, 238*, R333–R339.

Thrasher, T. N., Jones, R. G., Keil, L. C., Brown, C. J., & Ramsay, D. J. (1980). Drinking and vasopressin release during ventricular infusions of hypertonic solutions. *American Journal of Physiology, 238,* R340–R345.

Thrasher, T. N., & Keil, L. C. (1987). Regulation of drinking and vasopressin secretion: Role of the organum vasculosum laminae terminalis. *American Journal of Physiology, 253,* R108–R120.

Thrasher, T. N., Keil, L. C., & Ramsay, D. J. (1982a). Lesions of the organum vasculosum of the lamina terminalis (OVLT) attenuate osmotically-induced drinking and vasopressin secretion in the dog. *Endocrinology, 110,* 1837–1839.

Thrasher, T. N., Keil, L. C., & Ramsay, D. J. (1982b). Hemodynamic, hormonal and drinking responses to reduced venous return in the dog. *American Journal of Physiology, 243.* R354–R362.

Thrasher, T. N., Keil, L. C., & Ramsay, D. J. (1983). Altered responses to dehydration in dogs with lesions of the OVLT. *Proceedings of the International Congress of Physiological Sciences, 49.*

Thrasher, T. N., Keil, L. C., & Ramsay, D. J. (1987). Drinking, oropharyngeal signals and inhibition of vasopressin secretion in dogs. *American Journal of Physiology, 253,* R509–R515.

Thrasher, T. N., Nistal-Herrera, J. F., Keil, L. C., & Ramsay, D. J. (1981). Satiety and inhibition of vasopressin secretion after drinking in dogs. *American Journal of Physiology, 240,* E394–E401.

Thrasher, T. N., Simpson, J. B., & Ramsay, D. J. (1982). Lesions of the subfornical organ block angiotensin induced drinking in the dog. *Neuroendocrinology, 35,* 68–72.

Thrasher, T. N., Wade, C. E., Keil, L. C., & Ramsay, D. J. (1984). Sodium balance and aldosterone during dehydration and rehydration in the dog. *American Journal of Physiology, 247,* R76–R83.

Verney, E. B. (1947). The antidiuretic hormone and the factors which determine its release. *Proceedings of the Royal Society of London, Series B, 135,* 25–106.

Vincent, J. D., Amauld, E., & Biolac, B. (1972). Activity of osmosensitive single cells in the hypothalamus of the behaving monkey during drinking. *Brain Research, 44,* 371–384.

Vokes, T. J., & Robertson, G. L. (1988). Disorders of antidiuretic hormone. *Endocrinology and Metabolism Clinics of North America, 17,* 280–289.

Wade, C. E., Bie, P., Keil, L. C., & Ramsay, D. J. (1982a). Osmotic control of vasopressin in the dog. *American Journal of Physiology, 243,* E287–E292.

Wade, C. E., Bie, P., Keil, L. C., & Ramsay, D. J. (1982b). Effect of hypertonic intracarotid infusions on plasma vasopressin concentration. *American Journal of Physiology, 243,* E522–E526.

Wade, C. E., Keil, L. C., & Ramsay, D. J. (1983). Role of volume and osmolality in the control of plasma vasopressin in dehydrated dogs. *Neuroendocrinology, 37,* 349–353.

Weindl, A., (1973). Neuroendocrine aspects of circumventricular organs. In W. F. Ganong & L. Martini (Eds.), *Frontiers in neuroendocrinology, 1973* (pp. 3–32). New York: Oxford University Press.

Weisinger, R. S., Considine, P., Denton, D. A., Leksell, L. G., McKinley, M. J., Mouw, D., Muller, A., & Tarjan, E. (1982). Role of sodium concentration in the cerebro-spinal fluid in the salt appetite of sheep. *American Journal of Physiology, 242,* R51–R60.

Wolf, A. V. (1950). Osmometric analysis of thirst in man and dog. *American Journal of Physiology, 161,* 75–86.

Wood, R. J., Rolls, B. J., & Ramsay, D. J. (1977). Drinking following intracarotid infusions of hypertonic solutions in dogs. *American Journal of Physiology, 232,* R93–R100.

Yamashita, H. (1977). Effect of baro- and chemoreceptor activation on supraoptic nuclei neurones in the hypothalamus. *Brain Research, 126,* 551–556.

Yannet, H. (1940). Changes in the brain resulting from depletion of extracellular electrolytes. *American Journal of Physiology, 128,* 683–689.

Zucker, A., Gleason, S. D., & Schneider, E. G. (1982). Renal and endocrine responses to water deprivation in dog. *American Journal of Physiology, 242,* R296–R302.

15

Sodium Appetite

Edward M. Stricker and Joseph G. Verbalis

Introduction

Sodium is the backbone of extracellular fluid. Because cell membranes are functionally impermeable to sodium, it causes osmotic retention of water in the extracellular space and thereby enables the vascular system to expand. Given the importance of blood circulation in mammals, it is hardly hyperbole to conclude that the maintenance of adequate supplies of sodium in the extracellular fluid is essential to life in higher species.

Not surprisingly, there are a variety of physiological controls to preserve body sodium. Prominent among them is the adrenocortical hormone aldosterone, which promotes the conservation of sodium in the kidneys. A major factor controlling aldosterone secretion is the hormone angiotensin, formed when the renal enzyme renin acts on an otherwise inert peptide, angiotensinogen, in the blood. Angiotensin also supports blood pressure directly via pressor effects on blood vessels, especially during sodium deficiency. However, complementing these and other physiological adjustments to sodium deficiency is the specific appetite for NaCl.

Sodium appetite represents a strong motivation of animals to seek, obtain, and consume salty tasting fluids and foods. It is an innate motivation in rats (Handal, 1965a; Krieckhaus & Wolf, 1968; Nachman, 1962), analogous to thirst, and can induce them to work to obtain salt (Quartermain, Miller, & Wolf, 1967; Wagman, 1963); indeed, the desire for NaCl may be so strong that it can even overcome previously established aversions to salty tasting fluids (Frumkin, 1971; Stricker & Wilson, 1970). The taste of salt guides and maintains NaCl consumption (e.g., Morrison & Young, 1972; Smith, Holman, & Fortune, 1968), and

Edward M. Stricker   Department of Behavioral Neuroscience, University of Pittsburgh, Pittsburgh, Pennsylvania 15260.   Joseph G. Verbalis   Departments of Medicine and Behavioral Neuroscience, University of Pittsburgh, Pittsburgh, Pennsylvania 15261.

sodium-deficient rats consume NaCl even when it cannot be absorbed from the gastrointestinal tract (Epstein & Stellar, 1955; Mook, 1969), evidently because it tastes very good to them (Berridge, Flynn, Schulkin, & Grill, 1984; Jacobs, Mark, & Scott, 1988; also see Chapter 10).

The significant variables in the control of sodium homeostasis parallel those involved in body water homeostasis. However, whereas the regulation of thirst and secretion of the antidiuretic hormone, arginine vasopressin (AVP), are well established (see Chapters 14 and 16), as are the factors that control secretion of aldosterone and angiotensin, the potential regulation of sodium appetite remains quite controversial. This chapter addresses that issue. We focus exclusively on sodium appetite in rats and emphasize research reported during the 7 years since Denton's (1982) masterful, comprehensive monograph on this subject. We begin with a description of various models that have been used in the study of sodium appetite in rats, but the remainder of the chapter is devoted to the presentation and consideration of a new hypothesis to explain the biological bases of this behavior.

MODELS OF SODIUM APPETITE

Numerous models of sodium appetite in rats have been described, but in each case NaCl ingestion usually is seen in association with sodium deficiency. Such deficiency may occur whether or not animals are in negative sodium balance (i.e., total body sodium deficit). The oldest and most familiar model of sodium appetite derives from the uncontrolled loss of sodium in urine after adrenalectomy. Rats die of sodium deficiency within a few days after adrenalectomy when they are not given either mineralocorticoid therapy or replacement amounts of sodium. However, Richter (1936) demonstrated that adrenalectomized animals would voluntarily replace the lost sodium by appropriate ingestive behavior if given the opportunity to do so, and thereby maintain themselves indefinitely. Specifically, the rats consumed approximately 20 mL of 0.5 M NaCl per day, containing 10 mEq of sodium, an amount equivalent to the total sodium content in the extracellular fluid of adult rats. This impressive intake is analogous to the drinking behavior of rats with diabetes insipidus, which lose water in urine uncontrollably because of an absence of AVP and in compensation consume water in amounts equivalent to their total body water content each day (Richter, 1938). Other models for producing negative sodium balance in rats involve the administration of a natriuretic drug (Jalowiec, 1974), thermal dehydration (Yawata, Okuno, Nose, & Morimoto, 1987), or intraperitoneal glucose dialysis (Falk, 1965), each of which causes much smaller and more acute sodium deficits.

Note that studies of sodium appetite in rats typically employ concentrated NaCl solutions, which appear to be so unpalatable that they would not be consumed otherwise, much as the ingestion of bitter-tasting fluids has been used to signify strong motivation in studies of thirst. Furthermore, because the rat cannot excrete urine containing as much as 500 mEq Na per liter, solutions of this concentration are used as drinking fluid to eliminate the possibility that rats consume saline as a result of thirst.

A second group of models involves deficits in plasma volume (hypovolemia) achieved by an internal redistribution of extracellular fluid in the absence of total body sodium deficits. The most frequently studied example of such sodium appetite in rats is produced by subcutaneous injection of a hyperoncotic colloidal solution. The colloid gradually draws isosmotic plasma fluid across the capillary membrane to the interstitium in amounts proportional to the volume and concentration of the injected solution (Fitzsimons, 1961; Stricker, 1968). The loss of protein-free plasma fluid increases the plasma protein concentration, which then pulls interstitial fluid into the circulation from other tissues, but that fluid also is withdrawn and sequestered as subcutaneous edema. Thus, the Starling equilibrium forces that normally serve to sustain plasma volume after blood loss are overwhelmed, and hypovolemia develops. Fluid accumulation increases progressively over 12–16 hours, and its magnitude can be quite considerable: 5 mL of 30% PEG solution injected into an adult rat causes up to 30% reduction of plasma volume, and twice that dose can produce so much hemoconcentration that the kidneys fail. The induced plasma volume deficits linger for at least 24 hours, until the lymphatic system drains the injected colloid and sequestered fluid from the subcutaneous tissue.

Appropriate to their need for a dilute saline solution isosmotic with plasma, rats given a subcutaneous injection of polyethylene glycol (PEG) solution increase their intakes of both water and 0.5 M NaCl solution (Stricker & Jalowiec, 1970; Stricker & Wolf, 1966). Close analysis of these changes is illuminating (Stricker, 1981). At first, the PEG-treated rats drink water exclusively while excreting small amounts of sodium-poor urine; the renal conservation of water and sodium results from decreases in renal blood flow (Brenner & Berliner, 1969) and increases in the secretion of AVP (Dunn, Brennan, Nelson, & Robertson, 1973; Stricker & Verbalis, 1986) and aldosterone (Stricker, Vagnucci, McDonald, & Leenen, 1979). After 5 hours, the rats begin to drink the concentrated saline solution in rapid alternation with water consumption and thereby steadily increase the cumulative intakes of each fluid (Figure 1). Nevertheless, the renal conservation of water and sodium persists because most of the ingested fluid also is sequestered in the edema at the injection site. However, ultimately

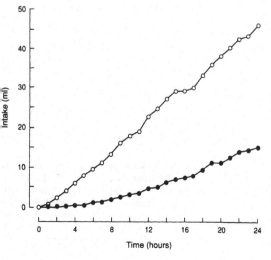

Figure 1. Cumulative mean volumes of water (open circles) and 0.5 M NaCl solution (closed circles) ingested by rats after 30% PEG treatment. Rats were deprived of food during the 24-hour test period. Noninjected control animals drank 20 mL of water and 5 mL of saline in 24 hours. (Adapted from Stricker & Jalowiec, 1970.)

EDWARD M.
STRICKER AND
JOSEPH G.
VERBALIS

tissue turgor limits further fluid accumulation at that site, and thereafter ingested fluid remains largely in the circulation and serves to repair plasma volume. Renal conservation of sodium ends at this time, approximately 12–18 hours after injection of the colloid. It is noteworthy that the PEG-treated rats do not stop drinking saline when the hypovolemia is abolished but continue to consume both NaCl solution and water throughout the 24-hour test period (Figure 1). In consequence, by the end of 24 hours they often consume as much as 10 mEq of sodium, an amount far greater than their intravascular deficit.

The onset of sodium appetite is seen more clearly when the intakes of water and saline are replotted as a function of one another rather than as a function of time (Figure 2). The initial period of exclusive water drinking (phase 1) is shown as the regression line paralleling the Y axis, with each point on the line representing a cumulative hourly reading of water and saline intakes. This line abruptly deflects toward the horizontal at 5 hours, indicating the onset of sodium appetite (phase 2). Many hours later, when the rats have consumed and retained sufficient volumes of water and saline so that they no longer are hypovolemic, the line suddenly deflects further toward the horizontal, but nonetheless the animals continue to drink both saline and water at a new, steady rate (phase 3).

It should be noted that Figure 2 makes clear what was not conveyed in Figure 1, namely, that there are three distinct phases in the ingestive behavior of rats after subcutaneous PEG treatment and that the ratio of water to saline consumption is quite different across the phases but constant within each phase. This regular, triphasic pattern of ingestion does not merely result from the rats' inexperience with hypovolemia; three weekly injections of 30% PEG solution

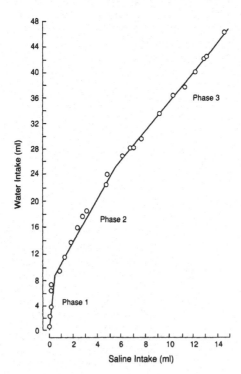

Figure 2. Mean intakes of water and 0.5 M NaCl solution by rats after 30% PEG treatment, redrawn from Figure 1. Symbols represent cumulative values for each hour of the 24-hour test period. Three distinct phases of drinking were observed, as indicated. (Adapted from Stricker, 1981.)

elicited comparable amounts and patterns of water and saline drinking each time (E. M. Stricker & J. G. Verbalis, unpublished observations; although see also Sakai, Fine, Epstein, & Frankmann, 1987). Nor is it an artifact of the highly concentrated saline solution that is offered to the rats. An initial thirst and delayed appearance of sodium appetite also are obtained when either 0.15 M or 0.30 M NaCl solution is available in a two-bottle test with water, and the net fluid mixtures consumed during the second phase of behavior after PEG treatment are identical to the mixture of 0.5 M NaCl and water that is ingested (Stricker, 1981). This mixture is equivalent to a NaCl solution whose sodium concentration is approximately 120 mEq/L, well below the normal Na^+ concentration of 150 mEq/L plasma water in rats; the possible significance of this relatively dilute intake is discussed below.

A similar pattern of alternating saline and water drinking also is seen when hypovolemia results from subcutaneous injection of a small volume of dilute formalin solution (Jalowiec & Stricker, 1970a; see also Braun-Menendez & Brandt, 1952; Wolf & Steinbaum, 1965). The induced damage to the capillary walls permits the extravasation of protein-rich plasma fluid, and that loss upsets the local Starling equilibrium of forces analogous to the effects of subcutaneous colloid treatment. Moreover, the decrease in plasma protein concentration swells interstitial fluid elsewhere, thus reducing plasma volume without producing total body sodium deficits. As with PEG treatments (Stricker, 1981), the induced sodium appetite is proportional to the administered dose of formalin (Handal, 1965b). Unlike the PEG-treated rat, however, after formalin treatment there is no initial period of exclusive water drinking, and rats drink saline and water in alternation immediately (Figure 3). In contrast, after tissue damage by tourniquet injury to a hind limb, thirst appears immediately, but sodium appetite is delayed for 8 hours (Figure 3; Stricker, 1980). Possible explanations for the differential effects of these two methods of tissue injury also are discussed below.

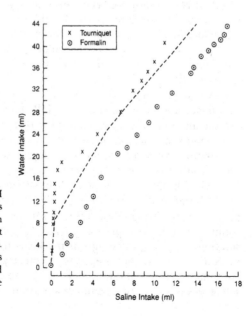

Figure 3. Mean intakes of water and 0.5 M NaCl solution by rats after subcutaneous formalin treatment (data replotted from Jalowiec & Stricker, 1970a) or tourniquet injury (data replotted from Stricker, 1980). Symbols as in Figure 2. Regression lines computed from the intakes of rats treated with 30% PEG solution, from Figure 2, are presented for purposes of comparison.

Still another model of sodium appetite in rats involves systemic administration of desoxycorticosterone (DOC), a precursor in the biosynthesis of aldosterone. Desoxycorticosterone has mineralocorticoid properties of its own and thus promotes renal sodium retention, which leads to marked osmoregulatory thirst when rats are fed standard sodium-rich diet (Rice & Richter, 1943). The DOC-treated rats also consume 15–25 mL of 0.5 M NaCl solution per day, amounts comparable to those ingested by adrenalectomized or PEG-treated rats.

Although the large NaCl intake elicited by DOC treatment in rats seems to be a pharmacological phenomenon that occurs paradoxically in association with excess body sodium, its existence allows the possibility that high physiological levels of circulating endogenous mineralocorticoid may play an important role in mediating the sodium appetite that occurs during body sodium deficiency or hypovolemia. Consistent with this possibility are the findings that administration of aldosterone itself increases NaCl intake in rats (Wolf & Handal, 1966) and that glucocorticoids, which also are secreted during hypovolemia, potentiate NaCl consumption when coadministered with DOC in rats (Braun-Menendez, 1952; Wolf, 1965; see also Fregly, 1967). Thus, it has been suggested that a U-shaped curve describes the relationship of blood mineralocorticoid levels to sodium appetite, with high intakes of NaCl seen either when blood levels are very low, as after adrenalectomy, or when they are very high (Fregly & Waters, 1966).

Each of the various models of sodium appetite in rats that was described above represents an unnatural laboratory occurrence resulting from pathological (adrenalectomy), pathophysiological (PEG treatment or tissue damage), or pharmacological conditions (DOC treatment). Thus, it is important to recognize that sodium appetite also has been observed in laboratory rats under more physiological circumstances. For example, prolonged deprivation of dietary sodium will produce sodium appetite in rats. The intakes observed after 4 days of deprivation were modest, but after 8 days the sodium-deprived rats drink 20 mL of 0.5 M NaCl solution in 7 hours (Figure 4; E. M. Stricker, E. Thiels, & J. G. Verbalis, unpublished observations), amounts comparable to the largest volumes of 0.5 M NaCl that have been consumed by rats under any circumstance. Note that increased thirst also occurred in this situation but followed the initial bout of saline ingestion, and the fluid mixture produced by these intakes is equivalent to a solution with a sodium concentration of 163 mEq/L; this value resembles the hypertonic fluid mixture ingested by rats during phase 3 after PEG treatment (Stricker, 1981).

A natural sodium appetite in rats also has been reported to occur during pregnancy and lactation (Barelare & Richter, 1938; Richter & Barelare, 1938), but the amounts of sodium consumed from a self-selection "cafeteria" array of pure food substances in these studies were less than 5 mEq per day during the last 10 days of gestation and first 15 days post-partum. A recent series of experiments confirmed that there is little spontaneous sodium appetite during the first 15 days of lactation when primiparous lactating rats were maintained on standard laboratory diets of differing sodium content (Thiels, Verbalis, & Stricker, in press; see also Braun-Menendez, 1953). The estimated daily loss of 1 to 2 mEq sodium in milk was replaced by basal daily intakes of 2 to 5 mL of 0.5 M NaCl solution. However, lactating rats markedly increased their spontaneous intakes of NaCl solution when deprived of dietary sodium for 4 days because they continued to lose sodium in milk and became sodium deficient; when access to saline was restored, these animals also drank 20 mL of 0.5 M NaCl solution in 7

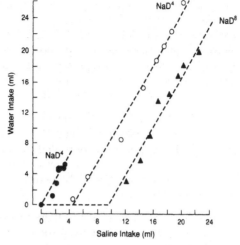

Figure 4. Mean intakes of water and 0.5 M NaCl solution by rats during a 7-hour test after dietary sodium deprivation. Rats either had been maintained on sodium-deficient diet for 4 days (NaD[4], closed circles) or for 8 days (NaD[8], closed triangles) before the test or were lactating and nursed their litters during a 4-day period of dietary sodium deprivation (open circles). Symbols represent cumulative values at 30 minutes and each hour thereafter. From E. M. Stricker, E. Thiels, & J. G. Verbalis, unpublished observations, and Thiels *et al.*, in press).

hours (Figure 4). Thus, markedly sodium-deprived rats and sodium-deficient lactating rats provide two robust models for the investigation of sodium appetite under physiological conditions, and any hypothesis for the bases of sodium appetite in rats must take into account these important situations as well.

HYPOTHESIS: BIOLOGICAL BASES OF SODIUM APPETITE

A useful perspective on the origin of sodium appetite in these various models may be obtained by considering the biological bases of other ingestive behaviors such as hunger and thirst (see Chapter 3). Briefly, it appears that ingestive behavior generally is controlled by multiple excitatory and inhibitory stimuli. Thus, as schematized in Figure 5, some excitatory signal stimulates sodium appetite, and in the presence of a salty-tasting food or fluid, NaCl ingestion occurs; however, that ingestion limits further intake both by removing the excitatory signal and by generating an inhibitory signal.

An extended discussion of the possible stimuli that might influence the

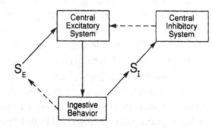

Figure 5. Schematic representation of central mechanisms by which sodium appetite may be controlled. Intake of hypertonic NaCl solution generates osmoregulatory inhibitory stimuli (S_I), which activate central inhibitory systems (perhaps located in parvocellular neurons of the hypothalamic paraventricular nucleus) and thereby suppress further NaCl consumption despite the continued presence of excitatory stimuli (S_E) derived from hypovolemia (i.e., a combination of angiotensin and neural signals from vascular baroreceptors) and the taste of NaCl. Osmotic dilution buffers that inhibitory effect. Unbroken arrows denote stimulatory influences, broken arrows inhibitory influences. Note the similarity between this scheme and that proposed for volume-regulatory thirst (Figure 2 in Chapter 3).

arousal of sodium appetite has been presented previously (Stricker, 1973, 1980; Stricker & Verbalis, 1988, in press). However, those discussions merely represent early stages in the development of the current perspective. Rather than recount those arguments here, we now propose a new hypothesis for the biological bases of sodium appetite that represents a synthesis and extension of the previous proposals and appears to account more adequately for the diverse findings to date. This hypothesis has three critical features:

1. There is a distinct mode of central nervous system activity in which sodium appetite is facilitated.
2. Sodium appetite is controlled by both excitatory and inhibitory stimuli.
3. The paraventricular nucleus of the hypothalamus is likely to be an important integrative structure in mediating inhibition of sodium appetite.

Evidence regarding each of these features is now reviewed, after which we reconsider the various models of sodium appetite from this perspective.

There is a distinct mode of central nervous system activity in which sodium appetite is facilitated. This mode is associated with hypovolemia and/or sodium deficiency and reflects a heightened sensitivity of the brain to signals of sodium appetite that would not elicit NaCl intake under conditions of sodium sufficiency. Note that we postulate a graded spectrum rather than a bimodal separation of such sensitivities and use the phrase "sodium-deficient mode of central nervous activity" in this chapter merely as a convenience to suggest the salt-seeking behavior of animals at an extreme point along that continuum.

The general view that brain function is influenced by the availability of body sodium is similar conceptually to the notion of a set point for sodium homeostasis (Hollenberg, 1980; Strauss, Lamdin, Smith, & Bleifer, 1958). Its application to issues of sodium appetite can be appreciated by considering the NaCl ingestion that occurs in hypovolemic rats. One of the most intriguing mysteries about this phenomenon was the 5-hour delay in NaCl consumption after PEG treatment. Whereas water consumption alone has relatively little effect to repair plasma volume deficits, because water distributes intracellularly as well as extracellularly, intake of NaCl is critical in repairing the volume deficit. Consequently, it seems paradoxical that thirst should appear first while sodium appetite is delayed. That mystery was solved by switching the maintenance diet from standard laboratory chow to sodium-deficient food pellets; under these conditions the rats drank saline solution within 30 minutes after PEG treatment (i.e., a different phase 1), well before any thirst was evident (Stricker, 1981; Figure 6). These observations suggest that the delay in sodium appetite normally seen after PEG treatment is an artifact of the common practice of feeding rats sodium-rich laboratory chow. Such maintenance not only retards the appropriate behavioral response to hypovolemia but blunts aldosterone secretion as well (Aguilera, Hauger, & Catt, 1978; Stricker et al., 1979).

Conversion to the sodium-deficient mode of central nervous system activity is achieved when rats are deprived of dietary sodium for only 1–2 days. Alternatively, such conversion can occur more rapidly if rats become acutely hypovolemic, hence the 5-hour delay in the appearance of sodium appetite after PEG treatment. In contrast, the adrenalectomized rat not supported by mineralocorticoid replacement therapy always is in the sodium-deficient mode and therefore is primed to consume NaCl when any stimulus for sodium appetite occurs. In

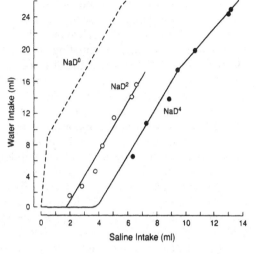

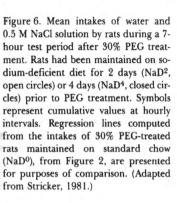

Figure 6. Mean intakes of water and 0.5 M NaCl solution by rats during a 7-hour test period after 30% PEG treatment. Rats had been maintained on sodium-deficient diet for 2 days (NaD[2], open circles) or 4 days (NaD[4], closed circles) prior to PEG treatment. Symbols represent cumulative values at hourly intervals. Regression lines computed from the intakes of 30% PEG-treated rats maintained on standard chow (NaD[0]), from Figure 2, are presented for purposes of comparison. (Adapted from Stricker, 1981.)

each case, it is their heightened sensitivity to signals of sodium deficiency that allows the animals to differentiate between dilutional hyponatremia (a reduction in plasma sodium concentration resulting from an excess of body water) and sodium deficiency (a reduction in plasma sodium concentration resulting from a loss of body sodium); in response to the former, the excess water is excreted in urine, but sodium appetite does not develop, nor is aldosterone secreted, whereas in response to the latter, sodium appetite and aldosterone secretion are prominent, but a water diuresis does not occur. Note the marked contrast between this distinction and the generally similar effects on water drinking and AVP secretion of hypernatremia resulting from excess body sodium or from loss of body water (see Chapters 14 and 16).

Sodium appetite is controlled by both excitatory and inhibitory stimuli. As mentioned previously, the taste of salt is a permissive excitatory stimulus for sodium appetite. But similar to thirst and hunger, there presumably are a variety of signals derived from changes in substrate levels, neural inputs to the brain, and endocrine actions that provide specific excitatory or inhibitory stimulation of NaCl consumption.

If analogous factors controlled sodium appetite and thirst, then NaCl ingestion should be stimulated by manifestations of sodium deficiency just as thirst is stimulated by manifestations of water deficiency; these would include hypovolemia, osmotic dilution, and elevated blood levels of angiotensin. However, sodium appetite is not stimulated shortly after PEG treatment, whereas thirst appears prominently, presumably largely because of the neural signals arising from low-pressure baroreceptors (Kaufman, 1984; Stricker, 1966; see Chapter 14). Sodium appetite might conceivably result from the same neural signal of hypovolemia when it reached some higher threshold, but there is no experimental support for this conjecture. The appearance of sodium appetite is delayed to a comparable extent in rats regardless of the rate at which hypovolemia develops (Stricker, 1981). Moreover, ligation of the inferior vena cava, which abruptly reduces venous return to heart, elicits thirst rapidly but not sodium appetite (Stricker, 1971b). Similarly, an abrupt pharmacological decrease in peripheral resistance lowers arterial blood pressure acutely and stimulates thirst but not sodium appetite (Fitzsimons & Stricker, 1971). Indeed, cava ligation and acute drug-induced hypotension each abolish the expected sodium appetite after PEG treatment in rats while potentiating thirst (Hosutt & Stricker, 1981; Stricker, 1971a). It seems unlikely that the absence of sodium appetite under those circumstances results from the mere presence of thirst as a competing drive because sodium appetite does not emerge even when the induced thirst is inhibited by osmotic dilution (Stricker, 1971a, 1971b).

On the other hand, it is important to note that in the experiments just cited rats were maintained on standard sodium-rich laboratory chow before the treatments. When animals were prefed sodium-deficient diet instead, stimulation of NaCl intake after PEG treatment was as prompt as the appearance of thirst when rats had been fed standard chow (Figure 6). The fact that a similar potentiation of sodium appetite was not obtained after cava ligation (Stricker et al., 1979) may not signify the absence of an appropriate excitatory stimulus but rather the presence of an overriding inhibition (see below).

In rats fed standard chow, hyponatremia does not appear to provide a stimulus for sodium appetite that is symmetrical to the well-known stimulation of thirst by hypernatremia resulting from NaCl loads. For example, water loads did not elicit NaCl consumption when body fluids were diluted by 5% and dilution was maintained for 8 hours (Stricker & Wolf, 1966), and diluting sodium concentration in the cerebrospinal fluid similarly was ineffective in stimulating sodium appetite (Epstein et al., 1984; Osborne, Denton, & Weisinger, 1987). Although rats in these studies were maintained on standard laboratory chow before the treatments, there is little evidence that hyponatremia is a necessary stimulus for sodium appetite when rats are presumed to be in the sodium-deficient mode of central nervous system activity because they consume hypertonic NaCl solutions after colloid treatment (Stricker, 1981), tissue damage (Jalowiec & Stricker, 1970b), or adrenalectomy (Wolf & Stricker, 1967), even when plasma sodium concentrations are normal. Moreover, PEG-treated rats did not drink more 0.5 M NaCl solution when hyponatremic after a 24-hour period in which they had been allowed access to drinking water than when they were water-deprived and normonatremic (Stricker, 1981). Although it remains possible that extreme osmotic dilution can provoke sodium appetite when the rat is in the sodium-deficient mode, it seems more likely that hyponatremia acts simply to blunt the inhibition of sodium appetite resulting from the postingestional consequences of consuming hypertonic NaCl solution.

In contrast, angiotensin probably provides an important excitatory stimulus for sodium appetite. Initial findings in rats that had been prefed standard sodium-rich chow did not support this possibility. For example, although intracerebroventricular injection of angiotensin increased NaCl intake in rats (Avrith & Fitzsimons, 1980; Bryant, Epstein, Fitzsimons, & Fluharty, 1980; Buggy & Fisher, 1974), it appeared to do so in substantial amounts only after an induced natriuresis had caused body sodium deficits (Fluharty & Manaker, 1983). The results obtained from such studies are quite different when rats begin the test in the sodium-deficient mode of central nervous system activity. Thus, intraventricular injection of angiotensin enhanced sodium appetite in rats deprived of dietary sodium (Buggy & Fisher, 1974) or pretreated with PEG solution (Fitts, Thunhorst, & Simpson, 1985a). Moreover, systemic injection of the drug captopril, which blocks the peripheral conversion of angiotensin I to angiotensin II but increases the delivery of angiotensin I to the brain for its conversion there

(Schiffrin & Genest, 1982), also precipitated the abrupt appearance of sodium
appetite in PEG-treated rats maintained previously on standard chow and greatly potentiated their NaCl intakes when they had been fed sodium-deficient diet
(Figure 7; Stricker, 1983; see also Elfont & Fitzsimons, 1985; Moe, Weiss, &
Epstein, 1984; Weisinger, Denton, Di Nicolantonio, & McKinley, 1988; Weiss,
Moe, & Epstein, 1986). Similarly, sodium appetite elicited by PEG treatment was
augmented in rats with septal lesions, especially when they had been maintained
on sodium-deficient diet previously (Stricker, 1984b); such rats are known to be
unusually sensitive to the central actions of angiotensin (Blass, Nussbaum, &
Hanson, 1974). Conversely, pharmacological blockade of the renin–angiotensin
system in the brain has been shown to blunt NaCl intake in PEG-treated or
sodium-deficient rats (Buggy & Janklaas, 1984; Elfont & Fitzsimons, 1985; Moe
et al., 1984; see also Sakai & Epstein, 1990). These and other findings have
demonstrated the significance of angiotensin as an excitatory stimulus of sodium
appetite in rats (Avrith & Fitzsimons, 1983; Di Nicolantonio, Hutchinson, &
Mendelsohn, 1982; Elfont, Epstein, & Fitzsimons, 1984; Fregly, 1980).

Consideration of putative excitatory stimuli does not provide a full account
of the possible factors that are involved in the control of sodium appetite. As
mentioned, inhibitory factors are known to be influential in the control of water
and food intakes, and there is clear indication that inhibitory factors play a role
in the control of NaCl ingestion too. The most compelling evidence in this
regard is the complete loss of sodium appetite that occurs when sodium-deficient
or hypovolemic rats are made anuric. This result occurs regardless of whether
anuria is produced in rats by bilateral nephrectomy or ureter ligation, by punc-
turing the bladder, or by severe hypovolemia and arterial hypotension (Stricker,
1971b, 1973; see also Fitzsimons & Stricker, 1971; Fitzsimons & Wirth, 1978;
Mills & Rodbard, 1950). There is no induced sodium appetite under these condi-

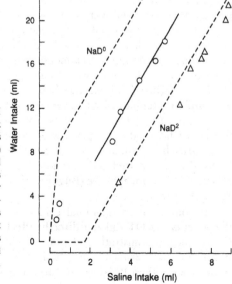

Figure 7. Mean intakes of water and 0.5 M
NaCl solution by rats during a 7-hour test
period after 30% PEG treatment. Some rats
were maintained on standard chow (NaD⁰)
prior to the test while others had been fed
sodium-deficient diet for 2 days (NaD²). Rats
received captopril (5 mg/kg, ip) 2 hours or
immediately after PEG treatment, respectively. Symbols represent cumulative values
at hourly intervals. Dashed lines represent
regression lines computed from the intakes
of 30% PEG-treated rats, from Figure 6, and
are presented for purposes of comparison.
(Adapted from Stricker, 1983.)

tions even when the anuric animals are allowed to drink palatable 0.15 M NaCl instead of a more concentrated solution (although see Chiaraviglio, 1976). The possibility of nonspecific disruption of drinking is not consistent with continued normal water intakes in such animals. Thus, whatever factors might be responsible for the stimulation of sodium appetite in rats during sodium deficiency and/or hypovolemia, it is evident that NaCl intake nevertheless may be suppressed under many circumstances. A possible basis for that inhibition is considered in the next section.

The paraventricular nucleus of the hypothalamus is likely to be an important integrative structure in mediating inhibition of sodium appetite. Sodium appetite almost invariably appears in PEG-treated rats after an initial period of exclusive water drinking (Stricker, 1981). The renal retention of ingested water dilutes body fluids and thereby inhibits secretion of the neurohypophyseal peptide hormones AVP and oxytocin (OT), both of which had been increased by the induced hypovolemia (Stricker & Verbalis, 1986). A synthesis of these observations leads to the hypothesis that sodium appetite may be inhibited by the activity of a subset of neurons in the supraoptic (SON) or paraventricular nuclei (PVN) or both, the location of virtually all vasopressinergic and oxytocinergic neurons (Sofroniew, 1983), and conversely that sodium appetite may emerge in response to appropriate stimuli only when the activity of those neurons is either suppressed or reduced to basal, unstimulated levels.

This hypothesis has been tested by measuring blood levels of AVP and OT under various circumstances in which sodium appetite was prominent: plasma levels of OT, but not AVP, were notably low in sodium-deficient adrenalectomized rats, in PEG-treated rats in phase 3 of drinking behavior, and in DOC-treated rats (Stricker & Verbalis, 1987). Such suppressed levels of plasma OT represented the first common biological factor known to be present in these very dissimilar models of sodium appetite. Moreover, OT secretion induced by hypovolemia was blunted by prior maintenance on sodium-deficient diet in association with enhanced sodium appetite (Stricker, Hosutt, & Verbalis, 1987). Thus, it seemed possible that OT secretion reflected inhibition of NaCl solution intake and that sodium appetite was expressed only when circulating levels of the peptide hormone were very low.

To evaluate this hypothesis further, plasma OT levels were examined when rats were hypovolemic and/or sodium deficient yet did not have a sodium appetite because of additional treatments causing anuria or arterial hypotension. In each case plasma OT levels were found to be substantially elevated (Stricker *et al.*, 1987; see Figure 8). Similarly, when sodium appetite in adrenalectomized rats was inhibited temporarily by injection of hypertonic NaCl, there was an acute increase in plasma OT levels whose duration varied in proportion to the duration over which sodium appetite was inhibited (Stricker & Verbalis, 1987). Moreover, carbachol (a cholinergic drug not attacked by cholinesterase) given into the brain of rats increased OT secretion (Kuhn & McCann, 1971) and inhibited an established sodium appetite (Buggy & Fisher, 1974; Fitts *et al.*, 1985a; Fitzsimons & Wirth, 1978).

Despite such correlational evidence, other experiments indicated clearly that circulating OT did not directly affect the intake of NaCl solution in rats. Continuous administration of synthetic OT from an implanted osmotic minipump, raising plasma OT to physiological levels and beyond, did not inhibit sodium appetite in PEG-treated rats, nor did systemic administration of an OT

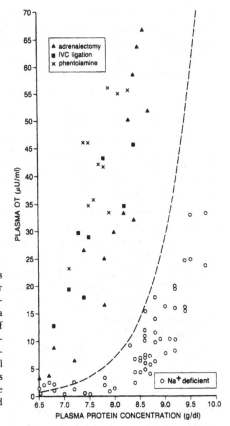

Figure 8. Plasma levels of oxytocin (OT) as function of plasma protein concentration after 30% PEG treatment in rats that had been bilaterally adrenalectomized, with inferior vena cava (IVC) ligation, made hypotensive by injection of phentolamine (10 mg/kg, i.p.), or prefed sodium-deficient diet for 2 days. Symbols represent data from individual animals. Exponential line was computed from data collected in rats maintained on standard laboratory chow before receiving 30% PEG treatments alone. (Adapted from Stricker *et al.*, 1987.)

receptor antagonist enhance NaCl intake in PEG-treated rats (Stricker & Verbalis, 1987). These results therefore indicate that increased circulating levels of OT do not, by themselves, provide an inhibitory stimulus of sodium appetite. Instead, plasma levels of OT may best be viewed as a peripheral marker of the parallel activity of some system of brain neurons that are intimately involved in the inhibitory control of sodium appetite (Stricker & Verbalis, 1988).

Oxytocin secreted from the neurohypophysis is synthesized in magnocellular neurons in the SON and PVN of the hypothalamus. Adjacent to the magnocellular neurons in the PVN, but not the SON, are parvocellular OT-containing neurons, which project widely throughout the brain to sites including the limbic system and the dorsal motor nucleus of the vagus in the brainstem (Swanson & Sawchenko, 1983). Thus, central oxytocinergic projections appear to be advantageously located to influence ingestive behavior and associated autonomic functions. Perhaps these centrally projecting hypothalamic nuclei are stimulated simultaneously with the magnocellular neurons projecting to the pituitary so that the induced activity somewhere in the brain serves to inhibit sodium appetite; conversely, sodium appetite may develop when activity is suppressed in this subset of PVN neurons.

We adopted two strategies to evaluate this hypothesis. One was to identify some measurable response that was influenced by the parvocellular oxytocinergic neurons and to determine whether sodium appetite was inhibited when that response was stimulated. In doing so, we made use of a recent report by Rogers and Hermann (1987) in which an oxytocinergic projection from the PVN to the dorsomotor nucleus of the vagus in the brainstem was observed to decrease gastric motility and emptying. In our studies, two chemical agents, lithium chloride and copper sulfate, both were found to inhibit gastric emptying in rats and to increase pituitary OT secretion as well (McCann, Verbalis, & Stricker, 1989; Verbalis, McHale, Gardiner, & Stricker, 1986). These agents also decreased NaCl consumption in sodium-deprived adrenalectomized rats and in PEG-treated rats, in the latter case without affecting the induced thirst (Stricker & Verbalis, 1987). Thus, specific stimulation of centrally projecting oxytocinergic neurons in rats also parallels the inhibition of NaCl intake and therefore may mediate that phenomenon.

The second approach used the reverse strategy and determined whether sodium appetite was potentiated when stimulation of the PVN was blocked. Osmoreceptors or osmoreceptor pathways in the ventral portion of nucleus medianus (vNM) are believed to be involved with stimulating the secretion of AVP and OT during osmotic dehydration. Accordingly, lesions of this brain area attenuated neurohypophyseal hormone secretion after injection of hypertonic NaCl solution, whereas the responses to hypovolemia were relatively unimpaired (Gardiner, Verbalis, & Stricker, 1985; Mangiapane, Thrasher, Keil, Simpson, & Ganong, 1983). Moreover, rats with lesions of vNM showed a marked spontaneous appetite for concentrated NaCl solution and drank 15–35 mL of 0.5 M NaCl daily for at least several months post-surgery (Gardiner, Jolley, Vagnucci, & Stricker, 1986). Examination of these animals indicated that they did not become sodium deficient after the brain lesions, nor was the renin–angiotensin system activated. Instead, we suggest that such lesions blocked the normal inhibitory effects on sodium appetite that occur when rats consume concentrated NaCl solution, thus permitting the ingestion of saline to persist unchecked.

A putative role of PVN neurons in the inhibitory control of sodium appetite would appear to be inconsistent with the sodium appetite seen in lactating rats (Richter & Barelare, 1938; Thiels et al., in press), inasmuch as suckling is an established stimulus of magnocellular PVN activity and pituitary OT secretion. However, recent experiments have indicated that parvocellular activity may not increase during suckling (Helmreich, Thiels, Verbalis, & Stricker, 1988; Lawrence & Pittman, 1985). Thus, suckling would not be expected to provide a stimulus that directly affected the proposed central control of NaCl consumption. Even if it did, parvocellular PVN neurons might still mediate inhibition of sodium appetite because lactating animals are inactive when nursing, and therefore suckling-induced OT secretion and saline ingestion necessarily occur at different times.

The hypothesis we have proposed is diagrammed schematically in Figure 9. In this figure the excitatory stimuli for drinking arise from hypovolemia and involve a baroreceptor signal plus angiotensin. These excitatory stimuli act in the brain to promote fluid intake. However, whether water, saline, or both fluids are consumed as a consequence of these stimuli depends as well on the switching function of a brain system that controls sodium appetite. Activation of that system sets the "switch" in the direction of water intake rather than NaCl intake. Because the presence of sodium appetite has been documented to vary inversely with plasma OT levels in rats, we have presumed that this switching system is

Figure 9. Patterns of water and NaCl solution intake in response to stimuli of fluid ingestion. During hypovolemia, neural signals from vascular baroreceptors act together with an endocrine signal
provided by angiotensin as excitatory stimuli (S_E)
that elicit drinking. How much drinking occurs
depends in part on the intensity of these signals.
Which fluids are consumed depends in part on a
graded switching mechanism that seems to be influenced by activity in a subset of parvocellular
neurons in the paraventricular nucleus (PVN) of
the hypothalamus. Predominant water intake occurs when this PVN activity is elevated (A), as during phase 1 after PEG treatment in rats maintained on standard chow, when the S_E are intense,
or when animals are anuric. Conversely, predominant NaCl intake occurs when this PVN activity is
suppressed (B), as in adrenalectomized animals or
intact lactating rats after a period of imposed dietary sodium deprivation or during hypovolemia
shortly after PEG treatment in rats prefed sodium-deficient diet. (C) Alternating intakes of
water and NaCl solution occur when the volume-
regulatory signals associated with hypovolemia

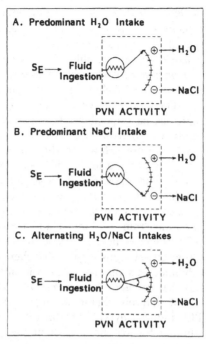

(i.e., S_E) occur in the presence of signals that tend to decrease PVN activity (e.g., osmotic dilution), as
during phase 2 after PEG treatment, after intraventricular injection of angiotensin in DOC-treated
rats, or after 8 days of dietary sodium deprivation.

likely to be located in some subset of oxytocinergic neurons in the PVN of the
hypothalamus. However, the PVN is an extraordinarily complex structure containing many different neuropeptidergic cells projecting widely throughout the
brain, any of which might be activated concurrently with magnocellular OT
neurons, and thus the switch obviously could be located in non-OT-containing
cells or even in some other area in the brain whose function also was inversely
related to that of magnocellular OT neurons in the PVN.

Under standard maintenance conditions in which laboratory rats are fed a
high-sodium diet, stimuli for drinking are directed away from NaCl intake and
toward water ingestion (Figure 9A). In contrast, when rats are sodium deprived
or made sodium deficient, the same stimuli for drinking now are directed toward NaCl intake and away from water ingestion (Figure 9B). However, under
certain conditions of competition between the osmo- and volume-regulatory
systems, in which osmotic dilution and hypovolemia are present concurrently,
the switching mechanism can oscillate between these extremes, first directing
behavior in one direction and then in the other (Figure 9C). In consequence,
both water and saline are consumed alternately, and plasma volume deficits are
corrected.

The set of the switch is influenced by some consequence of the reduced
availability of sodium in extracellular fluid. We presume that the interstitial
fluid, which contains a large and labile store of sodium, protects the brain from
sodium deficiency, acting as a buffer that, when depleted, finally permits the
impact of the sodium loss to affect the brain. How might such sodium deficiency

be detected? Rather than sensing interstitial fluid reserves or total body sodium, it seems more probable that when animals are sodium deficient, the excitatory neural system that controls sodium appetite becomes more responsive to appropriate excitatory stimuli. This change then establishes a new mode of brain neuronal activity whereby volume-regulatory signals that had stimulated thirst now are more likely to stimulate sodium appetite as well.

The set of the switch also can be influenced by the osmoregulatory system; high plasma osmolality sets the switch toward water intake, whereas osmotic dilution or hyponatremia, as occurs during sodium deficiency, sets it toward NaCl ingestion. However, there are other important influences on the switch that have nothing to do with osmoregulation, or indeed even with sodium or water homeostasis. For example, some nonspecific factors seem to involve certain kinds of traumatic stressors, especially those accompanied by severe hypovolemia and/or anuria. Such conditions always set the switch toward water intake and away from NaCl intake and are so influential that sodium appetite will not occur then despite the presence of frank hypovolemia and/or sodium deficiency. Other observations suggest that there may be a circadian rhythm in the central inhibition of sodium appetite, with inhibition normally being much stronger by day than by night (Gardiner et al., 1986; Rowland, Bellush, & Fregly, 1985).

Within this scheme, we suggest that there are four changes that occur after PEG-induced hypovolemia in rats. First, the plasma volume deficits provide a stimulus for drinking. As mentioned, the signal arises from low-pressure baroreceptors (Kaufman, 1984; Toth, Stelfox, & Kaufman, 1987; see also Fitzsimons & Moore-Gillon, 1980; Moore-Gillon & Fitzsimons, 1982; Zimmerman, Blaine, & Stricker, 1981) and stimulates intake in proportion to the volume deficit (Fitzsimons, 1961; Stricker, 1966, 1968). Second, hypovolemia elicits release of renin from the kidneys, which also is proportional to the induced volume deficit (Johnson, Mann, Rascher, Johnson, & Ganten, 1981; Leenen & Stricker, 1974). Angiotensin provides a second stimulus of drinking, in proportion to the induced increase in blood levels of the hormone (Fitzsimons & Simons, 1969; Mann, Johnson, & Ganten, 1980). It seems clear that these two stimuli of drinking are independent of one another; exogenous angiotensin elicits drinking in the absence of hypovolemia (Fitzsimons, 1969), whereas hypovolemia can elicit drinking in the absence of angiotensin (Fitzsimons, 1961; Stricker, 1973). There has been some lingering controversy as to how much contribution is made to drinking during hypovolemia by these separate neural and endocrine signals, but based on available evidence we believe that the neural signal has a predominant influence under normal physiological circumstances (Stricker, 1978a, 1978b; see Chapter 14; see also Mann et al., 1988).

Third, the plasma volume deficits help to establish the sodium-deficient mode of central nervous activity. This effect permits the hypovolemic signals to stimulate NaCl intake. They will do so rapidly when the brain already is in the sodium-deficient mode, as when the rats have been prefed sodium-deficient diet, but otherwise they will stimulate water intake because of the fourth effect of hypovolemia: it increases PVN activity and thereby sets the switch toward water intake. When that happens, the animal must reduce PVN activity before NaCl intake can appear. That reduction usually is accomplished by osmotic dilution resulting from renal retention of ingested water, unless the induced hypovolemia is very severe (Stricker et al., 1987; Stricker & Verbalis, 1986).

These four putative effects do not occur simultaneously. The stimulation of renin secretion occurs relatively rapidly, perhaps because angiotensin is an important pressor hormone and additionally promotes renal sodium conservation by eliciting aldosterone secretion. The stimulation of drinking also occurs relatively rapidly, perhaps because the induced NaCl intake (which occurs when animals are in the sodium-deficient mode) corrects their plasma volume deficits so effectively. In contrast, the increase in PVN activity is a much less sensitive response to hypovolemia than the others, and a more substantial hypovolemia is needed to cause marked secretion of the neurohypophyseal hormones (Stricker & Verbalis, 1986). Thus, there appears to be a window of time during which a progressively developing hypovolemia can stimulate sodium appetite while PVN activity remains at low basal levels, a window that is opened wider when rats are prefed sodium-deficient diet.

The various influences of angiotensin on fluid consumption are likely to be comparably complex. Thus, angiotensin also stimulates drinking and helps to establish a sodium-deficient mode of central nervous system activity, but when present in large amounts it appears to increase PVN activity (Ferguson & Kasting, 1988; Ferguson & Renaud, 1986; see also Reid, Brooks, Rudolph, & Keil, 1982) and thereby direct drinking toward water intake. Thus, for example, exclusive water ingestion is obtained when a relatively large dose of angiotensin is administered directly into the brain (Fluharty & Epstein, 1983), as also is observed to occur in rats when hypovolemia is very pronounced (Stricker, 1973). Conversely, when PEG-treated rats are given systemic captopril, the increased amount of angiotensin produced in the brain stimulates increased consumption of water when it is the only drinking fluid available (Evered & Robinson, 1984; Lehr, Goldman, & Casner, 1973), and it elicits increased drinking of water and NaCl solution when both fluids are available (Stricker, 1983). Both water and saline also are consumed when PVN activity is suppressed before central injection of angiotensin, as by prefeeding rats sodium-deficient diet (Buggy & Fisher, 1974) or by treating them with DOC (Fluharty & Epstein, 1983).

In short, the control of sodium appetite involves multiple stimuli known to affect other aspects of sodium homeostasis, some of which also are involved in the control of thirst and water homeostasis. We do not believe this overlap is merely coincidental. Instead, by considering these two systems together, a coherent picture emerges of the integrated central control of both drives and their interaction.

Models of Sodium Appetite: A Reconsideration

From the perspective offered by this new hypothesis, we now reconsider the possible biological bases of the three most common models of sodium appetite in rats: hypovolemia in the absence of total body sodium deficits, adrenalectomy, and DOC treatment.

Hypovolemia

Rats with plasma volume deficits need water and saline to restore circulatory volume. Accordingly, they increase their intakes of both fluids. A schema summarizing the interrelated controls of thirst and sodium appetite is presented in Figure 10. Thirst during hypovolemia appears to be controlled by some com-

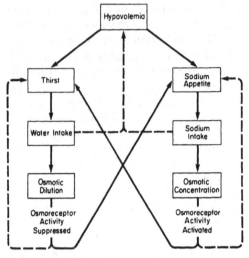

Figure 10. Schematic representation of the physiological mechanisms controlling thirst and sodium appetite during hypovolemia. Solid arrows indicate stimulation; broken arrows indicate inhibition. Hypovolemic rats alternately drink water and concentrated NaCl solution, depending on the current plasma osmolality, and ultimately consume sufficient fluid at an isotonic concentration to repair the volume deficit. However, the water and NaCl intakes are limited when animals have access to only one of the drinking fluids because of activation of the respective inhibitory osmoregulatory pathways. Conversely, neither inhibitory pathway is activated when the rats drink isotonic NaCl solution, and consequently intakes proceed unabated in response to the hypovolemic stimulus. (Adapted from Stricker & Verbalis, 1988.)

bination of excitatory neural signals from low-pressure cardiovascular baroreceptors and from angiotensin and inhibitory neural signals from cerebral osmoreceptors. Thus, water intake after PEG treatment is stimulated by hypovolemia but is not maintained despite continued plasma volume deficits because of the progressive osmotic dilution resulting from renal retention of the ingested water (Stricker, 1969). These animals need sodium both to expand plasma volume and to avoid excessive osmotic dilution, and appropriate to this need hypovolemic animals also manifest an appetite for NaCl. Sodium appetite appears to be controlled by the same excitatory stimuli that elicit thirst and by an inhibitory stimulus that results from activity in neurons projecting centrally from parvocellular subdivisions of the PVN. Both hypovolemia and increased blood levels of angiotensin appear to increase activity in those central inhibitory neurons, and thus PEG-treated animals do not begin to consume NaCl until osmotic dilution blunts their activity. Consumption of sufficient concentrated NaCl solution then raises plasma osmolality, thereby restoring the inhibition of sodium appetite while disinhibiting thirst; subsequent water intake then continues until the diluted plasma osmolality again inhibits thirst and, by quieting those critical PVN neurons, disinhibits sodium appetite. This scheme explains nicely the alternating pattern of drinking behaviors observed in PEG-treated rats. By such alternation of water and saline intakes, the hypovolemic animals ultimately consume sufficient amounts of both fluids to restore plasma volume.

Thus, the observed delay in the onset of sodium appetite after PEG treatment in rats results from three factors: the gradual development of hypovolemia, which allows time for conversion of the brain to the sodium-deficient mode of central nervous system activity; an inhibition of some aspect of PVN activity, which had been initially stimulated by the hypovolemia, as a result of water drinking and osmotic dilution; and progressively increasing plasma volume deficits and blood angiotensin levels, the associated stimuli for sodium appetite. Of these three factors, the first two can be obtained simply by maintaining rats on sodium-deficient diet before PEG treatment. They then begin the test in the sodium-deficient mode of brain function and with mild degrees of hyponatremia (6–8 mEq Na^+/L below normal; Stricker, 1981) and require only excitatory stimuli for sodium appetite to prompt NaCl ingestion. The fact that PEG treatment elicits saline consumption within 30 minutes under these conditions indicates how rapidly those stimuli are produced. However, because hypovolemia and angiotensin both increase pituitary OT secretion and therefore presumably increase activity of the subset of PVN neurons involved in the central inhibition of sodium appetite, one can suppress the immediate sodium appetite of PEG-treated rats prefed sodium-deficient diet simply by withholding access to all drinking fluids for 2–4 hours, during which time the plasma volume deficits grow progressively larger and provide an increasingly larger stimulus for water drinking. With PVN activity thereby elevated, sodium appetite is now inhibited, and thirst emerges instead; sodium appetite does not then reappear until PVN activity is again reduced by a greater osmotic dilution resulting from water consumption (Stricker, 1981).

The significant issue is not simply the removal of thirst as a competing drive, because under certain conditions osmotic dilution inhibits thirst without necessarily eliminating pituitary OT secretion, and therefore presumably without affecting central parvocellular PVN projections either. For example, sodium appetite can be inhibited in sodium-deprived, PEG-treated rats indefinitely by bilateral nephrectomy or ureter ligation, by cava ligation, and more temporarily by arterial hypotension, by toxic chemical agents such as LiCl, and by hypertonic NaCl solution, in each case in association with elevated levels of circulating OT and presumably PVN activity (Stricker *et al.*, 1987; Verbalis *et al.*, 1986). Thus, activity in these PVN neurons appears to provide a final critical pathway for the inhibition of sodium appetite, and NaCl ingestion in response to specific stimuli occurs only when that activity is reduced in sodium-deficient and/or hypovolemic rats.

The drinking responses of rats after tissue trauma can be understood in the same way. Although not yet measured after tourniquet injury, a large increase in pituitary OT secretion is likely to occur in association with the observed early absence of sodium appetite because hypovolemia is abrupt and pronounced (Stricker, 1980). However, because enhanced NaCl intake develops in those animals some hours later (Figure 3), perhaps after the operative trauma subsides, osmotic dilution resulting from the initial period of exclusive water drinking ultimately may inhibit OT secretion and with it activity in the subset of PVN neurons that presumably mediate inhibition of sodium appetite. Conversely, subcutaneous injection of small doses of formalin solution should cause tissue damage but without great trauma or increases in pituitary OT secretion; the rapid development of significant hyponatremia (as a result of induced cellular damage and the consequent release of sodium-poor intracellular water) in conjunction with the induced hypovolemia and hyperreninemia should act to elicit sodium appetite promptly, as has indeed been observed (Jalowiec & Stricker, 1970a).

The above considerations address one of the three paradoxical findings that are observed in the drinking behavior of rats after PEG treatment, namely, the delayed appearance of sodium appetite when rats had been maintained on standard sodium-rich chow. However, this perspective also can be used to help explain two other paradoxes observed following PEG treatment. One is the continued osmotic dilution of body fluids even after rats begin to consume hypertonic NaCl solution (see also Sugimoto, 1988). The emergent sodium appetite is appropriate to osmoregulation, of course, but it does not restore normal plasma osmolality, as water drinking does after osmotic dehydration. Thus, during phase 2 of drinking behavior after PEG treatment, each bout of NaCl consumption must rapidly increase activity in the PVN neurons that inhibit further NaCl intake while disinhibiting thirst, and these two effects occur before plasma sodium concentrations can return to 150 mEq/L of plasma water. Although hyponatremia may not directly stimulate NaCl ingestion in rats by itself, nevertheless there may be an osmoregulatory influence on sodium appetite within the context of hypovolemia and sodium deficiency; during this phase the animals seem to maintain a temporarily lowered osmotic set point that becomes the fulcrum around which, like a see-saw, thirst and sodium appetite alternate stimulation depending on current plasma osmolality or sodium concentration.

The third paradox in the drinking behavior of PEG-treated rats is the persistence of sodium appetite after sufficient water and saline have been consumed to repair the induced plasma volume deficits. The bases of thirst and sodium appetite then (phase 3) appear to be quite different from those that existed during hypovolemia (phase 2) because excitatory stimuli of sodium appetite no longer are present. However, it is of interest that plasma OT levels are unstimulated, presumably again reflecting reduced activity in the subset of PVN neurons that normally inhibit sodium appetite. This effect may result from an intrinsic delay in returning the brain from the sodium-deficient mode of central nervous system activity to the mode of brain function seen when body sodium stores are replete. Whatever the explanation, sodium appetite persists for some time while plasma sodium concentrations now return to normal levels (Stricker, 1981), and osmoregulatory stimuli of thirst are generated by the ingestion of hypertonic NaCl solution.

Note that the ratio of water to saline intake in phase 3 produces a fluid mixture whose sodium concentration is approximately 150–160 mEq/L, which is close to an isotonic solution. These values suggest that the PEG-treated rats rely on water drinking rather than sodium excretion for osmoregulation during this phase. Plasma aldosterone returns to low basal levels in the absence of hypovolemia (Stricker et al., 1979), so a persistence of mineralocorticoid hormone in blood is not responsible for the observed sodium retention. Perhaps an inhibited secretion of pituitary OT is significant in this regard. Although in this chapter pituitary OT secretion has been used simply as a peripheral marker of increased neuronal activity in the PVN, circulating OT is known to have natriuretic properties in rats (Balment, Brimble, & Forsling, 1980; Verbalis, Mangione, & Stricker, 1988). Thus, it seems possible that after an osmotic load in rats, sodium excretion (mediated in part by pituitary OT secretion) and the inhibition of sodium appetite are complementary functions much like renal water conservation (mediated by pituitary AVP secretion) and the stimulation of thirst. These dual effects of NaCl loads on sodium balance appear to be lost in phase 3 of drinking behavior after PEG treatment, and sodium appetite persists without rapid sodium excretion.

ADRENALECTOMY

Adrenalectomized rats allowed free access to saline solution lose 0.5 to 1.0 mEq of sodium every 1–2 hours, each time they excrete urine. Appropriate to this need, they repeatedly ingest NaCl in amounts that equal their urinary sodium losses (Jalowiec & Stricker, 1973; Stricker, 1981). By doing so they turn over large amounts of sodium, more than their total body sodium content each day, and their daily intake represents an accumulation of these numerous episodes of compensatory intake.

According to the current perspective, sodium appetite should occur whenever rats are in the sodium-deficient mode of brain function, providing that appropriate stimuli of sodium appetite are present and PVN activity is not elevated. This constellation of features in fact characterizes the rat after adrenalectomy. Thus, teleological issues aside, from this consideration of the possible biological bases of sodium appetite there should be little wonder that adrenalectomized rats have substantial NaCl intakes.

Given the chronic sodium deficiency of adrenalectomized rats, it also should not be surprising that their blood levels of OT, a natriuretic hormone, are very low in the basal state (Stricker & Verbalis, 1987). These levels presumably reflect a decrease in the activity of those PVN neurons that mediate inhibition of sodium appetite. Under the circumstances, it seems likely that the sustained hypovolemia and elevated blood levels of angiotensin associated with chronic sodium deficiency each contribute to the excitation of sodium appetite. Although the plasma volume deficits of adrenalectomized rats probably never become too large when the animals are maintained on *ad libitum* access to NaCl solutions because of their frequent drinking, these rats always are likely to be in a sodium-deficient mode of brain function and therefore should be very sensitive to the sodium-appetite-eliciting effects of hypovolemia.

Because of their absence of adrenal medullary and cortical hormones, sodium-deficient adrenalectomized rats are very vulnerable to hypovolemic challenges and other stressors. Thus, treatments that reduce blood pressure provoke unusually large increases in pituitary OT secretion in association with inhibition of sodium appetite (Stricker et al., 1987). Indeed, even when adrenalectomized rats were injected with 10% or 20% solutions of PEG instead of 30% PEG, to deplete plasma volume more slowly, half the animals showed thirst but little sodium appetite despite receiving two treatments (i.e., injection of PEG and adrenalectomy) that are potent stimuli of NaCl intake when given individually (Stricker, 1981). The other rats treated with the lower dose of PEG did show a potentiated sodium appetite, as would be expected of animals in a sodium-deficient mode of central nervous system activity (Stricker, 1981); presumably in these rats the sodium appetite became manifest before the induced hypovolemia was of sufficient magnitude to stimulate activity in the PVN neurons that mediate inhibition of sodium appetite. This is analogous to the case when intact rats are maintained on sodium-deficient diet before PEG treatment.

However, the prominent NaCl intakes of adrenalectomized rats likely do not simply reflect ongoing excitatory stimulation. After a 24-hour period of imposed sodium deprivation, adrenalectomized rats drank much more NaCl than is required to replace their sodium deficits but did not also increase water intake then (notably unlike PEG-treated rats in phase 3 of drinking behavior) despite a temporary excursion into positive sodium balance (Stricker, 1981). A plausible explanation for this phenomenon is that they had retained water during the period of sodium deprivation, and the osmotic dilution resulting from this water retention in combination with their urinary sodium losses had the effect of blunting inhibitory feedback during the initial period of exclusive saline drinking and thereby allowed the observed overconsumption of NaCl when saline solutions became available again.

Even when adrenalectomized rats were given by injection more than enough NaCl to repair their sodium losses and thereby provoke thirst, their NaCl consumption was interrupted only temporarily for osmoregulatory water drinking, and subsequently the animals ingested as much 0.5 M NaCl solution as adrenalectomized rats not given the injection (Stricker, 1981; also, Fitzsimons & Wirth, 1976). Similarly, after a 48-hour period of imposed sodium deprivation, adrenalectomized rats consumed 14 mL of 0.5 M NaCl solution in the first 60 minutes, more than three times their sodium deficits, yet for several hours afterwards still showed alternating water and saline intakes analogous to phase 2 drinking behavior after PEG treatment (Stricker, 1981). Similar observations have been made in intact rats that had been maintained on sodium-deficient diet for a prolonged period of time (Figure 4) or for shorter periods during lactation (Figure 4). These and other findings (Nachman & Valentino, 1966; Wolf, Schulkin, & Simson, 1984) therefore provide further evidence that the negative feedback system in the satiation of sodium appetite is very slow.

DEOXYCORTICOSTERONE TREATMENT

The increased NaCl intake by rats receiving daily treatments with DOC occurs despite the presence of excess body sodium and the clear absence of need. These animals certainly are not in a sodium-deficient mode of brain function; indeed, with elevated plasma volume, reduced plasma angiotensin levels, and incipient hypernatremia, their features are the opposite of those that characterize PEG-treated or adrenalectomized rats, and they do not include any apparent physiological excitatory stimulus of sodium appetite.

Despite these differences, the NaCl intakes elicited by the three familiar models of sodium appetite each are associated with reduced or unstimulated levels of plasma OT (Stricker & Verbalis, 1987). Whereas osmotic dilution and sodium deficiency seem to be responsible for that effect in the colloid-treatment and adrenalectomy models of sodium appetite, neither is present in DOC-treated rats, and therefore some other factor must be responsible for blunting NaCl-intake-inhibitory PVN activity in this preparation. One possibility is that the induced expansion of plasma volume, which is known to inhibit secretion of AVP (e.g., Menninger, 1981; Quillen & Cowley, 1983; see also Moore-Gillon & Fitzsimons, 1982; Toth *et al.*, 1987), has a similar effect on OT secretion. Regardless of its cause, the suppression of PVN activity in that subset of neurons mediating inhibition of NaCl intake, as reflected by the unstimulated pituitary OT secretion, should make it easier for all stimuli of sodium appetite to promote NaCl ingestion.

This may account for the enhanced NaCl intakes seen in response to intraventricular administration of angiotensin in rats that were pretreated with DOC (Fluharty and Epstein, 1983). Those important observations provide the foundation for the well-known proposal that angiotensin and mineralocorticoids act directly and synergetically in the brain to stimulate sodium appetite (see Chapter 2). However, reanalysis of the intakes of water and 0.5 M NaCl solution by animals given these two treatments indicates that water was consumed first, followed by an alternation in intake of both fluids; in other words, they drank in

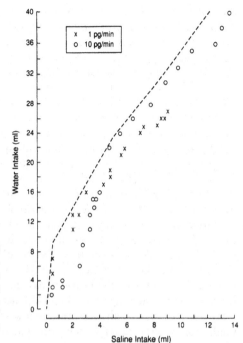

Figure 11. Mean intakes of water and 0.5 M NaCl solution by rats given intracerebroventricular infusion of angiotensin at the rate of 1 or 10 pg/min. Animals had been pretreated daily with 2 mg DOC acetate. Symbols represent cumulative values for each hour of the 24-hour test period. (Data replotted from Fluharty & Epstein, 1983.)

exactly the same pattern as do rats after subcutaneous PEG treatment (Figure 11; replotted from Fluharty & Epstein, 1983). Even when a large dose of angiotensin given by intraventricular injection stimulated 5 mL of saline ingestion in 30 minutes, it is noteworthy that close to 25 mL of water also was consumed during this period (Figure 7 in Fluharty and Epstein, 1983); these amounts fall on the regression lines summarizing the fluid intakes of PEG-treated rats (see Figure 11), as did the early intakes of rats drinking similarly large amounts of saline and water after injection of hog renin into the preoptic region (Avrith & Fitzsimons, 1983), thereby suggesting that the same pattern of water and saline consumption again was obtained.

According to this analysis, the observations of Fluharty and Epstein (1983) may be interpreted somewhat differently than the synergy hypothesis: in the DOC-treated rat, intracranial angiotensin may simply provide a signal for thirst and sodium appetite analogous to that produced by PEG treatment in rats maintained on standard chow. What permits sodium appetite to emerge under these circumstances is the inhibition of PVN activity caused by the DOC treatment, the same critical factor that allows NaCl ingestion to appear when osmotic dilution is present in conjunction with hypovolemia or sodium deficiency. Consistent with this hypothesis are observations that when rats were not pretreated with DOC, intracranial administration of relatively large amounts of angiotensin increased activity in PVN oxytocinergic neurons (Lang *et al.*, 1981) and stimulated thirst but not sodium appetite (Buggy & Fisher, 1974).

Recent observations suggest the presence of another important element in the NaCl ingestion of DOC-treated rats. If the observed intake of saline results from suppression of the inhibition of sodium appetite rather than from the presence of some unsuspected excitatory stimulus, then anything that would additionally suppress the PVN oxytocinergic neurons should further promote NaCl intake. One such factor may be the obligate polydipsia in DOC-treated rats. The oropharyngeal stimulation from water drinking is known to decrease AVP secretion in dehydrated dogs, humans, and other animals (see Chapter 14). We presume that water drinking similarly has a rapid inhibitory effect on the secretion of AVP and OT in dehydrated rats, and that parallel effects occur in parvocellular PVN neurons that inhibit sodium appetite. These assumptions allow the possibility that DOC-treated rats drink saline initially because activity is very low in those PVN cells that inhibit sodium appetite, and they reduce PVN activity further by consuming large amounts of water secondary to the renal retention of ingested NaCl. Although this speculation obviously requires further experimental attention, it should be noted that when food and fluids are present *ad libitum*, the intakes of saline by DOC-treated rats always are nested in their water intakes (E. M. Stricker, K. S. Gannon, & J. C. Smith, unpublished observations).

We are not saying that water drinking inevitably leads to NaCl ingestion in rats but that water drinking itself may do so when inhibitory signals to NaCl ingestion (and the corresponding PVN activity) are reduced, as reflected by plasma OT levels that already are very low. If so, then these same considerations might apply to the rapidly alternating intakes of saline and water that occur in phase 2 after PEG treatment (Figure 2), after tissue damage (Figure 3), after imposed sodium deprivation in male rats or lactating rats (Figure 4), after systemic injection of captopril in PEG-treated rats (Figure 7), after intraventricular injection of angiotensin in rats pretreated with DOC (Figure 11), and in any other condition of sodium deficiency or hypovolemia in which stimuli of sodium appetite are present together with osmotic dilution.

Concluding Remarks

Sodium appetite in rats was first reported by Richter (1936) more than 50 years ago, and for the subsequent 30 years the phenomenon received surprisingly little attention. But more recently there has been a renaissance of interest in the brain's control of sodium appetite, especially in the past decade, as behavioral neurobiologists have brought the sophisticated techniques of modern neuroscience to bear on this subject. This welcome flurry of research activity has generated many of the new observations that contributed substantially to the foundational base of the present proposal.

Our hypothesis rests on observations made in three different but well-studied models of sodium appetite in rats, each providing a state that differs in certain fundamental features from the others. The fact that a single hypothesis can provide a rational biological basis for the sodium appetite observed under these very dissimilar conditions encourages us to believe that it might have integrative value both in considering the burgeoning literature on sodium appetite and, even more importantly, in planning future experimental inquiries. However, this proposal differs from those that have been made heretofore, and while considering it the reader should be alerted to four significant departures we have taken from traditional approaches to this issue.

Models of thirst and sodium appetite in rats involving hypovolemia in the absence of total body sodium deficits have been studied for 25 years. However, the relationship between these drives never was clear, in part because studies of thirst and sodium appetite often were considered to be separate disciplinary issues. Accordingly, data regarding water and saline intakes were plotted separately as a function of time after experimental treatment so that the two responses could be evaluated independently of one another. Unfortunately, if these two responses are not regulated independently of one another but instead are functionally linked, then such conventional ways of presenting and evaluating data actually might obscure important relationships between the responses as well as confuse interpretation of each individual response.

As mentioned earlier, by replotting water and saline intakes together as a function of one another over time, it became apparent that there were distinct phases in the elicited drinking behavior following PEG-induced hypovolemia (Stricker, 1981). These findings helped to define the onset of sodium appetite and also served to identify and distinguish the changing bases for water intake and its influence on saline ingestion that occur in the posttreatment period. In addition, they provided a basis for comparing the thirst and sodium appetite produced in rats by various treatments and testing conditions. The invariant relationship between water and saline intakes during phase 2, displayed in numerous figures in this chapter, indicates the extraordinary consistency of drinking behavior that occurs whenever rats are made hypovolemic.

IMPORTANCE OF CENTRAL INHIBITORY SYSTEMS IN THE CONTROL OF SODIUM APPETITE

Studies of hypovolemic thirst indicated that important inhibitory factors influenced water ingestion, not just excitatory stimuli (Stricker, 1969). Le Magnen (1969) and others emphasized the same point in considering the controls of food intake, and that view has reoriented the field (Stricker, 1984a; see Chapter 3). With regard to sodium appetite, it has long been evident that there were many circumstances in which NaCl ingestion did not appear despite hypovolemia and/or sodium deficiency. Whereas much attention had been focused on the cardinal signs of sodium deficiency that represented excitatory stimuli to sodium appetite, little work sought to identify the manifestations that characterize the various circumstances in which inhibition of sodium appetite occurred.

The path of logic by which we were led to consider a possible role for a subset of PVN neurons in the central inhibitory control of sodium appetite has been discussed elsewhere (Stricker & Verbalis, 1988, in press). The availability of plasma OT levels as a convenient marker for such activity has been crucial to the development of these concepts, and the results obtained to date indicate a clear inverse relationship between sodium appetite and plasma OT levels. Indeed, we have yet to find a condition in which sodium appetite was elicited at a time of stimulated OT secretion, and that fact has encouraged us to turn and return to the PVN in our speculation.

Although a possible role for the PVN in the inhibition of sodium appetite had not been suspected previously, involvement of this structure in mediating satiety for food has been proposed (Leibowitz, 1978; Stricker, McCann, Flanagan, & Verbalis, 1988), and intraventricular injection of OT has been shown to reduce food intake in rats (Arletti, Benelli, & Bertolini, 1989; Olson *et al.*, 1989). Thus, it seems significant that many of the issues proposed to influence the control of sodium appetite apply as well to the control of food intake. For example, it is well recognized that factors relating to caloric homeostasis are not the only ones to influence food intake and that signals such as nausea and dehydration, which have nothing obvious to do with caloric flux, have potent inhibitory effects on food intake that appear to be mediated by parvocellular oxytocinergic neurons in the PVN (Stricker & Verbalis, 1990). Central pathways involved in inhibiting food and NaCl consumption could overlap under certain conditions if increased PVN activity acts to reduce the appetite for osmoles in general. Such activity then would be expressed both as a decreased intake of food, the usual source of osmotic particles, and as a decreased consumption of NaCl solution. More specific control of food and NaCl intakes could be obtained if the intakes were primarily influenced by separate signals, perhaps related to the availability of metabolic fuels on the one hand and to the presence of hypovolemia and activity in the renin–angiotensin system on the other, but still shared some component of a common inhibitory system. This speculation awaits further investigation.

Attention must be given in this regard to recent observations suggesting that peptides other than angiotensin and OT play an important role in the control of sodium appetite in rats. These include atrial natriuretic peptide (Antunes-Rodrigues, McCann, & Samson, 1986; Fitts, Thunhorst, & Simpson, 1985b) and peptides in the tachykinin family (Massi & Epstein, 1989; Massi, Perfumi, de Caro, & Epstein, 1988; Massi, Polidori, *et al.*, 1988). However, in each case an inhibition of NaCl intake was produced by chemical stimulation of the brain. Additional work clearly is required to determine how these results relate to normal physiology and whether similar conclusions would be reached by measuring the release of these peptides in the brain or by blocking their endogenous activity.

ALDOSTERONE LIKELY DOES NOT STIMULATE SODIUM APPETITE DIRECTLY BUT INDIRECTLY

An obvious and attractive way to integrate the paradoxical findings of sodium appetite in the DOC-treated rats with the sodium appetite expected in intact rats during sodium deficiency or hypovolemia is to assume that endogenous mineralocorticoids normally provide an excitatory stimulus of sodium appetite. This hypothesis is parsimonious in attributing to the same chemical agent the important functions of sodium intake and sodium retention during sodium deficiency and is reminiscent of the way in which gonadal steroid hormones have been proposed to influence both mating behavior and reproductive physiology. But is the hypothesis true?

The present considerations do not lead us to propose that mineralocorticoids increase NaCl ingestion because of some direct stimulatory effect they may have in the brain. Certainly the marked NaCl intakes of adrenalectomized rats always have indicated that aldosterone itself is not necessary for the expression of robust sodium appetite, and the same conclusion was reached when the effects of PEG treatment on sodium appetite were investigated. In adrenalectomized rats given DOC therapy so that they could tolerate the induced hypovolemia, completely normal patterns and volumes of water and saline intake were observed after injection of colloid (Stricker, 1983; see also Wolf & Steinbaum, 1965; Wolf & Stricker, 1967). Moreover, systemic administration of captopril provoked an abrupt increase in NaCl intake by the adrenalectomized PEG-treated rats, just as it did in intact PEG-treated rats (Stricker, 1983; see also Elfont & Fitzsimons, 1985). On the basis of these findings, it seems inescapable to conclude that elevated levels of circulating mineralocorticoids neither play a necessary role in converting the brain to the sodium-deficient mode after PEG treatment, nor do they influence the amounts of saline that are consumed in response to the hypovolemia. Instead, we have proposed that mineralocorticoids increase NaCl intake in rats not by directly providing an excitatory stimulus of sodium appetite but by indirectly decreasing activity in a central inhibitory system. These conclusions are compatible with recent observations that intraventricular injection of an aldosterone receptor blocker reduces NaCl ingestion in sodium-deficient rats (Sakai, Nicolaidis, & Epstein, 1986) but not in rats treated with systemic aldosterone (Gomez-Sanchez, Fort, & Gomez-Sanchez, 1990).

SPECIES DIFFERENCES

In this chapter we have focused exclusively on sodium appetite in rats. However, sodium appetite is a well-established phenomenon in many species (see review by Rowland & Fregly, 1988). For example, in sheep and other ruminants, controlled losses of parotid saliva through an externalized catheter usually are employed to produce large deficits in sodium bicarbonate (Denton & Sabine, 1961), but urinary sodium loss produced by adrenalectomy (Denton, Orchard, & Weller, 1969) or treatment with natriuretic drugs also has been used successfully (Zimmerman, Stricker, & Blaine, 1978). With these models the animals have been shown to have a strong motivation to ingest sodium-containing fluids and, like rats, will work hard to obtain them (Abraham et al., 1973; Baldwin, 1976; Bell & Sly, 1979). More physiological experiments indicate important roles for angiotensin and osmotic dilution in the stimulation of sodium appetite but not for mineralocorticoids (Denton, Nelson, Orchard, & Weller, 1969; Weisinger et al., 1982, 1987), as in rats, so that there may not be fundamental differences in the central regulatory systems controlling sodium appetite across these species as usually is supposed. In comparison, note that quite analogous control mechanisms for thirst and AVP secretion appear to have evolved in all mammalian species studied in the laboratory to date, including humans. As well, the importance of the renin–angiotensin–aldosterone system for blood pressure support and sodium conservation appears to be universal among mammalian species.

Even if the general features of sodium appetite prove to be comparable across species, the specific mechanisms by which central control of this behavior is mediated may be different. In this regard, too little is definitely known about the brain pathways that are involved in the control of sodium appetite in rats or any other species. Whether the PVN or some related brain structure has a critical role in mediating inhibition of sodium appetite in sheep and other ruminants is an important question that remains to be determined. In addressing this question, note that plasma levels of AVP may be useful as a marker of PVN activity because AVP, not OT, seems to be the neurohypophyseal peptide hormone that is associated with hypothalamic parvocellular neurons mediating inhibition of food intake in sheep and many other mammals, including man (Ebenezer, Thornton, & Parrott, 1989; Miaskiewicz, Stricker, & Verbalis, 1989; Verbalis, Richardson, & Stricker, 1987). Thus, more comparative studies of sodium appetite, especially including human subjects and nonhuman primates (see Schulkin, Leibman, Ehrman, Norton, & Ternes, 1984), are required before the present considerations of sodium appetite in rats can be extended to other mammals including humans (see Chapter 16).

REFERENCES

Abraham, S., Baker, R., Denton, D. A., Kraintz, F., Kraintz, L., & Purser, L. (1973). Components in the regulation of salt balance: Salt appetite studied by operant behaviour. *Australian Journal of Experimental Biology and Medical Science, 51*, 65–81.

Aguilera, G., Hauger, R. L., & Catt, K. J. (1978). Control of aldosterone secretion during sodium restriction: Adrenal receptor regulation and increased adrenal sensitivity to angiotensin II. *Proceedings of the National Academy of Sciences of the United States of America, 75*, 975–979.

Antunes-Rodrigues, J., McCann, S. M., & Samson, W. K. (1986). Central administration of atrial natriuretic factor inhibits saline preference in the rat. *Endocrinology, 118*, 1726–1728.

Arletti, R., Benelli, A., & Bertolini, A. (1989). Influence of oxytocin on feeding behavior in the rat. *Peptides, 10*, 89–93.

Avrith, D. B., & Fitzsimons, J. T. (1980). Increased sodium appetite in the rat induced by intracranial administration of components of the renin–angiotensin system. *Journal of Physiology (London), 301*, 349–364.

Avrith, D. B., & Fitzsimons, J. T. (1983). Renin-induced sodium appetite: Effects on sodium balance and mediation by angiotensin in the rat. *Journal of Physiology (London), 337*, 479–496.

Baldwin, B. A. (1976). Effects of intracarotid or intraruminal injections of NaCl or $NaHCO_3$ on sodium appetite in goats. *Physiology and Behavior, 16*, 59–66.

Balment, R. J., Brimble, M. J., & Forsling, M. L. (1980). Release of oxytocin induced by salt loading and its influence on renal excretion in the male rat. *Journal of Physiology (London), 308*, 439–449.

Barelare, B., Jr., & Richter, C. P. (1938). Increased sodium chloride appetite in pregnant rats. *American Journal of Physiology, 121*, 185–188.

Bell, F. R., & Sly, J. (1979). The metabolic effects of sodium depletion in calves on salt appetite assessed by operant methods. *Journal of Physiology (London), 295*, 431–443.

Berridge, K. C., Flynn, F. W., Schulkin, J., & Grill, H. J. (1984). Sodium depletion enhances salt palatability in rats. *Behavioral Neuroscience, 98*, 652–660.

Blass, E. M., Nussbaum, A. I., & Hanson, D. G. (1974). Septal hyperdipsia: Specific enhancement of drinking to angiotensin in rats. *Journal of Comparative and Physiological Psychology, 87*, 422–439.

Braun-Menendez, E. (1952). Aumento del apetito especifico para la sal provocado por la desoxicorticosterona. II. Sustancias que potencian o inhiben esta accion. *Revista Sociedad Argentina de Biologia, 28*, 23–32.

Braun-Menendez, E. (1953). Modificadores del apetito especifico para la sal en ratas blancas. *Revista Sociedad Argentina de Biologia, 29*, 92–103.

Braun-Menendez, E., & Brandt, P. (1952). Aumento del apetito especifico para la sal provocado por la desoxicorticosterona. I. Caracteristicas. *Revista Sociedad Argentina de Biologia, 28*, 15–23.

Brenner, B. M., & Berliner, R. W. (1969). Relationship between extracellular volume and fluid reabsorption by the rat nephron. *American Journal of Physiology, 217*, 6–12.

Bryant, R. W., Epstein, A. N., Fitzsimons, J. T., & Fluharty, S. J. (1980). Arousal of a specific and persistent sodium appetite in the rat with continuous intracerebroventricular infusion of angiotensin II. *Journal of Physiology (London), 301*, 365–382.

Buggy, J., & Fisher, A. E. (1974). Evidence for a dual role for angiotensin in water and sodium intake. *Nature, 250*, 733–735.

Buggy, J., & Janklaas, J. (1984). Sodium appetite decreased by central angiotensin blockade. *Physiology and Behavior, 32*, 737–742.

Chiaraviglio, E. (1976). Effect of renin–angiotensin system on sodium intake. *Journal of Physiology (London), 255*, 57–66.

Denton, D. (1982). *The hunger for salt: An anthropological, physiological and medical analysis.* Berlin: Springer-Verlag.

Denton, D. A., Nelson, J. F., Orchard, E., & Weller, S. (1969). The role of adrenocortical hormone secretion in salt appetite. In C. Pfaffmann (Ed.), *Olfaction and taste* (vol. 3, pp. 535–547). New York: Rockefeller University Press.

Denton, D. A., Orchard, E., & Weller, S. (1969). The relation between voluntary sodium intake and body sodium balance in normal and adrenalectomized sheep. *Communications in Behavioral Biology, 3,* 213–221.

Denton, D. A., & Sabine, J. R. (1961). The selective appetite for Na$^+$ shown by Na$^+$-deficient sheep. *Journal of Physiology (London), 157,* 97–116.

Di Nicolantonio, R., Hutchinson, J. S., & Mendelsohn, F. A. O. (1982). Exaggerated salt appetite of spontaneously hypertensive rats is decreased by central angiotensin-converting enzyme blockade. *Nature, 298,* 846–848.

Dunn, F. L., Brennan, T. J., Nelson, A. E., & Robertson, G. L. (1973). The role of blood osmolality and volume in regulating vasopressin secretion in the rat. *Journal of Clinical Investigation, 52,* 3212–3219.

Ebenezer, I. S., Thornton, S. N., & Parrott, R. F. (1989). Anterior and posterior pituitary hormone release induced in sheep by cholecystokinin. *American Journal of Physiology, 256,* R1355–R1357.

Elfont, R. M., Epstein, A. N., & Fitzsimons, J. T. (1984). Involvement of the renin–angiotensin system in captopril-induced sodium appetite in the rat. *Journal of Physiology (London), 354,* 11–27.

Elfont, R. M., & Fitzsimons, J. T. (1985). The effect of captopril on sodium appetite in adrenalectomized and deoxycorticosterone-treated rats. *Journal of Physiology (London), 365,* 1–12.

Epstein, A. N., & Stellar, E. (1955). The control of salt preference in the adrenalectomized rat. *Journal of Comparative and Physiological Psychology, 48,* 167–172.

Epstein, A. N., Zhang, D. M., Schultz, J., Rosenberg, M., Kupsha, P., & Stellar, E. (1984). The failure of ventricular sodium to control sodium appetite in the rat. *Physiology and Behavior, 32,* 683–686.

Evered, M., & Robinson, M. M. (1984). Increased or decreased thirst caused by inhibition of angiotensin-converting enzyme in the rat. *Journal of Physiology (London), 348,* 573–588.

Falk, J. L. (1965). Water intake and NaCl appetite in sodium depletion. *Psychology Reports, 16,* 315–325.

Ferguson, A. V., & Kasting, N. W. (1988). Angiotensin acts at the subfornical organ to increase plasma oxytocin concentrations in the rat. *Regulatory Peptides, 23,* 343–352.

Ferguson, A. V., & Renaud, L. P. (1986). Systemic angiotensin acts at subfornical organ to facilitate activity of neurohypophysial neurons. *American Journal of Physiology, 251,* R712–R717.

Fitts, D. A., Thunhorst, R. L., & Simpson, J. B. (1985a). Modulation of salt appetite by lateral ventricular infusions of angiotensin II and carbachol during sodium depletion. *Brain Research, 346,* 273–280.

Fitts, D. A., Thunhorst, R. L., & Simpson, J. B. (1985b). Diuresis and reduction of salt appetite by lateral ventricular infusions of atriopeptin II. *Brain Research, 348,* 118–124.

Fitzsimons, J. T. (1961). Drinking by rats depleted of body fluid without increase in osmotic pressure. *Journal of Physiology (London), 159,* 297–309.

Fitzsimons, J. T. (1969). The role of a renal thirst factor in drinking induced by extracellular stimuli. *Journal of Physiology (London), 201,* 349–368.

Fitzsimons, J. T., & Moore-Gillon, M. J. (1980). Drinking and antidiuresis in response to reductions in venous return in the dog: Neural and endocrine mechanisms. *Journal of Physiology (London), 308,* 403–416.

Fitzsimons, J. T., & Simons, B. J. (1969). The effect on drinking in the rat of intravenous infusion of angiotensin, given alone or in combination with other stimuli of thirst. *Journal of Physiology (London), 203,* 45–57.

Fitzsimons, J. T., & Stricker, E. M. (1971). Sodium appetite and the renin–angiotensin system. *Nature, New Biology, 231,* 58–60.

Fitzsimons, J. T., & Wirth, J. B. (1976). The neuroendocrinology of thirst and sodium appetite. In W. Kaufmann & D. K. Krause (Eds.), *Central nervous control of Na$^+$ balance—Relations to the renin–angiotensin system* (pp. 80–93). Stuttgart: Georg Thieme.

Fitzsimons, J. T., & Wirth, J. B. (1978). The renin–angiotensin system and sodium appetite. *Journal of Physiology (London), 274,* 63–80.

Fluharty, S. J., & Epstein, A. N. (1983). Sodium appetite elicited by intracerebroventricular infusion of angiotensin II in the rat: II. Synergistic interaction with systemic mineralocorticoids. *Behavioral Neuroscience, 97,* 746–758.

Fluharty, S. J., & Manaker, S. (1983). Sodium appetite elicited by intracerebroventricular infusion of angiotensin II in the rat: I. Relation to urinary sodium excretion. *Behavioral Neuroscience, 97,* 738–745.

Fregly, M. J. (1967). Effect of 9-a-fluorocortisol on spontaneous NaCl intake by adrenalectomized rats. *Physiology and Behavior, 2,* 127–129.

Fregly, M. J. (1980). Effect of the angiotensin converting enzyme inhibitor, captopril, on NaCl appetite of rats. *Journal of Pharmacology and Experimental Therapeutics, 215,* 407–412.

Fregly, M. J., & Waters, I. W. (1966). Effect of mineralocorticoids on spontaneous sodium chloride appetite of adrenalectomized rats. *Physiology and Behavior, 1,* 65–74.

Frumkin, K. (1971). Interaction of LiCl aversion and sodium-specific hunger in the adrenalectomized rat. *Journal of Comparative and Physiological Psychology, 75,* 32–40.

Gardiner, T. W., Jolley, J. R., Vagnucci, A. H., & Stricker, E. M. (1986). Enhanced sodium appetite in rats with lesions centered upon nucleus medianus. *Behavioral Neuroscience, 100,* 531–535.

Gardiner, T. W., Verbalis, J. G., & Stricker, E. M. (1985). Impaired secretion of vasopressin and oxytocin in rats after lesions of nucleus medianus. *American Journal of Physiology, 249,* R681–R688.

Gomez-Sanchez, E. P., Fort, C. M., and Gomez-Sanchez, C. E. (1990). Intracerebroventricular infusion of RU28318 blocks aldosterone-salt hypertension. *American Journal of Physiology, 258,* E482–E484.

Handal, P. J. (1965a). Immediate acceptance of sodium salts by sodium deficient rats. *Psychonomic Science, 3,* 315–316.

Handal, P. J. (1965b). Formalin induced sodium appetite: Dose–response relationships. *Psychonomic Science, 3,* 511–512.

Helmreich, D. L., Thiels, E., Verbalis, J. G., & Stricker, E. M. (1988). Suckling does not affect gastric motility in rats. *Society for Neuroscience Abstracts, 14,* 629.

Hollenberg, N. K. (1980). Set point for sodium homeostasis: Surfeit, deficit, and their implications. *Kidney International, 17,* 423–429.

Hosutt, J. A., & Stricker, E. M. (1981). Hypotension and thirst after phentolamine treatment. *Physiology and Behavior, 27,* 463–468.

Jacobs, K. M., Mark, G. P., & Scott, T. R. (1988). Taste responses in the nucleus tractus solitarius of sodium-deprived rats. *Journal of Physiology (London), 406,* 393–410.

Jalowiec, J. E. (1974). Sodium appetite elicited by furosemide: Effects of differential dietary maintenance. *Behavioral Biology, 10,* 313–327.

Jalowiec, J. E., & Stricker, E. M. (1970a). Restoration of body fluid balance following acute sodium deficiency in rats. *Journal of Comparative and Physiological Psychology, 70,* 94–102.

Jalowiec, J. E., & Stricker, E. M. (1970b). Sodium appetite in rats after apparent recovery from acute sodium deficiency. *Journal of Comparative and Physiological Psychology, 73,* 238–244.

Jalowiec, J. E., & Stricker, E. M. (1973). Sodium appetite in adrenalectomized rats following dietary sodium deprivation. *Journal of Comparative and Physiological Psychology, 83,* 66–77.

Johnson, A. K., Mann, J. F. E., Rascher, W., Johnson, J. K., & Ganten, D. (1981). Plasma angiotensin II concentrations and experimentally induced thirst. *American Journal of Physiology, 240,* R229–R234.

Kaufman, S. (1984). Role of right atrial receptors in the control of drinking in the rat. *Journal of Physiology (London), 349,* 389–396.

Krieckhaus, E. E., & Wolf, G. (1968). Acquisition of sodium by rats: Interaction of innate mechanisms and latent learning. *Journal of Comparative and Physiological Psychology, 65,* 197–201.

Kuhn, E. R., & McCann, S. M. (1971). Release of oxytocin and vasopressin in lactating rats after injection of carbachol into the third ventricle. *Neuroendocrinology, 8,* 48–58.

Lang, R. E., Rascher, W., Heil, J., Unger, T., Wiedmann, G., & Ganten, D. (1981). Angiotensin stimulates oxytocin release. *Life Sciences, 29,* 1425–1428.

Lawrence, D., & Pittman, Q. J. (1985). Response of rat paraventricular neurones with central projections to suckling, haemorrhage or osmotic stimuli. *Brain Research, 341,* 176–183.

Leenen, F. H., & Stricker, E. M. (1974). Plasma renin activity and thirst following hypovolemia or caval ligation in rats. *American Journal of Physiology, 226,* 1238–1242.

Lehr, D., Goldman, H. W., & Casner, P. (1973). Renin–angiotensin role in thirst: Paradoxical enhancement of drinking by angiotensin converting enzyme inhibitor. *Science, 182,* 1031–1034.

Leibowitz, S. F. (1978). Paraventricular nucleus: A primary site mediating adrenergic stimulation of feeding and drinking. *Pharmacology, Biochemistry and Behavior, 8,* 163–175.

Le Magnen, J. (1969). Peripheral and systemic actions of food in the caloric regulation of intake. *Annals of the New York Academy of Sciences, 157,* 1126–1157.

Mangiapane, M. L., Thrasher, T. N., Keil, L. C., Simpson, J. B., & Ganong, W. F. (1983). Deficits in drinking and vasopressin secretion after lesions of nucleus medianus. *Neuroendocrinology, 37,* 73–77.

Mann, J. F. E., Eisele, S., Rettig, R., Unger, T., Johnson, A. K., Ganten, D., & Ritz, E. (1988). Renin-dependent water intake in hypovolemia. *Pflugers Archiv, European Journal of Physiology, 412,* 574–578.

Mann, J. F. E., Johnson, A. K., & Ganten, D. (1980). Plasma angiotensin II: Dipsogenic levels and angiotensin-generating capacity of renin. *American Journal of Physiology, 238,* R372–R377.

Massi, M., & Epstein, A. N. (1989). Suppression of salt intake in the rat by neurokinin A: Comparison with the effect of kassinin. *Regulatory Peptides, 24,* 233–244.

Massi, M., Perfumi, M., de Caro, G. & Epstein, A. N. (1988). Inhibitory effect of kassinin on salt intake induced by different natrorexigenic treatments in the rat. *Brain Research, 440,* 232–242.

Massi, M., Polidori, C., Gentili, L., Perfumi, M., de Caro, G., & Maggi, C. A. (1988). The tachykinin NH$_2$-senktide, a selective neurokinin B receptor agonist, is a very potent inhibitor of salt appetite in the rat. *Neuroscience Letters, 92,* 341–346.

McCann, M. J., Verbalis, J. G., & Stricker, E. M. (1989). LiCl and CCK inhibit gastric emptying and feeding and stimulate OT secretion in rats. *American Journal of Physiology, 256,* R463–R468.

Menninger, R. P. (1981). Right atrial stretch decreases supraoptic neurosecretory activity and plasma vasopressin. *American Journal of Physiology, 241,* R44–R49.

Miaskiewicz, S. L., Stricker, E. M., & Verbalis, J. G. (1989). Neurohypophyseal secretion in response to cholecystokinin but not meal-induced gastric distention in humans. *Journal of Clinical Endocrinology and Metabolism, 68,* 837–843.

Mills, G. Y., & Rodbard, S. (1950). Voluntary fluid and salt intake in the normal and the nephrectomized rat receiving desoxycorticosterone. *Journal of Urology, 63,* 492–495.

Moe, K. E., Weiss, M. L., & Epstein, A. N. (1984). Sodium appetite during captopril blockade of endogenous angiotensin II formation. *American Journal of Physiology, 247,* R356–R365.

Mook, D. G. (1969). Some determinants of preference and aversion in the rat. *Annals of the New York Academy of Sciences, 157,* 1158–1174.

Moore-Gillon, M. J., & Fitzsimons, J. T. (1982). Pulmonary vein–atrial junction stretch receptors and the inhibition of drinking. *American Journal of Physiology, 242,* R452–R457.

Morrison, G. R., & Young, J. C. (1972). Taste control over sodium intake in sodium deficient rats. *Physiology and Behavior, 8,* 29–32.

Nachman, M. (1962). Taste preferences for sodium salts by adrenalectomized rats. *Journal of Comparative and Physiological Psychology, 55,* 1124–1129.

Nachman, M., & Valentino, D. A. (1966). Roles of taste and postingestional factors in the satiation of sodium appetite in rats. *Journal of Comparative and Physiological Psychology, 62,* 280–283.

Olson, B. R., Drutarosky, M. D., Chow, M. S., Hruby, V. J., Stricker, E. M., & Verbalis, J. G. (1989). Oxytocin agonist administered centrally decreases food intake in rats. *Society for Neuroscience Abstracts, 15,* 217.

Osborne, P. G., Denton, D. A., & Weisinger, R. S. (1987). Effect of variation of the composition of CSF in the rat upon drinking of water and hypertonic NaCl solutions. *Behavioral Neuroscience, 101,* 371–377.

Quartermain, D., Miller, N. E., & Wolf, G. (1967). Role of experience in relationship between sodium deficiency and rate of bar pressing for salt. *Journal of Comparative and Physiological Psychology, 63,* 417–420.

Quillen, E. W., Jr., & Cowley, A. W., Jr. (1983). Influence of volume changes on osmolality–vasopressin relationships in conscious dogs. *American Journal of Physiology, 244,* H73–H79.

Reid, I. A., Brooks, V. L., Rudolph, C. D., & Keil, L. C. (1982). Analysis of the actions of angiotensin on the central nervous system of conscious dogs. *American Journal of Physiology, 243,* R82–R91.

Rice, K. K., & Richter, C. P. (1943). Increased sodium chloride and water intake of normal rats treated with desoxycorticosterone acetate. *Endocrinology, 33,* 106–115.

Richter, C. P. (1936). Increased salt appetite in adrenalectomized rats. *American Journal of Physiology, 115,* 155–161.

Richter, C. P. (1938). Factors determining voluntary ingestion of water in normals and in individuals with maximum diabetes insipidus. *American Journal of Physiology, 122,* 668–675.

Richter, C. P., & Barelare, B., Jr. (1938). Nutritional requirements of pregnant and lactating rats studied by the self-selection method. *Endocrinology, 23,* 15–24.

Rogers, R. C., & Hermann, G. E. (1987). Oxytocin, oxytocin antagonist, TRH, and hypothalamic paraventricular nucleus stimulation: Effects on gastric motility. *Peptides, 8,* 505–513.

Rowland, N. E., Bellush, L. L., & Fregly, M. J. (1985). Nycthermeral rhythms and sodium chloride appetite in rats. *American Journal of Physiology, 249,* R375–R378.

Rowland, N. E., & Fregly, M. J. (1988). Sodium appetite: Species and strain differences and role of renin–angiotensin–aldosterone system. *Appetite, 11,* 143–178.

Sakai, R. R., & Epstein, A. N. (1990). The dependence of adrenalectomy-induced sodium appetite on the action of angiotensin II in the brain of the rat. *Behavioral Neuroscience, 104,* 167–176.

Sakai, R. R., Fine, W. B., Epstein, A. N., & Frankmann, S. P. (1987). Salt appetite is enhanced by one prior episode of sodium depletion in the rat. *Behavioral Neuroscience, 101,* 724–731.

Sakai, R. R., Nicolaidis, S., & Epstein, A. N. (1986). Salt appetite is suppressed by interference with angiotensin II and aldosterone. *American Journal of Physiology, 251,* R762–R768.

Schiffrin, E. L., & Genest, J. (1982). Mechanism of captopril-induced drinking. *American Journal of Physiology, 242,* R136–R140.

Schulkin, J., Leibman, D., Ehrman, R. N., Norton, N. W., & Ternes, J. W. (1984). Salt hunger in the rhesus monkey. *Behavioral Neuroscience, 98,* 753–756.

Smith, M. H., Jr., Holman, G. L., & Fortune, K. H. (1968). Sodium need and sodium consumption. *Journal of Comparative and Physiological Psychology, 65,* 33–37.

Sofroniew, M. V. (1983). Morphology of vasopressin and oxytocin neurones and their central and vascular projections. In B. A. Cross & G. Leng (Eds.), *The neurohypophysis: Structure, function and control* (pp. 101–114). Amsterdam: Elsevier.

Strauss, M. B., Lamdin, E., Smith, W. P., & Bleifer, S. J. (1958). Surfeit and deficit of sodium. *Archives of Internal Medicine, 102,* 527–536.

Stricker, E. M. (1966). Extracellular fluid volume and thirst. *American Journal of Physiology, 211,* 232–238.

Stricker, E. M. (1968). Some physiological and motivational properties of the hypovolemic stimulus for thirst. *Physiology and Behavior, 3,* 379–385.

Stricker, E. M. (1969). Osmoregulation and volume regulation in rats: Inhibition of hypovolemic thirst by water. *American Journal of Physiology, 217,* 98–105.

Stricker, E. M. (1971a). Inhibition of thirst in rats following hypovolemia and/or caval ligation. *Physiology and Behavior, 6,* 293–298.

Stricker, E. M. (1971b). Effects of hypovolemia and/or caval ligation on water and NaCl solution drinking by rats. *Physiology and Behavior, 6,* 299–305.

Stricker, E. M. (1973). Thirst, sodium appetite, and complementary physiological contributions to the regulation of intravascular fluid volume. In A. N. Epstein, H. R. Kissileff, & E. Stellar (Eds.), *The neuropsychology of thirst* (pp. 73–98). New York: H. V. Winston & Sons.

Stricker, E. M. (1978a). Excessive drinking by rats with septal lesions during hypovolemia induced by subcutaneous colloid treatment. *Physiology and Behavior, 21,* 905–907.

Stricker, E. M. (1978b). The renin–angiotensin and thirst: Some unanswered questions. *Federation Proceedings, 37,* 2704–2710.

Stricker, E. M. (1980). The physiological basis of sodium appetite: A new look at the "depletion–repletion" model. In M. Kare, R. Bernard, & M. Fregly (Eds.), *Biological and behavioral aspects of salt intake* (pp. 185–204). New York: Academic Press.

Stricker, E. M. (1981). Thirst and sodium appetite after colloid treatment in rats. *Journal of Comparative and Physiological Psychology, 95,* 1–25.

Stricker, E. M. (1983). Thirst and sodium appetite after colloid treatment in rats: Role of the renin–angiotensin–aldosterone system. *Behavioral Neuroscience, 97,* 725–737.

Stricker, E. M. (1984a). Biological bases of hunger and satiety: Therapeutic implications. *Nutrition Reviews, 42,* 333–340.

Stricker, E. M. (1984b). Thirst and sodium appetite after colloid treatment in rats with septal lesions. *Behavioral Neuroscience, 98,* 356–360.

Stricker, E. M., Hosutt, J. A., & Verbalis, J. G. (1987). Neurohypophyseal secretion in hypovolemic rats: Inverse relation to sodium appetite. *American Journal of Physiology, 252,* R889–R896.

Stricker, E. M., & Jalowiec, J. E. (1970). Restoration of intravascular fluid volume following acute hypovolemia in rats. *American Journal of Physiology, 218,* 191–196.

Stricker, E. M., McCann, M. J., Flanagan, L. M., & Verbalis, J. G. (1988). Neurohypophyseal secretion and gastric function: Biological correlates of nausea. In H. Takagi, Y. Oomura, M. Ito, & M. Otsuka (Eds.), *Biowarning systems in the brain* (pp. 295–307). Tokyo: University of Tokyo Press.

Stricker, E. M., Vagnucci, A. H., McDonald, R. H., Jr., & Leenen, F. H. (1979). Renin and aldosterone secretions during hypovolemia in rats: Relation to NaCl intake. *American Journal of Physiology, 237,* R45–R51.

Stricker, E. M., & Verbalis, J. G. (1986). Interaction of osmotic and volume stimuli in regulation of neurohypophyseal secretion in rats. *American Journal of Physiology, 250,* R267–R275.

Stricker, E. M., & Verbalis, J. G. (1987). Central inhibitory control of sodium appetite in rats: Correlation with pituitary oxytocin secretion. *Behavioral Neuroscience, 101,* 560–567.

Stricker, E. M., & Verbalis, J. G. (1988). Hormones and behavior: The biology of thirst and sodium appetite. *American Scientist, 76,* 261–267.

Stricker, E. M., & Verbalis, J. G. (in press). Biological bases of sodium appetite: Excitatory and inhibitory factors. In M. I. Friedman, M. G. Tordoff, & M. R. Kare (Eds.), *Chemical senses: Appetite and nutrition.* New York: Marcel Dekker.

Stricker, E. M., & Verbalis, J. G. (1990). Control of appetite and satiety: Insights from biologic and behavioral studies. *Nutrition Reviews, 48,* 49–56.

Stricker, E. M., & Wilson, N. E. (1970). Salt seeking behavior in rats following acute sodium deficiency. *Journal of Comparative and Physiological Psychology, 72,* 416–420.

Stricker, E. M., & Wolf, G. (1966). Blood volume and tonicity in relation to sodium appetite. *Journal of Comparative and Physiological Psychology, 62,* 275–279.

Sugimoto, E. (1988). Analysis of salt and water intake by continuous determination of blood volume and plasma sodium concentration. *Japanese Journal of Physiology, 38,* 519–529.

Swanson, L. W., & Sawchenko, P. E. (1983). Hypothalamic integration: Organization of the paraventricular and supraoptic nuclei. *Annual Reviews of Neuroscience, 6,* 269–324.

Thiels, E., Verbalis, J. G., & Stricker, E. M. (in press). Sodium appetite in lactating rats. *Behavioral Neuroscience.*

Toth, E., Stelfox, J., & Kaufman, S. (1987). Cardiac control of salt appetite. *American Journal of Physiology, 252,* R925–R929.

Verbalis, J. G., Mangione, M., & Stricker, E. M. (1988). Oxytocin is natriuretic at physiological plasma concentrations. *Society for Neuroscience Abstracts, 14,* 628.

Verbalis, J. G., McHale, C. M., Gardiner, T. W., & Stricker, E. M. (1986). Oxytocin and vasopressin secretion in response to stimuli producing learned taste aversions in rats. *Behavioral Neuroscience, 100,* 466–475.

Verbalis, J. G., Richardson, D. W., & Stricker, E. M. (1987). Vasopressin release in response to nausea-producing agents and cholecystokinin in monkeys. *American Journal of Physiology, 252,* R749–R753.

Wagman, W. (1963). Sodium chloride deprivation: Development of sodium chloride as a reinforcement. *Science, 140,* 1403–1404.

Weisinger, R. S., Considine, P., Denton, D. A., Leksell, L., McKinley, M. J., Mouw, D. R., Muller, A. F., & Tarjan, E. (1982). Role of sodium concentration of the cerebrospinal fluid in the salt appetite of sheep. *American Journal of Physiology, 242,* R51–R63.

Weisinger, R. S., Denton, D. A., Di Nicolantonio, R., & McKinley, M. J. (1988). The effect of captopril or enalaprilic acid on the Na appetite of Na-deplete rats. *Clinical and Experimental Pharmacology and Physiology, 15,* 55–65.

Weisinger, R. S., Denton, D. A., Di Nicolantonio, R., McKinley, M. J., Muller, A. F., & Tarjan, E. (1987). Role of angiotensin in sodium appetite of sodium-deplete sheep. *American Journal of Physiology, 253,* R482–R488.

Weiss, M. L., Moe, K. E., & Epstein, A. N. (1986). Interference with central actions of angiotensin II suppresses sodium appetite. *American Journal of Physiology, 250,* R250–R259.

Wolf, G. (1965). Effect of deoxycorticosterone on sodium appetite of intact and adrenalectomized rats. *American Journal of Physiology, 208,* 1281–1285.

Wolf, G., & Handal, P. J. (1966). Aldosterone-induced sodium appetite: Dose–response and specificity. *Endocrinology, 78,* 1120–1124.

Wolf, G., Schulkin, J., & Simson, P. E. (1984). Multiple factors in the satiation of salt appetite. *Behavioral Neuroscience, 98,* 661–673.

Wolf, G., & Steinbaum, E. A. (1965). Sodium appetite elicited by subcutaneous formalin: Mechanism of action. *Journal of Comparative and Physiological Psychology, 59,* 335–339.

Wolf, G., & Stricker, E. M. (1967). Sodium appetite elicited by hypovolemia in adrenalectomized rats: Re-evaluation of the "reservoir" hypothesis. *Journal of Comparative and Physiological Psychology, 63,* 252–257.

Yawata, T., Okuno, T., Nose, H., & Morimoto, T. (1987). Change in salt appetite due to rehydration level in rats. *Physiology and Behavior, 40,* 363–368.

Zimmerman, M. B., Blaine, E. H., & Stricker, E. M. (1981). Water intake in hypovolemic sheep: Effects of crushing the left atrial appendage. *Science, 211,* 489–491.

Zimmerman, M. B., Stricker, E. M., & Blaine, E. H. (1978). Water and NaCl intake after furosemide treatment in sheep (*Ovis aires*). *Journal of Comparative and Physiological Psychology, 92,* 501–510.

16

Clinical Aspects of Body Fluid Homeostasis in Humans

Joseph G. Verbalis

Introduction

Disorders of body fluid homeostasis represent one of the most commonly encountered problems in clinical medicine. This is because many different disease states can potentially disrupt the normal finely balanced system of intakes and outputs that act to maintain water and solute homeostasis. Consequently, a clear understanding of the mechanisms that control body fluid homeostasis in man is valuable for optimizing therapy of numerous clinical pathological states. This chapter first briefly summarizes what is presently known about normal mechanisms of water and solute homeostasis in man and then discusses in greater detail specific disease states involving disruptions of these normal regulatory mechanisms. Rather than attempting to provide a detailed review of all previous research in this field, I emphasize those areas where the most recent advances in our understanding of underlying pathophysiological mechanisms have occurred and also those clinical issues still enmeshed in controversy where additional work will be required to achieve a more complete understanding of derangements of body fluid homeostasis in man. Interested readers are referred to several excellent recent clinical reviews of this area for additional details (Baylis & Thompson, 1988; Robertson, 1987a; Schrier, 1988).

Normal Physiology

Water Balance

Water metabolism in man, as in all animals, represents a balance between intake and excretion of water. Although many factors affect both water intake and

Joseph G. Verbalis   Departments of Medicine and Behavioral Neuroscience, University of Pittsburgh, Pittsburgh, Pennsylvania 15261.

excretion, each side of this equation can be considered to consist of a "regulated" and an "unregulated" component, the magnitudes of which can vary markedly under different physiological and pathological conditions. The unregulated component of water intake occurs via the intrinsic water content of ingested foods, the consumption of beverages primarily for reasons of palatability or desired secondary effects (e.g., caffeine in coffee), or for social or habitual reasons (e.g., alcoholic beverages), whereas the regulated component of water intake can be best described as consisting of fluids consumed primarily in response to a perceived sensation of thirst. Similarly, the unregulated component of water excretion occurs via insensible water losses from a variety of sources (cutaneous losses in sweat, evaporative respiratory losses in exhaled air, gastrointestinal losses through diarrhea and fecal water content) as well as the obligate amount of water that the kidneys must excrete in order to eliminate solutes generated by body metabolism, whereas the regulated component of water excretion can be best described as consisting of the renal excretion of free water above the obligate amount necessary to excrete body solutes. This division is necessarily overly simplified, partly because the regulated and unregulated components often overlap. In addition, it is to some degree semantic, since all physiological processes are regulated in some way. Although some have advocated using the terminology of "primary," or need-induced, and "secondary," or need-free, thirst because of this point (Fitzsimons, 1976, 1979), these terms fail to emphasize the basic physiological difference between these various processes, namely, that the "unregulated" components occur for the most part independently of the major control systems for maintaining body fluid homeostasis, whereas the "regulated" components represent those which act specifically to maintain water balance and compensate for whatever perturbations may be caused by unregulated water losses or gains. Within this conceptual framework it is clear that the two prime factors that regulate water balance in man are thirst and pituitary secretion of arginine vasopressin.

THIRST. As described in previous chapters, drinking can be stimulated in animals either by intracellular dehydration caused by increases in the effective osmolality of the extracellular fluid or by extracellular dehydration caused by losses of extracellular fluid (Andersson, 1978; Fitzsimons, 1972). Substantial evidence to date has supported mediation of the former by osmoreceptors located in the anterior hypothalamus (Ramsay, 1985; Thrasher, 1985; Zimmerman, Ma, & Nilaver, 1987), whereas the latter appears to be stimulated primarily via activation of low- and high-pressure baroreceptors (Quillen, Reid, & Keil, 1988; Ramsay & Thrasher, 1986), with a possible contribution of circulating angiotensin II during more severe degrees of intravascular hypovolemia and hypotension (Mann, Johnson, Ganten, & Ritz, 1987). Controlled studies in animals have consistently reported thresholds for osmotically induced drinking ranging from 1% to 4% increases in plasma osmolality above basal levels (Fitzsimons, 1963; Szczepanska-Sadowska & Kozlowski, 1975; Wood, Rolls, & Ramsay, 1977; Wood, Rolls, & Rolls, 1982). Analogous studies in humans using quantitative estimates of subjective symptoms of thirst have confirmed that increases in plasma osmolality of similar magnitudes are necessary to produce an unequivocal sensation described as "thirst" in humans (Phillips, Rolls, Ledingham, Forsling, & Morton, 1985; Robertson, 1983; Wolf, 1950).

Conversely, the threshold for producing hypovolemic or extracellular thirst appears to be significantly greater in both animals and humans. Studies in a variety of species have suggested that sustained decreases in plasma volume or blood pressure of at least 4–8%, and in some species 10–15%, are necessary to stimulate drinking consistently (Fitzsimons, 1961b; Stricker, 1966; Thrasher, Keil, & Ramsay, 1982a). In humans it has been difficult to demonstrate any effects of mild to moderate hypovolemia to stimulate thirst independently of osmotic changes occurring with dehydration (Holmes & Montgomery, 1953; Robertson, 1984). Some have hypothesized that this blunted sensitivity to changes in extracellular fluid volume or blood pressure in humans may represent an adaptation that occurred as a result of the erect posture of primates, which predisposes them to wider fluctuations in blood and atrial filling pressures as a result of orthostatic pooling of blood in the lower body; stimulation of thirst and AVP secretion by such transient postural changes in blood pressure might lead to overdrinking and inappropriate antidiuresis in situations where extracellular fluid volume actually was normal but only abnormally distributed (Goldsmith, Cowley, Francis, & Cohn, 1984). Consistent with a blunted response to baroreceptor activation, recent studies have shown that systemic infusion of angiotensin II to pharmacological levels is a much less potent dipsogen in humans than in animals (Phillips, Rolls, Ledingham, Morton, & Forsling, 1985). Nonetheless, this basic stimulus is not completely absent in humans, as demonstrated by rare cases of polydipsia in patients with pathological causes of hyperreninemia (Rogers & Kurtzman, 1973). Consequently, the blunted thirst response to both vascular and hormonal manifestations of hypovolemia in man likely represents an example of how similar regulatory systems are utilized differently across species as a result of adaptation to environmental pressures.

Although osmotic factors clearly appear to be more effective stimulants of thirst in humans, it remains unanswered in man, as in other species, whether relatively small changes in plasma osmolality are to any significant degree responsible for day-to-day fluid intakes. Similar to animals, most humans consume the majority of their ingested water as a result of the unregulated components of fluid intake discussed previously and generally ingest volumes in excess of what can be considered to be actual "need" (Fitzsimons, 1979; Phillips, Rolls, Ledingham, & Morton, 1984). Consistent with this observation is the fact that under most conditions plasma osmolalities in man remain within 1–2% of basal levels, and these relatively small changes in plasma osmolality are generally below the threshold levels necessary to stimulate the sensation of thirst in most individuals (Figure 1). These considerations therefore suggest that despite the obvious vital importance of thirst during pathological situations of hyperosmolality and hypovolemia, under normal physiological conditions in man water balance is accomplished more by regulated free water excretion than by regulated water intake.

Before leaving thirst, it is appropriate to consider the question of the origin of the stimulus responsible for thirst and fluid ingestion in man. Historically debate has revolved around the question of whether thirst originates in the oropharynx as a result of a dry mouth (Cannon, 1919; Wolf, 1958) or is a more general sensation originating in the brain as a direct response to changes in ECF osmolality and volume. The last review of clinical disorders of thirst and fluid intake in the *Handbook of Physiology* summarized data correlating thirst in man to changes in salivary flow rates causing oropharyngeal dryness (Holmes, 1967). However, it has become abundantly clear from multiple experimental results in animals that although a dry mouth obviously accompanies states of dehydration

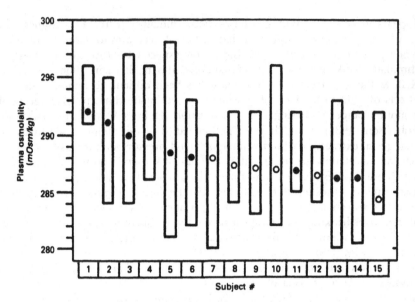

Figure 1. Osmotic thresholds for subjective thirst (top of bar) and AVP secretion (bottom of bar) in relation to basal plasma osmolality (open circles, female; closed circles, male) in 15 individual subjects studied with hypertonic saline infusion. (Reproduced with permission from Robertson, 1977.)

and therefore is a prominent manifestation of the dehydrated state, it cannot represent the primary stimulus for the sensation of thirst resulting from the body's need for additional water. The reader is referred to Fitzsimons (1979) for an extensive review of this area, which is beyond the scope of this chapter.

Nonetheless, it is still of interest to consider the influence of oropharyngeal factors on fluid intake in man. These can be divided into three main categories. First is the role of oropharyngeal factors in regulated, or need-induced, thirst. There is little question that mouth dryness is a prominent manifestation of thirst in man, and quantitation of this sensation correlates well with ratings of thirst as well as subsequent fluid ingestion (Phillips, Rolls, Ledingham, Forsling, & Morton, 1985). Consequently, mouth dryness represents the main subjective sensation that humans associate with a need to ingest fluids. Second is the role of oropharyngeal factors in unregulated, or need-free, thirst. Because of the strong association between a dry mouth and dehydration, it follows that situations causing a dry mouth even in the absence of intracellular or extracellular dehydration, such as ingestion of dry food, mouth breathing, talking, etc., also can lead to fluid ingestion. Although in these cases the dry mouth usually can be abolished simply by irrigation of the oropharynx rather than actual drinking, for a variety of reasons fluid is ingested to relieve this symptom in most situations. This unregulated ingestion can result in a significant amount of daily fluid intake in man, with potential pathological consequences that will be discussed further in the section on hypoosmolality. Finally, oropharyngeal factors have been shown to be of importance for satiation of thirst (Thrasher, Nistal-Herrera, Keil, & Ramsay, 1981), and clinical implications of this are discussed in the section on dehydration. Thus, although a dry mouth represents only a symptom rather than a primary cause of thirst, oropharyngeal factors clearly play an important role in various aspects of clinical disorders of body fluid homeostasis.

Vasopressin Secretion. As in animals, the prime determinant of free water excretion in man is the antidiuresis produced by circulating levels of arginine vasopressin (AVP) in plasma. With the advent of radioimmunoassays for AVP, the unique sensitivity of this hormone to small changes in osmolality, as well as the corresponding sensitivity of the kidney to small changes in plasma AVP levels, have become apparent. Although debate still exists with regard to the exact pattern of osmotically stimulated AVP secretion (Robertson, 1987b), most studies to date have supported the concept of a discrete osmotic threshold for AVP secretion above which a linear relationship between plasma osmolality and AVP levels occurs (Figure 2). The slope of the regression line relating AVP levels to osmolality defines the sensitivity of the response. Although this slope can vary significantly across individual human subjects, in part because of genetic factors (Zerbe, 1985), in general each 1 mosm/kg increase in plasma osmolality causes an increase in plasma AVP level from 0.4 to 0.8 pg/mL (Baylis, 1987; Robertson, 1977). The renal response to circulating AVP is similarly linear, with urinary concentration that is directly proportional to AVP levels from 0.5 to 4–5 pg/mL, after which urinary osmolality is maximal and cannot increase further despite additional increases in AVP levels. Thus, changes of 1% or less in plasma osmolality are sufficient to cause significant increases in plasma AVP levels with proportional increases in urine concentration, and maximal antidiuresis is already achieved after increases in plasma osmolality of only 6 to 12 mosm/kg (2–4%) above the threshold for AVP secretion.

However, even this analysis underestimates the sensitivity of this system to regulate free water excretion for the following reason. Although urine osmolality is directly proportional to plasma AVP levels, urine volume is *inversely* related to urine osmolality (Figure 3). Thus, an increase in plasma AVP level from 0.5 to 2 pg/mL has a far greater effect on urine volume than does a subsequent increase in AVP level from 2 to 5 pg/mL, in effect further magnifying the physiological effects of small initial changes in plasma AVP levels. The net result of these relations is a finely tuned regulatory system that adjusts the rate of free water excretion accurately to the ambient plasma osmolality via changes in pituitary AVP secretion. Furthermore, the rapid response of pituitary

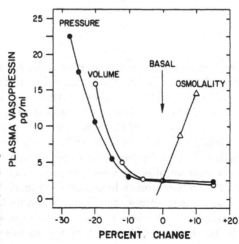

Figure 2. Comparative sensitivity of AVP secretion in response to changes in plasma osmolality (triangles) versus changes in blood volume (open circles) and pressure (closed circles) in human subjects. (Reproduced with permission from Robertson, 1987a.)

JOSEPH G.
VERBALIS

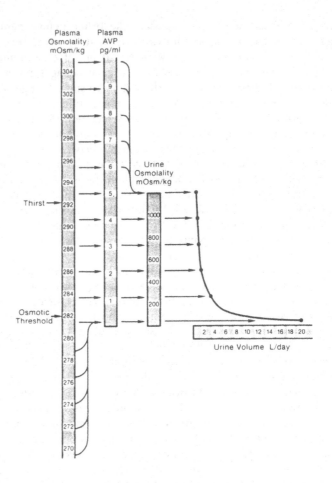

Figure 3. Schematic representation of normal physiological relationships among plasma osmolality, plasma AVP level, urine osmolality, and urine volume in man. Note the inverse relationship between urine osmolality and urine volume, resulting in disproportionate effects on urine volume of small changes in plasma AVP levels at lower concentrations. (Reproduced with permission from Robinson, 1985.)

AVP secretion to changes in plasma osmolality coupled with the short half-life (10–20 minutes) of AVP in human plasma (Engel, Rowe, Minaker, & Robertson, 1984; Lauson, 1974) enables this regulatory system to adjust renal water excretion to changes in plasma osmolality on a minute-to-minute basis.

As in the case of thirst, hypovolemia also is a well-known stimulus for AVP secretion in man (Robertson, 1977; Schrier, Berl, & Anderson, 1979); the appropriate physiological response to volume depletion obviously should include urinary concentration and renal water conservation. But again, similar to thirst, AVP secretion is much less sensitive to small changes in blood volume and blood pressure than osmolality (Figure 2). Some have even suggested that the AVP response to decreases in blood volume is absent in man (Goldsmith, 1988), though this most likely is simply a manifestation of the significantly higher threshold for AVP secretion to volemic stimuli. Such marked differences in AVP responses therefore represent additional corroborative evidence that osmolality represents a more sensitive regulatory system for water balance in man than does blood and ECF volume.

Nonetheless, modest changes in blood volume and pressure still can influence AVP secretion indirectly, even though they are weak stimuli by themselves. This occurs via shifting the sensitivity of AVP secretion to osmotic stimuli so that a given increase in osmolality will cause a greater secretion of AVP during hypovolemic conditions than during euvolemic states. Although this effect has been demonstrated in human (Moses & Miller, 1971; Robertson & Athar, 1976; Robertson, Aycinena, & Zerbe, 1982) as well as in multiple animal studies (Quillen & Cowley, 1983; Robertson, 1977), it has only clearly been shown with significant degrees of hypovolemia, and the magnitude of this effect during mild degrees of volume depletion remains conjectural. Nonetheless, it represents an additional mechanism to enhance the response of an already exquisitely sensitive osmoregulatory system for regulation of water excretion. Interestingly, a similar effect of hypovolemia to lower the osmotic threshold for drinking has been noted in dogs (Kozlowski & Szczepanska-Sadowska, 1975), although no human studies have yet addressed this issue. Consequently, the major effects of moderate degrees of hypovolemia on both AVP secretion and thirst may be to modulate the gain of the osmoregulatory responses, with direct effects on thirst and AVP secretion occurring only during more severe degrees of hypovolemia.

INTEGRATION. A synthesis of what is presently known about the regulation of thirst and AVP secretion in man leads to a relatively simple but elegant system to maintain water balance, illustrated for a single subject in Figure 4. Under normal physiological conditions, the sensitivity of the osmoregulatory system for AVP secretion accounts for maintenance of plasma osmolality within narrow limits by adjusting renal water excretion to small changes in osmolality. Stimulated thirst does not represent a major regulatory mechanism under these conditions, and unregulated water ingestion supplies adequate amounts of water in excess of true "need," which are then retained or excreted in response to osmoregulated pituitary AVP secretion. However, when this combination of unregulated water intake cannot, for whatever reason, adequately supply body needs even in the presence of plasma AVP levels sufficient to cause maximal antidiuresis, then plasma osmolality eventually rises to levels that stimulate unequivocal thirst and cause regulated water intake that is proportional to the elevation of osmolality above this threshold. In such a system thirst therefore in effect represents a backup mechanism that is called into play when pituitary–renal mechanisms prove insufficient to maintain plasma osmolality within a few percent of basal levels. This arrangement has the advantage of freeing man from frequent episodes of thirst that would require a diversion of focused activities toward behaviors oriented to seeking water during those times when water deficiency is sufficiently mild to be compensated for adequately by renal water conservation but would appropriately stimulate behavior toward water ingestion once water deficiency reaches more severe and potentially harmful levels. This system of differential effective thresholds for thirst and AVP secretion nicely complements studies demonstrating excess unregulated, or need-free, drinking in animals. This is summarized by Fitzsimons (1979): "There is normally a reserve of body water between an upper satiety level and a lower threshold level.

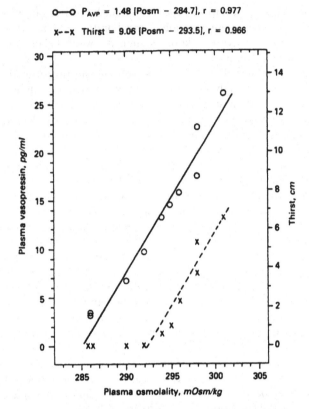

$P_{AVP} = 1.48 [Posm - 284.7], r = 0.977$

$Thirst = 9.06 [Posm - 293.5], r = 0.966$

Figure 4. Relationship of plasma AVP levels (pg/mL, open circles) and thirst (cm on a geometric rating scale, X's) to changes in plasma osmolality induced by hypertonic saline infusion in an individual human subject. Calculated regression functions indicate osmotic thresholds for AVP secretion (284.7 mOsm/kg) and thirst (293.5 mOsm/kg). (Reproduced with permission from Robertson, 1984.)

So long as this reserve exists thirst is not experienced." It is now apparent that the stimulation of AVP secretion prior to reaching plasma osmolalities causing subjective thirst acts as an important mechanism to maintain this "reserve" of body water, thus further freeing man and animals from the necessity for continual drinking.

Despite the intrinsic appeal of this theory and the presence of a substantial body of data in support of it (Robertson, 1983, 1984), some recent studies have questioned whether the threshold for thirst really is set higher than that for AVP secretion and renal water conservation in humans (Thompson, Bland, Burd, & Baylis, 1986; see also Chapter 14). The basis for this argument rests on distinguishing whether thirst is considered to be present or absent at basal levels of plasma osmolality. By allowing their subjects to differentiate between some degree of thirst rather than defining thirst to be absent under conditions of ad libitum fluid intake, these investigators have presented data suggesting similar thresholds for thirst and AVP secretion in man. This raises a crucial question concerning neural regulation of water ingestion, namely, whether central pathways that stimulate fluid intake can be activated in response to small changes in osmolality even before the recognition of the subjective sensation of thirst. If true, this would suggest a more prominent role for regulated water ingestion in the maintenance of water homeostasis even under physiological conditions.

However, several observations argue against this interpretation. First, basal plasma osmolality in man and animals is maintained at levels associated with easily measurable plasma AVP levels and significantly concentrated urinary osmolalities, thereby clearly indicating stimulated AVP secretion below the threshold for stimulation of water intake. Although increases in plasma osmolality of only 1–4% are necessary for stimulation of thirst, it must be recognized that the threshold for AVP secretion actually lies below basal plasma osmolality (Figure 1), accounting for the well-known observation that basal urine osmolalities in man and animals under conditions of *ad libitum* fluid intake are typically concentrated above isotonicity rather than maximally dilute. For example, in a study of spontaneous drinking in human volunteers, urine osmolalities were noted to be 958 ± 63 mosm/kg H_2O, demonstrating near maximal antidiuresis. Despite this, drinking was modest (1010 ± 151 mL over 10 hours) and occurred mainly in conjunction with meals and notably in the absence of any significant increases in plasma osmolality (Phillips, Rolls, Ledingham, & Morton, 1984). These results are consistent with maintenance of body water homeostasis via unregulated, or need-free, water intake in association with AVP-induced antidiuresis, all of which occurs well before the threshold is reached for osmotically stimulated, or regulated, thirst. Second, studies in animals have confirmed that increases in plasma osmolality stimulate drinking after approximately the same relative increase in osmolality that produces subjective thirst in man, arguing against activation of neural mechanisms stimulating water intake prior to production of the sensation of thirst. It should also be emphasized that animal studies suggesting similar thresholds for thirst and AVP secretion have generally utilized infusions of hypertonic saline starting from basal plasma osmolalities. Under these conditions, the observation that 1–2% increases in plasma osmolality are necessary for "significant" increases in plasma AVP levels is an artifact of the difficulty in accurately measuring small increases in plasma AVP levels that nonetheless are able to exert profound effects on the kidney (Figure 3). That osmoregulatory AVP secretion is already occurring during the basal state is amply demonstrated by the already elevated urine osmolalities and measurable plasma AVP levels; only by utilization of water loading to achieve osmolalities that suppress plasma AVP levels below limits of detectability prior to infusing hypertonic saline can any meaningful assessment of the relative increases in osmolality necessary to stimulate AVP secretion and urinary concentration versus thirst be made. Finally, pathological situations in which the ability of the kidneys to regulate free water excretion is impaired, such as chronic renal failure or diabetes insipidus, are often characterized by derangements of body fluid homeostasis, suggesting that water intake alone is not regulated finely enough to maintain effective plasma osmolality within the narrow limits that characterize the intact organism. Consequently, although the magnitude of the difference in effective osmotic set points for thirst and AVP secretion remains debatable and clearly is subject to substantial individual variability, the bulk of experimental results to date support the concept of a higher effective osmotic threshold for thirst than for AVP secretion in man.

JOSEPH G.
VERBALIS

The maintenance of body fluid volumes at normal levels, as well as the appropriate distribution of fluids between the intracellular and extracellular fluid compartments, obviously requires physiological mechanisms to ensure a balance of solutes as well as water. Because the bulk of body solutes that are effective osmotic particles are electrolytes, specifically Na^+, K^+, Cl^-, and HCO_3^-, for the most part discussions of solute metabolism appropriately focus on electrolyte balances. Except where noted, these two terms are used interchangeably in this chapter.

As in the case of water homeostasis, solute homeostasis requires a simple balance between intake and excretion of solutes, and particularly electrolytes. Also as in the case of water metabolism, it is possible to define regulated and unregulated components for both solute intake and solute excretion. However, unlike water intake, in humans there is little evidence to support a significant role for regulated solute intake in terms of specific appetitive behaviors for nonnutritive solutes, with the possible exception of only one pathological condition. Consequently, it follows that there must be an even greater dependence on mechanisms for regulated renal excretion of solutes than is the case for excretion of water. Whether for this reason or not, the specific mechanisms for renal excretion of solutes are more numerous and substantially more complex than the relatively simple, though quite efficient, system for AVP-controlled water excretion.

SODIUM APPETITE. The only solute for which any specific appetite has ever been clearly demonstrated in man is sodium (although, as with animals, this is generally expressed as an appetite for the chloride salt of sodium, so it should more appropriately be called NaCl appetite). Because of the importance of Na^+ for ensuring maintenance of the extracellular fluid (ECF) volume, which in turn directly supports blood volume and pressure, its uniqueness insofar as meriting a specific mechanism for regulated intake seems appropriate. However, despite abundant evidence demonstrating a salt appetite in rats that is proportionately related to Na^+ losses induced by various methods of sodium and extracellular fluid depletion (Denton, 1967; Jalowiec & Stricker, 1973; Stricker, 1981), there has been only one pathological condition in which a specific stimulated sodium appetite has been unequivocally observed in humans, namely, Addison's disease caused by adrenal insufficiency. Since the initial discovery of this disorder, salt craving has remained one of the well-known manifestations of Addison's disease (Baxter & Tyrrell, 1987), analogous to the well-known stimulation of sodium appetite in adrenalectomized rats (Richter, 1936). However, despite the production of sodium deficiency in most patients with untreated Addison's disease, unexplainably only 15–20% of such patients manifest this behavior (Henkin, Gill, & Bartter, 1963; Thorn, Dorrance, & Day, 1942).

Even more striking is the absence, or at least the lack of any prominence, of salt appetite during a variety of other disorders causing equivalent degrees of sodium and extracellular fluid volume depletion in humans. Patients with hemorrhagic blood loss, diuretic-induced hypovolemia, or hypotension of any etiology become thirsty when intravascular deficits are marked but almost never express any pronounced desire for salty foods or fluids. Yet, as with thirst, one might consider the possibility of subclinical activation of neural mechanisms stimulating salt intake without a conscious subjective sensation of salt "hunger." However, this possibility cannot be supported either, because many such patients actually become hyponatremic as a result of continued ingestion of only water or osmotically dilute fluids in response to their volume depletion (Alvis, Gehab, & Cox, 1985; Verbalis, 1990a). In fact, it is interesting that athletes must be taught to ingest sodium in the form of NaCl tablets or electrolyte solutions during periods of solute losses from profuse sweating rather than exhibiting a spontaneous salt appetite.

Analogous to the infrequency of stimulated sodium appetite in man, there is no evidence to support spontaneous inhibition of sodium intake under conditions of sodium and extracellular fluid volume excess; rather, every clinician is acutely aware of the difficulty of maintaining even moderate degrees of sodium restriction in patients with edema-forming states such as congestive heart failure and in patients with hypertension.

The absence of a prominent sodium appetite in man is not unique, however, and appears to be a characteristic of carnivores in general. This has been hypothesized to reflect an evolutionary adaptation in response to the naturally occurring high sodium intake of flesh-eating animals, as opposed to the low salt intake of herbivores, possibly accounting for the presence of specific mechanisms to seek sodium in the latter (Denton, 1982). Nonetheless, the fact that conditions such as adrenal insufficiency are associated with sodium appetite even in carnivores such as man again reinforces the view that the basic regulatory mechanisms for body fluid homeostasis are likely analogous in all mammalian species, but their relative importance and utilization vary across species in response to prior evolutionary pressures.

NATRIURESIS. Although specific mechanisms exist for regulated renal excretion of all major electrolytes, for reasons already mentioned this section concentrates on those mechanisms controlling excretion of the major extracellular solute, sodium. The classical mechanisms known to regulate renal sodium excretion are renal hemodynamic factors, specifically renal perfusion pressure and glomerular filtration rate, which influence sodium reabsorption in the proximal nephron, and adrenal aldosterone secretion, which affects sodium–potassium exchange in the distal nephron. Although these factors alone can account for much of the observed variation in renal sodium excretion, it has long been known that even together they still could not completely explain the occurrence of natriuresis in the absence of measurable changes in renal hemodynamic parameters or aldosterone secretion, such as during a modest isotonic saline volume expansion. This led to the postulation of the existence of a "third factor" for regulation of sodium excretion.

Several humoral factors have since been proposed as potential candidates for additional natriuretic hormones. Arginine vasopressin itself can stimulate natriuresis, presumably via receptor-mediated mechanisms similar to those by which both AVP (Balment, Brimble, Forsling, & Musabayane, 1984; Johnson, Kinter, & Beeuwkes, 1979) and oxytocin (Forsling & Brimble, 1985) stimulate sodium excretion in animals. However, the physiological situations during which AVP-induced natriuresis might be expected to play any significant role are quite limited and are discussed in following sections.

A better candidate for volume-induced natriuresis would seem to be the recently described atrial natriuretic peptide (ANP) secreted from the atria of the heart in direct proportion to increases in atrial stretch, whether in response to true volume expansion or to any stimuli causing increased atrial pressure (Cantin & Genest, 1985; Needleman & Greenwald, 1986). Many of the early studies of ANP-induced natriuresis in animals and man utilized supraphysiological infusates of this peptide (Maack *et al.*, 1984; Weidmann *et al.*, 1986). However, more recent studies using much lower infusion rates that reproduce plasma ANP concentrations within measured physiological ranges have also succeeded in demonstrating increases in renal sodium and water excretion (Richards *et al.*, 1988). Unfortunately, the magnitude of stimulated natriuresis occurring at more physiological levels appears to be quite small and insufficient to account for much of the sodium excretion observed in response to stimuli such as isotonic volume expansion. Furthermore, ANP is unlikely to represent the hypothesized "third factor," because it appears to be exerting its predominant effects either via changes in renal blood flow, specifically by increasing glomerular perfusion pressure via efferent arteriolar vasoconstriction in combination with afferent arteriolar vasodilatation (Marin-Grez, Fleming, & Steinhausen, 1986), or via a direct inhibition of adrenal aldosterone secretion (Atarashi, Mulrow, & Franco-Saenz, 1985). Thus, present data would suggest that in man ANP acts more as a modulator of the other known major mechanisms of renal sodium excretion than as a primary regulator of natriuresis. Nonetheless, this peptide remains particularly attractive in view of its spectrum of potential activities in conjunction with its pattern of release in response to appropriate stimuli such as volume expansion, and only well-controlled future studies with an effective receptor antagonist, when this becomes available, will finally define the actual role of ANP both in physiological and pathophysiological situations.

INTEGRATION. A synthesis of what is presently known about the regulation of sodium appetite and renal sodium excretion leads to the inescapable conclusion that achievement of sodium homeostasis in humans occurs almost exclusively via regulated sodium excretion. Normal dietary intakes in general ensure ingestion of excess amounts of the major body solutes, and the multiple renal controls of solute excretion then regulate the proper amount to be retained depending on physiological conditions. Sodium excretion in particular is closely regulated, with multiple levels of renal control ranging from the glomerulus (ANP) to the proximal tubule (renal hemodynamic factors) and finally the distal tubule (aldosterone). Interestingly, virtually all of the major known controls of sodium excretion in man are in some way controlled by ECF volume and blood pressure; other than the potential contribution of AVP to natriuresis and the still unknown factor associated with dehydration natriuresis, most of the major controls of solute excretion seem to be relatively insensitive to physiological changes in ECF osmolality.

The ability of the kidneys to virtually stop all sodium excretion allows maintenance of sodium homeostasis even during periods of prolonged sodium deprivation. However, the relative absence of a specific sodium appetite leaves humans particularly vulnerable to more severe degrees of volume depletion following pathological states of extracellular sodium and fluid losses. Although to some degree this is compensated for by thirst-induced water consumption, ingested water by itself is a relatively ineffective mechanism to expand plasma and ECF volume because of its distribution throughout all body fluid compartments. It therefore remains surprising that the more sensitive mechanisms to stimulate sodium intake in situations of ECF volume depletion that are so clearly present in many other species are apparently lacking in humans.

Osmotic and Volume Homeostasis

The previous sections have summarized the various mechanisms by which water and solute balances are maintained in man. However, an understanding of the physiology of body fluids also requires knowledge of how these two regulatory processes interact to achieve homeostasis of all body fluids. Body fluid homeostasis is directed at achieving stability of the two major functions of body fluids: (1) maintenance of body osmolality within narrow limits, or osmotic homeostasis, and (2) maintenance of extracellular and plasma volume at adequate levels, or volume homeostasis (Andersson, 1971). Osmotic homeostasis is important to prevent large osmotic shifts of water into and out of cells, which would interfere with normal cell function, while volume homeostasis is important to allow normal cardiovascular and circulatory function.

The previous analysis has suggested that in most animals, and to a substantially greater degree in man, water balance is more finely regulated by changes in osmolality while sodium balance is regulated to a greater degree by changes in ECF volume and blood pressure. Although this is an obvious oversimplification of exceedingly complex regulatory processes (and in particular fails to hold true for more severe degrees of volume depletion where water intake and excretion also to some degree contribute to preservation of volume homeostasis), nonetheless, under most conditions it represents an accurate summary of normal home-

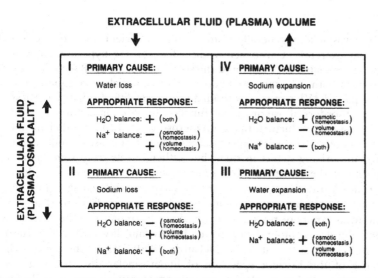

Figure 5. Summary of all potential combinations of derangements in both extracellular fluid (or plasma) volume and osmolality. For each category is shown the predominant clinical mechanism associated with that combination of altered ECF volume and osmolality, along with the effects on water and sodium balance predicted in response to the derangement under normal physiological conditions. Note that within each category, opposite responses for water or sodium balance are predicted depending on whether regulatory mechanisms for maintenance of osmotic homeostasis or volume homeostasis prevail.

ostatic mechanisms and provides a useful framework for understanding maintenance of body fluid homeostasis during normal physiological conditions as well as in pathological disorders of body fluid homeostasis.

The remainder of this chapter addresses specific pathological states. In general, these are divided into disorders of osmotic homeostasis, which for the most part are caused by abnormalities of water balance, and disorders of volume homeostasis, which for the most part result from abnormalities of solute, and specifically sodium, balance. Nonetheless, it is quite apparent that multiple interactions between osmotic and volume homeostasis must be considered for each disorder. When requirements for water and sodium intake and excretion are complementary, then both of these homeostatic systems act in concert to regulate body fluid physiology. However, multiple occasions arise clinically where body requirements for osmotic and volume homeostasis conflict, and these cases are particularly useful for understanding the overall integration of the many controls of body fluid homeostasis. As an example, Figure 5 illustrates four hypothetical pathological states arising from various possible combinations of deranged osmotic and volume homeostasis. For each category, the theoretical effects on water and sodium balances that would be predicted to occur for maintenance of osmotic versus volume homeostasis are shown. As each disorder is discussed in turn, it will be instructive to consider what is now known about how such regulatory conflicts are resolved and the subsequent implications this has for overall body fluid homeostasis in each case.

Disorders of Osmotic Homeostasis.

Hyperosmolality

Hyperosmolality generally results from insufficient water intake or excessive water excretion. Consequently, disorders causing hyperosmolality usually are those associated with impaired thirst or inadequate AVP secretion, or combinations of both. The most well-known of these is diabetes insipidus, resulting from impaired AVP secretion without any abnormality of thirst. Much less common are disorders of osmoreceptor function resulting in abnormalities of both AVP secretion and thirst. Although hyperosmolality from insufficient water intake is seen very frequently in clinical medicine, this is usually not caused by a specific underlying defect in thirst but rather results from some generalized incapacity of the patient to obtain and/or ingest fluids, such as a depressed sensorium. A prominent example of this is hyperosmolar coma caused by renal water losses from hyperglycemia-induced osmotic diuresis in patients who are unable for a variety of reasons to drink fluids in response to the diuresis. Such patients develop profound degrees of hyperosmolality with a high mortality (Gerich, Martin, & Recant, 1971; McCurdy, 1970). Despite the obvious clinical relevance of this broader spectrum of hyperosmolar disorders, this section focuses only on those disorders resulting from specific defects in AVP secretion and/or thirst.

DIABETES INSIPIDUS. Diabetes insipidus can result from either inadequate AVP secretion (central or neurogenic form) or inadequate renal sensitivity to circulating AVP levels (nephrogenic form). In either case the kidneys are unable to concentrate urine appropriately, leading to continued free water diuresis. Because renal mechanisms for sodium conservation are unimpaired, there is no accompanying sodium deficiency, and so this represents a disorder predominantly of water loss. As indicated in Figure 5, untreated syndromes of water loss (category I) lead to both hyperosmolality and volume depletion. However, until the water losses become severe, volume depletion is minimized by osmotic shifts of water from the intracellular fluid space to the more concentrated ECF. This phenomenon is not as evident following increases in the ECF sodium concentration, since in this case it simply results in a slower increase in the plasma Na^+ concentration than would be otherwise apparent. However, when nonsodium effective solutes such as mannitol are infused intravenously, the magnitude of this effect is much more readily appreciated by the progressive dilutional decrease in plasma Na^+ concentration caused by the translocation of intracellular water to the ECF.

Because uncontrolled diabetes insipidus results in intracellular dehydration to a far greater degree than extracellular dehydration, it therefore would be expected to activate primarily the regulatory mechanisms for maintenance of osmotic homeostasis. For water balance this consists of stimulation of thirst and AVP secretion to whatever degree the neurohypophysis is still able to secrete AVP. In cases where AVP secretion is totally absent, called complete diabetes insipidus, patients are dependent entirely on water intake for maintenance of water balance. However, in cases where some residual capacity to secrete AVP remains, called partial diabetes insipidus, plasma osmolality can eventually reach levels that allow moderate degrees of urinary concentration (recall from Figure 3 that even small concentrations of AVP can have substantial effects to limit urine volume). Although this represents a situation where water balance would be expected to be, and is, regulated almost exclusively by osmotic factors, a positive water balance obviously is appropriate for maintenance of both osmotic and volume homeostasis.

On the other hand, opposing forces might be expected to influence sodium balance, since a negative sodium balance would assist with maintenance of osmotic homeostasis by blunting increases in plasma osmolality caused by water losses, whereas a positive sodium balance would be necessary for maintenance of volume homeostasis and prevention of greater degrees of hypovolemia. Because sodium balance in man is largely a function of sodium excretion, and because this is controlled by mechanisms that for the most part are sensitive to volume rather than osmotic stimuli, one would expect volume homeostasis to prevail in limiting urinary Na^+ excretion in this case. This is true for more severe degrees of water loss, but with milder degrees of hyperosmolality not enough ECF volume depletion occurs, for reasons discussed previously, to significantly activate renal conservation mechanisms. Consequently, mild diabetes insipidus is characterized by a urinary Na^+ excretion that for the most part reflects sodium intakes without marked renal Na^+ conservation unless large unreplaced water losses occur to produce sufficient ECF volume depletion to activate sodium-retaining mechanisms. The interesting question of whether an osmoregulatory natriuresis producing a negative sodium balance occurs with mild hyperosmolality has not been adequately studied in humans with diabetes insipidus and is discussed in greater detail in later sections dealing with dehydration. However, if this phenomenon does occur, it does not appear to be of sufficient magnitude to produce clinically significant losses in body sodium or other solutes in these patients.

Although use of the long-acting AVP antidiuretic agonist 1-desamino-8-D-arginine vasopressin (DDAVP) has rendered therapy of diabetes insipidus relatively easy and allows most patients with this disorder to lead a normal life style (Richardson & Robinson, 1985), several interesting issues remain unanswered regarding this disease. First is the nature of the injury necessary to produce clinically significant diabetes insipidus, and the ability of the neurohypophysis to recover from, or compensate for, the injury. Studies in animals (Heinbecker & White, 1941; Laszlo & DeWied, 1966) have demonstrated that only 10–15% of hypothalamic vasopressin neurons are necessary to prevent symptomatic diabetes insipidus. This is consistent with the marked sensitivity of the kidney to small circulating concentrations of AVP. Similar human studies have not been done, but this is undoubtedly the case in man as well in view of the longstanding observation that simple hypophysectomy generally does not result in permanent diabetes insipidus; rather, a sufficient number of cells remain viable to allow relatively normal urinary concentration (Lipsett, MacLean, West, Li, & Pearson, 1956; Sharkey, Perry, & Ehni, 1961). Although recovery of function is generally the rule after damage in the area of the neural lobe, higher (and therefore more proximal) lesions cause increased frequencies of permanent diabetes insipidus (Ikkos, Luft, & Oliverocrona, 1955; Lipsett et al., 1956). This is presumably a result of greater amounts of retrograde neuronal degeneration following more proximal axonal damage (Verbalis, Robinson, & Moses, 1984). Interestingly, recent studies in rats have shown that treatment with DDAVP following pituitary stalk damage actually decreases the likelihood of functional recovery of AVP secretion afterwards (Herman, Marciano, & Gash, 1986), suggesting the possibility that ongoing stimulation as a result of hyperosmolality may actually be beneficial in terms of eventual cell survival. Since current therapy of posttraumatic or postsurgical diabetes insipidus entails treatment with DDAVP to ameliorate symptoms of polyuria and polydipsia (Verbalis et al., 1984), confirmation of these results in humans could modify the therapy of such patients in the immediate postinjury period.

Another issue of interest is that of thirst in patients with diabetes insipidus. Because magnocellular neurons are not themselves involved with thirst regulation, there is no theoretical reason to expect that diabetes insipidus should be associated with abnormalities of thirst. This has now been experimentally verified in human patients undergoing infusions of hypertonic saline (Thompson & Baylis, 1987a). Figure 6 shows that despite markedly impaired AVP secretion in patients with diabetes insipidus, thirst ratings using a visual analogue scale were equivalent to those of normal subjects. As further proof of normality of thirst regulation in these patients, note that thirst satiation began immediately on drinking and before any significant decrease in plasma osmolality in both groups. This represents a further extension of the well-established observation that AVP secretion is abruptly inhibited by oropharyngeal factors associated with the act of drinking before changes in plasma osmolality in dogs (Thrasher et al., 1981), which has subsequently been reproduced in several human studies (Geelen et al., 1984; Thompson, Burd, & Baylis, 1987), including those involving drinking of hypertonic solutions (Seckl, Williams, & Lightman, 1986) or simply irrigating the mouth with ice chips (Salata, Verbalis, & Robinson, 1987). Given the tight linkage that normally exists between thirst and AVP secretion, a similar effect on subjective thirst sensation would be expected, and the fact that this was found to be the case represents further evidence in support of the normality of

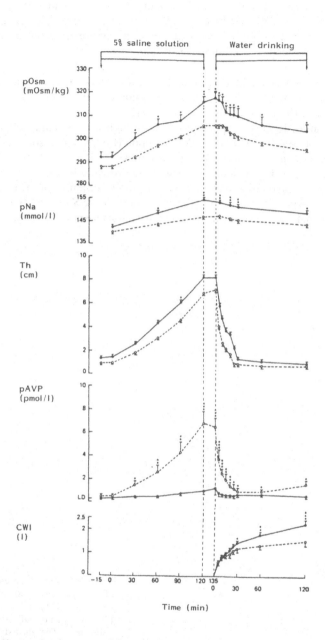

Figure 6. Changes in plasma osmolality (pOsm), plasma sodium (pNa), thirst (Th), plasma vas-opressin (pAVP), and cumulative water intake (CWI) in 13 patients with diabetes insipidus (closed circles) compared to five normal controls (open circles). All subjects were infused with hypertonic saline and then allowed to drink water *ad libitum*. (Modified with permission from Thompson & Baylis, 1987a.)

thirst mechanisms in patients with diabetes insipidus. Final proof that a normal thirst mechanism is present in such patients is that treatment with antidiuretic agents such as DDAVP results in complete normalization of plasma osmolality, which generally does not occur in patients with additional defects in thirst regulation as described in the following section.

OSMORECEPTOR DYSFUNCTION. There is an extensive literature in animals supporting the concept that the primary osmoreceptors that control both AVP secretion and thirst are located in the anterior hypothalamus. Lesions of this region in animals cause hyperosmolality through a combination of impaired thirst and osmotically stimulated AVP secretion (Buggy & Johnson, 1977; Leng, Blackburn, Dyball, & Russell, 1989; Thrasher, Keil, & Ramsay, 1982b). That such lesions do not result in primary damage to neurohypophyseal vasopressinergic neurons was confirmed by normal pituitary AVP contents and a normal AVP response to hypovolemic stimuli (Gardiner, Verbalis, & Stricker, 1985). It is not surprising that patients manifesting similar characteristics were described even

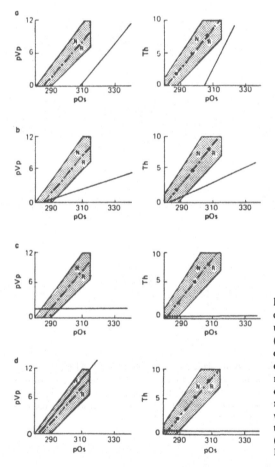

Figure 7. Schematic representation of different patterns of osmotically stimulated thirst (Th) and AVP secretion (pVp) in patients with osmoreceptor dysfunction. The shaded areas and dashed lines (—·—) represent normal responses (NR) to increases in plasma osmolality (pOs), and the solid lines represent the responses of patients with osmoreceptor dysfunction. See text for discussion of each pattern. (Reproduced with permission from Baylis & Thompson, 1988.)

prior to these studies in animals. Initial reports described this syndrome as "essential hypernatremia" (DeRubertis, Michelis, Beck, Field, & Davis, 1971; DeRubertis, Michelis, & Davis, 1974), and subsequent studies used the term "adipsic hypernatremia" in recognition of the profound thirst deficits found in most of the patients (Halter, Goldberg, Robertson, & Porte, 1977; Robertson *et al.*, 1982). Rather than focus on semantic issues, in many ways it makes more sense to group all of these syndromes as various types of disorders of osmoreceptor dysfunction.

Figure 7 schematically illustrates the four major patterns of osmoreceptor dysfunction as characterized by their thirst and AVP secretory responses: (a) upward resetting of the osmostat for both thirst and AVP secretion, (b) partial osmoreceptor destruction, (c) total osmoreceptor destruction, and (d) selective dysfunction of thirst osmoregulation. Most of the cases reported to date have represented various degrees of osmoreceptor destruction caused by a variety of brain lesions (Table 1). As opposed to lesions causing classical diabetes insipidus, these lesions usually occur more anteriorly in the hypothalamus, consistent with a presumed anterior hypothalamic location of the primary osmoreceptor cells. Whether some patients also have an inability to suppress as well as stimulate AVP secretion remains conjectural. Although some patients have been followed in whom water retention and hypoosmolality was produced by subsequent water loading (Robertson *et al.*, 1982), it cannot be proven with certainty that their apparently inappropriate antidiuresis was a result of destruction of inhibitory osmoreceptor neurons as opposed to some other effect of the primary brain lesion. However, given the very low frequency of this disorder, this question will likely have to be addressed via future animal studies.

TABLE 1. SPECIFIC CAUSES OF ADIPSIC HYPERNATREMIA[a]

Vascular (15%)
    Anterior communicating artery
        Aneurysms
        Ligation
    Intrahypothalamic hemorrhage
    Internal carotid ligation
Neoplastic (50%)
    Primary
        Craniopharyngioma
        Pinealoma
        Meningioma
        Chromophobe
    Metastatic
        Lung
        Breast
Granulomatous (20%)
    Histiocytosis
    Sarcoidosis
Miscellaneous (15%)
    Hydrocephalus
    Ventricular cyst
    Trauma
    Idiopathic

[a]From Robertson *et al.* (1982), with permission.

A second interesting point involves the stimulation of small amounts of AVP secretion at high plasma osmolalities in some patients (Figure 7b). As noted previously, this has been felt to reflect partial rather than total destruction of anterior hypothalamic osmoreceptor cells. However, it is interesting to note that virtually all animal studies involving anterior hypothalamic lesions have also suggested some residual AVP secretion. Several lines of evidence have demonstrated that magnocellular neurons themselves possess osmosensitive properties (Leng, Dyball, & Mason, 1985; Mason, 1980), although in general this response appears to require larger increases in osmolality than normally is necessary to cause AVP secretion. It is tempting to speculate that in some cases the residual AVP secretion seen in this syndrome may therefore represent the inherent osmosensitivity of the intact magnocellular neurons themselves rather than the presence of residual osmoreceptor neurons. Whatever the cause, it is unlikely that this pattern reflects a nonosmotic hypovolemic stimulus to AVP secretion, because in many cases hypertonic saline infusions causing secondary volume expansion have been used to increase osmolality.

Upward resetting of the osmostats for both AVP secretion and thirst constitutes what has been described as "essential hypernatremia," largely because these patients appeared to regulate their plasma osmolalities around a high set point despite clinically normal ECF volumes (Figure 7a). However, most such patients were not adequately studied in the past to ascertain whether they really manifested a relatively normal pattern of AVP secretion and thirst at some higher threshold of plasma osmolality. More recently cases have been described (Gill, Baylis, & Burn, 1985) in which the upward resetting clearly occurred in the absence of any hypervolemic state, which could also cause an elevation of the osmostat for AVP secretion (Ganguly & Robertson, 1980). Nonetheless, this disorder remains a relatively rare entity at this time. Whether it reflects a primary abnormality of osmoreceptor cell function or is secondary to altered synaptic influences on osmoreceptors or their interneurons is uncertain and likely will remain so, since no animal model for this syndrome has yet been described.

The final disorder of absent or deficient thirst despite normal AVP secretion (Figure 7d) is perhaps the most interesting pattern. It has been well studied in only one patient to date (Hammond, Moll, Robertson, & Chelmicka-Schorr, 1986). Although this could represent evidence that the osmoreceptors controlling thirst and AVP secretions are different even though tightly linked, this need not necessarily be true; such a pattern can be explained as well by a lesion distal to the osmoreceptors in areas more specifically involved with the cognitive perception of thirst. This possibility is discussed further in later sections.

Regardless of the exact nature of AVP secretion in these syndromes of osmoreceptor dysfunction, it is clear that the absence of a normal thirst mechanism represents a profound threat to these patients. Even in patients with substantial renal concentrating ability, insufficient intake of water invariably results in continued hyperosmolality. Mild or moderate hyperosmolality can generally be tolerated reasonably well, but in some cases more severe degrees of hypernatremia have been associated with significant neurological dysfunction (Arieff, 1984; Arieff & Guisado, 1976). Rapid changes in osmolality appear to be especially detrimental in this regard. Such patients generally must be treated with a fixed schedule of fluid intake in combination with DDAVP to enable retention of the ingested water (Robertson, 1984). Even with this conceptually simple therapy, they invariably become unstable when faced with even minor challenges to body fluid homeostasis, such as diarrhea or excessive sweating, and represent an exceedingly difficult management problem. Such cases therefore clearly illustrate the crucial role played by thirst as well as AVP secretion in the regulation of osmotic homeostasis during pathological conditions in man.

Hypoosmolality

Hypoosmolality of clinical significance almost always results from disorders causing impaired water excretion. Prominent among these are disorders characterized by abnormally regulated AVP secretion, although a variety of other cardiovascular and renal conditions can also lead to impaired water excretion with subsequent development of hypoosmolality (Schrier, 1988). Just as hyperosmolality can be produced by inadequate thirst in addition to insufficient AVP secretion, it might be predicted that excessive fluid intake from inappropriate thirst might also lead to hypoosmolality. Although excessive water intake from primary polydipsia does occur as a result of a variety of psychiatric, neurological, and metabolic disorders (Barlow & deWardener, 1959; Robertson et al., 1982), this actually only rarely causes hypoosmolality because of the substantial capacity of the kidneys for free water excretion. In the presence of maximally suppressed AVP levels, the normal human kidney can excrete as much as 20–25 L/day before significant free water retention occurs. However, the combination of polydipsia with sometimes subtle impairments of renal water excretion can and does cause significant derangements of osmotic homeostasis in man (Goldman, Luchins, & Robertson, 1988).

Syndrome of Inappropriate Antidiuresis. The syndrome of inappropriate antidiuresis (SIAD) causes hypoosmolality by impairing renal water excretion during conditions of plasma dilution that normally would be accompanied by maximal free water diuresis. As a result, hypoosmolality occurs because of dilution of plasma Na^+ and other body solutes. Although many factors can cause impaired renal water excretion, inappropriate AVP secretion is among the most prevalent causes of the syndrome (Anderson, Chung, Kluge, & Schrier, 1985) and is the focus of this section. Despite some intrinsic natriuretic properties of AVP (Johnson et al., 1979; Balment et al., 1984), the major physiological effect of excessive AVP secretion is antidiuresis. Thus, the hypoosmolality in this disorder is primarily caused by water retention. As depicted in Figure 5, syndromes of water excess (category III) cause not only hypoosmolality but also volume expansion. However, because the retained water is distributed throughout total body water, thereby expanding the intracellular as well as the extracellular fluid compartments, the actual expansion of ECF and plasma volume will not be as marked as might otherwise be imagined from the volume of water retained. Nonetheless, some degree of plasma volume expansion inevitably occurs, at least initially in the course of SIAD.

Based on the previously summarized mechanisms for normal physiological regulation of water balance, one might expect inhibition of thirst to occur in order to prevent further water intake, which would subserve maintenance of both osmotic and volume homeostasis. This certainly is true of animal models of hypoosmolality; animals infused with antidiuretic concentrations of AVP or DDAVP do not become hypoosmolar because they decrease water intake to levels that just balance insensible water losses (Verbalis, 1984). This has represented a major problem in animal studies, since in order to produce hyponatremia and hypoosmolality it has been necessary to infuse or inject water loads continually (Chan, 1973; Smith, Cowley, Guyton, & Manning, 1979; Gross & Anderson, 1982). Alternatively, it is possible to override the hypoosmolar-induced inhibition of water intake by offering more palatable fluids (Verbalis, 1984) or by feeding animals a liquid diet, thus obligating them to a large unregulated free water intake (Verbalis & Drutarosky, 1988). The degree to which such maneuvers have been necessary to produce sustained hypoosmolality serves to demonstrate that inhibition of thirst by osmotic dilution is a prominent mechanism in many animals. However, if this were true in humans as well, then the syndrome of inappropriate antidiuresis would not be as much of a clinical problem as it is.

The fact that many patients continue to drink fairly normal amounts of fluids despite profound degrees of hypoosmolality is striking and suggests that osmotic mechanisms that inhibit thirst are either absent or at least relatively weak in humans. It should be emphasized that rarely does stimulation of thirst occur with SIAD, but rather such patients manifest a failure to shut off normal fluid intake. One possible explanation for this is that osmotically regulated thirst may in fact be inhibited, but not the unregulated component of water ingestion that accounts for much of the daily water ingestion in humans. In this regard it is interesting to recall studies of animals in which infusions of water at rates sufficient to eliminate any regulated, or need-induced, drinking nonetheless still was associated with significant continued water intake in rats (Nicolaidis & Rowland, 1974). Based on these and other studies, the authors concluded that "oral drinking represents more than correction of homeostatic imbalance" and suggested that such excess need-free drinking actually could be considered "antiregulatory consumption" (Nicolaidis & Rowland, 1975). It would appear that the relative component of unregulated need-free oral ingestion might be even greater in humans to account for the continued fluid consumption that occurs despite the sometimes profound hyponatremia seen clinically in cases of SIAD (Verbalis, 1990b). Alternatively, some of these patients may also have defects in thirst regulation as part of their disease process, specifically a downward resetting of the osmostat for AVP secretion and thirst. This abnormality clearly occurs in some patients with SIAD (Defronzo, Goldberg, & Agus, 1976; Michelis, Fusco, Bragdon, & Davis, 1974; Zerbe, Stropes, & Robertson, 1980) and also is well known to occur physiologically during pregnancy in both animals (Durr, Stamoutsos, & Lindheimer, 1981) and man (Davison, Gilmore, Durr, Robertson, & Lindheimer, 1984). In such cases the reset thresholds for both thirst and AVP secretion result in a new steady state whereby all normal osmoregulatory controls remain in place but act to maintain osmotic homeostasis around a lower basal plasma osmolality. However, such resetting of osmotic thresholds has not been demonstrated in the majority of patients with SIAD (Zerbe et al., 1980) and consequently cannot explain the continued fluid intakes despite hypoosmolality of many patients in this latter group.

With regard to sodium balance, water excess similarly poses a conflict between osmotic and volume homeostasis, as summarized in Figure 5. In this case sodium conservation would assist in maintenance of osmotic homeostasis, whereas sodium excretion would represent the appropriate response for maintenance of volume homeostasis. In SIAD this conflict appears to be settled entirely in favor of volume homeostasis, again consistent with the major regulation of renal sodium excretion by mechanisms responding predominantly to blood volume and pressure. Increased sodium excretion has long been noted in SIAD, and studies in water-loaded human subjects demonstrated that the renal sodium excretion is directly related to the volume of water loading (Leaf, Bartter, Santos, & Wrong, 1953). Patients with SIAD who are water restricted normalize urine sodium excretion and do not continue to demonstrate renal salt wasting despite continued high plasma AVP levels.

The most interesting aspect of this natriuresis in SIAD is not that it occurs, but the degree to which it may contribute to the hypoosmolality by causing secondary sodium depletion. Some animal studies of chronic hyponatremia have suggested that sodium depletion may represent a major component of the hyponatremia that ensues (Smith *et al.*, 1979). Although this is an extreme view, there is little question that the sodium depletion induced by volume regulatory mechanisms does to some degree contribute to the hyponatremia of SIAD in both animal models (Verbalis & Drutarosky, 1988) and human disease (Cooke, Turin, & Walker, 1979; Schwartz, Bennett, Curelop, & Bartter, 1957; Stormont & Waterhouse, 1961). Consequently, chronic therapy of such patients consists not only of restriction of water intake but also of liberalization of NaCl intake to minimize total body sodium depletion. Nonetheless, sodium repletion alone will not have that much impact on the hypoosmolality, since it is well known that therapeutic infusions of isotonic saline in patients with SIAD simply cause increased excretion of Na^+ without any significant improvement of the plasma Na^+ concentration (Schwartz *et al.*, 1957). SIAD should therefore continue to be considered a disorder of primary water retention that is accompanied by secondary sodium depletion of variable magnitude and significance.

However, some limits must exist with regard to the degree of water expansion that is tolerable and compatible with life. Continued water loading in antidiuretic animals produces a phenomenon that has been called "escape" from antidiuresis. This is manifested by a decrease in urine osmolality and increase in urine volume despite continued AVP infusions (Chan, 1973). Studies in dogs have clearly demonstrated that escape does not occur because of decreased AVP levels or significant tolerance to AVP effects but rather as a result of the volume expansion produced by water retention (Cowley, Merrill, Quillen, & Skelton, 1984). This appears to be mediated via increases in renal artery perfusion pressure, because sustained antidiuresis occurs in AVP- and water-infused animals in whom renal artery pressure is maintained constant (Hall, Montani, Woods, & Mizelle, 1986). However, escape does not completely eliminate antidiuresis, nor does it cause correction of the hypoosmolality via a free water diuresis; rather, it simply prevents further water retention beyond critical levels of tolerance. Preliminary data in rats suggest that these "tolerance" limits are quite flexible and still allow a substantial degree of water retention and volume expansion prior to precipitation of renal escape from antidiuresis (Verbalis, Drutarosky, Ertel, & Vollmer, 1989). Therefore, this phenomenon can best be viewed as establishing a new steady state of water balance in which *additional* water retention is balanced by compensatory mechanisms for renal excretion of the excess water. Although this effect has been demonstrated reproducibly in animal models, the degree to which it occurs in SIAD in man is debatable, since most such patients demonstrate continued antidiuresis, low urine volumes, and persistently elevated urine osmolalities (DeTroyer & Demanet, 1976). Nonetheless, it likely does account for some of the variability often observed in the presentation of patients with SIAD (Verbalis *et al.*, 1989).

One of the most interesting and perhaps important questions to arise from animal and human studies of hypoosmolality is that of the cellular mechanisms of osmoreceptor function. Since the original demonstration of cerebral osmoreceptors by Verney (1947), osmoreceptive neurons have been hypothesized to respond to changes in cellular volume caused by osmotic shifts of water into or out of the cells. This proposed mechanism of osmoreception is illustrated in Figure 8. Because the actual osmoreceptors regulating AVP secretion and thirst have never been unequivocally identified, it has been impossible to evaluate this hypothesis other than by indirect means. One such indirect method results from studies of correction of hypoosmolality in animals. Exposure of the brain to hypoosmolar conditions initially causes brain edema through osmotic water shifts into the brain (Figure 9,2). However, this is then quickly followed by active loss of both electrolyte and nonelectrolyte solutes from brain tissue, thereby allowing reequilibration of brain volume (Figure 9,3). This process has been well documented in numerous animal studies (Arieff, Llach, & Massry, 1976; Holliday, Kalayci, & Harrah, 1968; Melton, Patlak, Pettigrew, & Cserr, 1987; Rymer & Fishman, 1973; Yannet, 1940) as well as in a variety of cell types *in vitro* (Hoffman, 1977) and unquestionably accounts for the ability of some patients with severe degrees of hypoosmolality to remain without symptoms attributable to brain edema (Arieff, 1984).

However, the occurrence of cellular volume regulation in cerebral osmoreceptors would have significant implications with regard to their function during subsequent alterations in ECF osmolality. Namely, relative increases in

NORMAL ECF EFFECTIVE OSMOLALITY

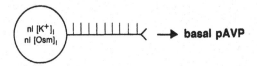

ACUTELY INCREASED ECF EFFECTIVE OSMOLALITY

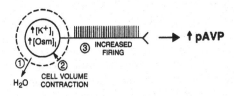

ACUTELY DECREASED ECF EFFECTIVE OSMOLALITY

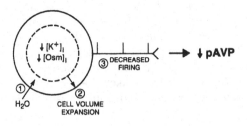

Figure 8. Schematic representation of the cell-volume theory of osmoreceptor function. Increases in extracellular fluid effective osmolality (middle panel) are hypothesized to stimulate AVP secretion as a result of osmotic shrinkage of the osmoreceptors, whereas decreases in extracellular fluid effective osmolality (lower panel) are hypothesized to inhibit AVP secretion as a result of osmotic swelling of the osmoreceptors. Note that in each case intracellular osmolality is obligatorily altered in parallel with changes in extracellular osmolality.

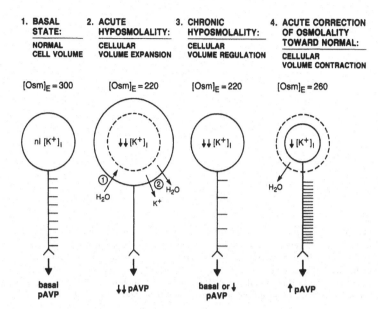

Figure 9. Resetting of the osmotic threshold for AVP secretion predicted by the cell-volume theory of osmoreceptor function. Note that after the initial osmotic swelling in response to hypoosmolality, osmoreceptors (similar to the rest of the brain) likely manifest a volume-regulatory decrease in cell volume as a result of extrusion of intracellular solute, mostly K^+. In this volume-regulated hypoosmolar state, subsequent increases in ECF osmolality, even below the normal osmotic threshold for AVP secretion, would be expected to cause osmotic shrinkage of the osmoreceptors. If cell volume is the main determinant of osmoreceptor activity, this should in turn cause AVP secretion in response to the relative increases in plasma osmolality despite low absolute levels of osmolality, as seen clinically in the reset-osmostat variant of the syndrome of inappropriate antidiuresis.

plasma osmolality should then cause premature osmoreceptor activity, thereby leading to AVP secretion at lower than normal levels of plasma osmolality (Figure 9,4). This would in effect cause a downward resetting of the osmotic threshold for AVP secretion and thirst and has been a mechanism postulated to account for the clinical observation of a reset osmostat in some cases of SIAD. This type of study would be impossible to do in humans because of the inherent risks from both hyponatremia itself (Arieff et al., 1976) and the subsequent correction of the hyponatremia (Sterns, Riggs, & Schochet, 1986). However, evaluation of this possibility in chronically hypoosmolar animals revealed no significant resetting of the osmotic threshold for AVP and oxytocin secretion (Figure 10) despite documented verification of near-total brain volume equilibration (Verbalis, Baldwin, & Robinson, 1986).

Although one interpretation of this result could be that the osmoreceptors do not volume regulate as does the rest of the brain, another is that the transduction of osmotic information by osmoreceptors does not occur via relative changes in cell volume but rather is a function of changes in other parameters, such as resting membrane potential. This would better allow for the observation of an apparent absolute threshold for AVP secretion and thirst independently of changes in cell volume during hyponatremia. In this context the occurrence of downward or upward resetting of the osmotic threshold for AVP secretion and

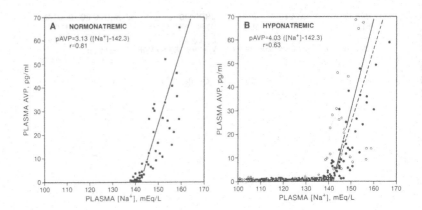

Figure 10. Plasma AVP levels as a function of plasma sodium concentration ([Na$^+$]) in control rats (panel A, squares) and rats with chronic sustained hyponatremia (panel B) for 2 (solid circles) or 5 days (open circles). Note the absence of any significant shift in the osmotic threshold for AVP secretion in the hyponatremic animals (solid regression line, hyponatremic; dashed line, normonatremic) and the lack of stimulated AVP secretion in response to relative changes in plasma [Na$^+$] below 140 meq/L. (Reproduced with permission from Verbalis et al., 1986.)

thirst in patients could then best be viewed as a result of alterations of afferent inputs to osmoreceptors, possibly resulting in basic changes in membrane potentials and excitability. Alternatively, because these mechanisms are not necessarily exclusive, changes in both osmoreceptor volume and electrolyte-induced membrane potential may regulate osmoreceptor activity under different conditions. For example, at normal or high plasma osmolalities changes in osmoreceptor cell volume may be the major determinant of neuronal activity, but during hypoosmolality membrane potential changes resulting from reduced intracellular potassium concentrations may be sufficient to limit any cell-volume-induced neuronal activity. Thus, although much is now known about the effects of changes in plasma osmolality on thirst and AVP secretion, it is clear that we still lack even a rudimentary understanding of the neurophysiology of the osmoreceptors themselves.

PRIMARY POLYDIPSIA. Primary polydipsia in humans is usually a result of psychiatric disease rather than intrinsic disorders of thirst (Barlow & deWardener, 1959; Robertson et al., 1982). Such patients ingest large amounts of fluids for a variety of reasons, but generally not because of true sensations of thirst. Most of these patients do not become hypoosmolar because of the already mentioned ability of the kidneys to excrete large amounts of free water. In fact, the presentation of most such patients is one of polyuria and polydipsia with a normal plasma osmolality, and evaluation of these patients is directed at differentiating them from patients with similar symptoms caused by partial diabetes insipidus (Miller, Dalakos, Moses, & Fellerman, 1970; Zerbe & Robertson, 1981). However, a subset of such patients do become hypoosmolar, and recent studies have suggested that this occurs when the high fluid intakes occur in combination with impairments of free water excretion resulting from both a small degree of inappropriate AVP secretion as well as a poorly defined renal defect (Goldman et *al.*, 1988). Neither of these diluting abnormalities would be clinically apparent by itself, but they become significant in the setting of unusually high fluid intakes.

Primary polydipsia also occurs in patients with a variety of brain lesions, and consideration of this possibility is mandatory in all patients before assuming they have "psychogenic" or psychiatrically induced polydipsia. Many such patients likely have a true alteration in their osmotic threshold for thirst; a downward shift of this threshold in the absence of any change in the threshold for AVP secretion would cause excessive thirst at plasma osmolalities less than that needed for AVP secretion and urinary concentration, thereby producing continuous polyuria and polydipsia. The occurrence of this phenomenon, as in the case of selective upward resetting of the thirst osmostat in some cases of "essential hypernatremia," has been considered to represent evidence for separate osmoreceptor populations subserving thirst and AVP secretion. Although this is obviously a possibility, such cases do not by themselves represent proof of this. Just as osmoreceptors constitute only the first step in activation of vasopressin secretion from magnocellular neurons, so do they likely also represent only the first step in the production of thirst. As schematically depicted in Figure 11, present clinical and experimental observations could also be explained by retaining the concept of a single population of osmoreceptive neurons. In this case, lesions or intrinsic abnormalities of the primary osmoreceptors (I) would cause parallel shifts of the thresholds for both thirst and AVP secretion, while more distal lesions or alterations in afferent inputs to either system (II or III) would cause more selective inhibition or stimulation of either thirst or AVP secretion. Additional clinical and experimental studies will be required to ascertain the exact anatomic relationship between osmoregulation of thirst and AVP secretion, but at the present time the number of situations where they are functionally tightly linked is far more striking than the relatively small number of pathological situations in which they appear to be dissociated.

DISORDERS OF VOLUME HOMEOSTASIS

HYPOVOLEMIA

Hypovolemia occurs when there is insufficient extracellular fluid to maintain blood volume and blood pressure at normal levels. For the reasons outlined

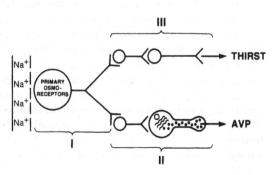

Figure 11. Schematic representation of the model with a single osmoreceptor for thirst and AVP secretion. Dysfunction of the osmoreceptors themselves (I) would cause parallel changes in thirst and AVP secretion, but more distal dysfunction in either magnocellular regulation (II) or thirst pathways (III) would cause more selective abnormalities of AVP secretion or drinking. Note that dysfunction at any point could potentially result in either excessive stimulation or inhibition of the final response.

in the previous section, pure water loss or inadequate intake produces hypovolemia, but the ECF volume deficits will be compensated for to a large degree by osmotic water shifts from the intracellular fluid space. The degrees of hyperosmolality that can be achieved in dehydrated patients with relatively mild effects on circulating blood volume and pressure are striking. This is obviously also in part because of the multiple mechanisms that maintain blood pressure homeostasis, and in particular the sympathetic nervous system. Clinically significant hypovolemia develops much more quickly and to a much more pathological degree when water losses are coupled with solute losses. In this case sodium balance becomes of crucial importance for maintenance of body fluid homeostasis. Because of the relative absence of regulation of sodium intake in humans, renal sodium as well as water conservation represents the main defense for maintenance of volume homeostasis during clinical hypovolemic states. It is therefore instructive to contrast hypovolemic states in which renal function is unimpaired, such as dehydration from nonrenal causes, from those in which an impairment of renal sodium conservation mechanisms itself either causes or exacerbates the hypovolemia, such as Addison's disease.

DEHYDRATION. By definition, dehydration indicates a deficiency of water in the body. If the deficiency is a result of pure water losses, as in diabetes insipidus, or lack of water intake, as in primary thirst disorders, then intracellular dehydration will be more marked than extracellular, and osmotic homeostatic mechanisms will be activated to a greater degree than the renal mechanisms subserving volume homeostasis. The opposite situation is seen with conditions of isotonic hypovolemia, typically produced by large amounts of blood loss from hemorrhage. This represents a situation where regulatory mechanisms for water and sodium balances will be activated purely by mechanisms for volume homeostasis without any contribution from osmotic regulatory systems. In clinical practice, however, most cases of dehydration lie somewhere between these extremes. This is because most pathological body fluid losses, whether from excessive sweating, gastrointestinal excretions, or diuretic use, result in losses of both water and sodium, but most of the fluid lost in such situations is relatively hypotonic (generally in the range of 0.45% NaCl, or half-isotonic). As a result, most clinical cases of dehydration are characterized by both hyperosmolality and hypovolemia. Consequently, although such cases still for the most part fall into category I of Figure 5, volume homeostatic mechanisms will tend to be more activated than for cases of pure water losses such as diabetes insipidus. It is hardly surprising, therefore, that renal sodium conservation occurs prominently in cases of dehydration in patients with normal renal function.

However, because both osmotic and volume homeostatic mechanisms are activated in most dehydrated patients other than those with pure isotonic dehydration, and because osmoregulatory mechanisms are generally more sensitive to small increases in plasma osmolality than are volume regulatory mechanisms to small decreases in ECF volume, it might be predicted that very early in the process of becoming dehydrated sodium excretion might actually be increased in response to osmoregulatory mechanisms, and only during more advanced stages of volume depletion would volume homeostatic mechanisms producing renal sodium conservation prevail. Although not yet well studied in humans, this sequence of events has long been known to occur in multiple animal species as the phenomenon of dehydration natriuresis (Bianca, Findlay,

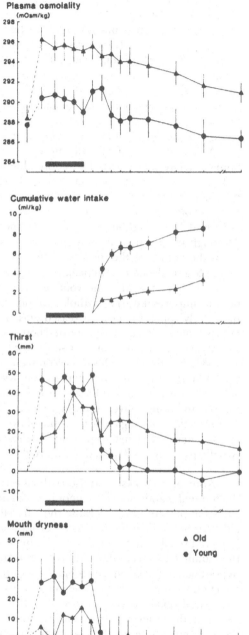

Figure 12. Plasma osmolality, thirst, cumulative water intake, and plasma AVP levels in elderly (triangles) and young subjects (circles) in response to a 24-hour period of water deprivation. Following a 20-minute equilibration period (solid bar), the subjects were allowed to drink *ad libitum* for 60 minutes and were then followed for another 60 minutes. (Modified with permission from Phillips, Rolls, Ledingham, Forsling, Morton, Crowe & Wollner, 1984.)

& McLean, 1965; Elkinton & Taffel, 1942; Luke, 1973; McKinley, Denton, Nelson, & Weisinger, 1983). The physiological mechanisms underlying the osmotically induced natriuresis remain unclear at this time (see Chapter 14), but it is likely that osmotically stimulated AVP, and in some species possibly oxytocin, secretion plays at least some role in this response (Edwards & LaRochelle, 1984). Whatever the mechanism, the apparent paradox of natriuresis occurring during early dehydration can be more readily understood as being not really paradoxical at all by taking into account an analysis of the relative degrees of activation of osmotic versus volume homeostatic mechanisms at various times in the progression of the dehydration from mild to more severe degrees of hypovolemia.

A second issue of clinical interest is that of age-related changes in the homeostatic responses to dehydration. Elderly patients are known to be particularly prone to dehydration regardless of the cause. This suggests the presence of defects in some or all of the various mechanisms subserving both osmotic and volume homeostasis. Studies in humans have consistently demonstrated intact or enhanced AVP secretion in response to osmotic stimuli (Helderman *et al.*, 1978). Although elderly patients generally have a decreased renal response to AVP, given the nature of the urine volume–urine osmolality relationship shown in Figure 3, a small decrease in urinary concentrating ability generally will not cause significantly increased urine volume losses, although this factor can certainly become important during pathological situations. The most significant defect in the elderly appears to be a decreased sensation of thirst. After 24 hours of fluid deprivation, elderly subjects reported both lesser degrees of subjective thirst and a less dry mouth than did younger subjects (Phillips, Rolls, Ledingham, Forsling, *et al.*, 1984). Furthermore, this was accompanied by less spontaneous drinking after the 24-hour period of water deprivation, with the effect that plasma osmolality and Na^+ concentration did not return to completely normal levels in the elderly subjects following rehydration (Figure 12). This dissociation between thirst and AVP secretion could again be the result of two separate populations of osmoreceptors or alternatively may represent the presence of more distal defects in thirst specific pathways as depicted in Figure 8.

ADDISON'S DISEASE. Addison's disease is caused by bilateral destruction of the adrenal glands, more commonly on an autoimmune basis than from tuberculosis in present times. Because the adrenal glands are the site of synthesis of aldosterone in addition to glucocorticoids, patients with Addison's disease have impaired renal sodium conservation. This leads to renal salt wasting in excess of the sodium losses seen with various other forms of dehydration, resulting in both hypoosmolality and hypovolemia. As shown in Figure 5 (category II), the net effect on sodium balance will be one of activating mechanisms for both sodium intake and renal sodium conservation. Because aldosterone acts distally in the nephron, the other more proximal sodium-conserving mechanisms cannot adequately compensate for the salt wasting in this case. However, as has been mentioned previously, this represents the only clinical disorder in which a stimulated sodium appetite has been consistently seen, even though only in a fraction of all patients. It is intriguing to speculate why this should be the case.

Studies in rats have shown that osmotic dilution is necessary for full expression of sodium appetite and that hyperosmolar conditions are inhibitory to sodium appetite (Stricker, 1981; Stricker & Verbalis, 1988). In this regard, it is interesting that Addison's disease in humans also represents a disorder in which hypoosmolality and profound hypovolemia coexist. Thus, it is tempting to explain the unique occurrence of sodium appetite in Addison's disease on the basis of simultaneous inhibition of osmoreceptor activity. Unfortunately, a consideration of other hypoosmolar hypovolemic states does not support this possibility. For example, patients on diuretic therapy frequently become hyponatremic and sometimes markedly hypovolemic as a result of renal Na^+ losses (Fichman, Vorherr, Kleeman, & Telfer, 1971) yet do not express any sodium appetite. Another relevant group are patients with isotonic fluid losses, such as hemorrhage, who then drink only water to replace their losses and thereby become hyponatremic. These patients clearly are hypovolemic, yet again a pronounced sodium appetite has never been noted clinically. Thus, Addison's disease appears to be unique with regard to expression of sodium appetite in man and represents an extremely interesting area for further clinical investigation.

Although responses producing a positive Na^+ balance would be appropriate for both volume and osmotic homeostasis in patients with Addison's disease, water intake again poses a potential conflict, since a positive water balance would help to promote volume homeostasis but at the expense of worsening osmotic homeostasis. It appears that volume regulatory mechanisms prevail in most cases, and such patients generally express stimulated thirst and AVP secretion. However, this issue has not been studied critically enough in this group of patients to know whether, as a result of the hypoosmolality, their thirst and AVP secretion are blunted relative to their degree of hypovolemia and also whether with lesser degrees of hypovolemia osmoregulatory mechanisms favoring a negative water balance might prevail, analogous to the opposite situation of early dehydration. If this were the case, it could exacerbate the occurrence of clinically significant volume depletion in patients with Addison's disease. However, differentiation of osmoregulatory free water excretion from the obligate water excretion accompanying the solute diuresis as a result of salt wasting would require carefully controlled studies. Whatever the resolution of this potential conflict in untreated patients, it is clear that treatment of Addisonian patients with corticosteroid therapy (both glucocorticoids and mineralocorticoids) completely reverses all the abnormalities of this disorder.

HYPERVOLEMIA

Hypervolemia is the result of excess ECF and plasma volume. Because water by itself distributes across all body fluid compartments, sodium excess represents a much more effective mechanism for volume expansion because of the predominant extracellular distribution of this solute. However, it should be remembered that hyperosmolality from excess solutes does not invariably lead to volume expansion. Patients with renal failure have high levels of blood urea that often increase total plasma osmolality by as much as 20–30 mosm/kg, yet this causes neither intracellular dehydration nor ECF and plasma volume expansion. This is of course because urea freely distributes across cell membranes and thus is not an "effective" solute (Gilman, 1937; Zerbe & Robertson, 1983). Consequently, only increases in concentrations of effective solutes are accompanied by ECF volume expansion, and such clinical disorders are caused almost exclusively in man by sodium retention.

HYPERALDOSTERONISM. Hyperaldosteronism in many ways represents the classic disease of primary sodium retention and subsequent volume expansion. High plasma aldosterone levels may result from an adrenal tumor (primary hyperaldosteronism), from excess activity of the renin–angiotensin syndrome with resultant angiotensin II stimulation of the adrenal (secondary hyperaldosteronism), or from pharmacological therapy with synthetic mineralocorticoids. Regardless of the etiology of the high plasma aldosterone activity, the net effect at the kidneys is continued sodium retention and potassium wasting. Thus, as in patients with SIAD, hyperactivity of a normal regulatory system leads to an abnormality of body fluid homeostasis. As might be expected, such patients are characteristically mildly hypernatremic and hypokalemic and manifest varying degrees of hypertension and subcutaneous edema from sodium and water retention.

As shown in Figure 5 (category IV), a primary sodium expansion should activate various mechanisms to produce a negative sodium balance, and these consist almost exclusively of renal mechanisms to enhance sodium excretion. But because aldosterone-induced sodium retention occurs distally in the nephron, the ability of other more proximal mechanisms to cause sufficient sodium excretion to produce a negative sodium balance is limited. However, as discussed previously with regard to the water retention of SIAD, there are limits to the degree of volume expansion that can be tolerated, regardless of whether this occurs as a result of water or sodium excess. Just as a pressure-mediated renal escape from antidiuresis occurs when water loading becomes excessive in some patients with SIAD, a similar "escape" from the antinatriuretic effects of aldosterone is well known to occur in response to the chronic volume expansion in this disorder (Knox, Burnett, Kohan, Spielman, & Strand, 1980). The mechanism by which aldosterone escape occurs has long been studied and debated, and recently investigations have focused on the potential role of increased plasma ANP levels in contributing to this effect (Ballermann, Bolch, Seidman, & Brenner, 1986; Granger et al., 1987). Although it is likely that such hormonal mechanisms, as well as probable aldosterone receptor down-regulation, do contribute to aldosterone escape, as in the case of renal escape from antidiuresis the major cause that has been identified in well-controlled animal studies appears to be a direct renal effect mediated by increased renal perfusion pressure with subsequent increased fractional sodium excretion (Hall, Granger, Smith, & Premen, 1984). In effect this represents a safety-valve type of mechanism; when renal artery pressure rises to unacceptable levels as a result of volume expansion, more sodium is excreted into the glomerular filtrate in amounts sufficient to overwhelm the aldosterone-mediated distal sodium resorption. Note that the term "escape" represents somewhat of a misnomer in this case, as well as in the case of escape from antidiuresis. Aldosterone effects are still present, but a new steady state of volume expansion has been reached in which no *additional* sodium retention occurs because of activation of compensatory mechanisms for sodium excretion. Consequently, although a new steady state is reached, a substantial degree of volume expansion persists.

Given this likely mechanism for escape from aldosterone, it would be predicted that escape occurs only in patients with primary hyperaldosteronism or mineralocorticoid therapy. Patients with secondary hyperaldosteronism for the most part have renal hypoperfusion from a decreased "effective" circulating volume (Schrier, 1988). Despite considerable edema from sodium retention, these patients continue to retain more sodium and fluid because of persistent aldosterone effects in addition to intrarenal hemodynamic mechanisms causing sodium retention. Thus, the edema and total body volume expansion with disorders such as congestive heart failure and cirrhosis is much more marked than with primary hyperaldosteronism. These patients lack the mechanisms to limit or "escape" from continued volume expansion and consequently can and often do present with massive amounts of subcutaneous edema from fluid retention. Interestingly, plasma ANP levels are quite high in patients with secondary hyperaldosteronism, substantially higher than following physiological volume expansion in normal subjects (Burnett et al., 1986; Shenker, Sider, Ostafin, & Grekin, 1985). This would suggest that although ANP may contribute to the natriuresis occurring during escape from hyperaldosteronism, nonetheless such effects are not sufficiently strong to significantly override intrinsic intrarenal hemodynamic mechanisms causing sodium conservation.

Once again, however, a potential conflict arises with regard to water balance because of opposite needs for maintenance of osmotic versus volume homeostasis in this case. The increased osmolality should stimulate thirst and AVP secretion in order to produce a positive water balance and thereby blunt further increases in osmolality, whereas the volume expansion should inhibit thirst and enhance water excretion. In this case the net result again depends on the chronicity as well as the severity of the salt loading. With an acute sodium excess, as would be the case with hypertonic saline infusions in normal subjects, increases in thirst and AVP secretion clearly occur (Phillips, Rolls, Ledingham, Forsling, & Morton, 1985; Robertson, 1983) and cause a positive water balance. However, with the more chronic and severe volume expansion that occurs with hyperaldosteronism, both pituitary and renal responses have the effect of allowing increased excretion of water. Although AVP is still secreted in response to increases in osmolality in patients with hyperaldosteronism, the chronic volume expansion tends to shift the threshold and sensitivity of the response so that less AVP is secreted per unit increase in osmolality (Ganguly & Robertson, 1980; Thompson & Baylis, 1987b). Modulation of the AVP response is likely mediated by the well-known effects of baroreceptor inhibition of AVP secretion (Epstein, Preston, & Weitzman, 1981; Menninger, 1981) and allows a decreased level of antidiuresis despite hyperosmolality in volume-expanded states. In addition to this pituitary effect, the renal mechanisms producing aldosterone escape discussed above also result in an increased water diuresis even in the presence of AVP activity at the collecting tubules, analogous to renal escape in SIAD. However, corresponding effects are not apparent with regard to thirst, and patients with primary hyperaldosteronism are persistently polydipsic. This indicates that, similar to the lack of thirst inhibition as a result of hyponatremia in SIAD, hypervolemia also does not appear to be a very potent inhibitor of thirst. This is consistent with earlier studies in the rat clearly demonstrating that volume expansion with isotonic saline failed to inhibit to any degree subsequent drinking in response to increases in osmolality (Fitzsimons, 1961a).

Multiple mechanisms probably account for the further increase in fluid intake above normal levels in patients with hyperaldosteronism. First, the mild hyperosmolality caused by sodium retention acts as a persistent osmotic stimulus. Second, when sufficienty severe, hypokalemia can cause both a nephrogenic form of diabetes insipidus (Rubini, 1961; Rutecki, Cox, Robertson, Francisco, & Ferris, 1982) and primary polydipsia (Berl & Teitelbaum, 1985; Berl, Linas,

Aisenbrey, & Anderson, 1977), the latter effect presumably as a result of cellular dehydration from total body potassium depletion. Finally, preliminary data on a single patient are suggestive that the osmotic threshold for thirst may actually be lower than normal in patients with primary hyperaldosteronism (Thompson & Baylis, 1987b), though this finding must be confirmed in a larger number of patients before assessing its general applicability to this disorder.

One might expect that yet another homeostatic mechanism in chronic hyperaldosteronism would be inhibition of sodium intake. However, this does not appear to occur, and patients with hyperaldosteronism require forced sodium restriction since they do not spontaneously decrease their salt ingestion. Although pharmacological doses of mineralocorticoids are well known to stimulate sodium appetite in animals (Rice & Richter, 1943; Wolf, 1965), this effect has not been observed in man, probably because the aldosterone levels achieved in most pathological states in man are still far lower than the pharmacological levels reached in animal studies. These observations therefore again provide evidence against any significant regulation of sodium intake in man under most physiological and pathological circumstances, other than the intriguing incidence in selected cases of Addison's disease.

SUMMARY

Much has been learned about the physiology of body fluid homeostasis in humans and the pathophysiology of clinical disorders of body fluid homeostasis since the last summary of this topic 20 years ago in the *Handbook of Physiology* (Holmes, 1967). Much still remains to be learned, particularly concerning osmoreceptor function and its relationship to specific neural pathways controlling thirst and AVP secretion, the role of peripheral and central ANP secretion in solute homeostasis, the mechanisms mediating dehydration natriuresis and its significance during clinical states of dehydration, and the neural basis of sodium appetite in patients with Addison's disease. Additional studies concerning these and other areas undoubtedly will provide new insights into the regulation of body fluid homeostasis in man and in turn will likely lead to entirely new areas of investigation outside of our present understanding in this field. This chapter has attempted to summarize some of our present concepts and to integrate characteristics of specific clinical disorders of water and solute metabolism into a more general theory of body fluid homeostasis. From this approach several general conclusions seem warranted.

First, it quite clear that total body fluid homeostasis is regulated in accordance with needs for maintenance of both osmotic and volume homeostasis. Sometimes these needs require similar effects on water and solute balances, but at other times they conflict. A complete and comprehensive understanding of clinical disorders of fluid homeostasis is not possible without taking into account the interactions between the body's needs for maintenance of both osmotic and volume homeostasis in each disorder. Although generalizations must always be made with caution, it appears from present data that under most physiological conditions osmotic homeostasis is a more sensitive regulator of both water and solute balance, and volume homeostatic mechanisms become more prominent only under pathological conditions of more marked degrees of extracellular fluid and plasma volume depletion.

Second, studies over the last two decades have confirmed the tight linkage between thirst and AVP secretion in most circumstances. Although at present we know neither the exact location of osmoreceptors nor the neurobiological basis of osmoreceptor function, the characteristics shared by thirst and AVP secretion strongly implicate closely related mechanisms. Whether there prove to be separate osmoreceptors for thirst and AVP secretion, as some instances of dissociation of these two factors might suggest, or a single set of osmoreceptors, in which case dissociation of thirst and AVP secretion could be accounted for by abnormalities distal to primary osmoreception, remains to be ascertained. Regardless of the answer to this question, it seems clear that thirst and AVP secretion are regulated quite similarly. Consequently, research in both fields will continue to benefit from attempts to relate and integrate these two complementary processes.

Finally, the intimate relationship between our knowledge of basic concepts of the physiology of water and solute metabolism in animals and in man has been emphasized throughout this chapter. It is obvious to clinical investigators that much of our present understanding of human physiology and pathophysiology in this area has been a direct consequence of many of the basic investigations performed in animals, some of which have been described in previous chapters of this *Handbook*. Despite a variety of important species differences with regard to specific aspects of body fluid homeostasis, for the most part these points appear to represent quantitative rather than qualitative differences. The basic physiological mechanisms by which organisms maintain constancy of their internal milieu of water and solutes has proven to be strikingly similar in all mammals; rather than completely different control systems, what differs across species are the finer points of the similar neural and renal mechanisms that regulate water and solute intake and excretion, and in particular the relative degrees of emphasis each species places on the multiple mechanisms that act to maintain body fluid homeostasis.

However, the crucial role that basic physiological studies in animals have played in understanding human physiology should not be construed as being indicative of a unidirectional flow of information in this field. Clinical studies have also provided important insights into normal mammalian body fluid homeostasis, in part as a result of two attributes that are unique to clinical studies. First, pathological states of water and solute imbalance have been much more intensively studied and better characterized in man than in animals (e.g., hypoosmolar states), and such naturally occurring derangements of body fluid homeostasis have in turn provided an enhanced understanding of normal physiological regulatory mechanisms in man as well as in other animals. Second, subjective sensations such as thirst can only be accurately ascertained in humans able to verbalize these sensations, thereby allowing clinical studies to provide a more exact elucidation of the relationship between behavioral and physiological mechanisms that act in a coordinated manner to achieve and maintain water and solute balance. Consequently, body fluid homeostasis represents a field in which clinical investigators and basic physiologists have learned much from each other's studies. I hope that this chapter will serve to further stimulate this historically mutually beneficial relationship.

JOSEPH G.
VERBALIS

Alvis, R., Geheb, M., & Cox, M. (1985). Hypo- and hyperosmolar states: Diagnostic approaches. In A. I. Arieff & R. A. DeFronzo (eds.), *Fluid, electrolyte, and acid–base disorders* (pp. 185–219). New York: Churchill Livingstone.

Anderson, R. J., Chung, H.-M., Kluge, R., & Schrier, R. W. (1985). Hyponatremia: A prospective analysis of its epidemiology and the pathogenetic role of vasopressin. *Annals of Internal Medicine, 102,* 164–168.

Andersson, B. (1971). Thirst—and brain control of water balance. *American Scientist, 59,* 408–415.

Andersson, B. (1978). Regulation of water intake. *Physiological Reviews, 58,* 582–603.

Arieff, A. I. (1984). Central nervous system manifestations of disordered sodium metabolism. *Clinics in Endocrinology and Metabolism, 13,* 269–294.

Arieff, A. I., & Guisado, R. (1976). Effects on the central nervous system of hypernatremic and hyponatremic states. *Kidney International, 10,* 104–116.

Arieff, A. I., Llach, F., & Massry, S. G. (1976). Neurological manifestations and morbidity of hyponatremia: Correlation with brain water and electrolytes. *Medicine, 55,* 121–129.

Atarashi, K., Mulrow, P. J., & Franco-Saenz, R. (1985). Effect of atrial peptides on aldosterone production. *Journal of Clinical Investigation, 76,* 1807–1811.

Ballermann, B. J., Bloch, K. D., Seidman, J. G., & Brenner, B. M. (1986). Atrial natriuretic peptide transcription, secretion and glomerular receptor activity during mineralocorticoid escape in the rat. *Journal of Clinical Investigation, 78,* 840–843.

Balment, R. J., Brimble, M. J., Forsling, M. L., & Musabayane, C. T. (1984). Natriuretic response of the rat to plasma concentrations of arginine vasopressin within the physiologic range. *Journal of Physiology, 352,* 517–526.

Barlow, E. D., & deWardener, H. E. (1959). Compulsive water drinking. *Quarterly Journal of Medicine, 28,* 235–258.

Baxter, J. D., & Tyrrell, J. B. (1987). The adrenal cortex. In P. Felig, J. D. Baxter, A. E. Broadus, & L. A. Frohman (Eds.), *Endocrinology and metabolism* (pp. 511–650). New York: McGraw-Hill.

Baylis, P. H. (1987). Osmoregulation and control of vasopressin secretion in healthy humans. *American Journal of Physiology, 253,* R671–R678.

Baylis, P. H., & Thompson, C. J. (1988). Osmoregulation of vasopressin secretion and thirst in health and disease. *Clinical Endocrinology, 29,* 549–576.

Berl, T., Linas, S. L., Aisenbrey, G. A., & Anderson, R. J. (1977). On the mechanism of polyuria in potassium depletion. *Journal of Clinical Investigation, 60,* 620–625.

Berl, T., & Teitelbaum, I. (1985). Effects of hypokalemia and hypercalcemia on water metabolism. In R. W. Schrier (ed.), *Vasopressin* (pp. 543–552). New York: Raven Press.

Bianca, W., Findlay, J. D., & McLean, J. A. (1965). Responses of steers to water restriction. *Research in Veterinary Science, 6,* 38–55.

Buggy, J., & Johnson, A. K. (1977). Preoptic–hypothalamic periventricular lesions: Thirst deficits and hypernatremia. *American Journal of Physiology, 233,* R44–R52.

Burnett, J. C., Jr., Kao, P. C., Hu, D. C., Hesser, D. W., Heublein, D., Granger, J. P., Opgenorth, T. J., & Reeder, G. S. (1986). Atrial natriuretic peptide elevation in congestive heart failure in the human. *Science, 231,* 1145–1147.

Cannon, W. B. (1919). The physiological basis of thirst. *Proceedings of the Royal Society, 90,* 283–301.

Cantin, M., & Genest, J. (1985). The heart and the atrial natriuretic factor. *Endocrine Reviews, 6,* 107–127.

Chan, W. Y. (1973). A study of the mechanism of vasopressin escape: Effects of chronic vasopressin and overhydration on renal tissue osmolality and electrolytes in dogs. *Journal of Pharmacological and Experimental Therapeutics, 184,* 244–252.

Cooke, C. R., Turin, M. D., & Walker, W. G. (1979). The syndrome of inappropriate antidiuretic hormone secretion (SIADH): Pathophysiologic mechanisms in solute and volume regulation. *Medicine, 58,* 240–251.

Cowley, A. W., Jr., Merrill, D. C., Quillen, E. N., & Skelton, M. M. (1984). Long-term blood pressure and metabolic effects of vasopressin with servo-controlled fluid volume. *American Journal of Physiology, 247,* R537–R545.

Davison, J. M., Gilmore, E. A., Durr, J., Robertson, G. L., & Lindheimer, M. D. (1984). Altered osmotic thresholds for vasopressin secretion and thirst in human pregnancy. *American Journal of Physiology, 246,* F105–F109.

Defronzo, R. A., Goldberg, M., & Agus, Z. S. (1976). Normal diluting capacity in hyponatremic patients: Reset osmostat or variant of SIADH. *Annals of Internal Medicine, 84,* 538–542.

Denton, D. A. (1967). Salt appetite. In C. F. Code (Ed.), *Handbook of physiology: Alimentary canal I* (pp. 433–459). Washington, DC: American Physiological Society.

Denton, D. A. (1982). *The hunger for salt.* Berlin: Springer-Verlag.

DeRubertis, F. R., Michelis, M. F., Beck, N., Field, J. B., & Davis, B. B. (1971). "Essential" hypernatremia due to ineffective osmotic and intact volume regulation of vasopressin secretion. *Journal of Clinical Investigation, 50,* 97–111.

DeRubertis, F. R., Michelis, M. F., & Davis, B. B. (1974). "Essential" hypernatremia. *Archives of Internal Medicine, 134,* 889–894.

DeTroyer, A., & Demanet, J. C. (1976). Clinical, biological and pathogenic features of the syndrome of inappropriate secretion of antidiuretic hormone. *Quarterly Journal of Medicine, 55,* 521–531.

Durr, J. A., Stamoutsos, B. A., & Lindheimer, M. D. (1981). Osmoregulation during pregnancy in the rat. Evidence for resetting of the threshold for vasopressin during gestation. *Journal of Clinical Investigation, 68,* 337–346.

Edwards, B. R., & LaRochelle, F. T., Jr. (1984). Antidiuretic effect of endogenous oxytocin in dehydrated Brattleboro homozygous rats. *American Journal of Physiology, 247,* F453–F465.

Elkinton, J. R., & Taffel, M. (1942). Prolonged water deprivation in the dog. *Journal of Clinical Investigation, 21,* 787–794.

Engel, P., Rowe, J., Minaker, K., & Robertson, G. L. (1984). Effect of exogenous vasopressin on vasopressin release. *American Journal of Physiology, 246,* E202–E207.

Epstein, M., Preston, S., & Weitzman, R. (1981). Isoosmotic central blood volume expansion suppresses plasma arginine vasopressin in normal man. *Journal of Clinical Endocrinology and Metabolism, 52,* 256–262.

Fichman, M. P., Vorherr, H., Kleeman, C. R., & Telfer, N. (1971). Diuretic-induced hyponatremia. *Annals of Internal Medicine, 75,* 853–863.

Fitzsimons, J. T. (1961a). Drinking by nephrectomised rats injected with various substances. *Journal of Physiology, 155,* 563–579.

Fitzsimons, J. T. (1961b). Drinking by rats depleted of body fluid without increases in osmotic pressure. *Journal of Physiology, 159,* 297–309.

Fitzsimons, J. T. (1963). The effects of slow infusions of hypertonic solutions on drinking and drinking thresholds in rats. *Journal of Physiology, 167,* 344–354.

Fitzsimons, J. T. (1972). Thirst. *Physiological Reviews, 52,* 468–561.

Fitzsimons, J. T. (1976). The physiological basis of thirst. *Kidney International, 10,* 3–11.

Fitzsimons, J. T. (1979). *The physiology of thirst and sodium appetite.* Cambridge: Cambridge University Press.

Forsling, M. L., & Brimble, M. J. (1985). The role of oxytocin in salt and water balance. In J. A. Amico & A. G. Robinson (Eds.), *Oxytocin: Clinical and laboratory studies* (pp. 167–175). Amsterdam: Elsevier.

Ganguly, A., & Robertson, G. L. (1980). Elevated threshold for vasopressin secretion in primary aldosteronism. *Clinical Research, 23,* 330.

Gardiner, T. W., Verbalis, J. G., & Stricker, E. M. (1985). Impaired secretion of vasopressin and oxytocin in rats after lesions of nucleus medianus. *American Journal of Physiology, 249,* R681–R688.

Geelen, G. L., Keil, L. C., Kravik, S. E., Wade, C. E., Thrasher, T. N., Barnes, P. R., Pyka, G., Nesvig, C., & Greenleaf, J. E. (1984). Inhibition of plasma vasopressin after drinking in dehydrated humans. *American Journal of Physiology, 247,* R968–R971.

Gerich, J. E., Martin, M. M., & Recant, L. (1971). Clinical and metabolic characteristics of hyperosmolar nonketotic coma. *Diabetes, 20,* 228–238.

Gill, G., Baylis, P., & Burn, J. (1985). A case of "essential" hypernatremia due to resetting of the osmostat. *Clinical Endocrinology, 22,* 545–551.

Gilman, A. (1937). The relation between blood osmotic pressure, fluid distribution and voluntary water intake. *American Journal of Physiology, 120,* 323–328.

Goldman, M. D., Luchins, D. J., & Robertson, G. L. (1988). Mechanisms of altered water metabolism in psychotic patients with polydipsia and hyponatremia. *New England Journal of Medicine, 318,* 397–403.

Goldsmith, S. R. (1988). Baroreflex control of vasopressin secretion in normal humans. In A. W. Cowley, Jr., J.-F. Liard, & D. A. Ausiello (Eds.), *Vasopressin: Cellular and integrative functions* (pp. 389–397). New York: Raven Press.

Goldsmith, S. R., Cowley, A. W., Jr., Francis, G. S., & Cohn, J. N. (1984). Effect of increased intracardiac and arterial pressure on plasma vasopressin in humans. *American Journal of Physiology, 246,* H647–H651.

Granger, J. P., Burnett, J. C., Jr., Romero, J. C., Opgenorth, T. J., Salazar, J., & Joyce, M. (1987). Elevated levels of atrial natriuretic peptide during aldosterone escape. *American Journal of Physiology, 252*, R878–R882.

Gross, P. A., & Anderson, R. J. (1982). Effects of DDAVP and AVP on sodium and water balance in conscious rats. *American Journal of Physiology, 243*, R512–R519.

Hall, J. E., Granger, J. P., Smith, M. J., & Premen, A. J. (1984). Role of renal hemodynamics and arterial pressure in aldosterone "escape." *Hypertension, 6*, I183–I192.

Hall, J. E., Montani, J. P., Woods, L. L., & Mizelle, H. L. (1986). Renal escape from vasopressin: Role of pressure diuresis. *American Journal of Physiology, 250*, F907–F916.

Halter, J. B., Goldberg, A. P., Robertson, G. L. & Porte, D. (1977). Selective osmoreceptor dysfunction in the syndrome of chronic hypernatremia. *Journal of Clinical Endocrinology and Metabolism, 44*, 609–616.

Hammond, D. N., Moll, G. W., Robertson, G. L., & Chelmicka-Schorr, E. (1986). Hypodipsic hypernatremia with normal osmoregulation of vasopressin. *New England Journal of Medicine, 315*, 433–436.

Heinbecker, P., & White, H. L. (1941). Hypothalamico-hypophyseal system and its relation to water balance in the dog. *American Journal of Physiology, 133*, 582–593.

Helderman, J. H., Vestal, R. E., Rowe, R. W., Tobin, J. D., Andres, R., & Robertson, G. L. (1978). The response of arginine vasopressin to intravenous ethanol and hypertonic saline in man: The impact of aging. *Journal of Gerontology, 33*, 39–47.

Henkin, R. I., Gill, J. R., Jr., & Bartter, F. C. (1963). Studies of taste thresholds in normal man and patients with adrenal cortical insufficiency. The role of adrenal cortical steroids and of serum sodium concentration. *Journal of Clinical Investigation, 42*, 727–735.

Herman, J. P., Marciano, F. F., & Gash, D. M. (1986). Vasopressin administration prevents functional recovery of the vasopressinergic neurosecretory system following neurohypophysectomy. *Neuroscience Letters, 72*, 239–246.

Hoffman, E. K. (1977). Control of cell volume. In B. L. Gupta, R. B. Moreton, J. Oschman, & B. J. Wall (Eds.), *Transport of ions and water in animals* (pp. 285–332). London: Academic Press.

Holliday, M. A., Kalayci, M. N., & Harrah, J. (1968). Factors that limit brain volume changes in response to acute and sustained hyper- and hyponatremia. *Journal of Clinical Investigation, 47*, 1916–1928.

Holmes, J. H. (1967). Thirst and fluid intake as clinical problems. In C. F. Code (Ed.), *Handbook of physiology: Alimentary canal I* (pp. 147–162). Washington, DC: American Physiological Society.

Holmes, J. H., & Montgomery, A. V. (1953). Thirst as a symptom. *American Journal of the Medical Sciences, 225*, 281–286.

Ikkos, D., Luft, R., & Oliverocrona, H. (1955). Hypophysectomy in man: Effects on water excretion during the first two postoperative months. *Journal of Clinical Endocrinology and Metabolism, 15*, 553–567.

Jalowiec, J. E., & Stricker, E. M. (1973). Sodium appetite in adrenalectomized rats following dietary sodium deprivation. *Journal of Comparative and Physiological Psychology, 83*, 66–77.

Johnson, M. D., Kinter, L. B., & Beeuwkes, R. III. (1979). Effect of AVP and DDAVP on plasma renin activity and electrolyte excretion in conscious dogs. *American Journal of Physiology, 236*, F66–F70.

Knox, F. G., Burnett, J. C., Jr., Kohan, D. E., Spielman, W. S., & Strand, J. C. (1980). Escape from the sodium retaining effects of mineralocorticoids. *Kidney International, 17*, 263–276.

Kozlowski, S., & Szczepanska-Sadowska, E. (1975). Mechanisms of hypovolaemic thirst and interactions between hypovolaemia, hyperosmolality and the antidiuretic system. In G. Peters, J. T. Fitzsimons, & L. Peters-Haefeli (Eds.), *Control mechanisms of drinking* (pp. 25–35). Berlin: Springer-Verlag.

Laszlo, F. A., & DeWied, D. (1966). Antidiuretic hormone content of the hypothalamo-neuro-hypophyseal system and urinary excretion of antidiuretic hormone in rats during the development of diabetes insipidus after lesions in the pituitary stalk. *Journal of Endocrinology, 36*, 125–137.

Lauson, H. D. (1974). Metabolism of the neurohypophysial hormones. In E. Knobil & W. H. Sawyer (Eds.), *Handbook of physiology: The pituitary gland and its neuroendocrine control* (pp. 287–393). Washington, DC: American Physiological Society.

Leaf, A., Bartter, F. C., Santos, R. F., & Wrong, O. (1953). Evidence in man that urinary electrolyte loss induced by Pitressin is a function of water retention. *Journal of Clinical Investigation, 32*, 868–878.

Leng, G., Blackburn, R. E., Dyball, R. E. J., & Russell, J. A. (1989). Role of anterior peri-third ventricular structures in the regulation of supraoptic neuronal activity and neurohypophysial hormone secretion in the rat. *Journal of Neuroendocrinology, 1*, 35–46.

Leng, G., Dyball, R. E. J., & Mason, W. T. (1985). Electrophysiology of osmoreceptors. In R. W. Schrier (Ed.), *Vasopressin* (pp. 333–342). New York: Raven Press.

Lipsett, M. B., MacLean, I. P., West, C. D., Li, M. C., & Pearson, O. H. (1956). An analysis of the polyuria induced by hypophysectomy in man. *Journal of Clinical Endocrinology and Metabolism, 16,* 183–195.

Luke, R. G. (1973). Natriuresis and chloruresis during hydropenia in the rat. *American Journal of Physiology, 224,* 13–20.

Maack, T., Marion, D. N., Camargo, M. J. F., Kleinert, H. D., Laragh, J. H., Vaughan, E. D., Jr., & Atlas, S. A. (1984). Effects of auriculin (atrial natriuretic factor) on blood pressure, renal function, and the renin–aldosterone system in dogs. *The American Journal of Medicine, 77,* 1069–1075.

Mann, J. F. E., Johnson, A. K., Ganten, D., & Ritz, E. (1987). Thirst and the renin angiotensin system. *Kidney International, 32,* S27–S34.

Marin-Grez, M. M., Fleming, J. T., & Steinhausen, M. (1986). Atrial natriuretic peptide causes preglomerular vasodilation and post-glomerular vasoconstriction in rat kidney. *Nature, 324,* 473–476.

Mason, W. T. (1980). Supraoptic neurons of rat hypothalamus are osmosensitive. *Nature, 287,* 154–157.

McCurdy, D. K. (1970). Hyperosmolar hyperglycemic nonketotic coma. *Medical Clinics of North America, 54,* 683–699.

McKinley, M. J., Denton, D. A., Nelson, J. F., & Weisinger, R. S. (1983). Dehydration induces sodium depletion in rats, rabbits, and sheep. *American Journal of Physiology, 245,* R287–R292.

Melton, J. E., Patlak, C. S., Pettigrew, K. D., & Cserr, H. F. (1987). Volume regulatory loss of Na, Cl, and K from rat brain during acute hyponatremia. *American Journal of Physiology, 252,* F661–F669.

Menninger, R. P. (1981). Right atrial stretch decreases supraoptic neurosecretory activity and plasma vasopressin. *American Journal of Physiology, 241,* R44–R49.

Michelis, M. F., Fusco, R. O., Bragdon, R. W., & Davis, B. D. (1974). Reset of osmoreceptors in association with normovolemic hyponatremia. *American Journal of Medical Science, 267,* 267–273.

Miller, M., Dalakos, T., Moses, A. M., & Fellerman, H. (1970). Recognition of partial defects in antidiuretic hormone secretion. *Annals of Internal Medicine, 73,* 721–729.

Moses, A. M., & Miller, M. (1971). Osmotic threshold for vasopressin release as determined by saline infusion and by dehydration. *Neuroendocrinology, 7,* 219–226.

Needleman, P., & Greenwald, J. E. (1986). Atriopeptin: A cardiac hormone intimately involved in fluid, electrolyte, and blood pressure homeostasis. *New England Journal of Medicine, 314,* 828–834.

Nicolaidis, S., & Rowland, N. (1974). Long-term self-intravenous "drinking" in the rat. *Journal of Comparative and Physiological Psychiatry, 87,* 1–15.

Nicolaidis, S., & Rowland, N. (1975). Systemic versus oral and gastrointestinal metering of fluid intake. In G. Peters, J. T. Fitzsimons, & L. Peters-Haefeli (Eds.), *Control mechanisms of drinking* (pp. 14–21). Berlin: Springer-Verlag.

Phillips, P. A., Rolls, B. J., Ledingham, J. G. G., Forsling, M. L., Morton, J. J., Crowe, M. H., & Wollner L. (1984). Reduced thirst after water deprivation in healthy elderly man. *New England Journal of Medicine, 311,* 753–759.

Phillips, P. A., Rolls, B. J., Ledingham, J. G. G., & Morton, J. J. (1984). Body fluid changes, thirst and drinking in man during free access to water. *Physiology and Behavior, 33,* 357–363.

Phillips, P. A., Rolls, B. J., Ledingham, J. G. G., Forsling, M. L., & Morton, J. J. (1985). Osmotic thirst and vasopressin release in humans: A double blind cross-over study. *American Journal of Physiology, 248,* R645–R650.

Phillips, P. A., Rolls, B. J., Ledingham, J. G. G., Morton, J. J., & Forsling, M. L. (1985). Angiotensin II-induced thirst and vasopressin release in man. *Clinical Science, 68,* 669–674.

Quillen, E., & Cowley, A. W. (1983). Influence of volume changes on osmolality–vasopressin relationships in conscious dogs. *American Journal of Physiology, 244,* H73–H79.

Quillen, E. W., Jr., Reid, I. A., & Keil, L. C. (1988). Cardiac and arterial baroreceptor influences on plasma vasopressin and drinking. In A. W. Cowley, Jr., J.-F. Liard, & D. A. Ausiello (Eds.), *Vasopressin: Cellular and integrative functions* (pp. 405–411). New York: Raven Press.

Ramsay, D. J. (1985). Osmoreceptors subserving vasopressin secretion and drinking—an overview. In R. W. Schrier (Ed.), *Vasopressin* (pp. 291–298). New York: Raven Press.

Ramsay, D. J., & Thrasher, T. N. (1986). Hyperosmotic and hypovolemic thirst. In G. de Caro, A. N. Epstein, & M. Massi (Eds.), *The physiology of thirst and sodium appetite* (pp. 83–96). New York: Plenum Press.

Rice, K. K., & Richter, C. P. (1943). Increased sodium chloride and water intake of normal rats treated with desoxycorticosterone acetate. *Endocrinology, 33,* 106–115.

Richards, A. M., McDonald, D., Fitzpatrick, M. A., Nicholls, M. G., Espiner, E. A., Ikram, H., Jans, S., Grant, S., & Yandle, T. (1988). Atrial natriuretic hormone has biological effects in man at physiological plasma concentrations. *Journal of Clinical Endocrinology and Metabolism, 67,* 1134–1139.

Richardson, D. W., & Robinson, A. G. (1985). Desmopressin. *Annals of Internal Medicine, 103,* 228–239.

Richter, C. P. (1936). Increased salt appetite in adrenalectomized rats. *American Journal of Physiology, 115,* 155–161.

Robertson, G. L. (1977). Vasopressin function in health and disease. *Recent Progress in Hormone Research, 33,* 333–385.

Robertson, G. L. (1983). Thirst and vasopressin function in normal and disordered states of water balance. *Journal of Laboratory and Clinical Medicine, 101,* 351–371.

Robertson, G. L. (1984). Abnormalities of thirst regulation. *Kidney International, 25,* 460–469.

Robertson, G. L. (1987a). Posterior pituitary. In P. Felig, J. D. Baxter, A. E. Broadus, & L. A. Frohman (Eds.), *Endocrinology and metabolism* (pp. 338–385). New York: McGraw-Hill.

Robertson, G. L. (1987b). Physiology of ADH secretion. *Kidney International, 32,* S20–S26.

Robertson, G. L., & Athar, S. (1976). The interaction of blood osmolality and blood volume in regulating plasma vasopressin in man. *Journal of Clinical Endocrinology and Metabolism, 42,* 613–620.

Robertson, G. L., Aycinena, P., & Zerbe, R. L. (1982). Neurogenic disorders of osmoregulation. *American Journal of Medicine, 72,* 339–353.

Robinson, A. G. (1985). Disorders of antidiuretic hormone secretion. *Clinics in Endocrinology and Metabolism, 14,* 55–58.

Rogers, P. W., & Kurtzman, N. A. (1973). Renal failure, uncontrollable thirst and hyperreninemia; cessation of thirst with bilateral nephrectomy. *Journal of the American Medical Association, 225,* 1236–1238.

Rubini, M. E. (1961). Water excretion in potassium deficient man. *Journal of Clinical Investigation, 40,* 2215–2224.

Rutecki, G. W., Cox, J. W., Robertson, G. W., Francisco, L. L., & Ferris, T. F. (1982). Urinary concentrating ability and antidiuretic hormone responsiveness in the potassium-depleted dog. *Journal of Laboratory and Clinical Medicine, 100,* 53–60.

Rymer, M. M., & Fishman, R. A. (1973). Protective adaptation of brain to water intoxication. *Archives of Neurology, 28,* 49–54.

Salata, R. A., Verbalis, J. G., & Robinson, A. G. (1987). Cold water stimulation of oropharyngeal receptors in man inhibits release of vasopressin. *Journal of Clinical Endocrinology and Metabolism, 65,* 561–567.

Schrier, R. W. (1988). Pathogenesis of sodium and water retention in high-output and low-output cardiac failure, nephrotic syndrome, cirrhosis and pregnancy. *New England Journal of Medicine, 319,* 1065–1072, 1127–1134.

Schrier, R. W., Berl, T., & Anderson, R. J. (1979). Osmotic and non-osmotic control of vasopressin release. *American Journal of Physiology, 236,* F321–F332.

Schwartz, W. B., Bennett, W., Curelop, S., & Bartter, F. C. (1957). A syndrome of renal sodium loss and hyponatremia probably resulting from inappropriate secretion of antidiuretic hormone. *American Journal of Medicine, 23,* 529–542.

Seckl, J. R., Williams, T. D. M., & Lightman, S. L. (1986). Oral hypertonic saline causes transient fall of vasopressin in humans. *American Journal of Physiology, 251,* R214–R217.

Sharkey, P. C., Perry, J. H., & Ehni, G. (1961). Diabetes insipidus following section of hypophyseal stalk. *Journal of Neurosurgery, 18,* 445–460.

Shenker, Y., Sider, R. S., Ostafin, E. A., & Grekin, R. J. (1985). Plasma levels of immunoreactive atrial natriuretic factor in healthy subjects and in patients with edema. *Journal of Clinical Investigation, 76,* 1684–1698.

Smith, M. J., Jr., Cowley, A. W., Jr., Guyton, A. C., & Manning, R. D., Jr. (1979). Acute and chronic effects of vasopressin on blood pressure, electrolytes and fluid volumes. *American Journal of Physiology, 237,* F233–F240.

Sterns, R. H., Riggs, J., & Schochet, S. S. (1986). Osmotic demyelination syndrome following correction of hyponatremia. *New England Journal of Medicine, 314,* 1535–1542.

Stormont, J. M., & Waterhouse, C. (1961). The genesis of hyponatremia associated with marked overhydration and water intoxication. *Circulation, 24,* 191–203.

Stricker, E. M. (1966). Extracellular fluid volume and thirst. *American Journal of Physiology, 211,* 232–238.

Stricker, E. M. (1981). Thirst and sodium appetite after colloid treatment in rats. *Journal of Comparative Physiology and Psychology, 95,* 1–25.

Stricker, E. M., & Verbalis, J. G. (1988). Hormones and behavior: The biology of thirst and sodium appetite. *American Scientist, 76,* 261–267.

Szczepanska-Sadowska, E., & Kozlowski, S. (1975). Equipotency of hypertonic solutions of mannitol and sodium chloride in eliciting thirst in the dog. *Pflugers Archiv, 358,* 259–264.

Thompson, C. J., & Baylis, P. H. (1987a). Thirst in diabetes insipidus: Clinical relevance of quantitative assessment. *Quarterly Journal of Medicine, 240,* 853–862.

Thompson, C. J., & Baylis, P. H. (1987b). Mechanisms responsible for thirst and polyuria associated with primary hyperaldosteronism. *British Medical Journal, 295,* 578–579.

Thompson, C. J., Bland, J., Burd, J., & Baylis, P. H. (1986). The osmotic thresholds for thirst and vasopressin release are similar in healthy man. *Clinical Science, 71,* 651–656.

Thompson, C. J., Burd, J. M., & Baylis, P. H. (1987). Acute suppression of plasma vasopressin and thirst after drinking in hypernatremic humans. *American Journal of Physiology, 252,* R1138–R1142.

Thorn, G. W., Dorrance, S. S., & Day, E. (1942). Addison's disease: Evaluation of synthetic desoxycorticosterone acetate therapy in 158 patients. *Annals of Internal Medicine, 16,* 1053–1096.

Thrasher, T. N. (1985). Circumventricular organs, thirst and vasopressin secretion. In R. W. Schrier (Ed.), *Vasopressin* (pp. 311–318). New York: Raven Press.

Thrasher, T. N., Keil, L. C., & Ramsay, D. J. (1982a). Hemodynamic, hormonal and drinking responses to reduced venous return in the dog. *American Journal of Physiology, 243,* R354–R362.

Thrasher, T. N., Keil, L. C., & Ramsay, D. J. (1982b). Lesions of the organum vasculosum of the lamina terminalis (OVLT) attenuate osmotically induced drinking and vasopressin secretion in the dog. *Endocrinology, 110,* 1837–1841.

Thrasher, T. N., Nistal-Herrera, J. F., Keil, L. C., & Ramsay, D. J. (1981). Satiety and inhibition of vasopressin secretion after drinking in dehydrated dogs. *American Journal of Physiology, 240,* E394–E401.

Verbalis, J. G. (1984). An experimental model of the syndrome of inappropriate antidiuretic hormone secretion in the rat. *American Journal of Physiology, 247,* E540–E553.

Verbalis, J. G. (1990a). Inappropriate antidiuresis and other hypo-osmolar states. In K. L. Becker (Ed.), *Principles and practice of endocrinology and metabolism* (pp. 237–247). Philadelphia: J. B. Lippincott.

Verbalis, J. G. (1990b). Inhibitory controls of drinking. In D. J. Ramsay & D. A. Booth (Eds.), *Thirst: Physiological and psychological aspects*. Berlin: Springer-Verlag (in press).

Verbalis, J. G., Baldwin, E. F., & Robinson, A. G. (1986). Osmotic regulation of plasma vasopressin and oxytocin after sustained hyponatremia. *American Journal of Physiology, 250,* R444–R451.

Verbalis, J. G., & Drutarosky, M. D. (1988). Adaptation to chronic hypoosmolality in rats. *Kidney International, 34,* 351–360.

Verbalis, J. G., Drutarosky, M. D., Ertel, R. J., & Vollmer, R. R. (1989). Adaptive responses to sustained volume expansion in hyponatremic rats. *Journal of Endocrinology, 122,* 421–431.

Verbalis, J. G., Robinson, A. G., & Moses, A. M. (1984). Postoperative and post-traumatic diabetes insipidus. *Frontiers in Hormone Research, 13,* 247–265.

Verney, E. B. (1947). The antidiuretic hormone and the factors which determine its release. *Proceedings of the Royal Society, London Series B, 135,* 25–106.

Weidmann, P., Hasler, L., Gnadinger, M. P., Lang, R. E., Uehlinger, D. E., Shaw, S., Rascher, W., & Reubi, F. C. (1986). Blood levels and renal effects of atrial natriuretic peptide in normal man. *Journal of Clinical Investigation, 77,* 734–742.

Wolf, A. V. (1950). Osmometric analysis of thirst in man and dog. *American Journal of Physiology, 161,* 75–86.

Wolf, A. V. (1958). *Thirst. Physiology of the urge to drink and problems of water lack.* Springfield, IL: Charles C. Thomas.

Wolf, G. (1965). Effect of deoxycorticosterone on sodium appetite of intact and adrenalectomized rats. *American Journal of Physiology, 208,* 1281–1285.

Wood, R. J., Rolls, B. J., & Ramsay, D. J. (1977). Drinking following intracarotid infusions of hypertonic solutions in dogs. *American Journal of Physiology, 232,* R88–R92.

Wood, R. J., Rolls, E. T., & Rolls, B. J. (1982). Physiological mechanisms for thirst in the nonhuman primate. *American Journal of Physiology, 242,* R423–R428.

Yannett, H. (1940). Changes in the brain resulting from depletion of extracellular electrolytes. *American Journal of Physiology, 128,* 683–689.

Zerbe, R. L. (1985). Genetic factors in normal and abnormal regulation of vasopressin secretion. In R. W. Schrier (Ed.), *Vasopressin* (pp. 213–220). New York: Raven Press.

Zerbe, R. L., & Robertson, G. L. (1981). A comparison of plasma vasopressin measurements with a standard indirect test in the differential diagnosis of polyuria. *New England Journal of Medicine, 305,* 1539–1546.

Zerbe, R. L., & Robertson, G. L. (1983). Osmoregulation of thirst and vasopressin secretion in human subjects: Effect of various solutes. *American Journal of Physiology, 244,* E607–E614.

Zerbe, R. L., Stropes, L., & Robertson, G. L. (1980). Vasopressin function in the syndrome of inappropriate antidiuresis. *Annual Review of Medicine, 31,* 315–327.

Zimmerman, E. A., Ma, L.-Y., & Nilaver, G. (1987). Anatomical basis of thirst and vasopressin secretion. *Kidney International, 32,* S14–S19.

PART V
Prospective Essays

17

The Behavioral and Neural Sciences of Ingestion

David A. Booth

Ingestive Behavior and Neural Processes

Intake and Ingestive Behavior

Behavior is information processing by the whole organism. Thus, ingestive behavior is a transformation of incoming information into phasic and selective transfers of materials into the body. The organization within eating and drinking behavior, or ingestive appetite, is a set of causal processes that accounts for intake as an effect of sensing the characteristics of diets and of other aspects of the external and internal environments.

Therefore, in order to build a neuroscience of ingestive behavior, we must do more than observe neural manipulations or correlates of intakes whose immediate controlling factors are unspecified. For example, an effect of a brain lesion or a drug's action on the size of meal eaten from a highly palatable diet is not sufficient to identify an effect of that manipulation on either satiation or palatability. A brain site whose destruction causes overeating is not thereby shown to be a satiety center. A drug that reduces food intake is not necessarily an anorexigenic agent, nor, if it acts centrally, need its target be receptors involved specifically in appetite. To justify such or any other behavioral interpretations, we have to characterize what processes within the ingestive behavior are related to the neural processes manipulated or measured.

That can be done qualitatively by showing that the specific neural intervention or correlate affects only one sort of influence on intake and not others that are also operative. This is the classic double-dissociation design; to reach a sound

David A. Booth School of Psychology, University of Birmingham, Birmingham B15 2TT, England.

conclusion, it must be shown that both the lesion site, the neurotransmitter, or the unit firing pattern and the aspect of behavior are associated exclusively with each other and not with any other neural or behavioral candidates.

Double dissociation can also be quantitated, sometimes even without experimental control (Booth, 1990). The strength of a causal process is measured by parameters of the statistical function of levels of the causal input on levels of the output effect, such as the sensitivity of the output to the input (Booth & Blair, 1988).

Such quantitative dissociation is familiar in pharmacology in the form of differences in dose–response functions between drugs thereby concluded to act on different receptors. It is well recognized also in physiology and in engineering. Some intervention (e.g., administering a hormone) modulates the slope and/or intercept of an input–output function (e.g., the dose response of another hormone's effect), while other influences on that output are unaffected. The behavioral sciences are badly weakened by unfamiliarity with the dissociation of psychophysical functions, i.e., stimulus–response relationships for overall functions of the whole organism.

One sort of ingestive psychophysical function is the effect of different strengths of a sensory characteristic on preference for a diet, e.g., calorically conditioned hedonic functions for sweetness in familiar foods (Booth, Lovett, & McSherry, 1972; Booth, Conner, & Marie, 1987). Quantitative evidence that a neural manipulation affects learned sweet preference depends on observing an effect on that psychophysical relationship but not on preference for another taste or on another effect of sweetness.

Other examples of ingestive psychophysics would be functions of meal initiation timing on variations from a learned cue complex (Booth & Davis, 1973; Weingarten, 1985), of nutrient-specific selection on phase of the lighting cycle (Leibowitz, 1988), or of an inhibition of food but not water ingestion on extent of gastric distension and on hepatic oxidation rate (Booth, 1972a, 1972b, 1976; Gibson & Booth, 1989). Gastric perceptivity in satiety (Robinson, 1989), for example, is the sensitivity of inhibition of eating to differences in gastric volume, unconfounded by parallel physiological changes or by the eater's expectations of the time course of satiety.

In short, dietary intake is not ingestive behavior. Whether the intake is of one material or of a selection of solid or liquid diets, the behavioral organization is not captured merely by measurement of food intakes by themselves or in response to some neural manipulation. To permit any analysis of behavior (and hence any behavioral neuroscience), there must also be control or measurement of the influences on the intake, be they from the diet(s), from the tissues of the eater's body, or from the external environment of other organisms or of inanimate objects. For behavior is the relationship between observable outputs from and observable inputs to the organism, not just the movements put out or their physical effect (in this case, transfer of material down the gullet).

As this method is brought into use in the neuroscience of ingestive behavior, so this research field will begin to attain full-fledged causal analysis of the functioning of the systems that concern it.

Prospects for ingestive neuroscience are therefore illustrated here by examples where intake research has the opportunity to coordinate analyses of both neural and behavioral causation.

From the above it follows that, even if we are primarily interested in explaining daily intakes of water and a maintenance food *ad libitum*, we cannot analyze the behavioral causation, let alone its neural bases, without controlling the sensory characteristics of the ingestates separately from their visceral impact and ecological contexts (Booth, 1988).

Chow and water present difficulties of opposite sorts in meeting this requirement.

The sources of energy and of other nutrients, such as vitamins and essential amino acids, fatty acids, and minerals, typically occur in differing forms in different foodstuffs. Each form of a nutrient has many sensory characteristics, usually not distinctive of the nutrient and yet, because of inherent molecular structures, difficult or impossible to mimic by some form of another nutrient. These difficulties are not reduced by using isolated forms of the nutrients. The separation techniques for proteins, for example, may introduce a taste or an odor that then is common among the protein preparations. The source of a starch preparation, such as maltodextrin (Polycose®), can carry over in a maize or potato odor. Such factors make it difficult to establish conclusively the existence of, for example, a starch taste or an innate protein appetite by behavioral experiments alone (Booth, 1987a; Booth, Conner, & Gibson, 1989).

Distilled water seems much simpler; it has only one texture, a taste, and some temperature or other. Yet the texture (and the thermal capacity) is inherent to water and to all the dilute aqueous solutions that organisms encounter naturally. This has so far prevented effective analysis of the origins of water recognition and preference and hence of the causal structure of thirsty behavior (Booth, 1979, 1990b).

Salt appetite is in a much better position. Qualitative variations can be rung on the sodium ion (and the anions). A qualitative demonstration of salt taste preference, and hence of salt appetite in the sense of a sodium-deficit-dependent preference, can be made by presenting a short-term choice between water and one salt solution of suitable concentration (or presenting just the saline if the water is never consumed). For analysis of physiological modulations of this avidity for the taste of salt, the fully quantitative measure would be a function of (relative) intakes on sodium chloride concentrations, dissociated from a similar short-term saccharin preference function. However, it should be noted that exposure to any particular level of a taste, including saltiness (Harris & Booth, 1987) and sweetness (Booth *et al.*, 1972, 1987), tends to induce greater avidity for that taste, and so these concentration–intake functions may not be stable unless differential exposures to the taste levels are carefully controlled.

FUNDAMENTAL BEHAVIORAL ISSUES FOR INGESTIVE NEUROBIOLOGY

In the 1970s, the basic issues for the physiological psychology of ingestion were commonly defined as selection, starting, and stopping. Unfortunately, this crisp and cogent agenda has been widely misunderstood, and so the same basic questions remain for the future.

The agenda became set in the form of ingestive initiation and termination factors because of a widening appreciation of the fact that the organisms on which most research was done ate in meals. Yet, as just argued above, questions about the influences on ingestive behavior, even those inspired by patterns in the temporal pattern of eating (Le Magnen & Tallon, 1966; Booth, 1978; Simpson & Bernays, 1983), cannot be answered by measuring intake alone, however refined the temporal analysis or the motor pattern recording.

The gross parameters of meal size and interval between meals can be used to obtain dose–response functions of physiological (Booth, 1972a) or ecological manipulations (Collier, 1980). The questions remain, however, which of the behavioral processes initiating, maintaining, or terminating meals are affected by the physiology or the ecology and in what manner. The neural underpinnings are yet a further issue that can only be addressed once we have identified how the control of behavior via the external and internal senses mediates dehydration anorexia or metabolic satiety and foraging work or fear of predation.

Effects on the finer temporal parameters of intake during a meal leave all the same questions open. If their controlling inputs were identified, though, intake parameters measured at the second-to-second level are more likely to relate to the dynamics of a neurophysiological network such as forebrain modulation of a brainstem masticatory pattern generator. Several examples of this are given later.

So, the meal initiation and termination issues would be productively recast as questions about the differences in inputs to the nervous system between occasions to eat and occasions to stop.

Also, it remains a fundamental behavioral issue how any organism even distinguishes food from water.

Answers to these basic questions may in fact emerge from resolutions of the superficially more complex issues of nutrient selection and its ecological, dietary, and visceral sensory bases. This is likely because even the intake of a single maintenance diet is the cumulative resultant of an ongoing series of choices suited to the changing circumstances of the moment (Booth, 1972b, 1980).

What nutrients are distinguished then? Are there alimentary appetites only for water, sodium, calories, and protein plus nonspecific aversive effects of some other nutritional deficiencies or excesses? How specific are the satieties to sensory characteristics, nutrient repletions, and associations between them? By what dietary discriminations and orientations of approach and consumption can selection of a particular nutrient be achieved in a given species? How specific and precise is the selection? Which of its mechanisms are organized genetically and by exposure to need or supply, and which are learned?

CONTROL OF INTAKE PATTERN

Distributions of intake over time, whether in meals or bouts and intervals between them or in a finer-grained "microstructural" analysis closer to the timing of neuromotor organization, at best identify output patterns that require causal explanation. These data cannot identify those causes, namely, the inhibitory or facilitatory influences on ingestion and its various subcomponents. So it is arbitrary to define meal size as satiation (or indeed as appetite) and meal-to-meal interval as satiety (or hunger). In thirst, for example, this move has led to use of bout sizes as measures of factors in both termination and initiation of drinking (Rolls & Rolls, 1983).

The meal pattern, rather, is an output measure on which the effectiveness of identified influences on ingestive behavior can be measured. Once that is done, the tasks can be characterized that neuronal processing has to achieve.

IDENTIFICATION OF INTAKE STRUCTURE. First, though, the measure extracted from the raw data must be realistic to the phenomena. The data can hardly be lumped across many mouthfuls before the analysis starts. As we will see when considering texture, even taking a mouthful as a unit is a poor interim compromise for a realistic neuroscience of sensorimotor control of ingestion. Hence, microstructural parameters must be built on time resolutions and preferably act resolutions no less than that given by the devices that record removal of powdered diets or small pellets. By using time intervals of the order of 1 second, different types of temporally stable processes controlling intake can often be identified by distinctive rates of events per opportunity and readily visualized by the logarithmic survivor plots recommended for this purpose by Allison and Castellan (1970) in drinking and Booth and Pain (1970) and Booth (1972c) for eating and since widely applied following Slater (1974) (cf. Clifton, 1987).

When one such family of mouthful frequencies has a constant central tendency throughout much of a meal, as is often the case for rats eating chow *ad libitum* for example (Booth, 1978), then motor control might usefully be characterized by an average eating rate within a bout or even across a whole meal. Within-meal bout sizes and interbout intervals may also be realistic characterizations of the modulating influences on the brainstem oscillators. Even meal sizes and intervals under constant conditions could be mechanistically meaningful epiphenomena, although it is hard to imagine how an organism might measure either parameter without reading weighing scales or a clock. The only ways to tell whether such parameters are significant is to relate meal sizes, meal-to-meal intervals, and their proportions to potential influences on them or at least to find neural manipulations that dissociate some parameters.

The causal importance of an influence will be measured as the sensitivity of meal size, postprandial ratio, within-bout ingestion rate, or some such to a range of normal strengths of that influence. Variation in this sensitivity or other parameters of the causal relationship as a result of a neural manipulation then provides evidence on how the process is executed by the neuronal machinery.

CONTROL OF MEALS. Such causal analysis of appetite in the meal sizes and meal-to-meal intervals of individual animals was used to obtain the first clear demonstration that food ingestion was inhibited by food carbohydrate energy while and after it was being absorbed (Booth, 1972a); in virtually freely feeding rats, both latency to initiate a meal and the size of that meal were sensitive to metabolic satiety. By varying the metabolic substrate and testing on the size of a meal at a fixed short delay, this conclusion was found to generalize to all sources of energy via hepatic oxidation (Booth, 1972d; Booth & Jarman, 1976).

The same test on the size of a briefly delayed meal was in turn used to show that rats with ventromedial hypothalamic lesions had normal metabolic satiety (Booth, Toates, & Platt, 1976; Duggan & Booth, 1986). On this basis, the abnormalities in the meal pattern of ventromedial hypothalamic rats implied, in calculations of energy flows around the body (Booth & Toates, 1974), that the main cause of ventromedial obesity is release of gastric emptying from the normal central inhibition during the rat's resting phase (Newman & Booth, 1981). This theory was subsequently confirmed by further simulations that also included abnormal autonomic control of insulin secretion (Campfield, Smith, & Fung, 1982) and by measurements of gastric emptying rate from immediately after placement of the lesion (Duggan & Booth, 1986; Duggan, Storlien, Kraegen, & Booth, 1988). These results have been complemented by the implication of slowed gastric emptying in the satiating actions both of the serotoninergic drug fenfluramine (Baker, Duggan, Barber, & Booth, 1988) and of the gut hormone cholecystokinin when exogenously administered (Moran & McHugh, 1988). This evidence strengthens the proposal that the metabolic effects of the gastric emptying pattern between meals and its central modulation over the 24-hour cycle have a more predominant role in the rat's meal pattern than the neuroendocrine mechanisms commonly invoked (Le Magnen, 1981).

Such progress in the neuroscience of satiety depends on interpreting the experimental control of variations in meal sizes and intermeal intervals as a measure of the operation of those satiating mechanisms that were activated by the manipulation. This contrasts with many proposals (Blundell, 1979; Smith & Gibbs, 1979; Le Magnen, 1985) such as that a decrease in meal size measures some entity called satiation and increase in latency to the next meal is a direct measure of something (different?) called satiety. The state of satiety and its creation in the process of satiation are input–output relations and therefore not definable as any aspect of intake in itself. Intake reductions are but one part of the evidence for the operation of causal processes by which normal effects of ingestion inhibit the tendency to ingest some or all foods. The size of the intake change is not the measure of the strength of those satiating influences; the strength of a satiety mechanism operative within a meal is the proportion of the variation in meal size that is under the control of that influence; the satiating effect after a meal is not the time to the next meal but that time relative to the size of the meal. A neural intervention is acting on satiety only if it alters the measured control of intake by one or more specifically activated mechanisms—an induced change of meal size need have nothing to do with any satiety mechanism. Moreover, it is arbitrary to call a lowering of intake an increase in satiety rather than reduced hunger, or a shorter latency to eat an increase in hunger rather than reduced satiety; until a lot more is known about the mechanisms controlling the tendency to ingest food in normal circumstances, there are no criteria to distinguish increased inhibition from decreased excitation.

CONTROL OF MICROSTRUCTURE. Sensory control of intake patterning within meals was early addressed by a learned dissociation of meal size from initial rate of eating (Booth, 1972b) and from choice preference at the start of the meal (Booth & Davis, 1973). These criteria of sensory control of meal size were subsequently elaborated to choice tests repeatedly during the meal (Booth, 1977, 1980) on the grounds that relative intakes in a choice are likely to be more sensitive than rate of intake of a single stimulus to the momentary facilitation or inhibition of intake response to a stimulus.

Nevertheless, much work has been done on the microstructure of intake of the maintenance diet or a single test diet. Study of the effects of drugs on intake patterning within meals is becoming increasingly productive, as dissociated effects on microstructural parameters are identified (Davis, 1989; Schneider, 1989). To advance behavioral neuroscience, such results should be related to theories of the brainstem controls of ingestatory movements rather than to conceptual labels that belong to the overall disposition, such as appetite or satiation.

The same applies to fine-grain analysis of sensory and postingestional control of bout sizes, such as that pursued by Mook and his students using saccharin solutions and sugars in rats. They have shown that concepts from control theory, such as feedforward and reference value, can help to make sense of otherwise complex data (Mook & Wagner, 1989). A more mechanistic style of behavioral interpretation will be needed if such phenomena are to be coordinated to the neural processing of ingestion. For example, they suggest that saccharin concentration provides a feedforward signal that sets the amount of lapping required to end a bout. Maybe they have evidence that strength of sweet taste provides tonic facilitation to a lapping pattern generator that is gated out by cumulative reafference, something like the inspiratory off switch (Cohen, 1979).

From Phrenology to Network Engineering

The elucidation of neural mechanisms by which an aspect of eating or drinking is generated thus puts great demands on psychological as well as neurobiological expertise. To contribute to definite progress in ingestive neuroscience, the observations have to be clearly interpretable in terms of both behavioral processes and neuronal system processes.

Until the 1960s, the only neural observations that were generally practicable were postmortem anatomy and *in vivo* manipulations limited to more or less localized physical damage to neural tissue. Therefore "physiological" psychology then consisted largely of sometimes quite exquisite analysis of behavior following brain lesions and interpretation of the results in terms of known neural connections.

Unfortunately, whether from lack of attention to the known interconnections or because the theory of functioning networks was so open-ended and therefore untestable, the behavior of the brain-damaged experimental animals was often not attributed to functioning of the remaining parts of the brain but to loss of the function of the damaged parts. With no empirically founded theory of how the neural processes in a destroyed region normally contribute to the organization of behavior, this move becomes a fallacious localization of function. Such phrenology (Booth, 1976) of hunger, thirst, and their satieties still persists in most textbooks and even in some of the rationales for current research.

Increasingly over recent decades, however, cellular and integrative neurophysiology and biochemical, pharmacological, and microanatomic data have accumulated that are sometimes adequate to inform realistic construction of theoretical neuronal networks that can perform specified subprocesses within behavioral and neurovisceral functioning. The best prospects for ingestive neuroscience are the testing of such "wetware" engineering theories of appetite organization by critical experiments with lesions, drug administration, and electronic and biochemical recording of cellular activity. Some examples built on personal experience are given below.

Neuroanatomy of Appetite

Some decades back (and sometimes still now), the approach was to measure intake and body weight (and perhaps blood and body composition, etc.) after making lesions in the brain or cutting peripheral nerves or "stimulating" tissue more or less locally by passing current through an electrode or by administering a chemical through a cannula. Decrease in intake or weight was taken as evidence of the location of an appetite center, and increase in intake or weight a satiety center. The outcome was the hypothalamic dual-center doctrine, which still has its advocates.

This lesion-and-stimulation phrenology of dietary intake has two major weaknesses that have not been equally well appreciated. The better known is that the critical tissue may not be where the intervention is made. Most workers now recognize that the neural basis of ingestive behavior is widely distributed over networks in the forebrain and brainstem as well as in the diencephalon. The more serious weakness is that saying where the organization is done, whether localized or distributed, explains nothing whatsoever even if true; the problem remains as to how the neuronal network achieves the behaviorally observable processing of information. To show that a satiety signal or a palatable stimulus is transmitted to a part of the brain poses the question; it does not start to answer how the brain integrates that input with others for decision when or what to eat. Each tract and each synaptic field around the subsequent network is transmitting or transforming some aspect of the information processing. We must at least characterize the behavioral changes produced by interference at any place in the network.

There was some resistance to the (il)logic of localization of overall ingestive function, even in its heyday. Lesions (e.g., Gold, 1973; Morgane, 1961) and chemical interventions (Booth, 1967, 1968) showed the importance of regions other than the lateral hypothalamus (LH) and ventromedial region of the hypothalamus (VMH) in food intake control. Also, there were early behavioral fractionations of the VMH (Graff & Stellar, 1964) and LH (Teitelbaum & Cytawa, 1965) lesion syndromes. However, this approach finally flowered in descriptive and experimental sensorimotor analyses (Turner, 1973) only at the time the neuroanatomic and behavioral fallacies of the dual-center approach were being comprehensively exposed by aphagic effects of destroying catecholamine transmitter pathways in and out of the LH (Ungerstedt, 1971; Stricker & Zigmond, 1976).

Even if a discrete lesion were shown to disrupt a specific behavioral mechanism, however, that still would not locate the neuronal engineering of that process at that place in the brain. Even a convergence of evidence from stimulation and recordings as well as lesions does not identify a brain center. This was early evident in the neuroscience of air hunger, a sadly neglected source of what one might call inspiration for ingestive neurobiology (Marler & Hamilton, 1966). The medullary regions postulated to be inspiratory and expiratory centers on the basis of opposite effects of transections and electrical stimulation lesions were later shown by unit recording to be merely efferent pathways from the neuronal networks organizing normal automatic breathing that extend up though the pons (Cohen, 1970).

Similarly to central respiratory control, Ungerstedt (1971) showed that destruction of nigrostriatal pathways outside the LH lesion/stimulation "appetite center" caused the same pattern of aphagia with recovery. This largely vindicated Morgane's (1961) early distinction between motivational and motor effects of LH lesions. Some of the integration of both excitatory and inhibitory influences on eating behavior in the normal mammalian brain does indeed seem to be located in and near lateral parts of the hypothalamus, but this is alongside and in interaction with several other regions such as neocortex, amygdala, and the brainstem. Furthermore, much of the involvement of the LH region in ingestion and in the control of instrumental behavior ("reward" effects) is probably as part of the reticular arousal system (Booth, 1976). Lateral hypothalamic lesions not only destroy dopaminergic tone for sensorimotor sequencing by the basal ganglia (Ungerstedt, 1971); they also destroy noradrenergic and serotoninergic afferents that optimize functions for perceptual discrimination, cognitive integration, and differentiation of responses carried out in cerebral neo- and archicortex. Le Magnen (1985) plausibly ascribes functions in eating and its learned control to the lateral hypothalamus that are those of Hull's general drive, i.e., nonspecific activation of behavior by tissue needs (Booth, 1987b). The subtleties of drive, reinforcement, and motivational selection among learned responses continue to defeat analyses of the laboratory phenomenon of electrical self-stimulation of the brain (Booth, 1987c); the recent chemical extensions of reward neuroscience have yet to receive the necessary behavioral analysis to address these issues. Either reinforcement or performance effects of dopaminergic activity in the nucleus accumbens may depend on specificity provided by input from the amygdala (Yim & Mogenson, 1980).

Also reminiscent of the respiratory centers, Gold (1973) showed that the obesity from VMH lesions could be produced by cutting pathways lateral and dorsal to the ventromedial nucleus of the hypothalamus. It had already been shown that VMH rats responded normally or supernormally to various satiety tests (Rabin, 1972), and so there has long been no justification for interpreting effects in the ventromedial nucleus of the hypothalamus as relevant to satiety (Booth *et al.*, 1976). This nucleus has impressive sensitivities to local and distant nutritional signals. Yet the evidence is that the VMH obesity is based on excessive deposition of fat resulting from autonomic disinhibition of insulin secretion (Le Magnen, Devos, Gaudilliere, Louis-Sylvestre, & Tallon, 1973; Powley, 1975) and/or of gastric emptying (Duggan & Booth, 1986; Duggan *et al.*, 1988), sometimes releasing hyperphagia and thereby fueling frank obesity but having nothing to do with the direct control of ingestive musculature.

The most productive use of behavioral or physiological effects of brain or nerve lesions, however small and however specific to a subnucleus, tract, or transmitter, would be to test preformulated theories of network connections and processing that have been based on synaptic and transsynaptic anatomic and electrophysiological evidence. This has been eminently practicable in those invertebrates that have ingestion-controlling ganglia containing large, functionally identifiable neurons. In mammalian brains, the specialization of pathways and regions can sometimes be similarly exploited. For example, the coordination of neurophysiological recordings, connection anatomy, and discrete lesions or complementing ablations has proved useful in narrowing down the search for the mammalian memory trace in the favorable case of a learning task such as auditory–airpuff nictitating membrane conditioning in the rabbit (Thompson, 1986).

It may be easier to make progress in mammalian neuroscience if the task is "simplified" by limiting the control of behavior to a part of the brain (Teitelbaum, 1977). Taste reflexes and their satiability (or at least their dearousal) have been demonstrated in decerebrate rats, for example (Grill & Norgren, 1978). This tactic will carry forward our understanding of the neural substrates of behavior if the tests of input–output function are designed to distinguish alternative ideas of how the networks are processing in this remainder of the brain.

DAVID A. BOOTH

The most common current variant of the inadequate logic of localization of function is explicitly or implicitly to seek a "chemical coding of behavior." In this sort of design, drugs are injected, and dietary intakes are measured. Two lines of work in which I have been involved illustrate how much more productive it can be to carry out behavioral analyses and to interpret the results in the light of neural processing around the brain and indeed the periphery. Instead of thinking of the neurotransmitter or its receptor unmechanistically, e.g., as augmenting or blocking an entity called satiety (or carbohydrate craving), the question asked is whether the subsynaptic neurons or their immediate connections are carrying out a detectable particular piece of information processing that explains the effect of the transmitter agonist or antagonist on overall ingestive behavior (Booth, 1982).

SEROTONIN, SATIETY, AND SELF-SELECTION. When the relevant behavioral and physiological analyses are done on the effects of the intake-suppressant serotoninergic (5-HT) drug fenfluramine, for example (Booth, Gibson, & Baker, 1986), it turns out that the drug's suppression of food intake in the freely feeding rat may not be via action in any central hunger-specific pathway (Booth & Baker, 1988). Much of the decrease in intake is attributable to disruption of brainstem sensorimotor reflexes (Baker & Booth, 1990) and to a prolongation of the appetite-suppressant effect of the previous meal by peripheral action on gastric emptying (Baker et al., 1988).

The same lack of attention to the behavioral processes involved in the effects of drugs on dietary intakes has vitiated recent attempts to study the neurotransmitter mediation of nutrient self-selection in rats or people (Booth, 1987d). Experimenters have neglected to allow for the basic fact that the immediate control of dietary intake is vested in the sensory characteristics of the diet(s). If the organism is really selecting a diet for its nutrient content, some sensory quality has to be used to predict the nutritional effect. When, for example, genuinely protein-specific dietary selection is measured, by bringing the learned sensory cue under the experimenter's control, then neither fenfluramine at the originally claimed doses nor indeed the old catecholaminergic (CA) appetite-suppressant amphetamine was found to alter relative intakes in rats (Gibson & Booth, 1988; Booth & Baker, 1990). Rather, the effects of fenfluramine on selection between casein and dextrin (the nutrient preparations generally used in the experiments with rats) were attributable to the drug's effects on textural preferences, presumably arising from changes in oral sensorimotor control.

Exclusive pursuit of the 5-HT and CA central transmitter hypotheses of carbohydrate and protein selection continues to delay advance in this field. Progress depends on further widening of renewed interest in the older and much better evidenced hypotheses of learned sensory control established by glucoreceptor and amino acid balance-detecting mechanisms in the brain (if not also by chemospecific receptors in the duodenal wall).

For solid advance in the neurobiology of nutrient selection, it will also be necessary for investigators to design their experiments to make sure that they are measuring effects on or of ingestive behavior that is directed by dietary stimuli that the eater is using to predict the nutritional effects of interest and not by other irrelevant stimuli.

In short, elementary behavioral and physiological analyses of the effects of a serotoninergic drug on dietary intakes have comprehensively refuted the notion that 5-HT has a predominant role either in central mechanisms of satiety or in macronutrient self-selection.

NORADRENERGIC SATIETY GATING IN ANTEROMEDIAL HYPOTHALAMUS. Another famous piece of ingestive neuropharmacology is the noradrenergic feeding effect discovered in the rat by Grossman (1960). It provided the first piece of chemical neurobiology to undermine the dogma of a lateral hypothalamic feeding center (Booth, 1967).

From the increase in size of a meal being eaten by the rat when very low doses of norepinephrine (NE) were infused into the hypothalamus (Ritter & Epstein, 1975) and little or no intake-stimulating effects in food-deprived rats, it is widely thought that NE action at this site blocks satiety. Yet it says nothing mechanistic to dub NE there an antisatiety transmitter; the question whose answer could take us forward is what the neurons under these noradrenergic synapses are normally doing that affects ingestive behavior and is altered by NE action.

My first publication in psychology was a primitive version of the relevant behavioral analysis of the NE feeding effect (Booth & Quartermain, 1965). This subsequently proved to be flawed (Sclafani & Toris, 1981) but still permits the conclusion that aversive tastes block the effect and hence that the centrally administered NE cannot be facilitating intake by interfering with an unpalatability of the test diets. Later, more detailed behavioral analysis (Matthews, Gibson, & Booth, 1985) also showed that the injected NE could not be stimulating intake by increasing the palatability of a sweet taste or of a calorically conditioned taste or smell preference. Also, NE injection cannot override the immediate duodenal bloat produced by ingestion of a large liquid meal (Gibson & Booth, 1986). Given no stimulation of hunger, this leaves only a blockade of one or more normal satiety signals to explain the NE effect on intake.

Effects of an NE injection that elicits chow intake on a choice between carbohydrate-conditioned flavors are consistent with a disruption of conditioned satiation (Matthews et al., 1985). The paraventricular nucleus and adjacent anteromedial hypothalamus, where NE injection elicits most chow intake (Leibowitz, 1978; Matthews, Booth, & Stolerman, 1978), has projections from the medullary visceral and taste relays (as well as to them and to autonomic motor nuclei) and also from ventral forebrain olfactory and learned preference systems; indeed, PVN neurons respond to both gastric and olfactory stimuli (Guevera-Aguilar, Jimenenez-Montufar, Garcias-Diez, Wayner, & Armstrong, 1988), as a network processing conditioned aroma satiation would have to. Thus, it is plausible to suppose that NE synapses gate out learned satiety after the convergence of the conditioned dietary and gastrointestinal cues.

It has been claimed that the stimulation of intake by PVN NE administration is carbohydrate specific. On the above theory, this would have to be because the somatic cues for the gated conditioned satiation arise from glucoreceptors such as those in the intestinal wall. The nutrient specificity of carbohydrate-conditioned satiation has yet to be determined. The latency of the feeding effect may be long enough to permit an insulin secretory loop (Gold, Jones, Sawchenko, & Kapatos, 1977) to create a carbohydrate need, disrupting the repletion signaling on which carbohydrate-specific conditioned satiation would depend.

Such a mechanism would not be in the NE projection but in the neural network on which it synapses in that particular region of the hypothalamus. Blockade of conditioned satiety could be one of many effects of a diffuse noradrenergic arm of an ascending arousal system that becomes particularly active in the rat as dusk approaches, for example (when the NE feeding effect seems to be greatest, and also carbohydrate intake: Leibowitz, Weiss, Yee, & Tretter, 1985). On this analysis, there is little basis to ascribe any broad metabolic function to the central NE projections. The behavioral and autonomic functions may be a general activation.

CENTRAL PEPTIDES, PHARMACOLOGY, AND INGESTIVE NEUROBIOLOGY. With the development of peptidergic pathway-tracing techniques over the last decade, peptides have come to seem to many to be the wave of the future for the neurobiology of ingestion and many other brain functions. If the same mistakes are made as with the monoamines, though, the peptides wave will also break on the beach of chemophrenology (Booth, 1976) and trickle into its sands. To switch the marine metaphor, the peptide (or the monoamine) should not just float in the head on verbal seaweed such as palatability, satiety, stress, or vagal activity or the notion that peptides have the same function in the brain as they do in the viscera (maybe similar only because of the same embryological origin). Rather, any central transmitter should be part of some specified machinery of a boat demonstrably on course to an observable destination such as ingestion.

This is a bright prospect for neuropharmacological and neurochemical contributions to ingestive neurobiology. Jointly behavioral and pharmacological or biochemical analyses would be related to a particular field of synaptic receptors involved in an electrophysiologically identifiable transformation of information. The old but persisting assumption that a particular transmitter predominates in the network controlling one behavior and not in the networks controlling any other behavior would finally be abandoned; it has never been plausible in the face of basic neuroanatomy and integrative neurophysiology, let alone the mapping of transmitters and receptors.

We might advance pharmacology by simply testing effects of drugs on intake, but not behavioral neurobiology. Similarly, neurochemical measurements made in different states of deprivation or refeeding, or even during sensory or postingestional stimulation or the emission of particular behavior, do not provide mechanistic information unless there is enough else known for the result to resolve alternative hypotheses as to what is going on in that bit of the brain. The challenge is to extend the experiments to include both behavioral and physiological analyses that distinguish alternative mechanistic hypotheses about the observed effects (Booth *et al.*, 1986), as illustrated above. Then, dose–response functions could be used to dissociate different receptor-specific actions on discrete processes within ingestive behavior (Booth, 1989, 1990a). This requires us to vary the normal operation of a distinctly localized and functionally characterized class of synapses, either pharmacologically, measuring the change in the behavioral transform carried out by that synaptic field, or behaviorally, observing the transmitter activity biochemically. In contrast, it cannot logically tell us anything about the normal way in which neural networks mediate behavior if we confine our interpretation to the intake or even the behavior observed when the networks are disrupted by blockade or destruction of a class of receptors or by stimulating them outside their normal range of functioning.

Opioid psychopharmacology illustrates the point well. The analgesias were found to have conditioned and unconditioned mechanisms, and some of each type were disrupted by opiate antagonists while others were not (Terman, Shavit, Lewis, Cannon, & Liebeskind, 1984). So the effects of drugs on a standard test for analgesia are neurobiologically uninformative; we need tests on discrete controlled mechanisms of analgesia. Enkephalinergic synapses might also be involved in stress-induced ingestion (McCarron & Tierney, 1989) or emotional overeating. Close behavioral analysis will be equally necessary to elucidate the neural mechanisms of these phenomena.

Neurophysiology of Appetite

Crucial to all future advances are unit recordings of neurons with known functional and anatomic connections in relation to measured brain inputs and analyzed ingestive behavior. Only such data can establish theories of the inter-neuronal transformations that "engineer" the organization of ingestive behavior toward a diet in the physiological and external context. Yet unit neurophysiology by itself or in mere correlation to behavior, when it is not designed to test a definite theory of the role of brain processes in behavior processes, becomes mere "electrophrenology" (Booth, 1976).

Tastes. Gustatory neuroanatomy and neurophysiology have been and will continue to be one of the major routes into the brain mechanisms of ingestion. The true tastes may not be as overwhelmingly important in eating and drinking as generally assumed. Nevertheless, they are easier dietary stimuli to control than aromas, oral textures, many aspects of visual appearance, and even the stimuli of astringency, mouthburn, etc. Also, there are innate reflexive motor patterns in response to tastants that are readily observable both to direct visual inspection and in their effects on maintained fluid consumption. As noted later, there are also powerful innate reflexes to mechanical stimulation within the mouth that have major effects on intake, but these are often harder to observe by eye, and their stimuli are harder to control, usually being modified or even generated by mastication.

Recent progress along the taste "pathway" will be accelerated by avoiding overrestrictive conceptual definitions and premature functional hypotheses.

For example, palatability and unpalatability should not be defined as ingestive and egestive mouth movements any more than they should be operationalized as composition of the diet, such as its content of glucose or of sucrose octaacetate. The palatability of a diet is its facilitatory influence on dietary choice and consumption. The role in palatability of the fixed action pattern elicited by a taste of sucrose solution in the rat or the newborn human baby and of its substrate in brainstem network activity will have to be determined empirically for each important instance of ingestive facilitation.

Furthermore, the generality of the facilitatory effect on ingestion of a dietary characteristic across a species or across situations within an organism should not be merely assumed, as it conventionally has been. The stability of the "palatability" of a diet, even of its taste, is also a wholly empirical matter. Indeed, when examined, facilitation has been found to be highly context dependent. In the specific hungers, sensory preferences among diets depend on nutritional need state (Baker *et al.*, 1988; Denton, 1985). In conditioned satiation, the learned flavor preference depends entirely on the learned level of gastrointestinal fill (Booth, 1972b; Gibson & Booth, 1989). One short-term mechanism of satiation is a decrease in palatability that is specific to the stimulus characteristics of recently ingested diets (Le Magnen, 1956; Rolls, 1976; Rolls, Rowe, & Rolls, 1982; Rolls & Wiggins, 1989).

The plasticity of the palatability of particular tastants is well demonstrated by the readily observed ingestive movements in response to saccharin in the rat. They can be replaced partly by egestive movements after the saccharin taste has been paired with the effects of injecting a lithium salt (Berridge, Grill, & Norgren, 1981); taste preference conditioning by morphine is also measurable in these motor patterns, and caloric conditioning of the sweetness preference peak (Booth et al., 1972; Conner, Haddon, Pickering, & Booth, 1988) should also be tested. The dependence on salt deficiency of these reactions to the taste of salt solutions has also been shown (Berridge, Flynn, Schulkin, & Grill, 1984), but the learned state dependence (on empty or full gut) and food habituation satiety remain to be measured in these movements. In each case that depends on forebrain modulation of the brainstem pattern generator activity, the microanatomic neurophysiology of the more complex circuitry will have to be elucidated, which would be a major advance in ingestive neurobiology.

This direction of advance is well illustrated by recent neocortical unit work in monkeys on the phenomenon of taste-specific satiation (Rolls, 1976), a part of so-called alliesthesia (Wooley, Wooley, & Dunham, 1972). There is, in fact, no change in the sensation of sweetness but an attenuation of the ingestive response to the sweetness repeatedly presented shortly beforehand, a habituationlike phenomenon (Booth, 1976) that would be independent of the general or nutrient-specific postingestional influences on intake or choice. Thus, one would not expect a change in the neuronal basis for taste discrimination during taste-specific satiation but rather a change in some taste-specific activity controlling the motor output or the attitude to food that guides it. The first taste relay showing the satiation is in fact a higher-order (association) cortical region (Rolls & Wiggins, 1989).

Unfortunately, unrealistic functional hypotheses about taste have sometimes diverted research from analysis of precisely what discriminations, preferences, habituation, associative learning, etc. the organism has in fact achieved and at the level of timing and afferent and efferent selectivity that is directly relevant to the neuroscience. A major example is the hedonics of sweetness.

A great deal of human appetite research has been taken up with the most implausible suggestion that plain sweetened water generates a pleasure that is suppressed by the postingestional effects specifically of carbohydrates, that this has the biological function of regulating energy balance, and that a weakness in such a mechanism causes obesity (Cabanac, 1971). Of course, it has always been believed that the suppression of interest in food by eating (satiety) has a negative feedback function in moderating caloric intake. Sweet water is not a food for most people, however, nor even for the rat until it has learned that a level of sweetness signals a supply of energy (Booth et al., 1972).

There is in addition a simpler and almost universal assumption that has been in danger of distorting design and interpretation of the behavioral neurophysiology. This is the idea that the sweet preference exists because sweetness has helped species to prosper by signaling sources of calories to them. Sugars simply are not abundant or critical enough in nature for this intuition to make scientific sense. A hungry primate does not need a sweet preference to encourage it to eat soft, colorful, ripe fruit. Rather few mammalian species can have survived by raiding bees' nests. It has been suggested that explanation be sought rather in the need to be averse to the taste of nitrogenous plant toxins but not to reject the taste of proteins, peptides, and amino acids, and particularly not to spit out mother's milk (Booth *et al.*, 1987; Booth, 1990c)!

This caloric signaling assumption, coupled with an expectation of finding solely receptor integration at the first sensory relay, has affected the interpretation of taste projections to the nucleus of the solitary tract (NTS). In the gustatorily naive rat, both the location and firing pattern of sweet units are distinct from those of bitter, sour, and salty units (Scott & Mark, 1986). This has been attributed to nutritional functions of the sweet taste that have been questioned above. There is a sounder, simpler, and mechanistic alternative. There are very few synapses in the shortest reflex pathways involved in the ingestive and egestive (expulsive) movement patterns in response to tastes. Both those motor pattern generators and the more broadly tuned motor networks in the reticular formation (Siegel, 1979) are likely to have phasic and trophic influences on the activities and connection patterns within the NTS. Hence, the grouping of NTS units by receptor type may simply be efferent control of sensory processing by motor reafference, the sort of network elaboration that should be expected in the mammalian medulla. Indeed, when an aversion is conditioned to a sweet stimulus, some sweet units then appear that are grouped with the innately aversive stimuli (Scott & Mark, 1986). Any satiating or appetizing mechanism would be expected to affect the taste units if they are innervated by reafference from in-/egestive networks. Therefore neurophysiological observations of this sort are not evidence that sensory processing at the first relay contributes to the control of ingestion.

VISCERAL AFFERENTS. Afferents to this same large nucleus come from the chemoreceptor zone just dorsal to it on the floor of the fourth ventricle and also from visceral branches of the vagus, including the vasculature, the lungs, the wall of the gastrointestinal tract, and the liver. Since the primary relays are topographically distributed through the nucleus, there is not the simple neural basis that some have assumed for the visceral conditioning of taste aversions. More relevant to normal ingestion is the possibility that visceral afferents contribute to the caloric reinforcement of taste and odor preferences (Baker & Booth, 1990; Booth *et al.*, 1972; Tordoff & Friedman, 1986) and vagal contributions to the somatic elements in food-specific learned meal-size control (Booth, 1972b; Deutsch, 1983; Gibson & Booth, 1989).

TEXTURES. The reflexes to taste are adapted to oral uptake or expulsion of material. This makes them rather obvious to external observation. The oral reflexes to food texture concern the preparation of fluid and particularly solid material for swallowing and digestion. Movements of the jaw are externally visible, but their coordination to movements of the tongue and indeed to the complicated changing pattern of mechanical stimuli generated by the food cannot be directly seen even when the lips are parted. Electromyography and fiberoptic probes gain better access, but physiological models of mastication based on a good understanding of both neuromotor organization and food physics will be necessary for theoretically productive interpretation of the data.

It is clear from neurophysiological and pharmacological studies of the sensorimotor networks in the brainstem that the mechanoreceptor reflexes are stronger and more refined than the taste reflexes. Several times in each mouthful, the muscles in the check snap tight to push food between the open teeth, normally without risking a bite on the cheek itself or indeed on the tongue. Tuning of the powerful inhibition on jaw closing from increasing pressure on the teeth presumably contributes to sensations of food crunchiness and rubberiness while regulating the breakdown of a great variety of solids into swallowable fragments or boli. The tongue wipes the food–saliva mixture on the palate, giving mechanoreceptor afferent patterns that indicate volume, slipperiness, and viscosity, including inhomogeneities and nonlinear stress–strain relations, from which are constructed sensations of fullness, smoothness, thickness, graininess, and "body."

Zeigler, Jacquin, and Miller (1985) justly accused the neuroscientists of food and fluid intake of "sensorimotor neglect," a neatly barbed pun on one of the few systematic programs of behavioral analysis that this research field has seen, the neurological testing of lesion-induced aphagia referred to earlier. Again, it would be no wiser to invest tactile fixed action patterns than taste reflexes with palatability or appetite, but they cannot be relegated to mere automaticities. Activity within the brainstem pattern generators is likely to be not just modulated by but highly integrated with midbrain and forebrain activity, not to say cerebellar processing.

The dangers of neglecting texture are shown by work on 5-HT mentioned earlier. The reduction in meal size induced by fenfluramine is likely to arise in part from interference with mastication, sometimes evident as a slowing of eating rate (Burton, Cooper, & Popplewell, 1981) or even a broad sedative effect (Booth et al., 1986). The same effect on serotoninergic synapses in ingestatory pattern generators explains the fenfluramine-induced shift in intake from dextrin to casein, since that shift interacts with purely textural differences between diets (Baker & Booth, 1990) and fails to occur with genuinely nutrient-selective intake, even when texture cued (Booth & Baker, 1990). Only if we are measuring the influence of texture on intake can we determine whether or not the neural manipulation is affecting that part of appetite and not having some other, perhaps extraneous, effect on intake.

AROMAS. Study of the smell of food is also technically more difficult than that of its taste. The greatest difficulty, as with texture, is control of the stimulus. The relative complexity of the olfactory world can be exaggerated, since even in gustation the labeled-line approach to four basic tastes has had to give way to pattern approaches, plastic and nonlinear. In addition, the dynamic nature of olfaction is insufficiently appreciated, or rather is incorrectly focused on the sniff (orthonasal stimulation) rather than the swallow (retronasal stimulation). As with texture (and taste), the neurobiology of olfaction might advance faster if the naturally produced stimuli that are important in ingestive behavior were studied as well as high-precision artificial stimuli.

Unit recordings are accumulating that reflect the well-known behavioral phenomena of integration of aroma with taste in flavor. Macroelectrode recording in olfactory bulb as well as hypothalamus showed that neural activity discriminating between odors was modulated by nutritional state according to the learned metabolic significance of the aroma (Pager, 1984); as it happens, much of the food-specific appetite and satiety conditioning in rats has used olfactory

cues (Baker *et al.*, 1988; Booth, 1972b, 1977, 1980; Gibson & Booth, 1988, 1989).
Anatomic and neurophysiological analyses of olfaction may be advanced by a
convergence with the conventional sensory approach from a movement "top
down" from behavioral evidence of how aroma is integrated into the control of
ingestion.

VISUAL APPEARANCE. Visual science is perhaps the most advanced field of
behavioral neurobiology, and so it would be good for ingestive neuroscience if
bridges could be built. However, the neuroscience and indeed the cognitive
science of vision are least advanced in just that area where it overlaps directly
with ingestion, namely, object recognition and recognition of a food by its
appearance.

Rolls' (1984) program of primate neurophysiology is indeed focused on this
overlap, spanning recognition of visual objects such as blocks and faces and the
multimodal perceptual control of ingestion. Foraging and ingestive choice involve the general problems of learned object recognition; unlike faces, natural
food objects are so diverse in form and color that they are unlikely to have innate
discriminative factors. Roll's group has found diencephalic and cortical units that
respond to visible differences in food shapes and/or colors and to combinations
(additive) or configurations (nonlinear) of visual and olfactory or gustatory features. Somatosensory units are common but cannot be analyzed in relation to
visual texture or any other modality without some understanding of the tactile
patterns produced by food in the mouth. Substantial progress in this area will
presumably depend on development of experimentally useful principles of form
perception and on ability to track neural changes involved in learning the combination of color and other cues that distinguishes a food.

PHYSIOLOGICAL FUNCTIONS OF NEURAL CONTROL

FUNCTION AND MECHANISM

We have seen some limitations of intuitions about the adaptive value of the
sweet preference, general roles for gut–brain peptides, and other aspects of
ingestive behavior. A scientifically and mathematically more disciplined approach to evolutionary functions has been developed in behavioral ecology.
These functional analyses concern the consequences of a behavior for the reproductive fitness of members of a species within past and current ecologies.
This approach to ingestive and other behavior has had to distinguish itself from
both psychological and physiological interests in how organisms work at the
moment. Fighting for the distinction has given some functional analysts an anti-
mechanistic attitude that will have to be reversed before their theories about the
processes of natural selection can be useful in guiding behavioral neurobiology
in any concrete detail.

For example, the costs of time or energy spent acquiring food or water and
the risks during foraging of becoming prey oneself clearly play major roles in the
ongoing control of ingestion in some circumstances (Collier, 1980). This, however, does not preclude parallel influences from—or even mediation of the
ecological effects by—physiological need and its repair. Furthermore, functional
behavior, like any other behavior, must rely on neural processing. Thus, for such

functional analyses to be relevant to ingestive neuroscience, a mechanistic analysis has to be provided of the stimulus configurations that are perceived as work or as risk in ways that affect meal sizes or intervals or dietary preferences as observed.

Another intuitive evolutionary hypothesis that has been widely held in ingestion research is that behavior is ruled by homeostatic targets in the internal milieu that exist because of their contribution to fitness of the species in past environments. Yet it is by no means obvious that such containment of physiological parameters within the ranges in which tissues work well has had benefits that are worth the costs. Yet the presumption of homeostasis hangs on that.

ROLES OF FOOD SELECTION IN HOMEOSTASIS

Making the usual assumptions, nevertheless, that there are some regulated physiological variables, to what extent does ingestive behavior contribute negative feedback to moderation of their deficiency or excess? The assumed or demonstrated answer to this question will influence the research strategies, such as whether and how satiety and hunger signals are sought.

Useful answers to this question are not provided by mere observation of compensatory responses to regulatory challenges. Signs of negative feedback control may or may not be seen, depending on what set of mechanisms a particular challenge has activated. Positive or negative results are not illuminating unless the mechanisms by which the effects of the challenge were detected and responded to are analyzed as well (Booth, 1972c).

As argued earlier, similar controls are likely to be affecting decisions to start and to stop eating a single diet when it is the only one available as well as to influence the selection among diets when two or more are present. Nonetheless, a considerable variety of CNS mechanisms could decide the activation and orientation of ingestive movements and the activities preparatory and consequent to the intake.

If the ecologically valid dietary selection in the face of a somatic deviation can be sensorially specified, then the ingestive behavior can be "hard wired," e.g., an osmotic threshold for facilitation of fluid ingestion. The almost universal assumption that responses to hunger signals are equally fixed is rather implausible, however, because both the dietary and the somatic situations for nutrients other than water, even just for energy from any nutrient, are so much more complicated than for water. This may be the main reason why our understanding of hunger signals seems so far behind that of signals from water deficit to drinking behavior. Also, for both drinking and eating, satiety signals cannot be fixed to the regulated variable. A suitably flexible system would connect the signal of deficit or excess to dietary and environmental cues in a configuration that facilitates ingestion when those dietary cues have been followed by repair of the deficit or reduction in the excess and inhibits ingestion when no such correction followed (Booth, 1972b, 1977, 1980). Such learning of somatic-state-dependent sensory preferences or aversions would adapt behavior to exploit the ecology to help sustain homeostasis of any detected variable capable of accessing the appropriate reinforcement system. The behavioral neuroscience of eating should be oriented to brain organization that could achieve such learned regulatory behavior.

Nevertheless, such deviation-correcting mechanisms are likely to be only some of the influences on ingestive behavior. Even on evolutionary grounds, ingestion would be expected to serve other functions beside homeostasis, and not every aspect of behavior or any other phenotype has to have had survival value. Furthermore, unlearned negative feedbacks and learnable ones cannot always act perfectly adaptively and healthily in all environments. This is least likely of all in biologically unprecedented environments such as those that humankind has created for itself through urbanization and then industrialization.

Intake and individual meals result from the cumulative operation of negative feedbacks, positive feedbacks, feedforward, and stochastic processes. It therefore makes no sense to try to categorize particular motor patterns or proportions of intake or of a meal as homeostatic and the remainder as non-homeostatic. It is a particular mechanism, not the outcome of its operation, that may in defined circumstances generally have or not have a regulatory role.

NUTRITION AND NEUROSCIENCE

The development of both the neurosciences of ingestion and the nutritional sciences has been retarded by confusions about homeostasis of metabolism and body composition and by unrealistically reductionist attitudes to behavior. Unfortunately, such muddying of the waters still persists where these fields of biomedical research and professional practice overlap.

This chapter and this volume illustrate the many fine opportunities provided by the neurobiology of ingestive behavior, if approached realistically, to advance integrative neurophysiology and, through it, quite possibly some aspects of molecular neuroscience such as pharmacology and genetics. However, the objective of advancing neuroscience is not necessarily coterminous with objective of advancing human or animal health. It cannot be assumed that the pharmacology of an anorexia or the genetics or neuroanatomy of an animal obesity will have any theoretical bearing, let alone any practical implications, for human weight control.

To have a right to expect ingestive neuroscience to contribute to understanding in the nutritional sciences, the aspects of ingestive behavior and of their physiological underpinning must to some extent be directed toward transferability to the health-relevant situation. The normal role of the human brain in everyday ingestion will be most relevant to applied nutrition and even much clinical nutrition. Hence, for nutritional relevance, animal models and experimental designs should illuminate normal mechanisms. Even psychiatric disorder often may not involve CNS abnormalities so much as functionally extreme conjunctions of normal operation. In any case, rather few of the problems of nutritional concern arise from abnormally disturbed mental states, even taking fully into account the eating disorders and the more dramatic weight changes during the course of depression (Booth, 1989).

Thus, normal central involvement in satiety and in the learned control of ingestion would therefore be the most obvious area in which to seek to develop contributions from neuroscience to nutrition. Within that framework, we should be alert to potential relevance in even the most obscure phenomenon.

For example, the idea that rats prefer to eat carbohydrate as they begin their active period was mentioned earlier, with some pharmacological interventions. One possibility was that noradrenergic arousal disinhibits the declining desire to eat as a meal progresses. This suggests a novel purely behavioral mechanism for emotional bingeing, a common problem among dieters, even without the severe symptoms of bulimia nervosa. Also of course, the relevant experimental paradigms in the rat suggest some systemic and central processes by which this disinhibition of satiation by normal arousal could be mediated in people. The pharmacology or the anatomy is not the point; they are tools that might be useful to dissect out the brain mechanisms of health-relevant behavior.

CAUSAL ANALYSIS IN INGESTIVE NEUROSCIENCE

Recapitulating then, in conclusion, the manner in which a complex system such as an organism relates to its environment can be described as a structure of causal processes relating input to output, regardless of the internal physical workings of the system. In the case of behavior, this "black box" causal organization has been known traditionally as the mental processes in the acts and reactions of the organism and more recently as behavioral and cognitive processes (which differ not in kind, only in complexity, with no agreed dividing line). Selection of a food is behavior, and so its physical bases in neuronal interactions and their connections to the environment cannot be addressed without analyzing the exact causal organization of choosing among foodstuffs. Measurement of output alone, or of input alone, clearly cannot provide the minimum data required for elucidating the causal structure of input–output relationships.

Behavioral neurobiology can advance only by gathering evidence that relates the organization of behavior to the causal processes within the nervous system. This means that the manipulation or the measurement of a single neural factor—a region of the brain, a class of transmitter receptors, or the firing of a neuron, for example—must be designed to test theory about the neuronal network processes mediating the aspect of behavior of interest. Likewise, behavioral observations and interpretation must relate some inputs to the nervous system to some outputs from it at the level of detail, specificity, and timing that bears on the neural processing being considered. Only the conjunction of those physiological and psychological sciences is capable of advancing our understanding of how observable relationships within behavior are neurally engineered.

Thus, neither the physiological organization of neuronal networks nor the causal processes within the mind of the organism can be neglected. Some of the neuroscience of ingestive behavior meets the demanding criteria. The more studies and research programs take an approach that is truly physiological and integrative, and properly psychological and behavior-analytic, the brighter are the prospects for ingestive neurobiology.

REFERENCES

Allison, J., & Castellan, N. J. (1970). Temporal characteristics of nutritive drinking in rats and humans. *Journal of Comparative and Physiological Psychology, 70*, 116–122.

Baker, B. J., & Booth, D. A. (1990). Effects of *dl*-fenfluramine on dextrin and casein intakes influenced by textural preferences. *Behavioral Neuroscience, 104*, 153–159.

Baker, B. J., Booth, D. A., Duggan, J. P., & Gibson, E. L. (1987). Protein appetite demonstrated. Learned specificity of protein-cue preference to protein need in adult rats. *Nutrition Research, 7*, 481–487.

Baker, B. J., Duggan, J. P., Barber, D. J., & Booth, D. A. (1988). Effects of *dl*-fenfluramine and xylamidine on gastric emptying of maintenance diet in freely feeding rats. *European Journal of Pharmacology, 150*, 137–142.

Berridge, K. C., Flynn, F. W., Schulkin, J., & Grill, H. J. (1984). Sodium depletion enhances salt palatability in rats. *Behavioral Neuroscience, 98*, 652–660.

Berridge, K. C., Grill, H. J., & Norgren, R. (1981). Relation of consummatory responses and preabsorptive insulin release to palatability and learned taste aversions. *Journal of Comparative and Physiological Psychology, 95*, 363–382.

Blundell, J. E. (1979). Is there a role for serotonin (5-hydroxytryptamine) in feeding? *International Journal of Obesity, 1*, 15–42.

Booth, D. A. (1967). Localization of the adrenergic feeding system in the rat diencephalon. *Science, 158*, 515–517.

Booth, D. A. (1968). Amphetamine anorexia by direct action on the adrenergic feeding system of rat hypothalamus. *Nature, 217*, 869–870.

Booth, D. A. (1972a). Satiety and behavioral caloric compensation following intragastric glucose loads in the rat. *Journal of Comparative and Physiological Psychology, 78*, 412–432.

Booth, D. A. (1972b). Conditioned satiety in the rat. *Journal of Comparative and Physiological Psychology, 81*, 457–471.

Booth, D. A. (1972c). Some characteristics of feeding during streptozotocin-induced diabetes in the rat. *Journal of Comparative and Physiological Psychology, 80*, 238–249.

Booth, D. A. (1972d). Postabsorptively induced suppression of appetite and the energostatic control of feeding. *Physiology and Behavior, 9*, 199–202.

Booth, D. A. (1976). Approaches to feeding control. In T. Silverstone (Ed.), *Appetite and food intake* (pp. 417–478). West Berlin: Abakon Verlagsgesellschaft/Dahlem Konferenzen.

Booth, D. A. (1977). Appetite and satiety as metabolic expectancies. In Y. Katsuki, M. Sato, S. F. Takagi, & Y. Oomura (Eds.), *Food intake and chemical senses* (pp. 317–330). Tokyo: University of Tokyo Press.

Booth, D. A. (1978). Prediction of feeding behaviour from energy flows in the rat. In D. A. Booth (Ed.), *Hunger models: Computable theory of feeding control* (pp. 227–278). London: Academic Press.

Booth, D. A. (1979). Is thirst largely an acquired specific appetite? *Behavioral and Brain Sciences, 2*, 103–104.

Booth, D. A. (1980). Conditioned reactions in motivation. In F. M. Toates & T. R. Halliday (Eds.), *Analysis of motivational processes* (pp. 77–102). London: Academic Press.

Booth, D. A. (1982). Book review: *Drugs and appetite*, T. Silverstone (Ed.), *Appetite, 3*, 377–378.

Booth, D. A. (1987a). How to measure learned control of food or water intake. In F. M. Toates & N. E. Rowland (Eds.), *Feeding and drinking. Techniques in the behavioral and neural sciences* (vol. 1, pp. 111–149). Amsterdam: Elsevier.

Booth, D. A. (1987b). Book review: J. Le Magnen, *Hunger. Animal Behaviour, 35*, 311–312.

Booth, D. A. (1987c). Book review: J. R. Stellar & E. Stellar, *The neurobiology of motivation and reward. Psychological Medicine, 17*, 521–522.

Booth, D. A. (1987d). Central dietary "feedback onto nutrient selection": Not even a scientific hypothesis. *Appetite, 8*, 195–201.

Booth, D. A. (1988). Objective measurement of determinants of food acceptance: Sensory, physiological and psychosocial. In J. Solms, D. A. Booth, R. M. Pangborn, & O. Raunhardt (Eds.), *Food acceptance and nutrition* (pp. 1–27). London: Academic Press.

Booth, D. A. (1989). Causal analysis of eating behavior in the eating disorders. *Annals of New York Academy of Sciences, 575*, 466–471.

Booth, D. A. (1990a). Measuring mechanisms of ingestive behavior. *Appetite, 14*, 74–76.

Booth, D. A. (1990b). Influences on human fluid consumption. In D. J. Ramsey & D. A. Booth (Eds.), *Thirst: its physiology and psychology* (pp. 1–12). London: Springer-Verlag.

Booth, D. A. (1990c). Learned role of tastes in eating motivation. In E. D. Capoldi & T. C. Powley (Eds.), *Taste, experience and feeding* (pp. 104–126). Washington, DC: American Psychological Association.

Booth, D. A., & Baker, B. J. (1988). Main serotoninergic effects of fenfluramine on food intake may not involve central appetite pathways. *Society for Neuroscience Abstracts, 14*, 613.

Booth, D. A., & Baker, B. J. (1990). *dl*-Fenfluramine challenge to nutrient-specific textural preference conditioned by concurrent presentation of two diets. *Behavioral Neuroscience, 104*, 226–229.

Booth, D. A., & Blair, A. J. (1988). Objective factors in the appeal of a brand during use by the individual consumer. In D. M. H. Thomson (Ed.), *Food acceptability* (pp. 329–346). London: Elsevier Applied Science.

Booth, D. A., Conner, M. T., & Gibson, E. L. (1989). Measurement of food perception, food preference, and nutrient selection. *Annals of New York Academy of Sciences, 561,* 226–242.

Booth, D. A., Conner, M. T., & Marie, S. (1987). Sweetness and food selection: Measurement of sweeteners' effects on acceptance. In J. Dobbing (Ed.), *Sweetness* (pp. 143–160). London: Springer-Verlag.

Booth, D. A., & Davis, J. D. (1973). Gastrointestinal factors in the acquisition of oral sensory control of satiation. *Physiology and Behavior, 11,* 23–29.

Booth, D. A., Gibson, E. L., & Baker, B. J. (1986). Gastromotor mechanism of fenfluramine anorexia. In S. Nicolaidis (Ed.), *Serotoninergic system, feeding and body weight regulation* (pp. 57–69). London: Academic Press.

Booth, D. A., & Jarman, S. P. (1976). Inhibition of food intake in the rat following complete absorption of glucose delivered into the stomach, intestine or liver. *Journal of Physiology, 259,* 501–522.

Booth, D. A., Lovett, D., & McSherry, G. M. (1972). Postingestive modulation of the sweetness preference gradient in the rat. *Journal of Comparative and Physiological Psychology, 78,* 485–512.

Booth, D. A., & Pain, J. F. (1970). Effects of a single insulin injection on approaches to food and on the temporal pattern of feeding. *Psychonomic Science, 21,* 17–19.

Booth, D. A., & Quartermain, D. (1965). Taste sensitivity of eating elicited by chemical stimulation of the rat hypothalamus. *Psychonomic Science, 3,* 525–526.

Booth, D. A., & Toates, F. M. (1974). A physiological control theory of food intake in the rat: Mark 1. *Bulletin of the Psychonomic Society, 3,* 442–444.

Booth, D. A., Toates, F. M., & Platt, S. V. (1976). Control system for hunger and its implications in animals and man. In D. Novin, W. Wyrwicka, & G. A. Bray (Eds.), *Hunger: Basic mechanisms and clinical implications* (pp. 127–142). New York: Raven Press.

Burton, M. J., Cooper, S. C., & Popplewell, D. A. (1981). The effect of fenfluramine on the microstructure of feeding and drinking. *British Journal of Pharmacology, 72,* 621–633.

Cabanac, M. (1971). Physiological role of pleasure. *Science, 173,* 1103–1107.

Campfield, L. A., Smith, F. J., & Fung, K. F. (1982). VMH hyperphagia and obesity. Role of autonomic neural control of insulin secretion. In B. G. Hoebel & D. Novin (Eds.), *Neural basis of feeding and reward* (pp. 203–220). Brunswick, ME: Haer Institute.

Clifton, P. C. (1987). Analysis of feeding and drinking patterns. In F. M. Toates & N. E. Rowland (Eds.), *Feeding and drinking. Techniques in the behavioral and neural sciences* (vol. 1, pp. 19–35). Amsterdam: Elsevier.

Cohen, M. I. (1979). Neurogenesis of respiratory rhythm in the mammal. *Physiological Reviews, 59,* 1105–1173.

Collier, G. H. (1980). An ecological analysis of motivation. In F. M. Toates & T. R. Halliday (Eds.), *Analysis of motivational processes* (pp. 125–151). London: Academic Press.

Conner, M. T., Haddon, A. V., Pickering, E. S., & Booth, D. A. (1988). Sweet tooth demonstrated: Individual differences in preference for both sweet foods and foods highly sweetened. *Journal of Applied Psychology, 73,* 275–280.

Davis, J. D. (1989). The microstructure of ingestive behavior. *Annals of New York Academy of Sciences, 575,* 106–121.

Denton, D. (1985). *The hunger for salt.* London: Springer-Verlag.

Deutsch, J. A. (1983). Dietary control and the stomach. *Progress in Neurobiology, 20,* 313–332.

Duggan, J. P., & Booth, D. A. (1986). Obesity, overeating and rapid gastric emptying in rats with ventromedial hypothalamic lesions. *Science, 231,* 609–611.

Duggan, J. P., Storlien, L. H., Kraegen, E. W., & Booth, D. A. (1988). Effect of procaine injection into the ventromedial hypothalamic area (VMH) of the rat on serum insulin, glucose and corticosterone and gastric emptying rate. *Physiology and Behavior, 43,* 29–33.

Gibson, E. L., & Booth, D. A. (1986). Feeding induced by injection of norepinephrine near the paraventricular nucleus is suppressed specifically by the early stages of strong postingestional satiety in the rat. *Physiological Psychology, 14,* 98–103.

Gibson, E. L., & Booth, D. A. (1988). Fenfluramine and amphetamine suppress dietary intake without affecting learned preferences for protein or carbohydrate cues. *Behavioural Brain Research, 30,* 25–29.

Gibson, E. L., & Booth, D. A. (1989). Dependence of carbohydrate-conditioned flavor preference on internal state in rats. *Learning and Motivation, 20,* 36–47.

Gold, R. M. (1973). Hypothalamic obesity: The myth of the ventromedial nucleus. *Science, 182,* 488–489.

Gold, R. M., Jones, A. P., Sawchenko, P. E., & Kapatos, G. (1977). Paraventricular area: Critical focus of a longitudinal neurocircuitry mediating food intake. *Physiology and Behavior, 18,* 1111–1120.

Graff, H., & Stellar, E. (1964). Hyperphagia, obesity and finickiness. *Journal of Comparative and Physiological Psychology, 55,* 418–424.

Grill, H. J., & Norgren, R. (1978). Chronically decerebrate rats demonstrate satiation but not bait-shyness. *Science, 201,* 267–269.

Grossman, S. P. (1960). Eating or drinking elicited by direct adrenergic or cholinergic stimulation of hypothalamus. *Science, 132,* 301–302.

Guevara-Aguilar, R., Jimenez-Montufar, L. L., Garcias-Diez, D. E., Wayner, M. J., & Armstrong, D. L. (1988). Olfactory and visceral projections to the paraventricular nucleus. *Brain Research Bulletin, 20,* 799–801.

Harris, G., & Booth, D. A. (1987). Infants' preference for salt in food: Its dependence upon recent dietary experience. *Journal of Reproductive and Infant Psychology, 5,* 97–104.

Leibowitz, S. F. (1978). Paraventricular nucleus: A primary site mediating adrenergic stimulation of feeding and drinking. *Pharmacology Biochemistry and Behavior, 8,* 163–175.

Leibowitz, S. F. (1988). Low-dose fenfluramine in carbohydrate selection in the early phase in the rat. Paper presented at the Third Benjamin Franklin/Lafayette Symposium, La Napoule.

Leibowitz, S. F., Weiss, G. H., Yee, F., & Tretter, J. B. (1985). Noradrenergic innervation of the paraventricular nucleus: Specific role in control of carbohydrate ingestion. *Brain Research Bulletin, 14,* 561–567.

Le Magnen, J. (1956). Hyperphagie provoquée chez le rat blanc par alteration du mécanisme de satiété peripherique. *Comptes Rendues de la Société de la Biologie, Paris, 150,* 136–139.

Le Magnen, J. (1981). The metabolic basis of dual periodicity of feeding in rats. *Behavioral and Brain Sciences, 4,* 561–607.

Le Magnen, J. (1985). *Hunger.* London: Cambridge University Press.

Le Magnen, J., Devos, M., Gaudilliere, J. P., Louis-Sylvestre, J., & Tallon, S. (1973). Role of a lipostatic mechanism in regulation by feeding of energy balance in rats. *Journal of Comparative and Physiological Psychology, 84,* 1–23.

Le Magnen, J., & Tallon, S. (1966). La periodicité spontanée de la prise d'aliments ad libitum du rat blanc. *Journal de Physiologie, 58,* 323–349.

Marler, P., & Hamilton, W. J. (1966). *Mechanisms of animal behavior.* New York: John Wiley & Sons.

Matthews, J. W., Booth, D. A., & Stolerman, I. P. (1978). Factors influencing feeding elicited by intracranial noradrenaline in rats. *Brain Research, 141,* 119–128.

Matthews, J. W., Gibson, E. L., & Booth, D. A. (1985). Norepinephrine-facilitated eating: Reduction in saccharin preference and conditioned flavor preferences with increase in quinine aversion. *Pharmacology Biochemistry and Behavior, 22,* 1045–1052.

McCarron, A., & Tierney, K. J. (1989). The effect of auditory stimulation on the consumption of soft drinks. *Appetite, 13,* 155–159.

Mook, D. G., & Wagner, S. (1989). Orosensory suppression of saccharin drinking in rat: The response, not the taste. *Appetite, 13,* 1–13.

Moran, T. H., and McHugh, P. R. (1988). Gastric and nongastric mechanisms for satiety action of cholecystokinin. *American Journal of Physiology, 225,* R628–R632.

Morgane, P. J. (1961). Medial forebrain bundle and "feeding centers" of the hypothalamus. *Journal of Comparative Neurology, 117,* 1–26.

Newman, J. C., & Booth, D. A. (1981). Gastrointestinal and metabolic consequences of a rat's meal on maintenance diet ad libitum. *Physiology and Behavior, 27,* 929–939.

Pager, J. (1984). A selective modulation of olfactory bulb electrical activity in relation to the learning of palatability in hungry and satiated rats. *Physiology and Behavior, 12,* 189–195.

Powley, T. L. (1975). The ventromedial hypothalamic syndrome, satiety, and a cephalic phase hypothesis. *Psychological Review, 84,* 89–126.

Rabin, B. M. (1972). Ventromedial hypothalamic control of food intake and satiety—a reappraisal. *Brain Research, 43,* 317–323.

Ritter, R. C., & Epstein, A. N. (1975). Control of meal size by central noradrenergic action. *Proceedings of National Academy of Sciences, U.S.A., 75,* 3740–3743.

Robinson, P. H. (1989). Gastric function in eating disorders. *Annals of the New York Academy of Sciences, 575,* 456–465.

Rolls, B. J., & Rolls, E. T. (1983). *Thirst.* Cambridge: Cambridge University Press.

Rolls, B. J., Rowe, E. A., & Rolls, E. T. (1982). How sensory properties of foods affect human feeding behavior. *Physiology and Behavior, 29,* 409–417.

Rolls, E. T. (1976). The neurophysiology of appetite. In T. Silverstone (Ed.), *Appetite and food intake* (pp. 1–16). Berlin: Dahlem Konferenzen.

Rolls, E. T. (1984). The neurophysiology of feeding. *International Journal of Obesity, 8 (Supplement 1),* 139–150.

Rolls, E. T., & Wiggins, L. L. (1989). The taste area of the caudolateral orbitofrontal cortex of the primate. *Chemical Senses, 13,* 747.

Schneider, L. (1989). Orosensory self-stimulation by sucrose involves brain dopaminergic mechanisms. *Annals of New York Academy of Sciences, 575,* 307–320.

Sclafani, A., & Toris, J. (1981). Influence of diet palatability on the noradrenergic feeding response in the rat. *Pharmacology Biochemistry and Behavior, 15,* 15–19.

Scott, T. R., & Mark, G. P. (1986). Feeding and taste. *Progress in Neurobiology, 27,* 293–317.

Siegel, J. M. (1979). Behavioral functions of the reticular formation. *Brain Research Reviews, 1,* 69–105.

Simpson, S. J., & Bernays, E. A. (1983). The regulation of feeding: Locusts and blowflies are not so different from mammals. *Appetite, 4,* 311–344.

Slater, P. J. B. (1974). The temporal pattern of feeding in the zebra finch. *Animal Behaviour, 22,* 506–515.

Smith, G. P., & Gibbs, J. (1979). Postprandial satiety. *Progress in Psychobiology and Physiological Psychology, 10,* 179–242.

Stricker, E. M., & Zigmond, M. J. (1976). Brain catecholamines and the lateral hypothalamic syndrome. In D. Novin, W. Wyrwicka, & G. A. Bray (Eds.), *Hunger* (pp. 19–32). New York: Raven Press.

Teitelbaum, P. (1977). Levels of integration of the operant. In W. Honig & J. E. R. Staddon (Eds.), *Handbook of operant behavior* (pp. 7–27). Englewood Cliffs, NJ: Prentice-Hall.

Teitelbaum, P., & Cytawa, J. (1965). Spreading depression and recovery from lateral hypothalamic damage. *Science, 147,* 61–63.

Terman, G. W., Shavit, Y., Lewis, J. W., Cannon, J. T., & Liebeskind, J. C. (1984). Intrinsic mechanisms of pain inhibition: Activation by stress. *Science, 226,* 1270–1277.

Thompson, R. F. (1986). The neurobiology of learning and memory. *Science, 233,* 941–947.

Tordoff, M. G., & Friedman, M. I. (1986). Hepatic portal glucose infusions decrease food intake and increase food preference. *American Journal of Physiology, 251,* R192–R196.

Turner, B. H. (1973). Sensorimotor syndrome produced by lesions of the amygdala and lateral hypothalamus. *Journal of Comparative and Physiological Psychology, 82,* 37–47.

Ungerstedt, U. (1971). Adipsia and aphagia after 6-hydroxydopamine induced degeneration of the nigro-striatal dopamine system. *Acta Physiologia Scandinavica [Supplement], 367,* 95–122.

Weingarten, H. P. (1985). Stimulus control of eating: Implications for a two-factor theory of hunger. *Appetite, 6,* 387–401.

Wooley, S. C., Wooley, O. W., & Dunham, R. B. (1972). Calories and sweet taste: Effects on sugar preference in the obese and nonobese. *Physiology and Behavior, 9,* 765–768.

Yim, C. Y., & Mogenson, G. J. (1980). Effect of iontophoretically applied dopamine on the responses of nucleus accumbens neurons to electrical stimulation of the amygdala. *Society for Neuroscience Abstracts, 6,* 292.

Zeigler, H. P., Jacquin, M. F., & Miller, M. G. (1985). Trigeminal orosensation and ingestive behavior in the rat. *Progress in Psychobiology and Physiological Psychology, 11,* 63–196.

18

Prospectus: Thirst and Salt Appetite

Alan N. Epstein

Thirst and Salt Intake

Those of you who are seeking a complete understanding of the behavioral neuroscience of thirst will be disappointed by these chapters. You will learn from them that we are still far from this goal. We do not, in other words, have a complete neurobiological account of the specific state (or states) within the brain that creates the urge to drink water, that generates an anticipation of water, that emits the behaviors that lead an animal to it, that allows water to be selected from among other ingestible commodities, that makes thirst and water drinking inherently reinforcing and memorable, and that controls the consumption of it until satiation is achieved, all with accompanying hedonic states that include the distress of water depletion, the excitement of its anticipation, and the pleasure of thirst satiation.

If we knew how the brain does all of this, this chapter would be a eulogy for an exhausted field. But it is not. We are still ignorant of most of what we need to know about the neuroscience of thirst, and I can make the following remarks and prophecies about a research venture that is diverse, exciting, and vigorous but largely unfinished.

The Nature of Thirst

First things first. What are we investigating? Thirst is the specific, central, motivational state of readiness to consume water.

By specific I mean that when it arises, water (its sensory qualities and its location in the environment) is somehow represented in the brain as the specific object of the animal's intentions and that this neural representation or image of

Alan N. Epstein   Department of Biology, University of Pennsylvania, Philadelphia, Pennsylvania 19104.

water governs the behavior until the animal contacts a commodity and accepts it as sufficiently waterlike for ingestion.

By central I mean, of course, that it goes on in the brain. It is modified, certainly, by events in the periphery, but what we are seeking to understand is essentially a problem of brain function, and we will not know what thirst is until we have understood its neuroscience.

And by motivational I mean two things. First I mean that thirst often arises in the absence of water, and drinking behavior often occurs in the absence of depletion signals. It cannot therefore be merely reflexive. And, second, I mean that the behavior is complex. It is divided into appetitive and consummatory phases, is guided by an anticipation of water, and is accompanied by affect. The appetitive behaviors of thirst will be individuated in ways that depend on the animal's past history of drinking behavior and on the ecological problems that it must solve in order to reach the water. Even the caged animal must move to its water reservoir and must emit behaviors that are appropriate to the problem of bringing the water into its mouth. The animal, in other words, uses operants to gain access to the water (Teitelbaum, 1977). And, remember, from its onset, which often occurs in the absence of water, the behavior is guided by an image of water, by an expectancy of it that is satisfied (or frustrated) only when the behavioral sequence is completed and consumption (or rejection) begins. And all of this is accompanied by affect. Feelings and emotional expressions are an integral part of thirst as they are of all motivated behaviors (Epstein, 1982a).

If this concept of the nature of thirst is accepted, we can understand why we are not just investigating drinking behavior. Drinking is not thirst. Although licking and swallowing are the common consummatory behaviors for water ingestion, there is no mandatory fixed action pattern for thirst in mammals. A mammal can "eat" water if it is a succulent plant (Milgram, Krames, & Thompson, 1974) or, more demonstrably, a solid mixture of nonnutritive cellulose and water (Hamilton, 1969), and it can "drink" its food if it is a liquid. Licking and swallowing are often an expression in behavior of the central state of thirst, and we use them to study the central state, but they are not thirst. They can, in fact, be dispensed with if the thirsty animal is provided with an alternate means for delivery of water into its stomach (Epstein, 1960) or veins (Rowland & Nicolaidis, 1974).

We can also understand why we are not investigating just a sensation. Thirst generates a percept that is represented in consciousness, but it is only the sensory aspect of thirst and can only be studied in humans. It is not the major issue of our science.

And, lastly, we can understand why we should not limit ourselves just to the study of the signals that control the arousal and satiation of drinking behavior. The brain does indeed process signals of several kinds to begin and to terminate water drinking, and the study of these signals (neural afferents, hydrational deficits, hormones) is one of the most feasible and informative of our current research strategies. But the brain is more than a processor of information. It is a self-activating organ that issues commands to somatic, autonomic, and endocrine action systems. These are, of course, modified by signals, but the signals do not produce the behavior. The brain does, and signals are integrated into neural programs for action that are both innate and acquired. If we limit our concept of thirst and the other motivated behaviors to the signal-processing idea, we risk misunderstanding them as kinds of Sherringtonian reflexes occurring in an essentially reactive brain.

Sleep is a better model for our thinking, and the neuroscience of sleep suggests a better program for the future of our work. What we know about sleep illustrates the brain's self-activating capabilities and its command functions within constraints. Sleep is not imposed on the brain by afferent inputs or other signals. It is generated by the brain and modulated by signals. And the sleeping brain sleeps in the several qualitatively different stages of sleep because of the constraints of its innate endowment of structure and function. I am suggesting that thirst is also a product of the brain's self-activation, that the state (or states) of thirst is inherent to the brain, modulated, of course, by signals but essentially endogenous. And I am also suggesting that it is diversified into several qualitatively different kinds or states of thirst, and these are discussed below.

We are confident of sleep's endogenous and varied nature because the EEG and other forms of electrical recording give us the signatures of sleep (Kleitman, 1963). The brain sleeps both behaviorally and electrically, and our colleagues in sleep research use the electrical signs of sleep to do their work. We do not have this advantage. There is no recordable index of the brain activities of thirst. In order to obtain it we will need a technology that we do not have, one that reveals multifocal activities that are both chemical and electrical and that occur dynamically in neuron assemblies throughout the neuraxis over extended periods of time. But I am confident that there is a specific state of activity under way inside the brain when animals seek water and drink it, and I believe that one of the major turning points in the future of thirst research will be the discovery of its signature.

THE BIOLOGICAL DIVERSITY OF THIRST

Thirst has evolved in the fully terrestial animals—the insects and the vertebrates beginning with the reptiles. But its evolution has not been simple. Water swallowing occurs in eels and other fishes (Hirano, 1974; Balment & Carrick, 1985), and, because it is a form of water ingestion, may represent an evolutionary prelude to true thirst. Where thirst is surely present, there is biological diversity in the behaviors of water drinking. Lapping, sucking, and pecking are all used to ingest water, and some animals bring it to their mouths with their forepaws or hands. The temporal patterns of drinking behavior vary. Some animals are frequent small-draft drinkers, others drink large drafts infrequently, and the behavior may occur either nocturnally or diurnally. There is even diversity in whether or not thirst is included in the animal's behavioral repertory. There are desert rodents that do not drink even when water is available (Schmidt-Nielsen, 1979), and the fully marine vertebrates (the marine turtles, birds, and mammals) have virtually no access to fresh water and rely on salt glands or on the isotonic flesh of their prey for water balance. Other vertebrates that live in fresh water or prey on fresh-water species (turtles, many birds, and mammals such as the beaver and otter) must solve the problem not of water ingestion but of reducing their water intake and of excreting the excess water that enters through permeable membranes.

With all this diversity we should not be surprised by species differences in the neurological mechanisms of thirst among animals, even among those that are closely related. There are, for example, differences in the responsiveness of rodents to the dipsogenic and natriorexigenic actions of angiotensin II (Wright, Morseth, Fairley, Petersen, & Harding, 1987; Rowland & Fregly, 1988). Differences will be even greater across taxons as shown by the fact that, when given into the brain ventricles, the tachykinins are dipsogenic in rats but antidipsogenic in pigeons (De Caro, 1986). One of the best examples we have of this diversity in neurological mechanisms is in the insects. Thirst has been studied in only a few species in this enormous order of animals and in them is controlled by mechanical distention of the vascular space (the hemocoele) rather than by the effective osmotic pressure and volume of the body fluids (Dethier & Evans, 1961). Desiccated flies can be made indifferent to water by filling their hemocoele with mineral oil or concentrated glucose solution (as well as water), and sated flies can be made thirsty by bleeding.

We still have much to learn about species differences in the neurobiology of thirst, and it is safe to predict that they will be found to be even greater in number and even more surprising as we study thirst and its brain mechanisms in more animal species. After all, most of our research uses only a handful of species (rat, sheep, dog, and to some extent the pigeon, monkeys, and man), and even among this small number we are confronted with the following facts: (1) sheep, unlike rats, dogs, the opossum, and the pigeon, continue to drink to intravenous angiotensin after ablation of the subfornical organ (Simpson, 1981) and (2) the organum vasculosum of the lamina terminalis must be intact in sheep and dogs in order for them to drink to cellular hydration (Thrasher, Keil, and Ramsay, 1982), but in the rat the necessary tissue is in the lateral preoptic area and nearby portions of the anterior hypothalamus (Blass & Epstein, 1971; Peck & Novin, 1971; Almli & Weiss, 1974).

BIOLOGICAL DIVERSITY WITHIN SPECIES

There are three other kinds of biological diversity that are not as readily acknowledged for thirst, in particular, and for other areas of behavioral neuroscience in general. All three occur *within* rather than across species, and all of them have been only poorly studied. These are the diversities that arise, first from differences in developmental stage and aging, second, from differences in sex, and, third, from differences in the states of thirst themselves.

DEVELOPMENT AND DIVERSITY. Animals that drink do so differently as they progress through their ontogeny, and the brain mechanisms for the behavior change as they mature (Epstein, 1984). We know something about this problem from studies of the onset of water drinking in rats. Young rats begin to drink in their third postnatal week (Babicky, Pavlik, Pavlik, Ostadalova, & Kolar, 1972) and then drink more frequently during weaning, but without the careful segregation of feeding and drinking that is characteristic of the adult (Kissileff, 1969a, 1971). These are the only published studies of the natural onset of drinking, and both use the rat, and there are none of drinking in the aged animal.

Changes in the brain mechanisms for thirst with development are illustrated by Leshem's work on the ontogeny of the dipsogenic effect of angiotensin (Leshem & Epstein, 1988; Leshem, Boggan, & Epstein, 1988). Very young rat pups drink in response to both systemic and intracranial angiotensin, but they accept either water or milk. Leshem showed that the selective water drinking that is diagnostic of the adult behavior begins at 8 days of age, when the pups begin to drink more water than milk in response to intracranial angiotensin, and reaches its developmental climax at 16 days, when they drink only water. Throughout this sequence of developmental stages angiotensin arouses drinking only when the pups are expressing nascent forms of adult ingestion while they are away from their dams and not suckling. When they are reunited with her and treated with angiotensin, their behavior is unchanged. They do not increase their intake of suckled milk despite its high water content, as if the thirst evoked by angiotensin were somehow uncoupled from suckling and, therefore, not expressed in their behavior (for further discussion, see ONTOGENY OF THIRST, below). Again, the work has been done only in the rat, and longitudinal developmental studies of the other determinants of thirst have not been published.

SEX DIFFERENCES. In the rat, differences in ingestive behavior between the sexes are common and conspicuous. They occur for water and food (Tarttelin & Gorski, 1971; Vijande, Costales, Schiaffini, & Marin, 1978) intake, for salt consumption (Krecek, 1973), and for the intake of other sapid solutions (Zucker, 1969). They are often expressions of gender and can be altered by castration or hormone administration in the neonatal period and in adulthood. For example, female rats drink more salt than males, but the difference disappears if the males are castrated shortly after birth (Krecek, 1973). And the decreased water intake that occurs during estrus in the adult female is eliminated by ovariectomy and reinstated by estrogen administration (Findlay, Fitzsimons, & Kucharczyk, 1979).

Except for Jonklaas and Buggy's intriguing work, the neurobiological aspects of this problem have not attracted the interest they deserve. They first confirmed their earlier work (Jonklaas & Buggy, 1984) in which intracranial implants of estrogen reduced the water intake that is evoked by central injection of angiotensin II and then showed (Jonklaas & Buggy, 1985) that this occurs only in female rats and is most effective when the estrogen is administered to the medial preoptic area. They also demonstrated, with regional binding studies, that angiotensin II binding was decreased in the medial peroptic region of estrogen-treated animals. And, lastly, they found that the antidipsogenic effect of estrogen occurs only in animals whose neonatal hormonal history was normal. That is, genetic females that were masculinized by neonatal androgen did not reduce their water intake when treated with estrogen in adulthood, and males that were castrated at birth were feminized. They drank less while receiving estrogen. Sex differences in angiotensin-induced water drinking and in salt intake (see NEED-FREE SALT INTAKE, below) appear, therefore, to be expressions in adulthood of the sexual differentiation of the brain that is produced in the perinatal period by the central actions of the gonadal steriods.

THE SEVERAL STATES OF THIRST. It should not be necessary to insist on the diversity of the states or kinds of thirst, but much of our published work implies otherwise. We often write about thirst as if it were qualitatively singular, one kind of activity in the brain that underlies all water-drinking behavior and that does not vary except quantitatively among episodes of drinking. I do not believe this. The drinking behavior of our most frequently studied animal argues against it. Rats that have free access to water and are eating dry food drink most often just before or just after they eat (food-associated drinking occurring pre- or postprandially), but they also drink, albeit rarely, between meals, usually in association with grooming (Kissileff, 1969b). When their salivary production is impaired they drink repeatedly within meals (prandial drinking), often excessively (Epstein, Spector, Samman, & Goldblum, 1964). They also drink, of course, after they have been deprived of water for periods in excess of 6 to 8 hours and when they are defending themselves against heat stress by saliva-spreading (Hainsworth, Stricker, & Epstein, 1968) or when they are moved abruptly from the cold to a neutral environment (Katovich, Barney, Fregly, Tyler, & Dasler, 1979). They can also be required to use water drinking as an operant, entirely outside of its natural context (Williams & Teitelbaum, 1956), and will drink water in excess when they are forced to wait while hungry for the delivery of successive morsels of dry food (Falk, 1967).

This diversity in the kinds of water-drinking behavior warns us that there are kinds of thirst just as there are kinds or states of sleep. An animal that is drinking preprandially cannot be expressing a thirst that is the same in its biological determinants and neurological mechanism as one that drinks after eating a meal, and both of these must be different from the drinking that is done after a bout of grooming or saliva spreading. Conditions within the oral cavity will be major determinants of the drinking that is associated with grooming and thermolysis, whereas water depletions and expectancies of the need for water are the most likely determinants of the drinking that precedes a meal. And this kind of drinking must be different, again in its biological determinants and neurological mechanisms, from the drinking that occurs postprandially. After a meal, but not before, isotonic serum will have shifted into the gut without net water loss, and meal taking has neurochemical and hormonal consequences that simply cannot operate to control preprandial drinking.

Most of us assume the singularity of the neural mechanisms of thirst because we provoke it experimentally and study drinking behavior out of its natural context. Scott Kraly's recent analysis of meal-associated water drinking in the rat is a welcome counterexample (Kraly, 1989). In a careful analysis of the drinking done by the rat before, during, and after eating, he finds that water intake that is food associated (done while the animal is eating) is reduced by one-third when either angiotensin or histamine actions are blocked, whereas water intake before or after meals is unaffected by the same treatments. In the sham-feeding rat, in which pregastric food-associated events predominate, histamine or angiotensin blockage abolishes the animal's robust drinking.

We should not let the sameness of water-drinking behavior deceive us into believing that there is only one kind of thirst. We should, instead, be asking how many kinds are there? Or, more precisely, we should be asking how many qualitatively different states of brain activity there are that can generate drinking behavior in response to different sets of biobehavioral determinants.

ANGIOTENSIN AS A HORMONE OF THIRST

Kraly's demonstration of a role for angiotensin in the drinking done while rats are eating (pregastric food-associated drinking) is the most recent evidence of a role for the hormone in thirst. From the first demonstrations of the effects of its injection it was clear that angiotensin was an exceptionally powerful dipsogen (Fitzsimons, 1969). Now, more than 20 years later, we know that the hormone is potent when infused intravenously (Fitzsimons & Simons, 1969; Hsiao, Epstein, & Camardo, 1977; Evered & Robinson, 1981) and that the circulating levels of the hormone that are achieved by its intravenous infusion match the endogenous levels that are produced by several dipsogenic treatments including those that occur naturally (Johnson, Mann, Rascher, Johnson, & Ganten, 1981). We also know now that several experimental thirsts are entirely angiotensin dependent (Fitzsimons, 1964; Fitzsimons & Elfont, 1982; Evered & Robinson, 1981; Katovich *et al.*, 1979), and it is becoming increasingly likely that the hormone plays a role in deprivation-induced and food-associated drinking behavior.

Angiotensin remains the most potent dipsogenic agent known when given into the anterior cerebral ventricles, and especially when it is injected directly into the subfornical organ, where it can arouse drinking behavior in the satiated rat at doses in the femtomole range (Simpson, Epstein, & Camardo, 1978). Its potency as a blood-borne hormone of thirst is also impressive. This was underestimated in the original studies of the problem (Fitzsimons & Simons, 1969; Hsiao *et al.*, 1977) in which artifacts of the method of intravenous infusion were not avoided. Recent studies that have abolished the pressor effect of intravenous angiotensin II (Evered & Robinson, 1981) demonstrate a 100% increase in the dipsogenic potency of the blood-borne hormone across a broad range of doses (1 to 100 ng/min), and they show that the threshold for elicitation of drinking is the same as that for the pressor response. Other studies that suppress prostaglandin production with oral indomethacin yield the same results (Kenney & Moe, 1981). Presumably, the chronic catheter that is necessary for intravenous administration of the hormone somehow activates a prostaglandin reaction that is a suppressor of angiotensin-induced drinking. A similar enhancement of the dipsogenic potency of systemic angiotensin II is produced by parabrachial (Ohman & Johnson, 1986) or area postrema (Edwards & Ritter, 1982) lesions. And, lastly, infusions of the hormone into the carotid circulation of the dog, which give it direct access to the brain and do not provoke a marked pressor response, elicit drinking at low physiological doses (Fitzsimons, Kucharczyk, & Richards, 1978).

The work of Johnson and his colleagues (Mann, Johnson, & Ganten, 1980; Johnson *et al.*, 1981) is especially relevant. They showed that thirst-provoking treatments such as 48 hours of water deprivation, caval ligation, moderate doses of isoproterenol, and hyperoncotic colloid dialysis all produce blood levels of endogenous angiotensin II that are well above the amounts (approximately 200 fmol/mL of plasma) that are produced by intravenous infusion of the hormone at the dipsogenic threshold (approx 10 ng/min) that was determined by Hsiao *et al.*, (1977). And this threshold was measured before any of us were aware of the artifacts of the inhibitory effect of the pressor response and suppression of drinking by the prostaglandins. When the dipsogenic threshold for circulating angiotensin II is reevaluated with precautions against these and other antidipsogenic artifacts, it will undoubtedly be lower and closer to, if not below, the plasma angiotensin II levels that are produced by everyday dipsogenic conditions such as consumption of a dry meal and 12 to 24 hours of water deprivation.

It must also be remembered that judging the dipsogenic potency of blood levels of infused angiotensin places an abnormal burden on it as a cause of drinking. Angiotensin never acts alone during instances of physiological thirst. After water deprivation, for example, angiotensin is always accompanied by other dipsogenic preconditions, most frequently hypovolemia and cellular dehydration (Hatton & Almli, 1969; Ramsay, Rolls, & Wood, 1977; Epstein, 1982b). It is remarkable that exogenous angiotensin is as potent as it is even when it is taken out of its normal physiological context and is administered as the sole dipsogen.

Lastly, angiotensin's role in thirst has been confirmed by use of specific blockers (Malvin, Mouw, & Vander, 1977; Barney, Threatte, & Fregly, 1983), either analogues of the natural octapeptide that are competitive inhibitors of receptor occupancy or other compounds such a captopril that inhibit the angiotensin-converting enzyme and thereby prevent the biosynthesis of endogenous angiotensin II. Experimental thirsts such as those that are induced by isoproterenol (Houpt & Epstein, 1971) or by caval ligation (Fitzsimons, 1964) are completely abolished by angiotensin receptor blockers and by captopril (Rettig, Ganten, & Johnson, 1981), and the water intake that is produced by 48 hours of water deprivation is markedly reduced by the same treatments (Barney et al., 1983).

Remarkable progress has been made in understanding the neural circuit that mediates the thirst that is aroused by angiotensin. It derives from two discoveries. These were (1) the discovery by Ganten and his colleagues (Ganten, Fuxe, Phillips, Mann, & Ganten, 1978) of the cerebral renin–angiotensin system and their demonstration of its independence from the blood-borne system of renal origin, and (2) the identification by Simpson and his colleagues (Simpson, 1981; Simpson & Routtenberg, 1973) of the subfornical organ (SFO) as the essential site at which blood-borne angiotensin II acts to arouse thirst.

These new findings set the stage for two concurrent lines of work that focused attention on the nucleus medianus of the preoptic area (MnPO) as the crucial site within the brain for the dipsogenic action of angiotensin. These were the mapping of the neural connections of the SFO by Miselis and his students (Miselis, 1981; Miselis, Shapiro, & Hand, 1979), which includes a reciprocal pathway between SFO and MnPO, and the findings of Johnson and his colleagues (Buggy & Johnson, 1977; Johnson & Buggy, 1977; Johnson & Wilkin, 1987) of the important role of the tissues of the ventral lamina terminals (their "AV3V"), which includes the MnPO, in thirst, ADH release, and control of blood pressure. The work of Johnson and Lind (Lind & Johnson, 1982; Lind, Swanson, & Ganten, 1985) was especially important because it demonstrated, on the one hand, that the MnPO was rich in both angiotensinergic terminals and angiotensin-sensitive cells and that it is, on the other hand, essential for thirst aroused by both blood-borne angiotensin and angiotensin that acts within the brain.

As a result of this remarkable series of investigations, we know now that both peripherally and centrally generated angiotensin II contribute to the water drinking that is evoked by the hormone. They do so through an angiotensin-sensitive circuit that includes the SFO, which lacks a blood–brain barrier, and the MnPO, which is within the barrier. Drinking is evoked by activation of angiotensin receptors at both sites, and they are linked by angiotensinergic neurons of the SFO that project to the MnPO. We also know that noradrenergic projections to the MnPO from the brainstem are necessary for its role as the mediator of the dipsogenic effect of angiotensin (Bellin, Bhatnagar, & Johnson, 1987).

Much still needs to be learned about the role of angiotensin in the generation of drinking behavior. We need to know how the two renin–angiotensin systems (the peripheral renin–angiotensin–aldosterone system and the renin–angiotensin system that is endogenous to the brain) cooperate to produce the behavior, how the hormone is integrated into the complex of conditions that lead to normal drinking, and how it may participate in thirst satiety. But, despite what still needs to be understood, the evidence that we now have justifies the prediction made 20 years ago by Fitzsimons (1969) when he discovered a renal dipsogen and suggested that angiotensin II is a hormone of thirst.

Progress is being made in this important problem, and you will have read about it in the other chapters in this volume. We do know that thirst is mediated by the forebrain. The chronic decerebrate rat does not seek water and will not ingest it in response to a variety of thirst challenges even when it is infused directly into its mouth (Grill & Miselis, 1981). Where thirst determinants such as cellular dehydration and angiotensin have been identified, neural circuits for their action are being described in the circumventricular organs and the anterior forebrain (Miselis, 1981; Johnson & Wilkin, 1987). Global deficits in several kinds of thirst are produced by ablations of the lateral hypothalamus (Epstein, 1971), of the ventral lamina terminalis (the "AV3V"; Buggy & Johnson, 1977; Johnson, 1985), and of the zona inserta (Grossman, 1984), suggesting that they are convergence areas for information in the thirst circuit. And Mogenson's (1987) recent synthesis of mechanisms for motivated behavior reminds us of the equal importance of brainstem circuitry for the organization of efferent commands and sequential action.

Much remains to be done, and it will be a formidable task. First, as I emphasize above, we are not searching for the neural substrate of behaviors like jaw and tongue movements or of swallowing, which are problems of motor control. We are instead attempting to describe a neural apparatus that uses the motor programs for jaw and tongue action and for swallowing while it mediates a cluster of central states whose common property is the intention to drink water. Moreover, we will have to understand how the several states of thirst arise within the neural apparatus and how the special combination of biological determinants of each kind of thirst gain control of it while it governs the expression of drinking behavior.

Secondly, the task is formidable because all kinds of thirst are complex behaviors. They occur in appetitive sequences, are characterized by expectancy and expression of affect, and have endocrinological, perceptual, motivational, cognitive, memorial, and hedonic aspects whose mediation requires coherent activity throughout the cerebrum, brainstem, and cord. We should expect the neural apparatus for thirst to be distributed throughout the anterior forebrain, the limbic system, and the amygdala, and we should expect it to use the brainstem and cord to generate not only the motor sequences of the appetitive and consummatory behaviors of water ingestion but also the more evanescent expectancies and affects that are equally important aspects of the behavior.

SPONTANEOUS OR FREELY EMITTED DRINKING BEHAVIOR

Thanks to the masterful work of Ramsay, B. Rolls, and Wood published more than a decade ago (1977), we now have a physiological account of the causes of the drinking that is induced by water deprivation. It is aroused by the additive effects of deficits in both the cellular and extracellular water compartments. Earlier work had shown that the dipsogenic effect of cellular dehydration depends on osmosensors in the lateral preoptic area and immediately adjacent portions of the anterior hypothalamus (Blass & Epstein, 1971; Almli & Weiss, 1974) and that the drinking evoked by hypovolemia is mediated by a combination of neural afferents from the great vessels of the low-pressure circulation and of angiotensin II acting on the subfornical organ (Fitzsimons, 1979). In the Ramsay, Rolls, and Wood experiments, rats, dogs, and monkeys that had been prevented from drinking overnight were found to be dehydrated by water losses from both their cellular and extracellular water compartments, and their thirsts were satiated by selective and partial restoration of both deficits. Rehydration of their forebrain osmosensors was achieved with intravascular infusion of small volumes of water, and their extracellular volume was restored with intravenous isotonic saline. When offered access to water after these treatments, animals that had been deprived of it overnight drank only trivial volumes despite the fact that they were still in negative water balance. These experiments were a welcome and convincing confirmation of the double-depletion hypothesis, which predicted that thirst would be understood as the sum of the dipsogenic effects of deficits in both water compartments (Epstein, 1973), and they have the added importance of making clear that the thirst that is aroused by double depletion is the thirst of water deprivation.

We also have a satisfactory account of the drinking that is done, at least by rats, while they eat. As you will recall, Kraly's recent work demonstrates that water intake that is associated with active food ingestion is controlled by an angiotensin-mediated histaminergic mechanism (Kraly & Corneilson, 1988).

But we are still left with no account of the mechanism of the drinking that is done by nondeprived animals just prior to meals or at times that are unrelated to eating. We simply do not know why animals drink at these times, and the several names that have been invented for this kind of drinking are symptomatic of our ignorance. It has been called secondary, nonhomeostatic (or nonregulatory), and spontaneous. I prefer to call it "freely emitted" drinking, which does not imply, as does spontaneous, that it has no cause; which does not carry the implication that it is somehow less important than primary or deficit-induced drinking; and which does not define it as what remains when homeostatic drinking is accounted for.

The only thing we can say with certainty about freely emitted drinking (and, again, the reference animal is our most common subject, the domestic rat) is that it occurs most frequently at night when the animal is performing its other freely emitted behaviors (locomotion and investigation, elimination of urine and feces, yawning and stretching, grooming and scratching, and, of course, eating). These occur in bouts that interrupt periods of sleep that are briefer than in the daytime. The behaviors are episodic and associated with each other (De Castro, 1989). The animal wakes and performs several of these freely emitted behaviors in a cluster or bout that often includes water drinking, and this fact suggests how freely emitted drinking may be generated.

Could these bouts of freely emitted behaviors be habitual domestic routines that have been acquired over the lifetime of the animal? Could they be based on a combination of causes that have recurred on a daily (or, rather, nightly) basis since the animal was weaned and that have provided the comfort of relief from minor distresses and peripheral irritants, the pleasure of the performance of benign behaviors, and the opportunity to forestall nutrient deficits by the consumption of small amounts of water (and food) *before* it is needed in the homeostatic sense? Freely emitted drinking behavior may, in other words, be one of the behaviors in the well-practiced domestic routines of behavior that afford the rat comfort, pleasure, and the avoidance of nutrient deficits.

This suggestion rejects the idea that freely emitted drinking behavior can be understood as an instance of classic feedback homeostasis. It predicts that water deficits will not be found in animals expressing this behavior and that they are not performing it to restore water losses. Behavior serves homeostasis but is not its slave. There is a richness in the natural behaviors of animals and a complexity in the capabilities of their nervous systems that goes beyond homeostasis. We should be looking elsewhere for our hypotheses about the mechanism of such phenomena as freely emitted drinking and should take more seriously the suggestion, made by the much-quoted work of Fitzsimons and LeMagnen (1969), that drinking can anticipate water need. Although they did not use the term, they studied freely emitted feeding and drinking behaviors in rats that were switched from a high-carbohydrate to a high-protein diet that required more water for its metabolism. The animals increased their water intakes during the first day of high-protein feeding, but they did not consume the excess water while they ate until several days later. That is, the animals adjusted to the greater need for water in two stages. First they responded to it by drinking after they had eaten. But then they reorganized the extra water intake into temporal register with their food intake, thereby anticipating the greater need for it that was created by the metabolism of the added protein.

The Ontogeny of Thirst, Especially in Mammals

Everything that is alive, both plants and animals, has had an individual developmental history, from the most lowly of creatures to animals like ourselves, and no aspect of biology can be completely understood without an understanding of its ontogeny. Research on the ontogeny of thirst began only in the early and mid-1970s (Babicky, Pavlik, Pavlik, Ostadalova, & Kolar, 1972; Almli, 1973; Wirth & Epstein, 1976), and it has already taught us several things of value.

First, we know from research on the suckling rat that thirst is precocious. Neonatal rat pups that are still entirely dependent on mother's milk for all of their food and water will drink water that is delivered directly into their mouths or is available from puddles at their feet when they are made thirsty by cell dehydration, deprivation (removal from the dam), hypovolemia, or by systemic or intracranial angiotensin, and they do so before the end of the first week of postnatal life (Wirth & Epstein, 1976). We also know that they do not begin to ingest water freely until they are 16 or 17 days old (Babicky et al., 1972; Almli, 1973), demonstrating that the neural mechanisms for water drinking and for thirst are mature well before the animal begins to drink water.

Second, we know that the water drinking that is elicited from the neonatal rat by all of these dipsogens is an early expression of adultlike ingestive behavior that is mediated by thirst mechanisms that are nascent in the brain of the pup. It occurs only when they are away from their dam and not while they are suckling. In fact, dipsogens either have no effect on pups that are suckling from their dam or decrease their intake of mother's milk despite its high water content (Leshem & Epstein, 1988). Moreover, pups that are expressing this precocious form of adult thirst make the movements of drinking (repeated mouth opening, lapping, and swallowing) and do not attempt to suck the water they are offered (Almli, 1973; Wirth & Epstein, 1976).

This distinction between suckling on the one hand and feeding and drinking (or independent ingestion) on the other and their underlying neurological mechanisms arose from studies of fluid intake in neonatal rats that were either sucking mother's milk from their dam or were being fed milk or water, usually by direct infusion into their mouths, while they were away from her. It was first made by Drewett (1978) and has been discussed recently by several authors (Epstein, 1984; Hall & Williams, 1983). It is nicely illustrated by Leshem's recent work, cited earlier in this chapter, which describes the developmental timetable for the maturation of the brain mechanisms for angiotensin-induced water intake (Leshem, Boggan, & Epstein, 1988). He activated brain angiotensin by injecting pups intracranially with renin and demonstrated that the selective intake of water that is characteristic of adult angiotensin-induced drinking behavior matures in the neonatal rat at 16 days of age. This was shown in pups that had been removed from their dam and were offered water or milk that was infused directly into their mouths while they rested in a heated and humidified chamber. Activation of their brain angiotensin resulted in an increase in both water and milk intake in pups that were between 8 and 15 days of age. Sixteen-day-old pups and pups that were older increased only their water intake. But when the experiment was repeated in pups that were reunited with their dam and were suckling in a natural litter, intracranial renin had no effect on their intake of mother's milk even when they were less than 16 days old and could be induced by the brain renin injections to consume both water and milk while they were in the test chamber and away from their dams.

The newborn rat, and presumably newborn mammals in general, is born with several separate neurological systems for ingestion; one is for suckling, which is the most recent ingestive behavior to have evolved among mammals, and which is expressed when the pup is with its dam, and the others are for independent feeding and drinking. These latter mechanisms are nascent in the neonate's brain and are somehow suppressed whenever the pup is with its dam. At weaning suckling is suppressed, and the neurological systems for independent feeding and drinking gain complete control of ingestive behavior.

To understand this, think of locomotion and of how common it is among animals like amphibians and birds to have two kinds of locomotory behavior, each with its own neurological mechanism and each with its own developmental calendar. Many birds, for example, both walk and fly. They usually walk before they fly, and different neurological mechanisms obviously are employed for the two behaviors. Similarly, the mammalian brain contains neurological mechanisms for different kinds of ingestive behavior, and much remains to be done in future research to understand both the nature of these neural mechanisms and how their succession is controlled.

A third way in which this work on the ontogeny of the ingestive behaviors has been valuable is the clarity with which it demonstrates that these behaviors are innate. Animals that have never encountered water and that have not previously been deprived of it (recall that the high water content of milk and the pups' frequent ingestion of it assure that they are not dehydrated) drink it selectively when treated for the first time with a variety of dipsogens, and this is true of birds (Stricker & Sterritt, 1967) as well as mammals (Wirth & Epstein, 1976). The neural mechanisms for thirst are products of the animal's genome that have been realized in its normal development and are preformed in its brain. Thirst and the acts of water ingestion are not acquired. Learning, especially of the places in its environment at which fluids will be found and of the operants that can be employed to gain access to it, is, of course, important once weaning has begun (Kriekhaus & Wolf, 1968; Teitelbaum, 1977; Paulus, Eng, &

Schulkin, 1984), but the urge to drink water and the motor programs for its ingestion once it has passed the lips are part of the animal's phenotype.

Fourthly, the ontogenetic research has revealed that the behaviors of independent ingestion mature abruptly and sequentially (Wirth & Epstein, 1976; Ellis, Axt, & Epstein, 1984; Leshem, Boggan, & Epstein, 1988). The animal does not become competent for feeding and drinking slowly over several days, and it does not express all aspects of both behaviors at once. Instead, they are either absent from or, once a critical age has been reached, present in the animal's behavioral repertoire. For example, there is a clear sequence in the ages at which the several kinds of thirst make their debuts during the first week of the rat pup's life. Immediately after birth the pup can not be made to drink water by any known dipsogen, but at 3 days of age cellular dehydration thirst matures, followed 2 days later by that of hypovolemia and then by the drinking that is elicited by angiotensin, which, as discussed above, becomes selective for water abruptly at 16 days of age.

The same principles are illustrated by the ontogeny of feeding. Distension of the upper GI tract inhibits milk intake at birth (Houpt & Epstein, 1973), but decreases in intracellular fuel utilization (glucoprivation and ketoprivation) increase food intake only after weaning (Houpt & Epstein, 1973; Gisel & Henning, 1980; Leshem, Flynn, & Epstein, 1990). Again, there is an invitation here for future work. What is happening in the brain of the pup to provide for these successive steps toward fully mature ingestive behaviors? What are the developmental events in growth of neural circuitry or in maturity of receptor systems or effector mechanisms that occur in what could be a matter of hours to suddenly increase the animal's behavioral competence?

And lastly, what should we make of the precocity of independent ingestive behaviors in general and of thirst in particular? Is the fact that the neonatal rat has neural mechanisms for responses to dipsogens and for the drinking of water long before they are required nothing more than an expression of nascent adult behaviors? Or are these early capabilities for adult ingestion important for the optimal development of adult ingestive behaviors? There are at least two ways in which their precocity may be important. First, the arousal of the several kinds of thirst and the performance of the behaviors of water ingestion early in life may be necessary for their full development, as is the case for pattern vision (Wiesel & Hubel, 1963). Mammals are endowed by their genome with the programs for the development of the appropriate neural mechanisms for species-specific behaviors, and the optimal ontogeny of the mechanisms may require that they be used in the neonatal period. We have an indication that a process of this kind may operate in the development of suckling. Newborn rat pups that are deprived of contact with their dam for the first 12 hours after birth never suckle as effectively as their normally reared peers (Dollinger, Holloway, & Denenberg, 1978).

The early performance of the independent ingestive behaviors may be important for their insertion into the routines of freely emitted behaviors that the animal will express later in life. The rat pup suckles, locomotes, grooms, scratches, urinates, and defecates before it eats and drinks. When weaning begins it nibbles at food and samples water. It does so while performing its other behaviors, and these may be added to the clusters of behaviors it expresses while it is awake because, as discussed above, they are comforting and pleasurable, and because they forestall nutrient deficits. Later, when weaning is completed, meals and drafts of water may occur as parts of habitual domestic routines of behaviors that, as I suggest above, are the settings in which freely emitted drinking (and eating) will be understood.

This makes weaning a crucial period for future research. How do animals like rats make the transition from suckling (which occurs most often in daytime because the dam is away from her litter more often at night) to adult feeding and drinking? And how do they organize the bouts or clusters of activities in which their freely emitted domestic behaviors will occur for the rest of their lives? We know that in the rat the process of weaning begins at the end of the second week of suckling and continues for at least another week or 10 days. We know something about the decline in the dam's attractiveness to her pups as her production of ceacotroph declines (Leon, 1974), about the adoption by the pups of their nocturnal way of life (Levin & Stern, 1975), and about the tutoring the pups receive from adult females in the location and acceptability of foods (Galef & Clark, 1977). But I know of only one study (Kissileff, 1971) of both the feeding and drinking behaviors of pups at weaning (and it shows, interestingly, that they have a high incidence of prandial drinking), and of no studies of the ontogeny of all the freely emitted behaviors as they are expressed by the developing mammal.

Salt Intake

Salt intake (or natriorexia) is similar to thirst in important respects. It is innate, it serves homeostasis, and it can also occur when the animal is replete and is therefore, on occasion, another of the freely emitted behaviors. Salt appetite or salt hunger is the form of the behavior that serves homeostasis. It is the increase in the intake of salty commodities that occurs when the animals is sodium deficient. The freely emitted form of the behavior has been called salt preference, but I prefer to think of it as "need-free" salt intake. It is common among mammals and occurs in birds that have access to dilute salty solutions and, of course, is one of man's most interesting and medically relevant ingestive behaviors (Fregly & Kare, 1982).

Salt Appetite

Future research on salt appetite will be shaped by the two contrasting proposals for its physiological mechanism that have been made in the past decade. Denton and his colleagues in Australia, on the one hand (Weisinger et al., 1982; Blair-West et al., 1987), have demonstrated in sheep and cows that salt appetite can be aroused and satiated by decreases or increases in brain sodium, and they believe that the behavior is governed by sodium sensors in the walls of the third ventricle. My colleagues and I, on the other hand (Epstein, 1985), can arouse or satiate the behavior in the rat and pigeon by either administering or by pharmacologically blocking angiotensin II and aldosterone, which are the hormones of renal sodium conservation, and we propose that they are also the hormones of salt appetite. We believe that they generate the behavior by acting in synergy on the brain. In both proposals, sodium deficiency in the periphery triggers events in the brain (activity in brain sodium sensors in the Australian proposal, activation of brain angiotensin and release of aldosterone and its subsequent action on the brain in ours) that then generate the behavior, but the mechanisms by which these are produced are unknown.

There are several other proposed mechanisms for salt appetite. Two of them, Fregly's and Contreras', focus on changes in salt sensitivity within the oral cavity and are, therefore, in the tradition of Richter. He proposed, when he discovered the phenomenon, that the adrenalectomized rat drank more NaCl solution because of a lowered taste threshold for sodium (Richter, 1936). These periperal sensory theories of the behavior are attractively simple but conceptually inadequate. Their explanation of the behavior begins only after the salty commodity has contacted the taste receptors. They have nothing to say about the events that precede that contact, the events within the brain by which the urge to consume salt is aroused, by which it is expected and sought in appetitive behavior, and by which the animal's hedonic reaction to the salt is altered.

Fregly's concept has the virtue of having been the first to propose a hormonal basis for the appetite (Fregly & Waters, 1967). Although he acknowledges a role for angiotensin in his most recent discussion of his ideas (Fregly & Rowland, 1986), the mechanism he proposes is based on changes in salivary sodium and in sodium taste sensitivity that are secondary to changes in circulating aldosterone or exogenous mineralocorticoid. According to Fregly, intake of salt increases both when aldosterone levels rise and when they fall. Intake is at a nadir in the sodium-replete rat, whose blood aldosterone and salivary sodium are at normal levels, and increased both when the concentration of the hormone falls or, as in the adrenalectomized rat, is absent and when the concentration of the hormone rises, as in sodium deficiency.

In addition to being entirely peripheralist, these ideas suffer from an apparent internal contradiction. Increases in salt intake on either side of the normal aldosterone levels and normal salivary sodium of the replete rat are assumed to occur both when aldosterone is absent and salivary sodium is high and when salivary sodium is low because of the high aldosterone level of sodium deficiency (Fregly, 1968). In addition, changes in salivary sodium are not necessary for expression of the appetite because it occurs in the DOCA-treated (Vance, 1965) and sodium-depleted (Wong & Krantz, 1977) salivarectomized rat. Increased salt intake and an apparent salt appetite reported by Fregly after salivarectomy and no other treatment is an apparent artifact of prandial drinking because it disappeared when the animals were switched from dry to wet food (Fregly, 1968). Lastly, any suggestion that bases the arousal of the appetite on the peripheral action of aldosterone, or of angiotensin for that matter, is weakened by the fact that the appetite of both the adrenalectomized and the sodium-deficient rat is entirely unaffected by peripheral blockade of either aldosterone or angiotensin II receptors with specific competitive analogues of each hormone (Sakai, Nicolaidis, & Epstein, 1986; Sakai & Epstein, 1990).

After making the interesting discovery that sodium-deficient rats have fewer salt-best fibers in their chorda tympani nerves, Contreras proposed the most recent of the peripheralist theories of the appetite (Contreras & Frank, 1979). He suggested that the animals increase their intake of sodium-containing commodities in order to bring the frequency of the afferent discharge in the peripheral sodium channel up to optimum levels. Again, this concept does not address the issue of the nature of the neural events that bring the animal to the salt and that precede its contact with the oral surface. And it does not explain the qualitative changes in the evaluation of the sodium stimulus that are expressed in the facial reactions of the deficient animal. The report of Schulkin and colleagues (Berridge, Flynn, Schulkin, & Grill, 1984), that the purely aversive facial reactions to the oral infusion of small volumes of strong NaCl solution that are elicited from the sodium-replete rat are converted to a mixture of acceptance and aversion reactions by an episode of sodium deficiency suggests that the sensory events that are associated with the appetite are not just changes in magnitude. They are, instead, changes in kind, and the changes are in the direction of positive hedonics. This is shown nicely by Scott's recent report that sodium deficiency shifts the overall activity of solitary nucleus gustatory units from salt-best to sugar-best cells (Jacobs, Gregory, & Scott, 1988), in the right direction, in other words, for a more positive hedonic evaluation.

Two other proposals have been made for the mechanism of the appetite, but only one of these is the focus of current work. Wolf and Stricker (1967) some years ago suggested that a reservoir somewhere in the body measured its own sodium content and somehow exercised control over salt ingestion. It has never been found and is not discussed in Stricker's most recent review of the mechanisms of salt appetite (Stricker & Verbalis, 1988). Lastly and most recently, Stricker and Verbalis (1987) suggest that endogenous brain oxytocin inhibits salt intake and that expressions of the appetite are the result of the release of behavioral mechanisms within the brain from this inhibition. Suppression of the appetite with intracranial oxytocin or, more convincingly, arousal of it with oxytocin blockers administered within the brain would support this idea, but these data have not yet been reported.

We are therefore left with two proposed mechanisms for salt appetite that are frameworks for future research. The Australian sodium receptors of the third ventricle that work in sheep and cattle and the Philadelphia synergy of angiotensin and aldosterone that works in rats and pigeons (Massi & Epstein, in press). The two proposals may be the result of real species differences. But that is less interesting than the possibility that they are extremes of a common mechanism. Increased salt intake can be aroused in sheep by pharmacological doses of DOCA (a mimic of aldosterone: Hamlin, Webb, Ling, & Bohr, 1988) and by angiotensin II, especially if it is given at high doses into the cerebral ventricles (Coghlan et al., 1981). But the hormones have not been given at the same time and at lower doses, which would test for the operation of a synergy mechanism. And the rat has not been fully evaluated for the possibility that sodium deficiency acts directly on the brain to trigger the activation of cerebral angiotensin in parallel with its well-known direct action on the kidney. Decreases in brain sodium may activate brain angiotensin in both sheep and rats, and when this is accompanied by peripheral sodium deficit (as it will certainly be during instances of the arousal of the appetite by natural causes), aldosterone will be released, and a synergy of the two hormones in the brain may cause the behavior.

Increases in salt intake can also be aroused by pregnancy (Pike & Yao, 1971) and by the hormones of reproduction (Schultes, Covelli, Denton, & Nelson, 1972) and ACTH (Blaine, Covelli, Denton, Nelson, & Schultes, 1975; Denton, 1984), but these have been demonstrated only in the rabbit. Intracranial carbachol, which is a pharacological mimic of acetylcholine, suppresses salt intake (Fitts, Thunhorst, & Simpson, 1987), as do the tachykinins (Massi, Micossi, De Caro, & Epstein, 1986) and the atrial natriuretic peptides (Fitts, Thunhorst, & Simpson, 1985) when they are given into the cerebral ventricles. The amphibian peptide kassinin, which is a ligand for tachykininlike receptors in the mammalian brain, is an especially interesting agent. It suppresses salt intake (both salt appetite and need-free intake) at very low doses without interfering with food intake or with angiotensin-induced drinking (Massi, Perfumi, De Caro, & Ep-

tomy, ureteric ligation, or bladder puncture, produce a profound aversion to
salty commodities that has not yet been analyzed (Fitzsimons & Stricker, 1971).
There may be an interesting and complex set of endogenous agents that arouse
and satiate salt appetite. We can look forward to learning more about them and
their interactions as well as about the role that sodium deficiency plays in trigger-
ing their actions.

ADDITIONAL ASPECTS OF THE HORMONAL SYNERGY

The angiotensin–aldosterone synergy proposal assumes that in the intact,
sodium-deficient animal the hormones act on separate receptor systems within
the brain for each hormone. This seems likely. Salt appetite can be produced by
each hormone acting alone (Rice & Richter, 1943; Braun-Mendenez & Brandt,
1952; Chiaraviglio, 1976; Avrith & Fitzsimons, 1980; Bryant, Epstein,
Fitzsimons, & Fluharty, 1980), pharmacological blockade of each hormone alone
does not reduce the appetite aroused by administration of the other hormone
(Sakai et al., 1986; Sakai & Epstein, 1990), and medial amygdala damage to the
rat brain selectively abolishes the salt appetite aroused by the mineralocorticoids
(Schulkin, Marini, & Epstein, 1989). Such animals do not increase their salt
intake in response to aldosterone or DOCA but remain responsive to the natriorexigenic effect of angiotensin.

Research on the physiology of salt appetite began with Richter's discovery of
the increase in salt intake that is produced by adrenalectomy (Richter, 1936) and
with his and Braun-Mendenez's demonstrations that pharmacological doses of a
mineralocorticoid also produce the behavior in the sodium-replete rat (Rice &
Richter, 1943; Braun-Mendenez & Brandt, 1952). Both of these classic phenomena can now be understood as instances of the abnormal operation of the
synergy mechanism. The mechanism is activated by both hormones when they
are acting within their physiological ranges, but it can also be activated by each
hormone acting alone if it is in excess. The increased salt intake that is induced
by large doses of DOCA or aldosterone is an obvious example. Adrenalectomy is
its counterpart. Removal of the adrenal removes the source of aldosterone. In its
absence sodium loss in the urine cannot be controlled. This leads to release of
renal renin and, somehow, to activation of brain angiotensin, which is the cause
of the salt appetite of the adrenalectomized rat. Activation of brain angiotensin
with intracranial renin or renin substrate (Fitzsimons, 1979) leads to increases in
NaCl as well as water intake, and blockade of angiotensin receptors within the
brain of the adrenalectomized rat with competitive analogues of angiotensin II
completely suppresses the adrenalectomized rat's avidity for salt (Sakai & Epstein, 1990) without interfering with its other ingestive behaviors.

And, lastly, studies of the synergy mechanism have revealed what may be an
organizational effect of the hormones on the brain mechanisms for salt appetite.
Rats that have had prior sodium depletions or that have been exposed to brain
angiotensin and aldosterone while sodium replete drink salt solutions more
rapidly and in greater volume during subsequent expressions of salt appetite
(Sakai, Fine, Frankmann, & Epstein, 1987). This results in a lifelong enhancement of depletion-induced salt intake. The enhancement is not gradual but
quantal, and it does not depend on the drinking of the salt solution during the
initial expression of the appetite. It appears that the hormones have two effects

during the first occasion of their synergistic action. They arouse an innate brain circuit for salt intake, and they somehow produce irreversible changes in the brain that make the animal thereafter more responsive to the combined actions of angiotensin and aldosterone.

NEED-FREE SALT INTAKE

Rats and other animals drink salt that they do not need. That is, if NaCl solutions are available in addition to water and commercial diets (which usually contain ten times more sodium than is required by the rat for its daily nutrition), rats that are in good health will consume small volumes of it nightly even if its concentration is as high as 3%. This need-free intake is greater in females, which may be an expression of a suppressive effect of androgen (Krecek, 1973). This kind of salt intake is also enhanced by prior sodium depletions. Several such depletions are required for maximum effect, but fewer are needed in females, and their intakes escalate to higher asymptotic levels (Sakai, Frankmann, Fine, & Epstein, 1989). Four prior depletions at weekly intervals produce final levels of average daily 3% NaCl intake that are similar in males to those that are consumed by many adrenalectomized rats (10–12 mL/night) and that approach 20–25 mL/night in females. This is voluntary salt intake by rats that have restored their prior deficits, that have no renal pathology, and that do not have elevated levels of angiotensin and aldosterone. It is, therefore, a model of human salt overconsumption.

It should attract research interest in the future, first because of its medical relevance and second because of its ontogenetic implications: can the enhancement of need-free intake be produced in adult animals that have had sodium depletion in infancy? Third, it will attract interest because of its biological implication. What role does it play in the life of the rat, and why have the brain mechanisms for it evolved? Does it promote high avidity for salt in animals that do not need it in order to assure that future needs will be avoided? Is it greater in females because they must donate sodium to their offspring?

And, lastly, this phenomenon of enhanced daily need-free salt intake in the multidepleted rat will attract attention because it, like the related phenomenon of increased need-induced salt intake in animals with similar depletion histories, is an irreversible change in brain function that is somehow linked to the actions of hormones. This means that all the power and precision of modern biological technology is available for its cellular, molecular, and genetic analysis, and this analysis can be done with continuous reference to an innate, specific, sexually dimorphic, and medically relevant motivated behavior.

SUMMARY

The prospects for research on the behavioral neuroscience of thirst are discussed, and the following issues are emphasized. First, the research problem for students of thirst is an understanding of the central neural mechanism that creates the state of readiness to drink water, and it is suggested that sleep is a better model for our thinking than are Sherringtonian reflexes. Second, the research problem is complicated, but made more interesting, by the biological diversity of thirst, which is revealed in species (or rather taxonic) differences as well as in differences in thirst within a single species. These are differences that arise from the sex and the developmental stage of the animal or from its advanced age and from differences in the biobehavioral determinants of thirst, making it likely that there are qualitatively different central states of thirst. Third, the nature of the neural mechanism itself is identified as a major problem for the future of the science, and much needs to be done to understand how the physiological determinants of thirst (cellular dehydration, hypovolemia, and angiotensin II) use it to arouse homeostatic drinking behavior. Fourth, non-homeostatic or "freely emitted" water drinking remains an intriguing and important problem and may be understood when well-practiced routines of domestic behaviors are more fully investigated. And, fifth, studies of the ontogeny of thirst will continue to be rewarding because they may yield an understanding of freely emitted drinking and because they offer opportunities for inquiry into the following issues: the precocity of the brain mechanisms for thirst in the infant mammal, the differences between suckling behavior and early independent ingestion, the innateness of the ingestive behaviors, and the sequential and quantal nature of the development of the behaviors and therefore of their neural substrates. Finally, ontogenetic studies should be pursued because they provide an opportunity for inquiry into whether the preformed neural mechanisms of thirst, like those for others species-specific behaviors, are perfected by use in the neonatal period.

Rats and other animals ingest salt both when they need it (need-induced salt appetite or salt hunger) and when they do not (need-free salt intake). Future research on the behavioral neuroscience of need-induced salt appetite will be shaped by the contrasting results of work on sheep and cows, in which the behavior is controlled by changes in brain sodium, and of work on rats and pigeons, in which it is controlled by a synergy of angiotensin II and aldosterone. Need-free salt intake is incompletely understood, and additional research on it is urgently needed because it is the most common and most medically relevant form of the behavior.

Lastly, attention is called, first, to the fact that both need-induced and need-free salt intake are enhanced by prior episodes of sodium depletion, suggesting that the brain mechanisms for them are organized by the hormones of sodium conservation and, second, to the fact that the enhancement of need-free intake provides an experimental model for salt overconsumption in human populations.

Acknowledgments

I am grateful to Jay Schulkin, Sandra Frankmann, Eliot Stellar, and especially to Carl Thompson for careful criticisms of an earlier draft of this chapter. The writing of this chapter and the personal research discussed here were supported by NS 03469 and HD 25857.

REFERENCES

Almli, R. C. (1973). Ontogeny of onset of drinking and plasma osmotic pressure regulation. *Developmental Psychobiology, 6,* 147–158.

Almli, R. C., & Weiss, C. R. (1974). Drinking behaviors: Effects of lateral preoptic and lateral hypothalamic destruction. *Physiology and Behavior, 13,* 527–538.

Avrith, D., & Fitzsimons, J. T. (1980). Increased sodium appetite in the rat induced by intracranial administration of components of renin angiotensin system. *Journal of Physiology, (London), 301,* 349–364.

Babicky, A., Pavlik, L., Pavlik, J., Ostadalova, I., & Kolar, J. (1972). Determination of the onset of spontaneous water intake in infant rat. *Physiologia Bohemoslovaca, 21,* 467–471.

Balment, R. J., & Carrick, S. (1985). Endogenous renin–angiotensin system and drinking behavior in flounder. *American Journal of Physiology, 248,* R157–R160.

Barney, C. C., Threatte, R. M., & Fregly, M. J. (1983). Water deprivation-induced drinking in rats: Role of angiotensin II. *American Journal of Physiology, 244,* R244–R248.

Bellin, S. I., Bhatnagar, R. K., & Johnson, A. K. (1987). Perinventricular noradrenergic system are critical for angiotensin-induced drinking and blood pressure responses. *Brain Research, 403,* 105–112.

Berridge, K. C., Flynn, F. W., Schulkin, J., & Grill, H. (1984). Sodium depletion enhances palatability in rats. *Behavioral Neurosciences, 98,* 652–660.

Blaine, E., Covelli, M., Denton, D., Nelson, J., & Schultes, A. (1975). The role of ACTH and adrenal glucocorticoids in the salt appetite of wild rabbits (*Oryctolagus cuniculus* (L). *Endocrinology, 97,* 793–801.

Blair-West, J. R., Denton, D. A., Gellatly, D. R., McKinley, M. J., Nelson, J. F., & Weisinger, R. S. (1987). Changes in sodium appetite in cattle induced by changes in CSF sodium concentration and osmolality. *Physiology and Behavior, 39,* 465–469.

Blass, E. M., & Epstein, A. N. (1971). A lateral preoptic osmosensitive zone for thirst in the rat. *Journal of Comparative and Physiological Psychology, 76,* 378–394.

Braun-Menendez, E., & Brandt, P. (1952). Augmentation de l'appetit specifique pour le chlorure de sodium provoquee par le desoxycorticosterone: Caracteristiques. *Comptes Rendus de la Societe de Biologie, 146,* 1980–1982.

Bryant, R. W., Epstein, A. N., Fitzsimons, J. T., & Fluharty, S. J. (1980). Arousal of a specific and persistent sodium appetite in the rat with continuous intracerebroventricular infusion of angiotensin II. *Journal of Physiology (London), 301,* 365–382.

Buggy, J., & Johnson, A. K. (1977). Preoptic–hypothalamic periventricular lesions: Thirst deficits and hypernatremia. *American Journal of Physiology, 233,* R44–R52.

Chiaraviglio, E. (1976). Effect of renin–angiotensin system on sodium intake. *Journal of Physiology (London), 255,* 57–66.

Coghlan, J., Considine, P., Denton, D., Fei, D., Leksell, L., McKinley, M., Muller, A., Tarjan, E., Weisinger, R., & Bradshaw, R. (1981). Sodium appetite in sheep induced by cerebral ventricular infusion of angiotensin: Comparison with sodium deficiency. *Science, 214,* 195–197.

Contreras, R. J., & Frank, M. E. (1979). Sodium deprivation alters neural responses to gustatory stimuli. *Journal of General Physiology, 73,* 569–578.

De Caro, G. (1986). Effects of peptides of the "gut–brain–skin triangle" on drinking behavior of rats and birds. In G. De Caro, A. N. Epstein, & M. Massi (Eds.), *The physiology of thirst and sodium appetite* (pp. 213–226). New York: Plenum Press.

De Castro, J. M. (1989). The interactions of fluid and food intake in the spontaneous feeding and drinking patterns of rats. *Physiology and Behavior, 45,* 861–870.

Denton, D. A. (1984). *The hunger for salt.* New York: Springer-Verlag.

Dethier, V. G., & Evans, D. R. (1961). Physiological control of water ingestion in the blowfly. *Biology Bulletin, 121,* 108–116.

Dollinger, M. J., Holloway, W. R., & Denenberg, V. H. (1978). Nipple attachment in rats during the first 24 hours of life. *Journal of Comparative and Physiological Psychology, 92,* 619–626.

Drewett, R. F. (1978). The development of motivational systems. *Progress in Brain Research, 48,* 407–417.

Edwards, G., & Ritter, R. (1982). Area postrema lesions increase drinking to angiotensin and extracellular dehydration. *Physiology and Behavior, 29,* 943–950.

Ellis, S., Axt, K., & Epstein, A. N. (1984). The arousal of ingestive behaviors by chemical injection into the brain of the suckling rat. *Journal of Neuroscience, 4,* 945–955.

Epstein, A. N. (1960). Water intake without the act of drinking. *Science, 131,* 497–498.

Epstein, A. N. (1971). The lateral hypothalamic syndrome: Its implications for the physiological psychology of hunger and thirst. In E. Stellar & J. M. Sprague (Eds.), *Progress in physiological psychology* (vol. 4, 263–311). New York: Academic Press.

Epstein, A. N. (1973). Epilogue: Retrospect and prognosis. In A. N. Epstein, H. R. Kissileff, & E. Stellar (Eds.), *The neuropsychology of thirst* (pp. 315–332). New York: Winston.

Epstein, A. N. (1982a). Instinct and motivation as explanations for complex behavior. In D. W. Pfaff (Ed.), *Physiological mechanisms of motivation* (pp. 25–55). New York: Springer-Verlag.

Epstein, A. N. (1982b). The physiology of thirst. In D. W. Pfaff (Ed.), *Physiological mechanisms of motivated behavior* (pp. 315–332). New York: Springer-Verlag.

Epstein, A. N. (1984). The ontogeny of neurochemical systems for feeding and drinking. *Proceedings of the Society for Experimental Biology and Medicine, 175,* 127–134.

Epstein, A. N. (1985). The dependence of the salt appetite of the rat on the hormonal consequences of sodium deficiency. *Journal de Physiologie (Paris), 79,* 495–498.

Epstein, A. N., Spector, D., Samman, A., & Goldblum, C. (1964). Exaggerated prandial drinking in the rat without salivary glands. *Nature, 201,* 1342–1343.

Evered, M. D., & Robinson, M. M. (1981). The renin–angiotensin system in drinking and cardiovascular responses to isoprenaline in the rat. *Journal of Physiology (London), 316,* 357–362.

Falk, J. L. (1967). Control of schedule-induced polydipsia: Type, size, and spacing of meals. *Journal of Experimental Analysis of Behavior, 10,* 199–206.

Findlay, A. L. R., Fitzsimons, J. T., & Kucharczyk, J. (1979). Dependence of spontaneous and angiotensin-induced drinking upon the estrous cycle and ovarian hormones. *Journal of Endocrinology, 82,* 215–225.

Fitts, D. A., Thunhorst, R. L., & Simpson, J. B. (1985). Diuresis and reduction of salt appetite by lateral ventricular infusions of atriopeptin II. *Brain Research, 348,* 118–124.

Fitts, D. A., Thunhorst, R. L., & Simpson, J. B. (1987). Modulation of salt appetite by lateral ventricular infusions of angiotensin II and carbachol. *Brain Research, 346,* 273–280.

Fitzsimons, J. T. (1964). Drinking caused by contraction of the inferior vena cava in the rat. *Nature, 204,* 479–480.

Fitzsimons, J. T. (1969). The role of a renal thirst factor in drinking induced by extracellular stimuli. *Journal of Physiology (London), 201,* 349–368.

Fitzsimons, J. T. (1979). *The physiology of thirst and sodium appetite.* Cambridge: Cambridge University Press.

Fitzsimons, J. T., & Elfont, R. M. (1982). Angiotensin does contribute to drinking induced by caval ligation in rat. *American Journal of Physiology, 243,* R558–R562.

Fitzsimons, J. T., Kucharczyk, J., & Richards, G. (1978). Systemic angiotensin-induced drinking in the dog: A physiological phenomenon. *Journal of Physiology (London), 276,* 435–448.

Fitzsimons, J. T., & Le Magnen, J. (1969). Eating as a regulatory control of drinking in the rat. *Journal of Comparative and Physiological Psychology, 67,* 273–283.

Fitzsimons, J. T., & Simons, B. J. (1969). The effect on drinking in the rat of intravenous angiotensin, given alone or in combination with other stimuli of thirst. *Journal of Physiology (London), 203* 45–57.

Fitzsimons, J. T., & Stricker, E. M. (1971). Sodium appetite and the renin–angiotensin system. *Nature New Biology, 231,* 58–60.

Fregly, M. J. (1968). The role of hormones in the regulation of salt intake in rats. In M. R. Kare & O. Maller (Eds.), *The chemical senses and nutrition* (pp. 115–138). Johns Hopkins University Press.

Fregly, M. J., & Kare, M. R. (1982). *The role of salt in cardiovascular hypertension.* New York: Academic Press.

Fregly, M. J., & Rowland, N. E. (1986). Hormonal and neural mechanisms of sodium appetite. *News in Physiological Sciences, 1,* 51–54.

Fregly, M. J., & Waters, I. W. (1967). Hormonal regulation of the spontaneous sodium chloride appetite of rats. In T. Hayashi (Ed.), *Olfaction and taste II* (pp. 439–458). New York: Pergamon Press.

Galef, B. G., Jr., & Clark, M. M. (1977). Mother's milk and adult presence: Two factors determining initial dietary selection by weanling pups. *Journal of Comparative and Physiological Psychology, 78,* 220–225.

Ganten, D., Fuxe, K., Phillips, M. I., Mann, J. F. E., & Ganten, V. (1978). The brain isorenin-angiotensin system: Biochemistry, localization, and possible role in drinking and blood pressure regulation. In D. Ganong & L. Martini (Eds.), *Frontiers in Neuroendocrinology* (vol. 5, pp. 61–99). New York: Raven Press.

Gisel, E. G., & Henning, S. J. (1980). Appearance of glucoprivic control of feeding behavior in the developing rat. *Physiology and Behavior, 24,* 313–318.

Grill, H. J., & Miselis, R. R. (1981). Lack of ingestive compensation to dehydrational stimuli in decerebrates. *American Journal of Physiology, 240,* R81–R86.

Grossman, S. P. (1984). A reassessment of the brain mechanisms that control thirst. *Neuroscience and Biobehavioral Reviews, 8,* 95–104.

Hainsworth, F. R., Stricker, E. M., & Epstein, A. N. (1968). The water metabolism of the rat in the heat: Dehydration and drinking. *American Journal of Physiology, 214,* 983–989.

Hall, W. G., & Williams, C. L. (1983). Suckling isn't feeding, or is it? A search for developmental continuities. *Advances in the Study of Behavior, 13,* 219–254.

Hamlin, M. N., Webb, R. C., Ling, W. D., & Bohr, D. F. (1988). Parallel effects of DOCA on salt appetite, thirst, and blood pressure in sheep. *Proceedings of the Society for Experimental Biology and Medicine, 188,* 46–51.

Hamilton, C. L. (1969). Problems of refeeding after starvation in the rat. *Annals of the New York Academy of Sciences, 157,* 1004–1017.

Hatton, G. I., & Almli, C. R. (1969). Plasma osmotic pressure and volume changes as determinants of drinking thresholds. *Physiology and Behavior, 4,* 207–214, 1969.

Hirano, T. (1974). Some factors regulating water intake by the eel *Anguilla japonica. Journal of Experimental Biology, 61,* 737–747.

Houpt, K. A., & Epstein, A. N. (1971). The complete dependence of beta-adrenegic drinking on the renal dipsogen. *Physiology and Behavior, 7,* 897–902.

Houpt, K. A., & Epstein, A. N. (1973). The ontogeny of the controls of food intake in the rat: GI fill and glucoprivation. *American Journal of Physiology, 225,* 58–66.

Hsiao, S., Epstein, A. N., & Camardo, J. S. (1977). The dipsogenic potency of peripheral angiotensin II. *Hormones and Behavior, 8,* 129–140.

Jacobs, K. M., Gregory, M. P., & Scott, T. T. (1988). Taste responses in nucleus tractus solitarius of sodium-deprived rats. *Journal of Physiology (London), 406,* 393–410.

Johnson, A. K. (1985). The periventricular anteroventral third ventricle (AV3V): Its relationship with the subfornical organ and neural systems involved in maintaining body fluid homeostasis. *Brain Research Bulletin, 15,* 595–601.

Johnson, A. K., & Buggy, J. (1977). A critical analysis of the site of action for the dipsogenic effect of angiotensin II. In J. Buckley & C. M. Ferrario (Eds.), *Central actions of angiotensin and related hormones* (pp. 357–386). New York: Pergamon Press.

Johnson, A. K., Mann, J. E. F., Rascher, W., Johnson, J. K., & Ganten, D. (1981). Plasma angiotensin II concentrations and experimentally induced thirst. *American Journal of Physiology, 240,* 229–234.

Johnson, A. K., & Wilkin, L. D. (1987). The lamina terminalis. In P. M. Gross (Ed.), *Circumventricular organs and body fluids* (vol. 3, pp. 125–141). Boca Raton, FL: CRC Press.

Jonklaas, J., & Buggy, J. (1984). Angiotensin–estrogen interaction in female brain reduces drinking and pressor responses. *American Journal of Physiology, 247,* R167–R172.

Jonklaas, J., & Buggy, J. (1985). Angiotensin–estrogen central interaction: Localization and mechanism. *Brain Research, 326,* 239–249.

Katovich, M. J., Barney, C. C., Fregly, M. J., Tyler, P. E., & Dasler, R. (1979). Relationship between thermogenic drinking and plasma renin activity in the rat. *Aviation Space and Environmental Medicine, 50,* 721–724.

Kenney, N. J., & Moe, K. E. (1981). The role of endogenous prostaglandin E in angiotensin-II induced drinking. *Journal of Comparative and Physiological Psychology, 95,* 383–390.

Kissileff, H. (1969a). Food associated drinking in the rat. *Journal of Comparative and Physiological Psychology, 67,* 284–300.

Kissileff, H. (1969b). Oropharyngeal control of prandial drinking. *Journal of Comparative and Physiological Psychology, 67,* 309–319.

Kissileff, H. (1971). Acquisition of prandial drinking in weaning rats and in rats recovering from lateral hypothalamic lesions. *Journal of Comparative and Physiological Psychology, 77,* 97–109.

Kleitman, N. (1963). *Sleep and wakefulness.* Chicago: University of Chicago Press.

Kraly, S. (1989). Drinking elicited by eating. In A. N. Epstein & A. R. Morrison (Eds.), *Progress in psychobiology and physiological psychology* (pp. 315–332). New York: Winston.

Kraly, S., & Corneilson, R. (1988). Angiotensin II mediates drinking elicited by histamine in rats. *Society for Neuroscience Abstracts, 14,* 196.

Krecek, J. (1973). Sex differences in salt taste: The effect of testosterone. *Physiology and Behavior, 10,* 683–688.

Kriekhaus, E. E., & Wolf, G. (1968). Acquisition of sodium by rats: Interaction of innate mechanisms and latent learning. *Journal of Comparative and Physiological Psychology, 65,* 197–201.

Leon, M. (1974). Maternal pheromone. *Physiology and Behavior, 13,* 441–453.

eshem, M., Boggan, B., & Epstein, A. N. (1988). The ontogeny of drinking evoked by activation of brain angiotensin in the rat pup. *Developmental Psychology, 21*, 63–75.

eshem, M., & Epstein, A. N. (1988). Thirst-induced anorexias and the ontogeny of thirst in the rat. *Developmental Psychobiology, 21*, 651–662.

eshem, M., Flynn, F. W., & Epstein, A. N. (1990). The ontogeny of the metabolic controls of ingestion: Does brain energy privation control ingestion in the rat pup? *American Journal of Physiology, 258*, R365–R375.

evin, R., & Stern, J. M. (1975). Maternal influences on ontogeny of suckling and feeding rhythms in the rat. *Journal of Comparative and Physiological Psychology, 89*, 711–723, 1975.

ind, R. W., & Johnson, A. K. (1982). Central and peripheral mechanisms mediating angiotensin-induced thirst. In D. Ganten, M. Priuta, M. I. Phillips, & A. Scholkens (Eds.), *The renin angiotensin system in the brain* (pp. 353–364). New York: Springer-Verlag.

ind, W. R., Swanson, L. W., & Ganten, D. (1985). Organization at angiotensin II immunoreactive cells and fibers in the rat central nervous system. *Neuroendocrinology, 40*, 2–24.

Malvin, R. L., Mouw, D., & Vander, A. J. (1977). Angiotensin: Physiological role in water-deprivation-induced thirst in rats. *Science, 197*, 171–173.

Mann, J. F. E., Johnson, A. K., & Ganten, D. (1980). Plasma angiotensin II: Dipsogenic levels and angiotensin-generating capacity of renin. *American Journal of Physiology, 238*, R372–R378.

Massi, M., & Epstein, A. N. (in press). Angiotensin/aldosterone synergy governs the salt appetite of the pigeon. *Appetite*.

Massi, M., Micossi, L. G., De Caro, G., & Epstein, A. N. (1986). Suppression of drinking but not feeding by central eledoisin and physalaemin in the rat. *Appetite, 7*, 63–71.

Massi, M., Perfumi, M., De Caro, G., & Epstein, A. N. (1988). Inhibitory effect of kassinin on salt intake induced by different natriorexigenic treatments in the rat. *Brain Research, 440*, 232–242.

Milgram, N. W., Krames, L., & Thompson, R. (1974). Influence of drinking history on food-deprived drinking in the rat. *Journal of Comparative and Physiological Psychology, 87*, 126–133.

Miselis, R. R. (1981). The efferent projections of the subfornical organ of the rat: A circumventricular organ within a neural network subserving water balance. *Brain Research, 230*, 1–37.

Miselis, R. R., Shapiro, E. R., & Hand, P. J. (1979). Subfornical organ efferents to neural systems for control of body water. *Science, 205*, 1022–1025.

Mogenson, G. L. (1987). Limbic–motor integration. *Progress in Psychobiology and Physiological Psychology, 12*, 117–158.

Ohman, L., & Johnson, A. K. (1986). Lesions in lateral parabrachial nucleus enhance drinking to angiotensin II and isoproterenol, *American Journal of Physiology, 251*, R504–R511.

Paulus, R. A., Eng, R., & Schulkin, J. (1984). Preoperative latent place learning preserves salt appetite following damage to the central gustatory system. *Behavioral Neuroscience, 98*, 146–151.

Peck, J. W., & Novin, D. (1971). Evidence that osmoreceptors mediating drinking in rabbits are in the lateral preoptic area. *Journal of Comparative and Physiological Psychology, 74*, 134–147.

Pike, R. L., & Yao, C. (1971). Increased sodium chloride appetite during pregnancy in the rat. *Journal of Nutrition, 101*, 169–176.

Ramsay, D. J., Rolls, B. J., & Wood, R. J. (1977). Thirst following water deprivation in dogs. *American Journal of Physiology, 232*, R93–R100.

Rettig, R., Ganten, D., & Johnson, A. K. (1981). Isoproterenol-induced thirst: Renal and extrarenal mechanisms. *American Journal of Physiology, 241*, R152–R159.

Rice, K. K., & Richter, C. P. (1943). Increased sodium chloride and water intake in normals rats treated with desoxycorticosterone acetate. *Endocrinology, 33*, 106–115.

Richter, C. P. (1936). Increased salt appetite in adrenalectomized rats. *American Journal of Physiology, 115*, 155–161.

Rowland, N. E., & Fregly, M. (1988). Sodium appetite: Species and strain differences and role of renin–angiotensin–aldosterone system. *Appetite, 11*, 143–178.

Rowland, N., & Nicolaidis, S. (1974). Periprandial self-intravenous drinking in the rat. *Journal of Comparative and Physiological Psychology, 87*, 16–25.

Sakai, R. R., & Epstein, A. N. (1990). The dependence of adrenalectomy-induced sodium appetite on the action of angiotensin II in the brain of the rat. *Behavioral Neuroscience, 104*, 167–176.

Sakai, R. R., Fine, W. B., Frankmann, S. P., & Epstein, A. N. (1987). Salt appetite is enhanced by one prior episode of sodium depletion in the rat. *Behavioral Neuroscience, 101*, 724–731.

Sakai, R. R., Frankmann, S. P., Fine, W. B., & Epstein, A. N. (1989). Prior episodes of sodium depletion increase the need-free sodium intake of the rat. *Behavioral Neuroscience, 103*, 186–192.

Sakai, R. R., Nicolaidis, S., & Epstein, A. N. (1986). Salt appetite is completely suppressed by interference with angiotensin II and aldosterone. *American Journal of Physiology, 251*, R762–R768.

Schmidt-Nielsen, K. (1979). *Desert animals.* New York: Dover Press.

Schulkin, J., Marini, J., & Epstein, A. N. (1989). A role for the medial region of the amygdala in mineralocorticoid-induced salt hunger. *Behavioral Neuroscience, 103*, 724–731.

Schultes, A. A., Covelli, M. D., Denton, D. A., & Nelson, J. F. (1972). Hormonal factors influencing salt appetite in lactation. *Australian Journal of Experimental Biological and Medical Sciences, 50*, 819–826.

Simpson, J. B. (1981). The circumventricular organs and the central actions of angiotensin. *Neuroendocrinology, 32*, 248–256.

Simpson, J. B., Epstein, A. N., & Camardo, J. S. (1978). Localization of dipsogenic receptors for angiotensin in subfornical organ. *Journal of Comparative and Physiological Psychology, 92*, 768–795.

Simpson, J. B., & Routtenberg, A. (1973). The subfornical organ: Site of drinking elicitation by angiotensin II. *Science, 818*, 1172–1174.

Stricker, E. M., & Sterritt, G. M. (1967). Osmoregulation in the newly hatched domestic chick. *Physiology and Behavior, 2*, 117–119.

Stricker, E. M., & Verbalis, J. G. (1987). Central inhibitory control of sodium appetite in rats: Correlation with pituitary oxytocin secretion. *Behavioral Neuroscience, 101*, 560–567.

Stricker, E. M., & Verbalis, J. G. (1988). Hormones and behavior: The biology of thirst and sodium appetite. *American Scientist, 76*, 261–267.

Tarttelin, M. F., & Gorski, R. A. (1971). Variations in food and water intake in the normal and acyclic female rat. *Physiology and Behavior, 7*, 847–852.

Teitelbaum, P. (1977). Levels of integration of the operant. In W. K. Honig & J. E. R. Staddon (Eds.), *Handbook of operant behavior* (pp. 67–83). Engelwood Cliffs, NJ: Prentice-Hall.

Thrasher, T. N., Keil, L. C., & Ramsay, D. J. (1982). Lesions of organum vasculosum of the lamina terminalis (OVLT) attenuate osmotically-induced drinking and vasopressin secretion in the dog. *Endocrinology, 110*, 1837–1845.

Vance, W. B. (1965). Observations on the role of salivary secretions in the regulation of food and fluid intake in the white rat. *Psychology Monographs, 79*, 1–22.

Vijande, M., Costales, M., Schiaffini, O., & Marin, B. (1978). Angiotensin-induced drinking: Sexual differences. *Pharmacology Biochemistry and Behavior, 8*, 753–755.

Weisinger, R. S., Considine, P., Denton, D. A., Leksell, L. G., McKinley, M. J., Mouw, D., Muller, A., & Tarjan, E. (1982). Role of sodium concentration of the cerebrospinal fluid in the salt appetite of sheep. *American Journal of Physiology, 242*, R51–63.

Wiesel, T. N., & Hubel, D. H. (1963). Single-cell responses in striate cortex of kittens deprived of vision in one eye. *Journal of Neurophysiology, 26*, 1003–1017.

Williams, D. R., & Teitelbaum, P. (1956). Control of drinking behavior by means of an operant-conditioning technique. *Science, 124*, 1294–1296.

Wirth, J. B., & Epstein, A. N. (1976). Ontogeny of thirst in the infant rat. *American Journal of Physiology, 230*, 188–198.

Wolf, G., & Stricker, E. M. (1967). Sodium appetite elicited by hypovolemia in adrenalectomized rats: Re-evaluation of the "reservoir" hypothesis. *Journal of Comparative and Physiological Psychology, 63*, 252–257.

Wong, R. & Krantz, L. (1977). Desalivation and saline ingestion in rats. *Behavioral Biology, 19*, 130–134.

Wright, J. W., Morseth, S. L., Fairley, P. C., Petersen, E. P., & Harding, J. W. (1987). Angiotensin's contribution to dipsogenic additivity in several rodent species. *Behavioral Neuroscience, 101*, 361–370.

Zucker, I. (1969). Hormonal determinants of sex differences in saccharin preference, food intake, and body weight. *Physiology and Behavior, 4*, 595–602.

19

Making Sense Out of Calories

Mark I. Friedman

INTRODUCTION

It is axiomatic that body energy balance is equal to the difference between energy intake and expenditure. The simplicity of this formula, however, masks the enormous complexity of the biological processes that maintain caloric homeostasis. Elaborate physiological, neural, endocrine, and biochemical mechanisms maximize the efficient allocation and use of body energy resources. But because organisms continually expend energy to live, calories must also be brought in from external sources. This is achieved by eating food. The nervous system directs this behavior in response to signals that link food intake with the requirements of caloric homeostasis. How this control is accomplished is largely a mystery. Much is known already about the mechanisms that control energy expenditures. However, we will not have a full understanding of caloric homeostasis until we solve the behavioral side of the energy balance equation as well.

Control of food intake is usually characterized as a very complicated affair influenced by many factors. As more research is done, the list of controls grows. The complexity is daunting, and the conclusion, inevitably, is that food intake is under multifactorial control. It almost seems that there is the expectation that a clear understanding of food intake will materialize when enough controls are added to the list. The opposite, however, is probably true: the longer the list, the less we really know. The growing roster betrays a lack of integration and synthesis, of understanding and explanation. It is time to start paring the list.

The most urgent challenge in the years ahead is to develop theories of the control of food intake that bring order to the empirical observations that continue to mount. Without theoretical direction, we should expect a fragmentary understanding of the problem. Control of food intake has been studied from a variety of perspectives, including those of gastrointestinal physiology, neurology,

---

MARK I. FRIEDMAN   Monell Chemical Senses Center, Philadelphia, Pennsylvania 19104.

endocrinology, metabolism, and psychology. This pluralism, although providing a broad research front, no doubt has contributed to the proliferation of factor thought to govern feeding. All of these approaches, however, contend with the same basic issue of stimulus and receptor: What is the signal controlling food intake, and where is the sensor that detects it? Drawing on this common element—making the problem of food intake a sensory problem—is one way to organize the search for controls and start sorting the list.

Where we stand on the issue of stimulus and receptor for feeding as we approach the 21st century is in many ways reminiscent of the state of knowledge of vision, the model sensory system, at the time of the ancient Greeks (see Polyak 1957). They did not know that light is the stimulus and the eye the receptor organ for sight. Instead they believed that visual images resulted from the projection of visual rays out of the eyes onto objects in the world. As absurd as this "emanation" theory of vision seems now, it survived with few opponents for over 2,000 years until the beginning of the 17th century, when it was established that the eye is an optical instrument for collecting the light stimulus and projecting it onto retinal receptors. Solving the problem of stimulus and receptor will be just as crucial for progress in understanding control of food intake. We hope, given our substantial foundation of physiological and biochemical knowledge, that we are two decades, not two millennia, from a solution.

A Sensory Problem

Viewing the study of food intake as a problem in sensation turns the search for controls into a search for sensory stimuli. This by itself helps to shorten the roster of controls of food intake, since many of the factors that are often included in such lists are not sensory events. Many are mechanisms that translate the effects of sensory input; they serve but do not initiate, modulate, or direct. For example, according to this perspective, neurotransmitters are not controls of food intake, although they may mediate the effect of a controlling signal. Experimental treatments also are not controls. Thus, for example, the eating response after administration of an inhibitor of glucose utilization is "glucoprivic eating" only in an operational sense; it does not reflect a glucoprivic control of food intake unless the stimulus elicited by that treatment has been elucidated. It is tempting to attribute the property of a control or stimulus to intervening mechanisms and experimental treatments as biological and pharmacological agents that affect food intake are discovered and as new technologies are developed to measure more and more biological parameters. A strong theoretical orientation to the problem of food intake control would avoid this confusion, for it would show us which experimental treatments to use, which to measure, and where to look for mechanisms.

Brain versus Body

Physiological theories of the control of food intake have alternated between those emphasizing the central nervous system and those focusing on peripheral organs. In the last 10 years, the field has taken a decided turn from central toward peripheral explanations. In addition to a resurgence of interest in the role of the gastrointestinal tract, and the stomach in particular, the liver has received increasing attention as an organ involved in food consumption. Usually, when one theory overthrows another, scientific knowledge takes a leap forward. However, it is not clear that the shifts between central and peripheral theories have advanced our understanding of the control of food intake; a pendulum swings, but it really goes nowhere.

Seeing controls of food intake as sensory stimuli changes how the role of the brain and peripheral organs are conceptualized. The issue behind central versus peripheral theories of feeding is not whether the nervous system is involved—it obviously is or there would be no feeding. Nor is the question whether a theory based on the organization of the brain is better than one based on the function of a peripheral organ. Neither is better because the source of control is misplaced. Control of intake does not lie in a particular organ or place but in the sensory stimuli that are detected and processed by the nervous system, whether they originate or are sensed in a peripheral organ or the brain. Taking a sensory approach to food intake control thus breaks the brain–body dichotomy that has plagued the field for so long (see Le Magnen, 1971). It also eliminates anatomic sites from our list of controls.

Identification of the sensory signals directing food intake will change our experimental approach to the analysis of neural mechanisms for feeding. When the stimuli can be defined, the mechanism will follow; without the stimuli, the mechanism loses its function, and the purpose of the investigation fades. Imagine if vision were still thought to result from the projection of rays out of the eyes onto the external world. Early anatomic descriptions of the visual system were shaped by this theory, and we might still believe that the optic nerves were hollow tubes that served as a conduit for the visual rays. Conceivably, Hubel and Wiesel could have performed their electrophysiological studies of the visual cortex, except they would have thought they were studying an efferent system for projecting rays, not an afferent system for coding patterns of light. Knowing the stimuli controlling feeding will provide meaningful points of entry into the nervous system, whether they be in the periphery or brain, and a functional basis and rationale for the neuroscientific investigation of food consumption.

Short Term versus Long Term

There are many different facets to the control of feeding, and one might expect that different controlling signals are involved in each. Which sensory systems are singled out for study will depend on what it is we wish to explain. The controls on our list therefore can be organized and pared in different ways depending on which aspects of feeding are under consideration. Ordering the list like this also provides insight into the function and role of the various stimuli that control feeding.

It is common practice to distinguish between controls of food intake that operate over the long term, usually a day or more, and those acting in the short term, over minutes or hours. Long-term control is typically associated with maintenance of overall body energy balance, whereas short-term control is tied to the sporadic nature of feeding behavior and food intake. Whether this distinction is based in physiology or is merely an illusion of the time interval chosen for measurement is unclear. Food intake measured over long intervals is less variable than when it is measured over shorter intervals and may appear, therefore, more closely tied to a homeostatic function. However, this does not mean that controls of food intake linked to caloric homeostasis cannot operate on a moment-to-moment basis as well. We are simply not in the position to assign such functions to controls of food intake until the controlling stimuli have been identified.

How versus How Much

Instead of dividing controls along a continuum of time, it may be more useful to sort them according to their involvement in food intake as opposed to feeding behavior. This distinguishes the means (behavior), which can vary enormously depending on environmental and other constraints, from the end (intake), which is tied to regulatory processes of caloric homeostasis and varies less. In this case, "food intake" is the amount consumed, whereas "feeding behavior" means the pattern of feeding. It is a distinction between "how much" food is eaten (food intake) and "how" that quantity is consumed (feeding behavior). Feeding behavior here refers to the frequency and duration—the timing—of meals or eating bouts, not the discrete motor patterns associated with food ingestion, which may be governed by yet another set of stimuli (see Chapter 6).

Controls of feeding behavior commonly are separated into those for meal initiation and those for meal termination. It is questionable whether such start/stop functions can be assigned to physiological controls of feeding before the signals have been identified. Empirically, the distinction is difficult to justify because one treatment can both initiate eating and lengthen a feeding bout depending on relatively minor differences in testing procedures (e.g., Friedman & Tordoff, 1986). Meal initiation and termination may simply be different sides of the same coin. From a theoretical perspective, it is not clear what two hypothetical stimuli (or classes of stimuli) add that one cannot accomplish; one stimulus for feeding might wax and wane, and other unrelated events might also disrupt eating. Nor is it possible to say whether the stimulus is excitatory or inhibitory. Either a decrease in an inhibitory signal or increase in an excitatory one could start a meal, whereas a decline in an excitatory stimulus or an increase in an inhibitory one could cause a meal to stop. Any explanation will do. There do seem to be more experimental treatments that can stop feeding than start it (see Stricker, 1984), but this does not necessarily imply that feeding behavior is under a largely inhibitory control. Provoking a feeding bout may simply require a more specific stimulus than ending one does, and feeding may be more easily disrupted by nonspecific events or competing needs as the stimulus to eat declines.

We probably will not understand the determinants of meal size and frequency until the stimulus and receptor mechanism controlling food intake—the amount consumed—are identified. It is the amount of food eaten that defines the pattern of consumption, because the behavior operates in the service of caloric requirements. When access to food is limited, eating patterns adjust, sometimes very considerably, in order to maintain caloric intake (Collier, 1986). In contrast, inducing a marked change in eating patterns has no effect on caloric intake (Weingarten, 1984). Meals often are seen as the building blocks of food intake; the amount consumed is the sum of the behavior. However, since feeding behavior operates within the boundaries set by the amount of energy that needs to be ingested, the question is what proportion of the variability in meal taking is determined by caloric requirements, not vice versa.

Within the context of energy balance, it is the number of calories consumed that is crucial, not the pattern in which those calories are consumed. The link from the physiological to behavioral component of caloric homeostasis should therefore be sought in the controls, the sensory stimuli, governing how much food is consumed. What kinds of signals serve this important function? Where and how are the demands of caloric homeostasis turned into sensations the nervous system can recognize?

Gustatory and Olfactory Sensations

There is no doubt that the chemical senses are important in the recognition of food and in guiding its selection. Whether the taste and smell of food affect the amount consumed over anything but the short term, however, is another matter. It is widely believed that the hedonic response to food flavor, the palatability of food, drives food intake; when food tastes good it is overconsumed, when it tastes bad it is underconsumed. Increased food palatability has been offered as the explanation for the chronic overeating induced by feeding rats cafeteria diets, high-fat diets, or sucrose solutions. Indeed, the assumption that diet-induced hyperphagia is related to the palatability of the food is so ingrained that the term "palatable" is often used, much as "pelleted" or "liquid" are, to describe a physical property of the diet. What is remarkable about the notion that palatability drives intake, given its almost universal acceptance, is the paucity of evidence to support it. When one looks for an effect lasting beyond a few hours, there is no evidence (Ramirez, Tordoff, & Friedman, 1989).

Studies suggesting that diet palatability increases food intake are flawed for several reasons (see Ramirez *et al.*, 1989). In most cases they provide no independent evidence that the diets used were palatable. They were thought to be so because intake increased or because they were palatable to the experimenter. Additionally, in every case, changes in the sensory properties of the diets were confounded with changes in the nutritional composition, which alone can produce increases in food intake through effects on digestion and metabolism. There also is direct evidence that changes in diet palatability alone do not alter food intake (see Ramirez *et al.*, 1989). For example, rats do not overeat nutritionally controlled diets to which are added palatable flavors shown to be highly preferred in independent tests (Naim, Brand, Kare, & Carpenter, 1985). Overeating of a preferred liquid diet occurs even when the diet is made less palatable than the control diet (Ramirez, 1988).

It is possible that under some conditions changes in diet palatability *per se* could be shown to alter food intake for more than a brief period. As the evidence stands now, however, it seems unlikely that such conditions would be anything but extraordinary. In any case, it seems clear that a reevaluation of the role of the chemical senses and the hedonic aspect of food in the control of food intake is in order. If, as it appears, palatability can be eliminated as an explanation of diet-induced hyperphagia, then what remains to be accounted for is the mechanism by which modifications in diet composition increase caloric intake and produce obesity. Answering this nutritional question will require analysis of the dietary factors involved as well as understanding of the signals that control food intake.

MARK I.
FRIEDMAN

In the same work in which he outlined his emanation theory of vision, Plato proposed that the large capacity of the stomach and long length of the intestine prevent a hunger so persistent as to make philosophy and culture impossible (*Timaeus*). Since the ancient Greeks, an association between the gastrointestinal tract and appetite for food has been suggested many times. In this century, the most notable advocate was Cannon, who proposed in 1912 that gastric contractions were the cause of hunger. This theory was generally accepted for many years, even long after evidence refuting it had accumulated (see Bolles, 1975). Early on, Cannon was careful to distinguish hunger, which for him was the sensation associated with hunger pangs, from food ingestion, which was determined by other factors as well. The focus was on the experience of hunger (see Rosenzweig, 1962). In contrast, contemporary research is concerned with the role of the stomach in the control of food and caloric intake. It is Cannon's concept of homeostasis, not his theory of hunger, that informs this work.

GASTRIC SIGNALS

Interest in a gastric control of food intake has been revived in recent years largely through the studies of Deutsch (see Chapter 7) and McHugh and Moran (1985), which suggest that gastric distention inhibits food intake in a quantitative manner. Because the degree and duration of distention depend on the rate of gastric emptying, food intake is thought to change in order to compensate for variations in the rate of emptying, which could be determined by changes in pyloric function. When emptying is constant, intake adjusts to compensate for differences in the time required to empty gastric contents that vary in caloric value. Presumably, nerves from the stomach, or perhaps the pylorus, carry the gastric stimulus to the brain. Potentially, this signal could account for changes in meal size and frequency. For example, slow emptying would result in faster distention and a shorter, smaller meal, whereas fast gastric emptying might rapidly reduce the signal of distention and initiate a meal sooner.

Experiments suggesting a role for gastric distention and emptying on food intake usually employ short-term tests in animals adapted to restricted feeding schedules that limit the time they have to eat. Measures of gastric emptying are taken over a short time interval, usually associated with the scheduled feeding period. These and other special procedures are used to standardize food intake test periods, stabilize intakes, and conveniently measure gastric emptying. However, whether the results obtained in these experiments can be generalized to other, more normal conditions is an important unanswered question. A feeding schedule that demands a rapid rate of intake to meet caloric requirements may push the capacity for gastric emptying closer to its limits and make a signal of gastric fill especially salient. Such a situation might obtain during the early period of refeeding after the normal diurnal fast. On the other hand, when animals are allowed to eat *ad libitum*, the coupling between gastric fill (or clearance) and food intake may not be so striking (Edens & Friedman, 1988).

We need to clarify the relationship, if any, between alterations in gastric function and chronic changes in food intake. Specific surgical manipulations of the stomach or its innervation usually do not lead to changes in food intake although they might alter feeding patterns. It is difficult to attribute chronic changes in food intake to alterations in stomach emptying because the amount of food ingested drives intestinal adaptations that determine the absorptive capacity of the intestinal tract and in turn may affect rates of gastric clearance. Because postabsorptive events can modify gastric function (e.g., Granneman & Friedman, 1980), it is also possible that changes in intake and gastric emptying, which are functionally linked as part of a food delivery system, are controlled in parallel by mechanisms operating beyond the gastrointestinal tract.

GUT PEPTIDES

During the last 10 years there has been a burst of interest in the role of gut peptides, particularly cholecystokinin (CCK), in the control of feeding behavior. Injection of CCK has been shown to reduce short-term food intake in a variety of species (Baile, McLaughlin, & Della-Fera, 1986). Administration of the hormone also elicits a pattern of behavior—termed the "satiety sequence"—that is seen following a meal. The effect of CCK appears to be on meal-taking behavior rather than overall food intake, as administration of the peptide in a meal-contingent fashion alters *ad libitum* feeding patterns but not daily caloric intake (West, Fey, & Woods, 1984).

Because CCK is an endogenous substance, it is tempting to think that its effect on feeding is physiological, not pharmacological. But despite the amount of work done in this area, it is unknown whether inhibition of feeding behavior is a normal action of the peptide. The reduction of food intake after injection of CCK constitutes the primary evidence that the peptide inhibits feeding, yet it is not clear that CCK acts specifically; for example, sexual behavior is also suppressed by the same doses that depress food intake (Mendelson & Gorzalka, 1984). Similarly, the behaviors associated with the satiety sequence are also observed in the postcopulatory interval (Dewsbury, 1967). Growing evidence suggests that CCK has nonspecific adverse effects on feeding. Administration of CCK has been shown to produce learned taste and place aversions in animals (e.g., Deutsch & Hardy, 1977; Moore & Deutsch, 1985; Swerdlow, Van der Kooy, Koob, & Wenger, 1983), has been associated with reports of malaise in humans (e.g., Miaskiewicz, Stricker, & Verbalis, 1989), and has been found to elicit neuroendocrine responses associated with the emetic reflex (see Verbalis, Richardson, & Stricker, 1987). This issue of CCK-induced illness was raised in the first reports that the peptide decreased food intake and persists to this day. There are many fascinating aspects to CCK physiology. Until the issues of specificity and illness are resolved, we will not know whether control of feeding behavior is one of them.

METABOLIC SENSATIONS

The idea that metabolism generates stimuli that control food intake is a fairly recent one. Its most direct origins are in the work of Carlson (1916), who, in light of Cannon's prevailing gastric model, proposed that decreases in blood glucose level initiate the gastric contractions that cause hunger. Mayer (1955) later modified this mechanism by moving the critical event from the extracellular blood to the inside of cells and by transforming it from a change in level to a change in utilization. According to his "glucostatic" hypothesis, increases in the intracellular utilization of glucose in hypothalamic cells provide the stimulus for satiety, whereas decreases lead to hunger. About the same time, Kennedy (1953) shifted the question of stimulus by advocating a role of body fat stores and lipid metabolism in the control of food intake. As in the glucostatic model, Kennedy saw the hypothalamus as a receptor site in this "lipostatic" control of food intake. In 1963, Russek proposed that the liver was a site for glucoreceptors that controlled food intake by modifying activity in hypothalamic neurons. This relocated the receptor for feeding from the brain to a peripheral organ that is pivotal in metabolic regulation and caloric homeostasis. His later discussions of hepatic metabolic events that control food intake moved the stimulus again, this time down specific metabolic pathways (see Russek & Racotta, 1986).

Metabolic signals have been implicated in control of food intake—the amount consumed—under a variety of physiological, pathological, and experimental conditions (e.g., Friedman, in press; Friedman & Stricker, 1976; Le Magnen, 1981; Wade & Gray, 1979). There is good evidence for a metabolic control in many species, including humans (e.g., Anil & Forbes, 1988; Shurlock & Forbes, 1981; Thompson & Campbell, 1977; Woods, Stein, McKay, & Porte, 1984). Currently, liver and brain are still the best candidates for receptor sites that detect the metabolic sensations directing food intake, but the nature of the metabolic stimulus is still not well defined. By analogy to vision, we have a good idea that the eye is involved in sight but have yet to discover light.

A Metabolic Perspective

Energy balance is maintained by the machinery of metabolism, and it is at the level of metabolism that the link is made between the demands of caloric homeostasis and the behavior of eating. The metabolic stimulus and receptor controlling food intake will be difficult to identify and locate. The sensory events and the sensors that monitor them reside in the interior, cellular environment and are not readily accessible for study. The receptors may be hard to recognize and isolate because they do not have to have a specialized morphology but instead may simply have special neural connections. The stimulus may be hard to isolate as well, since metabolic pathways are overlapping and sequentially ordered. Also, fuel metabolism, like all homeostatic systems, has evolved to defy perturbation, and this capacity of the system to compensate for disturbances thwarts attempts to study it. It also makes it an enticing problem.

These and other problems will tax our ingenuity and technical prowess. However, the existing body of knowledge on fuel metabolism, along with the array of techniques that have been developed to study it, confers important advantages in an analysis of the metabolic control of food intake, advantages not unlike those that knowledge of physics and optics had for understanding the stimulus for vision and function of the eye.

The Stimulus

Traditionally, two metabolic stimuli have been implicated in the control of food intake: one generated by the utilization of glucose and another associated with the metabolism and storage of body fat. The mechanisms for feeding associated with these two stimuli have long been imagined to work together (Mayer, 1955). Such a coordinated control has been used to account for a variety of phenomena including diurnal rhythms of eating, maintenance of caloric intake, and the response to over and underfeeding (see Friedman & Tordoff, 1986; Le Magnen, 1981). Recently, direct evidence for a coordinated metabolic control of food intake has been provided by studies showing that simultaneous inhibition of fatty acid and glucose utilization produces a synergistic increase in food intake (Friedman & Tordoff, 1986; Friedman, Tordoff, & Ramirez, 1986). This interactive effect strongly suggests that an integrated signal from the metabolism of glucose and fat controls feeding. Integration could occur at a neural level, with inputs from different sensors detecting changes in glucose and fat metabolism. This has been the prevailing view since the early 1950s. Alternatively, integration may occur at a metabolic level, with some event common to the metabolism of glucose and fat generating the stimulus. This hypothesis, and the more general issue of neural versus metabolic integration, needs to be explored in the 1990s.

The idea that a common pathway in the processing of metabolic fuels produces a stimulus directing food intake has its origins in the work of Ugolev and Kassil (1961; see also Kassil, Ugolev, & Chernigovskii, 1970), who proposed that the signal emerges from the tricarboxylic acid cycle, where carbohydrates, fats, and amino acids are enzymatically degraded to carbon dioxide and hydrogen atoms. More recently, others have focused on the ensuing process of oxidative phosphorylation and electron transport (e.g., Friedman & Stricker, 1976; Langhans & Scharrer, 1987; Nicolaidis & Even, 1985), in which water is formed and adenosine triphosphate (ATP), a primary molecular energy source, is generated. The oxidative stimulus is thought to be generated in nonmuscular tissue. Because it is associated with oxidative phosphorylation, it occurs in mitochondria and is independent of the type of fuel and whether it derives from diet or endogenous stores. The changes in oxidation are inversely related to food intake; relative increases in oxidation produce decreases in food intake, whereas decreases result in increased intake. And the fluctuations in oxidation need not be extreme, at least not in the sense of a typical depletion–repletion model, but can occur within a normal range associated with feeding and fasting.

At a theoretical level, an oxidative stimulus for food intake is very attractive. It is a more parsimonious view than one based on separate signals from glucose and fat metabolism, and it accounts for a number of findings that are anomalous for traditional glucostatic and lipostatic models (Friedman & Stricker, 1976). It also helps to explain a number of phenomena related to control of food intake and feeding behavior, some of which are discussed below. Nevertheless, despite the theoretical appeal, the event providing the oxidative stimulus has not been identified, and the evidence for such a signal remains circumstantial. There is, of course, always the possibility that separate stimuli associated with glucose utilization, fat metabolism, and oxidation together control food intake. However, the relative lack of work on the oxidative stimulus and the potential insights that could derive from elucidating it suggest this is where efforts should be concentrated in the coming years.

Attempts to characterize the metabolic sensations controlling food intake eventually must deal with the issue of signal transduction—how processes of fuel metabolism are translated into an event that affects the nervous system. What speculation there has been about the mechanism for transduction has centered largely on changes in cellular potentials (Russek, 1975; Langhans & Scharrer, 1987) with the underlying assumption that the receptor is a nerve cell. This may not be the case, however; the sensor could be a specialized secondary receptor separate from the primary afferent nerve. There is precedent for such a sensory mechanism in taste cells, the carotid body, and portal vein osmoreceptors. Alternatively, the receptor may function as an endocrine cell and secrete an agent that acts at some distance. Any number of transduction mechanisms are possible—there certainly is a variety to choose from that have been identified in other sensory systems. For the oxidative stimulus, I tend to favor a mechanism based on intracellular calcium. Cytosolic calcium concentration can vary with mitochondrial oxidative metabolism (see Lehninger, Vercesi, & Bababunmi, 1978) and, as a second messenger, can initiate both secretory events and changes in nerve cell function (see Mandel & Eaton, 1987). Calcium controls intracellular fuel metabolism (see Lehninger *et al.*, 1978) and therefore could serve as an intermediary that coordinates cellular and behavioral mechanisms of caloric homeostasis.

THE RECEPTOR

Before the metabolic stimulus can be established and characterized in detail, and before any analysis of transduction mechanism can begin, we must know where to look. The liver and brain are generally considered the sense organs for detecting metabolic events that control food intake, although the specific receptor cells have not been identified or isolated. Going from a tissue to a cellular analysis of the metabolic receptors is an important goal. How quickly we get there will depend on how certain we are about the location of the receptor organ.

Identification of a stimulus and receptor are interrelated problems. Characterization of the stimulus requires that its effect on a specialized receptor be demonstrated, and identification of the receptor demands that it be activated or altered by its adequate stimulus. In the case of the metabolic control of food intake, the two problems are especially intertwined, since the metabolic organization that generates the stimulus also defines the tissue or cell type: different tissues have characteristic metabolic profiles. On balance, this interrelationship should be advantageous because work on the stimulus will inform the search for the receptor, and vice versa.

There are four lines of evidence that suggest that changes in hepatic metabolism modulate food intake (see Novin & VanderWeele, 1977; Russek, 1986). First, infusions of metabolic fuels into the hepatic portal vein reduce food intake more effectively than infusions into the systemic circulation (see Tordoff & Friedman, 1986, 1988). Second, food intake is altered by metabolic substrates, pharmacological agents, or substrate analogues that act primarily or exclusively in liver (e.g., see Russek, 1986; Tordoff & Friedman, 1988; Tordoff, Rafka, DiNovi, & Friedman, 1988). Third, changes in hepatic metabolism parallel the behavioral effects of nutrient infusions (Tordoff & Friedman, 1988). Fourth, hepatic nerve section alters food intake and the ingestive response to treatments that act specifically on liver (e.g., Anil & Forbes, 1988; Friedman & Granneman, 1983; Friedman & Sawchenko, 1984). That a hepatic control of food intake might operate under normal conditions is further suggested by the observation that the suppressive effect of portal substrate infusion is seen under *ad libitum* feeding conditions when substrate is infused at a rate and concentration that falls within the normal parameters of substrate delivery to the liver (e.g., Anil & Forbes, 1988; Tordoff & Friedman, 1986, 1988). Whether other factors associated with the prandial delivery of nutrients to the liver (e.g., increased blood flow, pancreatic secretions, release of gut peptides) alter the response to portal vein substrate infusion should be examined.

How information about hepatic metabolism reaches the brain to control food intake needs to be determined. To date there is no evidence for a humoral signal, although this is a logical possibility that should be explored. Hepatic nerve section studies point to a neural route, but it is not clear that the effects of nerve section result from interruption of sensory information, as hepatic nerves contain both afferent and efferent fibers (Friedman, 1988). Experiments showing that cutting hepatic nerves eliminates the effects on food intake of metabolic treatments specific to the liver inspire more confidence that sensory nerves are involved; however, in this case, section of efferent fibers could alter the effect of the metabolic manipulation. Also, hepatic nerves do not necessarily innervate only the liver (Prechtl & Powley, 1987). Thus, although nerve section studies can indicate that the liver is involved in food intake control, they do not establish a sensory function for the nerves that are cut. To do this, techniques for specific hepatic deafferentation must be developed.

At this time, evidence for a cerebral metabolic receptor controlling food intake is less convincing, largely because it is difficult to restrict experimentally induced changes in metabolism to the brain. Support for a cerebral site stems primarily from experiments showing that brain lesions alter the feeding response to metabolic manipulations and that administration of metabolic fuels, hormones, or inhibitors into the brain or its ventricles alters food intake. The specificity of brain lesion effects has been questioned before (Friedman & Stricker, 1976) and remains an issue that obscures interpretation of behavioral results and makes localization of function difficult. Direct manipulation of brain metabolism by cerebral injection is less problematic in this regard, especially when it produces increases in food intake instead of decreases, which can result from nonspecific effects. Such cerebral injections also cause neurally mediated perturbations in peripheral metabolism and gastrointestinal function; indeed, a hyperglycemic response is used to verify placement of cannulas for intracerebroventricular injection of metabolic inhibitors that stimulate feeding. It will be an important next step to determine whether a change in food intake after injection into the brain results directly from activation of cerebral metabolic receptors controlling food intake or is an indirect response to changes in metabolism and digestion that are detected by peripheral (hepatic or gastric) receptors. Is feeding the report or an echo from cerebral receptors?

In many ways this question about direct and indirect effects of cerebral metabolic receptors is reminiscent of the confusion about afferent and efferent functions of the eye, except, in the case of food intake, the mechanism behind the effects of brain manipulations may be efferent. Specifically, autonomic efferents under central control might modulate peripheral metabolism in the service of caloric homeostasis. Peripheral receptors, in turn, could assess the impact of these "emanations" to provide feedback to the brain. Under some circumstances, the neurally induced metabolic responses, like perturbations in metabolism produced by other means, would generate a peripheral stimulus that alters food intake and corrects the homeostatic imbalance, although under different conditions an adjustment in metabolism would suffice. In this way, the nervous system may coordinate physiological and behavioral mechanisms of caloric homeostasis.

MARK I.
FRIEDMAN

Although a picture of the sensory apparatus is just beginning to emerge, it is not too early to start exploring the ramifications a metabolic perspective holds for understanding the control of food intake, feeding behavior, and related issues. Indeed, the reach of a metabolic explanation of food intake tests and defines the power of this approach.

CALORIC INTAKE

The metabolic pathways provide a common ground for control of energy expenditures and energy intake. It seems reasonable that stimuli arising from these pathways would control caloric intake in response to energetic demands because it is through metabolism, and the oxidative pathway in particular, that biologically useful calories are realized. It is at the level of metabolism that calories are made into messages the nervous system can read. Although food intake may be controlled by a metabolic signal that carries information about calories, this stimulus is not just about calories in food but also about those from endogenous fuel sources. Because not all metabolic pathways are involved in energy production (i.e., in generating the stimulus), the fate of metabolic fuels from either diet or internal reserves becomes important information in interpreting regularities, changes, and disturbances in caloric intake. To link physiological states or experimental treatments with changes in intake, we have to know more about the metabolic processes that direct fuels into and away from oxidative pathways. We need a metabolic map of caloric intake, a map of the sensory field in which the metabolic stimulus lies.

EATING DISORDERS

Research on control of food intake has obvious ties to the study of obesity, and this relationship ought to be explored in terms of a metabolic control. In particular, the role of fuel partitioning between pathways of storage and oxidation should be examined in some depth. Typically, the connection between body fat stores and food intake has been considered in terms of a negative feedback humoral signal that reflects the degree of adiposity; however, despite an enormous amount of effort, there is no direct evidence that such a signal exists or that it is involved in hyperphagia (see Friedman, in press; Harris & Martin, 1984). Insulin has received a great deal of attention in this regard lately, but recent studies indicate that the hormone is not itself a signal controlling feeding. Rather, it appears that insulin affects food intake through its actions on fuel partitioning, which may vary depending on diet (Friedman, in press; Friedman & Ramirez, 1987).

Kennedy (1953) and others (see Friedman, in press) have suggested that fuel storage can influence food intake indirectly by altering the metabolic disposition of signals controlling food intake. From this point of view, hyperphagia might be a compensatory response to the "loss" of oxidizable fuels into storage; that is, overeating is secondary to increased fat deposition (Friedman, in press; Friedman & Stricker, 1976). This hypothesis, which ties in neatly with current studies showing decreased energy expenditure in obesity, should be examined in more detail. A metabolic approach to food intake control may be helpful as well in the study of appetite problems in other diseases associated with marked metabolic disturbances, such as cancer, diabetes, and liver disease. Such an analysis might provide insights into the nature of the stimulus for feeding as well as the etiology of the appetite disorders. Few treatment strategies for eating disorders are directed at the stimulus end of the control process. Identification of the metabolic stimulus and receptor for food intake control therefore might open other avenues for intervention.

Meal Patterns

Knowing more about the metabolic events that govern caloric intake should also tell us much more about feeding behavior. This would be true in general terms because, as discussed above, feeding patterns adjust to accommodate caloric requirements. However, studies by Le Magnen (1981) and Campfield and Smith (Chapter 8) and their colleagues showing that small decreases in blood glucose precede a spontaneous meal in rats raise the possibility that metabolic stimuli also can be tied empirically to normal meal taking. The prediction of feeding behavior from metabolic antecedents is an important goal because it provides a crucial test of any metabolic theory of food intake. However, the ability to make such predictions ultimately will depend on how well the metabolic stimulus can be isolated and specified. It is thus important to bear in mind that whereas the initiation of a meal may be related to fluctuations in blood glucose, this does not demonstrate that the measured change in glucose causes the meal. There may be other metabolic perturbations that occur concomitantly with those in blood glucose level, and earlier metabolic events that produce the fluctuations in glucose also could provoke a meal. For example, shifts in liver metabolism away from glucose production might cause the change in blood glucose as well as a feeding response. The simultaneous, on-line analysis of meal taking and metabolism is a powerful technique. Its application will surely be a challenge for the 21st century should the metabolic stimulus turn out to be an intracellular event rather than a change in circulating metabolic fuels.

Toward an Integrated Picture

We are beginning to see how chemosensory and metabolic stimuli operate together to control feeding behavior and food selection. Cephalic-phase digestive and endocrine reflexes elicited by the taste and smell of food have been suspected for many years of influencing feeding behavior. Recent experiments have shown that a nonnutritive sweet taste increases short-term food intake and creates a preference for the flavor of that food (Tordoff & Friedman, 1989a). Physiological studies indicate that cephalic-phase metabolic responses in liver may underlie these appetitive effects of sweet taste stimulation (Tordoff & Friedman, 1989b). Hepatic metabolism of ingested fuels, and perhaps hepatic fuel oxidation in particular, can produce unconditioned stimuli for the formation of a conditioned flavor preference (Tordoff & Friedman, 1986; Tordoff, Tepper, & Friedman, 1987). Such a mechanism could provide a metabolic basis for food selection or for learning about the caloric consequences of different foods. What the metabolic stimulus is, whether it is the same as that controlling intake or meal taking, and how the integration of metabolic and chemosensory events is accomplished are questions that will need to be addressed.

These kinds of studies linking the chemical (taste and smell) and metabolic senses in the control of feeding behavior point the way toward a broader goal: an integrated picture of the control of food intake and feeding behavior. The role of gastric signals should be included as well, and here too the relationship between gastrointestinal function and postabsorptive metabolism ought to be explored in more detail. For example, changes in liver metabolism during a meal could alter stomach-emptying rate (see Friedman, 1988) and thereby modulate the signal of gastric fill to reflect the metabolic value of the food being consumed. This postabsorptive assessment, when associated with the food's flavor, could, in turn, provide information on the degree of gastric fill needed to deliver a sufficient amount of calories from a food with those sensory properties. Other scenarios like this certainly are possible, and elucidating the metabolic stimulus and its receptor should play a central role in creating and testing them.

METABOLIC PERCEPTIONS

Theories of hunger at the beginning of this century and earlier were concerned with the experience—the perception—of hunger. The emphasis on behavior, both its regulatory aspects and patterning, came later. The terms "hunger" and "satiety" usually are used these days in an operational sense to describe changes in intake or food-related behaviors. Hunger and satiety are frequently assessed in studies of human food intake, but there is little attempt to determine the physiological origins of these subjective feelings. At the end of this century we should return again to this original problem.

In early studies on gastric contractions, the cause of the perception "hunger" was misplaced. This was shown clearly by experiments demonstrating the persistence of hunger after gastric nerve section or gastrectomy. The fact that the feeling of hunger could be referred to a stomach that was no longer there—a phantom stomach—indicated another source for the experience. Direct manipulations of metabolism have been known for many years to alter subjective feelings of hunger and satiety (e.g., Bulato & Carlson, 1924; Janowitz & Ivy, 1949; Thompson & Campbell, 1977). It may not be too farfetched to think that at the beginning of the next century we may have some insight into how metabolic sensations are transformed by the brain into the perceptions we call hunger and satiety.

CONCLUSION

How will we judge our progress in understanding the control of food intake? Success in alleviating clinical problems of food intake and appetite will provide one yardstick, but it is possible to be clinically effective without understanding the basic mechanisms involved. There was surgery for cataracts, and even eyeglasses, before the function of the eye was fully appreciated. The impact of our efforts on related problems will provide a gauge of progress. For example, elucidating the role of specific metabolic events in the control of feeding should inform the study of metabolism much as understanding vision taught early physicists about light. Most telling, however, will be our future perspective on the current state of affairs. Looking back in the year 2000, will we think that today's list of factors affecting feeding is too short and mistakenly believe that progress as been made? Or, will we see a relatively long list and view the end of the 20th century as the time when we started to make sense out of the control of food intake?

REFERENCES

Anil, M. H., & Forbes, J. M. (1988). The roles of hepatic nerves in the reduction of food intake as a consequence of intraportal sodium propionate administration in sheep. *Quarterly Journal of Experimental Physiology, 73*, 539–546.

Baile, C. A., McLaughlin, C. L., & Della-Fera, M. A. (1986). Role of cholecystokinin and opioid peptides in control of food intake. *Physiological Reviews, 66*, 172–234.

Bolles, R. C. (1975). *Theory of motivation* (pp. 110–113). New York: Harper & Row.

Bulato, E., & Carlson, A. J. (1924). Influence of experimental changes in blood sugar level on gastric hunger contractions. *American Journal of Physiology, 69*, 107–115.

Carlson, A. J. (1916). *The control of hunger in health and disease.* Chicago: University of Chicago Press.

Collier, G. (1986). The dialogue between the house economist and the resident physiologist. *Nutrition and Behavior, 3*, 9–26.

Deutsch, J. A., & Hardy, W. T. (1977). Cholecystokinin produces bait shyness in rats. *Nature, 266*, 196.

Dewsbury, A. A. (1967). A quantitative description of the behavior of rats during copulation. *Behavior, 29*, 154–178.

Edens, N. K., & Friedman, M. I. (1988). Satiating effect of fat in diabetic rats: Gastrointestinal and postabsorptive factors. *American Journal of Physiology, 255*, R123–R127.

Friedman, M. I. (1988). Hepatic nerve function. In I. M. Arias, W. B. Jakoby, H. Popper, D. Schachter, & D. A. Shafritz (Eds.), *The liver: Biology and pathobiology* (pp. 949–959). New York: Raven Press.

Friedman, M. I. (in press). Body fat and the metabolic control of food intake. *International Journal of Obesity.*

Friedman, M. I., & Granneman, J. (1983). Food intake and peripheral factors after recovery from insulin-induced hypoglycemia. *American Journal of Physiology, 244*, R374–R382.

Friedman, M. I., & Ramirez, I. (1987). Insulin counteracts the satiating effect of a fat meal in rats. *Physiology and Behavior, 40*, 655–659.

Friedman, M. I., & Sawchenko, P. E. (1984). Evidence for hepatic involvement in the control of ad libitum food intake in rats. *American Journal of Physiology, 247*, R106–R113.

Friedman, M. I., & Stricker, E. M. (1976). The physiological psychology of hunger: A physiological perspective. *Psychological Review, 83*, 409–431.

Friedman, M. I., & Tordoff, M. G. (1986). Fatty acid oxidation and glucose utilization interact to control food intake in rats. *American Journal of Physiology, 251*, R840–R845.

Friedman, M. I., Tordoff, M. G., & Ramirez, I. (1986). Integrated metabolic control of food intake. *Brain Research Bulletin, 17*, 855–859.

Granneman, J., & Friedman, M. I. (1980). Hepatic modulation of insulin-induced gastric acid secretion and EMG activity in rats. *American Journal of Physiology, 238*, R346–R352.

Harris, R. B. S., & Martin, R. J. (1984). Lipostatic theory of energy balance: Concepts and signals. *Nutrition and Behavior, 1*, 253–275.

Janowitz, H. D., & Ivy, A. C. (1949). Role of blood sugar levels in spontaneous and insulin-induced hunger in man. *Journal of Applied Physiology, 1*, 643–645.

Kassil, V. G., Ugolev, A. M., & Chernigovskii, V. N. (1970). Regulation of selection and consumption of food and metabolism. *Progress in Physiological Sciences, 1*, 387–404.

Kennedy, G. C. (1953). The role of depot fat in the hypothalamic control of food intake in the rat. *Proceedings of the Royal Society London, Series B, 140*, 578–592.

Langhans, W., & Scharrer, E. (1987). Evidence for the role of sodium pump of hepatocytes in the control of food intake. *Journal of the Autonomic Nervous System, 20*, 199–205.

Lehninger, A. L., Vercesi, A., & Bababunmi, E. A. (1978). Regulation of calcium release by the oxidoreduction state of pyridine nucleotides. *Proceeding of the National Academy of Sciences, 75*, 1690–1694.

Le Magnen, J. (1971). Advances in studies on the physiological control and regulation of food intake. In E. Stellar & J. M. Sprague (Eds.), *Progress in physiological psychology* (vol. 4, pp. 203–261). New York: Academic Press.

Le Magnen, J. (1981). The metabolic basis of dual periodicity of feeding in rats. *The Behavioral and Brain Sciences, 4,* 561–607.

Mandel, L. J., & Eaton, D. C. (Eds.). (1987). *Cell calcium and the control of membrane transport.* New York: Rockefeller University Press.

Mayer, J. (1955). Regulation of energy intake and the body weight: The glucostatic theory and the lipostatic hypothesis. *Annals of the New York Academy of Sciences, 63,* 15–42.

McHugh, P. R., & Moran, T. H. (1985). The stomach: A conception of its dynamic role in satiety. In J. M. Sprague & A. N. Epstein (Eds.), *Progress in psychobiology and physiological psychology* (pp. 197–232). New York: Academic Press.

Mendelson, S. D., & Gorzalka, B. B. (1984). Cholecystokinin-octapeptide produces inhibition of lordosis in the female rat. *Physiology and Behavior, 21,* 755–759.

Miaskiewicz, S. L., Stricker, E. M., & Verbalis, J. G. (1989). Neurohypophyseal secretion in response to cholecystokinin but not meal-induced gastric distention in humans. *Journal of Clinical Endocrinology and Metabolism, 68,* 837–843.

Moore, B. O., & Deutsch, J. A. (1985). An antiemetic is antidotal to the satiety effects of cholecystokinin. *Nature, 315,* 321–322.

Naim, M., Brand, J. G., Kare, M. R., & Carpenter, R. G. (1985). Energy intake, weight gain and fat deposition in rats fed flavored, nutritionally controlled diets in a multichoice ("cafeteria") design. *Journal of Nutrition, 115,* 1447–1458.

Nicolaidis, S., & Even, P. (1985). Physiological determinant of hunger, satiation and satiety. *American Journal of Clinical Nutrition, 42,* 1083–1092.

Novin, D., & VanderWeele, D. A. (1977). Visceral involvement in feeding: There is more to regulation than the hypothalamus. *Progress in Psychobiology and Physiological Psychology, 7,* 193–241.

Polyak, S. (1957). *Vertebrate visual system.* Chicago: University of Chicago Press.

Prechtl, J. C., & Powley, T. L. (1987). A light and electron microscopic examination of the vagal hepatic branch of the rat. *Anatomy and Embryology, 176,* 115–126.

Ramirez, I. (1988). Overeating, overweight and obesity induced by an unpreferred diet. *Physiology and Behavior, 43,* 501–506.

Ramirez, I., Tordoff, M. G., & Friedman, M. I. (1989). Dietary hyperphagia and obesity: What cause them? *Physiology and Behavior, 45,* 163–168.

Rosenzweig, M. R. (1962). The mechanisms of hunger and thirst. In L. Postman (Ed.), *Psychology in the making* (pp. 73–143). New York: Knopf.

Russek, M. (1963). Participation of hepatic glucoreceptors in the control of intake of food. *Nature, 197,* 79–80.

Russek, M. (1975). Current hypotheses in the control of feeding behaviour. In G. J. Mogenson & F. R. Calaresu (Eds.), *Neural integration of physiological mechanisms and behaviour* (pp. 128–147). Toronto: University of Toronto Press.

Russek, M. (1986). Possible participation of oro-, gastro-, and enterohepatic reflexes in preabsorptive satiation. In M. R. Kare & J. G. Brand (Eds.), *Interaction of the chemical senses with nutrition* (pp. 373–393). New York: Academic Press.

Russek, M., & Racotta, R. (1986). Possible participation of oro-, gastro-, and enterohepatic reflexes in preabsorptive satiation. In M. R. Kare & J. G. Brand (Eds.), *Interaction of the chemical senses with nutrition* (pp. 373–393). New York: Academic Press.

Shurlock, T. G. H., & Forbes, J. M. (1981). Evidence for hepatic glucostatic regulation of food intake in the domestic chicken and interaction with gastro-intestinal control. *British Poultry Science, 22,* 333–346.

Stricker, E. M. (1984). Biological bases of hunger and satiety: Therapeutic implications. *Nutrition Reviews, 42,* 333–340.

Swerdlow, N. R., Van der Kooy, D., Koob, G. F., & Wenger, J. R. (1983). Cholecystokinin produces conditioned place-aversions, not place preferences in food deprived rats: Evidence against involvement in satiety. *Life Sciences, 32,* 2087–2093.

Thompson, D. A., & Campbell, R. G. (1977). Hunger in humans induced by 2-deoxy-D-glucose: Glucoprivic control of taste preference and food intake. *Science, 198,* 1065–1067.

Tordoff, M. G., & Friedman, M. I. (1986). Hepatic portal glucose infusions decrease food intake and increase food preference. *American Journal of Physiology, 251,* R192–R196.

Tordoff, M. G., & Friedman, M. I. (1988). Hepatic control of feeding: Effect of glucose, fructose, and mannitol infusions. *American Journal of Physiology, 254,* R969–R976.

Tordoff, M. G., & Friedman, M. I. (1989a). Drinking saccharin increases food intake and preference I. Comparison with other drinks. *Appetite, 12,* 1–10.

Tordoff, M. G., & Friedman, M. I. (1989b). Drinking saccharin increases food intake and preference: IV. Cephalic phase and metabolic factors. *Appetite, 12,* 37–56.

Tordoff, M. G., Rafka, R., DiNovi, M. J., & Friedman, M. I. (1988). 2,5-Anhydro-D-mannitol: A fructose analogue that increases food intake in rats. *American Journal of Physiology, 254,* R150–R153.

Tordoff, M. G., Tepper, B. J., & Friedman, M. I. (1987). Food flavor preferences produced by drinking glucose and oil in normal and diabetic rats: Evidence for conditioning based on fuel oxidation. *Physiology and Behavior, 41,* 481–487.

Ugolev, A. M., & Kassil, V. G. (1961). [Physiology of appetite]. *Uspekhi Sovremennoi Biologii, 51,* 352–368.

Verbalis, J. G., Richardson, D. W., & Stricker, E. M. (1987). Vasopressin release in response to nausea-producing agents and cholecystokinin in monkeys. *American Journal of Physiology, 252,* R749–R753.

Wade, G. N., & Gray, J. M. (1979). Gonadal effects on food intake and adiposity: A metabolic hypothesis. *Physiology and Behavior, 22,* 583–593.

Weingarten, H. P. (1984). Meal initiation controlled by learned cues: Basic behavioral properties. *Appetite, 5,* 147–158.

West, D. B., Fey, D., & Woods, S. C. (1984). Cholecystokinin persistently suppresses meal size but not food intake in free-feeding rats. *American Journal of Physiology, 246,* R776–R787.

Woods, S. C., Stein, L. J., McKay, L. D., & Porte, D. (1984). Suppression of food intake by intravenous nutrients and insulin in the baboon. *American Journal of Physiology, 247,* R393–R401.

20
Clinical Issues in Food Ingestion and Body Weight Maintenance

Paul R. McHugh

Introduction

The alliance between clinician and basic scientist in the study of food ingestion and body weight control has been a sometime thing. Clinicians, in particular, concerned with management of obesity and anorexia nervosa, wax and wane in their patience with progress in basic science. Most opt for simple symptomatic treatments of such patients—dieting for the obese and renourishment for the anorexic—recognizing that the identification of the causes of these disorders and thus rational treatment and prevention is far from achieved. Basic scientists on the other hand intermittently look to the human resources of the clinician with enthusiasm, finding great interest in the "animal who talks," only eventually to return to laboratory animals when faced with the complicated problems of the "animal that deceives, denies, and distorts."

The purpose of this brief chapter is to lay out the natural linkages and some highlights of achievement of this alliance. I do this, as have others, in part to encourage what can be a faltering relationship but also to direct attention to distinctly different scientific domains in each of which clinicians and basic scientists can cooperate in the search for the causes of human disorders. It will be my point that it is the mutual responsibility of members of this alliance to appreciate that at present two rather distinct lines of investigation exist that can be culled to reveal details of causation in different human disorders. Eventually a union of these lines will be achieved, but this is not imminent, partly because both lines of investigation are still immature. An appreciation, though, of what each has

Paul R. McHugh   Department of Psychiatry and Behavioral Sciences, The Johns Hopkins University School of Medicine, Baltimore, Maryland 21205.   An early and abridged version of this chapter was presented at the Symposium on Psychobiology of Eating Disorders held at the New York Academy of Sciences on October 13–15, 1988.

proposed in the past and a focus on prototypic examples of contemporary study should reassure anyone that the clinical–basic science alliance remains lively and its future full of opportunity.

I can be more specific. Human feeding and weight control, like all the issues that rest on motivated behaviors, can be appreciated as having proximate and ultimate causes. The proximate or immediate causes of food intake and its control emerge from *experiments* revealing basic, regulatory physiology with its diversity of hierarchically organized mechanisms in the body comprehensible from the organismal to the molecular levels. Investigations to explain some feeding disorders have in the past and do in the present reveal how crucial aspects of these proximate controls can be injured. An important "two-way trade" between bedside and laboratory bench exists right now in this study.

On the other hand, the ultimate causes of any motivated behavior such as food intake and some of its uniquely human problems rest on progressively species-specific formative elements of the organism, that is, the human genome, its shaping over time by natural selection, and the successful emergence of a developmental program that depends on the interaction between the genetic instructions and the staged experiences imposed by the environment. As an ecological process, this program leads to the phenotypic expression of the genotypic direction. It is human epidemiology, the *prospective surveillance* of naturally occurring events, rather than experiment, that can bring to light the play of ultimate factors in the construction of some human disorders of food intake and weight control. This epidemiologic perspective broadly construed to grasp ethological, ecological, and genetic considerations has revealed equally intriguing results, as have experimental investigations into pathophysiology. It is the perspective that promises a synthesis of causation that may bring rational treatment and prevention together for many human disorders of body weight control.

Finally, to be still more specific and to introduce particular issues and our contemporary knowledge of them, human disorders of food intake and body weight control fall into two great and, at least at the moment, distinctive categories. One is based on crude pathology within the individual organism disrupting the normal functioning of the apparatus that controls the behavior, what might be called "spanner in the works" conditions. These represent the most natural and immediate sites for cooperation between clinicians and laboratory scientists. In the second category are those distortions or deviations in food intake and weight control specific to mankind, emerging with development and its capacity to direct, reinforce, or hinder behavioral expression. Such issues as genetic differences across populations, schedules of maturation, and variations in sociocultural processes that interact with intrapersonal factors all offer information to explain disorders of this second type and again represent opportunities for coordination of the work of clinicians and behavioral scientists.

PROXIMATE OR IMMEDIATE FACTOR COLLABORATION BETWEEN CLINICIANS AND BASIC SCIENTISTS

HYPOTHALAMIC OBESITY

The main reason for beginning with the proximate factors is that here successful collaboration between clinicians and basic scientists in the study of food intake and weight control first occurred. The very first recognition of an association between pathological lesions at the base of the brain and abnormal obesity was made in man by Fröhlich (Fröhlich, 1901; Bruch, 1939; Fulton, 1940) and was promptly followed up by Erdheim in 1904 who announced that it was injury to the hypothalamus and not to the pituitary gland that produced such obesity. These remarkable observations from the clinic can be considered the opening events in the collaboration of clinicians and basic scientists in obesity because they prompted Hetherington and Ranson (1940) to ask what specific structures in the hypothalamus must be destroyed to produce these effects. This question could best be addressed in animals and at the laboratory bench and required the redeployment of the Horsley–Clarke instrument for precise placement of lesions. The success at reproducibly evoking obesity with strategically placed lesions in the ventral and medial hypothalamus of the rat was a landmark discovery in every sense of the word. On the one hand, it promoted an enthusiasm for the Horsley–Clarke instrument throughout physiological psychology and, on the other, a sharp focus on the hypothalamus and obesity that led to Brobeck's demonstration (Brobeck, Tepperman, & Long, 1943) that the obesity was associated with overeating and not simply to passive metabolic factors turning calories into fat (Fulton, 1940; Erdheim, 1904; Hetherington & Ranson, 1940). The relationship of the overeating to autonomic imbalances and to the metabolic and endocrine responses provoked by the brain lesion is now the active focus of research (Bray & York, 1979; Friedman & Stricker, 1976; Powley, 1977) and is illuminated in other chapters in this volume.

It is not appropriate in this chapter to go into more detail about this great achievement, as I wish primarily to point it out as the classic example of work on the proximate or immediate causes in the control of food intake and body weight, where clinicians and basic scientists continue to collaborate to advantage. Suffice it to say that it has cultivated the general appreciation of the existence of physiological controls on body weight that generate their effects through behavior. The particular role of the hypothalamus itself remains to be enunciated. The preliminary view of "centers" for feeding and satiety has been replaced with concepts of a more dynamic character that offer promise in comprehending the stages in the behavioral responses of hypothalamic hyperphagia and obesity that rest on control theory. As well, an appreciation of an interactive role between the brain and the physiology and metabolism of the peripheral body has emerged. The concepts of transmitter-specific neural paths converging within the hypothalamus have opened a view that the behavioral outcomes from lesions depend on injury to both intrinsic hypothalamic mechanisms and fibers of passage. Hypothalamic obesity still remains an issue for clinicians and for bench scientists to study, but its pathophysiology is emerging, and the stimulus its discovery gave to the alliance remains powerful.

In fact, the ongoing examination of hypothalamic hyperphagia reveals the complicated interaction of behavioral, autonomic, and metabolic issues in the production of obesity. For example, with increasing obesity, fat cells become less responsive to insulin, an event that can be protective because it will tend to limit triglyceride synthesis and further fat deposition. However, if provocations to overeat persist, a vicious cycle will appear as more and more insulin is secreted to overcome the cellular resistance. The increasing insulin levels will enhance the other stimuli to hyperphagia. It is this interplay of behavioral and metabolic factors that often makes a confident analysis of the mechanisms behind an obese state complicated. The obvious clinical corollary is that the treatment of any form of obesity is likely to be difficult and subject to relapse if a primary lesion cannot be eliminated.

CANCER ANOREXIA

If hypothalamic obesity is the classic example of collaboration between clinicians and basic scientists over a proximate factor in the control of food intake, then cancer anorexia and cachexia are the contemporary issues that reveal the intrinsic liveliness in this collaboration.

What more clear example of the "spanner in the works" view is there than the behavioral and physiological effects of cancer? The patient with cancer may have a change in his food appetite as the very first sign of his disease. A prompt improvement in appetite may be the best indicator of successful oncologic treatment even before other objective evidence of tumor regression is apparent. The dramatic nature of this anorexia as it abruptly interrupts the eating patterns of a lifetime, often before any other sign of neoplasm is noted, the significant role that the anorexia and associated wasting of body fat and lean body mass plays in the survival of the patient (often leading to death long before the local or metastatic aspects of tumor growth would have done), and the poor quality of the survival period with the anorexia- and cachexia-induced malaise and distress have focused and sustained clinical attention on this condition.

The early assumptions that the explanation for this anorexia/cachexia might be sought from the growth characteristics of cancer, with functional or anatomic derangements to the gastrointestinal tract, cerebral metastases with increased intracranial pressure, pain-induced anorexia, or simply increased metabolic requirements from tumor growth soon led to the appreciation that these factors, though active in some patients, were not common in the majority. In fact the incidence and severity of anorexia and weight loss bore no precise relation to the size, site, stage, or histology of the neoplasm (Holroyde & Reichard, 1986).

What did seem clear was that inadequate food intake was a partial explanation. Secondary features tied to being a cancer patient such as depression of mood, surgical experience, or a side effect of chemotherapy did not explain this anorexia in more than a small minority. However, there have been relatively few attempts to measure the food intake of cancer patients or their energy expenditure. What data are available show that a mean daily caloric intake of 1,150-1,550 kcal is customary in adult weight-losing cancer patients (Moloney, Moriarty, & Daly, 1983; Walsh, Bowman, & Jackson, 1983) and that 75% of cancer patients are normo- or hypometabolic (Knox, Crosby, Feurer, Buzby, Miller, & Mullen, 1983). An adequate interpretation of these data is hindered by a lack of associated information needed to evaluate their importance, such as the time course of weight loss and the premorbid food intake of the same patients.

These inadequacies in the study of human cancer patients led to the sense that cancer anorexia/cachexia is a complex outcome of a multiplicity of factors, no one of them adequate to explain much of the phenomenon. Yet it is the dramatic nature of the syndrome in humans that sustains attention on the issue. It is tailor-made for investigation in the laboratory.

In the study of tumor-bearing experimental animals, it was readily possible to demonstrate the phenomena of anorexia and cachexia with the induction of cancer. Again, the condition was not explained by the simple effects of the umor, and although such standard laboratory approaches as seeking out earned food aversions have offered occasional demonstrations of a role in cancer anorexia, they have not held up with further study (Levine & Emory, 1987).

The most useful results from the behavioral studies in the laboratory were the clear signs that some circulating factor, either from the neoplasm or from the host, was active in cancer anorexia. The best indication was from parabiotic rats in which a tumor in one of the parabiotic pair induced anorexia and profound weight loss in both, without the spread of cancer to the other rat (Norton, Moley, Green, Carson, & Morrison, 1985). The evidence that diminished food intake and body weight depended on the area postrema in tumor-bearing rats (Bernstein, Treneer, & Kott, 1985) was another indication that some circulating factor noticed by the chemosensitive trigger zone of the brain was active in cancer anorexia.

It was with these suggestions that an intriguing set of discoveries has opened up inquiries into the biology of anorexia and of cancer itself. Cerami and coworkers (Cerami, Ikeda, Le Trang, Hotez, & Beutler, 1985; Rouzer & Cerami, 1980), in efforts to determine the cause of endotoxin-induced hypertriglyceridemia in infected animals, identified a circulating factor that suppressed the synthesis of lipoprotein lipase, the enzyme that promotes lipogensis in adipocytes. This factor was produced by the macrophages of the animals and could be garnered from cell cultures. It is one of the cytokines synthesized and released by blood monocytes and tissue macrophages in response to inflammation (Tracey, Lowry, & Cerami, 1987). These investigators, reasoning that this factor may be involved in the pathophysiology of progressive bodily wasting in disease, designated this factor "cachectin" and have demonstrated that the tissue loss it provokes is quite different from that seen in simple caloric restriction and starvation (Fong et al., in press).

Simultaneously, other workers (Carswell et al., 1975) had demonstrated that an antitumor factor could be induced in mice by bacterial endotoxin, an idea prompted in part by a set of clinical observations of tumor regression after infection made almost a century ago by a New York surgeon (Coley, 1893; Coley, 1906). This factor, which killed a variety of murine and human tumor cells in vitro and in vivo, was also produced by hematopoietic mononuclear cells including macrophages and was designated "tumor necrosis factor" (TNF). Many investigators working with cachectin and TNF demonstrated their comparable molecular weight. Beutler and associates (1985) documented recently that purified cachectin has a tumor necrosis factor activity in vitro and established from inspection of the amino acid sequences and molecular clones of TNF and cachectin that they are homologues of one protein, referred to now as "TNF/cachectin" Beutler & Cerami, 1986).

Recombinant TNF/cachectin is now available, and many laboratories have helped to uncover a long list of biological activities affected by TNF/cachectin Beutler & Cerami, 1987; Oliff, 1988). Crucial to the interest in body weight regulation is the demonstration that the factor suppresses the in vitro activity of several enzymes that contribute to lipogenesis in adipocytes. It is the suppression of these and other anabolic enzymes that gives the factor its ability to deplete body lipid stores and provokes peripheral protein wasting irrespective of caloric intake. Along with its other activities such as stimulation of interleukin-1 secretion, this may provide insight into an aspect of its tumor-killing capacity.

It is, however, also clear that TNF/cachectin does lead to reduced food intake, a phenomenon that is revealed best with intraperitoneal delivery of TNF/cachectin in animals. In a similar way, anorexia is observed in humans given TNF/cachectin as an oncologic therapy (Sherman *et al.*, 1988). Thus, with the discovery of this factor we have come around to new opportunities to appreciate many of the features represented by cancer anorexia and cachexia. This host-derived factor can mediate and provoke many of the metabolic changes associated with cachexia and as well lead to reduced caloric intake in many cancers. The weight loss is likely a result of a combination of these two aspects of TNF/cachectin's actions, and thus we understand the rather frequent observation of examples of cancer patients with cachexia in whom the degree of weight loss exceeds what would be expected from the reduction in caloric intake. As well, we can understand the appearance of anorexia in some patients early in their course of illness when their tumor is small and site-restricted. The anorexia may be a direct or indirect result of this circulating factor but probably depends in part on the biochemical activation of area postrema cells.

Many obvious questions remain to be answered in this new and emerging aspect of control of body weight and food intake. But the overriding question is how to make sense out of this combination of activities tied to a common factor released into the circulation by the body in response to a wide range of assaults, from bacterial and parasitic infection to neoplasia. Are the anorexic and cachexic features simply unfortunate side effects of a factor that offers the host some means of eliminating a tumor, or is there some biological utility to the suppression of caloric intake simultaneously with an attack on the tumor itself? It has been long known that caloric restriction in animals with advanced tumors will inhibit tumor growth and reduce the incidence of spontaneous malignancies in laboratory animals. As well, the metabolic factors that lead to depletion of tissue protein may lead to the preservation of hepatic protein, a phenomenon that is usually found during an acute illness and may have survival value but can be inappropriately sustained with continuing neoplasia.

There is much more to come in addressing the ultimate fitness of this combination of features, metabolic changes associated with cachexia and reduced caloric intake, in TNF/cachectin. It offers great opportunities for investigation to students of food intake and of disease of all kinds. Agonists and antagonists for this hormone/factor are presently being sought (Oliff, 1988) and will certainly prove useful to clinician–scientists concerned with nutritional abnormalities in general and cancer anorexia in particular. The entire realm of biology from genetic mechanisms to neural transmission has been opened up by this effort to study a remarkable behavioral phenomenon, cancer anorexia and cachexia, that links body weight control to disease.

In this section, I have chosen to discuss two research examples in food intake and weight control, one classic and the second new and emerging, that showed the lively interaction between clinicians and basic scientists evoked by recognition of pathological states ("spanner in the works"). These reveal both the excitement and the kind of steady progress that emerge from an appreciation of the proximate or immediate factors tied to the body's machinery that can evoke particular behavioral patterns. The analysis, prompted by human syndromes, quickly turned to basic experimental studies, moving along physiological and biochemical lines. The great advantages to this work on proximate mechanisms are not restricted to the disease focus that provoked the initial studies to illuminate pathophysiology but also lie in the revelation of the need to comprehend the normal mechanisms of physiological controls that are the paths to the home-ostasis in health represented by day-to-day regulations of behavior. This work thus naturally draws together scientists whose interest is on breakdown of the body and its relief with those whose fascination is found in the recognition of the "wisdom of the body" that must rest on homeostatic principles and integrative actions of biology. Disease as an "experiment of nature," like all experiments, opens much of life to study.

The Ultimate or Distal Factor Collaboration among Clinicians and Basic Scientists

It is my contention that all human disorders of food intake are not rooted in disease. Some are maladaptions. The search for the causes of maladaptive human behaviors leads us in other directions—away from the laboratory and the experiment and toward prospective surveys of naturally occurring events in human populations. This is a shift of focus, at least temporarily, from the study of the specific controls on food intake to a way of revealing how a misfit between these controls and emerging life situations can lead to disorder. An appreciation of factors more distal than a broken part within the bodily mechanisms can derive from epidemiologic surveys broadly construed to include genetics.

"Simple" or Idiopathic Obesity

As with the previous section, we begin this section with a classic line of research and focus on obesity first. Epidemiology, which looks at the distribution of disorders in populations and their associated risk factors, is fundamentally important in human studies and has its linkages to genetics, ethology, ecology, and behavioral science. The particular focus here is not on the "epidemics" of disease from which it got its name but on behavior within a population, the behavior of food intake.

Body weight in mankind, when not affected by some clear pathology of the body, as in hypothalamic disease, actually seems a smoothly distributed characteristic much as is height. There may be some extra numbers on the high side of the curve, but individuals are often called obese when they deviate to some arbitrary degree above the mean. The acceptance of the body mass index [BMI, weight (kg)/height (m) squared] as the best anthropometric indicator of obesity has eliminated some of the arbitrariness of definition because it both deemphasizes the effect of height on weight and correlates quite well with other measures of adiposity such as skinfold thickness. Body mass index is a graded, bell-shaped characteristic in the United States population (with perhaps some excess on the right-hand side), with obesity defined as a BMI greater than 27 (kg/m^2) for men and 25 for women.

The first indication that obesity has a crucial sociocultural aspect came from the epidemiologic Midtown Manhattan Study of mental disorders (Srole, Langner, & Michael, 1962; Langner & Michael, 1963) in the community. This study showed a clear relationship between social class and obesity. Obesity was much more prevalent among the lower classes. It seemed avoided for some reason by the upper classes. The fascinating point that subsequently emerged was that this class distinction is a feature of developed societies. Just the reverse relationship i found in underdeveloped nations, where the well-to-do tend to obesity and the poor, perhaps because of caloric starvation, are thin, often to the point of emaci ation. These observations have held up in many studies and beyond themselve offer intriguing speculations about the meaning of weight regulation acros classes and cultures. Such sociocultural interactions with food intake include sign of wealthy status if one is plump in a culture or country where food calorie are scarce and an opposite social meaning of thinness ("A woman cannot be too rich or too thin") in cultures where calories are abundant and evidence of the control of their consumption is valued.

An intriguing sociocultural obesity and its complications has been identified when a race of people who survived in famine are transported to or emerge in calorie-rich world. These last, exemplified best by the Pima Indians of the United States, may develop obesity and its complications because of a "thrifty genotype rendered detrimental by progress" (Neel, 1962), that is, a genetic en dowment for fat storage that would have the survival value of reducing the expenditure of energy during times of shortage but would overstore energy as fat when food became and remained plentiful. There is a considerable amoun of evidence (Leibel & Hirsch, 1984; Ravussin et al., 1988; Roberts, Savage, Cow ard, Chew, & Lucas, 1988) that many obese people do have a reduced energy expenditure as indicated by their metabolic rate at rest. This low 24-hour energy expenditure is a clear risk factor for obesity, appears to be a familial trait, and directs attention to the search for a genetic basis for human obesity.

A. J. Stunkard has provided two important studies that indicate a strong influence of heredity on human obesity by employing two standard techniques for revealing a genetic element, an adoption study and a comparison of mono zygotic and dizygotic twins. The Danish adoption study (Stunkard, Sorenson, e al., 1986) used the birth and health history registers available in that country Stunkard and colleagues demonstrated a clear correlation (using body mass index) between the weights of adoptees and their natural parents but no correla tion with the weight of the adoptive parents.

In his twin study, Stunkard's group (Stunkard, Foch, & Hrubec, 1986) dem onstrated a much higher correlation of BMI in monozygotic than in dizygotic twins. In fact, the implications of these data are that as much as 80% of the variance in BMI can be accounted for in some way by the genetic makeup. This is a remarkable figure and, as he points out, contrasts with data from twin studies in conditions assumed previously to be more strictly tied to heredity such a hypertension (57%) and epilepsy (50%).

The epidemiologic question that is emerging now in the study of human obesity is how to put these observations together, that is, the clear evidence of environmental and sociocultural influences on obesity with the equally clear evidence of a genetic contribution. Once again the nature–nurture issue seems to beset us. However, as in every behavior-based disorder (and obesity at its root depends at least in part on the behavior of food consumption), the ultimate and crucial issue to comprehend is not the genes alone, nor the environment alone but the interaction of the two factors. "The action is in the interaction," say Stunkard, and he bases his opinion directly on his understanding of the herita bility concept derived from monozygotic and dizygotic twin comparisons. This method, intended to discriminate the environmental factors that dizygotic and monozygotic twins share from the genetic factors that render monozygotic twin dentical, has the disadvantage of placing all the interactive causal factors of genes and environment into the genetic compartment and thus overestimates its contribution.

With behavior-driven disorders such as obesity, it is very likely that not only are there genetic controls on the basic metabolism (and those genes can and should be discovered) but also that there are genetic controls on the actions of individuals in a given environment. Thus, genes may influence such things as the individual's search for calories within the environment, the avoidance of protective activities such as exercise, and a preference for such sedentary behaviors as reading, writing, or viewing television.

Much of the excitement in obesity research today is in seeking major genes that can interact with the environment in this way. The simplest example is the intriguing genetic differences in sweetness tasting (Bartoshuk, 1979). People differ in their "sweet worlds"; some live in a "vivid and bright" sweet world because their capacity to appreciate sweetness is genetically enhanced. Others without this amplified capacity live in a "pastel" sweet world (these are Linda Bartoshuk's felicitous metaphors). The potential for consumption to be influenced by the faculty for taste interacting with environments that differ in the availability of refined sugar seems obvious.

Thus, it is the capacity for genetic analysis right down to the molecular level that has increased our explanatory power. As well, though, the ultimate explanations for obesity in given individuals and populations will be found in the interplay of genetic constitution with the reinforcing environment that is itself susceptible to change over time. These explanations will emerge from epidemiologically based studies and then join up with laboratory studies to illuminate multiple behavioral and metabolic features that lead to obesity.

Anorexia Nervosa

If obesity research is emerging into the daylight, the study of anorexia nervosa remains in the shadows. On first blush this seems surprising, given that obesity seems a "constitutional" issue defined simply by a deviation above the population mean for body weight, whereas anorexia nervosa seems to have a more categorical nature with symptomatic criteria like a disease. One might expect a search for proximate causes and mechanisms to have been more successful in anorexia nervosa than has proven true.

There remains, despite a degree of diagnostic agreement, not only great ignorance about the fundamental cause of this condition but real doubts about certain defining features, such as whether there are distinctly separate entities of abnormal weight control, restrictive anorexia nervosa, and binging bulimia nervosa, or whether these are simply variants of the same disorder. It is even debatable whether the food-restricting, perception-altered, emaciated anorexic patient is to be seen as an individual deviating quantitatively to an extreme in a commitment to the food fads and body weight consciousness of our society or whether she suffers from a clear qualitative aberration from normal as a syndromic definition tends to imply (Garner, Olmsted, & Garfinkel, 1983). There are champions for each of these points of view, and there are also those who would merge the differences by proposing that anorexia nervosa has both quantitative and qualitative aspects when viewed over its natural history. These fundamental disagreements, though, identify our ignorance about this condition. It is possible, however, to lay out the established information in a fashion that holds promise for future research and eventual comprehension.

To begin with, I think that the most useful criteria of anorexia nervosa are those of Russell (see Table 1). They are simple, derived from a long personal clinical experience with a broad range of patients, and indicate that whatever may be the cause for the condition, behavior is basic to it.

Whereas in obesity the increased body weight emerges gradually over the course of development because of some insensible imbalance between food intake and energy consumption, in anorexia nervosa the patient is conscious of a struggle to achieve a reduction in her body weight. This struggle is represented by her active efforts to avoid or eliminate caloric intake. This behavior and its sustaining mental state (the overvalued idea in Russell's terms) can often be given a clear onset in time and even an initiating stimulus. The search for the explanation of this behavior—the stubborn rejection of nourishment based on a resolve to be very thin—that could reveal its cause, explain what sustains its course and its variants, and might bring about its rational rather than symptomatic treatment has so far been unsuccessful.

Thus, for example, the last decades have seen a good deal of physiological, metabolic, and endocrine research in anorexia nervosa, prompted in part by the view embedded in Russell's criteria that there is a "specific endocrine disorder" in these patients. This research has attempted to identify bodily derangements that might represent either predispositions or precipitants to the behavior. These studies, often construed as attempts to reveal a proximate pathophysiology at the root of anorexia nervosa and thus to identify it as another disease expressing itself in behavior, include assessment of such hypothalamically derived functions as gonadotropin levels and body temperature responses but also extend to peripheral physiology in evaluations of gastrointestinal motility and changes in ovarian anatomy.

Every observation of a deviation of these features from normal in anorexia nervosa has proven on follow-up after refeeding to be a symptom of the food-restricted and starved state and thus likely to be a result rather than a cause or even a predisposing feature of the behavior. All anomalies disappear with renourishment and restoration of body weight. Even the underlying suspicion that anorexia nervosa represents a form of the lateral hypothalamic syndrome proved on a careful examination to be unlikely (Stricker & Andersen, 1980).

From the physiological point of view, anorexia nervosa has so far provided a model of human starvation and its consequences. This is not a trivial matter for the clinician because many of the features of chronic starvation are important to identify when a course of treatment and explanations for distressing symptom.

TABLE 1. RUSSELL CRITERIA FOR ANOREXIA NERVOSA[a]

1. Self-induced loss of weight (resulting mainly from the studied avoidance of foods considered by the patient to be fattening).
2. A characteristic psychopathology consisting of an overvalued idea that fatness is a dreadful state.
3. A specific endocrine disorder that in the postpubertal girl causes the cessation of menstruatio or a delay of events of puberty in the prepubertal or early pubertal female.

[a]From Russell (1979).

of the disorder are offered to the anorexic patient. Thus, it is helpful to be able to identify the early sense of fullness and even nausea during refeeding as a complication of reduced gastric emptying (Robinson, McHugh, Moran, & Stephenson, 1988). Information that there is a need for extra body weight recovery to restore normal menses and complete reproductive capacity (Frisch, 1985) is critical in defining for the patient the goal weight sought by the treatment team. Other discoveries such as the sonographic identification of ovarian atrophy and its restoration with refeeding have permitted a definition of stages of recovery (Treasure, Gordon, King, Wheeler, & Russell, 1985).

However, if the aim of research is the discovery of cause, these pathophysiological studies have revealed only the symptoms and consequences of the behavior of food restriction and little of its generation. We are still left with the question, what can lie behind this resolve—this drive to be thin—this insistent rejection of nourishment and its natural effects on the body? One turns to seek ultimate factors in frustration from the failures to find proximate ones.

For this reason, an epidemiologic approach with its potential to identify common factors that differentiate these patients from normals in broader terms than pathophysiology is promising and gives a sense of direction to contemporary studies. The time–place–person triad of epidemiology brings out several unique facts that may come to illuminate the syndromic criteria. These can be briefly summarized.

Anorexia nervosa is a behavioral disorder that primarily affects (1) youthful females of the (2) upper classes in (3) developed nations in the (4) contemporary era. This is the major set of features, the ones that deserve most attention, and whose limits are to be appreciated.

First, the female sex of the afflicted. This is not an absolute exclusion, but since females outnumber males ten to one in anorexia nervosa, we certainly should devote attention to aspects of women's lives and to see the male examples as anomalous and offering perhaps some opportunity to identify a feature that might point up an element provoking or sustaining the behavior in females. The high incidence of homosexuality (Herzog, Norman, Gordon, & Pepose, 1984) among the males with anorexia nervosa suggests, for example, an extra strength for concerns about sexuality at the heart of anorexia nervosa.

That it is youthful females, primarily those emerging from puberty, rather than an even distribution of the disorder across all ages promotes a focus of attention on the events of puberty and its immediate aftermath in women, not on all of the other psychosocial features that distinguish the lives of women from men across the life span. These events of puberty include the establishment of a cyclical program of gonadotropin release and the emergence of a female bodily contour tied to the capability and sustenance of procreation. These bodily changes in women are more than simply biological events, since they both influence the self-attitude of a young woman and draw attention to her as an individual emerging from girlhood into the phase where her attractiveness and suitability for marriage are being considered. The increased incidence of anorexia nervosa in young women who are in occupations where their appearance is a premium, such as ballet students and fashion models, tends to secure a view that a risk factor for the condition is attention to bodily appearance.

There is evidence that those who succumb to anorexia nervosa, though, have some constitutional vulnerability that makes these interpersonal and cultural features particularly burdensome to them. This evidence has emerged both from the epidemiologic and clinical studies. Thus, a genetic contribution to anorexia nervosa is secure. It rests on an increased concordance in monozygotic as compared to dyzygotic twins and on a higher rate of anorexia nervosa among sisters (Strober, Morrell, Burroughs, Salkin, & Jacobs, 1985). Exactly how this genetic contribution may function is uncertain, but it may emerge indirectly through the personality characteristics that derive in part from genetic instructions. As I shall discuss, the restricting anorexic tends to have a personality of introversion and self-reflection, and these features may enhance the influence of sociocultural pressures that provoke concerns about appearance.

Finally, the condition is increasing in its incidence. The evidence for a sharp rise in anorexia nervosa includes the observations of Theander (1970) who noted a fivefold increase in Malmo, Sweden between the 1930s and 1950s, of Jones, Fox, Babigan, & Hutton (1980), who demonstrated a doubling of incidence in Monroe County in New York State, and Szmukler (1985), who documented a rise in the annual incidence from 1.6 to 4.1 per 100,000 population in Northeast Scotland between the late 1960s and the early 1980s. In fact, as many as one in 250 schoolgirls in England (Crisp, Palmer, & Kalucy, 1976) have the condition, although many fewer suffer from it to the degree of needing hospitalization. All these studies of incidence considered and rejected the possibility that these increases might be more apparent than real and dependent on the increased diagnostic interest of physicians in anorexia nervosa or on families seeking medical attention for their daughters prompted by an enhanced awareness of the disorder gained through the media.

Along with this increased incidence, there is also evidence for a recent spread of the disorder outside its prime targets. Females both younger and older than the original 14- to 25-year-old range are now being reported. The appearance of anorexia nervosa in prepubescent girls can stunt bodily growth and development. Older women are often examples of relapses, but all specialists in anorexia nervosa can now report the initial appearance of the disorder in an occasional case over age 30.

There is clear evidence that anorexia nervosa is spreading beyond the confines of the upper classes of developed nations. Several examples have been reported in American blacks from impoverished backgrounds as well as among women in the upper classes of the "third world" nations (Buhrich, 1981). There is even a suggestion that there is an increased incidence of anorexia nervosa in males. Certainly most anorexia nervosa services now regularly have at least one male patient in the hospital at any time, a rare phenomenon a decade ago. The appearance of the apparent new disorder, bulimia nervosa (Russell, 1979), in which the weight loss is less and the morbid fear of fatness is associated with the behavior of gorging and purging rather than calorie restricting, suggests that whatever is provoking anorexia nervosa, it is spreading to engage other behavioral activities tied to maintaining a low body weight.

These then are the epidemiologic observations. The next question is how to use them to approach the cause or causes of anorexia nervosa. Thus, they do not suggest that this condition is a result of some bodily disease or internal chemical imbalance. At least, no such disorder can be proposed for such a relatively circumscribed group of victims embraced by class and culture. As with obesity, it is probable that internal physiological consequences of the nutritional state may sustain the behavior or encourage it once it becomes established.

But why should a behavior provoking a state of starvation break out among young women who live in the midst of plenty? The pathophysiological consequences do not explain the initiation. We thus turn to interpret this behavior by identifying goals promoted by the contemporary culture that attract particular individuals.

A plausible formulation of the epidemiologic data is to see this behavior as incited by a set of sociocultural attitudes that rains down on all in this era. Its effect is to pick out a vulnerable minority, mostly women, to induce, cultivate, and sustain a response that can vary in its strength depending on a number of other provocative or protective features only some of which we have identified.

The sociocultural force most often mentioned is the emerging and persisting enthusiasm for thinness as an ideal bodily shape. This enthusiasm certainly can be tied to the fear of fatness of anorexia nervosa patients and perhaps verified as a provocative factor by the increased incidence of the disorder among occupations where thinness is demanded. However, essentially synchronous with the emerging cultural enthusiasms for thinness has been an increasing sexualization of adolescence in the last several decades. A preoccupation with bodily appearance, shape, and size is a natural response to this influence. Anorexia nervosa may be for some young women a way of reacting to and even defending against the complicated nexus of choice and demand imposed on them at this time as they develop out of girlhood. This would be to view anorexia nervosa as a female ascetic response to the broadly construed sexual engrossments of this period.

Any construal that a cultural force induces a maladaptive behavior, though, must propose that there is some particular vulnerability in the people who respond to this pressure, since cultural forces act on all members of the society, only some of whom display the behavior in question. As has been mentioned above, anorexia nervosa appears much more frequently among individuals with an introverted temperament. That is, they often have the kind of affective constitution that tends to the self-reflections and sustained moods that generate concerns about the future.

A social setting that draws attention to her changing bodily appearance and may include some inklings about sexuality can be experienced more powerfully by an introverted than extraverted young woman as a challenge to her self-management. It may be met by a stance of restriction and a resolve to sustain thinness as a kind of conflicted search for control of her situation. Once fasting and restriction are established, then the physiological and physical changes that emerge, and even the bodily sensations associated with them, can aid and sustain the resolve by giving the individual evidence of success. These successive phenomena may enhance other reinforcements such as peer support to shift her from simply a weight consciousness that is almost universal in contemporary women into attitudes powerful enough to sustain a persisting, recurrent, and even life-threatening behavior.

This view of anorexia nervosa as a spiral of behavioral responses with the initial food restriction prompted by sociocultural pressures striking on the vulnerable and then being reinforced by consequences that encourage more restriction in a self-perpetuating fashion matches some of the clinician's experience with this condition and particularly the clear clinical observation that early interruption of restriction is the best treatment for the disorder (Szmukler & Eisler, 1985). As well, this view of an interactive process can explain the increasing incidence of the behavior as the cultural pressures continue, become locally amplified as in certain families or institutions, and through suggestion and example draw in individuals who are not so obviously predisposed as the initial victims.

Thus, the epidemiologic evidence can be combined with the clinical evidence to propose that anorexia nervosa is a maladaptive response based ultimately on the characteristics of human beings to see implications in life circumstances. This would fit the clinical opinion of Russell that anorexic patients have an "overvalued idea," i.e., a morbid devotion to an opinion shared by many others but provocative of disorder when it becomes a ruling passion.

Thus, most of the clinical and epidemiologic data suggest that the cause of anorexia nervosa is embedded in the combination of vulnerable personalities conflicted over the values of this era and adapting to them with a behavior that can become self-engrossing and self-sustaining. This formulation leaves several unanswered questions. Among them: Why *are* women so much more vulnerable to this disorder than men? What is the distortion of perception that can blind anorexics to their thinness? Why is it so hard to treat? These questions and others remain as provocative issues for further research to confirm or reject the meaningful interpretation set forward here. At the moment, however, it seems the best account we have for this strange condition.

In summary, from the consideration of both idiopathic obesity and anorexia nervosa, we can propose that disruptions in human feeding do not always derive from bodily disease and the injury to normal controls but rather can emerge from more ultimate factors such as human genetic variation, reinforcing life events, and the role of meaning tied to culturally based values that shape the expression of behaviors in our species. Although we might all agree that eventually these ultimate influences must work through the bodily mechanisms discerned by experiment, it is also clear that these superimposed anomalies of feeding are not so easily duplicated by animal models or adequately assessed at a tissue level. They seem tied to the biopsychology issues unique to mankind: a species with its own genetic endowment that has evoked a capacity to imagine a future, to represent it with an idea, and to find therein a meaning for behavior. These anomalies are illuminated by methods derived from epidemiology that permit a multiplicity of causes and sustaining mechanisms to be glimpsed perhaps first as disconnected risk factors and only with further careful, often crosscultural study put together into coherent testable explanations formulated as the outcome of constitutional factors, life experience, and shared values.

CONCLUSION

My major aim in this chapter was not to give an encyclopedic account of the human disorders of feeding and weight control. Such an achievement was not only outside my charge, it is outside my scope. Rather, I wished to emphasize a natural separation among these disorders seen in the clinic and the different contemporary logics that are being applied to their study. I hoped to impart the view that there are two major lines of investigation, on the one hand laboratory based, emphasizing biochemical and physiological experiment, and on the other epidemiologically based, emphasizing surveillance of human populations distinguished by genetic and sociocultural variations. These lines can be distinguished as the search for proximate and ultimate causal factors. There is, I believe, emerging a greater contact between these two lines of research, but they remain, by their nature, relatively separate and distinct. They evoke different questions and must be approached by experts in quite different methods.

However, it is the concept of motivated behaviors, directed toward goals tied to the vital needs of the organism, based on psychoneural systems that function to organize activity and its commerce with surrounds, that holds this work together. These systems are subject to pathological change from within the organism and to deviant influences from without. In this chapter I have looked at feeding behavior and some of its disorders; it should be obvious that a similar approach is also to be applied to disorders in any of the other motivated behaviors of mankind.

Acknowledgments

I want to thank Marie Killilea, Timothy Moran, and James Wirth for very helpful criticism and suggestions of earlier drafts of this chapter.

This work was supported by National Institute of Diabetes and Digestive and Kidney Diseases (DK-19302).

REFERENCES

Bartoshuk, L. M. (1979). Bitter taste of saccharin related to the genetic ability to taste the bitter substance 6-*n*-propylthiouracil. *Science, 205*, 934–935.

Bernstein, I. L., Treneer, C. M., & Kott, J. N. (1985). Area postrema mediates tumor effects on food intake, body weight, and learned aversions. *American Journal of Physiology, 249*, R296–R300.

Beutler, B., & Cerami, A. (1986). Cachectin and tumour necrosis factor as two sides of the same biological coin. *Nature, 320*, 584–588.

Beutler, B., & Cerami, A. (1987). Cachectin: More than a tumor necrosis factor. *The New England Journal of Medicine, 316*, 379–385.

Beutler, B., Greenwald, D., Hulmes, J. D., Chang, M., Pan, Y.-C. E., Mathison, J., Ulevitch, R., & Cerami, A. (1985). Identity of tumour necrosis factor and the macrophage-secreted factor cachectin. *Nature, 316*, 552–554.

Bray, G. A., & York, D. A. (1979). Hypothalamic and genetic obesity in experimental animals: An autonomic and endocrine hypothesis. *Physiological Reviews, 59*, 719–809.

Brobeck, J. R., Tepperman, J., & Long, C. N. H. (1943). Experimental hypothalamic hyperphagia in the albino rat. *Yale Journal of Biology and Medicine, 15*, 831–853.

Bruch, H. (1939). Progress in pediatrics. The Frohlich syndrome. *American Journal of Diseases of Children, 58*, 1282–1289.

Buhrich, N. (1981). Frequency of presentation of anorexia nervosa in Malaysia. *Australian and New Zealand Journal of Psychiatry, 15*, 153–155.

Carswell, E. A., Old, L. J., Kassel, R. L., Green, S., Fiore, N., & Williamson, B. (1975). An endotoxin-induced serum factor that causes necrosis of tumors. *Proceedings of the National Academy of Sciences, U.S.A., 72*, 3666–3670.

Cerami, A., Ikeda, Y., Le Trang, N., Hotez, P. J., & Beutler, B. (1985). Weight loss associated with an endotoxin-induced mediator from peritoneal macrophages: The role of cachectin (tumor necrosis factor). *Immunology Letters, 11*, 173–177.

Coley, W. B. (1893). The treatment of malignant tumors by repeated inoculations of erysipelas: With a report of ten original cases. *American Journal of Medical Sciences, 105*, 487–511.

Coley, W. B. (1906). Late results of the treatment of inoperable sarcoma by the mixed toxins of erysipelas and *Bacillus prodigiosus. American Journal of Medical Sciences, 131*, 375–430.

Crisp, A. H., Palmer, R. L., & Kalucy, R. S. (1976). How common is anorexia nervosa? A prevalence study. *British Journal of Psychiatry, 128*, 549–554.

Erdheim, J. (1904). Ueber Hypopysenganggeschwulste und Hirncholestratome. *Sitzungsb. d. Akad. d. Wissensch. Math.-naturw Kl., 113*, 537–726.

Fong, Y., Moldawer, L. L., Marano, M., Wei, H., Barber, A., Manogue, K., Tracey, K. J., Kuo, G., Fischman, D. A., Cerami, A., and Lowry, S. F. (1989). Cachectin/TNF or IL-la induces cachexia with redistribution of body proteins. *American Journal of Physiology, 256*, R659–R665.

Friedman, M. I., & Stricker, E. M. (1976). The physiological psychology of hunger: A physiological perspective. *Psychological Reviews, 83*, 409–431.

Frisch, R. E. (1985). Fatness, menarche and female fertility. *Perspectives in Biology and Medicine, 28*, 611–633.

Fröhlich, A. (1901). Ein Fall von Tumor der Hypophysis cerebri ohne Akromegalie. *Wiener Klinische Rundschau, 15*, 883–886, 906–908.

Fulton, J. F. (1940). Introduction: Historical resume. *Research Publications, Association for Research in Nervous and Mental Disease, 20*, xiii–xxx.

Garner, D. M., Olmsted, M. P., & Garfinkel, P. E. (1983). Does anorexia nervosa occur on a continuum? *International Journal of Eating Disorders, 2*, 11–20.

Herzog, D. B., Norman, D. K., Gordon, C., & Pepose, M. (1984). Sexual conflict and males. *American Journal of Psychiatry, 141*, 989–990.

Hetherington, A. W., & Ranson, S. W. (1940). Hypothalamic lesions and adiposity in the rat. *Anatomical Record, 78*, 149–172.

Holroyde, C. P., Reichard, G. A. (1986). General metabolic abnormalities in cancer patients: Anorexia and cachexia. *Surgical Clinics of North America, 66*, 947–956.

Jones, D. J., Fox, M. M., Babigan, H. M., & Hutton, H. E. (1980). Epidemiology of anorexia nervosa in Monroe County. *Psychosomatic Medicine, 42*, 551–558.

Knox, L. S., Crosby, L. O., Feurer, I. D., Buzby, G. P., Miller, C. L., & Mullen, J. L. (1983). Energy expenditure in malnourished cancer patients. *Annals of Surgery, 197*, 152–162.

Langner, T. S., & Michael, S. T. (1963). *Life stress and mental health.* Glencoe, IL: Free Press.

Leibel, R. L., & Hirsch, J. (1984). Diminished energy requirement in reduced-obese patients. *Metabolism, 33*, 164–170.

Levine, J. A., & Emory, P. W. (1987). The significance of learned food aversions in the aetiology of anorexia associated with cancer. *British Journal of Cancer, 56*, 73–78.

Moloney, M., Moriarty, M., & Daly, L. (1983). Controlled studies of nutritional intake in patients with malignant disease undergoing treatment. *Human Nutrition and Applied Nutrition, 37A*, 30–35.

Neel, J. V. (1962). Diabetes mellitus: A "thrifty" genotype rendered detrimental by "progress"? *American Journal of Human Genetics, 14*, 353–362.

Norton, J. A., Moley, J. F., Green, M. V., Carson, R. E., & Morrison, S. D. (1985). Parabiotic transfer of cancer anorexia/cachexia in male rats. *Cancer Research, 45*, 5547–5552.

Oliff, A. (1988). The role of tumor necrosis factor (cachectin) in cachexia. *Cell, 54*, 141–142.

Powley, T. L. (1977). The ventromedial hypothalamic syndrome, satiety, and a cephalic phase hypothesis. *Psychological Reviews, 84*, 89–126.

Ravussin, E., Lillioja, S., Knowler, W. C., Christin, L., Freymond, D., Abbott, W. G. H., Boyce, V., Howard, B. V., & Bogardus, C. (1988). Reduced rate of energy expenditure as a risk factor for body weight gain. *New England Journal of Medicine, 318*, 467–472.

Roberts, S. B., Savage, J., Coward, W. A., Chew, B., & Lucas, A. (1988). Energy expenditure and intake in infants born to lean and overweight mothers. *New England Journal of Medicine, 318*, 461–466.

Robinson, P. H., McHugh, P. R., Moran, T. H., & Stephenson, J. D. (1988). Gastric control of food intake. *Journal of Psychosomatic Research, 32*, 593–606.

Rouzer, C. A., & Cerami, A. (1980). Hypertriglyceridemia associated with *Trypanosoma brucei brucei* infection in rabbits. *Molecular and Biochemical Parasitology, 2*, 31–38.

Russell, G. F. M., (1979). Bulimia nervosa: An ominous variant of anorexia nervosa. *Psychology and Medicine, 9*, 429–448.

Sherman, M. L., Spriggs, D. R., Arthur, K. A., Imamura, K., Frei, E., & Kufe, D. W. (1988). Recombinant human tumor necrosis factor administered as a five day continuous infusion in cancer patients: Phase I toxicity and effects on lipid metabolism. *Journal of Clinical Oncology, 6*, 344–350.

Srole, L., Langner, T. S., & Michael, S. T. (1962). *Mental health in the metropolis: The Midtown Manhattan Study.* New York: McGraw-Hill.

Stricker, E. M., & Andersen, A. E. (1980). The lateral hypothalamic syndrome: Comparison with the syndrome of anorexia nervosa. *Life Sciences, 26*, 1927–1934.

Strober, M., Morrell, W., Burroughs, J., Salkin, B., & Jacobs, C. (1985). A controlled family study of anorexia nervosa. *Journal of Psychiatric Research, 19,* 239–246.

Stunkard, A. J., Foch, T. T., & Hrubec, Z. (1986). A twin study of human obesity. *Journal of the American Medical Association, 256,* 51–54.

Stunkard, A. J., Sorenson, T. I. A., Hanis, C., Teasdale, T. W., Chakraborty, R., Schull, W. J., & Schulsinger, F. (1986). An adoption study of human obesity. *New England Journal of Medicine, 314,* 193–198.

Szmukler, G. I. (1985). The epidemiology of anorexia nervosa and bulimia. *Journal of Psychiatric Research, 19,* 143–153.

Szmukler, G. I., & Eisler, I. (1985). The implications of anorexia nervosa in a ballet school. *Journal of Psychiatric Research, 19,* 177–181.

Theander, S. (1970). Anorexia nervosa: A psychiatric investigation of 94 female patients. *Acta Psychiatrica Scandinavica Supplementum, 214,* 1–194.

Tracey, K. J., Lowry, S. F., & Cerami, A. (1987). Physiological responses to cachectin. *Ciba Foundation Symposium, 131,* 88–108.

Treasure, J. L., Gordon, P. A., King, E. A., Wheeler, M., & Russell, G. F. (1985). Cystic ovaries: A phase of anorexia nervosa. *Lancet, 2,* 1379–1382.

Walsh, T. D., Bowman, K. B., & Jackson, G. P. (1983). Dietary intake of advanced cancer patients. *Human Nutrition and Applied Nutrition, 37A,* 41–45.

Index